A Textbook of *Entomology*

Herbert H. Ross
Late Professor of Entomology
University of Georgia

Charles A. Ross
Western Washington University

June R. P. Ross
Western Washington University

FOURTH EDITION

JOHN WILEY & SONS
New York Chichester Brisbane Toronto Singapore

Copyright © 1948, 1956, 1965, 1982, by John Wiley & Sons, Inc.

All rights reserved. Published simultaneously in Canada.

Reproduction or translation of any part of
this work beyond that permitted by Sections
107 and 108 of the 1976 United States Copyright
Act without the permission of the copyright
owner is unlawful. Requests for permission
or further information should be addressed to
the Permissions Department, John Wiley & Sons.

Library of Congress Cataloging in Publication Data:

Ross, Herbert Holdsworth, 1908–
A textbook of entomology.

Includes bibliographies and index.
1. Entomology. I. Ross, Charles Alexander.
II. Ross, June R. P. III. Title.
QL463.R68 1982 595.7 81-16097
ISBN 0-471-73694-5 AACR2

Printed in the United States of America

10 9 8 7 6 5 4 3 2 1

To Herbert Ross's many friends and colleagues who helped in so many ways to make the vast and diverse field of entomology so excitingly interesting to him.

Preface

Entomology is an ever-growing science that continues to expand in basic knowledge and perspectives through the publication of a large amount of research and ideas. In the 15 years since the third edition of this textbook was published, entomology has taken enormous steps forward in its understanding of the ecology, physiology, behavior, and interactions of insects. These new developments will continue to lead to additional and even more fundamental questions about insects, and new evidence will be needed in order to reach new answers and to test new ideas. The results will significantly change some of our current perceptions and ideas, and, in the long term, will lead us to a much better understanding of our science.

This book serves as an introduction to entomology by presenting the fundamental aspects of the field, and it is organized so that students will acquire a general view of the whole field. This edition preserves the major aims and balance of the earlier editions with emphasis on a knowledge of basic relationships between different groups of insects and between insects and their environment. The relationships between insects and their environments are emphasized in chapters on life processes, behavior, past history, ecology, beneficial insects, and harmful insects. Some chapters are new or nearly completely rewritten to reflect the significant changes that have taken place in entomological thought during recent years. Substantial portions of other chapters were rewritten, and the chapter on insect orders was reorganized to represent more closely the probable evolutionary relationships of the different groups.

As before, evolutionary relationships are stressed as a central theme. Preliminary family trees are included for the common families in a number of orders to visualize more readily their evolutionary relations. It is increasingly evident that our information in such areas as comparative biochemistry, behavior, and anatomy will not be fully understood until the relevant facts are fitted into systems of evolutionary progression. The largely rewritten chapter on the history of insects gives a geological perspective of the dynamic history of insect evolution and the forces surrounding it.

The keys to the orders and families are designed to accommodate only common members and hence are far from complete. They are intended to aid beginning students in realizing the differences used in delimiting orders and families and to give them practice in the usage of keys.

We wish to thank the many persons who have been of assistance in planning and writing the various editions of this textbook. In preparing this fourth edition we particularly wish to thank all of those who have offered encouragement, suggestions, and criticisms. Drs. Donald Ashdown of Texas Tech. University, Frank M. Carpenter of Harvard University, Robert W. Dicke of the University of Wisconsin, Jarmila Kukalova-Peck of Carleton University, Jack Lattin of Oregon State University, Fred A. Lawson of the University of Wyoming, W. P. McCafferty of Purdue University, Robert W. Matthews and Janice R. Matthews of the University of Georgia, T. M. Peters of the University of Massachusetts, and Robbin W. Thorp of the University of California at Davis read all or parts of the manuscript for this edition and made many valuable and thoughtful comments and suggestions. Special thanks to Douglas A. Craig of the University of Alberta, Alfred Dietz of the University of Georgia, K. G. A. Hamilton of Agriculture Canada, Preston Hunter of the University of Georgia, Alden Lea of the University of Georgia, and the publishers, John Wiley & Sons, for their help and thoughtful consideration. Mrs. Jean A. Ross provided helpful background information for this revision. We also thank the many organizations and persons who gave permission to use their materials for illustrations in this book. They are individually acknowledged under each illustration. Miss Patricia Combs, Mrs. Chris Moreland, and Mrs. Joan Roley, Western Washington University, also deserve our special gratitude for their care in typing various parts of the manuscript.

Charles A. Ross
June R. P. Ross

Contents

Chapter 1. GROWTH OF NORTH AMERICAN ENTOMOLOGY. **1–26**

Beginnings of modern biology. Progress in the eighteenth century. Development of North American entomology. American expansion period, roughly 1867–1900. Twentieth century developments.

Chapter 2. ARTHROPODA: INSECTS AND THEIR ALLIES. **27–58**

Ancestral relations. Synopsis of the phylum Arthropoda. Superclass Hexapoda.

Chapter 3. EXTERNAL ANATOMY. **59–106**

Body wall and exoskeleton. Body regions. The head. Cervix or neck. Development of the generalized insect segment. Thorax. Abdomen.

Chapter 4. INTERNAL ORGAN SYSTEMS. **107–125**

Digestive system. Excretory system. Circulatory system. Tracheal system. Nervous system. Musculature. Reproductive system. Specialized tissues.

Chapter 5. LIFE PROCESSES. **127–174**

Structure and function of the skeleton. Nutrition. Digestion. Excretion and salt and water balance. Respiration. The blood and circulation. Metabolism. Muscular systems and motion.

Chapter 6. RESPONSE AND BEHAVIOR. **175–222**

Sources of information: Receptors and sense organs. Nervous coordination and integration. Endocrine system and hormones. Communication. Responses and types of behavior. Organization of social insects.

Chapter 7. LIFE CYCLES, GROWTH, AND REPRODUCTION. 223–264

Development. Maturity. Food habits. Seasonal cycles. Reproduction. The role of hormones.

Chapter 8. THE ORDERS OF ENTOGNATHS AND INSECTS. 265–502

Class Entognatha. Class Insecta. Keys to the orders of common Entognatha and adult Insecta. Diplura. Protura. Collembola. Microcoryphia. Thysanura. Ephemeroptera. Odonata. Dictyoptera. Isoptera. Phasmatodea. Orthoptera. Dermaptera. Grylloblattodea. Embioptera. Plecoptera. Zoraptera. Psocoptera. Phthiraptera. Thysanoptera. Hemiptera. Raphidioptera. Megaloptera. Neuroptera. Coleoptera. Hymenoptera. Mecoptera. Siphonaptera. Diptera. Trichoptera. Lepidoptera.

Chapter 9. THE PAST HISTORY OF INSECTS. 503–549

Forms of evidence. Paleogeography. Extinct orders. Fossil insects and their relation to living orders. Insects and the history of life.

Chapter 10. ECOLOGICAL CONSIDERATIONS. 551–602

Biomes and communities. Ecosystems. Populations. Ecology of individuals; Environmental factors—weather, physical and chemical conditions of the habitat, food, enemies, protection against enemies, competition.

Chapter 11. USEFUL INSECTS. 603–630

Predatory and parasitic insects. Insect predators. Insect parasites. Weed management. Insect pollination. Makers of useful products. Other uses. Decomposers and nutrient recycling.

Chapter 12. INSECT PESTS AND THEIR CONTROL. 631–666

Insect pests of plants. Insect pests of humans and animals. Insect pest control: Biological control, Plant resistance, Physical control, Chemical control. Insect pest management. Integrated pest management. Challenges of the immediate future.

Index I-1–I-30

1
Growth of North American Entomology

Insects and their relatives, the subject matter of entomology, are a remarkable group of animals. They have evolved into the largest class of all living things in respect to the number of different kinds or species now inhabiting the earth. They occupy a fantastic number of ecological niches, ranging from living on other insects to eating either living or dead plants or living in or on higher animals. Often each species is adapted to live in only one narrow niche. Their body parts, tissues, and organs are in their own way as intricate as those of the higher vertebrates, even though many full-grown insects are only 1 or 2 mm. long. In most insects, the reproductive potential is enormous. Because they feed on all the natural foods eaten by humans and because they affect humans' health adversely in many ways, insects have been considered our most serious competitor for the food and fiber products. At times, it has seemed as if the insects were indeed about to reach an ascendancy in this competitive relationship, but to date we have achieved an uneasy balance with our little but innumerable foes. At present, the costs for this uneasy balance and for our successes over the insects are increasing constantly.

Humans have always had their troubles with insects. When humans first emerged, they already had fleas and lice and were fed on by mosquitoes and pestered by flies. In those early days, when human populations were scattered and sparse, the human struggle was on a primitive plane—to find natural food from day to day and to escape the onslaughts of predatory animals. At this period it is doubtful that insects and insect-borne diseases were nearly so important as enemies to humans as were other inimical factors of the environment. In fact, on the average, insects were probably of great help because termites, grasshoppers, grubs, and the like could be found and eaten when other foods were not obtainable.

From primeval conditions human progress has been based essentially on changing various factors of our environment and making it better suited for our own survival and increase. But every change that benefited us also benefited a host of insects. Gradually, as the starker enemies of primeval life, such as the leopard and tiger, ceased to be a great threat to primitive humans, insects became increasingly important as a challenge to their civilized status.

In the first place, increase in human populations allowed a great increase of such insect ectoparasites as lice and fleas. This was a result of the ready accessibility of additional host individuals for the insects and, therefore, better opportunities for dissemination and chances for reproduction. The same factors favored the increase and spread of pestilence, including insect-borne diseases. When large cities arose, they were repeatedly swept by outbreaks of these maladies, in the same way that Imperial Rome was decimated by bubonic plague in the second century A.D.

Insects became a real factor with food as with health. When humans began to store food, it was attacked by a host of insects that before had been of no significance in the human environment. In the tremendous food-storage organization of today, insects destroy millions of tons of food annually in spite of widespread and expensive control programs.

When human populations outstripped the food-producing capacity of natural surroundings, animals began to be domesticated. The concentration of these allowed an increase of their ectoparasites and diseases, thus partially nullifying the effort to enlarge the food supply. The cultivation of crops brought about the greatest change with regard to insects. Agriculture congregated plant hosts so that their insect attackers could build up extensive populations on them. The Egyptian writer in the time of Rameses II (1400 B.C.) commiserates with the peasant that "Worms have destroyed half of the wheat, and the hippopotami have eaten the rest; there are swarms of rats in the fields, and the grasshoppers alight there." In the more recent period of crop improvement, new varieties of plants developed for increased yield have frequently been more attractive to certain insects than original wild hosts, with a resultant influx of destructive species to the cultivated crops.

This situation has been made more serious by our development of transportation between all parts of the world. Insects of many species have been carried to continents new to them, where they have found favorable climates, succulent acceptable cultivated hosts, and a freedom from the natural enemies that had kept their numbers in check in their original homes. Sometimes the result has been disastrous, as, for example, the entry of the European corn borer and the Japanese beetle into North America. These two species are of little economic importance in their native ranges in Eurasia, but in the United States they have caused losses to crops in the magnitude of millions of dollars per year.

North America has been especially hard hit by losses from insects. This is the result of the cultivation here in recent centuries of many crops not indigenous to the area, to introduction of many new pests, and to changes wrought by agriculture that have favored many endemic insect species. Insect damage to agricultural products for 1974 was estimated to be over $25 billion, plus an estimated cost of control applications of approximately $5 billion, a total of $30 billion. These figures do not include losses to forest products, illnesses and deaths from insect-borne diseases, or work losses caused by the irritation of biting insects, all difficult to estimate, but conservatively in the neighborhood of $15 billion. This gives a grand total annual insect bill of $45 billion.

It is difficult to visualize a loss in the magnitude of billions of dollars. Let us put it another way. In 1974 each person in the United States, on the average, paid $225 toward insects. This means $900 for a family of four. Some of this sum was spent for insecticides and some for replacing damaged goods, but most of it was disguised as increased cost of commodities of plant or animal origin, such as lumber, clothing, and food.

A review of insect damage gives the entire group a sinister aspect. But the adage, "There is some good in everything," finds a real place even among this group of apparent despoilers. Many kinds of insects are definitely beneficial. In fact, less than 10% of the known insect species are pestiferous. The most conspicuous example of a beneficial insect is the honeybee, which not only produces a marketable crop of high cash value, but also pollinates many valuable plant species. Most of our fruits and legumes are dependent for pollination on a large number of insect groups such as bees, moths, flies, and beetles. Without these insects we would have no apples, pears, peas, squash, and seeds of other insect-pollinated plants.

Another group of economically beneficial insects embraces a large assemblage of predaceous and parasitic insects whose hosts are other insects. These include ichneumon wasps, parasitic wasps, parasitic flies, and ladybird beetles. The adults or larvae of these species prey on or parasitize many important insect pests. In some cases they can be used as an efficient control method. The vedalia ladybird beetle, for example, is potentially one of the chief means of combating the cottony cushion scale, an insect destructive to citrus orchards in California.

Integrated pest management is the field of activity that attempts to combat destructive insects and to encourage beneficial insects. It is comparable in many ways to the field of medicine, which has arisen out of the challenge to combat sickness and disease. In North America, integrated pest management involves a financial outlay of considerable proportions. Many thousands of people are employed primarily in the investigation of economic species and the development of control measures. Although the primary objective of pest management is the reduction of insect damage, it has long been evident that a wide knowledge of scientific information is necessary as a foundation for effective control. For this reason there has been an appreciation of basic entomological research in many directions. Some phases seemed of little importance when first started, yet later proved of inestimable value in control problems.

BEGINNINGS OF MODERN BIOLOGY

Because of their tremendous abundance, it might seem that insects would have been used a great deal in the early investigations in fundamental biology. The small size of the average insect, however, militated against this. Extremely delicate methods of dissection are required to study anatomy and physiology, and powerful microscopic equipment is necessary for taxonomic studies of almost any insect group. To a large extent, therefore, the fundamentals of biology were based on observations of larger animals. The early development of entomology was a process of transposing to insect studies the principles discovered in related fields. To gain a better appreciation of the growth of entomology in the New World, it is instructive to review the origin and evolution of its parent science, modern biology, which arose in the European theater of the Old World.

At the time of the discovery of America by the Spaniards, the progress of world science was barricaded by the "age of authority." In the literature of the times were heated arguments over such matters as the number of teeth in a horse; learned authors were quoted, but apparently no one thought to examine a horse and actually count its teeth.

The subsequent sixteenth and seventeeth centuries, which saw the exploration and early colonization of the Americas, witnessed also the overthrow of authority and the return of observation and experiment in science. Both exploration and scientific advances had their root in the same fundamental causes. In the centuries following the Renaissance, people developed again the desire to look and think for themselves.

In the field of the biological sciences, Vesalius' work on human anatomy (1543) rejuvenated observation, and Harvey's proof of arterial and venous circulation of blood (1628) introduced experiment. Together, as Locy says, "they stand at the beginning of biological science after the Renaissance." Introduction of the microscope in the seventeenth century led to the microanatomical works of Malpighi and Swammerdam, and to discoveries of microorganisms with which Leeuwenhoek astonished the scientific world.

During the seventeenth century entomology really started to develop. In fact, 1667 and 1668 may be considered almost its birth date, for in 1667 Redi used insects in demonstrations to test the theory of spontaneous creation. He exposed meat in jars, some covered by parchment, others by fine wire screen, and some not covered. The meat spoiled and attracted flies, which laid eggs in the exposed meat, resulting in a crop of maggots. Of the two covered jars, no eggs were laid on that covered by parchment, but the flies were attracted to the screen-covered one and laid eggs on the screen, since they could not reach the meat itself. Redi observed in this instance that, when the eggs hatched, maggots appeared on the screen instead of on the meat. He concluded, therefore, that maggots in meat resulted from the eggs of insects, and not from spontaneous generation as was previously supposed.

In 1668 Malpighi published anatomical studies of the silkworm, and Swammerdam published his first insect studies. These men produced the first accurate studies of insect anatomy, preparing skilled illustrations showing details of minute structures and organs (Fig. 1-1). These model works were the inspiration for later work in insect anatomy.

Another important phase of the biological sciences paralleled these advances. As people began to observe nature, interest awakened in natural history, and books on the subject appeared. Early treatises by Wotton (1552) and Gesner (1551–1556) were elemental and general, and were characterized by a lack of discrimination between different kinds of related animals. Imaginary animals of folklore were even given consideration as if they actually existed. Later works by Ray at the end of the seventeenth century were on a sounder basis and introduced a clear species concept of living organisms essentially similar to that understood today. Natural history museums (Fig 1-2) came into being, stimulated by the many bizarre and unfamiliar objects brought back to Europe by travelers and mariners. Many of these museums were operated as hobbies by

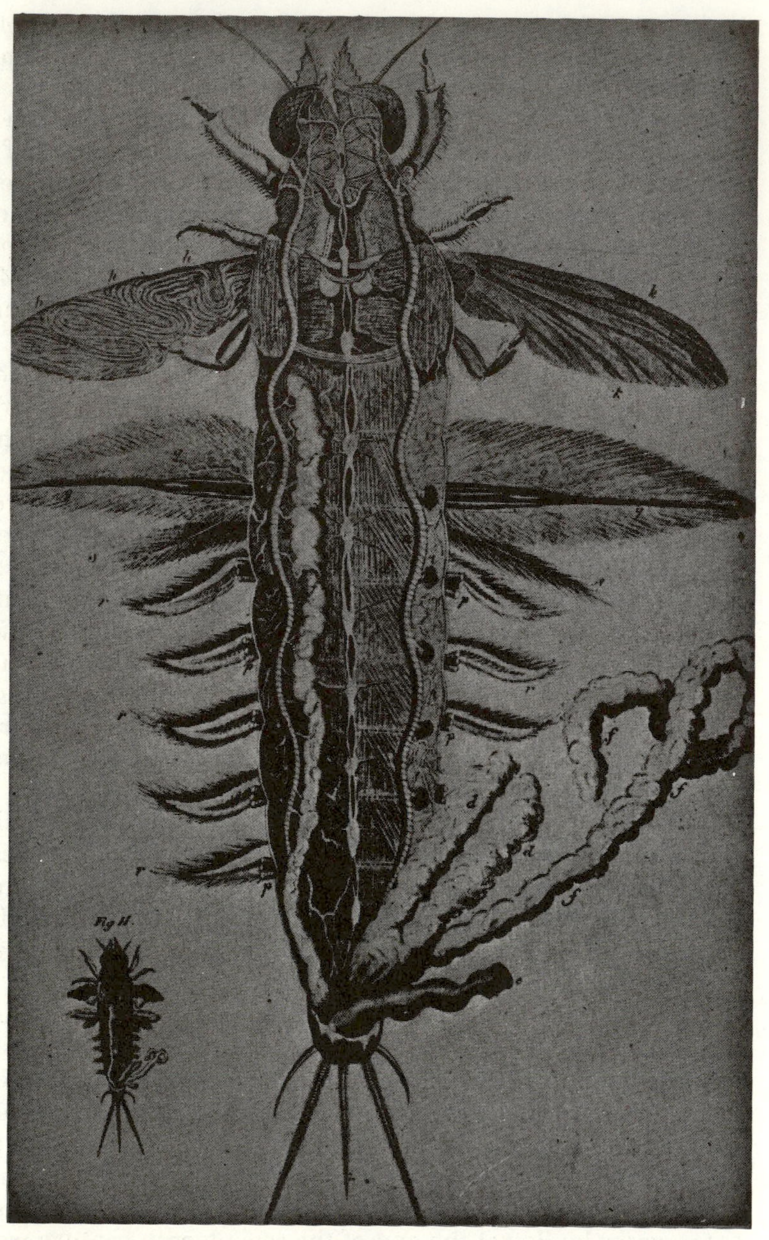

Fig. 1-1. Anatomy of a mayfly nymph, dissected and drawn by Swammerdam. One of the very early studies of insects, published about 1675. (From Essig, 1942, *College entomology*, by permission of Macmillan Co.)

Fig. 1-2. The famous museum of Olaus Worm, illustrated by its Swedish founder as it appeared in 1655. (Reproduction loaned by Waldo Shumway)

wealthy persons and were the forerunners of the extensive private collections which later played an important role in the development of taxonomy.

PROGRESS IN THE EIGHTEENTH CENTURY

Historians depict the eighteenth century as the climax of the revolt against authority, in which despotism and ecclesiasticism were to some extent replaced by individuals asserting their right to be ends in themselves. Historical expression of this tide of feeling is found in the American Revolution and the French Revolution. In the United States further expression is found in the great rise of both public and private education.

In this same century, undoubtedly manifesting the same individualistic trend, the biological sciences in Europe progressed to a new peak. Entomology especially attracted a large number of talented workers. Lyonet, a Hollander, contributed anatomical work of the finest detail, his first and best publication describing the anatomy of the larva of the willow moth (1750). More important in this period from the standpoint of arousing widespread public interest were the voluminous works of the German, Roesel; the Frenchman, Réaumur; and the Swede, DeGeer. All three of these authors published detailed well-illustrated observations on many insects, their life histories, habits, and characteristics.

About the middle of the eighteenth century occurred a movement of extreme importance to the entire field of natural science. It has already been mentioned that John Ray introduced the first clear concept of species. But the names for these species were phrases or descriptions (in Latin), often several lines long, cumbersome, and inconsistently used. In most cases the first name was a noun and corresponded to our present-day usage of a generic name; the remainder of the phrase was adjectival and modified this "generic" name. In John Ray's time, for example, one of the butterflies had the name *Papilio media, alis pronis, praefertim interioribus, maculis oblongis argenteis perbelle depictis*, and another had the name *Papilio parva nigra duplici in alis exterioribus macula alba insignis*. The first word in each name was the noun *Papilio*, the remainder was an adjectival description of the species. Students in field classes, attempting to keep up with recording lists of plants and animals being found, undoubtedly were the first to shorten these names. Gradually they were shortened to such names as *Papilio media* and *Papilio parva*, each name composed of the original noun and only one adjective originally referring to some distinctive feature of the species. This is known as binomial nomenclature.

In this period the naturalist Linnaeus (Fig. 1-3) was coming into prominence as a systematist, organizing the known plants and animals into one of the first comprehensive classifications. In 1758, the tenth edition of his work *Systema Naturae*, was published. In this, for the first time, the binomial method of names was employed uniformly throughout a large and comprehensive book. The method proved so successful that workers in all fields adopted it almost immediately. In fact, so profound was Linnaeus' influence on later workers that the tenth edition of his *Systema Naturae* has been designated as the official beginning point for zoological nomenclature. Although Latin is no longer the standard language of science, as it was until the end of the seventeenth century, the Latin names have been preserved and are used for scientific names throughout the world.

The stabilization of binomial nomenclature was of tremendous scientific advantage in two ways. First, it gave an easily designated and unambiguous "handle" to species, so that workers in different

Fig. 1-3. Carolus Linnaeus (1707–1778) at the age of 40 years. (After Schull)

fields and different countries were better able to identify the species with which others were working. This was of prime importance for integrating advances in comparative anatomy, physiology, and other fields of biology. Second, it provided a system of names that could be expanded indefinitely in a simple manner to accommodate additional genera and species. How necessary to future progress such a simple method was, may be seen at once by this tabulation: For the entire world Linnaeus recognized about 4500 species of animals, including 2000 species of insects; today over 1,250,000 species of animals are recognized, and of these roughly 900,000 are insects.

Thus taxonomists after Linnaeus were presented with an open invitation to describe and name the myriad species occurring in all parts of the world. Much of the early work was superficial and has been criticized by many, but it furnished the basis for analyses that led to the formation of the theory of evolution and to the organization of such fields as ecology and limnology.

The field of insect taxonomy in particular had been handicapped under the old system. After Linnaeus' work it began to emerge as a specialized subject. The first outstanding insect taxonomist was Fabricius, a Danish student of Linnaeus. Fabricius' first work, *Systema Entomologica*, appeared in 1775; others followed from 1782 to 1804. Fabricius treated the entire insect fauna of the world. By the end of his career it became apparent that this was too large a unit for intensive study by one person. As a result, many workers of the early nineteenth century following Fabricius studied either only one of the larger insect groups or the fauna of only one country.

The works of Réaumur, Linnaeus, DeGeer, and Fabricius stimulated a tremendous development of taxonomic study of insects among European entomologists. They also served as the most important basis for the beginning of entomology in North America.

DEVELOPMENT OF NORTH AMERICAN ENTOMOLOGY

American entomology came into existence about the beginning of the nineteenth century. For the first two thirds of the century, development was slow, witnessing the appearance of scattered pioneering works that form the backbone of further progress in science. But after the Civil War many factors contributed to a hastening of the growth of entomology in the United States. The resultant demand for entomological investigation found eager and able enthusiasts available, with the result that by the end of the century, American entomology had blossomed into a well-balanced science of wide practical and theoretical scope.

Pre-Nineteenth Century Work

Before the nineteenth century only small fragments were known about North American insects. Naturalist Mark Catesby (1679–1749) was possibly the first to illustrate North American insects in his book, *A Natural History of Carolina, Florida and the Bahama Islands, containing figures of the Birds, Beasts, Fishes, Insects and Plants*. Fabricius named some species, relying on specimens sent to him by various collectors or specimens that had been acquired by private collections in Europe. John Abbott, an Englishman who settled in Georgia, collected much material for European collectors in the period about 1780 and prepared many drawings of insects. The economic losses occasioned by insects were noticed with grave concern by Thomas Jefferson in 1782. He was particularly aware of the damage caused in stored grain and gave a few remarks on the problems of control, pointing out a need for further study. But until a few years before 1800 no concerted effort was evident by residents of the United States to investigate the native insect fauna.

Pioneering Period, Roughly 1800–1866

Work on American insects by American workers began about the turn of the nineteenth century. One of the first workers was W. D. Peck, who published many articles on the injurious insects of the New England states. These articles appeared from 1795 to 1819 in various agricultural journals. The pioneer work on North American entomology was *A Catalogue of Insects of Pennsylvania*, published in 1806 by F. V. Melsheimer. The chief value of this little 60-page book was its stimulating effect. Its author, his collection, and his association with later workers were a real aid in opening up the subject. His insect collection, incidentally, was the first comprehensive one to be built up in North America and was ultimately purchased many years later by the Harvard Museum of Comparative Zoology.

In 1812 a group of enthusiastic naturalists organized the Academy of Natural Sciences of Philadelphia. This nucleus of scientists was the cradle of serious descriptive work in many fields of American biology, and among them was entomology. Thomas Say (Fig. 1-4) was the outstanding entomologist of the group. He published the first useful classic in the field, three well-illustrated volumes (1817–1828), *American Entomology, or descriptions of the insects of North America*. The excellence of this work, together with his other papers on insects, has earned for Say the well-deserved title "Father of American Entomology." Say died at New Harmony, Indiana, in 1834.

In 1823 Dr. T. W. Harris (Fig 1-5) of Massachusetts published the first of a series of papers on the life history and economic importance of many insects. Harris was a student of Peck, who taught natural history at Harvard, from which Harris graduated in 1815. Harris collected and observed insects constantly, and the breadth of his published work increased. It culminated in 1841 in his monumental *Report on Insects Injurious to Vegetation*; this was twice reprinted and revised, the last time in 1852. Harris received $175 from the

Fig. 1-4. Thomas Say (1787–1834), the father of American entomology. (After Howard, 1930, courtesy of U.S.D.A., E.R.B.)

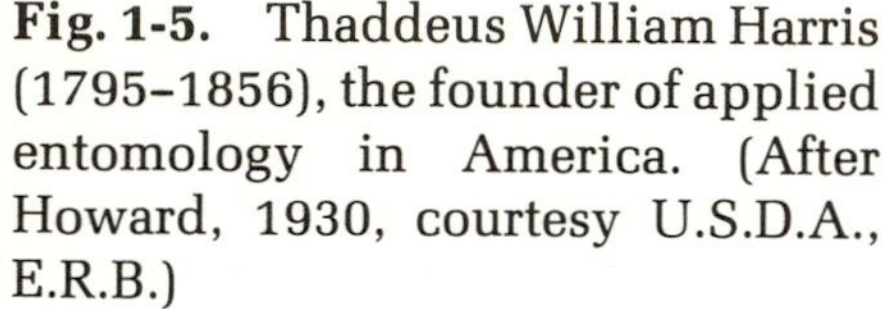
Fig. 1-5. Thaddeus William Harris (1795–1856), the founder of applied entomology in America. (After Howard, 1930, courtesy U.S.D.A., E.R.B.)

State of Massachusetts for this work; this was the first tax-supported entomological program in North America. It was also the first real textbook of economic entomology, and Harris is justly regarded as the founder of applied entomology in America.

The influence of Say and Harris took root immediately. In a few years a dozen authors published papers on the life history, habits, predations, and control measures of insects.

It is interesting to look back at the remedies in vogue for insect control during that period. It must be remembered that the arsenicals, pyrethrum, DDT, organophosphates, carbamates, and many other effective insecticides had not yet been discovered in the Western World, although both arsenicals and pyrethrum had been used for insect control for many centuries in parts of the Orient. A few of the standard remedies, to quote from Harris, included "Hand picking; sweeping into pans; spray with whitewash and glue; sulphur and Scotch snuff; fumigation with tobacco under a movable tent; syringe with whale-oil soap solution; soap and tobacco water," and many recommendations for cultural control. We can see in these the forerunners of many control measures recommended today.

Interest in agriculture led to the establishment in 1853 of a new Bureau of Agriculture in the federal government. This bureau appointed Townend Glover as Entomologist and Special Agent. Glover's duties were varied, including the preparation of exhibits of agricultural seeds, plants, and fruits, as well as insects. But in addition he did considerable investigational work, especially on the insects attacking orange trees and cotton. Glover had the belief that a picture of an insect is of much greater value than the prepared insect specimen. His greatest entomological efforts were consequently devoted to making copper etchings illustrating the insects of North America.

The farmers' losses caused by insect damage were attracting more and more attention. In response to this, the State of New York in 1854 appropriated $1000 for investigations on insects, especially those injurious to vegetation. Dr. Asa Fitch (Fig. 1-6) was chosen for this work, which he continued from 1854 to 1872. He wrote 14 fine reports, the result of a great amount of original observation performed with great care, which made available information on the life history of many insects. After Harris's work, these reports were the next great stimulus for further development of entomological work in the United States. Although such was not his title, Fitch was usually called State Entomologist; in his activities he was this in a very real sense and was the first one in the United States.

While Fitch was at the height of his career, several other entomologists were coming into prominence, including B. D. Walsh and C. V. Riley in Illinois, and E. T. Cresson and A. R. Grote in Philadelphia. Walsh's principal noneconomic work was done from 1860 to 1864, but his great contribution to economic entomology belongs to the account of the last third of the century. Riley, Cresson, and Grote also made their great contributions in this later period.

Fig. 1-6. Asa Fitch (1809–1878), the first of the state entomologists. (After Howard, 1930, courtesy U.S.D.A., E.R.B.)

The Civil War, which concluded the first two thirds of the century, seems to have had only slight effect on entomological work, most of which was being done north of the actual battle area. Its effects, however, had far-reaching entomological consequences in the years immediately following.

Science in Europe for this Period. While the foundation works of American entomology were being written by Say, Harris, Fitch, and Walsh, two very important series of events were taking place in Europe. These had only slight contemporaneous effect in America but were a great contributing factor to entomological development in the next period.

In the first place, European taxonomists were making needed strides in redefining taxonomic concepts, especially families and genera, to accommodate the huge tide of insect species being discovered in the world. Every large order received some attention, and creditably workable systems of classification were set up for them. European workers had access to libraries and collections far superior to any in America. They prepared keys and illustrated works, many of which were simply transposed by later American taxonomists to fit American species of insects.

The second circumstance is one that concerned all the biological sciences. In this period (1800–1866), there developed ideas that revolutionized the outlook in the entire field. Up to this point, work had been almost entirely descriptive, with scarcely any concept of fundamental laws. Now these came to light in rapid succession. Owen brought forward the idea of analogy and homology of parts; Cuvier and Lamarck founded comparative anatomy; Milne Edwards propounded the idea of division of physiological labor; Müller demonstrated the interrelationship of anatomy and physiology; Schwann and Schleiden demonstrated the cell theory; Bichat founded histology; Von Baer founded modern embryology; and Schultze defined protoplasm. To climax this galaxy of ideas, Darwin and Wallace established the practicality of the theory of organic evolution.

That these discoveries were made in such rapid succession is not strange. Scientists had been on the verge of seeing them for many years, and as soon as one fundamental was discovered it served as a key to unlock the next half-anticipated secret. Together with genetics and bacteriology (both discovered later), these discoveries gave a preliminary outline for the entire range of known biological laws. These, of course, were as fundamental to basic progress in the study of insects as in the study of any other group of living things.

First Entomology in American Colleges. In the early part of the pioneering period courses in natural history began to appear in various colleges in North America. Until the middle of the nineteenth century there were meager, mostly theoretical and classificatory. They were given chiefly by lecture, sometimes with demonstrations but with little or no field or laboratory work. This applied in large measure to chemistry and physics also. Louis Agassiz at Harvard was the first teacher in zoology to break away from this and introduce laboratory methods in teaching. The greatest impetus for the laboratory method of teaching, however, was the great upsurge of inquiry following the publication in 1859 of Darwin's *Origin of Species*. At about this same period such new institutions as Cornell and Johns Hopkins Universities emphasized the teaching of science. In the United States this was coincidental with the establishment of the land-grant colleges in 1862 by the Morrill Act of Congress. This promoted education in agriculture, mechanical arts, and natural sciences. Entomology was included only as a part of biology courses, but the foundations were being laid for the later development of its teaching.

AMERICAN EXPANSION PERIOD, ROUGHLY 1867-1900

In the two or three decades after the Civil War, American entomology expanded at a prodigious rate. Many events interacted to bring about this expansion, including the following, of great importance:

1. As a result of the westward migration of thousands of people following the Civil War, agriculture expanded in the Middle West and on the Pacific Coast. Devastating insect outbreaks occurred periodically. The farmers' demand for entomological assistance resulted in the rapid development of both state and federal organizations in economic entomology.
2. American insect collections and libraries had gradually improved and, with help obtainable from European literature, they opened the door for more extensive and better descriptive work. The fundamentals of biology, recently discovered, provided avenues for many lines of investigation with insects.
3. Demand for trained entomologists brought about teaching of entomology in colleges and universities.

Agricultural Entomology

Before the beginning of this period (1867-1900), New York State, through Asa Fitch, was the only state actively sponsoring entomology. In 1866, Illinois appointed a State Entomologist (although he did not become active until 1867), and in 1868 Missouri followed suit. In Illinois the appointment was given to B. D. Walsh, who had written many fine articles on taxonomic and economic entomology. Walsh (Fig. 1-7) died in an accident in 1869 but wrote three reports as State Entomologist before that tragedy. After him William LeBaron occupied the post for 5 years, followed by Cyrus Thomas from 1875 to 1882, and then by S. A. Forbes. In Missouri the appointment was given to C. V. Riley, who held the post from 1868 to 1876. Riley's annual reports were outstanding, in both scientific content and illustration, and were a tremendous stimulus to other authors.

From 1874 to 1876 the migratory locust invaded a number of the important grain-growing states. This outbreak was studied by Riley, who saw in it the need for action on a national scale against injurious insects. His efforts to secure national legislation persuaded Congress to establish the United States Entomological Commission. This was the first recognition in a broad way that applied entomology was of national importance and dealt with many problems whose thorough investigation transcended state lines. This commission had Riley as chief, and A. S. Packard, Jr., and Cyrus Thomas as the two other members. The commission was active officially for only 3 years, but did some excellent work and published several extremely useful reports and bulletins. These treated not only the migratory locust but also a wide variety of other economic insects.

For a short period in 1878 Riley (Fig. 1-8) succeeded Glover as Entomologist for the Federal Department of Agriculture. J. H. Comstock then held the office for 2 years, after which Riley again held it for 15 years. On Riley's return, the entomological work received such support that it was reorganized as a separate Division of En-

Fig. 1-7. Benjamin Dann Walsh (1808–1869), an early vigorous writer on various phases of entomology, later first State Entomologist of Illinois. (After Forbes)

Fig. 1-8. C. V. Riley (1843–1895), insect illustrator par excellence, who first built up the Federal Bureau of Entomology. (After Howard, 1930, courtesy U.S.D.A., E.R.B.)

tomology. Under Riley's leadership it rapidly developed into a large and useful organization with field stations in many parts of the country.

During these years several able entomologists in Canada began writing notable contributions on the insects of that country. The good work of two pioneers stand out conspicuously, that of the Rev. C. J. S. Bethune, who began publishing in 1867, and Dr. William Saunders, whose earliest paper appeared the following year. The efforts of these and other enthusiasts culminated in the founding of the Ontario Entomological Society in 1870 and the establishment of the office of Honorary Entomologist by the Department of Agriculture of Canada in 1884. This post was given to James Fletcher, who in 1887 was transferred to the staff on the Central Agricultural Experiment Station as Entomologist and Botanist. Fletcher had little help and a tremendous territory to cover; that he served the entomological needs of Canadian agriculture so well and so long is proof of his ability and industry. He died in 1908, following surgery.

Until 1887 organized work in economic entomology in the United States was being done only by the federal government and by New York, Illinois, and Missouri. Work was also being done by individuals in many other states, from Maine to California. But in 1887 the demands of an ever-expanding agriculture resulted in the Hatch Act, establishing agricultural experiment stations in all states. From their beginning, investigations of injurious insects were stressed, and the need for trained entomologists soon far exceeded the meager supply.

By the last decade of the century, applied entomology was an influential and producing concern. Outstanding work was being done by many workers in the Federal Division of Entomology, notably L. O. Howard, who in 1894 succeeded Riley as chief. The state organizations (many connected with the agricultural experiment stations) included several brilliant men in their roster. To mention only a few, S. A. Forbes in Illinois, John B. Smith in New Jersey, and E. P. Felt in New York contributed immense amounts of original research and wrote monumental reports, many of lasting value.

Two items of interest had a special effect on entomological thought and procedure in this last third of the century: (1) About 1869 Paris green was discovered to be an effective insecticide, and its success opened up the entire field of "stomach poisons" to be used against insects. (2) The cottony cushion scale, introduced about 1870 into California, had become such an abundant and serious pest of citrus trees in the 1880s that it threatened the extinction of the citrus-growing industry in the West. Known insecticides failed to deter the pest. Finally, natural insect enemies of the scale were imported from Australia. One of these, the vedalia ladybird beetle, destroyed the scale with such effectiveness that in a few years it ceased to be a problem. Such a wonder-working event established the importance of biological control as a possible means of combating injurious insects.

Insect Taxonomy, Morphology

During this latter third of the nineteenth century an almost complete foundation was laid for the classification of North American insects. A large number of workers contributed to this, including J. L. LeConte and G. H. Horn on Coleoptera; A. S. Packard, Henry Edwards, and A. R. Grote on Lepidoptera; E. T. Cresson, Edward Norton, and L. O. Howard on Hymenoptera; S. W. Williston, Osten Sacken, and D. W. Coquillet on Diptera; S. H. Scudder on Orthoptera; P. R. Uhler and O. Heidemann on Hemiptera; J. H. Comstock (Fig. 1-9) on Coccidae or scale insects; Herbert Osborn on ectoparasites and Homoptera. In Canada the Abbé L. Provancher was outstanding, especially for his work on Hymenoptera.

These are only a few of the "old masters" who described the first great bulk of the North American insect fauna and gave us our first working synopses. Many outstanding European entomologists also contributed to this literature. It is noteworthy that many of the most outstanding taxonomists of this and the succeeding era were amateur entomologists who made their great contributions as a hobby, without remuneration. To mention only a few: LeConte and Horn were practicing physicians; Edwards was an actor; Norton and

Fig. 1-9. John Henry Comstock (1849–1931), one of the "old masters" in the teaching of entomology in America. (After Howard, 1930, courtesy U.S.D.A., E.R.B.)

Cresson were businessmen; Williston a geologist; and Provancher a clergyman.

Following Linnaeus's establishment of a system of binomial classification, different points of view arose regarding many phases of the application of scientific names. As these differences became acute, inconsistent usage threatened to nullify the benefits of the binomial system, and taxonomists of all groups sought measures to bring about uniformity of practice. Success finally crowned their efforts at the International Zoological Congress held at Berlin in 1901, with the adoption by the zoological world of the International Rules of Zoological Nomenclature and the organization of the International Commission of Zoological Nomenclature. It was at this historic meeting that Linnaeus's tenth edition of *Systema Naturae* was designated as the beginning point for zoological scientific names.

Complementary to the development of better taxonomy was the origin and growth of large research collections. Until about 1865, North American insect collections were relatively small, usually consisting of, at most, a few thousand specimens. As the complexity of insect identification became obvious, the need for extensive collections, both to aid accurate identification of unknowns and to further progress in taxonomy became apparent. It was early recognized that accurate identification was essential for sound fundamental research in all fields and for consistent control recommendations. In the United States, the first serious effort in this direction was made by Louis Agassiz, who in 1867 appointed Hermann A. Hagen to build up a collection of insects in the Museum of Comparative Zoology at Harvard University. Since then many institutions, including various academies of sciences, universities, and other state and federal organizations, have amassed extensive insect collections. Many workers maintain personal collections of considerable size.

Teaching of Entomology

Until about 1867 entomology was taught in American colleges only as part of courses in biology or natural history. But in 1866 B. F. Mudge gave a course entitled "Insects Injurious to Vegetation" at Kansas State Agricultural College; in 1867, A. J. Cook gave a course in entomology at the Michigan Agricultural College; in 1870, Hagen gave informal courses in entomology at Harvard; in 1872, C. H. Fernald began teaching at Maine State College; in 1873, Comstock began teaching at Cornell University; and, in 1879, Herbert Osborn taught at Iowa State College of Agriculture. These men were the real founders of the teaching of entomology in the United States. They had little organized or general literature to use as a basis for teaching and, to quote from Osborn, "were feeling their way in the matter of both content and method for entomological instruction.

The task confronting these men was enormous: learning or discovering the multitude of details about insects, including life histories, morphology, development, and classification, and combining it with the then new concepts of general physiology, embryology, phylogeny, and evolution. The splendid and coherent courses that developed from the welding of all this material represent a triumph indeed for these pioneer teachers. Of special importance in this connection were the early textbooks written by A. S. Packard (Fig. 1-10), who was a trail blazer in this field.

Creation of the agricultural experiment stations in 1888 led to a tremendous demand for better-trained entomologists for economic positions and stimulated teaching in this field. By the end of the century, entomology courses had been organized in most of the leading universities and colleges stressing natural sciences. This was especially true of the land-grant colleges. Outstanding men, such as S. A. Forbes at Illinois, G. A. Dean at Kansas, and M. V. Slingerland at Cornell, set an early example of combining the fundamental and practical aspects in what may be called the first modern courses in applied entomology.

Fig. 1-10. A. S. Packard (1839–1905), who wrote some of the early entomology textbooks. (After Howard, 1930, courtesy U.S.D.A., E.R.B.)

TWENTIETH CENTURY DEVELOPMENTS

Outstanding new discoveries of the latter part of the nineteenth century and continuing into the twentieth century investigated the role of insects in relation to many major human and livestock diseases. In 1879, Patrick Manson in southern China discovered that mosquitoes are the carriers of the parasite that causes filariasis. About 1898, Ronald Ross in India proved the association of malaria and anopheline mosquitoes. In 1900, Walter Reed and co-workers proved that the mosquito *Aedes aegypti* carries yellow fever. This series of discoveries solved the transmission mystery of some of the world's worst diseases and established the important role that insects and other arthropods play in relation to human health. This was the birth of medical entomology. Continued investigation has shown that insects, mites, and ticks transmit an ever-increasing number of diseases, including bubonic plague, dengue, typhus fever, trench fever, Rocky Mountain spotted fever, and African sleeping sickness. The medical world has in many cases turned to a control of the arthropod disease carrier as a means of combating the disease because of no known immunization methods.

Further emphasis on the role of insects arose when many introduced insects of foreign origin became established in the United States and Canada and produced catastrophic damage to agriculture. Since 1889, the gypsy moth has threatened to wipe out fruit trees and other trees in the New England states; starting before 1900, the

San Jose scale became a countrywide fruit tree scourge; between about 1895 and 1920, the destructive cotton boll weevil invaded the entire Cotton Belt; and in the early 1930s, the European corn borer loomed as a serious threat to the Midwestern corn crop. In the 1930s, the Mediterranean fruit fly threatened the entire citrus industry of Florida. New accidental introductions still occur, such as the destructive alfalfa aphid, cereal leaf beetle, and khapra beetle.

One of the most spectacular episodes in insect control began in 1945 and 1946 with the commercial introduction of the new synthetic insecticide DDT. Control of insect pests by insecticides previously had been a relatively expensive undertaking, because the necessary dosages of insecticides were high and the control only moderately satisfactory. By contrast, DDT, a chlorinated hydrocarbon, was so effective in extremely small dosages and was so persistent that the cost of control was reduced drastically. Other newly synthesized chlorinated hydrocarbons, such as benzene hexachloride (BHC), aldrin, and dieldrin, gave equally good results in a wide variety of situations. These compounds appeared to open vast new possibilities for the widespread use of insecticides on crops with a relatively low yield per acre or low production value per year, such as some field crops, range animals, grassland, and forests. The medical world envisioned a great reduction or even the elimination of scourges, such as malaria and bubonic plague, by the eradication of their insect vectors. It seemed, in short, that these "magic insecticides" would finally give us the means to eliminate insects as competitors for food and fiber and as health hazards.

These optimistic views were challenged, however, in only a few years. Strains of insect pests soon began to appear that were increasingly resistant to DDT and similar insecticides. The immediate result was application of greater and greater concentrations of the insecticides to obtain the desired level of pest control. At about the same time, it was discovered that the insecticides affected a broad spectrum of organisms, killing not only their target organism but also a wide array of the target organism's natural predators and parasites. The net result was that the DDT-resistant pest now had fewer natural enemies and could reproduce in greater than usual numbers. Also, many species that had previously been at low population levels and considered nonpest insects and mites rapidly became numerous enough to become important new pests, resulting in the necessity for more frequent applications of even greater concentrations of insecticides.

Chlorinated-hydrocarbon insecticides also have proved extremely persistent in the natural environment, with half-lives ranging to several years. As a result, they have accumulated and quickly entered food chains where they were rapidly concentrated in the tissues of terrestrial and aquatic vertebrates. This included humans, to whom high concentrations of these insecticides are also toxic. Thus by the late 1960s, the valuable commercial sale of fishes from Lake Michigan was virtually prohibited because of the accumulation of chlorinated hydrocarbons in fish tissue.

Birds are extremely susceptible to the toxic effects of chlorinated hydrocarbons. Many birds eat nearly their own weight in food each day, and the principal food of a large number of species of birds is insects. Insects that had a high level of insecticide in or on them were easy catches for birds. Thus in the later part of the 1950s the populations of many birds, particularly insectivorous species, began to decline dramatically. Public awareness was focused on the seriousness of these problems in 1962 by Rachel Carson in her book, *Silent Spring*. There soon followed a number of state and federal laws that controlled, limited, and even prohibited the application of DDT and other chlorinated hydrocarbons.

The dilemma was real. Many insects were, and still are, extremely important and very destructive pests that need to be controlled. On the other hand, the early "magic insecticides" of chlorinated hydrocarbons were far from being a panacea, and their indiscriminate use within a few years resulted in even more serious types of problems. New approaches to insect pest problems were needed. These have yielded significant results and far better understanding of how insects and other pests can be controlled without also destroying their natural or biological controls at the same time. They include a wide range of approaches, such as:

1. Using detailed knowledge of the life cycles of pests and their predators in planning for a low but acceptable level of damage, so that the pest's predators are able to maintain their populations at effective levels;
2. Locating and developing bacterial, viral, and other insect infections;
3. Using sexual attractants to lure and capture pest insects;
4. Using hormones that cause reproductive or life-cycle failures.

Also, new short-lived insecticides are used, but as infrequently as possible and only at those times during the life cycles of pests when they are vulnerable and when it is relatively safe for their natural predators. Some insect pests can be effectively controlled by simply varying the planting time of a crop. Furthermore, plant geneticists have produced a number of crop strains that have a high resistance to pests. All these types of approaches and others must be carefully considered and variously combined to form integrated pest-control programs. The very appropriate term *integrated pest management* is applied to the effective combination of any and all control possibilities available to particular pest-control problems.

In the twentieth century, entomological research activity has been stimulated by the challenges of marked fluctuations of insect-control demands and by the search by curious individual investigators to advance entomological knowledge. In addition to improvements of earlier equipment, such as the light microscope, this research activity has progressed immeasurably by new advances with transmission and scanning electron microscopes, by microtechniques for chemical analysis, by sophisticated electronic measuring devices, and by computer applications to biological and chemical data. Advances in systematics and related fields have been immeasurably aided by the construction of modern roads and the availability of air travel, permitting collecting and sampling over a tremendous area of the earth. As a result, insects are providing increasing information of general biological interest, especially in biochemistry, physiology, behavior, ecology, biogeography, and evolution.

Sources of Entomological Information. As the science of entomology has grown, so has its literature. Information about insects is available in a large number of different types of publications. New information commonly appears in periodicals, such as journals, bulletins, and symposia that are published by regional, national, and international entomological groups. For example, the Entomological Society of America publishes a number of journals including a *Bulletin* and an *Annals* series, as well as the *Journal of Economic Entomology* and many books that examine special topics about insects. The Royal Entomological Society of London also publishes a diverse set of entomological publications. There are similar entomological societies in many countries of the world. Some fields within entomology are so active that they are able to support their own specialized journals, for example, *Journal of Insect Physiology*, *Journal of Applied Entomology*, *Environmental Entomology*, and *Mosquito News*.

A considerable amount of entomological literature is published by federal and state governments. In North America, both the U.S. Department of Agriculture and Agriculture Canada publish a broad range of information about insects that is of use to all aspects of agriculture. Translations of studies from foreign languages are also published by these organizations. State governments also publish extensive accounts on entomological research and investigations.

Periodicals, such as *American Midland Naturalist*, *Canadian Journal of Zoology*, *Hilgardia*, *Journal of Animal Ecology*, *Journal of Experimental Zoology*, *Nature*, *Oecologia*, *Parasitology*, *Science*, *Scientific American*, and many others, commonly have important articles on entomology. In addition, several publications review major fields within entomology on a frequent basis. These include *Review of Applied Entomology*, *Annual Review of Entomology*, and *Quarterly Review of Biology*.

Various museums, universities, and other organizations frequently publish monographic studies on particular insect topics. Bibliographical information may be found in *Biological Abstracts*, *Abstracts of Entomology*, *Science Citation Index*, and *Zoological Record*.

REFERENCES

Cloudsley-Thompson, J. L., 1976. *Insects and history*. New York: St. Martin's Press. 242 pp.

Decker, G. C., 1964. The past is prologue. *Ent. Soc. Amer. Bull.*, **10**(1):8–15.

Dupuis, C., 1974. Pierre Andre Latreille (1762–1833): The foremost entomologist of his time. *Annu. Rev. Entomol.*, **19**:1–14.

Essig, E. O., 1931. *A history of entomology*. New York: Macmillan. 1029 pp.
1942. *College entomology*. New York: Macmillan. 900 pp., illus.

Gilbert, P., 1977. A compendium of the biographic literature of deceased entomologists. *Brit. Mus. Nat. Hist. Publ.* XIV, No. 786, 455 pp.

Gill, T., 1908. Systematic zoology: Its progress and purpose. *Smithsonian Inst. Rep. for 1907*. pp. 449–472.

Howard, L. O., 1930. A history of applied entomology (somewhat anecdotal). *Smithsonian Misc. Collections*, **84**:1–564 pp., illus.

Leftwich, A. W., 1976. *A dictionary of entomology*. London: Constable, 360 pp.

Locy, W. A., 1910. *Biology and its makers*. New York: Henry Holt & Co. 477 pp.

Mallis, A., 1971. *American entomologists*. New Brunswick, N.J.: Rutgers Univ. Press. 549 pp.

Mickel, C. E., 1973. John Ray: Indefatigable student of nature. *Annu. Rev. Entomol.*, **18**:1–16.

Osborn, H., 1937. *Fragments of entomological history*. Columbus, Ohio, by the author.

Russell, L. M., 1978. Leland Ossian Howard: A historical review. *Annu. Rev. Entomol.*, **23**:1–15.

Smith, E. H., 1976. The Comstocks and Cornell: In the people's service. *Annu. Rev. Entomol.*, **21**:1–26.

Smith, R. F., T. E. Mittler, and C. N. Smith (Eds.), 1973. *History of entomology*. Palto Alto: Annual Reviews, Inc. 517 pp.

Spencer, G. J., 1964. A century of entomology in Canada. *Can. Entomol.*, **96**:33–59.

Weiss, H. B., 1936. *The pioneer century of American entomology*. New Brunswick, N.J., by the author.

2
Arthropoda: Insects and Their Allies

Insects belong in the great phylum of jointed-legged animals, the Arthropoda. In addition to insects, this phylum contains a tremendous variety of forms, such as crabs, sowbugs, spiders, mites, trilobites, centipedes, and many others. Basically these are segmented animals having segmentally arranged appendages that are also segmented; in different groups, however, a large number of divergencies from this basic plan occur, especially in the immature stages.

ANCESTRAL RELATIONS

The Arthropoda undoubtedly arose from a wormlike creature similar in general appearance to the Annelida or segmented worms. The body of this ancestor (Fig. 2-1*A*) consisted of a series of uniform segments, each a full ring of the body. The head was a simple structure, probably bearing sensory bristles. The mouth was situated on the ventral side between the head and the first ring or segment of the body. Because of its position in front of the mouth or stomodeal opening, the head region in this early stage is termed the prostomium.

As has been emphasized by Manton (1973), this wormlike ancestor differed from present-day Annelida in many important characteristics of fundamental body structure. In the Annelida each segment of the body has a separate body cavity or coelom, those of successive segments being separated by a wall or septum; each segment has its excretory system (paired nephridia), opening to the outside through individual ducts; and the segmentally arranged

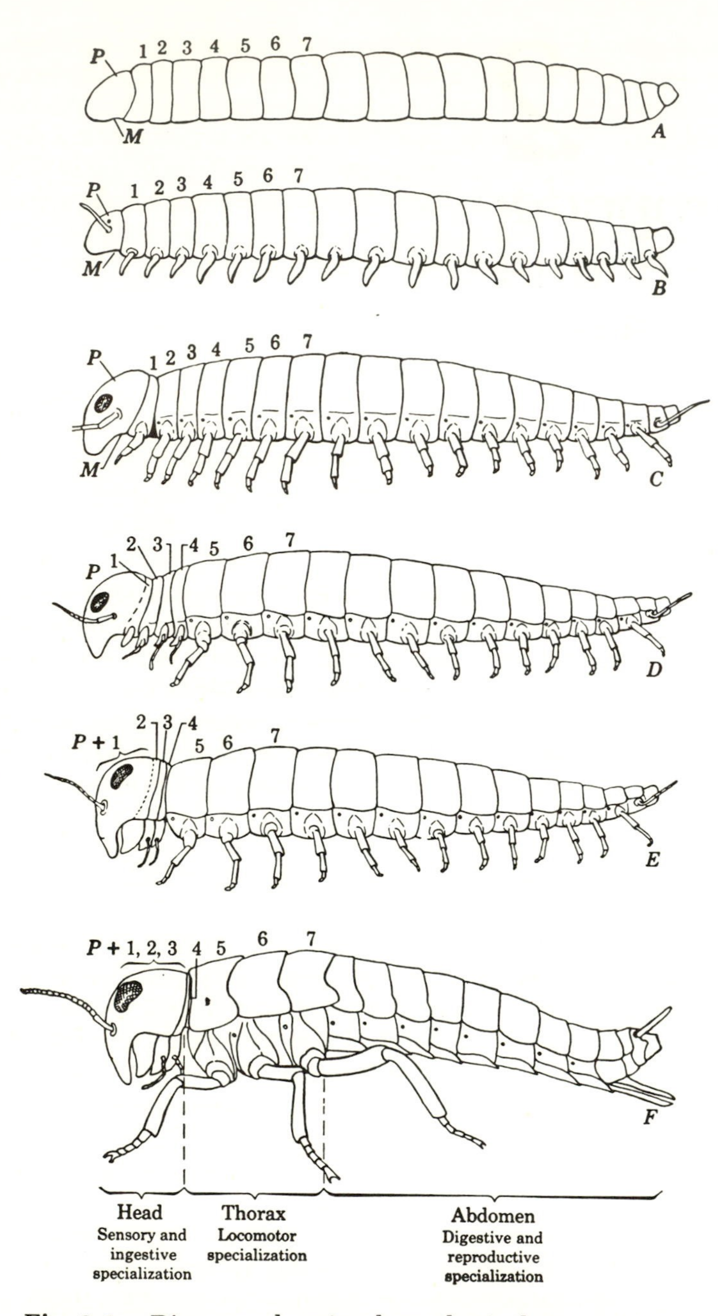

Fig. 2-1. Diagram showing hypothetical stages (*A* to *F*) in the development of body regions and appendages from a wormlike ancestor to an insect. *M*, Mouth; *P*, prostomium. (Modified from Snodgrass)

legs, if present, are of a primitive type. In postulated members of the line leading to the Arthropoda, there are no intersegmental divisions, hence the body cavity extends the full length of the body; the excretory system is composed of a single organ serving the entire animal; and legs, if present, are more specialized in many features of musculature than those of the Annelida. Hypothetical steps in the evolutionary progress beginning with this simple pre-arthropod stage and leading through generalized arthropods to insects are pictured in Fig. 2-1*A* to 2-1*E*.

The first great step was the development of a pair of ventral appendages or legs on each body segment, aiding in locomotion (Fig. 2-1*B*). Apparently the last segment, the periproct, bearing the anus, never had appendages.

Probably next after this, the appendages of the first body segment fused and formed the *labrum*, a ventral flap directing food into the mouth opening. The appendages of the second segment moved dorsally and became the *antennae*. The appendages of the third segment are visible in the early embryos of insects but apparently never evolved into persisting structures in stages of later development. One by one these first three segments, called the *labral*, *antennal*, and *intercalary segments*, respectively, fused with the ancestral *prostomium*, forming a head structure of compound origin from six segments, and their segmental ganglia fused with the primeval brain or *archicerebrum*, resulting in a composite arthropod brain.

These changes in head and brain structure were accompanied by another evolutionary event important in feeding. The mouth became situated first posterior to the labral segment, then behind the antennal segment, and finally behind the intercalary segment. Undoubtedly, by this time in pre-arthropod evolution, the appendages of the next few segments, although only slightly modified in shape, functioned in pushing food toward the mouth. This backward displacement of the mouth, together with the posterior extension of the labrum, would have provided a better pocket in which food could be concentrated before being swallowed. Because its definition as prostomium no longer applied after these changes in mouth position, the primeval head subsequent to these events has been termed the *acron*.

The displacement of the mouth opening from a position between the acron and segment 1 to a position between segments 3 and 4 requires some explanation. In the early stages of embryonic development, the ventral body structures were formed before the segment became full "rings" or complete segments. What eventually became paired nerve centers or ganglia first developed in each segment as paired but unconnected lateral masses of nerve tissue called *neuromeres* (nerve bodies). Early in embryonic development, these separate neuromeres may have moved anteriorly on each side of the mouth opening to fuse with the primeval brain.

Presumably about this period in the evolution of the arthropods, the appendages of segment 4 evolved into food-gathering and shredding structures, the *mandibles*. The neuromeres, however, did not migrate anteriorly, and the mouth opening remained between segments 3 and 4.

The three phyla, Onychophora, Pentastomida, and Tardigrada, may represent this stage in evolutionary development, especially

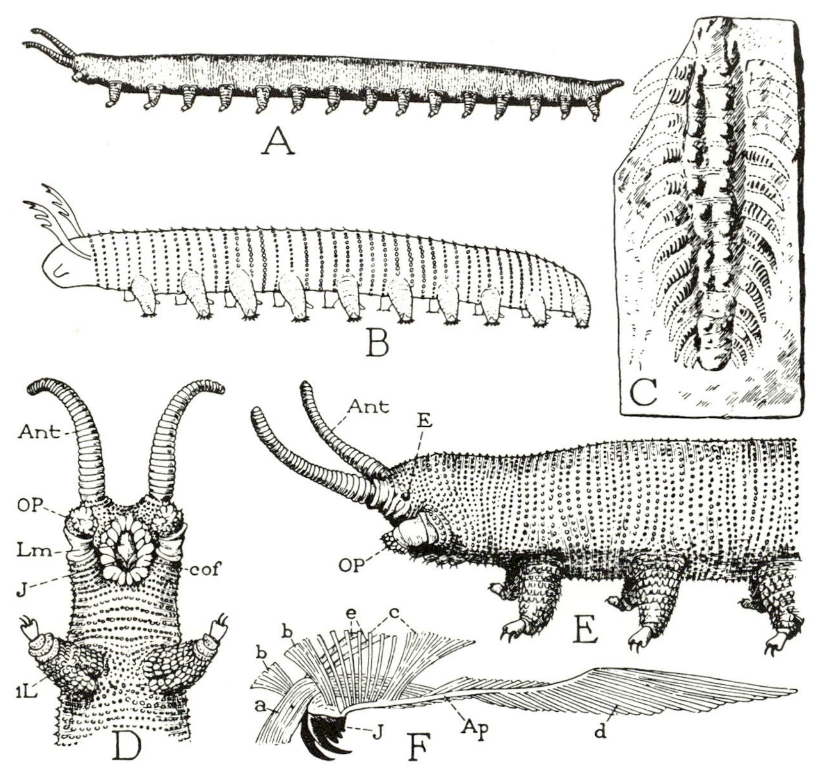

Fig. 2-2. Onychophora, ancient and modern. (*A*) *Peripatoides novae-zealandiae*; (*B*) restoration of the Middle Cambrian fossil *Aysheaia pedunculata*; (*C*) a supposed onychophoran *Xenusion auerswaldae* from Precambrian or Early Cambrian quartzite; (*D, E*) ventral and lateral views of anterior portion of *Peripatoides novae-zealandiae*; (*F*) right jaw of same, showing muscles. *a* to *d*, jaw muscles; *Ant*, antenna; *Ap*, apodeme; *cof*, circumoral fold; *E*, eye; *J*, jaw; *L*, leg; *Lm*, labrum; *Op*, oral papilla. (After Snodgrass, 1938)

living members of the phylum Onychophora (*Peripatus* and its allies, Fig. 2-2). Onychophorans are soft bodied but have what appear to be segmentally arranged pairs of appendages, although the body segmentation is not obvious externally. Furthermore, the legs (called *lobopods*) have specialized muscles and muscle-attachment bases such that segmental legs could readily have evolved from them (Manton, 1973). They also possess a pair of structures that resemble mandibles and are in about the right position for them, but whether or not they are homologues of arthropod mandibles is not certain. Even if they are, onychophorans are likely a separate and distinct branch of this evolutionary line.

Members of the two other phyla, the Tardigrada (Fig. 2-3) and the Pentastomida (Fig. 2-4), are so reduced in structure that their true relationships are open to question. The Tardigrada, or water bears (Fig. 2-3) are minute animals that live in wet moss and in both fresh and salt water. The body, never more than a millimeter long, has four pairs of lobiform legs terminating in claws; the head has neither apparent mouthparts nor other appendages. The Pentastomida, or linguatulids (Fig. 2-4) are a small group in which the adult is wormlike and the earliest stages are minute four-legged creatures in general appearance resembling mites. The linguatulids are internal parasites of a variety of vertebrates.

After these events, a new series of specializations occurred in the lineage leading to the arthropods.

1. The legs became segmented, resulting in improved locomotion.
2. Parts of the body wall became hardened and many internal flanges (called *apodemes* and *apophyses*) evolved; these flanges are attached to the ventral hardened parts and provide better attachments for the muscles causing movements of body parts and appendages (see also Fig. 5-1*C*, *D*).
3. The appendages of several postoral segments became spe-

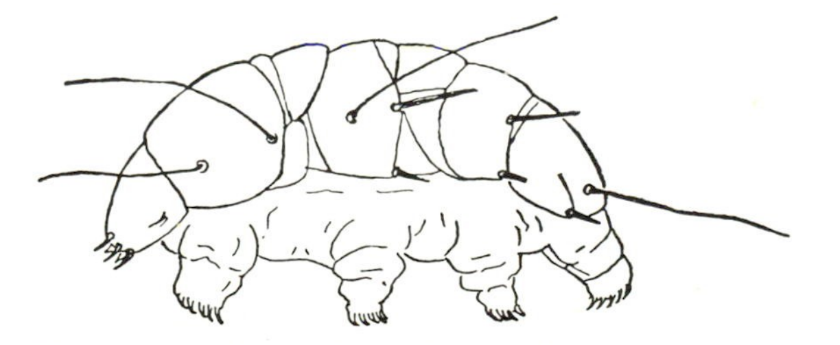

Fig. 2-3. A tardigrade. (After U.S.D.A., E.R.B.)

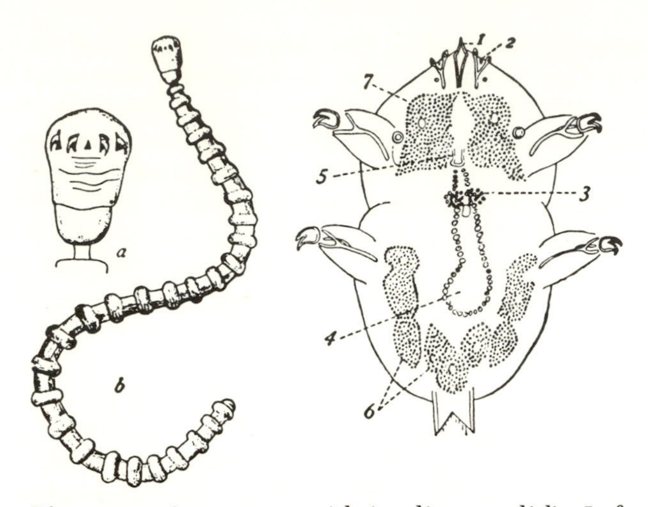

Fig. 2-4. A pentastomid (or linguatulid). Left, *Porocephalus annulatus*; (*a*) ventral view of head; (*b*) ventral view of entire animal. Right, larva of *Porocephalus proboscideus*, ventral view: (*1*) boring anterior end; (*2*) first pair of sclerotized processes seen between the forks of the second pair; (*3*) ventral nerve ganglion; (*4*) alimentary canal; (*5*) mouth; (*6* and *7*) gland cells. (After Stiles and Shipley)

cialized in function (although at first not in structure) for moving food forward to the vicinity of the mouth opening.

When these developments finally occurred, the result was a true member of the phylum Arthropoda.

The earliest known members of the arthropods, the jointed-legged animals, are the trilobites, constituting the marine subphylum Trilobita (Figs. 2-5, 2-6). They occurred in the Cambrian Period some 600 million years ago but have been extinct for about 200 million years. The body has a small anterior head with antennae and labrum, and posterior to the head a number of short legs that probably aided in feeding. These legs and their bases, however, were not fused with the head structure.

The trilobites evolved into a wide array of morphologically diverse groups. It is probable that two of these gave rise to the other two large groups of arthropods, the subphyla Chelicerata and Mandibulata.

In the Chelicerata, the antennae were lost and the body became

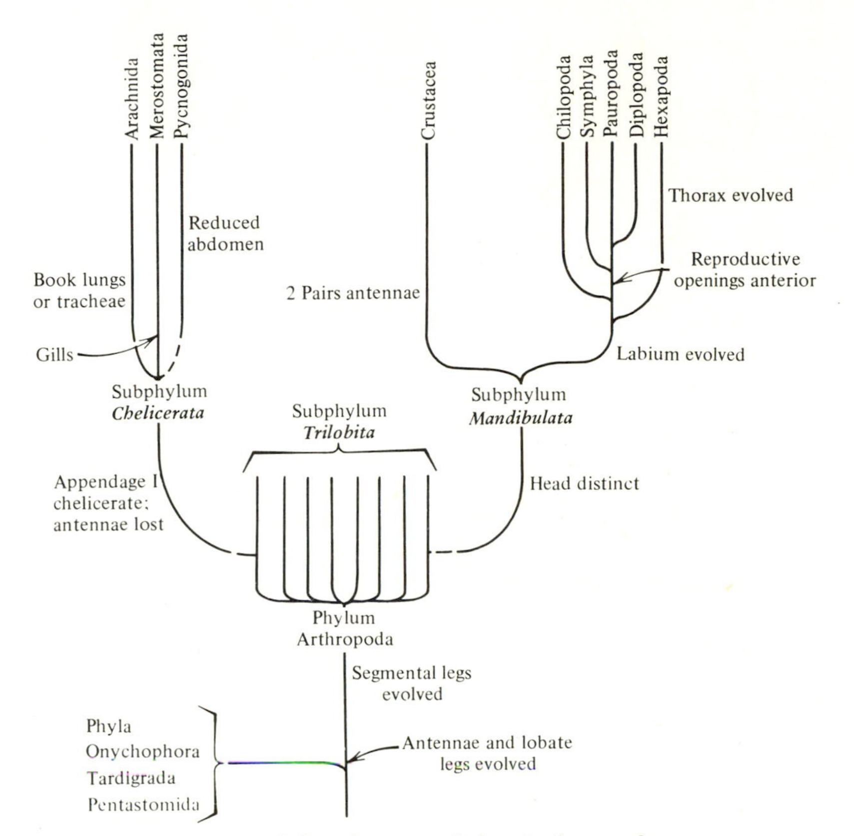

Fig. 2-5. Suggested family tree of the Arthropoda.

organized into two regions, the *cephalothorax* (specialized for feeding and locomotion) and *abdomen* (specialized for reproduction). This subphylum contains the scorpions, spiders, mites, and their allies.

In the Mandibulata, the stem leading to the hexapods, the first three postoral segments and their appendages fused with the head, which thus became a composite structure consisting of the primeval acron and six body segments. The appendages of the last three segments added to the head evolved into a variety of feeding structures that aid in gathering food and getting it into the mouth. The three segments bearing these mouthparts are called the *gnathal segments*. This fusion brought together in one functional unit all the external structures ultimately connected with feeding, a pattern that holds throughout the Mandibulata with few exceptions.

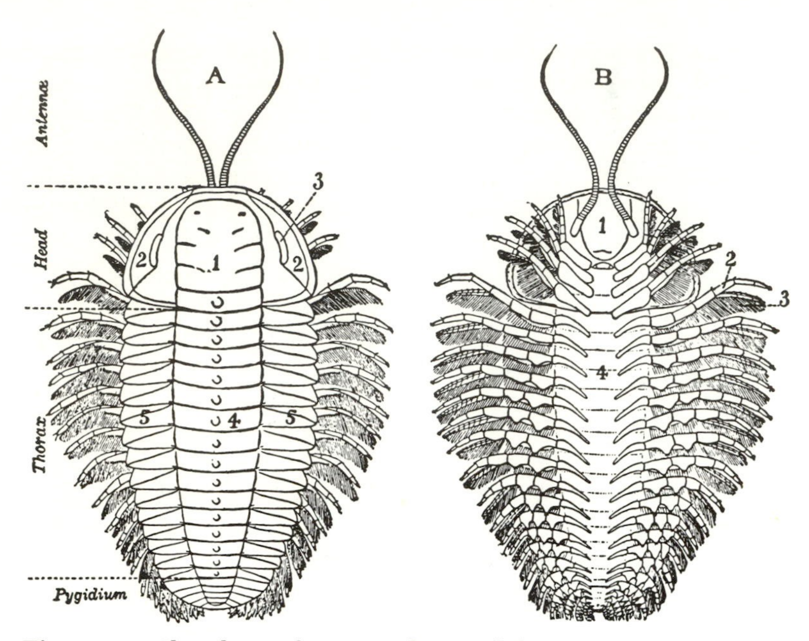

Fig. 2-6. Sketches of a complete trilobite *Triarthrus becki.* (*A*) Dorsal or upper side of carapace, showing three lobes, pleura (*5*), rachis or axis (*4*), glabella (*1*), and free cheeks (*2*), which bear the eyes (*3*). (*B*) Ventral or underside, showing biramous limbs (*2*, *3*) attached to rachis, and upper lip or hypostoma (*1*), which covers mouth. The biramous legs had a dual purpose: The upper feathered branch served for breathing gills and swimming paddles; the lower bare branch served for crawling. The short anterior appendages probably aided in feeding. (From Schuchert, after Beecher)

The ancestral mandibulate gave rise to two lines that are each represented today by a large diverse group, the Crustacea (the crabs, shrimps, and their allies), and the Myriapoda–Insecta group (the centipedes and their allies, and the insects and theirs). The Crustacea are predominantly aquatic and it is probable that the ancestral mandibulate was also. The ancestor of the Myriapoda–Insecta line, however, was probably terrestrial. It probably lived in the leaf litter along the edge of small shady ponds and streams, much like some existing isopods (Crustacea). Some zoologists consider that the Crustacea are more distantly related to the other mandibulates and that they show convergent evolution of feeding structures.

In this myriapod–insect line the third pair of gnathal appendages, called the *second maxillae* in the Crustacea, fused and formed a continuous transverse flap called the *labium*. This fusion completed the formation of a new and effective functional structure, namely, a preoral mouth cavity bounded front and back by the labrum and labium and at the sides by the mandibles and maxillae, the whole allowing great flexibility in food handling. Størmer (1977) considered this development a decisive adaptation for terrestrial life.

The terrestrial myriapod–insect ancestor gave rise to three lines that have persisted to the present, one represented by the class Chilopoda (centipedes); another by the three classes Symphyla, Pauropoda, and Diplopoda (millipeds); and the other by the class Insecta. In the first four classes, functional legs persisted on almost all segments of the trunk region.

In the insect branch a further body specialization evolved. The first three pairs of locomotor appendages became larger; the remainder became reduced and finally disappeared or became modified into nonlocomotor structures (Fig. 2-1*F*). This centralized the locomotor function in the first three segments, behind the head, which then formed a well-marked body region, or *thorax*. The posterior portion of the body containing most of the internal organs is called the *abdomen*. The posterior appendages of the abdomen became modified as organs for mating or oviposition. Some of the Crustacea have a distinct thorax and abdomen, but in these the thorax is usually composed of about eight segments.

Summarizing these developments from the primitive legless arthropod ancestor, it seems reasonable to suppose that (a) similar generalized appendages were developed on all postoral segments; and (b) these were continuously modified and became segregated into groups for specialized functions. In the insects this has resulted in the present distinctive body form composed of three regions: head, with sensory appendages and mouthparts; thorax, bearing three pairs of legs; and abdomen, containing most of the vital organs and having terminal appendages adapted for reproductive functions.

The relationships of the arthropods suggested here are outlined in Fig. 2-5. In this interpretation all hexapods arose from one ancestral arthropod lineage and they are, therefore, monophyletic in origin. As explained by Kristensen (1975), the scheme suggested here requires the least number of evolutionary parallels to explain the origin of insects. Other authors have presented different views and have proposed that not all insects arose from the same arthropod ancestors, so they consider insects to be polyphyletic in origin (Manton, 1949, 1964, 1977). Both monophyletic and polyphyletic interpretations are examined by authors in Gupta (1979).

SYNOPSIS OF THE PHYLUM ARTHROPODA

SUBPHYLUM TRILOBITA

The body was divided into head, thorax, and pygidium, the whole usually flattened and divided by two longitudinal furrows into three lobes (Fig. 2-6). The head was a loosely organized region consisting of the primeval head (bearing a pair of long segmented antennae) and four body segments each bearing a pair of biramous appendages. Over this structure was a shell-like carapace. Many species had a pair of well-developed eyes. Each segment of the remainder of the body bore a pair of biramous appendages, except for the last segment, the telson. The various groups of trilobites exhibited a great variety of shapes and sizes. Most of them were 4 to 6 centimeters long, but some were quite tiny (10 mm) and some were giants, attaining a length of over 60 centimeters.

Trilobites were an abundant marine group in the early Paleozoic era but became extinct at the close of that portion of geological time. In general actions most of them were probably similar to present-day isopods, swimming a little, running over the bottom, and feeding as scavengers. It is thought that some were carnivorous, others were pelagic and lived on plankton, and still others burrowed in the bottom and ingested mud and ooze.

SUBPHYLUM CHELICERATA

The body of a typical chelicerate is divided into two regions, the cephalothorax and the abdomen. The cephalothorax of the adult has no antennae. It usually bears six pairs of appendages: the anterior chelicerae (closely associated with the mouth); the pedipalps, which are often chelicerate; and four pairs of walking legs. The abdomen bears no jointed appendages; its segmentation may or may not be visible externally.

Members of the subphylum are of great interest to entomologists. Many, such as mites and ticks, transmit disease organisms in much the same manner as do insect vectors of disease. A great many kinds of mites feed on crops, livestock, and humans, causing heavy economic losses. The chelicerates usually occur in company with economic insects and are controlled in much the same manner and by use of the same techniques. As a result the mites and spiders have gradually become a part of the field of entomology.

SYNOPSIS OF CLASSES

1. Abdomen bearing pairs of external gills situated under platelike coverings (Fig. 2-7) .... **Merostomata**
   Abdomen without external gills 2
2. Abdomen consisting of a minute finger-like structure situated between bases of hind legs (Fig. 2-11) .... **Pycnogonida**
   Abdomen large, usually larger than cephalothorax (Fig. 2-8) or merged with it (Fig. 2-9) .... **Arachnida**

Class MEROSTOMATA (Gigantostraca)

The abdomen bears ventral pairs of appendages forming gills and platelike coverings and ends in a spikelike tailpiece or telson. The gills are used both for respiration and as swimming paddles. The extinct subclass Eurypterida lacked a carapace and looked somewhat like a scorpion. In the living marine subclass Xiphosura, the cephalothorax has a large horseshoe-shaped carapace; the horseshoe crab, *Limulus* (Fig. 2-7) is the only North American representative.

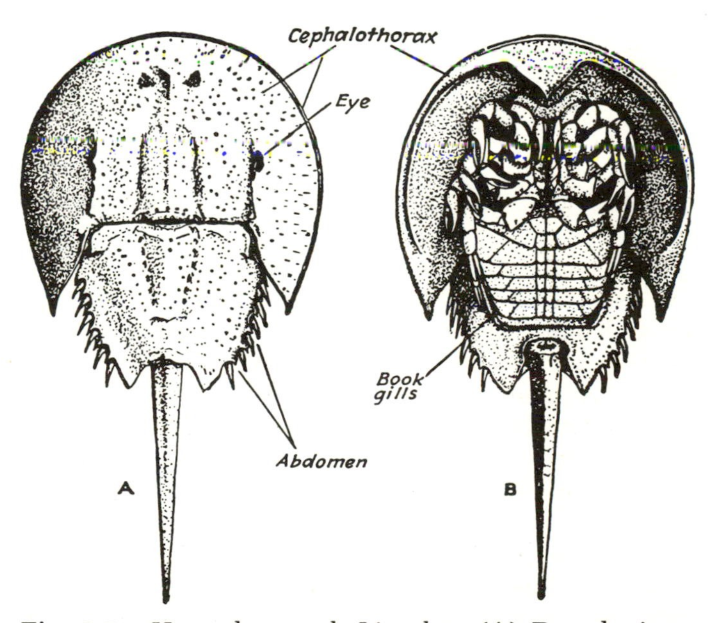

Fig. 2-7. Horseshoe crab *Limulus*. (*A*) Dorsal view; (*B*) ventral view. (From Wolcott, *Animal biology*, by permission of McGraw-Hill Book Co.)

The abdomen is large but has no external gills or locomotor organs. From front to back the cephalothorax bears a pair of chelicerae that are normally chelate, a pair of pedipalps that may be chelate, and four pairs of legs. In the ticks and mites the pedipalps and chelicerae are highly modified and closely associated with the mouth opening, this entire region having somewhat the appearance of a head.

This class is large and varied. Most of its groups are terrestrial animals but a few of the mites are aquatic.

SCORPIONIDA. This order is a small group containing only the true scorpions, which range from about 13 to 18 mm in length. In these arachnids, the end of the tail bears a poisonous sting (Fig. 2-8*A*), which the scorpion can use with great agility. The venoms of various species differ in the severity of their effects. In North America the group occurs chiefly in the Southwest, the range of one or two species extending as far northeastward as Illinois.

UROPYGIDA. A chiefly tropical or subtropical order, its species range in length from 2 to 65 mm. The large whiptail scorpion *Mastigoproctus giganteus* occurs in the southern United States from coast to coast. A few genera of small species also occur there; these have only short tails and are often placed in the separate order Schizomida.

PALPIGRADA. Another small order of minute species, the palpigrads are about 2.5 mm long. They live under stones and in North America are restricted to the Southwest.

AMBLYPYGIDA. This is a small tropical and subtropical group formerly included in the Uropygida. Its species range in length from 4 to 45 mm. The somewhat flattened body resembles that of spiders. Like the uropygids, they are nocturnal. They prefer a humid environment.

PSEUDOSCORPIONIDA. The pseudoscorpions (Fig. 2-8*E*) are an abundant, widespread order. They are small, the largest not exceeding 7 or 8 mm in length. Most of the species occur under bark and in leaf mold, but the genus *Chelifer* is also found in houses.

RICINULEIDA. A small group of rare, mostly tropical arachnids resembling spiders, but the anterior margin of the body bears a large movable flap that extends over the chelicerae. One small species *Cryptocellus dorotheae* has been collected in the Rio Grande Valley of Texas.

ARANEAE. This order comprises the spiders (Fig. 2-8*B*), which in North America alone number several thousand species. Neither legs nor pedipalps are chelicerate; the abdomen shows only traces of segmentation and is constricted at the base to form a thread-waisted joint with the cephalothorax. The pedipalps of the male are highly

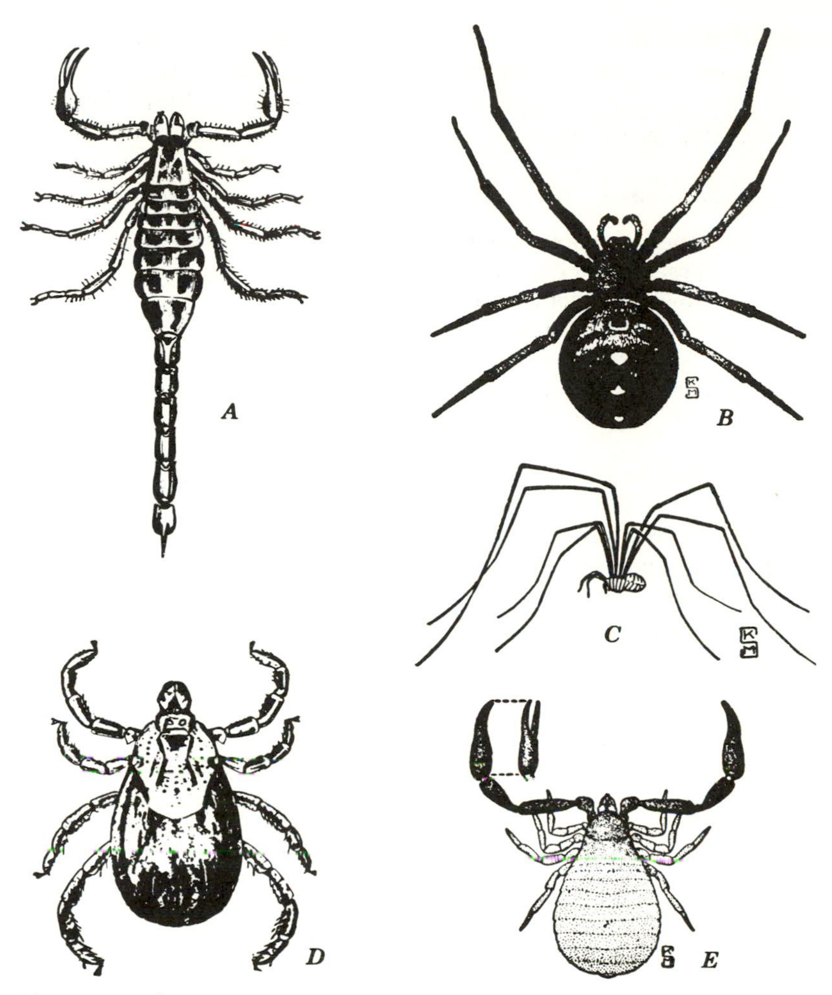

Fig. 2-8. Class Arachnida. (*A*) A Scorpion *Buthus carolinianus*; (*B*) the black widow spider *Latrodectus mactans*; (*C*) a harvestman or phalangid; (*D*) the eastern dog tick *Dermacentor variabilis*; (*E*) a pseudoscorpion *Larca granulata*. (*A* after Packard, *Guide to the study of insects*, Henry Holt & Co.; *D* from U.S.D.A.; *B*, *C*, and *E* from Illinois Natural History Survey)

modified to transfer sperm to the female genital organs; these modifications assume varied shapes and are used extensively in the taxonomy of the group. Spiders are predaceous on insects and other small animals, but aside from this they are extremely varied in habits. Some species hunt their prey, running it down or jumping on it; the crab spiders wait in flowers or other places for the prey to come within reach; many groups spin webs in which prey is snared. The family Theraphosidae comprises large hairy hunting spiders called tarantulas.

KEY TO ORDERS (BASED ON FREE-LIVING ADULTS) OF THE CLASS ARACHNIDA

1. Apex of abdomen ending in a narrowed, slender, tail-like portion composed of 4 to 15 segments 2
   Apex of abdomen without a tail 4
2. Tail stout, ending in a bulblike sting (Fig. 2-8*A*) .... **Scorpionida**
   Tail very slender, apex unarmed 3
3. Pedipalps stout, heavy, and relatively short, contrasting greatly with the slender legs; last 2 segments of pedipalps often forming a pincer .... **Uropygida**
   Pedipalps slender, never pincerlike, no thicker than the legs .... **Palpigrada**
4. Pedipalps chelate 5
   Pedipalps not chelate 7
5. First pair of legs extremely long and whiplike, with many subdivisions .... **Amblypygida**
   First pair of legs neither whiplike nor multisegmented 6
6. Chelae of pedipalps long, sometimes massive (Fig. 2-8*E*) .... **Pseudoscorpionida**
   Chelae of pedipalps much smaller .... **Ricinuleida**
7. Apex of abdomen with 3 or 4 pairs of spinnerets, best seen from lateral or ventral views; abdomen unsegmented in most forms ...... **Araneae**
   Apex of abdomen without spinnerets 8
8. Abdomen unsegmented .... **Acarina**
   Abdomen distinctly segmented 9
9. Abdomen as short as cephalothorax .... **Opilionida**
   Abdomen much longer than cephalothorax .... **Solpugida**

ACARINA. These are the mites and ticks, ranging from less than 1 to 15 mm in length. The cephalothorax and abdomen are fused into a single continuous body region devoid of external segmentation (Figs. 2-8*D*, 2-9, 2-10). Mites are extremely varied in structure and habits. Although there are fewer species of mites than spiders, the mites are the most important group economically in the entire class Arachnida. Several families, including ticks (Ixodidae, Fig. 2-8*D*), chiggers (Trombiculidae), mange or itch mites (Sarcoptidae), and follicle mites (Demodicidae), are ectoparasites of insects, birds, and mammals. Species of "red spiders" (Tetranychidae, Fig. 2-9*A*, *B*), attack leaves and cause defoliation of many crops. Various Tyroglyphidae, bulb mites, attack stored products and bulbs and roots of bulb crops. Members of the family Eriophyidae (Fig. 2-9*C*, *D*) produce blisters and galls on several commercial crops, notably pears.

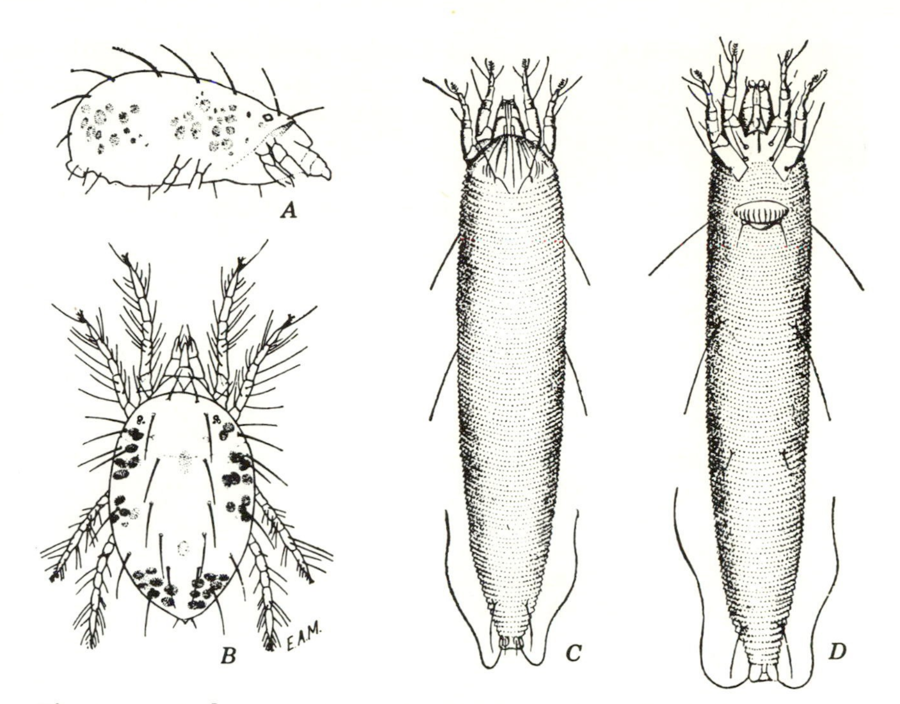

Fig. 2-9. Order Acarina. (*A*, *B*) A "red spider" *Tetranychus pacificus*, lateral and dorsal aspects; (*C*, *D*) a blister mite *Eriophyes pyri*, dorsal and ventral aspects. (From U.S.D.A., E.R.B.)

In addition to economic species, there are many mites living in soil and ground cover and as ectoparasites on many native birds and mammals. Two families, the Hydrachnellidae and Unionicolidae, are found in freshwater habitats, and the Halacaridae are marine water mites.

OPILIONIDA. The harvestmen, or daddy longlegs (Fig. 2-8C), have a broad abdomen, the entire body stout and oval or round in outline. In our larger species the legs are spindly, frequently five or more times as long as the body. The smallest species are minute, mitelike, and have short legs. These animals are common in damp shaded woods, where they move about on leaves, tree trunks, and in ground cover, seeking small insects and other food. In North America the group is widespread, represented by about 200 species.

SOLPUGIDA. This is a small order containing hairy, agile arachnids of moderate size. They are often called vinegarones or sunspiders. The pedipalps and legs are short, and the chelicerae are massive. Vinegarones are tropicopolitan, in North America occurring in the Southwest and northeastward into Kansas.

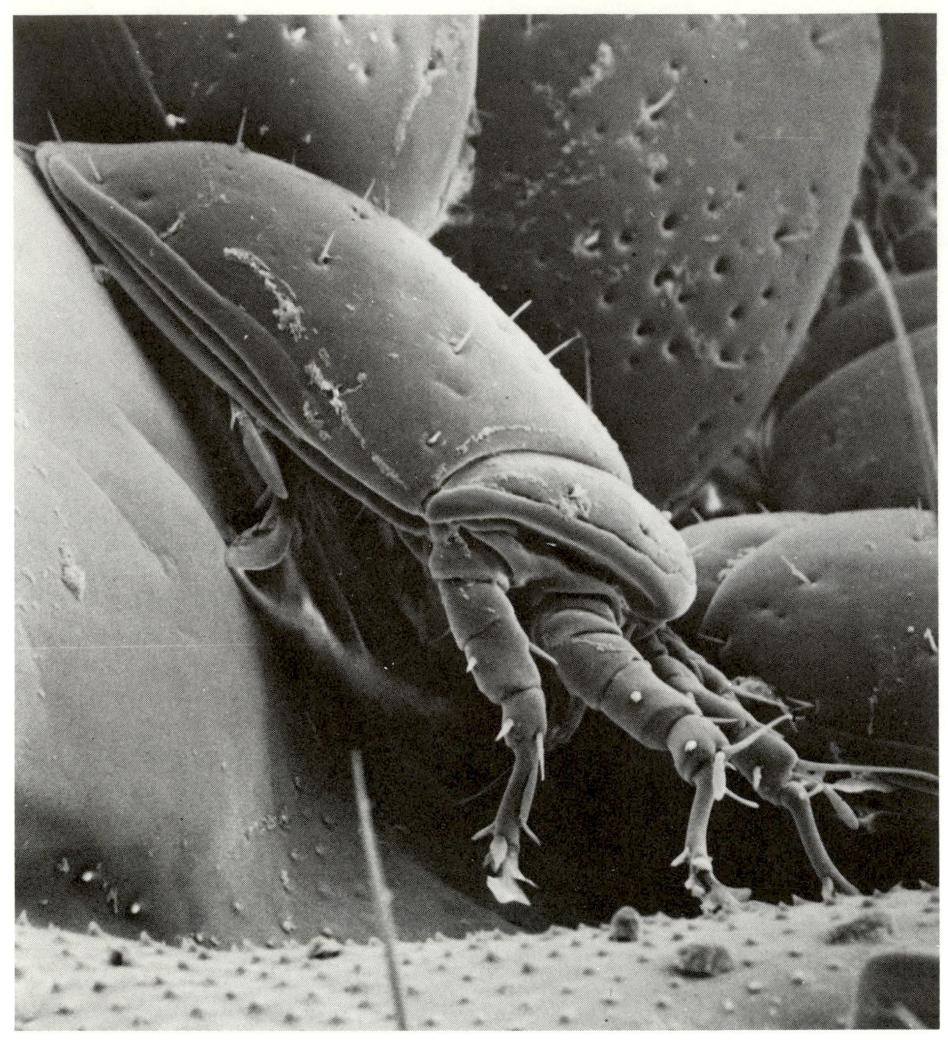

Fig. 2-10. Beetle mite on the mouthparts of larva of *Pericoptus truncatus* (Coleoptera); ×3700. (Courtesy Dept. of Entomology, Univ. of Alberta, and D. A. Craig)

Class PYCNOGONIDA

This class contains the sea spiders (Fig. 2-11), a small marine group of spiderlike forms having a minute peglike abdomen. Sea spiders are found mostly on hydroids and sea anemones, but occasionally on jellyfish.

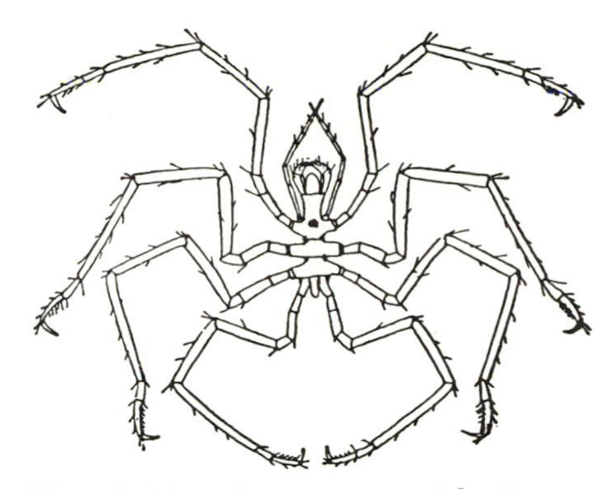

Fig. 2-11. A pycnogonid *Nymphon*. (From U.S.D.A., E.R.B.)

SUBPHYLUM MANDIBULATA

Adults of typical Mandibulata differ from those of the Chelicerata in having a head separate from the thorax, the head bearing one or two pairs of antennae and three or four pairs of feeding appendages or mouthparts. From the subphylum Trilobita, they differ in having these accessory feeding organs fused with the primeval head or prostomium to form a head of a compound origin.

Many forms of the mandibulate class Crustacea, such as the crayfish, share several characteristics with the trilobites, including a similar division of the body into head, thorax, and abdomen (the latter called a *pygidium* in the trilobites), the elongate thorax, and lateral or mesal appendages on at least the thoracic legs. In several classes of Mandibulata there are parasitic forms that are extremely difficult to diagnose. These are not included in the accompanying key.

KEY TO CLASSES (ADULT, FREE-LIVING FORMS ONLY) OF THE MANDIBULATA

1. Body having only 2 regions, the head and trunk, each trunk segment with similar segmental appendages 2
   Body having 3 regions, the head, the thorax, and the abdomen, the latter with the segmental appendages not leglike, or reduced or absent ... 5
2. Each trunk segment with 1 pair of appendages 3
   Each trunk segment with 2 pairs of appendages 4
3. First trunk segment with a pair of stout, curved poison claws (Fig. 2-16) .... **Chilopoda**
   Appendages of first trunk segment similar to those of other segments (Fig. 2-17) .... **Symphyla**

4. Trunk with about 9 pairs of appendages; antennae branched (Fig. 2-19) .... **Pauropoda**
   Trunk with over 50 pairs of appendages; antennae simple (Fig. 2-18) .... **Diplopoda**
5. Thorax with only 3 segments (Fig. 2-20) ..... **Insecta** and **Entognatha** (= superclass Hexapoda)
   Thorax with more than 3 segments (Fig. 2-12) .... **Crustacea**

Class CRUSTACEA

To this class, illustrated in Fig. 2-12, belongs such a varied assortment of forms that it is difficult to give a brief diagnosis that will apply to all. The majority have the following characteristics: body divided into head, thorax, and abdomen; head and thorax often closely joined and called the cephalothorax; head having two pairs of antennae, and, in a few groups, four pairs of accessory feeding appendages, including a pair of mandibles, two pairs of maxillae, and a pair of maxillipeds; thorax usually having 4 to 20 distinct segments, each with a pair of segmented appendages; abdomen having one to many segments, with short appendages or none. A few parasitic or sedentary groups have extreme reduction in both body segments and appendages, as in parasitic Copepoda, (Fig. 2-12*B*), and the barnacles. Several groups have a stout carapace covering much of the body, as in the crayfish (Fig. 2-12*C*); some others have a shell, bivalve in appearance, that encloses most of the body and appendages. In the sow bugs or pill bugs (Fig. 2-12*E*), comprising the order Isopoda, the body is somewhat flattened.

Most groups of Crustacea are aquatic, either marine or freshwater. A few are amphibious, such as certain species of crayfish. The only abundant terrestrial forms in North America are certain families of Isopoda, such as pill bugs or sow bugs. Species of these families occur chiefly in humid situations, such as in and under rotting logs, under stones, or in the soil. At night some of these species leave their shelters and wander about freely. Pill bugs are pests of stored vegetables and of moist organic-rich areas, such as soil in plant pots. They may reach pest levels, both outdoors and indoors.

Crustaceans rival insects in the variety of diverse forms developed in the class. They share another characteristic with the insects, that in may groups there is a succession of changes in form, or metamorphosis, in the life history of the individual. In the crustaceans this is well exemplified by the shrimp (Fig. 2-13), which passes through four quite different immature stages before attaining the adult stage, giving a total of five distinctive body forms in the life cycle of the species.

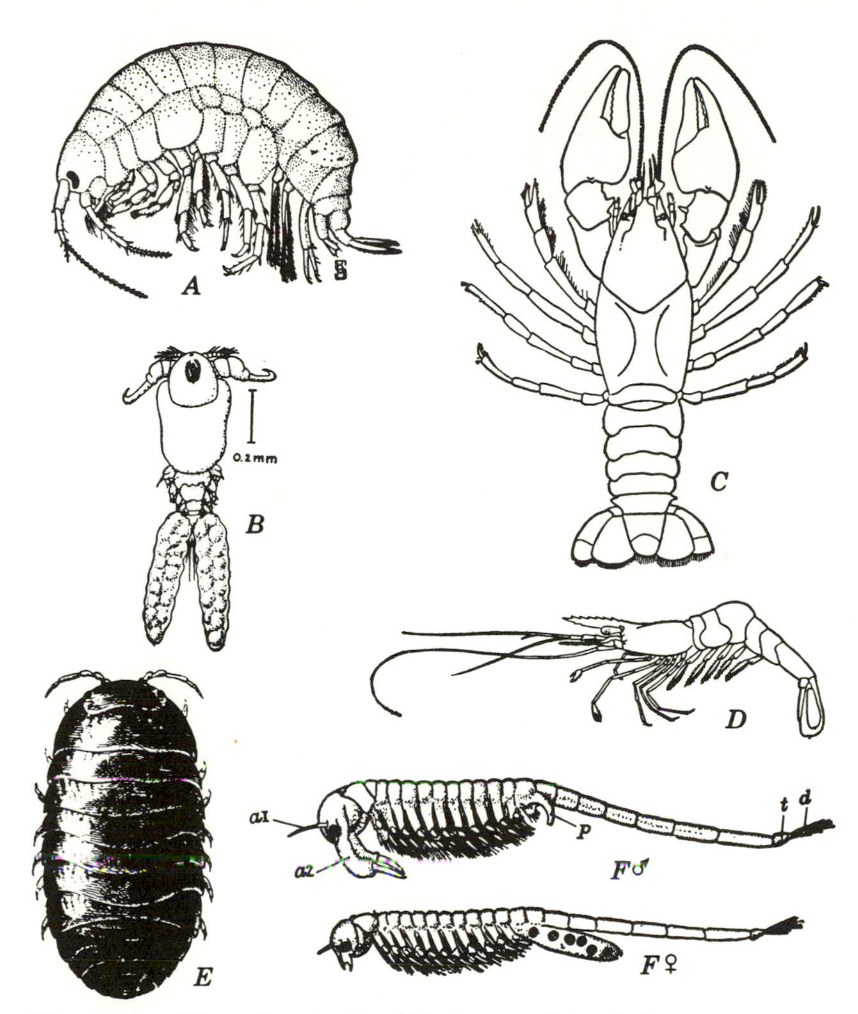

Fig. 2-12. Class Crustacea. (*A*) An amphipod *Gammarus* sp.; (*B*) a parasitic copepod *Ergasilus caerulus*; (*C*) a crayfish *Cambarus bartoni*; (*D*) a shrimp *Palaemonetes exilipes*; (*E*) an isopod *Armadillidium vulgare*; (*F*) a fairy shrimp *Branchinecta paludosa*. All are freshwater forms except *E*, which is terrestrial. (*A* from Illinois Natural History Survey; *E* from U.S.D.A.; others from Ward and Whipple)

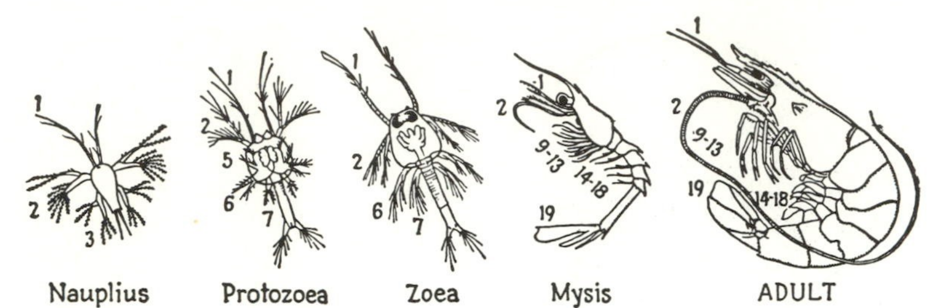

Fig. 2-13. Metamorphic stages of the shrimp *Penaeus*, showing the succession of changes in body form and in the appendages (1 to 19). (From Storer, *General zoology*, McGraw-Hill Book Co., after Muller and Huxley)

Five subclasses of Crustacea are recognized. Four of these have terrestrial or freshwater species; the fifth (barnacles) is exclusively marine.

MYRIAPOD GROUP

Four classes, the Diplopoda, Chilopoda, Symphyla, and Pauropoda, have centipede-like shapes and collectively are often termed the myriapods. They all have a distinct head (composed of the original prostomium fused with several body segments whose appendages form the mouthparts) and an elongate trunk region bearing segmental ambulatory or walking legs. In each of these four classes the antennae are present, sometimes well developed. Although these classes share many superficial resemblances, they exhibit many differences in basic structure.

Class CHILOPODA

Here belong the centipedes (Figs. 2-14, 2-15). They are elongate and have many segments, with a pair of legs on each segment, and with the reproductive openings on the penultimate body segment. The head bears long antennae and has eyes that are either compound or composed of single facets. The mouthparts (Fig. 2-16) consist of three pairs of appendages: the jawlike mandibles; the *first maxillae*, which are fused and resemble the insect labium; and the *second maxillae* or *palpognaths*, which are leglike, sometimes with their bases fused to form a bridge below the first maxillae. An interesting feature of the Chilopoda is the *poison claws* (Fig. 2-16C). These are appendages of the first trunk segment but are held beneath the head and superficially resemble mouthparts.

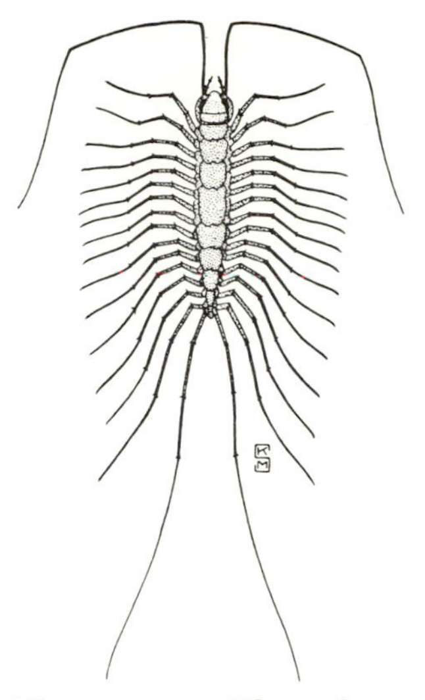

Fig. 2-14. The house centipede *Scutigera forceps*. (From Illinois Natural History Survey)

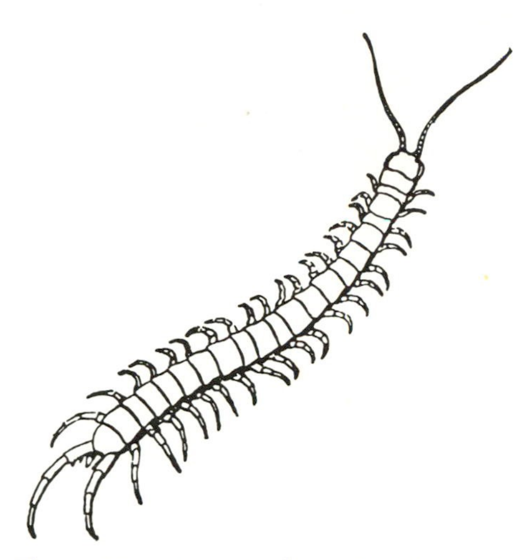

Fig. 2-15. A typical centipede. (After Snodgrass)

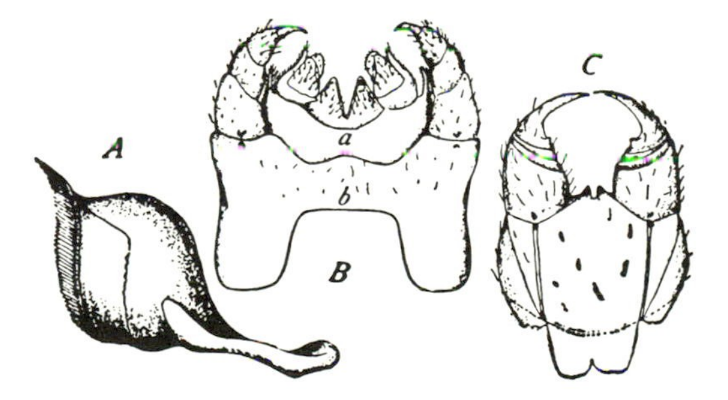

Fig. 2-16. Mouthparts and poison jaws of a centipede *Geophilus flavidus*. (*A*) Right mandible. (*B*) The two pairs of maxillae; (*a*) the united coxae of the first maxillae; (*b*) the united coxae of the second pair or palpognaths. (*C*) The poison claws or toxicognaths. (After Latzel)

Chilopoda are predaceous in habits. About 100 species are known from the United States. Most of them are nocturnal, moving about only at night in search of prey and hiding during the day in leaf mold, rotten logs, and galleries in soil. Species in temperate climates seldom exceed 4 cm in length, but a few tropical species 20 or 25 cm long occur in the extreme southern United States. One species (Fig. 2-14) having especially long legs is common in houses.

Class SYMPHYLA

Members of this class are about 5 mm long and centipede-like in form (Fig. 2-17) and have the trunk composed of about 15 segments (none fused in pairs) of which 11 or 12 bear legs. The reproductive openings are at the anterior end of the body. The head possesses long antennae, and the mouthparts consist of mandibles, maxillae, and labium. Most members of the group are uncommon, usually occurring in humus, but certain species that eat the roots of plants are pests in gardens and greenhouses. In the United States the genus *Scolopendrella* is sometimes found in ground-cover samples. On cursory examination, small individuals may be confused with Pauropoda, which occur in the same type of situation.

Class DIPLOPODA

This includes the millipedes or thousand-legged worms (Fig. 2-18). Except for a few segments at each end of the body, the body segments have fused into pairs so that each apparent segment has two pairs of legs. The mouthparts consist of a pair of mandibles and a platelike *gnathochilarium*, thought to be the fused maxillae. The labium is apparently atrophied. The reproductive organs open behind the second pair of legs.

Millipedes live in a leaf mold, rotten logs, and other humid places. About 100 species occur in the United States, especially in forested localities. Most species feed on decaying plant material. A few feed on living plants and become of local economic importance.

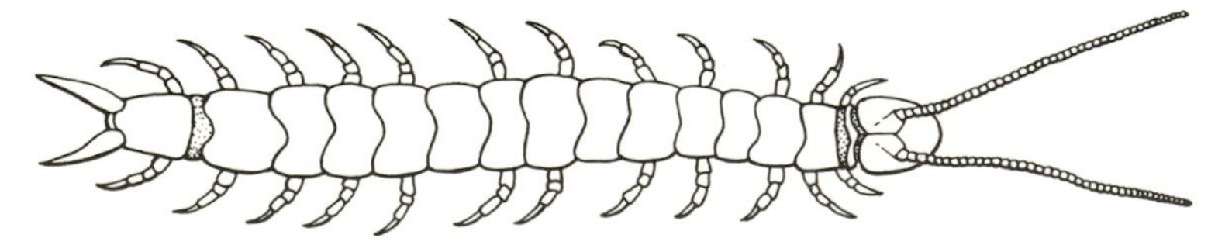

Fig. 2-17. A symphylid *Scutigerella immaculata*. (Adapted from Snodgrass)

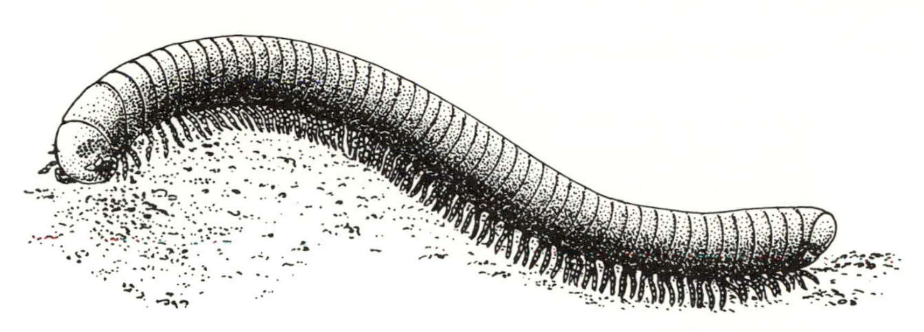

Fig. 2-18. A millipede *Parajulus* sp. (Original by C. O. Mohr)

Class PAUROPODA

Several genera of minute animals comprise this class. The body consists of 11 or 12 segments, of which the dorsal portions are fused in pairs; 8 or 9 segments each bear a pair of legs, each pair evenly spaced from the next. The antennae are biramous, unlike any of their relatives. Each eye is represented by only a small spot. The mouthparts consist of a pair of mandibles and a curious complex lower lip thought to be the same as the gnathochilarium in the Diplopoda. As in the Diplopoda, the openings of the reproductive organs are located in the anterior part of the body.

The known North American species number about 35 and range in length from about 0.5 to 2 mm. They occur on fallen, decayed twigs and logs, especially in situations where leaf mold has accumulated. Most of our species have slender white bodies, and look like small centipedes (Fig. 2-19). Members of the family Eurypauropodidae are short and stout, resembling minute sow bugs (isopods). The class has been recorded from only scattered localities in eastern and western North America. Much remains to be discovered about these curious animals.

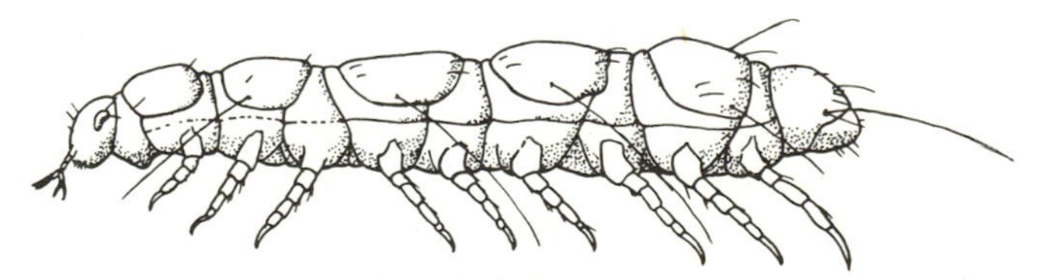

Fig. 2-19. A pauropod *Pauropus silvaticus*. (Adapted from Snodgrass)

SUPERCLASS HEXAPODA

All the insects and entognaths belong to the superclass Hexapoda. They are distinguished primarily in that they have a three-segmented thorax, each segment typically bearing a pair of legs. From this arrangement comes the term Hexapoda, meaning six-legged.

Class ENTOGNATHA

Until recently the orders Collembola, Protura, and Diplura were included in the apterygote (wingless) insects. However, these three orders differ from true insects in having *entognathous mouthparts*; that is, the mouthparts lie inside the head in a cavity formed by oral folds of the genae. Living forms are described in Chapter 8.

Class INSECTA

This class includes those hexapods that have *ectognathous mouthparts*; that is, the mouthparts are not enclosed by the head but are external to the head.

Characteristics. A typical adult insect has three body regions (Fig. 2-20). The anterior region is the head, which bears eyes, antennae, and three pairs of mouthparts. The next region is the thorax, which is composed of three segments, each usually bearing a pair of legs; in many groups the second and third segments each bear a pair of wings. The posterior portion of the body is the abdomen. It consists of as many as 11 segments and has no legs. The eighth, ninth, and tenth segments usually have appendages modified for mating activities or egg laying. The exoskeleton in insects, as in other arthropods, provides both the protection for the vital organs and the support that maintains the body shape. The chief internal organs consist of the following parts: (1) a tubular digestive tract; (2) a long valvular heart for pumping the blood; (3) a system of pipelike tracheae for respiration; (4) paired reproductive organs opening at the posterior end of the body; (5) an intricate muscular system; and (6) a nervous system consisting of a brain, paired segmental ganglia, and connectives.

Insects lay eggs, except for a few forms that bear living young. The young insects molt from time to time in their development to the mature or adult stage, and generally at each molt an increase in size or the development of special parts takes place.

Immature insects do not have wings. The only known exception to this is found in the order Ephemeroptera, mayflies, in which the last brief immature instar has functional wings. Immature insects

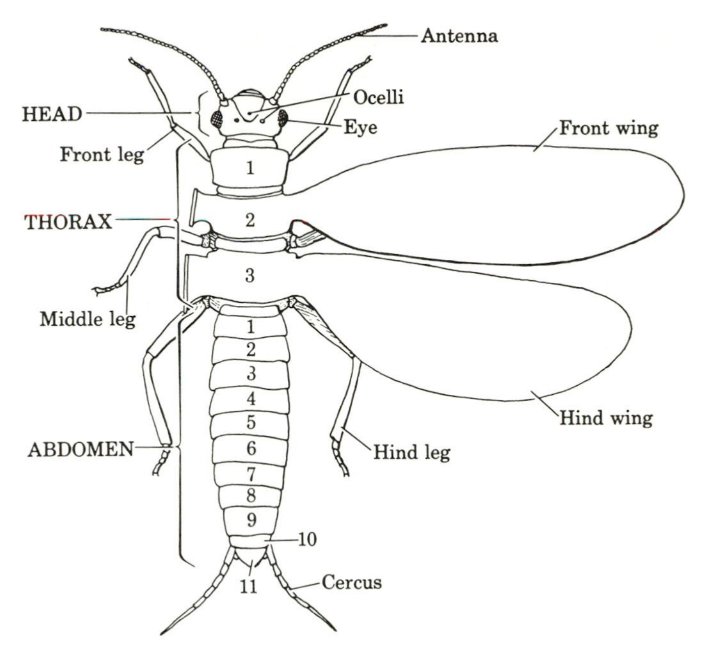

Fig. 2-20. Diagram of a typical adult winged insect. (Adapted from Snodgrass)

may be entirely unlike their adults in general appearance and may lack legs and many other structures typical not only of insects but also of arthropods.

Taxonomic Diversity. The insects have evolved into many strikingly different kinds of organisms. Living forms are classified in 27 orders, listed in Table 2-1 and described in Chapter 8.

Success as a Group. Insects have attained the largest number of kinds of any animal group with an estimated 900,000 described species (Fig. 2-21). The total number of described species for all animal groups is about 1.5 million. Sometimes insects occur in such numbers as to swarm in dark clouds or, attracted to lights, blanket the streets of a city a foot or more deep with their bodies.

In competition with other animals they have been able to fit into and populate almost every nook and cranny of the globe except the depths of the ocean. They abound throughout the tropics and also are one of the very few permanent animal inhabitants of the South Polar Region; aquatic insects may nearly pave the bottom of large rivers and lakes and also develop to maturity in the water in hoof-prints; in one case grasshoppers may range over miles of prairie, in

Table 2-1 Classification of Living Hexapod Orders

- Superclass Hexapoda
  - Class Entognatha:
    - Order Diplura—campodeans and japygids
    - Order Protura—proturans
    - Order Collembola—springtails
  - Class Insecta:
    - Subclass Apterygota—wingless insects
      - Order Microcoryphia—bristletails
      - Order Thysanura—silverfish
    - Subclass Pterygota—winged insects
      - Series Paleoptera—ancient winged insects
        - Order Ephemeroptera—mayflies
        - Order Odonota—dragonflies
      - Series Neoptera—modern or folding-wing insects
        - Order Dictyoptera—cockroaches
        - Order Isoptera—termites
        - Order Phasmatodea—walking sticks
        - Order Orthoptera—grasshoppers, crickets
        - Order Dermaptera—earwigs
        - Order Grylloblattodea—grylloblattids
        - Order Embioptera—embiids
        - Order Plecoptera—stoneflies
        - Order Zoraptera—zorapterans
        - Order Psocoptera—psocids, booklice
        - Order Phthiraptera—chewing and sucking lice
        - Order Thysanoptera—thrips
        - Order Hemiptera—bugs (Heteroptera and Homoptera)
        - Order Raphidioptera—snakeflies
        - Order Megaloptera—dobsonflies
        - Order Neuroptera—lacewings
        - Order Coleoptera—beetles
        - Order Hymenoptera—bees, ants, wasps, sawflies
        - Order Mecoptera—scorpionflies
        - Order Siphonaptera—fleas
        - Order Diptera—two-winged flies, midges, mosquitoes
        - Order Trichoptera—caddisflies
        - Order Lepidoptera—moths, butterflies

another a brood of maggots may feed and mature within a single rotting walnut husk, and in still another a wasp may mature within the tiny seed of a small plant.

These are but a few fragments of the evidence that insects are a remarkably successful group of organisms. With this in mind, it is interesting to speculate on some of the reasons why they have developed to such huge numbers of both species and individuals.

Adaptive Features. Among arthropods, insects represent the culmination of evolutionary development in terrestrial forms. They have exploited the mechanical advantages of an exoskeleton and used them as a basis on which to add specializations that give them still further advantages over their competitors. Chief advantages of an exoskeleton include (1) a large area for internal muscle attachment; (2) an excellent possibility of evaporation control, especially in small-bodied animals; and (3) almost complete protection of vital organs from external injury.

To this foundation other specializations have been added, some morphological and some physiological, which have been decided factors in assisting insects to attain their present development. The more outstanding of these specializations are enumerated in the following paragraphs.

Functional Wings. The power of flight greatly increased the chances of survival and dispersal, except on wind-swept islands. It increased feeding and breeding range and provided a new means of eluding enemies. Increased feeding range undoubtedly opened the way for adoption of foods of more specific limitation, especially in those cases in which the host or breeding medium occurred in small quantities and scattered situations. For example, it would allow a species to adopt carrion as a food, since the individual with functional wings could seek out and reach carcasses that would be not only many miles away, but also suitable as food for only a short period.

Small Size. In the main, insect evolution has followed the course of developing many small individuals rather than few large ones. This made available many new specific foods occurring in small quantity and has increased chances of hiding from and eluding enemies. Small size has the disadvantage that the total body surface area is disproportionately large in comparison to the body volume. This results in a high rate of evaporation, which makes terrestrial life virtually impossible for a thin-skinned animal. The sclerotized cuticular exoskeleton of insects provides a check to this high rate of evaporation, and its impermeable characteristics are undoubtedly one of the principal factors that has allowed insects to retain their small size.

Adaptability of Structures. The same structure has become adapted to perform different functions in different groups. For instance, the front legs of mantids and ambush bugs grasp and hold prey while it is being devoured, thus functioning as accessory mouthparts, rather than as ambulatory legs. In other instances essen-

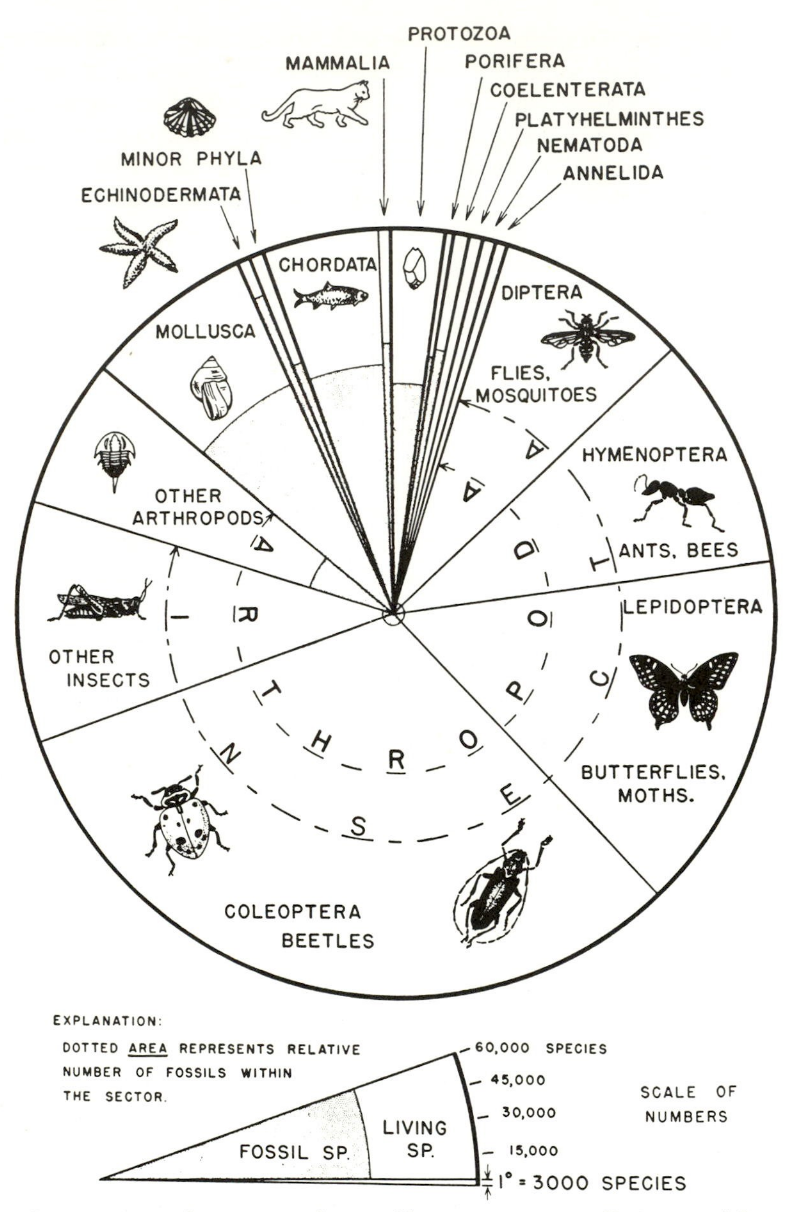

Fig. 2-21. Relative numbers of known species, living and fossil, of various phyla. (From Muller and Campbell, 1954, with permission of *Systematic Zoology*).

Table 2-2. Estimated Number of Living Animal Species of Major Groups Known at Present for the World

GROUP	NUMBER OF SPECIES
Chordata	55,000
Echinodermata	5700
Arthropoda exclusive of insects	80,000
Insecta	900,000
Mollusca	128,000
Annelida	8700
Nematoda	10,000
Platyhelminthes	12,700
Cnidaria	10,000
Porifera	5000

tially the same structure has become adapted to function under entirely different conditions, for example, the modifications of the respiratory system, which are adapted for many types of aquatic and terrestrial life.

Complete Metamorphosis. This is a specialization in which the life history is divided into four distinct parts: (1) the egg; (2) the larva or feeding stage; (3) the pupa, a quiescent transformation stage; and (4) the adult or reproductive stage. It occurs in the insect orders having the largest number of species, including the beetles and flies. In all but a few instances of this type of life history, all real growth is the result of larval feeding; the adult has only to maintain a more or less static metabolism and at the most provide sufficient food for maturation of sperms or eggs. This system has enabled the larva and adult to live in entirely different places and under different conditions, so that the larva has been able to take advantage of conditions most favorable for rapid growth and the adult to live in conditions best suited to fertilization, dispersal, and oviposition. Complete metamorphosis opened to the group an infinite variety of habitat and food possibilities. In connection with it there has often been a varied development of complex behavior. In addition, extremely short life cycles have frequently evolved, based on the extraordinary feeding and digestive ability of some of the larvae; a flesh-fly maggot, for instance, can develop from hatching to a full-grown larva in 3 days. In short, complete metamorphosis has enabled a species to combine the advantages of two entirely different ways of life and at the same time avoid many of the disadvantages of both.

Increase in Number of Species. Factors contributing to the success of insect populations would tend to assure the *persistence* of species, but would not cause an increase in the *number* of species. Yet the extraordinary number of insect species is one of the most important factors in their success as a group. Other factors must be invoked to account for the extraordinary number of different insect species. According to evolutionary theory the following circumstances are especially important in this connection:

1. Many insect species can live only within narrow limits of certain ecological factors such as host, temperature, and humidity, and their species ranges become broken into isolated segments by relatively small long-range changes in climate such as those accompanying an Ice Age.
2. Because of their power of flight, winged insects are commonly transported over large expanses of water or other barriers by moving air masses, such as in the Gulf of Mexico. As a result of this broad dispersal, populations of a species may colonize new sites that are geographically isolated from the main populations. Such founding and colonizing populations may evolve into new species if individuals in the populations are able to adapt to the new conditions.
3. Genetic incompatibility in mating and breeding between isolated populations may arise with unusual rapidity, in part because of the short generation time in insects, resulting in an accelerated rate of speciation.

No one of these factors can be considered the most important reason that insects have achieved their present diversity and numbers. The process has been most complex, with various combinations of these factors and undoubtedly others working together and producing the end result. It should be borne in mind that not every insect lineage has every one of these specializations. For example, entire orders of insects, such as the chewing lice, sucking lice, and fleas, have lost wings, correlated with a limited sphere of activity on or near the host. Complete metamorphosis does not occur in about half the orders; but in these, other features come into play. It must be remembered, too, that these specializations are only a few of the most important of the very large number that have evolved in the class Insecta.

REFERENCES

Anderson, D. T., 1973. *Embryology and phylogeny in annelids and arthropods*. Oxford: Pergamon. 495 pp.

Baker, E. W., and G. W. Wharton, 1952. *An introduction to acarology*. New York: Macmillan. 465 pp.

Barnes, R. D., 1980. *Invertebrate zoology*, 4th ed. Philadelphia: Saunders College/Holt Rinehart & Winston. 1089 pp.

Chamberlin, J. C., 1931. The arachnid order Chelonethida. *Stanford Univ. Publs., Univ. Ser. Biol. Sci.*, **7**:1–284.

Cloudsley-Thompson, J. L., 1968. *Spiders, scorpions, centipedes and mites*. Oxford: Pergamon. 278 pp.

Comstock, J. H., and W. J. Gertsch, 1940. *The spider book*. Ithaca, N.Y.: Comstock. 729 pp.

Cook, D. R., 1974. Water mite genera and subgenera. *Mem. Am. Entomol. Soc.*, No. 21. 860 pp.

Driver, E. C., 1950. *A guide to the identification of the common land and freshwater animals of the United States* (rev. ed.). Published by the author, 119 Prospect St., Northhampton, Mass. 558 pp.

Edmondson, W. T., H. B. Ward, and G. C. Whipple (Eds.), 1959, *Freshwater biology*, 2nd ed. New York: Wiley. 1248 pp.

Gertsch, W. J., 1979. *American spiders*, 2nd ed. New York: Van Nostrand Reinhold. 274 pp.

Gupta, A. P. (Ed.), 1979. *Arthropod phylogeny*. New York: Van Nostrand Reinhold. 762 pp.

Hoff, C. C., 1949. The pseudoscorpions of Illinois. *Bull. Ill. Nat. Hist. Surv.* **24**:413–498.

Kaston, B. J., 1948. Spiders of Connecticut. *Bull. Conn. Geol. Nat. Hist.* Surv. **70**:1–784.

——— 1978. *How to know spiders*, 3rd ed. Dubuque, Iowa: W. C. Brown.

King, P. E., 1973. *Pycnogonids*. London: Hutchinson. 144 pp.

Krantz, G. W., 1970. *A manual of acarology*. Oregon State Univ. Bookstores, Inc. 335 pp.

Kristensen, N. P., 1975. The phylogeny of hexapod "orders." A critical review of recent accounts. *Z. Zool. Syst. Evolutionsforsch*. **13**(1):1–44.

Levi, H. W., and L. R. Levi, 1968. *A guide to spiders and their kin*. New York: Golden Press. 160 pp.

MacSwain, J. W., and U. N. Lanham, 1948. New genera and species of Pauropoda from California. *Pan-Pacific Entomol.*, **24**:69–84.

Manton, S. M., 1949. Studies on the Onychophora VII. The early embryonic stages of *Peripatopsis*, and some general considerations concerning the morphology and phylogeny of arthropods. *Phil. Trans. R. Soc. B.*, **227**:411–464.

——— 1964. Mandibular mechanisms and the evolution of arthropods. *Phil. Trans. R. Soc. B.*, **247**:1–183.

——— 1973. The evolution of arthropodan locomotory mechanisms. Part II. Habits, morphology and evolution of the Uniramia (Onychophora, Myriapoda, Hexapoda) and comparisons with Arachnida, together with a functional review of uniramian musculature. *Zool. J. Linn. Soc.*, **53**:257–375.

——— 1977. *The Arthropoda: Habits, functional morphology and evolution*. Oxford: Oxford Univ. Press. 527 pp.

McGregor, E. A., 1950. Mites of the family Tetranychidae. *Am. Midl. Naturalist*, **44**:257–420.

Michelbacher, A. E., 1938. The biology of the garden centipede, *Scutigerella immaculata*. *Hilgardia*, **11**:55–148.

Pennak, R. W., 1978. *Freshwater invertebrates of the United States*, 2nd ed. New York: Wiley. 1248 pp.

Ross, H. H., 1962. *A synthesis of evolutionary theory*. Cliffside, N.J.: Prentice-Hall. 387 pp.

Snodgrass, R. E., 1938. Evolution of the Annelida, Onychophora, and Arthropoda. *Smithsonian Misc. Collections*, **97**(6):1–159.

Storer, T. I., R. L. Usinger, R. C. Stebbins, and J. W. Nybakken, 1979. *General zoology*, 6th ed. New York: McGraw-Hill. 902 pp.

Størmer, L., 1977. Arthropods from the Lower Devonian (Lower Emsian) of Alken an der Mosel, Germany. Part 5: Myriapoda and additional forms, with general remarks on fauna and problems regarding invasion of land by arthropods. *Senckenb. Lethaea*, **57**:87–183.

Waterhouse, D. F. et al., 1970. *The insects of Australia*. Carleton, Victoria: Melbourne Univ. Press. 1029 pp.; Suppl. 1974, 146 pp.

Wharton, G. W., 1952. A manual of the chiggers. *Entomol. Soc. Wash. Mem.*, **4**:1–185.

Williams, S. R., and R. A. Hefner, 1928. The millipedes and centipedes of Ohio. *Ohio Biol. Surv. Bull.* **18**, 4(3):91–146.

3
External Anatomy

This chapter deals primarily with the external parts and divisions of the body and its appendages. Before proceeding to the description of these parts, it is necessary to understand a few generalities regarding the body wall, from which the parts are formed.

BODY WALL AND EXOSKELETON

The body wall of insects secretes an exoskeleton that serves as a framework for the internal attachment of muscles. The exoskeleton gives rigidity to the insect body and support for the muscles comparable to that of the internal skeleton of vertebrates. To the exoskeleton are attached the principal muscles that give the body cohesion. The body wall may have considerable spring or flexibility; but it will not stretch, except for a short time after a molt.

Serving as both protection and a rigid attachment for muscles, various parts of the body wall are hardened or sclerotized. If all the body wall were uniformly hard, there would be no possibility of movement, whether for purposes of locomotion or for expansion to accommodate important activities such as food ingestion or development of eggs. Movement is possible because the hardened body areas form a series of plates, or *sclerites*, between which the body wall is soft and flexible, or membranous. This arrangement permits the development of hard exterior plates for protection and rigidity and at the same time allows many types of movement.

A simple example of how this works is found in the abdomen of the mosquito (Fig. 3-1). When the abdomen is not engorged, a cross section of the body wall of the abdomen is narrow and elliptical (Fig. 3-1*A*); the dorsal plate and ventral plate are connected at the sides by a strip of membrane that is very finely accordion pleated. As the abdomen enlarges during a blood meal, these membranes simply unfold, allowing the dorsal and ventral plates to be pushed farther apart by the increasing volume of food being pumped into the abdomen. At its greatest expansion, the body cross section is nearly circular (Fig. 3-1*B*).

Another common type of membranous connection, shown in Fig. 3-2, works on the principle of telescoping rings. When the body is retracted, as in *A*, the rings overlap, and the membrane is drawn in with the telescoped section. When it is extended, as in *B*, the sections may be pushed out to the limit of the length of their connecting membranes.

Figure 3-3 illustrates how a flexible membranous strip affords articulation of a leg joint. To the right, a narrow strip of membrane forms a hinge between the leg and one plate; to the left a roll of membrane connects the leg and the other plate. In *A* the leg is held straight out; note that the left membrane forms a series of pleats. In *B* the appendage has been pulled forward; this movement stretched

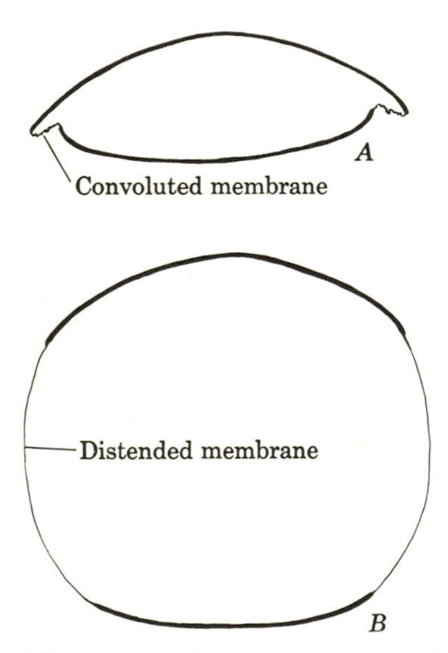

Fig. 3-1. Cross section of mosquito abdomen, diagrammatic. (*A*), Contracted; (*B*) expanded.

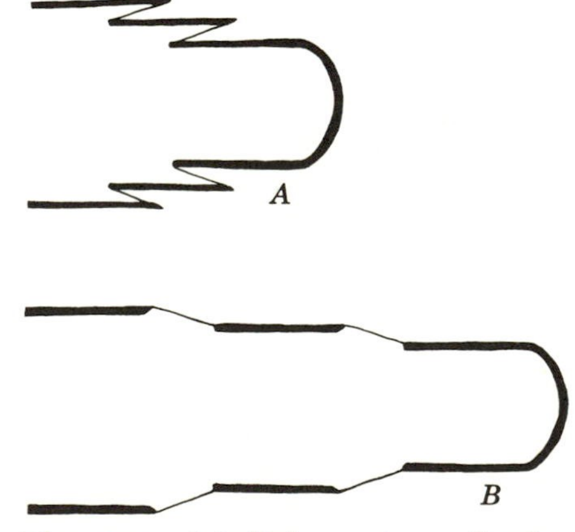

Fig. 3-2. Medial section of telescoping ring segments, diagrammatic. (*A*) Retracted; (*B*) extended. The thin portion represents membrane.

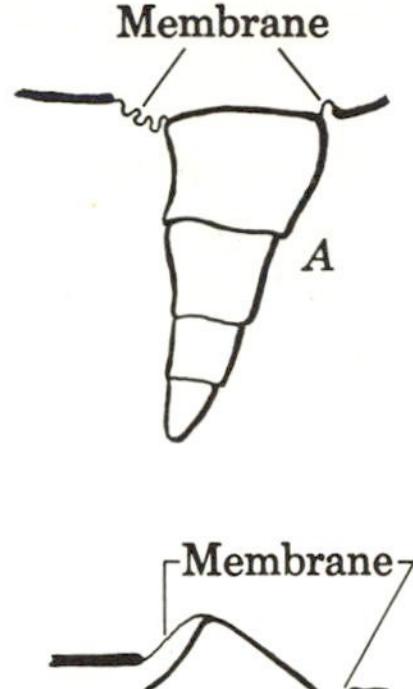

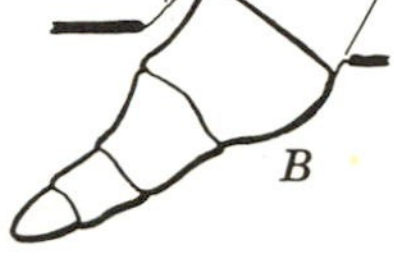

Fig. 3-3. Articulation of a membranous leg joint. (*A*) Held straight out; (*B*) pulled forward.

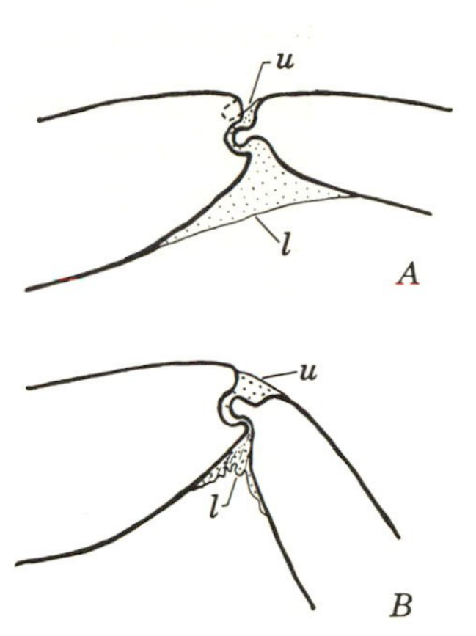

Fig. 3-4. Membranous connections of a ball-and-socket leg joint. (*A*) Held straight out; (*B*) pulled in. *u*, Upper membrane; *l*, lower membrane.

out the left membrane as the corner of the leg moved inward and back.

A common type of articulation found at leg joints of adult insects is illustrated in Fig. 3-4, in which the two segments are connected by a ball-and-socket joint. When the segment to the right is extended, as in *A*, the upper membrane *u* is folded, and the lower membrane *l* is stretched out. When the segment is pulled down, as in *B*, the upper membrane is pulled out, and the lower membrane is folded or pleated.

In all four of these examples movement has been made possible by flexibility, not elasticity. The illustrations do not show the muscles that actually *cause* the movements, but only the membranous connections that *permit* movement.

Sclerites. The hardened or sclerotized areas of the body are called sclerites. The major sclerites are usually separated by areas or lines of membrane. Many major sclerites may be subdivided by furrows or new lines of membrane into additional sclerites. Sclerites may also unite, usually with an evident line, furrow, or seam along the line of fusion. In entomological usage these types of demarcation—membrane strips, furrows, and fusion lines—are called sutures. The term sclerite is applied loosely to any sclerotized surface area, bounded by sutures of any type.

External Processes. The surface of the integument usually bears many kinds of processes, including wrinkles, spurs, scales, spines, and hair. These are outgrowths of the body wall. They are of only incidental interest to the general subject of external anatomy but are extremely important in functions such as sound production and sensing various types of stimuli.

Internal Processes. There are many internal processes that are formed by the invagination of the body wall. These are called *apodemes* (see also Fig. 5-1C). Their point or line of invagination is almost always indicated by an external pit or groove. These pits and grooves provide some of the most reliable landmarks for identifying their parent sclerites. The apodemes provide internal areas for muscle attachment.

BODY REGIONS

The adult insect body is divided into three parts: the head, thorax, and abdomen. The phylogenetic origin of these regions is discussed on page 63. The head is usually a solidly constructed capsule no longer having obvious segmentation. The thorax and abdomen have both preserved distinct segments of more or less ringlike form.

Legless immature instars of some insects have little differentiation between body regions. The head is usually distinct, but the segments of both thorax and abdomen may be identical in appearance and form a single uniform body region, the trunk.

Orientation. In describing the relative position of various parts of an insect, several sets of terms are used to indicate direction or position. Certain body regions are used as a basis for orientation, chiefly the following:

1. *Anterior Portion*, the portion of the body bearing the head; or that portion of any part that is toward the head end.
2. *Posterior Portion*, the portion of the body bearing the cauda, or "tail end" of the abdomen; or that portion of any part that is toward the posterior end.
3. *Dorsum*, the top or upper side of the body or one of its parts.
4. *Venter*, the underside or lower side of the body or one of its parts.
5. *Meson*, the longitudinal center line of the body, projected on either the dorsal or ventral aspect, or any point in between.

6. *Lateral Portion*, the side portion of the body or one of its parts.
7. *Base*, *Apex*; in appendages or outgrowths of the body, such as antennae or legs, the point or area of attachment is called the base; the tip or furthermost point from the attachment is called the apex. In parts of appendages, such as a segment of a leg, the same orientation is used; the part articulated nearest the body is the base or proximal portion; the part away from the body is the apex or distal portion.

THE HEAD

The head (Fig. 3-5) comprises the anterior body region of an insect. It is normally a capsule with a sclerotized upper portion, which contains the brain, and a membranous floor, in which is situated the oral opening or mouth.

Origin

As explained in Chap. 2, the insect head is a composite structure consisting of many parts. These have fused together so completely that only a few of the original areas can be identified in the immature and adult stages. We are, therefore, aware of the full composite nature of the head only through evidence found in embryos. There has been controversy over the units that comprise the head, clearly summarized by Rempel (1975), whose views are adopted here.

According to the Scholl-Rempel-Church theory, the head consists of an ancestral acron (called a prostomium in groups in which it is not fused with trunk segments) and six trunk segments (Fig. 3-6). In the embryo, each segment bears a pair of ventral appendages (Fig. 3-7) and apodemes. In later development these structures become

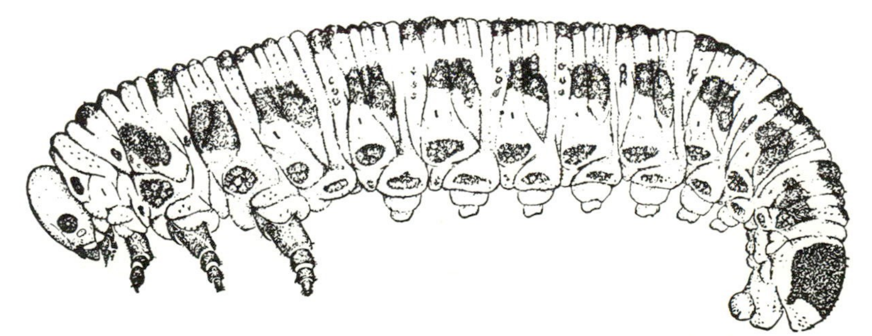

Fig. 3-5. Sawfly larva *Neodiprion lecontei*, illustrating a hypognathous head. (After Middleton)

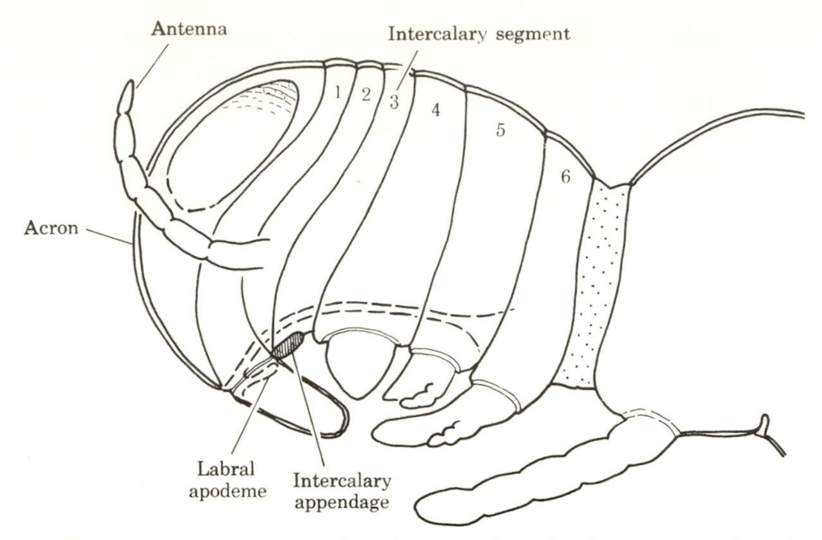

Fig. 3-6. Construction of embryonic head of an insect showing ancestral acron and six trunk segments. (Adapted from Rempel, 1975)

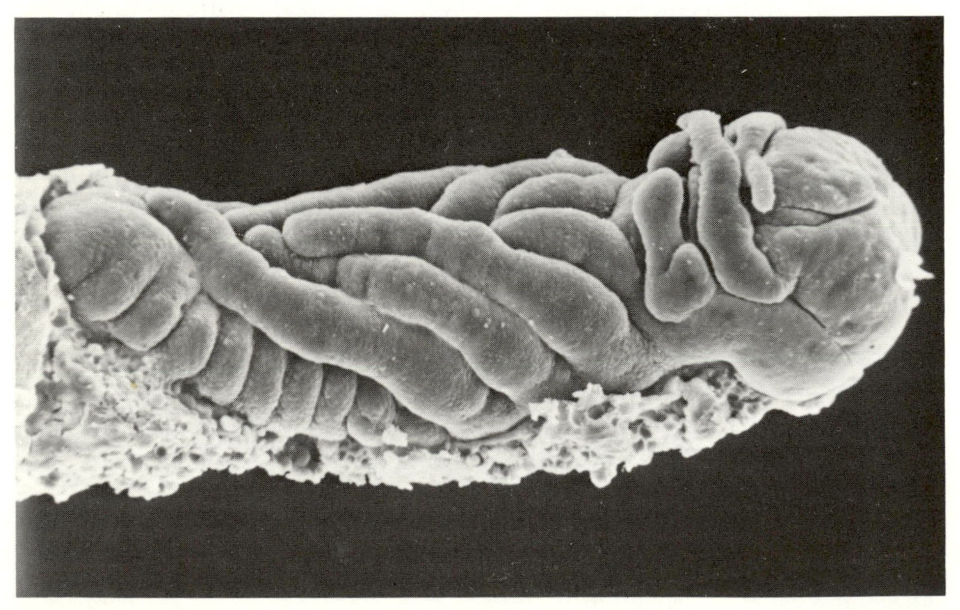

Fig. 3-7. Embryo of leafhopper *Microstelus focifrons*, ×75. (Courtesy Dept. of Entomology, Univ. of Alberta, and D. A. Craig)

highly modified. A diagram of the pre-insect head, showing the postulated arrangement of segments and appendages, is shown in Fig. 3-6.

Typical Insect Head

By the time insects had evolved, the various parts comprising the head had compressed into a structure very different from that of its progenitor. In a typical insect head (Fig. 3-8), the anterior region or face, the dorsal portion, and lateral portions form a continuous sclerotized capsule that is open beneath, like an inverted bowl. On this capsule are situated a pair of compound eyes, three ocelli, and a pair of antennae. The labrum hangs down from the lower front margin of the capsule to form a flap in front of the mouth. The ventral portion of the head forms a membranous floor posterior to the mouth; from this floor arises the hypopharynx, bearing the opening of the salivary duct. On each side of this floor hang down the three pairs of appendages forming the chewing organs or mouthparts, consisting of the mandibles, maxillae, and labium. These articulate with the ventral margin of the capsule. The posterior portion of the

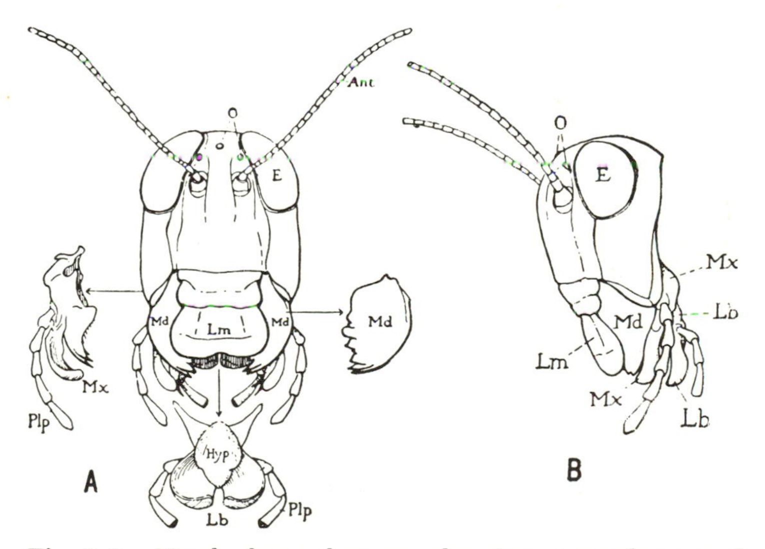

Fig. 3-8. Head of grasshopper, showing appendages and chief organs. (*A*) Anterior aspect; (*B*) lateral aspect. *Ant*, Antenna; *E*, compound eye; *Hyp*, hypopharynx; *Lb*, labium; *Lm*, labrum; *Md*, mandible; *Mx*, maxilla; *O*, ocelli; *Plp*, palpus. (After Snodgrass)

head is shaped like an inverted horseshoe, the capsule forming the dorsal and lateral portion, the labium closing the bottom of the shoe; the open center is called the occipital foramen, through which pass the esophagus, nerve cord, salivary duct, aorta, tracheae, and free blood. Inside the head is a series of braces called the tentorium (Fig. 3-9). The head also contains the brain and other associated structures. The capsule itself bears two sets of visual structures, the compound eyes and the ocelli (Fig. 3-10).

Compound Eyes. Situated one on each side of the head, these are usually large, many-faceted structures (Fig. 3-11). Each eye is

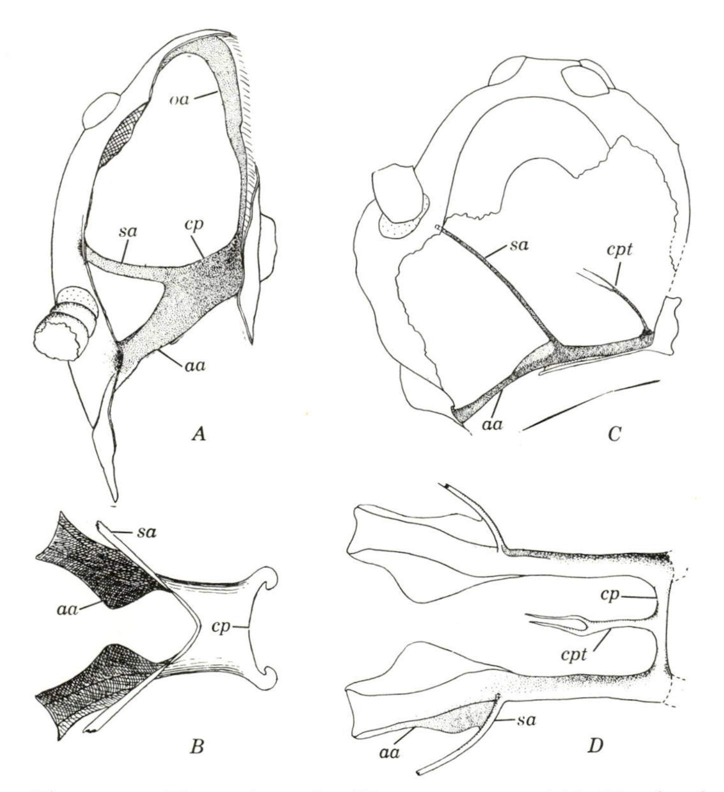

Fig. 3-9. Tentorium in Hymenoptera. (*A*) Head of *Macroxyela*, cut away to show tentorium, lateral aspect; (*B*) tentorium of same, dorsal aspect; (*C*) head of *Aleiodes* cut away to show tentorium, lateral aspect; (*D*) tentorium of same, dorsal aspect. *aa*, Anterior arms; *cp*, corporotentorium; *cpt*, corpotendon; *oa*, dorsal flange extending from the tentorium along the side of the head; *sa*, superior arms.

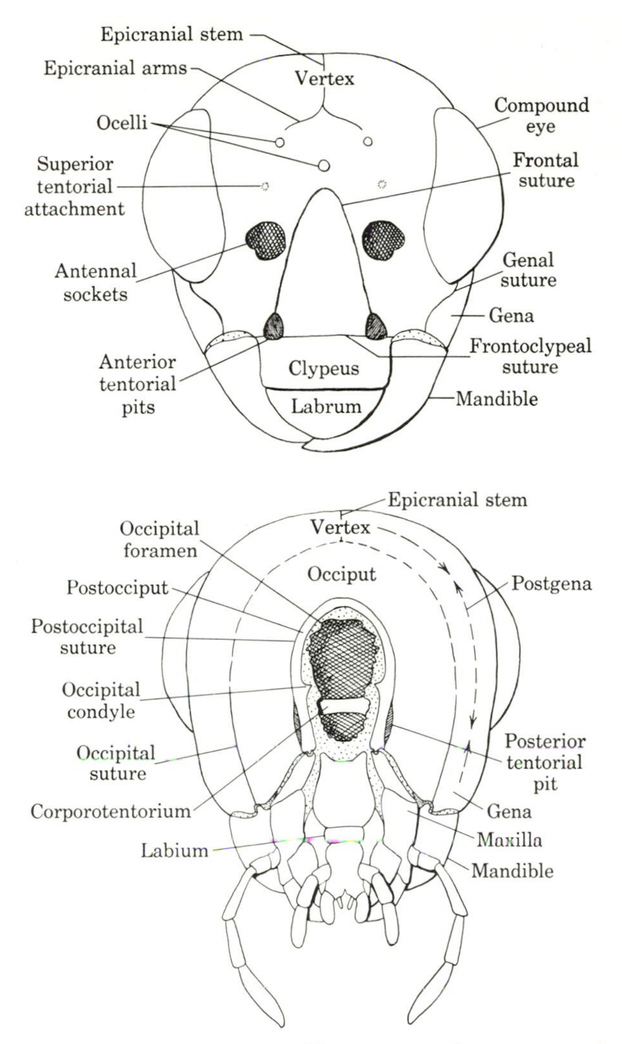

Fig. 3-10. Diagrams illustrating the principal sutures and areas of the head. (*A*) Anterior aspect (maxillae and labium omitted); (*B*) posterior aspect.

situated on or surrounded by a narrow ringlike or shelflike *ocular sclerite*. In the group of insects that have complete metamorphosis (holometabolous insects) the larval eyes, if present, are reduced to one to six single-faceted lenses (Fig. 3-12) called *stemmata*. The group of stemmata on each side of the head is called an *ocularium*. In adult insects the number of facets may be extremely large. The

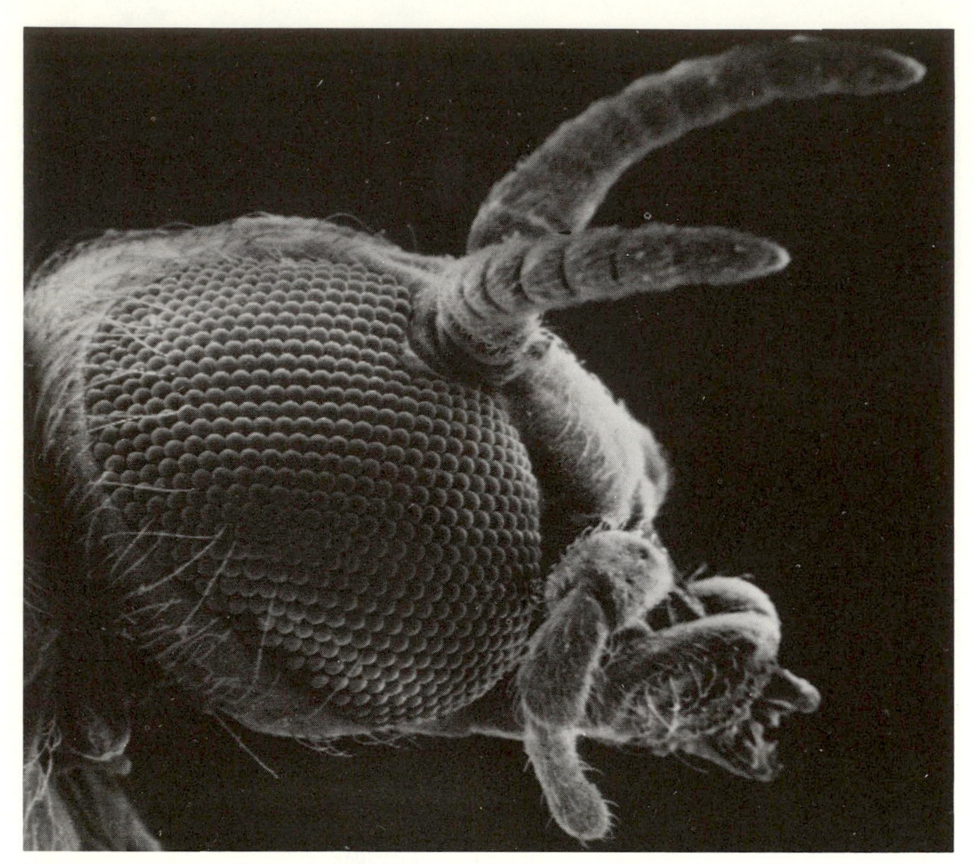

Fig. 3-11. Lateral view of head showing compound eye and mouthparts of adult female black fly *Simulium*, ×150. (Courtesy Dept. of Entomology, Univ. of Alberta, and D. A. Craig)

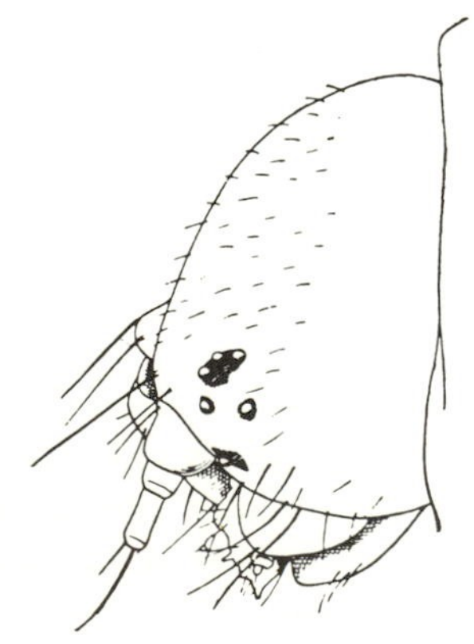

Fig. 3-12. Head of a caterpillar showing the ocularium. (From Folsom and Wardle, 1934, *Entomology*, by permission of The Blakiston Co.)

housefly has about 4000 facets to an eye, and some beetles about 25,000.

Ocelli. These are three single-faceted organs situated on the face and usually between the compound eyes (Fig. 3-10). The upper two are arranged as a pair, one on each side of the meson, and are called the *lateral ocelli*. The lower one is on the meson and is the *median ocellus*. Ocelli are missing or not apparent in many insects.

Clypeus. This is the liplike area between the frontoclypeal suture and the labrum. It never articulates with the frons but is joined solidly with it. The *labrum* hangs below it and articulates by means of the membranous connection between them.

Gena. This is the lower part of the head beneath the eyes and posterior to the frons. There is sometimes a *genal suture* on the anterior portion of the face between the frons and gena; if this suture is absent, the division between the frons and gena is indefinite.

Frons or Front. This is the area on the lower anterior face above the clypeus, bounded laterally by the *frontal* (sometimes called frontogenal) *sutures* and ventrally by the *frontoclypeal suture*. These both originate at the anterior tentorial pits. The frons is frequently absent.

Occiput or Occipital Arch. This is the area comprising most of the back of the head. It is divided from the vertex and genae by the *occipital suture*; in many groups this suture is either reduced to a crease or completely obliterated, in which case the occiput can be defined only as a general area merging anteriorly with vertex and gena. The ventral portions of the entire occipital arch area are sometimes called the *postgenae*.

Postocciput. This is the narrow ringlike sclerite that forms the margin of the occipital foramen. It is separated from the occiput by the *postoccipital suture*, almost universally present in adult insects. The postocciput bears the *occipital condyle*, on which the head articulates with the cervical sclerites of the neck region.

Tentorium (Fig. 3-9). The head is strengthened internally by a set of sclerotized apodemes or invaginations of the body wall that have evolved primarily as more rigid supports for the attachment of muscles connected with the mouthparts. In Collembola, Protura, Diplura, and their allies, the centipedes, the apodemes are more or less platelike or rodlike structures sometimes connected by ligamentous bridges. In the ancestors of the pterygote insects these structures enlarged, fused, and evolved into a strong internal skeleton of the head called the *tentorium*. Typically the tentorium (Fig. 3-9) is composed of four principal parts: the *anterior arms*, *posterior arms*, *corporotentorium* or central mass, and *dorsal arms*. The anterior arms are invaginated from the *anterior tentorial pits*, which usually are well defined externally as pits at each lower corner of the frons. The posterior arms are invaginated from the *posterior tentorial pits*, which almost always persist as external slits on the postoccipital suture. The corporotentorium represents the inward extension, meeting, and fusing of the anterior and posterior arms. The dorsal arms are considered as secondary outgrowths of the anterior arms, because there is no large or persistent pit associated with their point of attachment with the head capsule, usually located between the antennal sockets and lateral ocelli. The shape and relative position of the tentorial parts are extremely different in various groups of insects.

The Head Appendages

The paired appendages of the six trunk segments fused with the ancestral head or acron and became part of the present compound, consolidated structure we call the head.

As is also the case with the segmental appendages occurring on other body regions of insects, the head appendages evolved from simple legs typical of early arthropods. The trilobite leg may have been such a prototype (Fig. 3-13). It consisted of a basal segment or *coxopodite* that presumably articulated with a sclerous plate in the lateral wall of the body, and an apical segmented portion, the *telopodite*. The coxopodite had a long lateral process that probably served the dual function of a respiratory gill and a swimming organ, and often a hard mesal projection, the *endite*. In insects the parts of this prototype became modified in a variety of ways.

In the insect line the lateral process was lost. The result would

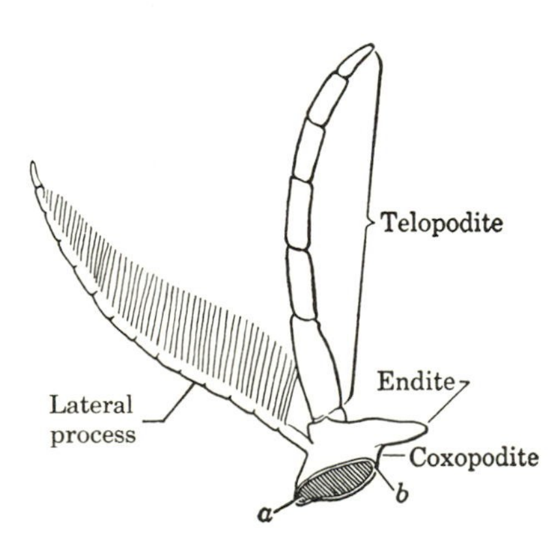

Fig. 3-13. Leg of a trilobite. *a* and *b*, points of articulation. (Redrawn from Snodgrass)

have been a structure similar to the insect antenna. The labrum lost any obvious evidence of segmentation. The intercalary appendages essentially atrophied. In the postoral appendages (mandibles, maxillae, and labium), the coxopodite and its endites appear to have become greatly enlarged and specialized for food handling and the telopodite reduced and specialized as a sensory organ. The mandible lost the telopodite completely, and evolved into a biting, crushing, or sucking structure.

Antennae

Typically in adults the antennae are a pair of movable segmented appendages that arise from the face, usually between the eyes. The first segment is termed the *scape*, the second the *pedicel*, and the remainder together the *flagellum*. They articulate in the antennal socket, which is sometimes surrounded by a narrow ringlike antennal sclerite. The periphery of the socket has a small projection on which the antenna articulates. Embryologically, the antennae are the appendages of head segment 2. Antennae are extremely varied in shape, and names have been applied to the more striking types. A few examples are listed here and illustrated in Fig. 3-14:

Filiform or threadlike	*Clavate* or clubbed
Setaceous or tapering	*Capitate* or having a head
Moniliform or beadlike	*Lamellate* or leaflike
Serrate or sawlike	*Pectinate* or comblike

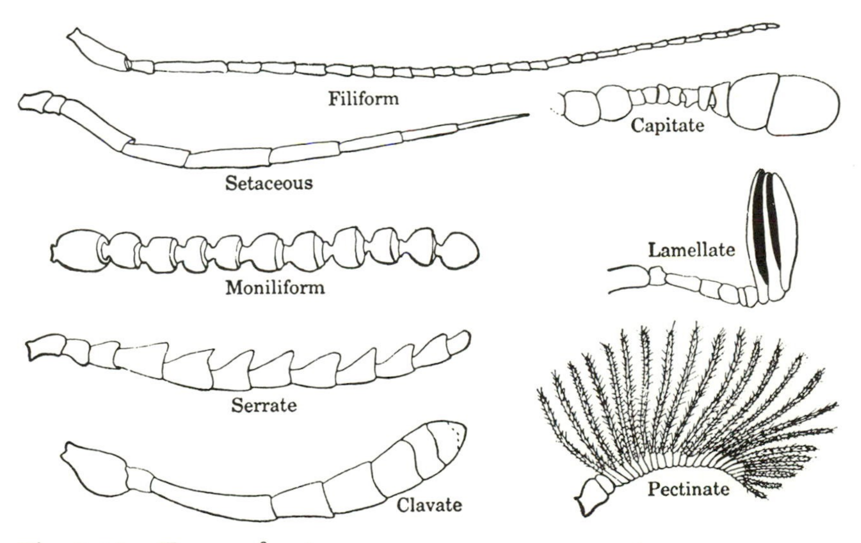

Fig. 3-14. Types of antennae.

In larvae, the antennae are usually greatly reduced both in length and number of segments.

The Mouthparts

The three most conspicuous elements of insect mouthparts are the mandibles, maxillae, and labium. These represent modifications of typical paired arthropod limbs. The shape of these parts in insects is so different from that in the original ancestral forms that evidence from other arthropod groups is necessary to demonstrate the relationship. A study of the appendages of fossil arthropods, together with an analysis of the comparative morphology of appendages of living forms, indicates that all present-day arthropod appendages arose from a simple generalized form.

Generalized Arthropod Appendage (Fig. 3-15). The basal segment or *coxopodite* is implanted in the side of the body wall. The apical segments form the *telopodite*. Each segment has potentialities for developing processes on both the lateral and mesal sides, the lateral processes called *exites*, and the mesal processes called *endites*. A primitive and early modification is illustrated by the leg of the trilobite (Fig. 3-13). Note that the coxopodite has a gill-like exite and a spurlike endite; the telopodite is simple and without processes.

Intercalary Appendages (Fig. 3-6). These are a pair of simple appendages that appear on head segment 3 early in embryological development in some insects, then disappear. The intercalary segment at an early stage of embryonic development apparently marks the position of the mouth.

Labrum. Typically this is a movable flap hanging down from the edge of the clypeus (Fig. 3-8, *Lm*), and covering the mouth. Its inner side forms the front of the pre-oral cavity and its called the *epipharynx*, which usually bears raised lobes and sensory setae of many types. The labrum is the fused pair of appendages of head segment 1.

Mandibles (Figs. 3-8, 3-14, 3-16). The mandibles are the paired appendages of head segment 4, which is immediately behind the mouth; they appear to be directly behind the labrum. Typically they are hard and sclerous, and have various sets of teeth and brushes. The palp has been lost. The mandibles therefore appear to be the modified coxopodites and their endites of the primeval crustacean segment. The entognathan orders have only a single point of articulation with the lateral margin of the head. In the Thysanura and

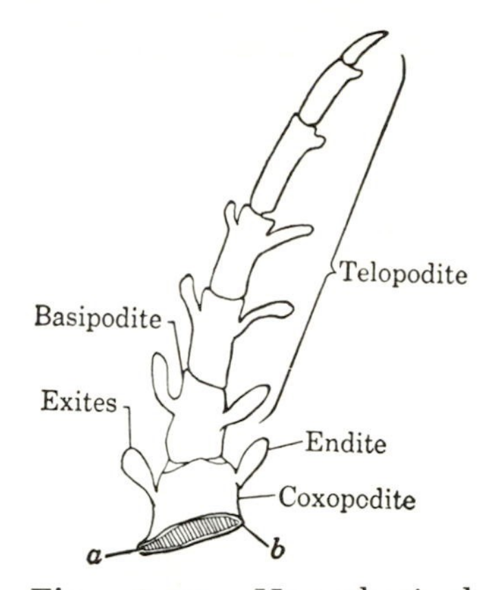

Fig. 3-15. Hypothetical arthropod appendage. *a* and *b*, points of articulation. (Modified from Snodgrass)

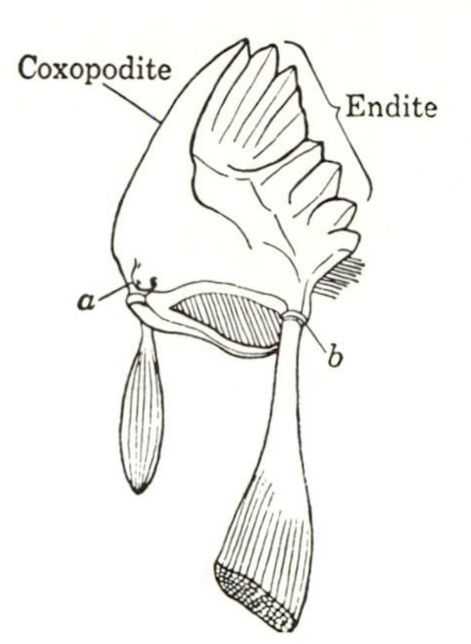

Fig. 3-16. Mandible of a grasshopper. *a* and *b*, points of articulation. (Redrawn from Snodgrass)

winged insects there is a second, more dorsal or mesal point of articulation that provides a more powerful opening and closing motion for these appendages. At each of these articulations arise strong tendons that project into the head and serve as attachments for the muscles that operate the mandibles.

Maxillae (Fig. 3-17). The maxillae lie directly behind the mandibles and are the paired appendages of head segment 5. They have only a lateral articulation with the base of the head capsule. The coxopodite is composed of two parts, the cardo and stipes, and bears the remaining parts of the maxilla.

In entomological literature, little reference is made to the terms coxopodite, telopodite, endite, and so forth, which refer to the fundamental divisions and processes of arthropod appendages. In this account they have been used until now to assist in correlating insect parts with those of other Arthropoda. From this point on, however, it is pertinent to make a change in terminology and employ those terms usually applied in entomological usage. These terms refer to parts that are in most instances differentiated only in insects, and the terms are therefore necessary for accurate identification of the part or area.

The generalized type of maxilla is a masticating structure that is divided into several well-marked parts (Fig. 3-17) as follows.

Cardo. The cardo is the triangular basal sclerite that is attached to the head capsule, and that serves as a hinge for the movement of the remainder of the maxilla.

Stipes. The central portion or body of the maxilla is called the stipes and is usually somewhat rectangular in shape. The stipes is situated above the cardo and is the basis for the remaining parts of the maxilla.

Galea. The galea is the outer (lateral) lobe articulating at the end of the stipes. It is frequently developed as a sensory pad or bears a cap of sense organs. The galea and lacinia are endites.

Lacinia. The inner (mesal) lobe articulating at the apex of the stipes, the lacinia, is usually mandible-like in general form with a series of spines or teeth along its mesal edge.

Palpus. The antenna-like segmented appendage that arises from the lateral side of the stipes, the palpus is commonly five-segmented. Presumably, it is entirely sensory in function.

Labium (Figs. 3-14, 3-18). This structure, representing the paired appendage of head segment 6, forms the lip posterior to the maxillae. It appears to be a single unit but really consists of a second pair of maxillae that have fused on the meson to form a single functional structure. The parts of the labium correspond closely to those of the

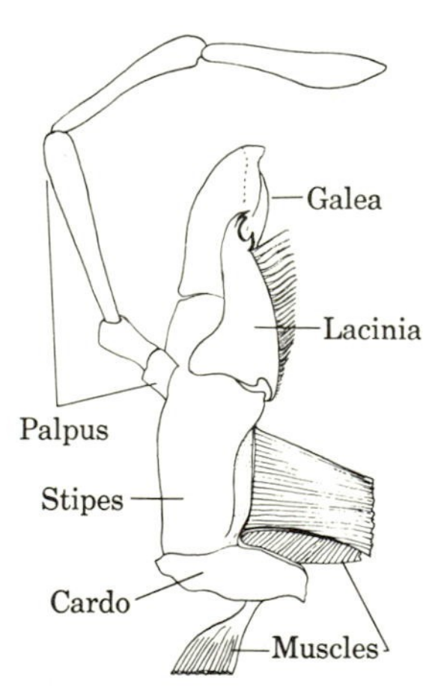

Fig. 3-17. Maxilla of a cockroach, illustrating a generalized type. (After Snodgrass)

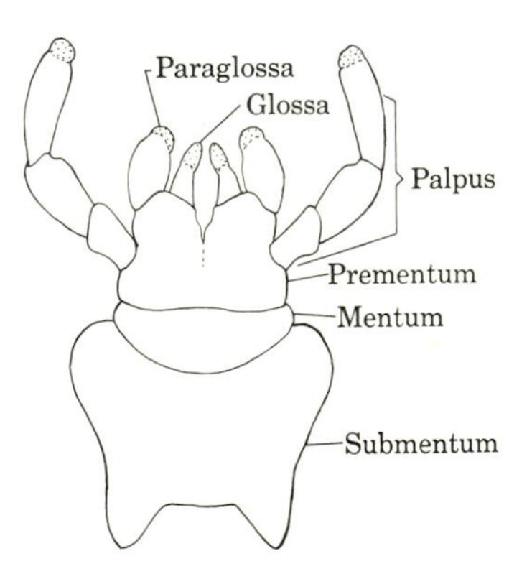

Fig. 3-18. Labium of a cockroach, illustrating a generalized type. (Adapted from Imms)

maxillae, and their homologies have been established by studies of muscles and their points of attachment.

The accompanying table illustrates the homologies of the corresponding parts of the maxillae and labium. When consulting this table, refer also to Figs. 3-17, and 3-18. A description of these parts follows.

Table of Homologies

MAXILLA		LABIUM
Cardo	corresponds to	Postlabium { submentum, mentum }
Stipes	corresponds to	Prementum or stipulae
Palpi	corresponds to	Palpi
Lacinia	corresponds to	Glossae
Galea	corresponds to	Paraglossae

Postlabium. The basal region of the labium, the postlabium, hinges with the head membranes. It is frequently divided into two parts: a basal *submentum* and an apical *mentum*. The postlabium represents the fused cardines of the maxillae.

Prelabium. The apical region of the labium, the prelabium includes various lobes and processes. The central portion or body is the *prementum* (sometimes also called *stipulae*), which bears a pair of *labial palpi*, one on each side of the prementum, and each usually three-segmented in generalized forms.

The apical portion of the prelabium frequently forms a sort of tongue and for this reason is called the *lingula*. It varies greatly in structure but usually is divided into two pairs of lobes: (1) the *glossae*, a pair of mesal lobes usually close together; and (2) the *paraglossae*, a pair of lateral lobes that usually parallel the glossae. In many groups such as the Hymenoptera, the glossae are fused to form an *alaglossa* (Fig. 3-19). In other cases the glossae and paraglossae may be fused together into a single solid lobe called a *totoglossa* (Fig. 3-20).

Hypopharynx. From the ventral membranous floor of the head arises the hypopharynx (Fig. 3-21). It usually forms a protruding lobe or mound. In generalized insects the hypopharynx is so closely associated with the base of the labium as to be considered a part of it. Unlike the other mouthparts, the hypopharynx is not an appendage but an unsegmented outgrowth of the body wall. On it is situated the opening of the salivary gland.

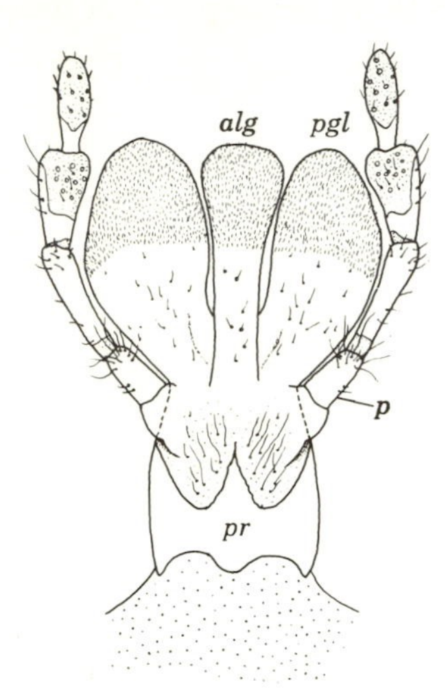

Fig. 3-19. Part of labium of hymenopteran *Trichiosoma triangulum*, illustrating an alaglossa. *alg*, Alaglossa; *p*, palpus; *pgl*, paraglossa; *pr*, prementum.

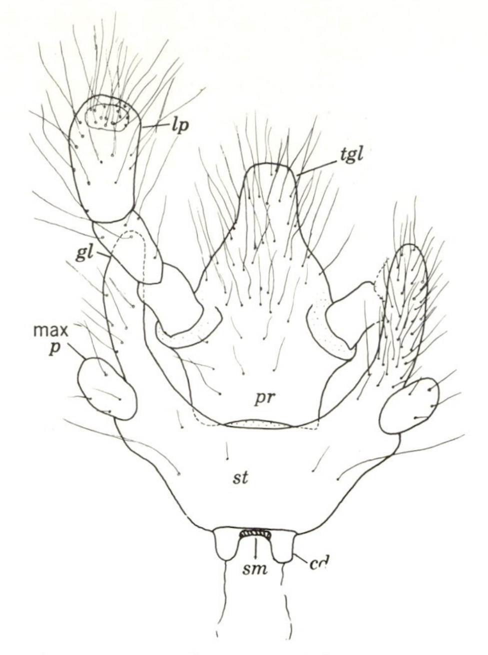

Fig. 3-20. Labium and fused maxillae of hymenopteran *Tremex columba*, illustrating a totoglossa. *cd*, Cardo; *gl*, fused galea and lacinia; *lp*, labial palpus; *max p*, maxillary palpus; *pr*, prementum; *sm*, submentum; *st*, fused stipites; *tgl*, totoglossa.

Superlinguae. In a few primitive insects a pair of simple lobes occur in close association with the hypopharynx. Embryologically they appear as lobes associated with the mandibles, but in advanced embryos they become more closely attached to the hypopharynx. These superlinguae occur in a few insect orders such as the Thysanura and Ephemeroptera, in some Crustacea, and in some members of the Symphyla. They apparently are a primitive arthropodan character that has atrophied in the more specialized insects.

Principal Types of Mouthparts

Insect mouthparts have become modified in various groups to perform the ingestion of different types of food and by different methods. The more diverse and interesting types are listed here, chosen to illustrate the varied shapes assumed by homologous parts,

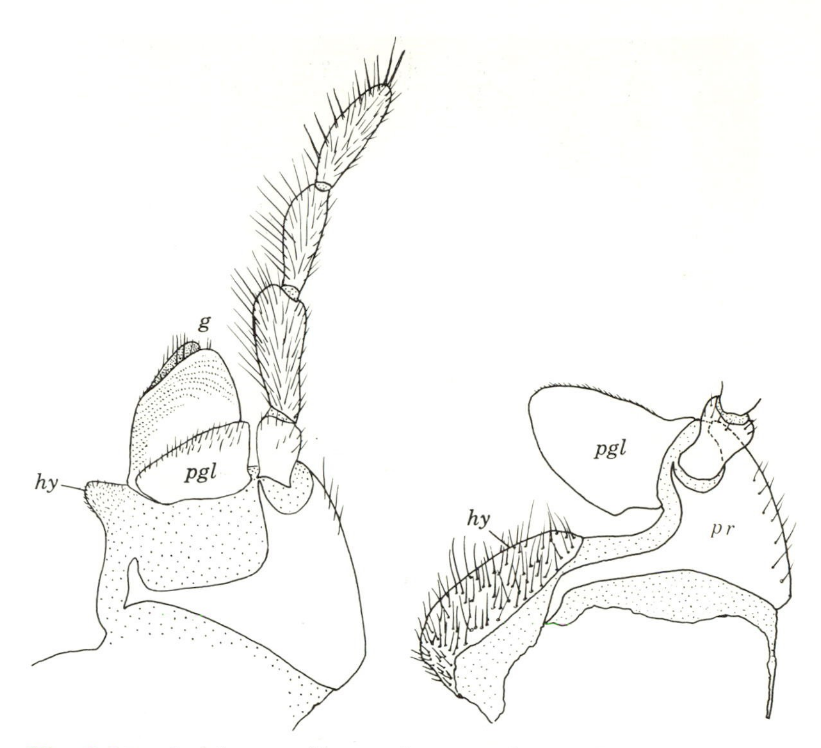

Fig. 3-21. Labium and hypopharynx of a sawfly *Arge pectoralis* (right), and braconid *Aleiodes terminalis* (left). g, Glossa; *hy*, hypopharynx; *pgl*, paraglossa; *pr*, prementum.

and the different uses to which they are put. Many other types exist, many of them representing intermediate stages between some of the types treated here.

Chewing Type. In this type the various appendages are essentially as in the preceding Figs. 3-16, 3-17 and 3-18; see also Figs. 3-22, 3-23. The mandibles cut off and grind solid food, and the maxillae and labium push it into the esophagus. Grasshoppers and lepidopterous larvae are common examples. It seems certain that the chewing type of mouthparts is the generalized one from which the other types developed. This view is upheld by important evidence of two kinds. In the first place, such mouthparts are most similar in structure to those of the centipedes and symphylids, which are the closest allies of the insects. In the second place, chewing mouthparts occur in almost all the generalized insect orders, such as the cockroaches, grasshoppers, and thysanurans; and they occur in the larvae of at least the primitive families of the neuropteroid orders. In many of

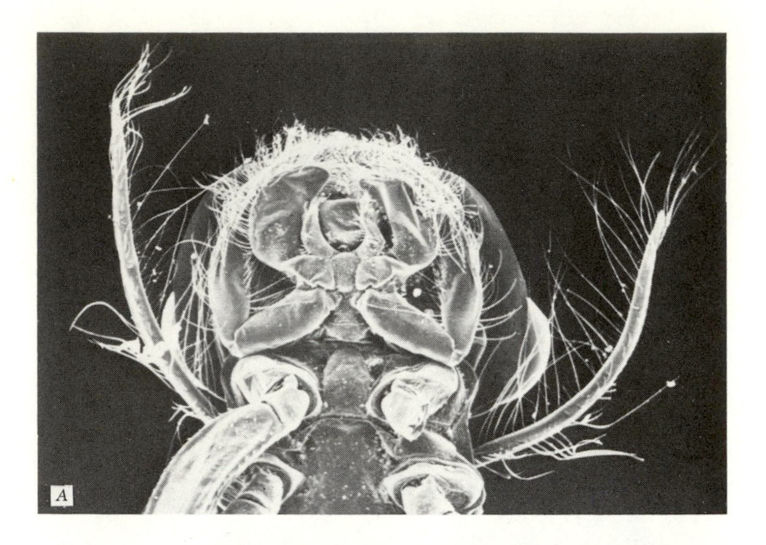

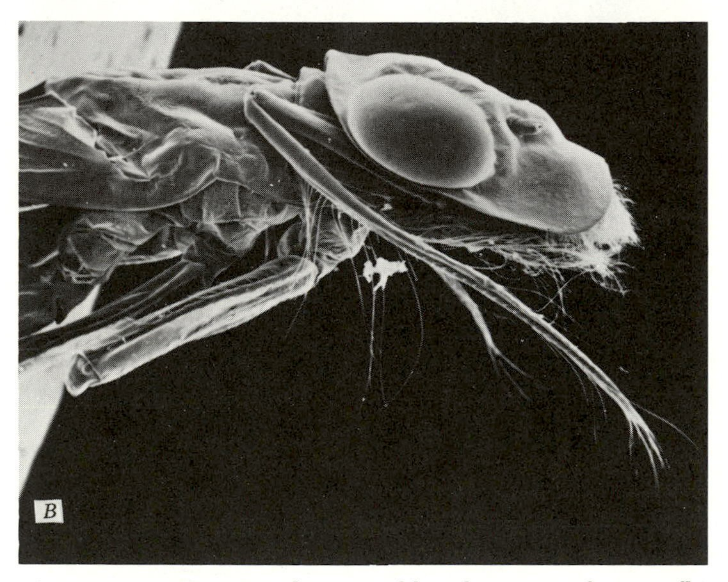

Fig. 3-22. (*A*) Ventral view of head region of a mayfly nymph *Arthroplea*. The maxillary palps are used as filtering structures. (*B*) Lateral view of same specimen; ×35. (Courtesy Dept. of Entomology, Univ. of Alberta, and D. A. Craig)

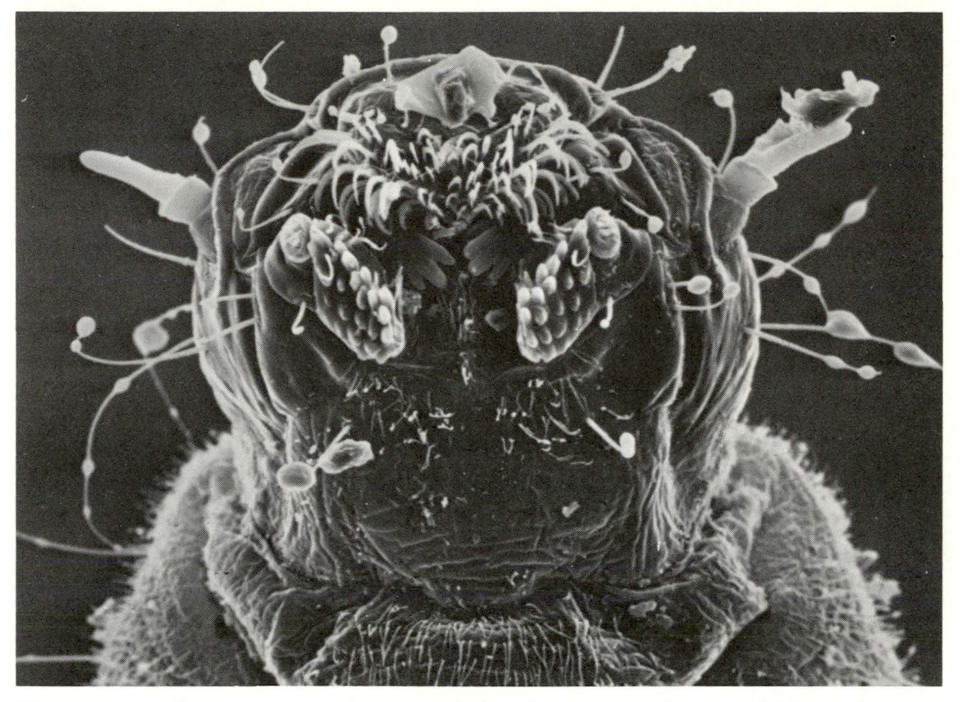

Fig. 3-23. Chewing mouthparts of the first instar larva of *Microchorista philpoti* (Mecoptera; Nannochoristidae); ×60. (Courtesy Dept. of Entomology, Univ. of Alberta, and D. A. Craig)

the neuropteroid orders the adults frequently also have the chewing type of mouthparts which is little changed from the primitive type: for example, the Coleoptera and most of the Hymenoptera.

Filtering Type. Some aquatic larvae, particularly some of the black flies (Simuliidae), have an elaborate cephalic fan used in filtering microorganisms (Fig. 3-24). Others are both filter feeders and browsers and the cephalic fans are less elaborate (Fig. 3-25).

Cutting-Sponging Type (Fig. 3-26). In horseflies (Tabanidae) and certain other Diptera, the mandibles are produced into sharp blades, and the maxillae into long probing styles. The two cut and tear the integument of a mammal, causing blood to flow from the wound. This blood is collected by the spongelike development of the labium and conveyed to the end of the hypopharynx. The hypopharynx and epipharynx fit together to form a tube through which the blood is sucked into the esophagus.

Fig. 3-24. Cephalic fan and mouth region of the black fly *Simulium tahitiensis*; ×90. (Courtesy Dept. of Entomology, Univ. of Alberta, and D. A. Craig)

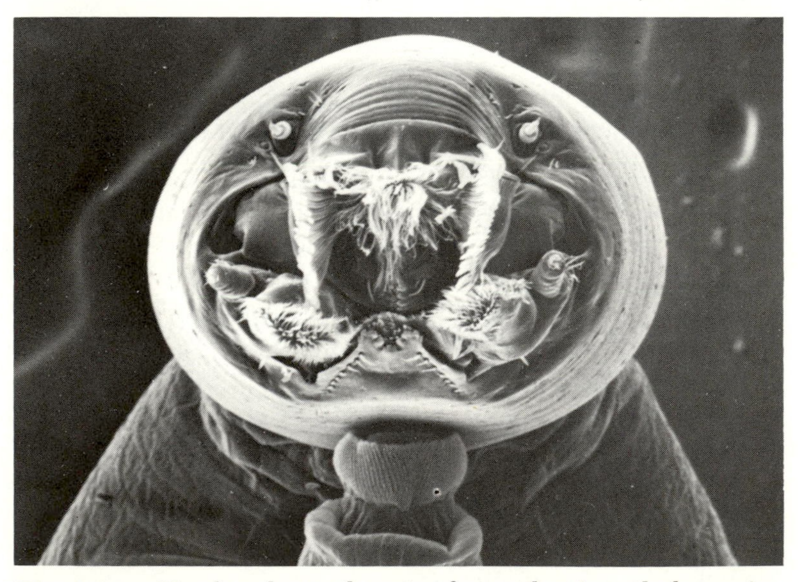

Fig. 3-25. Head and mouthparts of a predominantly browsing black fly larva *Crozetia crozetensis* with reduced cephalic fans; ×150. (Courtesy Dept. of Entomology, Univ. of Alberta, and D. A. Craig)

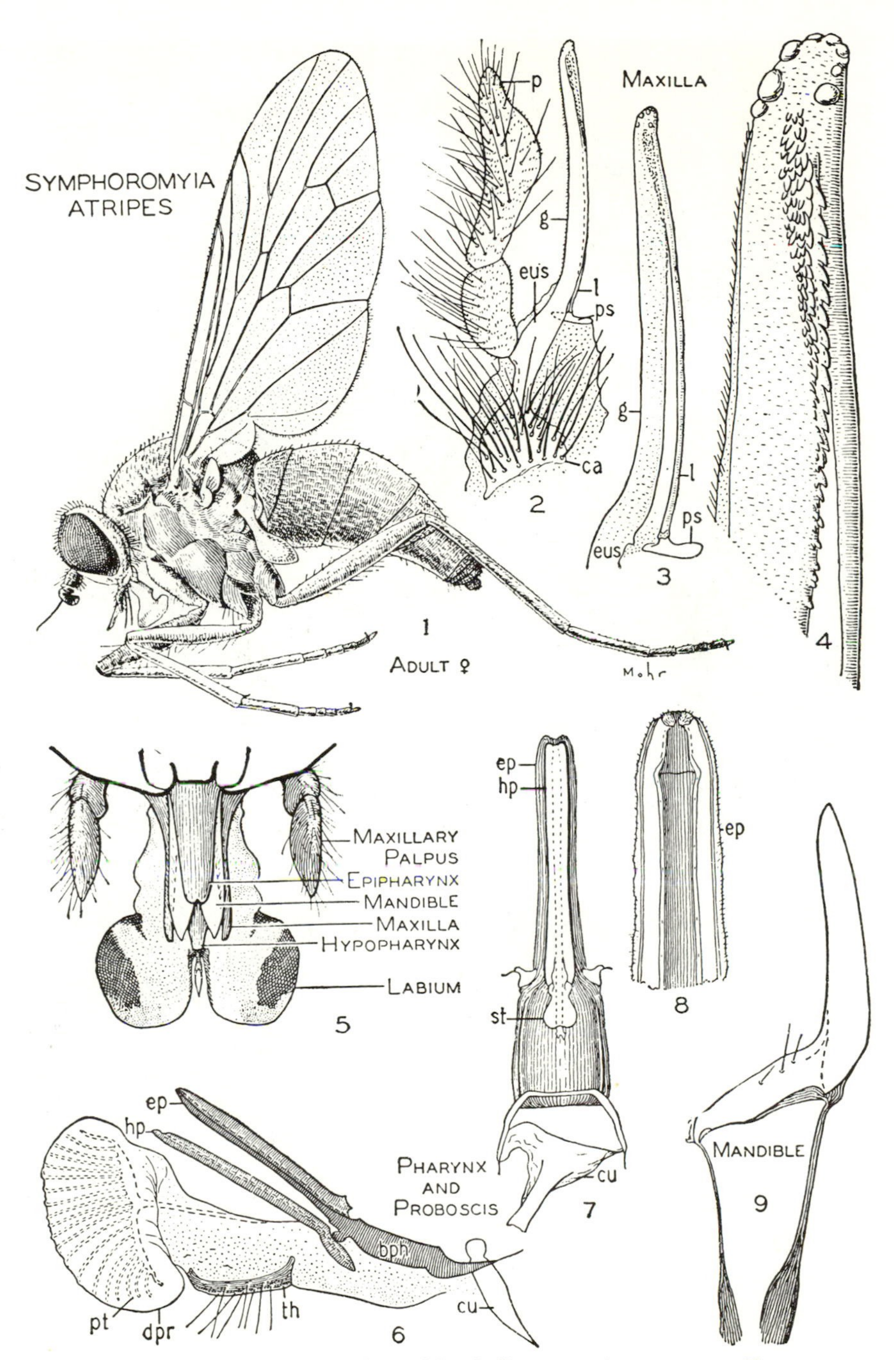

Fig. 3-26. Mouthparts of false black fly *Symphoromyia*, illustrating cutting-sponging type. (From Illinois Natural History Survey)

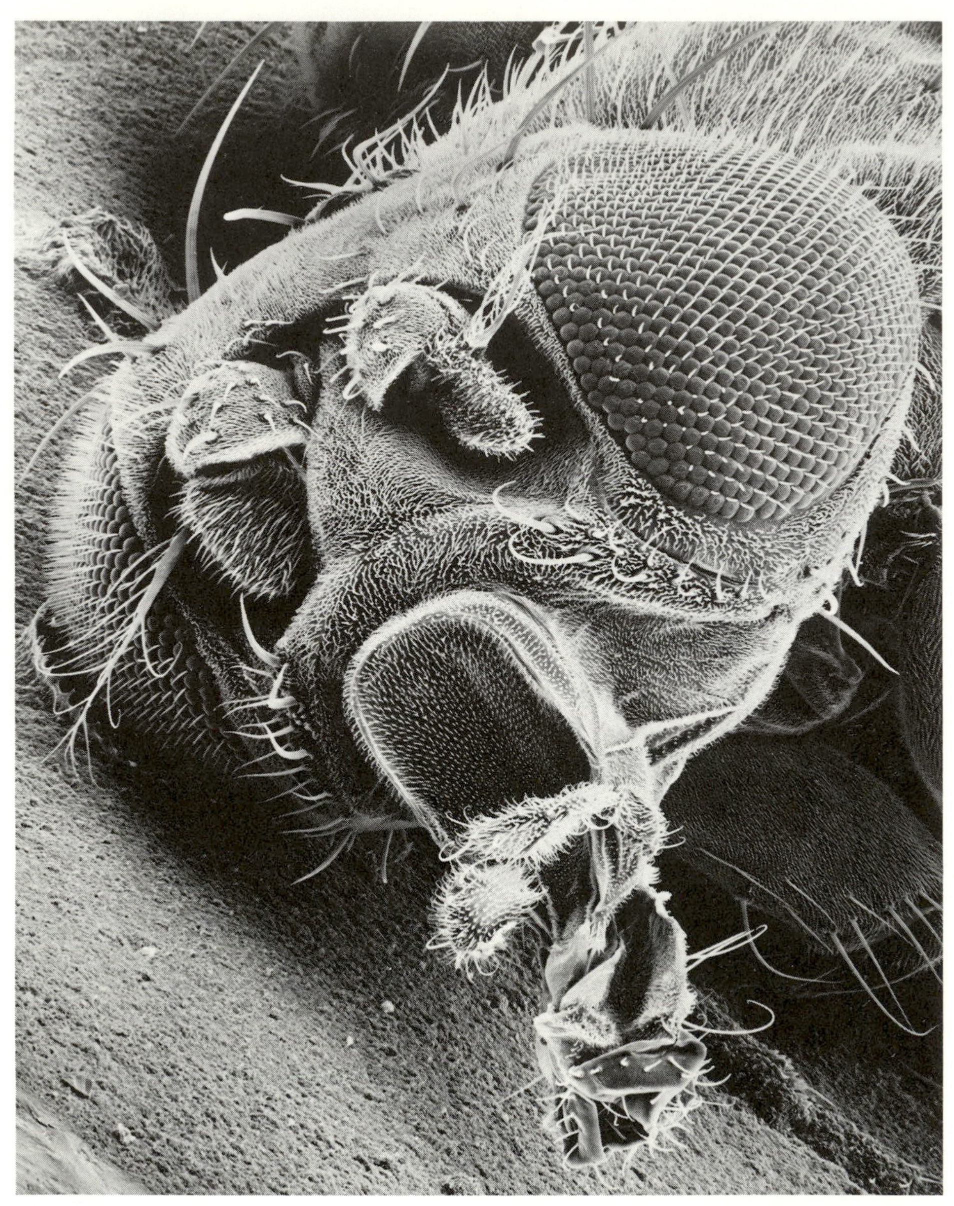

Fig. 3-27. View of head of fruit fly *Drosophila melanogaster* showing detail of appendages and mouthparts; ×180. (Courtesy of Nanometrics, Inc.)

Sponging Type (Fig. 3-27, 3-28). A large number of the nonbiting flies, including the housefly, have this type, fitted for using only foods that are either liquid or readily soluble in saliva. This type is most similar to the cutting-sponging type, but the mandibles and maxillae are nonfunctional, and the remaining parts form a proboscis with a spongelike apex, or labella. This is thrust into the liquid food, which is conveyed to the food channel by minute capillary channels on the surface of the labella. The food channel is formed in this type also by the interlocking elongate hypopharynx and epipharynx, which form a tube leading to the esophagus. Certain solid foods, such as sugar, are eaten by flies with these mouthparts. This is accomplished as follows: First, the fly extrudes a droplet of saliva onto the food, which dissolves in the saliva; this solution is then drawn up into the mouth as a liquid.

Chewing-Lapping Type. Another type of mouthparts for taking up liquid food is found in the bees and wasps, exemplified by the honeybee (Fig. 3-29). The mandibles and labrum are of the chewing type

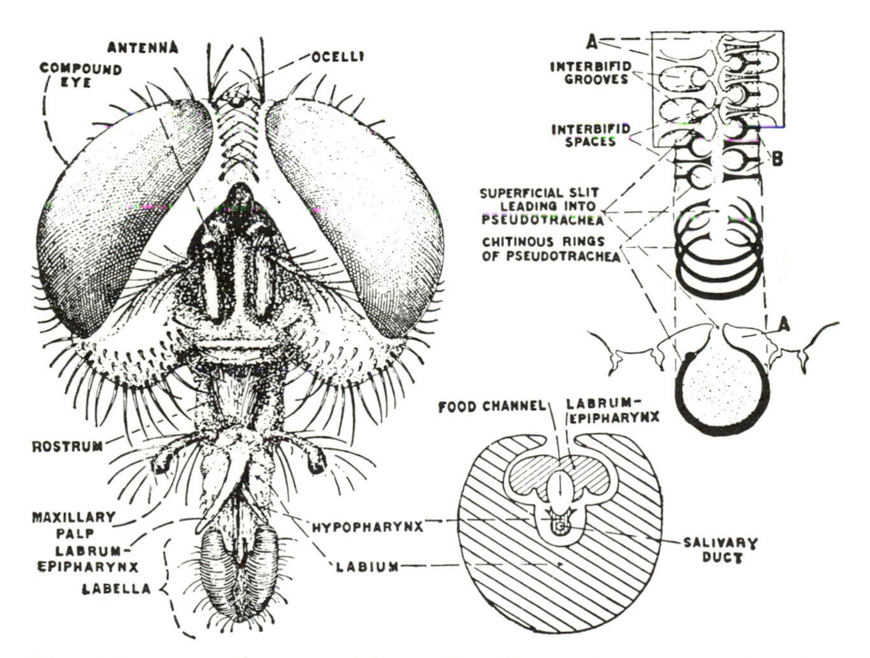

Fig. 3-28. Mouthparts of housefly, illustrating a sponging type. (From Metcalf and Flint, *Destructive and useful insects*, by permission of McGraw-Hill Book Co.)

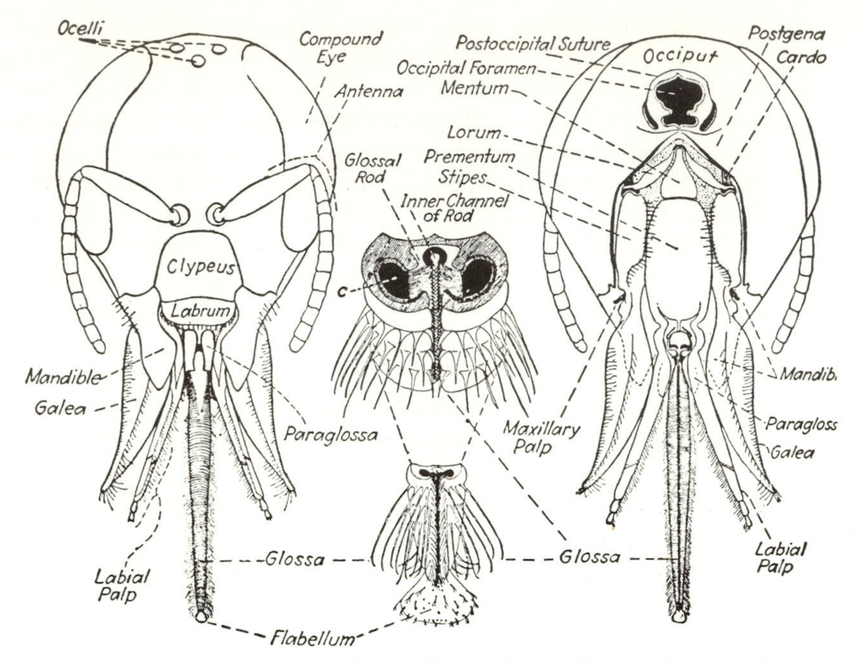

Fig. 3-29. Mouthparts of the honeybee, illustrating a chewing-lapping type. (From Metcalf and Flint, *Destructive and useful insects*, by permission of McGraw-Hill Book Co.)

and are used for grasping prey or molding wax or nest materials. The maxillae and labium are developed into a series of flattened elongate structures, of which the alaglossa (usually called the glossa) forms an extensile channeled organ. This latter is used to probe deep into nectaries of blossoms. The other flaps of the maxillae and labium fit up against the glossa and form a series of channels down which the saliva is discharged and up which food is drawn. There is some difference of opinion among observers as to the exact mechanics by which passage of the liquids is attained.

Piercing-Sucking Type (Fig. 3-30). The mouthparts of many groups of insects are modified to pierce tissues and suck juices from them. This includes aphids, cicadas, leafhoppers, scale insects, and others that suck juices from plants; assassin bugs, water striders, and predaceous forms of many sorts that suck juices from insects and other small animals; and mosquitoes, bedbugs, lice, and fleas, which suck blood from mammals and birds. In this group the labrum, mandibles, and maxillae (sometimes the hypopharynx also) are slender and long and fit together to form a delicate hollow needle (Figs. 3-31,

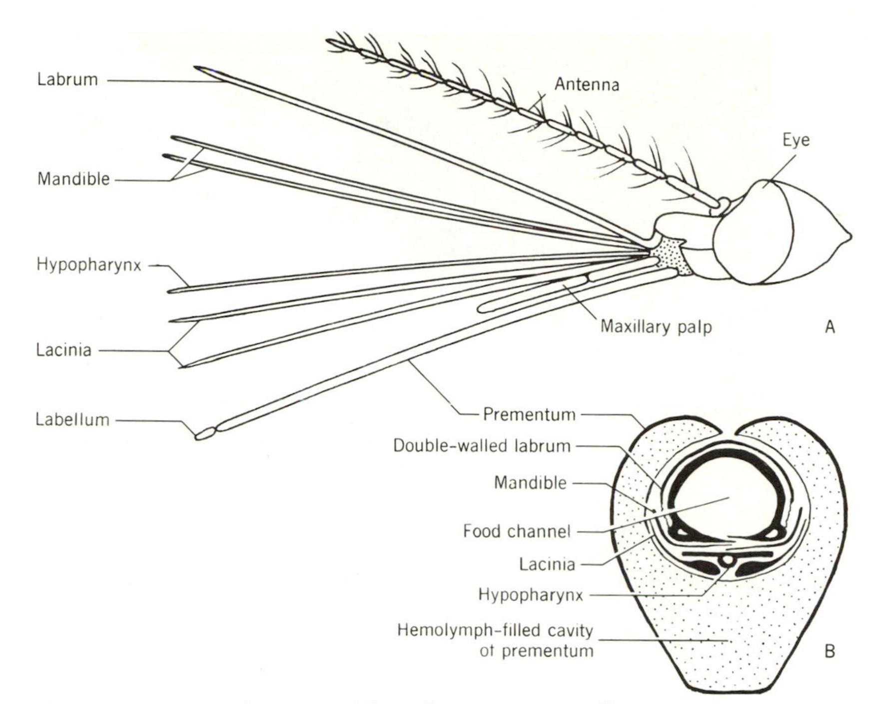

Fig. 3-30. Mouthparts of female mosquito, illustrating a piercing-sucking type. The double wall of the labrum permits this piece to be guided inside the host tissue. (*A*) Head and separated mouthparts; (*B*) cross section near middle of beak. (*A*, after Waldbauer; *B*, modified from Waldbauer)

3-32). The labium forms a stout sheath that holds this needle rigid. The entire structure is called a beak. To feed, the insect presses the entire beak against the host, then inserts the needle into the host tissues, and sucks the host juices through the needle into the esophagus.

An interesting and apparently generalized kind of this type of mouthparts is found in the thrips (see p. 331). Several components of the mouthparts are stylet-like but together form a rasping cone rather than a beak. On the extreme of complexity are the sucking lice or Anoplura, which have definite retractile beaks whose structure is so modified that only a few of the original parts remain (see page 327).

Siphoning-Tube Type (Fig. 3-33). Adult Lepidoptera feed on nectar and other liquid food. These are sucked up by means of a long proboscis, composed only of the united galea of each maxilla. These form a tube that opens into the esophagus.

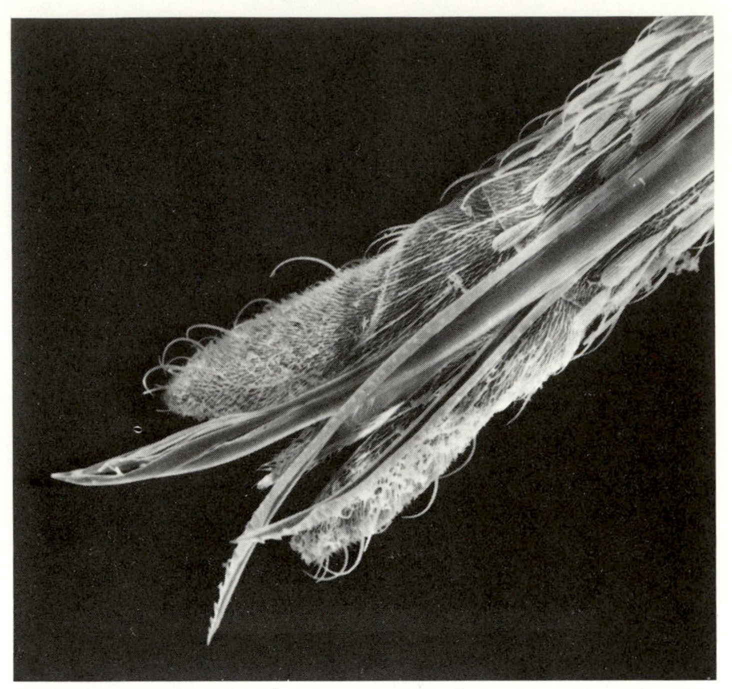

Fig. 3-31. Mouth of mosquito *Aedes aegypti*, showing sensory-covered labium and sawlike lacinia; ×300. (Courtesy Dept. of Entomology, Univ. of Alberta, and D. A. Craig)

CERVIX OR NECK

Between the head and trunk is a membranous region that forms a neck, or *cervix*. This has sometimes been regarded as a separate body segment, the microthorax, but little evidence has been found to support this view. It seems more likely that the cervix includes areas of both the labial head segment and the prothoracic segment, forming a flexible area between the two.

Embedded in the cervix are two pairs of *cervical sclerites* (Fig. 3-34), which serve as points of articulation for the head with the trunk. The two sclerites on each side are hinged with each other to form a single unit, which articulates anteriorly with the occipital condyle on the postocciput of the head and posteriorly with the prothorax. Frequently the cervical sclerites are fused with the pleurae of the prothorax.

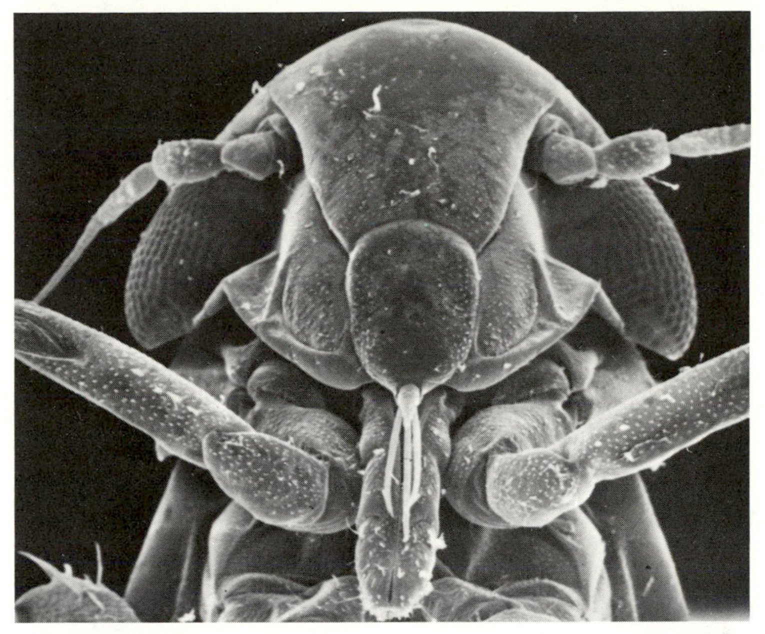

Fig. 3-32. Piercing-sucking mouthparts of first instar of the plant hopper *Macrosteles focifrons*; ×115. (Courtesy Dept. of Entomology, Univ. of Alberta, and D. A. Craig)

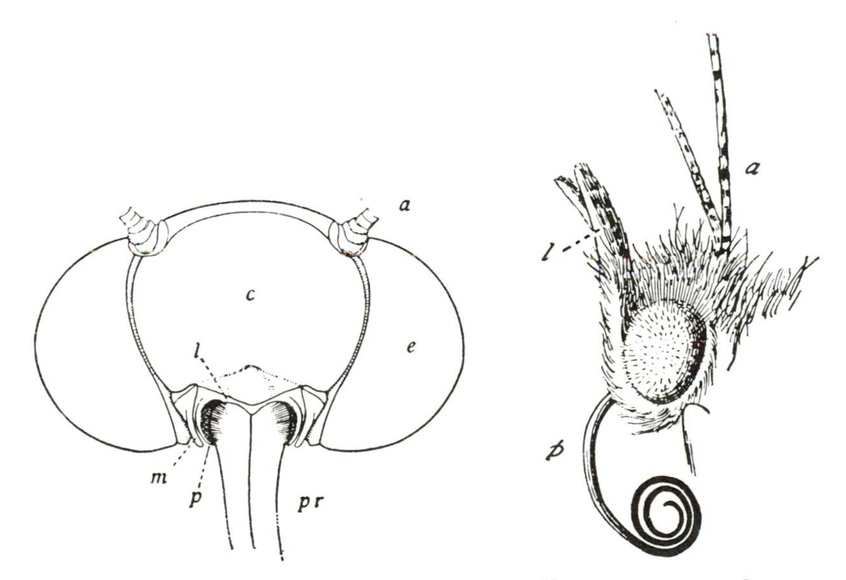

Fig. 3-33. Mouthparts of Lepidoptera, illustrating a siphoning type. Left, a moth: *a*, antenna; *c*, clypeus; *e*, eye; *l*, labrum; *m*, mandible; *p*, pilifer; *pr*, proboscis. Right, a butterfly: *a*, antennae; *l*, labial palpus; *p*, proboscis. (From Folsom and Wardle, 1934, *Entomology*, by permission of The Blakiston Co.)

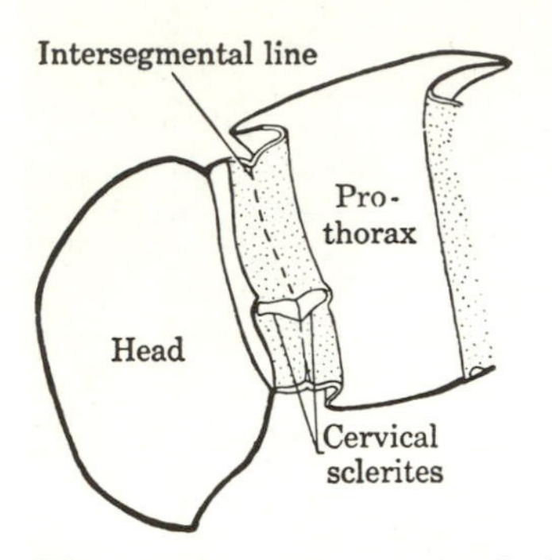

Fig. 3-34. Diagram of the cervical sclerites of an insect. (Modified from Snodgrass)

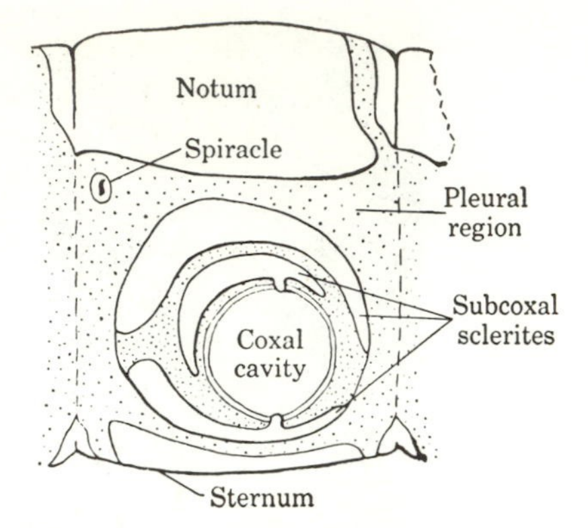

Fig. 3-35. Diagram of a simple insect body segment. (Redrawn from Snodgrass)

DEVELOPMENT OF THE GENERALIZED INSECT SEGMENT

The structure of existing Chilopoda, Entognatha, and primitive insects suggests that the body segments evolved from a very simple type (Fig. 3-35), composed of five elements:

1. The tergum or sclerotized dorsal plate, called the notum when referring to the thoracic tergum.
2. The sternum or sclerotized plate.
3. The pleural region connecting tergum and sternum; it is entirely membranous.
4. A pair of segmented legs; the basal segment or coxopodite of each leg is embedded in the membrane between the tergum and sternum. The coxopodite is divided into a basal portion (*subcoxa*) and an apical portion (*coxa*). In Fig. 3-35 the subcoxa is divided into three sclerites.
5. A pair of spiracles, one in the membrane above each leg.

In a few archaic groups, such as the proturans, and in the Chilopoda is found a type of segment (Fig. 3-36), which represents the simple prototype. The tergum and sternum are unchanged. The subcoxa is represented by cresentic sclerites, or areas, one mesad of the coxa, and two laterad of it. The latter two appear to be units between the coxa of the leg and the tergum. These detached subcoxal sclerites in the pleural region are the forerunners of the pleural sclerites. The coxa forms the functional articulating base of the leg.

In the next evolutionary development (Fig. 3-37), the subcoxal sclerites became immovably implanted in the segmental wall to form a solid base on which the functional leg articulates. The mesal subcoxal sclerites fused with the sternum, and the lateral subcoxal sclerites became flattened adjacent sidepieces together called the pleuron (pl., *pleura*). This condition is considered the generalized one from which developed both the specialized thoracic wing segments and the simplified abdominal segments. It is illustrated by many living forms, notably in the thorax of the immature stages of many such diverse groups as stoneflies, caddisflies, and lacewings (Fig. 3-37).

THORAX

The thorax is the body region between the head and abdomen. It is composed of three segments, the prothorax, mesothorax, and metathorax, respectively.

In those orders that never developed wings, the three segments are nearly alike in general structure. The tergum and sternum are platelike, and the pleural sclerites (the subcoxal arcs) are small or degenerate (Fig. 3-36).

In the orders of winged insects, the three thoracic segments are extremely dissimilar. The prothorax has essentially the same parts as

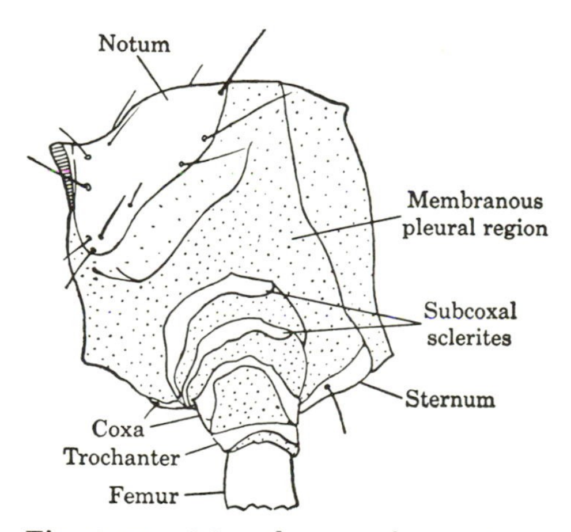

Fig. 3-36. Mesothorax of a proturan *Acerentomon*. (Redrawn from Snodgrass)

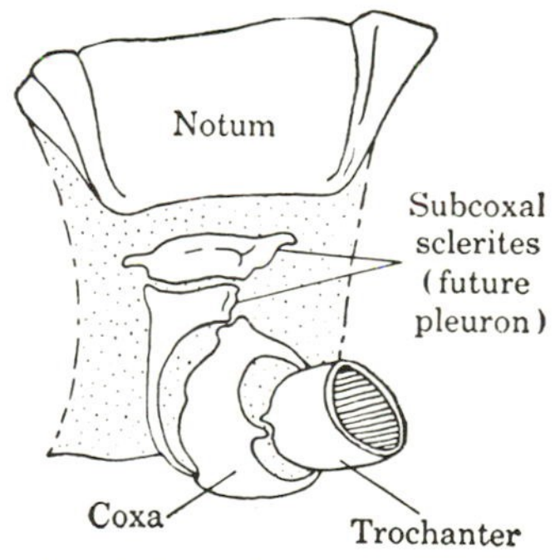

Fig. 3-37. Prothorax of a stonefly nymph *Perla*. (Redrawn from Snodgrass)

the basic condition (Fig. 3-37), although the various sclerites may be consolidated or recombined to such an extent that their exact interpretation may be difficult. The mesothorax and metathorax have undergone a veritable morphological revolution correlated with the musculature necessary to combine both running and flying mechanisms in one segment. Many new sclerites have been added, and many of them regrouped.

Generalized Winged Segment. There are three principal areas in this segment as in the general wingless one: the tergum (called the *notum* when applied to segments of the thorax), the sternum, and the pleura. Their derivation from the primitive body segment is shown diagramatically in Fig. 3-38. Each has many specializations, but those of the pleura are the most conspicuous external features accompanying the winged condition. In existing forms of winged insects the parts follow the generalized condition illustrated in Fig. 3-39.

Pleuron. This sclerite has become enlarged to form a conspicuous lateral plate. It has a ventral coxal process against which the leg articulates and a dorsal wing process against which the wing articulates. The pleuron is divided into an anterior portion, the *episternum*, and a posterior portion, the *epimeron*, by a *pleural suture* that extends from the coxal process to the wing process. This suture marks the line of invagination of the internal pleural apodeme, the *pleurodema*. Anteriorly and posteriorly the pleuron is fused with the sternum, the areas of fusion forming bridges before and behind the coxal cavities.

Notum. This area is divided into two principal sclerites, the anterior *alinotum* and the posterior *postnotum*. The alinotum has an anterior apodeme or *phragma* and also is the sclerite connecting

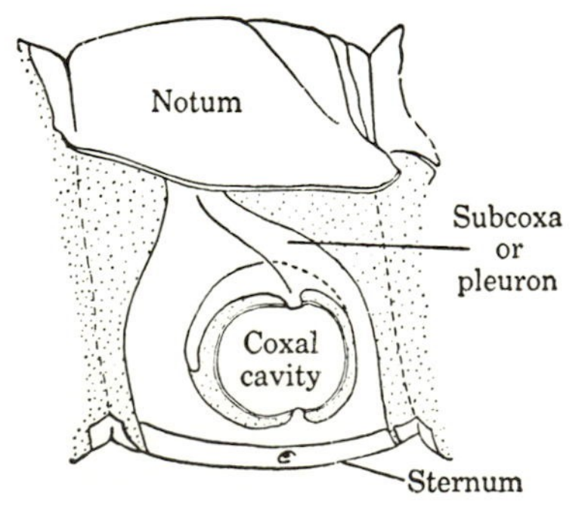

Fig. 3-38. Diagram of a hypothetical step in the development of a winged segment. (Redrawn from Snodgrass)

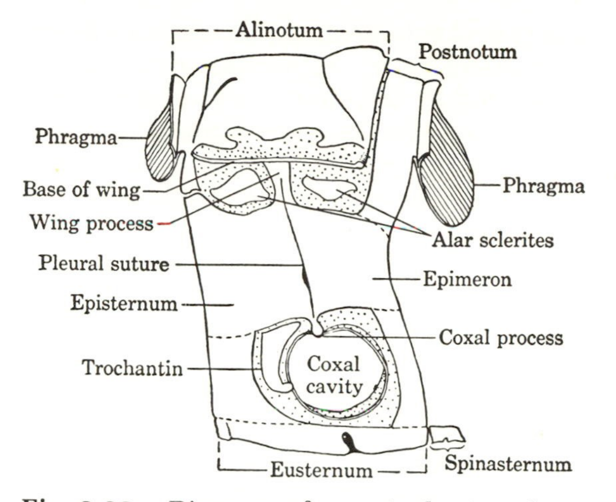

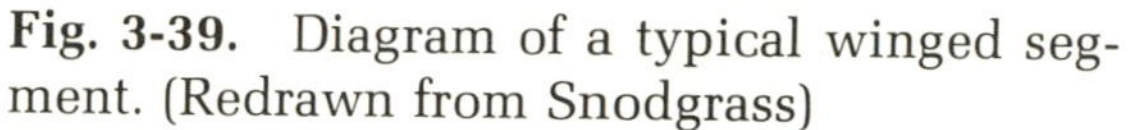

Fig. 3-39. Diagram of a typical winged segment. (Redrawn from Snodgrass)

directly with the wing. It is subdivided in various patterns in different groups. The postnotum also bears a phragma and is connected laterally not with the wings but with the epimeron of the pleura to form a bridge behind the wings. The postphragmal portion of the postnotum really is a part of the next posterior segment that has become a functional part of the one in front. Transpositions of this type are frequent in insects.

Sternum. This plate is joined by anterior and posterior bands to the pleura, thus forming the socket in which the coxa is situated. The central portion, called the *eusternum*, has a furrow that marks the line of invagination of the large internal apodeme called the *furca*, so named because it is forked or double at its apex (Fig. 3-40). Posterior

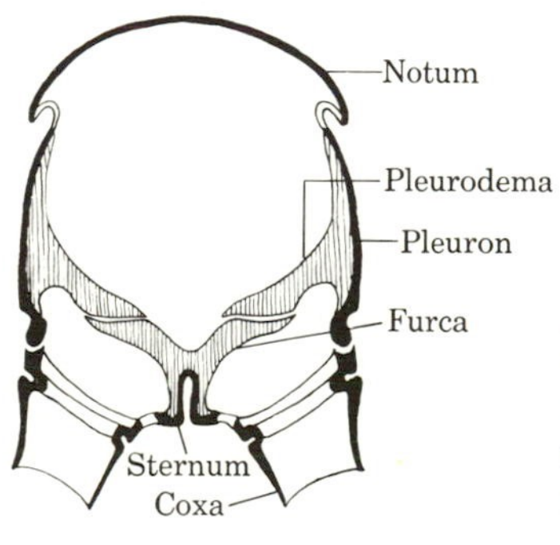

Fig. 3-40. Diagrammatic cross section of thoracic segments illustrating the furca and pleurodemae. (Redrawn from Snodgrass)

to the eusternum is a small sclerite, the *spinasternum*, bearing internally a single apodeme, the *spina*. This spinasternum had its origin in the membrane between the segments, but usually it is coalesced with the segment anterior to its place of origin.

Internal Skeleton. The various apodemes of the segments are frequently referred to collectively as the internal skeleton. They serve as areas of attachment for many of the large leg and wing muscles. The pleurodemae and furcae of a segment fit together closely as an almost continuous band (Fig. 3-40).

Existing Forms. In a general discussion of the thorax it is impractical to go into more detail than the preceding outline because of the almost endless variety in the thoracic structure of existing insects. Many orders exhibit a distinctive basic pattern, and in large orders such as the Coleoptera and Diptera there may be many extreme modifications within the same order.

Often there is little apparent similarity between some existing form and the generalized type. In such cases identification of the sclerites must be preceded by orientation in relation to stabilized features or landmarks. The apodemes and the sutures that are their external indications, plus articulation points for legs and wings, are the most reliable.

Legs. The typical thoracic leg consists of six parts, the *coxa*, *trochanter*, *femur*, *tibia*, *tarsus*, and *pretarsus* (Fig. 3-41). The coxa is the segment that articulates with the body; it may bear a posterior lobe called the *meron*. The tarsus of adult insects is usually subdivided into two to five segments. The pretarsus appears as a definite small end segment of the leg in larvae of several orders and in the Collembola. In almost all other insects the pretarsus is represented by a complex set of claws and minute sclerites set in the end of the tarsus. The Collembola and Protura are also unique in that the tibia and tarsus form a single tibiotarsus.

In general, insects have simple legs designed for walking or running (Fig. 3-41*A*). There have developed, however, a large number of modifications that fit the legs for other uses. These include jumping types with greatly enlarged femora, as in the grasshoppers (Fig. 3-41*B*); grasping types armed with sharp opposing spurs and spines (Fig. 3-41*C*), as in the praying mantis; swimming types, having long brushes of hair and flattened parts, as in the water boatmen; and digging types, with strong, scraperlike parts (Fig. 3-41*E*), such as found in the mole crickets.

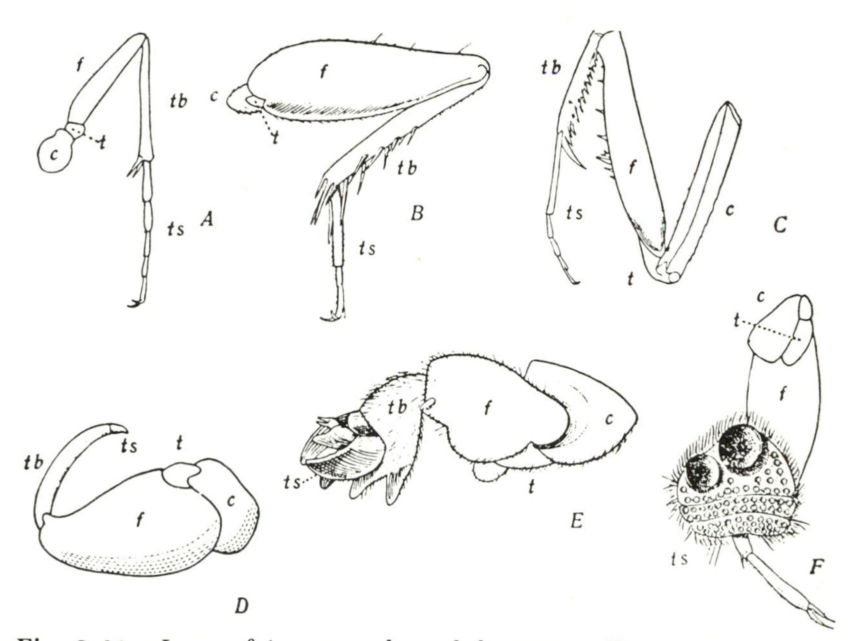

Fig. 3-41. Legs of insects adapted for (*A*) walking; (*B*) jumping (hind leg of cricket); (*C*) grasping (front leg of mantid); (*D*) clasping (front leg of bug); (*E*) digging (front leg of mole cricket); (*F*) holding fast by suction (front leg of male diving beetle). *c*, Coxa; *f*, femur; *t*, trochanter; *tb*, tibia; *ts*, tarsus. (From Folsom and Wardle, 1934, *Entomology*, by permission of The Blakiston Co.)

Wings

Insect wings are unique structures found in no other organisms. In such flying animals as bats and birds, the wings are highly modified front legs. In insects this is not the case. Their wings are outgrowths of the body wall along the lateral margins of the dorsal plate or notum. Unlike most other insect appendages, the wings have no muscles attached inside them. Wings, giving the power of flight, have been one of the most important reasons for the success of the insect group as a whole. No other invertebrate group has ever developed them.

Living winged insects, called *pterygote* insects (derived from *pteryx*, the Greek word for wing), typically have two pairs of wings, one pair each on the mesothorax and metathorax. The prothorax may have lateral foliaceous expansions but no articulated wings.

Structure. In basic design, an insect wing is very simple. It is a thin flaplike extension of the body wall, with an upper and lower membrane, and a set of strengthening supports called *veins* running more or less the length of the wing. The veins are connected by a series of *crossveins*. The pattern of veins and crossveins is termed the *venation*. Several small sclerites are associated with the narrow, flexible membranous strip between the notum and the functioning wing. Dorsally (Fig. 3-42*A*) there are three basal sclerites (marked 1, 2, and 3 in the figure), of which at least 1 and 3 have a close articulation with the notum; and three corresponding plates (s, m, and c), to which the veins attach. Ventrally (Fig. 3-39) the two alar sclerites lie one on each side of the wing process of the pronotum. Muscles attached to the alar sclerites control some movements of the wings.

Insect wings exhibit innumerable differences in venation that are of great value in the classification of orders, families, and genera. These different types of venation have apparently evolved from a single ancestral type whose basic features can be reconstructed from a comparison of fossil and living forms. These features are illustrated in Fig. 3-43.

Each main vein has a definite name; these are listed here in order of occurrence from the anterior to the posterior margin of the wing. Certain veins have definite typical branches. Standard abbreviations are indicated for the veins (Fig. 3-43).

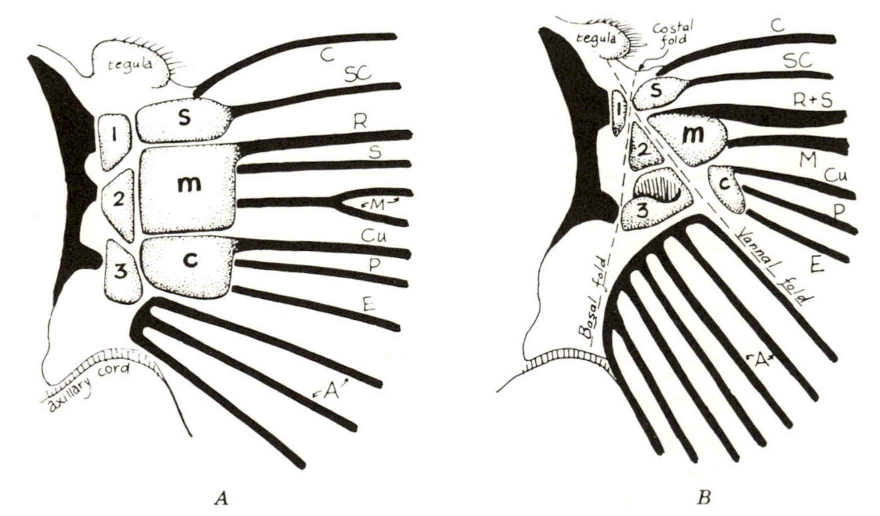

Fig. 3-42. Axillary sclerites (1, 2, 3) and associated plates (s, m, c) at base of wing. (*A*) Theoretical ancestral type; (*B*) generalized wing-folding type. c, Cubital plate; m, median plate; s, subcostal plate. See text for further explanation. (After Hamilton, 1971)

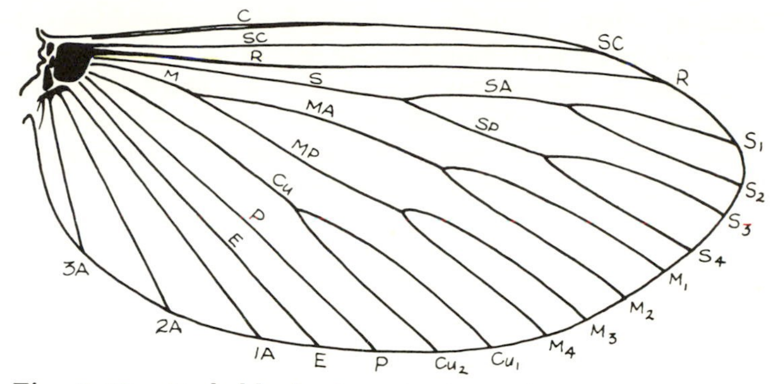

Fig. 3-43. Probable basic venation of a primitive insect wing, crossveins omitted. For explanation of abbreviations, see text. (After Hamilton, 1972)

Costa (C) usually forms the thickened anterior margin of the wing. It is unbranched and arises from the subcostal plate (Fig. 3-42).

Subcosta (Sc) runs immediately below the costa, always in the bottom of a trough and between the costa and radius. It also is usually unbranched and arises from the subcostal plate.

Radius (R) is the next main vein. It is typically a stout one and connects at the base with the median plate. It is usually undivided.

Sector (S) arises from the median plate and is typically four-branched in dichotomous fashion. The first two divisions are termed anterior and posterior sector, respectively (*SA* and *SP*).

Media (M) also arises from the median plate and is dichotomously divided into anterior and posterior media, (*MA* and *MP*), and each of these is usually divided into two branches.

Cubitus (Cu) arises from the cubital plate and is typically two-branched. It is almost invariably a strong convex vein running along the top of a ridge.

Plical Vein (P) arises from the cubital plate and is unbranched. It is concave; that is, it runs along the bottom of a furrow between *Cu* and the empusal vein.

Empusal Vein (E) also arises from the cubital plate and is unbranched. This is sometimes convex, running along the top of a ridge.

Anal Veins (1A, 2A, etc.) are a cluster of veins with a common base associated directly with the third axillary sclerite.

Names of Veins. Over the years, the homologies of the wing veins have been the subject of much difference of opinion and controversy. A summary of these ideas is given by Hamilton (1971–1972). The vein terminology outlined above differs in several particulars from that currently used by many authors. For ease of comparison with other terminology, these are outlined here.

1. Sector is the vein system previously termed *radial sector* by most authors. It is, however, a vein system in its own right and is not a branch of the radius.
2. The plical and empusal veins have been regarded variously as branches of the cubitus or as anal veins.
3. A pair of very short veinlike structures at the extreme base of the posterior wing margin have often been referred to as *jugal veins*. New evidence suggests that these are not veins but are new structures associated with wing folding. It is suggested that these be termed the *jugal bar*.

Crossveins. In general, crossveins are designated by the veins they connect. Thus crossveins connecting radius and the anterior branch of sector would be termed 1*r*–*s*, 2*r*–*s*, and so on; those between the posterior branch of media and cubitus would be termed 1*m*–*cu*, 2*m*–*cu*, and so on. Crossveins between branches of the same vein often have different terminologies in different taxonomic groups. There are a few standard exceptions. The crossvein between costa and subcosta at the base of the wing is called the humeral crossvein (*h*), and others from the costa to the next posterior vein are called simply *costal crossveins*.

Origin. There are several theories as to how these unique insect wings evolved. Because the two most primitive winged insect orders, the Ephemeroptera and Odonata, and also another fairly primitive order, the Plecoptera, have aquatic nymphs, it was first postulated that wings arose as swimming flaps that aided in propelling nymphs through the water. Making the transition from a swimming flap to an aerial wing, however, poses serious difficulties that detract from the probability of the theory. Another theory proposed by Alexander and Brown (1963) suggested that wings originated as lateral outgrowths of the thorax in connection with sexual activity. To date, sexual structures that might have been the progenitors of flight organs have not been observed on the thorax of either fossil or living insects. The most plausible theory seems to be that wings began as lateral flaplike extensions of the thoracic nota; perhaps they originally evolved as a camouflage function, or as a result of a flattening of the body as an adaptation to hiding in crevices or under scales or bark. If the insect was disturbed or alarmed, these lateral angulations would have aided the insect in alighting right side up and being ready to run when dropping from a higher to a lower level. Even so, it must have happened only under a particular set of circumstances. The insects must have frequented surfaces such as leaves a few meters from the ground or vertical surfaces such as rocks or trees, otherwise the insects would not have attained enough speed in the course of dropping for the angular sides of the notum to "bite" into the airstream. It was calculated by Flower (1964) that short-legged insects slightly under 1 cm in length and with only a slight flange along the edges of the thorax could have satisfied these requirements and have a stable, swooping glide at the end of their drop. The insects presumably were preyed on by swift predators, such as spiders and centipedes, when speed and aerial maneuverability would have been an asset to survival.

Of living insects, insects like the firebrat *Thermobia* would have satisfied these aerodynamic requirements. *Thermobia* and some other primitive wingless insects, such as *Machilis* (Fig. 8-7), have other features that would have been advantageous in such a gliding escape endeavor: The thorax is situated near the balance point of the organism, and the three pairs of large, strong thoracic legs would give it the thrust for a fast takeoff. It now seems fairly certain that primeval planing in an insect resembling *Thermobia* was the precursor of insect flight.

Another aspect to consider in deciphering the evolution of insect flight is that, in many of the oldest known fossils of pterygote insects, the prothorax had lateral flat expansions called *paranotal lobes* that are suggestive of planing structures. Furthermore, these lobes show an array of veins that appear to be homologous with the main veins of typical articulated functional wings. With this added information, it is possible to reconstruct the evolution of insect wings with a high degree of probability.

When stabilized dropping had been achieved, there would have come into play a selection pressure for the notal angles to enlarge into planing fins or lobes that would decrease the time required for the onset of gliding and increase the length of the glide. An early lobe was probably like Fig. 3-44*A*, enlarging through the stages shown in Figs. 3-44*B*, *C*, *D*, and *E*. Similar lobes were presumably present on all three thoracic segments. As the lobes increased in size, tracheal branches extended into them; in the adult, these became the centers of the thickenings of the wing integument or veins. Figure 3-44*F* illustrates the largest lobe known to occur on the prothorax. It

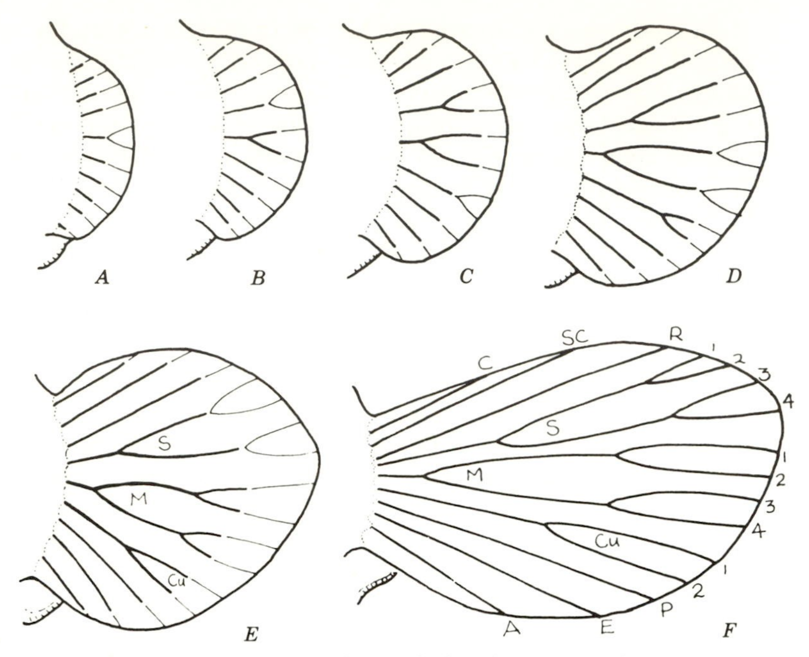

Fig. 3-44. (*A* to *E*) Hypothetical development of tracheation of paranotal lobes leading to the venation of a primitive wing, as shown in *F*. (From Hamilton, 1971)

is probable that at this stage a hinge had already evolved at the wing base, and from that time the meso- and metathoracic wings became larger and the lobes of the prothorax smaller.

The first small wings were probably relatively thick and ungainly. As they became larger, the veins became stronger and the wing membrane between them thin and light. The earliest known efficient wings were also pleated lengthwise like a fan with the veins running along the creases (Fig. 3-45), and had a system of crossveins connecting the veins. This combination of fanlike or fluted design and trusslike support enabled a tough but thin membrane to have sufficient rigidity for flight.

The next step in wing evolution was a realignment of the axillary sclerites and plates in such a way as to provide diagonal folding at the base of the wing (Figs. 3-42*B*, 3-46). With this arrangement the wings could be folded along the vannal and basal folds backward flat over the body. Along with this step much of the wing fluting was eliminated and the area between R and Cu became flat.

In various orders of insects, many changes of wing venation have evolved. In some, additional veins occur; in others, various veins

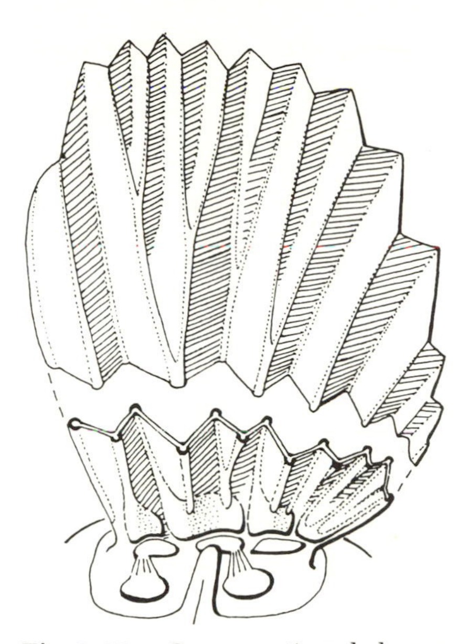

Fig. 3-45. Cross-sectional shape of a paleopteran wing. Strong fluting of alternating convex and concave veins provides added strength to the wing. (From Hamilton, 1971)

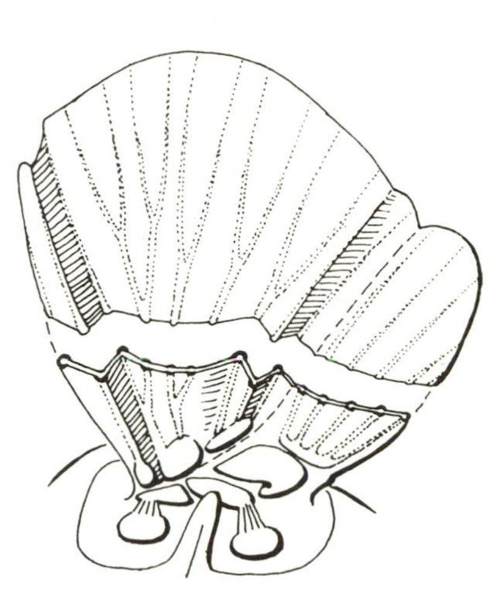

Fig. 3-46. Cross-sectional shape of a neoteran wing. Wing fluting is modified to retain strength and to accommodate wing folding. (From Hamilton, 1971)

have either fused or been lost. The same is true of crossveins. In larger insects, rapid flight is usually associated with long narrow wings; fluttering flight, as in butterflies, is associated with shorter, broader wings. In many small insects, such as aphids, the venation may be greatly reduced and the flight is weak, little more than drifting. Each type of wing modification is correlated with the habits of the adult insects. In many groups, the ability to fly has been completely lost, as in the camel crickets and fleas, which have no wings.

ABDOMEN

The abdomen is the third and posterior region of the insect body. It is relatively simple in structure, compared with the thorax, and in adults has no walking legs. Primitively it is 12-segmented but this condition is distinct only in Protura (Fig. 8-5), and certain embryonic stages. Usually the abdomen consists of 10 or 11 segments (Fig. 3-47*A*). In some forms much greater reduction occurs, as in the

Collembola, which have only 6 abdominal segments (Fig. 8-6). Many groups, such as the housefly group, have the last several segments developed into a copulatory structure or an ovipositor that is normally retracted within the preceding segments.

Segmental Structure. In adult insects, a typical segment consists of (1) a tergum or dorsal plate, (2) a sternum or ventral plate, (3) lateral areas of membrane connecting tergum and sternum, and (4) a spiracle on each side, usually situated in the lateral membrane. In some larval forms (Fig. 3-47*B*), and a few adults, there are sclerites in the lateral membrane; some of these undoubtedly represent vestiges of the subcoxal sclerites of the primitive appendages (see Fig. 3-36).

Appendages. These may be divided roughly into two groups: those not associated with reproduction, and those developed for reproductive activities such as mating or oviposition.

Nonreproductive Appendages. In most adult insects abdominal appendages are absent except on the terminal segments. A few primitive forms have retained degenerate legs represented by styli, as in the silverfish (Fig. 3-48). The appendages of the eleventh segment, the cerci, are present in most insects (Fig. 3-47). They are usually tactile organs and in such groups as caddisflies become part of the male genitalia. The cerci may appear to belong to the tenth or

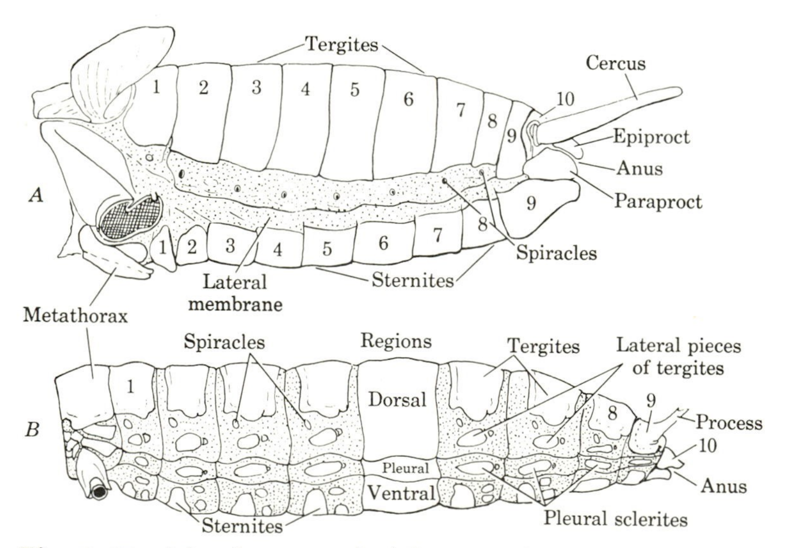

Fig. 3-47. Metathorax and abdomen of insects. (*A*) Adult cricket *Gryllus*; (*B*) larva of ground beetle *Calosoma*. (Redrawn from Snodgrass)

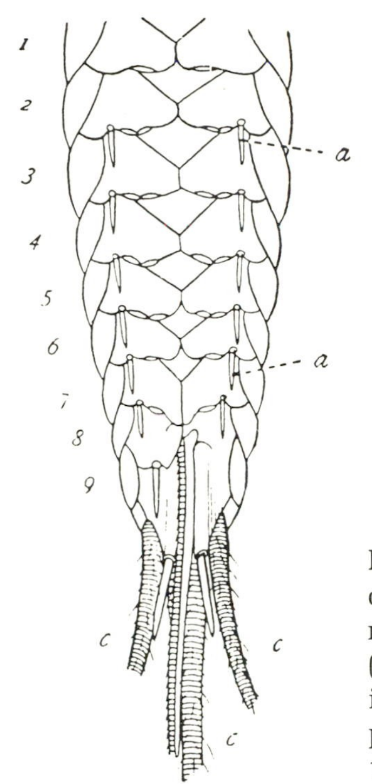

Fig. 3-48. Ventral aspect of the abdomen of a female *Mechilis maritima*, to show rudimentary limbs (a) of segments 2 to 9. (The left appendage of the eighth segment is omitted.) *c*, Lateral cerci and median pseudocercus. (From Folsom and Wardle, 1934, after Oudemans)

ninth segment if the eleventh or tenth segment is reduced. In larvae and nymphs a great variety of abdominal appendages are developed. Well-known examples are the larvapods of caterpillars (Fig. 3-5) and the segmental gills of mayfly nymphs (Fig. 8-10).

Reproductive Appendages. These generally include appendages of the eighth and ninth segments. Many entomologists believe that these appendages are homologous with true segmental appendages such as the mouthparts and legs. Others believe that the primeval appendages of the eighth and ninth segments have been lost and that the appendages associated with reproduction are special structures arising from more mesal outgrowths of the same segments.

Female. The ovipositor (Fig. 3-49) is composed chiefly of three pairs of blades, the first, second, and third valvulae, respectively. The *first valvulae* arise from a pair of plates, the *first valvifers* of the eighth segment. The valvifer and valvula probably correspond to the coxopodite and telopodite of the generalized arthropod segment (Fig. 3-15). The *second valvifers* bear a ventral pair of blades, the *second valvulae*, and a dorsal pair, the *third valvulae*. In most in-

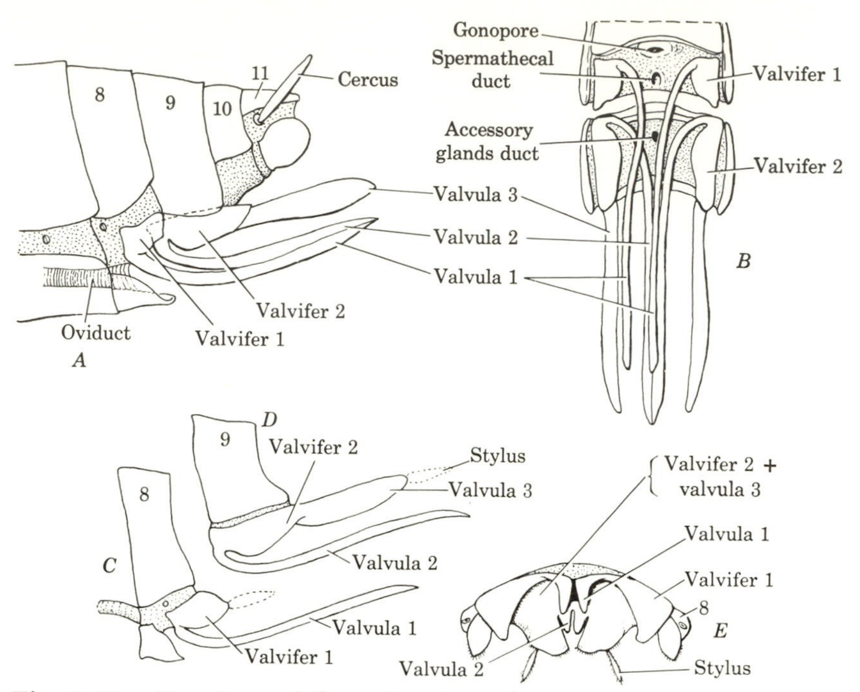

Fig. 3-49. Structure of the ovipositor of pterygote insects (*A* to *D*, diagrammatic). (*A*, *B*) Showing segmental relations of the parts of the ovipositor. (*C*, *D*) Lateral view of genital segments and parts of ovipositor dissociated. (*E*) Nymph of *Blatta orientalis*, ventral view of genital segments with lobes of ovipositor. (Redrawn from Snodgrass)

sects with a well-developed ovipositor, such as the sawflies (Fig. 3-50), the first and second valvulae form a cutting or piercing organ with an inner channel down which the eggs pass. The third valvulae form a scabbard or sheath into which the ovipositor folds when retracted. In Orthoptera, either all three pairs fit together to form the functional ovipositor, or the second valvulae form a short egg guide.

In many insects the valvulae are only poorly or not at all developed, in which case the apical segments of the abdomen generally form an extensile tube that functions as an ovipositor. This is exemplified by many of the Lepidoptera and Diptera (Fig. 3-51)

Male. In this sex the appendages of the ninth segment are usually combined with parts of the ninth segment proper and sometimes parts of the tenth to form a copulatory organ. In each order this organ usually displays fundamental peculiarities. It is extremely difficult

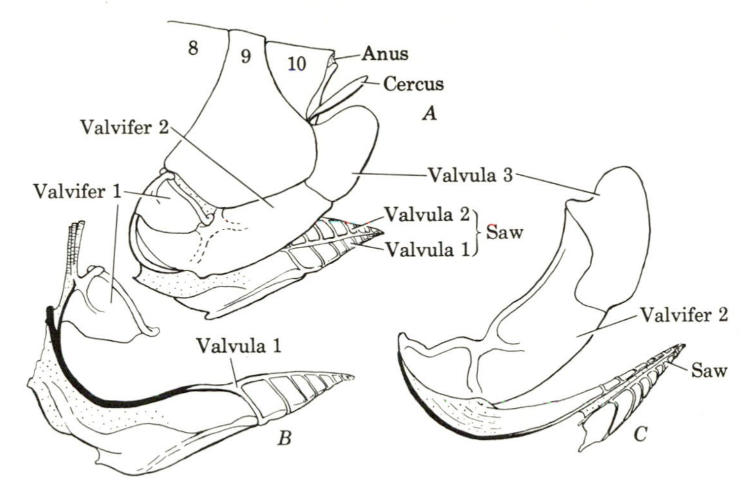

Fig. 3-50. The ovipositor of a sawfly. (*A*) Showing relation of basal parts to each other and to ninth tergum; (*B*) first valvifer and valvula; (*C*) second valvifer with second and third valvula. (Redrawn from Snodgrass)

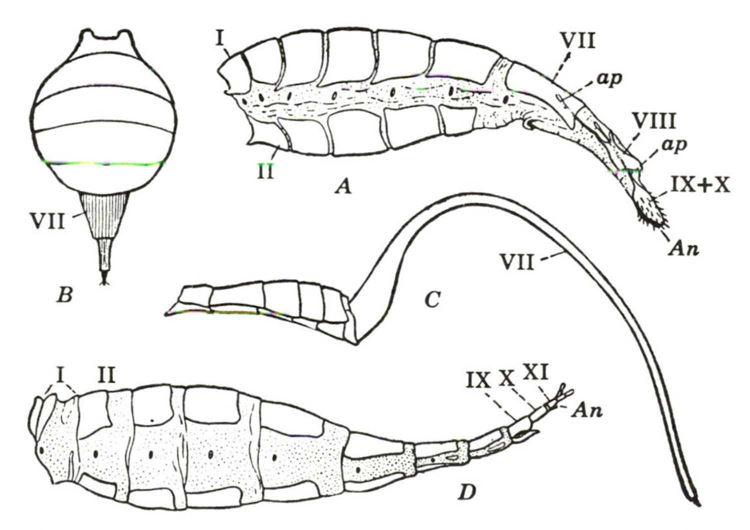

Fig. 3-51. Examples of an "ovipositor" formed of the terminal segments of the abdomen. (*A*) A moth *Lymantria monacha* (from Eidmann, 1929). (*B*) A fruit fly *Paracantha culta*. (*C*) A fruit fly *Toxotrypania curvicauda*. (*D*) A scorpionfly *Panorpa* sp. *ap*, Apodeme; *An*, anus. (After Snodgrass)

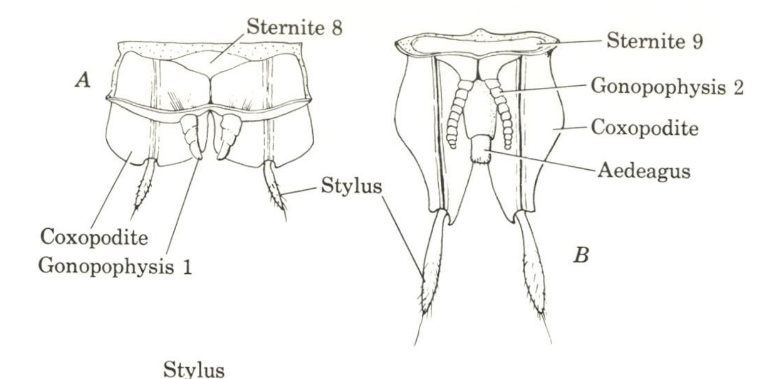

Fig. 3-52. Male genitalia of the microcoryphian *Machilis variabilis*, dorsal aspect. (*A*) First gonopods, showing gonapophyses of eighth segment; (*B*) second gonopods and median copulatory organ. (Modified from Snodgrass)

to homologize the individual parts of these copulatory organs throughout the insect orders to be certain of their relation to what must have been the simple parts and appendages from which they are derived.

Structural differences in the copulatory organs furnish excellent taxonomic characters in many groups of insects for the differentiation of families, genera, or species. In any one group the constituent parts of the organ are usually well marked, and in each group there is a clear terminology for the designation of these parts. Until closer agreement is reached regarding the homologies of these structures in different orders, it is more practical to employ the terminology in established usage for any particular group. A simple type is illus-

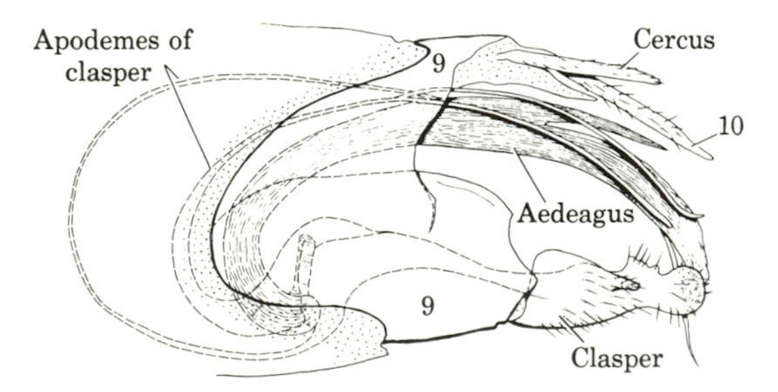

Fig. 3-53. Male genitalia of a caddisfly. (From Illinois Natural History Survey)

trated by Microcoryphia (Fig. 3-52), and a more complicated type occurs in the caddisflies (Fig. 3-53). These two examples by no means give the full range of different types among the insect orders but will serve to indicate some of the variety in structure that may be found.

REFERENCES

Alexander, R. D., and W. L. Brown, Jr., 1963. Mating behavior and the origin of insect wings. *Occ. Pap. Mus. Zool. Univ. Mich.* **628**:1-19.

Chapman, R. F., 1969. *The insects, structure and function.* New York: American Elsevier. 819 pp.

Edmunds, G. F., Jr., and J. R. Traver, 1954. The flight mechanics and evolution of the wings of Ephemeroptera, with notes on the archetype insect wing. *J. Wash. Acad. Sci.*, **44:**390-399.

Flower, J. W., 1964. On the origin of flight in insects. *J. Insect Physiol.*, **10**:81-88.

Folsom, J. W., and R. H. Wardle, 1934. *Entomology with special reference to its ecological aspects*, 4th ed. Philadelphia: Blakiston. 605 pp.

Hamilton, K. G. A., 1971. The insect wing, Part I. Origin and development of wings from notal lobes. *J. Kans. Entomol. Soc.*, **44**(4):421-433.

——— 1972. The insect wing, Part II. Vein homology and the archetypal insect wing. *J. Kans. Entomol. Soc.*, **45**(1):54-58.

——— 1972. The insect wing, Part III. Venation of the orders. *J. Kans. Entomol. Soc.*, **45**(2):145-162.

——— 1972. The insect wing, Part IV. Venational trends and the phylogeny of the winged orders. *J. Kans. Entomol. Soc.*, **45**(3):295-308.

Matsuda, R., 1958. On the origin of the external genitalia of insects. *Ann. Entomol.* Soc. Am., **51**:84-94.

——— 1963. Some evolutionary aspects of the insect thorax. *Annu. Rev. Entomol.*, **8:**59-76.

——— 1970. Morphology and evolution of the insect thorax. *Mem. Entomol. Soc. Can.*, **76:**1-431.

——— 1976. *Morphology and evolution of the insect abdomen.* Oxford: Pergamon. 532 pp.

Rempel, J. G., 1975. The evolution of the insect head: The endless dispute. *Quaest. Entomol.* **11**(1):7-25.

Richards, O. W., and R. G. Davies, 1977. *Imms' general textbook of entomology*, 10th ed. London: Chapman & Hall. Vols. 1 and 2.

Schmitt, J. B., 1938. The feeding mechanism of adult Lepidoptera. *Smithsonian Misc. Collections,* **97**(4):1-28.

Scudder, G. G. E., 1961. The comparative morphology of the insect ovipositor. *Trans. R. Entomol. Soc. London,* **113**:25–40.

——— 1971. Comparative morphology of insect genitalia. *Annu. Rev. Entomol.* **16:**379-406.

Snodgrass, R. E., 1935. *Principles of insect morphology.* New York: McGraw-Hill. 667 pp.

——— 1944. The feeding apparatus of biting and sucking insects affecting man and animals. *Smithsonian Misc. Collections,* **104:**(7):1-113.

——— 1950. Comparative studies on the jaws of mandibulate arthropods. *Smithsonian Misc. Collections,* **116**(1):1-85.

Waterhouse, D. F. et al., 1970. *The insects of Australia.* Carleton, Victoria: Melbourne Univ. Press. 1029 pp.

Weber, H., 1952. Morphologie, Histologie and Entwicklungsgeschichte der Articulaten. *Fortschr. Zool.,* N.F., **9:**18-231.

——— 1954. *Grundriss der Insektenkunde,* 3rd rev. ed. Stuttgart: Fischer. 428 pp.

4 Internal Organ Systems

The internal organ systems of insects and related hexapods perform many of the vital functions of life. These organs are protected from the external environment by the body wall. If parts of certain organs project as flaps or lobes beyond the body outline, they are encased by a thin mantle of body wall so that they lie within the exoskeleton.

DIGESTIVE SYSTEM

The digestive system is the food tract and its accessory parts. It is composed of the *alimentary canal* (*digestive tract*) and various glands connected with it either directly or indirectly. Typically these include the salivary glands, gastric caeca, and Malpighian tubules.

Alimentary Canal (Fig. 4-1). This organ is a tube passing through the central part of the body. Its anterior opening, the mouth, is situated at the base of the *preoral cavity* (the space enclosed by the mouthparts); its posterior opening, the *anus*, is on the posterior body segment. The alimentary canal is divided into three distinct parts: an anterior *foregut* (*stomodeum*); a middle *midgut* (*mesenteron*); and a posterior *hindgut* (*proctodeum*). Usually between the foregut and the midgut is the *stomodeal* or *cardiac valve*, and between the midgut and the hindgut is the *proctodeal* or *pyloric valve*. The foregut and hindgut result from embryonic infoldings of the ectoderm; the midgut is formed from the endoderm; this is discussed further in the section on embryology.

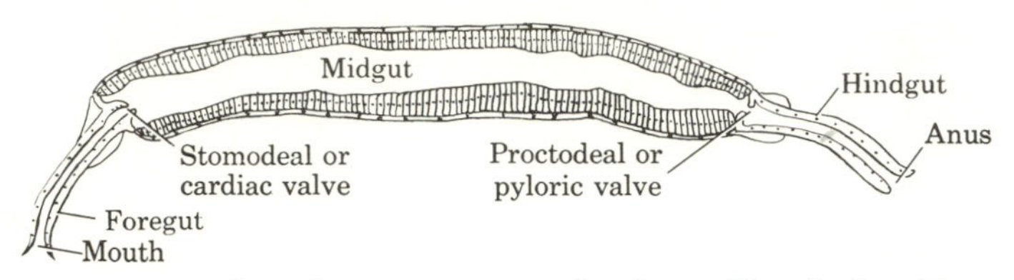

Fig. 4-1. The alimentary canal of a collembolan *Tomocerus niger*, showing the primary components of the food tract without secondary specializations. (Redrawn from Snodgrass)

In a few primitive hexapods, the three parts of the alimentary canal are simple and tubular in shape (Fig. 4-1). In most insects, however, each of these parts has become differentiated into functional subdivisions. The typical structure of these is as follows (Fig. 4-2).

Foregut or Stomodeum. This portion is usually divided into three main portions: (1) an anterior, more or less tubular portion, the *esophagus*; followed by (2) an enlarged portion, the *crop*, which narrows to (3) a valvelike *proventriculus* at the junction with the mesenteron. An indefinite portion of the esophagus at the mouth opening is frequently called the *pharynx* but is difficult to identify without a knowledge of the musculature. The boundary between esophagus and crop is frequently arbitrary, as in Fig. 4-2; in some insects such as certain moths (Fig. 4-3C), the crop is developed into a spherical chamber; this modification is carried still further by many flies (Fig. 4-4), and the crop forms a sack connected to the esophagus by a long lateral tube. The proventriculus may be a simple valve opening into the mesenteron; in insects that eat solid food, it bears a series of hooks for food shredding and is called the *gastric mill*.

Midgut or Mesenteron. This middle portion of the alimentary canal is the place where most digestion takes place. It may be called

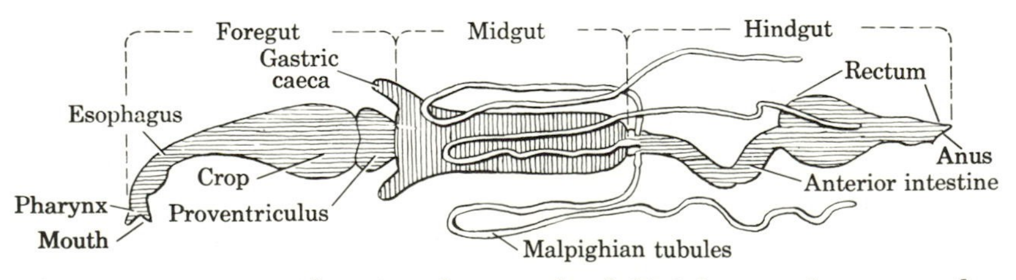

Fig. 4-2. Diagram showing the usual subdivisions and outgrowths of the alimentary canal. (Redrawn from Snodgrass)

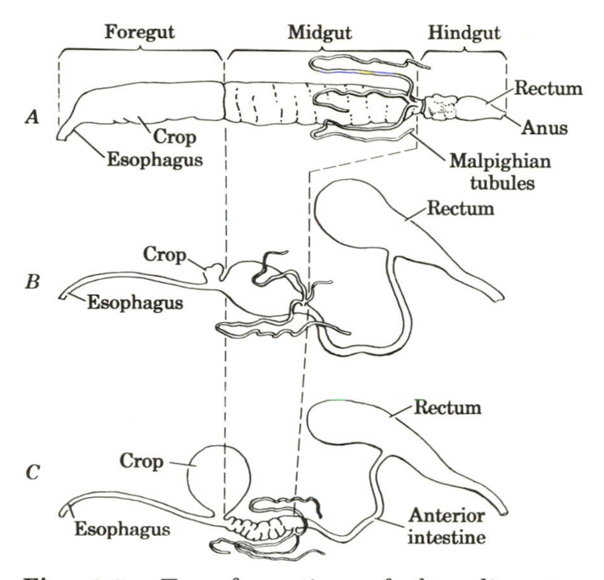

Fig. 4-3. Transformation of the alimentary canal of a moth *Malacosoma americana*, from the larva (*A*) through the pupa (*B*) to the adult (*C*). (Redrawn from Snodgrass)

the *ventriculus* or stomach. Usually it is tubular, but occasionally it is subdivided into definite parts. This subdivided condition is most pronounced in the Hemiptera, in which the midgut may have three or four sections. The midgut typically bears several fingerlike outgrowths, the *gastric caeca*. These usually occur at the anterior end of the stomach (Fig. 4-2), but may be situated on more posterior portions.

Hindgut or Proctodeum. This posterior portion of the alimentary canal varies greatly in different insects but is usually divided into a tubular *anterior intestine* and an enlarged *posterior intestine*. This latter is termed the *rectum* and is connected directly with the anus.

Labial or Salivary Glands. Most insects possess a pair of glands lying below the midgut (Fig. 4-4) and associated with the labium. Each gland has a duct running anteriorly. These unite, usually within the head, to form a single duct that opens into the preoral cavity between the labium and the hypopharynx. The function of these glands differs in various insects and in some has not been determined definitely. In most insect these labial glands secrete saliva, as in cockroaches. In lepidopterous and hymenopterous larvae these glands secrete silk, used in making larval nests and pupal

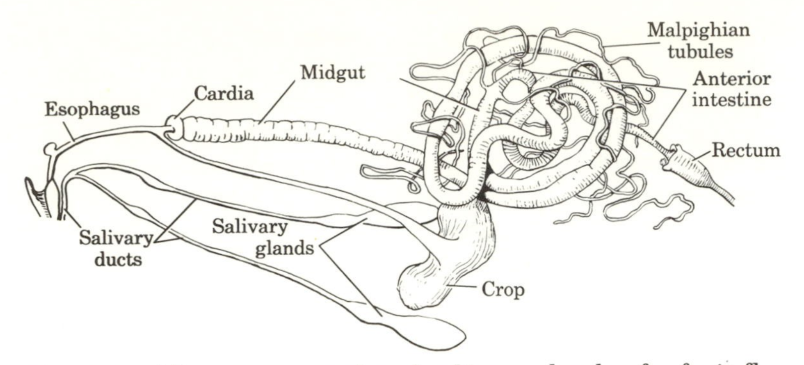

Fig. 4-4. Alimentary canal and salivary glands of a fruit fly *Rhagoletis pomonella*, showing the diverticular crop and the cardia of the midgut, characteristic of many Diptera. (Redrawn from Snodgrass)

cells. In blood-sucking insects the labial glands secrete an anticoagulant that keeps ingested blood in liquid form.

EXCRETORY SYSTEM

Malphighian Tubules. With few exceptions insects possess a group of long, slender tubules branching from the alimentary canal near the junction of the midgut with the hindgut (Fig. 4-2). These are the Malpighian tubules, which are excretory in function. The number of these tubules varies from 1 to 250 or more. When a large number is present, they are often grouped into bundles of equal size.

CIRCULATORY SYSTEM

The circulatory system comprises chiefly the blood, and tissues and organs that cause its circulation through the body. In many animals, such as the vertebrates, the blood travels only through special vessels (arteries, capillaries, and veins). This condition is called a closed system. In insects this is not the case. For most of its course the blood simply flows through the body cavity, irrigating the various tissues and organs. There is a special pumping organ or dorsal vessel situated dorsally in the insect body, which pumps blood from the posterior portion of the body and empties it into the internal cavity of the head. From this cavity the blood again flows back through the body, is drawn into the heart, and is again pumped forward, and so on. This kind of arrangement is called an *open system*, and the body cavity through which the blood flows is called a *hemocoel*. The circulatory system also has associated with it one or more small secondary pumping organs near or within the appendages, widely scattered absorbing or phagocytic cells, and the large storage and synthesizing fat body.

Blood. The fluid that circulates through the body cavity is called the blood or *hemolymph*. It consists of a liquid part, the *plasma*, and an assortment of free floating cells, called blood corpuscles or *hemocytes*. Study of the blood involves histology and physiology, and is discussed in the next chapter.

Dorsal Vessel. As its name implies, the dorsal vessel (Fig. 4-5*A*) lies directly beneath the dorsum, or dorsal wall. It extends the length of the body, from the posterior end of the abdomen into the head. It is the principal pulsating organ that causes the flow of the blood.

The dorsal vessel is divided into two parts: a posterior portion called the *heart*, and an anterior portion called the *aorta*. In general,

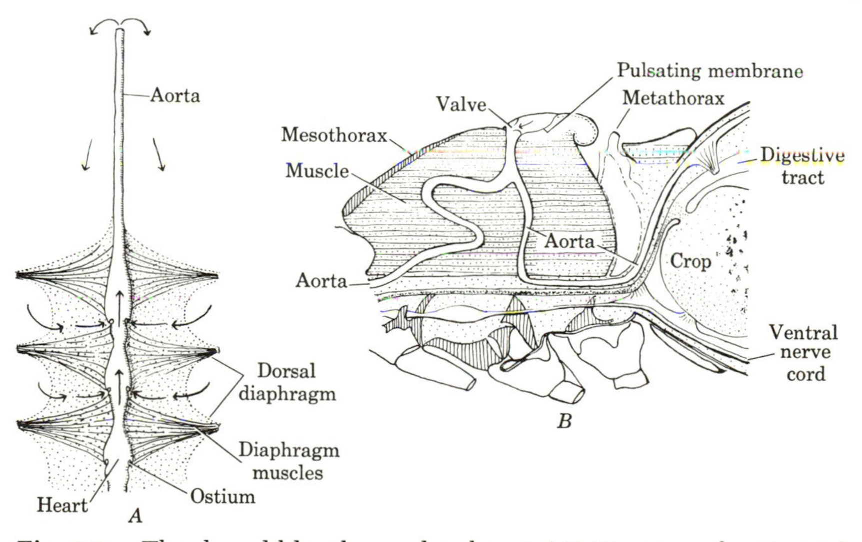

Fig. 4-5. The dorsal blood vessel or heart. (*A*) Diagram of aorta and three chambers of the heart with corresponding part of the dorsal diaphragm, dorsal view, arrows indicating the course of blood circulation. (*B*) Vertical section of thorax and base of abdomen of *Sphinx convolvuli* showing pulsating membrane in mesothorax. (Redrawn from Snodgrass)

the heart is the pulsating portion, and the aorta is the tube that carries the blood forward and discharges it into the head.

The *heart* is usually more or less swollen in each segment to form *chambers* separated by constrictions. This chambered portion typically consists of nine parts, occurring in the first nine segments of the abdomen. Each chamber has a pair of lateral openings or *ostia*, through which blood enters the chamber. In certain insects the heart may depart radically from this typical condition. In cockroaches and japygids, for example, the first two chambers occur in the meso- and metathorax. In the bug *Nezara* the heart consists of a single large chamber having three pairs of ostia.

The *aorta* is typically a simple tubular extension of the heart. In some forms, such as cranefly larvae and adult mosquitoes, the aorta also pulsates and thus is an accessory to the heart in causing circulation.

Dorsal Diaphragm and Sinus. Connected to the underside of the heart are pairs of muscle bands known as *wing muscles* or *alary muscles*. This name is applied because the muscles form flat fans or wings which connect the heart and the lateral portions of the terga. These aliform muscles, when well developed, form a fairly complete partition between the main body cavity and the region around the heart. In such cases the partition is called the *dorsal diaphragm*, and the segregated heart region is termed the *dorsal sinus*. The diaphragm and sinus extend only as far as the heart and are not continued forward in the region of the aorta.

Accessory Pulsating Organs. In addition to the heart there may occur other pulsating organs assisting in blood circulation. The two of most frequent occurrence are the thoracic pulsating organs and the ventral diaphragm. Other accessory structures are found only rarely.

Thoracic Pulsating Organs (Fig. 4-5*B*). In many insects, especially rapid fliers such as hawk moths, there is a pulsating organ that draws blood through the wings and discharges it into the aorta. The pulsating organ itself is a cavity in the scutellum provided with a flexible or pulsating membrane. The outlet of the structure is a tube called the aortic diverticulum that connects directly with the aorta.

Ventral Diaphragm. Many Orthoptera, Hymenoptera, and Lepidoptera have muscle bands developed *over* the ventral nerve cord in much the same manner as the aliform muscles form a diaphragm *under* the heart, which is dorsal in position. When such a muscle band is formed over the nerve cord, it is known as the ventral diaphragm. By expansion and contraction it produces a flow of blood posteriorly and laterally.

TRACHEAL SYSTEM

Most insects possess a system of internal tubes or *tracheae*, that conduct free air to the cells of the body. This system of tubes is the *tracheal system*, and it performs the function of *external respiration*. In addition to insects, a few groups of the Arthropoda possess a well-developed tracheal system. These groups include some of the Arachnida, a few Crustacea, and most of the Chilopoda. Rudimentary tracheal tubules are found in Onychophora and Diplopoda.

This tracheal system is of necessity highly complex, because it must branch into a myriad of fine tubules, each of which reaches intimately only a small group of cells. This intricate branching of tracheae in insects is analogous to that of the blood vessels and capillaries in the vertebrates.

Principal Components of Tracheal System

A common type of tracheal system is shown in Fig. 4-6. The tracheae form definite groups in each segment and receive air from the exterior by means of segmentally arranged pairs of openings called *spiracles*(*s*). The spiracles join more or less directly with a main *tracheal trunk* (*t*), a pair of which usually run the full length of the body. In each segment there arise from these trunks various branches (always paired, since one comes from each trunk) that aerate the tissues of the organs. The number and position of these branches vary greatly in different insects, but generally there are three large branches given off on each side in any one segment: (1) a dorsal branch aerating the dorsal vessel and dorsal muscles; (2) a ventral or visceral branch aerating the digestive and reproductive organs; and (3) a ventral branch aerating the ventral muscles and nerve cord.

The fine tips of the tracheae divide into minute capillary tubes, or tracheoles, usually 1 micron or less in diameter. These tracheoles ramify between and around cells of other tissues and are the functional part of the system through which oxygen diffuses into the body cells.

Tracheal Trunks. The segmental arrangement of clusters of tracheal branches indicates that originally insects had an independent tracheal system in each postoral segment, with no connection

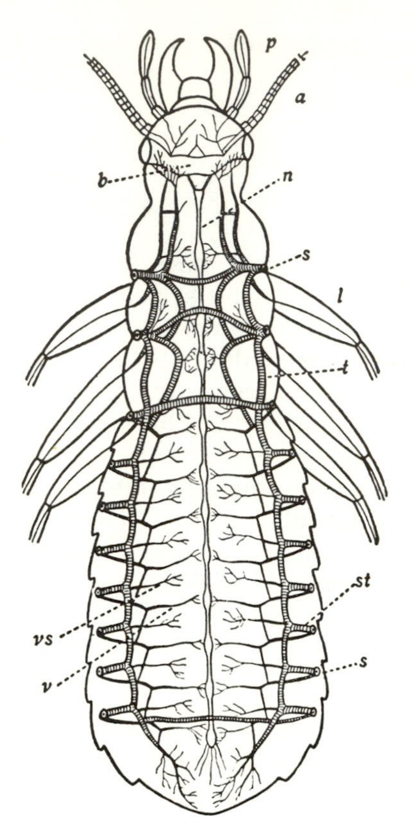

Fig. 4-6. Tracheal system of an insect. *a*, Antenna; *b*, brain; *l*, leg; *n*, nerve cord; *p*, palpus; *s*, spiracle; *st*, spiracular branch; *t*, main tracheal trunk; *v*, ventral branch; *vs*, visceral branch. (From Folsom, after Kolbe)

between tracheae of different segments. With only few exceptions, however, insects of the present have connections between the tracheae of adjoining segments if the tracheal system is developed. These connecting tubes form trunks. In many insects the main tracheal trunks are lateral in position and are called the *lateral tracheal trunks*. Frequently a second pair of *dorsal tracheal trunks* are found, one on each side of the heart. These are usually small in diameter and secondary to the lateral trunks. In most fly larvae the opposite is the case. There the dorsal trunks are greatly developed and are the chief respiratory passages (Fig. 4-7).

Tracheal Air Sacs. An important feature in many groups is the development of air sacs that serve as air-storage pockets. These are often enlargements of the tracheal trunks, as in Figs. 4-8 and 4-9. In many fast-flying insects, such as the housefly and the bees, the sacs fill a large part of the body cavity. These sacs can be squeezed and

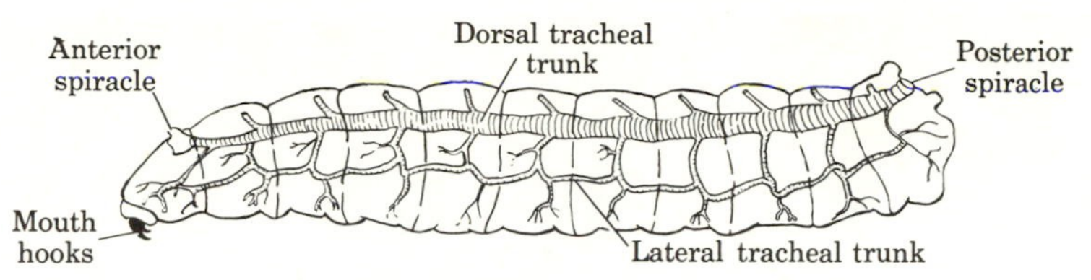

Fig. 4-7. Fly larva illustrating large dorsal tracheal trunks. (Redrawn from Snodgrass)

released by muscular contraction of the body to act like bellows and increase intake and expulsion of air.

Spiracles. When functional, spiracles have an important control over air intake. They are extremely varied in size, shape, and structure. If functional, they all have some sort of closing device. This device may be *external* (usually in the form of two opposed lips), or it may be *internal* (usually in the form of a clamp that pinches the trachea shut).

Open Tracheal Systems

Systems in which spiracles are open and functional are called open systems. The more generalized type has 10 pairs of spiracles, a pair on the mesothorax, metathorax, and each of the first eight abdominal segments. Many modifications of this type occur, including such examples as mosquito larvae, having spiracles on only the eighth abdominal segment; most of the maggots, which have only prothoracic spiracles and the terminal pair on the eighth abdominal segment; fly pupae, which have only the prothoracic spiracles; sev-

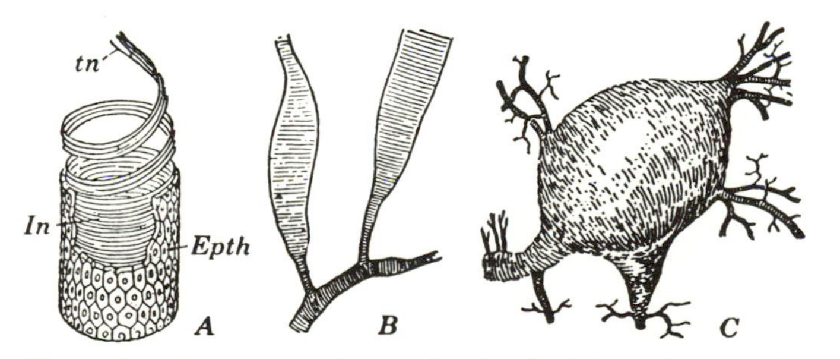

Fig. 4-8. Structure of a tracheal tube (*A*) and examples of tracheal air sacs (*B*, *C*) *Epth*, Epithelium; *In*, intima; *tn*, taenidium in spiral band of cuticular intima artificially separated. (After Snodgrass)

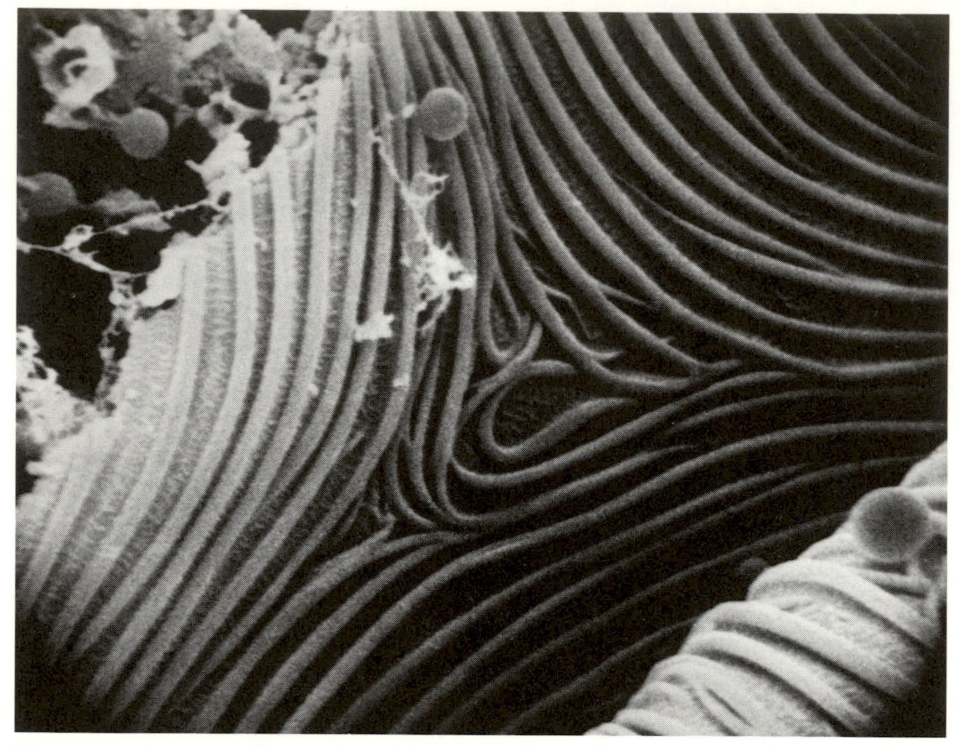

Fig. 4-9. Inner surface of an insect trachea. The curving ridges (taenidia) provide support and the membranous bands (intima) provide the flexibility for the tubules; approx. ×40,000. (Courtesy of Electron Microscope Laboratory, Univ. of Georgia; photographed by C. C. Dapples and A. O. Lea)

eral aquatic forms, such as the rat-tailed maggots, having the posterior spiracles on a long extensile tube that is exserted through the breeding medium into the air.

Closed Tracheal Systems

In many forms of insects the spiracles are either functionless or entirely absent. In these cases the tracheal system is termed *closed*. It is usually well developed otherwise, as concerns the tracheal trunks and their interior branches. In most closed systems the spiracles are replaced by a network of fine tracheoles that run under the skin or into gills. This is illustrated by nymphs and larvae of many aquatic insects, such as mayflies, stoneflies, damselflies, and midges.

An interesting modification among the aquatic insects occurs in dragonfly nymphs. In these the rectum contains internal gill-like folds. Fine tracheae extend throughout these folds. The nymph periodically draws water into the rectum, and expels it, bathing these *rectal gills* and thus aerating the tracheae in them.

NERVOUS SYTEM

The nervous system in insects is highly developed and consists of a central system and a stomodeal system. As in other animals, the nervous system serves to coordinate the activity of the insects with conditions both inside and outside the body.

Central Nervous System. The basic units of the central nervous system (Fig. 4-10), are essentially: (1) the brain, situated in the head, and (2) paired nerve centers or *ganglia*, one ganglion for each segment. The ganglia are connected by double fibers into a cord, and the anterior ganglion is connected with the brain. Concurrently with the fusion of body segments that occurred in the evolution of the insect group (Fig. 2-1), there occurred also a fusion of the ganglia belonging to each segment. For this reason the nerve centers in the head bear little apparent resemblance to the primitive condition, for the head is in reality composed of the archaic head, acron or prostomium, plus six body segments that join with it to make a solid mass.

Brain. The brain (Fig. 4-11) is situated in the head above the esophagus and for this reason is frequently referred to as the *supraesophageal ganglion*. It has three principal divisions: (1) the *protocerebrum*, which innervates the compound eyes and ocelli; (2) the

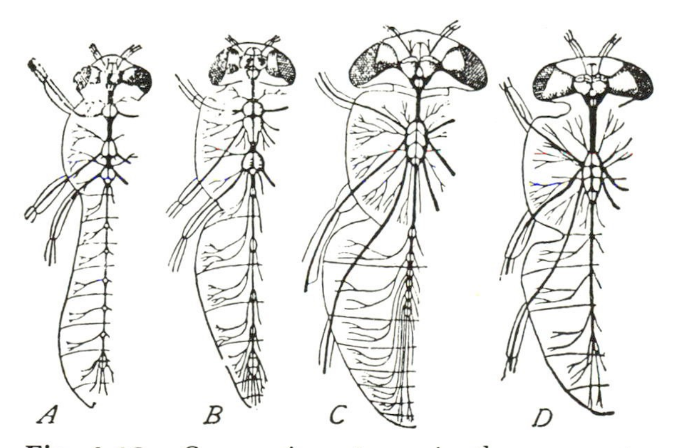

Fig. 4-10. Successive stages in the concentration of the central nervous system of Diptera. (A) *Chironomus;* (B) *Empis;* (C) *Tabanus;* (D) *Sarcophaga.* (From Folsom and Wardle, 1934, after Brandt)

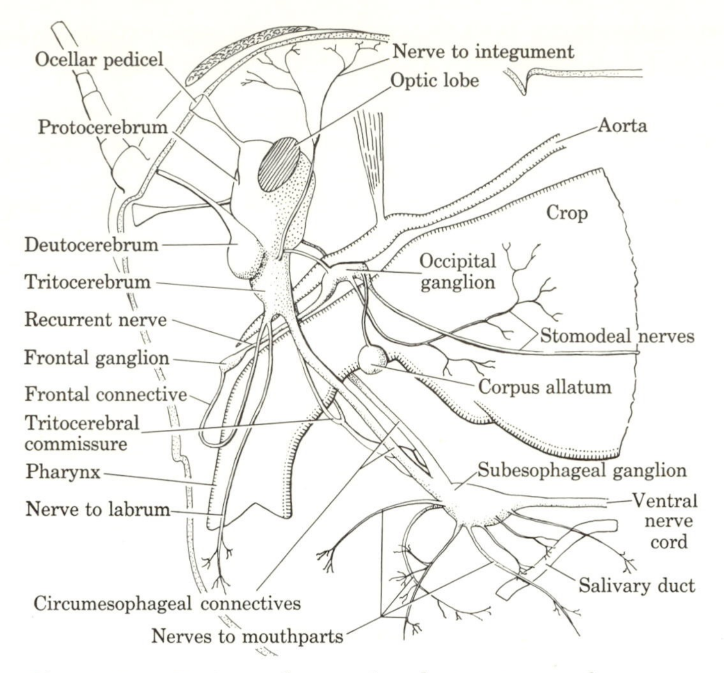

Fig. 4-11. Brain and associated structures of a grasshopper, lateral aspect. (Redrawn from Snodgrass)

deutocerebrum, which innervates the antennae; and (3) the *tritocerebrum*, which controls the major sympathetic nervous system. All three of these parts are definitely paired.

In its long evolutionary development, the various parts of the insect head have shifted somewhat in general orientation. Because of this shifting, the brain, which was originally *in front of* the mouth, is now *above* the mouth or esophagus. The protocerebrum and deutocerebrum are situated above the esophagus and for this reason are considered to be the outgrowth of the primitive prostomial brain such as is found in the annelids. The tritocerebrum is intimately joined with the deutocerebrum, but its two halves are connected by a commissure or connective fiber that passes *underneath* the esophagus. For this reason it is thought to be the ganglion of the first true body segment, now fused with the head.

Subesophageal Ganglion. Situated in the head, beneath the esophagus and joined to the brain by a pair of large connectives, is a large nerve center, the subesophageal ganglion. It is in reality the fused ganglia of the original mandibular, maxillary, and labial segments. This composite ganglion gives rise to the nerve trunks servicing the mouthparts. From this nerve center a pair of connectives pass through the neck into the thorax.

Ventral Nerve Cord. In the thorax and abdomen there is typically a nerve ganglion in the ventral portion of each segment. The ganglia of adjoining segments are joined by paired connectives, the whole forming a chain of nerve centers stretching posteriorly from the prothorax (Fig. 4-10). This chain is the ventral nerve cord. It is joined to the subesophageal ganglion by the connective passing through the neck. The thoracic ganglia give rise to the nerves controlling the legs and wings, and the abdominal ganglia have branches and fibers to the abdominal muscles and abdominal appendages.

The generalized ventral nerve cord is composed of a chain of well-separated ganglia. In various groups of insects certain of these may fuse to form a smaller number of larger units. This type of modification is demonstrated strikingly in the order Diptera (Fig. 4-10). Primitive members of this order possess a fairly generalized nerve cord; in more specialized families the thoracic ganglia fuse into a single large mass, and the abdominal ganglia become smaller and finally are scarcely discernible. The stages in this series of modifications are shown in Fig. 4-10, *A* to *D*.

Stomodeal Nervous System. Insects possess a so-called sympathetic nervous sytem (Fig. 4-12) that controls some of the "involuntary" motions of the anterior portions of the alimentary tract and dorsal blood vessel. There is considerable doubt, however, as to the exact function of some branches. It is more appropriate to term it the stomodeal system, because most of the parts are situated on the top or sides of the stomodeum (foregut). The central structure of this stomodeal system appears to be the *frontal ganglion*, which is

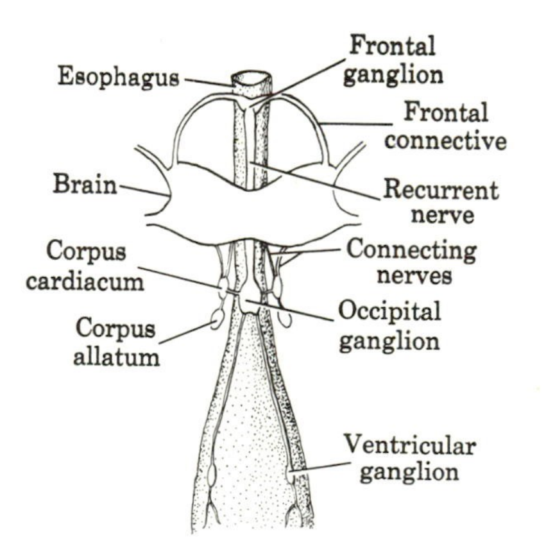

Fig. 4-12. Diagram of sympathetic nervous system of an insect. (Compiled from various sources)

situated in front of the brain and connected with the tritocerebrum by a pair of fibers. From the frontal ganglion a median recurrent nerve runs back beneath the brain and along the top of the esophagus, where it connects with a system of small ganglia and nerves. This group innervates the foregut, salivary ducts, the aorta, and apparently certain muscles of the mouthparts.

MUSCULATURE

The insect body is provided with an extremely complex system of muscles. These control almost all the movements of the body and its appendages. Some insects may possess over 2000 muscle bands.

In a gross body dissection, muscle tissue is one of the conspicuous features within the insect body. It does not form a continuous system but is distributed in different areas and enters into the composition of several organs. On the basis of distribution, muscle tissue may be grouped into three categories.

Visceral Muscles. The digestive tract and ducts of the reproductive system have an outer layer of muscle, which produces peristaltic movements. The muscle may be in circular, longitudinal, or oblique bands, or a combination of these. Special muscles occur in such places as the closing or opening mechanism of spiracles and in the mouth region. Muscles form pulsating bands that assist in the operation of the circulatory system.

Segmental Bands. The various segments of the body are connected by series of muscle bands that maintain body form (Fig. 4-13). In the abdomen, the terga are connected by longitudinal dorsal bands, and the sterna are connected by longitudinal ventral bands. The tergum and sternum of the same segment are connected by oblique or perpendicular tergosternal muscles. In the thorax the musculature appears entirely different. The most conspicuous muscles are large cordlike groups that operate the legs and wings; the other muscles are subordinate to these in size and prominence. In addition to these major muscle groups, there are many smaller bands that may be extremely complicated in pattern (Fig. 4-14). In both thorax and abdomen the exact muscle pattern differs markedly in various kinds of insects.

Muscles of the Appendages. The movable appendages have muscle bands of varying size and complexity. The mandibles of chewing insects have a few muscle groups that fill a large portion of the head capsule, but there are no muscles within the mandible itself (Fig. 4-15). On the other hand, appendages that are divided into

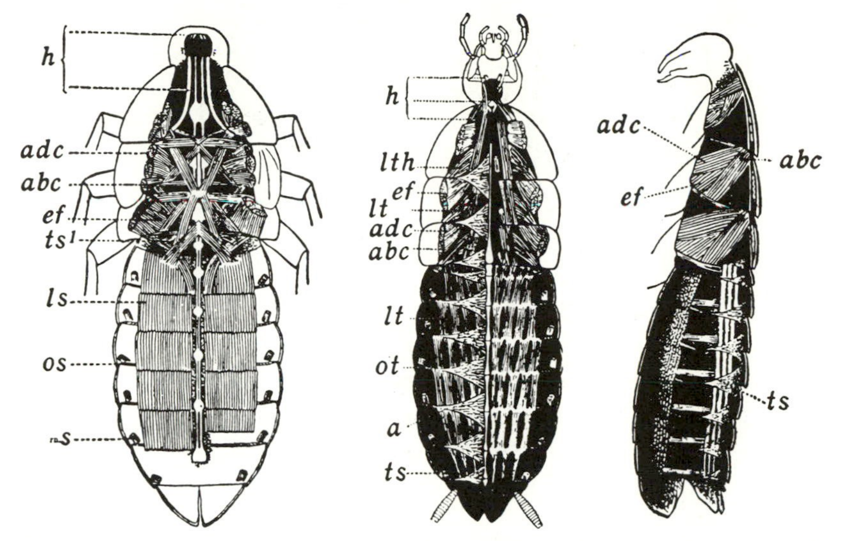

Fig. 4-13. Body musculature of a cockroach, showing ventral, dorsal, and lateral walls, respectively. *a*, Alary muscle; *abc*, abductor of coxa; *adc*, adductor of coxa; *ef*, extensor of femur; *h*, head muscles; *ls*, longitudinal sternal; *lt*, longitudinal tergal; *lth*, lateral thoracic; *os*, oblique sternal; *ot*, oblique tergal; *ts*, tergosternal; ts^1, first tergosternal. (From Folsom and Wardle, 1934, after Miall and Denny)

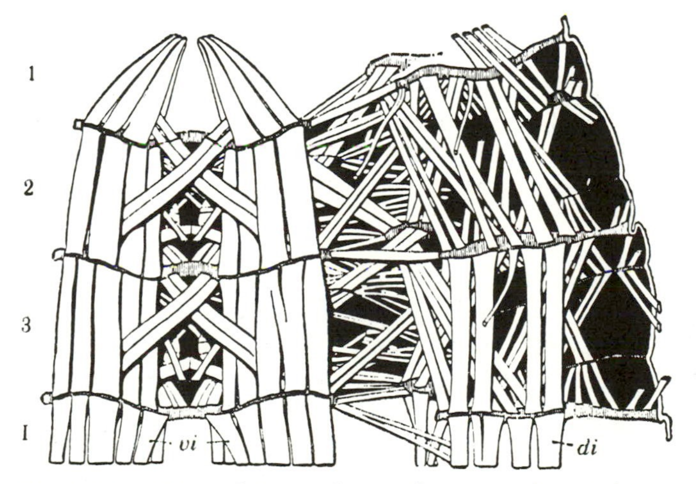

Fig. 4-14. Musculature of mesothorax and metathorax of a caterpillar. *di*, Dorsal bands; *vi*, ventral bands. (After Snodgrass)

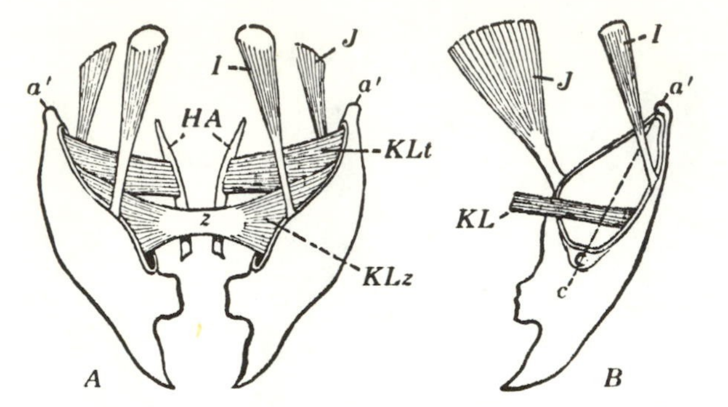

Fig. 4-15. Diagram of the mandibular muscles of insects. (*A*) Apterygote type with one articulation, *a'*; (*B*) pterygote type with two articulations, *a'*, *c*. (After Snodgrass)

segments, such as the maxillae and legs (Fig. 4-16), not only are activated by the large muscles inside the body, but in addition have muscles extending from segment to segment.

REPRODUCTIVE SYSTEM

Insects are primarily dioecious, in that normally only one sex is represented in any one individual. A few rare instances are known of hermaphroditic insects in which both sexes are represented in the same individual. The most notable case is the cottony cushion scale, *Icerya purchasi*.

In insects the reproductive system is a highly developed set of organs situated in the abdomen. There is a close parallel between the parts of the male and female systems, and most parts of both are bilaterally symmetrical.

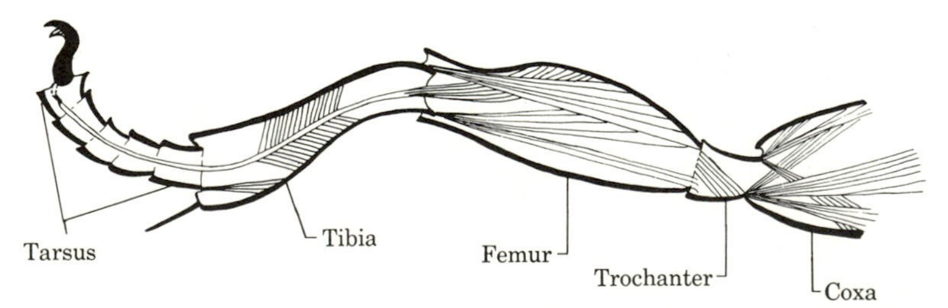

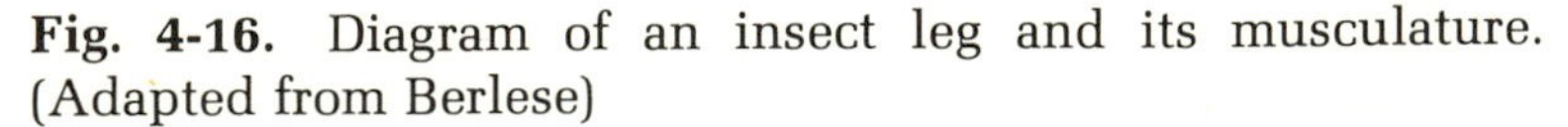

Fig. 4-16. Diagram of an insect leg and its musculature. (Adapted from Berlese)

Female Reproductive System. The female system consists essentially of (1) a group of ovarioles in which the eggs are produced, (2) a spermatheca in which sperms are stored, and (3) a duct arrangement through which the eggs are discharged outside the body. A typical system is illustrated in Fig. 4-17*B*, *C*. There are two ovaries, one on each side of the body. An *ovary* consists of several to many *ovarioles* or tubules. Each ovariole ends in an attachment thread, called the terminal filament; the upper part of the ovariole contains the forming eggs and the lower larger portion contains the more matured eggs; the bottom of the ovariole forms a small duct or pedicel. The pedicels of each group unite to form a *calyx*. Each calyx opens into a lateral oviduct. The oviducts of the two sides join to form a common *oviduct*. This opens into an egg-holding chamber, or *vagina*, which opens directly into the external ovipositor, or egg-laying mechanism.

Two structures are connected with the dorsal wall of the oviduct. One is the *spermatheca*, a single bulbous organ in which spermatozoa are stored and that has a gland attached to its duct. The other is a paired structure, the *accessory glands* or *colleterial glands*, which secrete adhesive material used in making a covering over egg masses or gluing eggs to a support.

In some of the more primitive groups such as the Orthoptera, the vagina may be only a pouchlike invagination of the eighth sternite.

Many deviations are found from the system just described, with

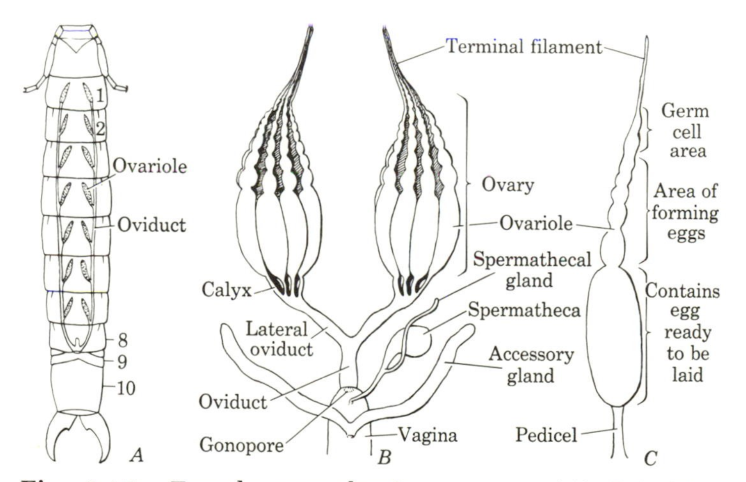

Fig. 4-17. Female reproductive system. (*A*) Primitive type of *Heterojapyx gallardi*; (*B*) diagram of common type found in many insects; (*C*) diagram of a single ovariole. (Redrawn from Snodgrass)

differences occurring in the number and shape of ovarioles and tubules, ducts, and glands. In many groups the spermatheca and vagina exhibit many shapes that are of considerable taxonomic value.

The primitive family Japygidae has a most interesting reproductive system. The ovarioles (Fig. 4-17*A*) are arranged segmentally, linked together by a pair of long lateral oviducts that fuse and form a single oviduct near the egg-laying aperture. This condition suggests that the ancestral insect groups possessed independent ovaries in each segment and that there has occurred a constant migration and consolidation of these to the posterior end of the body, evolving finally the typical system shown in Fig. 4-17*B*, *C*.

Male Reproductive System. In general organization the male system is similar to the female. It consists primarily of a pair of testes, associated ducts and sperm reservoirs, and outlets to the outside of the body. A common type is shown in Fig. 4-18.

Each testis consists of a group of *sperm tubes*, or *follicles*, in which the sperm are produced. The sperm tubes open into a common duct, the *vas deferens*, which in turn opens into a reservoir, the *seminal vesicle*. From each seminal vesicle proceeds a duct; the two ducts join to form a common ejaculatory duct. This duct runs through the *penis*, or *phallus*, at the end of which is the sperm escape opening. Structures usually associated with the penis form

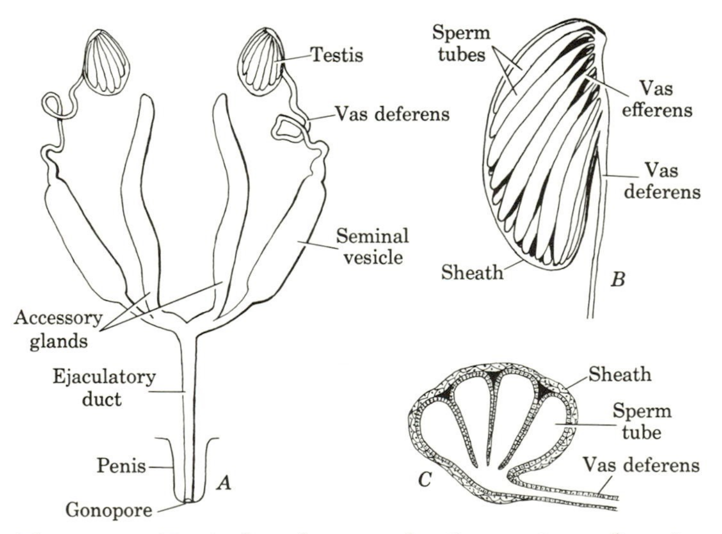

Fig. 4-18. Typical male reproductive system of an insect. (*A*) Entire system; (*B*) structure of a testis; (*C*) section of a testis and duct. (Redrawn from Snodgrass)

the male external genitalia, the *aedeagus*, and include a rigid sheath around the true membranous penis. Associated with the internal part of the ejaculatory duct are *accessory glands*, which may be single or paired.

SPECIALIZED TISSUES

In addition to the extensive systems outlined in the preceding pages, the insect body contains some other smaller or less definitely organized tissues. The most important are the *fat body* and the *corpora allata*.

The Fat Body. This is a loosely organized aggregation of cells that occurs throughout the insect body, especially in the later larval or nymphal instars. The cells of the fat body may be packed so tightly as to appear like organized tissue. The function of the fat body is to store food and to participate in intermediate metabolism.

Corpora Allata. These are a pair of ganglia-like bodies (Fig. 4-11), closely associated with the stomodeal nervous system. They secrete hormones important in regulating metamorphosis and the development of some adult tissues. They are further discussed with other structures of the nervous system in Chap. 6.

REFERENCES

Chapman, R. F., 1969. *The insects, structure and function.* New York: American Elsevier. 819 pp.

Richards, O. W., and R. G. Davies, 1977. *Imms' general textbook of entomology*, 10th ed. London: Chapman & Hall. Vols. 1 and 2.

Wigglesworth, V. B., 1972. *The principles of insect physiology*, 7th ed. London: Chapman & Hall. 827 pp.

See also references for Chapters on External Anatomy, Life Processes, Response and Behavior, and Life Cycles, Growth, and Reproduction.

5 Life Processes

In insects, as in all living organisms, persistence from generation to generation (in other words, survival of the species) depends on successful *growth* and *reproduction*. These are two of the basic functions or activities of life. An insect must extract from its surroundings those substances (nutrients) it needs for growth and use them to build a mature individual, and finally to reproduce. In this sense, an insect species, or any other species for that matter, may be considered as extremely complex, highly integrated processes of mechanical-physical-chemical systems. These various processes, which constitute the organism's *physiology*, are the result of the functional activities of the organs and tissues described in the previous two chapters. This chapter discusses life processes in several of the important functional systems and their interactions within an insect. The next chapter deals with coordination and behavior and examines the response of insects to stimulation and irritability that they experience in their surroundings. It includes the nervous and hormonal systems. Life cycles, growth, and reproduction are included in Chap. 7.

The basic physical and chemical processes involved in the functioning of an insect are the same as those occurring in other forms of animal life, and include movement, metabolism (chemical reactions carried on within the organism, tissues, and cells), respiration, circulation, homeostasis (maintenance of the constancy of the internal environment around the cells), transport of fluids (including waste fluids), adaptation, and irritability (certain responses to physical and chemical stimuli in the organism's environment).

STRUCTURE AND FUNCTION OF THE SKELETON

A distinctive feature of insects is the chitinous exoskeleton. This is an integument surrounding the body and appendages and is secreted by an underlying layer of cells, the epidermis (Fig. 5-1*A*). Most of the integument is composed of cuticle, but it also contains many kinds of external hairs and sense receptors, and its internal surface has processes (*apodemes* and *apophyses*, Fig. 5-1*C*, *D*) of many types for attachment of muscles. The integument is called an exoskeleton because of its exterior position and it serves as a covering to protect the organism not only from predators, but also from abrasion, heat loss, and desiccation. It also determines the form and shape of the insect's body. Particular layers form barriers to water, pathogens, and insecticides. Loss of water by evaporation is the greatest threat to terrestrial organisms, and nearly all insects are terrestrial or aerial for at least some portion of their lives. Evaporation is a function of surface area, not volume, and as size decreases, the ratio of surface area to volume increases. Thus, because insects are both small and terrestrial, they are faced with a major problem of protecting from excessive evaporation the small total amount of water contained in their bodies. The protection lies in the impermeable nature of the insect cuticle, which is remarkably resistant to the passage of water vapor. Without such efficient protection it is doubtful if an insect flying in the air, for even a short time, could escape fatal desiccation.

Epidermis. This is a single inner layer of epithelial cells below the cuticle and rests on an amorphous granular layer, the *basement membrane*. Most of the epithelial cells secrete the cuticle of the integument, which is the inert, nonliving outermost part of the insect. Various specialized epithelial cells associated with the epidermis form *sense organs* and *dermal glands* (Fig. 5-1*A*, *B*).

Cuticle. The insect cuticle performs several dissimilar functions. It protects, it supports, and it serves as a source of food during starvation and molting. It also is involved in the formation of sense organs and mouthparts and it lines the tracheae.

The cuticular exoskeleton of an insect is a thin, yet very strong shell. In its mechanical properties, cuticle has a tensile strength comparable to some metals, weight for weight. For example, cuticle has a tensile strength of about 10 kg/mm^2 compared to 100 kg/mm^2 for steel, but steel is seven times heavier. Aluminum is about as strong as cuticle but twice as heavy.

In addition, cuticle can be exceedingly hard. The mandibles of some insects are able to bore through metal, such as tin or copper.

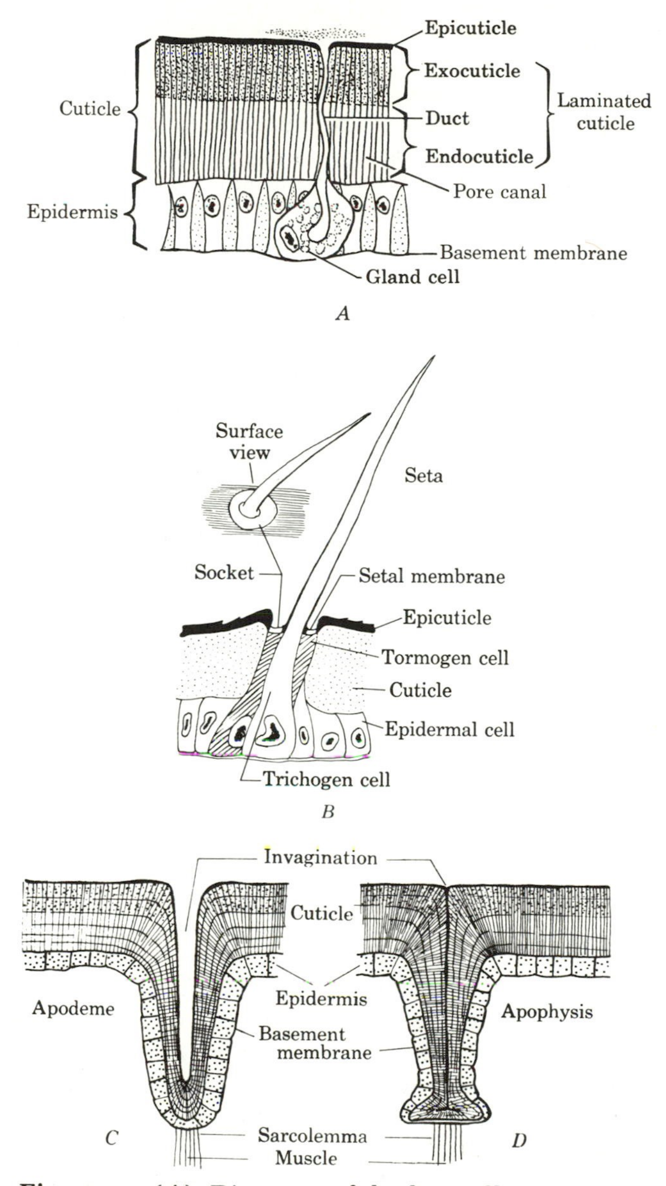

Fig. 5-1. (*A*) Diagram of body wall structure (Adapted from Wigglesworth). (*B*) A seta and its socket (Adapted from Snodgrass); (*C*) Cuticular invagination forming an apodeme; (*D*) Cuticular invagination forming an apophysis (C & D modified from Richards, 1951)

Typically the cuticle consists of a thick inner *laminate cuticle* and a thin outer *epicuticle*.

The laminate cuticle consists of an inner *endocuticle* and an outer *exocuticle* (Fig. 5-1*A*). Both are composed of chitin (a high-molecular-weight polysaccharide) and protein bound together to form a complex glycoprotein. The laminate cuticle is considered to be microfibers of chitin in a matrix of protein. The exocuticle is absent at joints and along sutures where the skeleton will split during molting.

The lamellae of the cuticle are deposited as sheets of microfibers, each of which is extremely thin and each of which may have the axes of the microfibers rotated a few degrees with respect to the ones above and below (Fig. 5-2). This type of construction, like plywood, has great strength. The outer part of the laminate cuticle (exocuticle) is chemically stable and is not dissolved by the molting fluid. It includes the laminate cuticle that is usually tanned with quinones (converted phenols), which change the microfibers into tough, inert compounds by the formation of cross-linkages between adjacent protein chains.

Extending through the laminate cuticle and across the lamellae are minute *pore canals*. They extend from the epidermis to the inner layers of the epicuticle and contain filamentous extensions from the epithelial cells (Figs. 5-1, 5-3). These pore canals function as ducts for the transport of chemical secretions, such as wax, tanning substances, and cuticular material from the epithelial cells. There may be 50 to 70 pore canals associated with each epidermal cell. In the cockroach, their density is very great, being about 1.2 million per square millimeter.

Epicuticle. The epicuticle, although an exceptionally thin region, is composed of four layers that are secreted by the epithelial cells and dermal glands (Fig. 5-3). It consists of an *outer layer of cement* (secreted by the dermal glands), underlain by *wax layers*, which are succeeded by an exceptionally thin layer of dense *cuticulin* and then a thicker inner layer of *homogenous cuticle* that is adjacent to the outer part of the exocuticle.

The *outer layer of cement* is present wherever dermal glands occur and varies greatly in thickness and extent. It is thickest in the exposed exoskeleton of beetles, and is absent in many adult insects that are covered with scales.

The *wax layer* consists of extraordinarily long hydrocarbon chains.

The *cuticulin* is a layer covering the whole insect except for some sensory structures. It is a double lipid layer, appearing in cross sec-

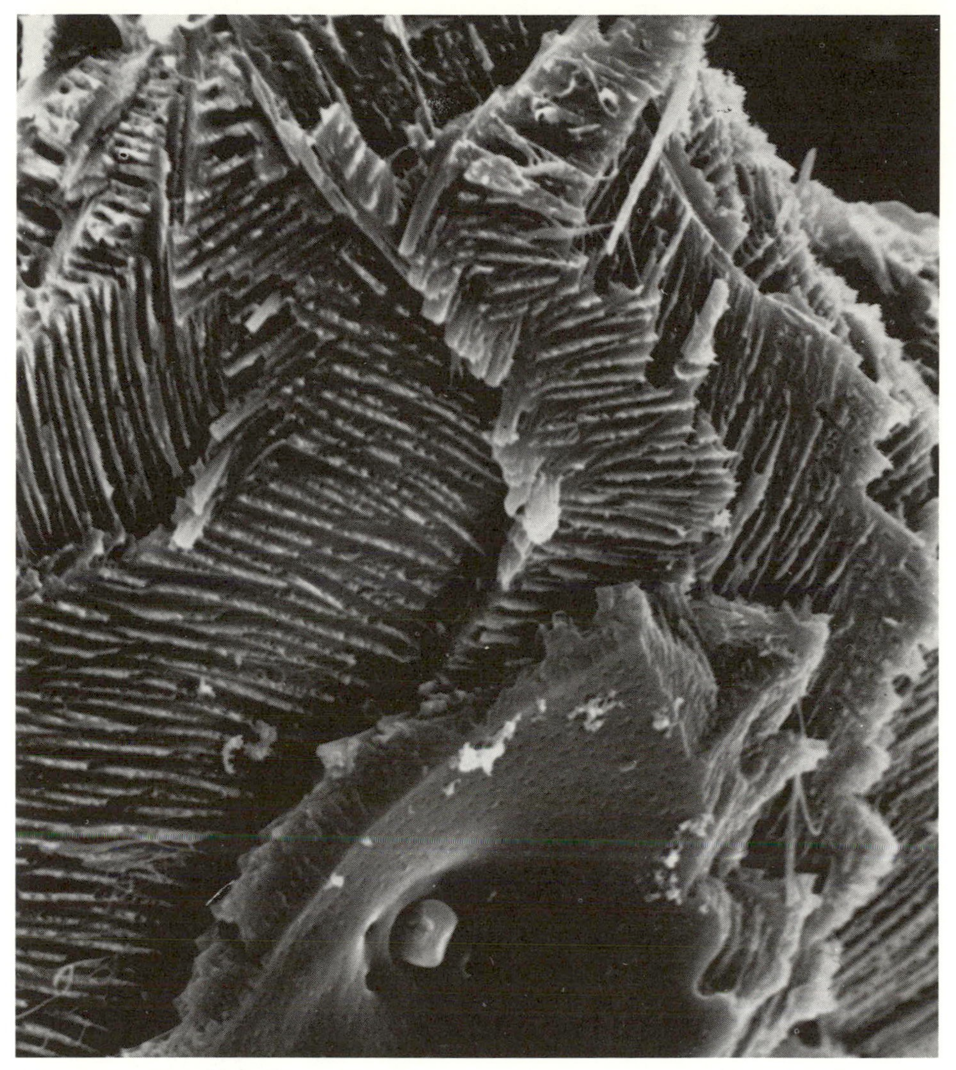

Fig. 5-2. Exoskeleton in head region of coleopteran *Haplorhynchites aeneus* showing different layers of lamellae of the cuticle penetrated by a sensory pore; ×7200. (Courtesy Dept. of Entomology, Univ. of Alberta, and D. A. Craig)

tion as a thin dense line about 100 to 200 Å thick. This exceptionally important layer serves several functions. It is permeable to chemicals activating molting but is itself impermeable to the molting fluid and resistant to digestion by that fluid. In some circumstances, it is permeable to wax and also water. In addition, the surface pattern and

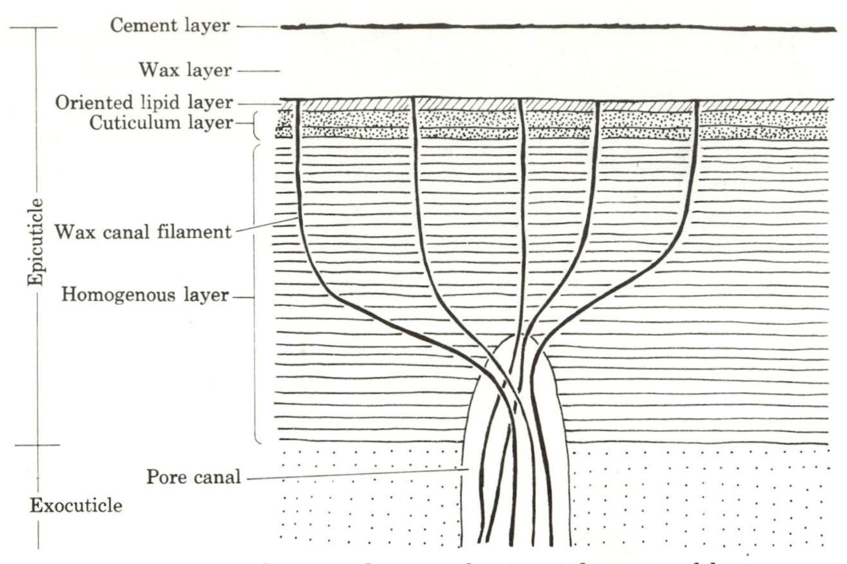

Fig. 5-3. Diagram showing layers of epicuticle crossed by wax canals. (After Gluud, 1968)

ornamentation of an insect's exoskeleton and the formation of hairs, scales, and mitochondria are closely associated with the deposition of the cuticulin layer, which provides support for the waxy layers.

The dense *homogenous layer* covers most of the insect, including tracheae (but not tracheoles) and parts of the linings of some glands and sense organs.

The epicuticle is traversed by elongate cylindrical filaments, 60 to 130 Å in diameter (Fig. 5-3). These filaments may be lipid–water liquid crystals. Because they are most abundant in regions where wax is secreted, they initially were called *wax canals*. They probably serve as the route for transporting the waxes that are deposited on top of the cuticulin. The epicuticle, then, is a route to the surface for waxes and a barrier to water loss and water penetration. Disruption of the wax layer may cause terrestrial insects to lose water.

Specialized cells. Some epidermal cells have special functions, either the secretion of fluids or the formation of definite structures such as hairs.

Dermal Glands. Some of the epidermal cells may become specialized to form either sense organs or glands. The dermal glands commonly consist of three cells: a secretory cell and two other cells forming the duct from the secretory cell through the epidermis and cuticle. The duct continues to the surface of the cuticle, and the products of the secretory cell are discharged onto the surface during cuticle formation, forming the cement layer.

Setae (Fig. 5-1B). Most of the flexible hairs or bristles of insects are formed by epidermal cells called *trichogen cells*. At the time of the actual formation of the hair, the trichogen cell is large and nucleated and has a duct that passes through the cuticle to the surface. From this point the products of the cell build up the hair. Closely associated with the trichogen cell is a *tormogen cell*, which forms a socket (usually flexible) around the base of the hair. A hair or bristle of this histological origin is called a *seta* (pl. *setae*). The parent cells may degenerate after the seta is formed.

Specialized setae originate in the same manner. These include scales, poison hairs, and sensory setae, discussed in Chap. 6.

Color. Many insect colors are located in the epidermis or exoskeleton. Insects colors are of two types: *pigments* and *structural colors*.

Pigments such as *carotenoids*, *sclerotin*, and *melanin* are deposited in the cuticle and produce different colors by selective absorption of different wavelengths of light. Carotenoids generally produce yellow, orange, or red colors in insects. Sclerotin and melanin are dark pigments, many times producing black coloration.

Tetrapyrrole pigments (bile pigments) are present in the hemolymph of many insects and are blue. This blue pigment when present with a carotenoid pigment produces a green color seen in many insects with a thin, transparent cuticle.

Pteridines are fluorescent pigments found in the wings of insects, eyes, and other tissues. Some of the white, yellow, orange, and red colors in insects are produced by these pigments.

Omnochromes are masking pigments in the accessory cells of the light-sensitive cells that form the compound eyes (*ommatidia*). Red, yellow, and brown are the colors found in the eyes. They are also present in many insect tissues including the epidermis, and are excretory and wing pigments in the Lepidoptera.

Structural colors are produced by extremely delicate and minute lamellae in the cuticle that break up light into various wavelengths by scattering, reflection, and interference. The individual lamellae are commonly separated by a chemical, such as uric acid in scarabaeid beetles. This grouping of lamellae, referred to as vanes, produces many of the metallic and iridescent colors of beetles. The most common example of this occurs in the moths and butterflies. In these the wings are covered with scales (modified setae), and the scales bear ribs running the length of the scale. Studies with the electron microscope have shown that each rib is composed of several parallel, extremely thin fenestrate groups of lamellae. Studies on the

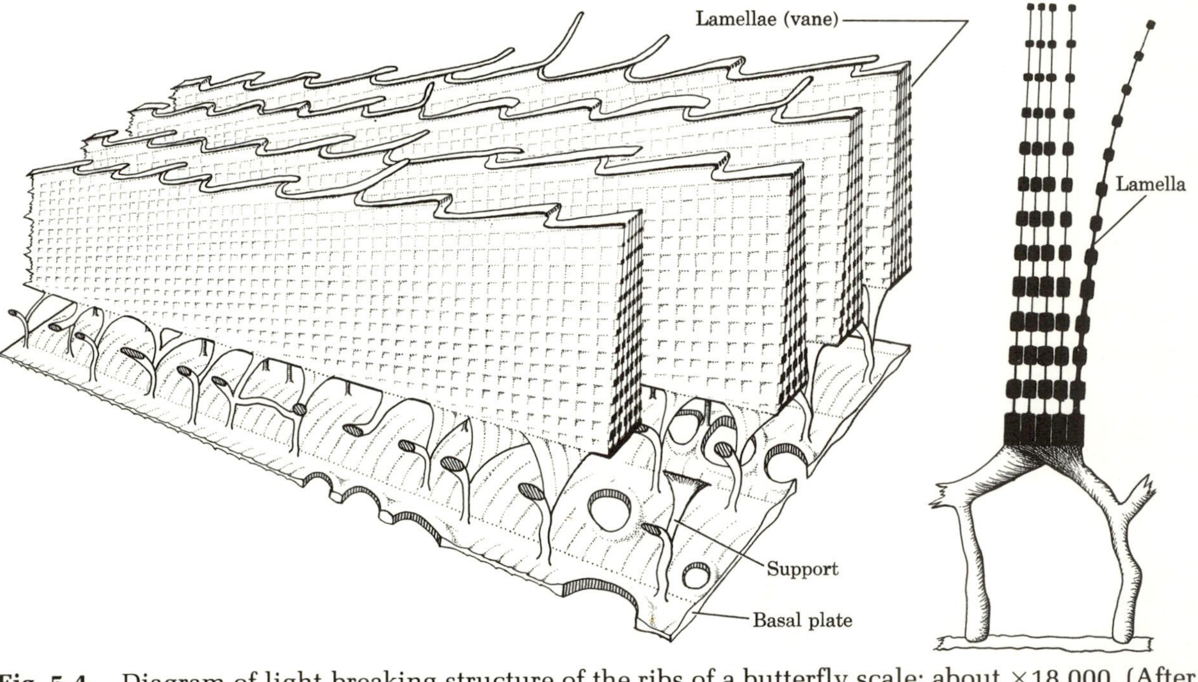

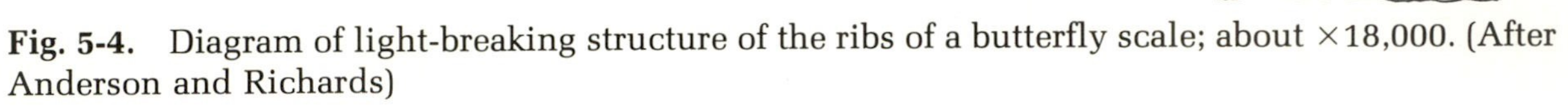

Fig. 5-4. Diagram of light-breaking structure of the ribs of a butterfly scale; about ×18,000. (After Anderson and Richards)

tropical *Morpho* butterflies indicate that ribs of more simple structure produce nonmetallic colors and that ribs of great complexity (Fig. 5-4) produce the dazzling iridescent colors for which these butterflies are famous.

Epidermal Growth: Molting. When an immature insect has reached the size permitted by the cuticle, the old cuticle must be replaced by a larger new one in order that continued growth is possible (Fig. 5-5). This periodic shedding of cuticle is called *molting*. It is initiated by the separation of old cuticle from the underlying epidermis (the process of *apolysis*) and terminated by the shedding of the remains of the old cuticle (the process of *ecdysis*). Molting, however, involves many changes before, between, and after apolysis and ecdysis. It is under hormonal control. In general, the following steps occur.

1. The epithelial cells may either divide mitotically and become closely packed or they may increase in size, and the cuticle separates from the epidermis.
2. A new epicuticle is formed on the newly exposed epidermis.
3. In the meantime, molting fluid in an inactive state is secreted into the space between the newly forming epicuticle and the old endocuticle. By the time the new epicuticle becomes impervious, the

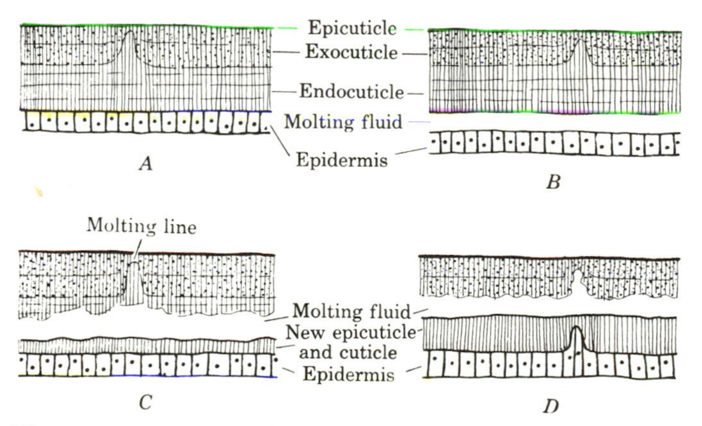

Fig. 5-5. Diagram showing molting of old cuticle and secretion of new cuticle. (*A*) Complete exoskeleton and underlying epidermis; (*B*) initiation of apolysis with the separation of the epidermis and endocuticle and secretion of molting fluid; (*C*) digestion of old endocuticle and secretion of new cuticle; (*D*) exoskeleton just before molting with both the old and the new exoskeleton. (Modified from Hackman, 1971)

molting fluid has changed to an active state containing the enzymes chitinase and proteinase that digest the endocuticle of the old cuticle. The products of digestion are resorbed by the epithelial cells and used to build up a new cuticle. Up to 90% of the old cuticle may be recycled in this fashion.

4. When the new endocuticle is about half completed, certain enlarged dermal glands discharge their contents over the outside of the new cuticle. This secretion forms the final waxy layer of the epicuticle.

5. When the new cuticle is fully formed, the insect has to break out of the old one. The initial rupture is made along a mesal line of weak cuticle that typically runs along the dorsum of the thorax and extends on to the head. This rupture is caused by the pressure of the blood. The insect contracts the abdomen, forcing the blood into the thorax and causing it to bulge until the cuticle breaks along the line of weakness. The insect may swallow air (or water, if aquatic) to aid in this process. The insect then wriggles and squirms free from the old exoskeleton. Before this time, the molting fluid is usually reabsorbed by the body, so that at the time of molting (ecdysis), the area between the old and new skins may be dry.

6. For a short period after molting the new cuticle can be stretched, at least in the nonsclerotized (membranous) portions. During this short period, the insect stretches the cuticle first by swallowing air or water, thus increasing its internal volume, and then by increasing the blood pressure in first one body region and then another. These actions "blow up" the regions and stretch the integument. When the blood pressure is reduced, the stretched integument *does not shrink again* but the membranous parts contract into a series of small folds or minute accordionlike pleats. In larvae with no sclerotized body areas, these folds may occur over the entire body (Fig. 5-6). In insects with definite sclerotized plates, the folds occur in the membrane between the sclerites. As the body increases

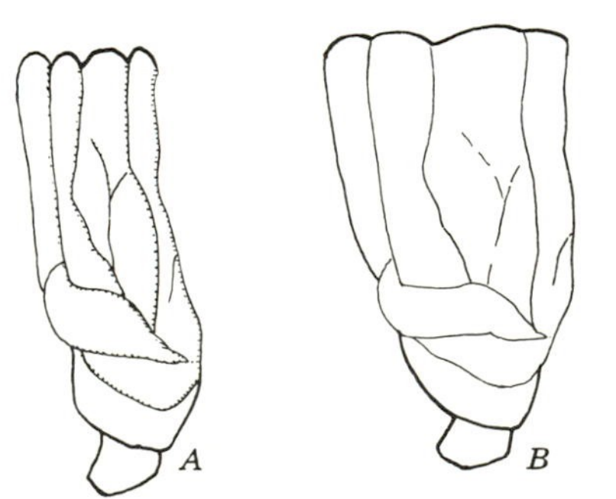

Fig. 5-6. Drawing of abdominal segment of a sawfly larva showing membranous folds. (*A*) Immediately after molting; (*B*) after growth and expansion.

in size with subsequent growth, the integument increases by a simple expansion of the folds, as in Figs. 3-1 and 3-2. When this avenue for increase is exhausted, the insect must molt again to allow further size increase. The act of shedding the old skin may take only a few seconds, or it may require an hour or more.

7. After molting, the dermal glands secrete a thin layer, the cement layer, over the integument. This acts as a waterproofing coat and protects the waxy layer of the epicuticle from abrasion. Material is added to the endocuticle by further secretion of the epithelial cells throughout the intermolt period.

8. After its complete formation, the new cuticle is impermeable to many substances, especially water, and is locally sclerotized and colored to assume its normal condition. In many groups, such as the grasshoppers, this occurs just after the stretching process that follows molting. In other cases, for instance, in adult Trichoptera, Lepidoptera, and many Hymenoptera, this occurs before molting while the adult is still encased in the pupal skin. Such an incased adult, with the old cuticle of epicuticle and exocuticle and also its new cuticle, is said to be in the *pharate* condition.

NUTRITION

In common with many living things, insects require water, a large selection of amino acids, some nucleic acids, a source of sterols, a number of vitamins, certain fatty acids, and a large number of minerals and inorganic substances including phosphorus, potassium, iron, copper, zinc, cobalt, and calcium. These are called the *minimum nutritional requirements*. Although foods such as carbohydrates and fats are eaten and used as energy sources by many insects, other organic compounds can be "burned" by the tissues for energy.

Even though watching a whole insect may obscure the fact, the ultimate basis of both growth and reproduction is cell division. Cell division itself requires duplication of the nuclear deoxyribonucleic acid (DNA) and its associated complex nuclear proteins. Certain cells, such as those producing setae, must be supplied with large quantities of nutrients for the cell to convert into extrusions that are highly elaborated chemical compounds. These circumstances help to explain the dietary needs for the amino acids and other organic compounds that are essential for the manufacture of proteins and nucleic acids (DNA, RNA).

Like other animals, insects cannot synthesize certain amino acids and many other organic compounds they need. That is why these compounds are nutritionally required. As a consequence, insects obtain them by eating other living or dead organisms that have either synthesized these compounds (green plants) or have themselves obtained these compounds directly or indirectly from green plants (other animals or parasitic plants). Insects synthesize nucleic acid. However, the rate of nucleic acid synthesis may be slower than the needs of the insect, particularly in some species of Diptera and Coleoptera, so the organism seeks nucleic acid sources in its diet.

These nutritional requirements are not daily but lifetime requirements. Insects to a high degree rely on nutrient reserves. They accumulate nutritive materials of different kinds in body tissue to use later in their life cycles. Thus a caterpillar eats green plants and stores large quantities of required nutrients that suffice for maturation and egg laying of the adult, which itself may feed primarily on carbohydrates, or may not feed at all. In other insects such as certain mosquitoes, however, the egg-laying capacity is dependent in large measure on the protein-rich blood meals of the adult female. The mosquitoes present another interesting phenomenon common in insects. The larva feeds chiefly on whole microorganisms, the adult male on nectar, the female (of most species) on vertebrate blood. Many insect species feed only on green plants (leaves, bark, roots, pollen, or heartwood); others only on dung, rotten wood, or animal carcasses; and still others on various kinds of living animals. Whatever its food, in each kind of insect there has evolved a system of digestion by which the insect can normally procure its nutritional requirements from its food.

Vitamins and Cholesterol. Vitamins play an important role in cell metabolism and the needs of various insect species differ. Vitamins A, C, D, E, and those of the B complex are used in different metabolic functions of insects. Insects that need vitamin C (ascorbic acid) generally feed on plants, but certain species, such as the roach *Blattella*, synthesize it. Some insects need vitamin E for reproductive processes.

Most insects do not synthesize sufficient sterols for their nutrient requirements and need cholesterol in their diet. Cholesterol is, therefore, like vitamins, a needed food suplemented in the diet.

Experiments on blowfly larvae have shown that some of the needed vitamins are obtained from symbionts (symbiotic microorganisms) and mixed with the normal diet, which may itself be deficient in these substances. The symbionts are present in the alimentary tract, in specialized organs, or intracellularly, and provide nutrients, including various vitamins and amino acids that are not available for nutrition of the host.

Water Requirements. Like other organisms, water is essential for metabolic processes in insects. It is a vital item of the insect diet. Most insects obtain sufficient water for their needs from foodstuffs with a fairly high water content, such as foliage and blood. They also have developed many structural and physiological specializations to conserve water. Some insects conserve water to such a degree that they are able to exist entirely on dry materials. In these instances the insect makes use of the water resulting from the oxidation of the foodstuffs. But even then the food generally must contain a small percentage of water to supplement the metabolic water. Some insects even absorb water from humid air, such as the mealworm and most shore-dwelling beetles.

DIGESTION

Digestion is the process of breaking down and changing food so that it is absorbed by the epithelium of the gut and passed into the blood that furnishes nutriments to the body. Because the food of different insects includes an array of diverse materials, many modifications are found in the digestive systems, each adapted to handling a particular type of food. The digestive system may be entirely different in larval and adult stages of the same insect (Fig. 4-3), especially in the Diptera, in which the food of the various stages is entirely different. Within an order there may be strikingly different digestive systems. The Hymenoptera, for instance, include such diverse forms as sawfly larvae, which are herbivorous, and wasp larvae, which are internal parasites—each having a different type of digestive system.

A generalized type of digestive system is found in herbivorous and omnivorous insects such as cockroaches, grasshoppers, and the larvae and adults of many beetles. This type is used as a basis for the following account. A few of the more conspicuous modifications from this type are discussed; the enormous number of other modifications are a specialized study of the subject on their own.

Salivation. In a great variety of insects saliva is mixed with the food before it is ingested. In chewing insects, the saliva is secreted into the mouth and mixed with the food there. In sucking insects the saliva sometimes is injected into the liquid foods, and the mixture is then sucked into the pharynx. The saliva is usually produced by the labial glands.

Typically each gland is like a long bunch of grapes, each "grape" a small cluster or *acinus* of secreting cells. Each acinus has its own duct; these join successively to form the large duct of the whole gland. The acinus may contain cells of different histological structure. The labial glands having a function connected with food may be segregated into two general groups, based on the principal substance they are known to secrete.

*1. **Digestive Group.*** In many insects the labial glands are the chief source of amylase, which digests starch. This is usually secreted into the food mass before it is swallowed, and the actual digestion takes place in the digestive tract. In adult Lepidoptera and bees the glands secrete invertase, which is exuded at the tip of the proboscis and drawn into the stomach with the nectar.

*2. **Anticoagulin Group.*** The labial glands of bloodsucking insects secrete no digestive enzymes but instead produce an anticoagulin. The anticoagulin prevents the ingested blood meal from clotting and plugging up the beak and digestive tract.

Extraintestinal Digestion. In special cases, digestive enzymes are extruded from the body onto or into the food and effect at least partial digestion before the food is taken into the digestive tract. This is called extraintestinal digestion. Plant lice, for instance, extrude saliva containing amylase from the beak into the host tissues and in this way digest starch in the host plant cells. Many predaceous beetles that lack salivary glands eject their intestinal enzymes through the mouth onto their prey. When digestion has occurred, the fluids produced are reabsorbed. Flesh-feeding maggots extrude proteolytic enzymes from the anus and effect extraintestinal digestion of the tissues in which they live and that form their food.

Ingestion. Insects take their food into the alimentary canal by way of the mouth. In insects with chewing mouthparts, the mandibles and maxillae cut off and shred the food. The closing together of these opposing structures presses the food to the back of the mouth or cibarium, at the base of the hypopharynx (Fig. 5-7). The hypopharynx is then pulled upward and forward, forcing the food into the pharynx, which is the anterior end of the esophagus. From this point the food is moved along the digestive tract by peristaltic action. In insects with sucking mouthparts (Fig. 5-8), the pharynx forms a bulblike pump, which expands and contracts by action of head muscles. The *pharyngeal pump*, as it is called, pulls the liquid food through the beak and into the region of peristaltic control. Digestive enzymes or other secretions may be mixed with the food before it is swallowed. Digestive enzymes commonly found in the salivary secretions and regions of the digestive tract are amylase, maltase, invertase, tryptase, peptidase, and lipase.

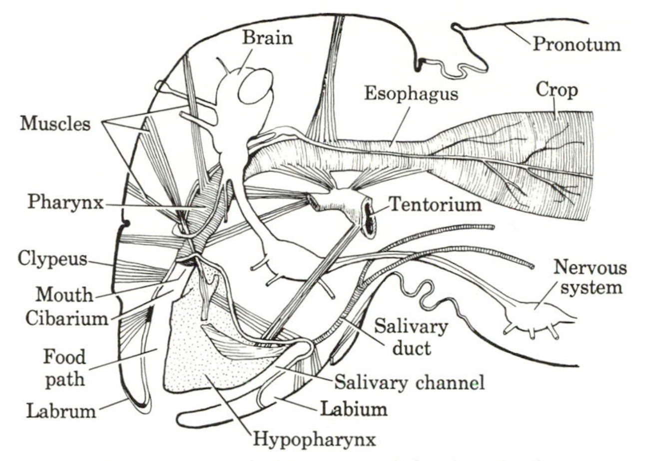

Fig. 5-7. Sectional diagram of the head of a chewing insect, showing the generalized parts, areas, and musculature used in swallowing. (Redrawn from Snodgrass)

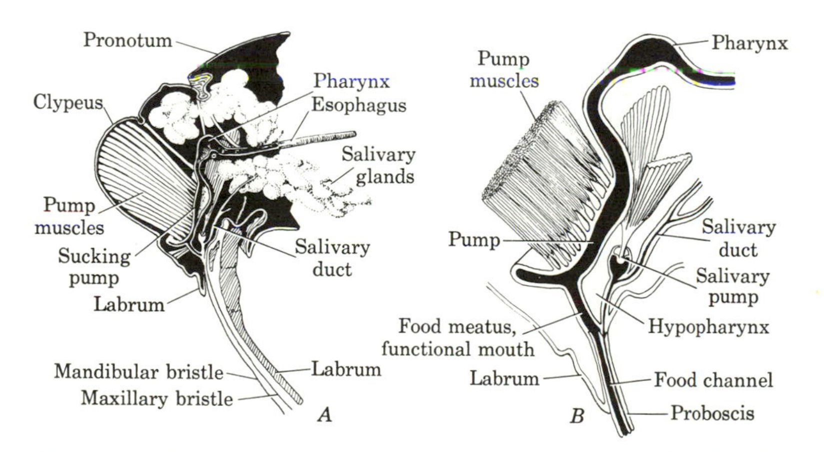

Fig. 5-8. The sucking pump and salivary syringe of a cicada. (*A*) Section of the head showing position of the sucking pump (cibarium) with dilator muscles arising on the clypeus. (*B*) Section through the mouth region, showing food meatus, suck pump, and salivary syringe. (Redrawn from Snodgrass)

An exception to oral ingestion occurs in the earliest larval stages of some internal parasites, which absorb their nutriment through the general body surface from the tissues or blood of their host.

Foregut or Stomodeum. The food is passed through the esophagus into the foregut. There is considerable variety in the functions of the foregut; it may serve simply as a passage into the midgut, or may form a capacious crop in which the food can be stored and partial digestion may take place. In some cases, as in the Orthoptera, digestive juices are passed forward from the midgut to the foregut.

The foregut typically consists of a layer of simple epithelial cells (Fig. 5-10*A*) that secrete a definite cuticle. It is believed that this cuticle is practically impermeable to enzymes and to the products of digestion and that little or no absorption takes place through it. The function of the cuticle is probably to prevent absorption of only partially digested compounds, because such premature absorption would interfere with complete digestion.

Proventriculus. Orthoptera and other groups that eat coarse food have a set of powerful shredding teeth in the proventriculus for dividing the food into smaller particles. A typical arrangement is found in the cockroach (Fig. 5-9), where six stout teeth do the shredding. Fleas employ a mass of sharp, needle-like teeth directed backward. During digestion these are driven backward at the same time that the blood meal in the midgut is thrust forward, and the fine teeth crush the blood corpuscles, causing them to disintegrate. These movements are caused by rhythmic and opposing muscle contractions. In other insects the proventriculus is simply the narrowed end of the foregut.

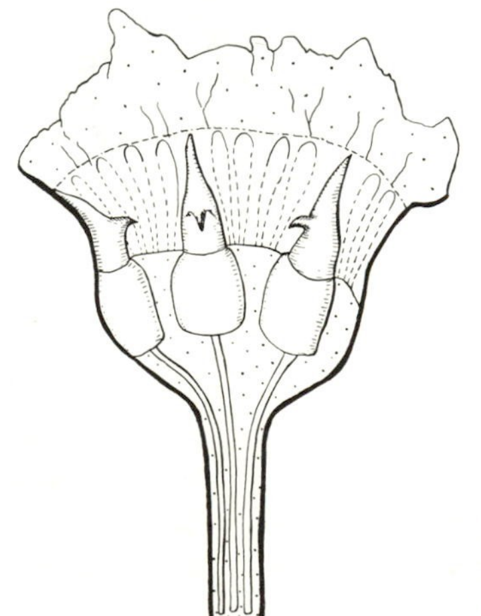

Fig. 5-9. Proventriculus of a cockroach laid open to show three of the six shredding teeth.

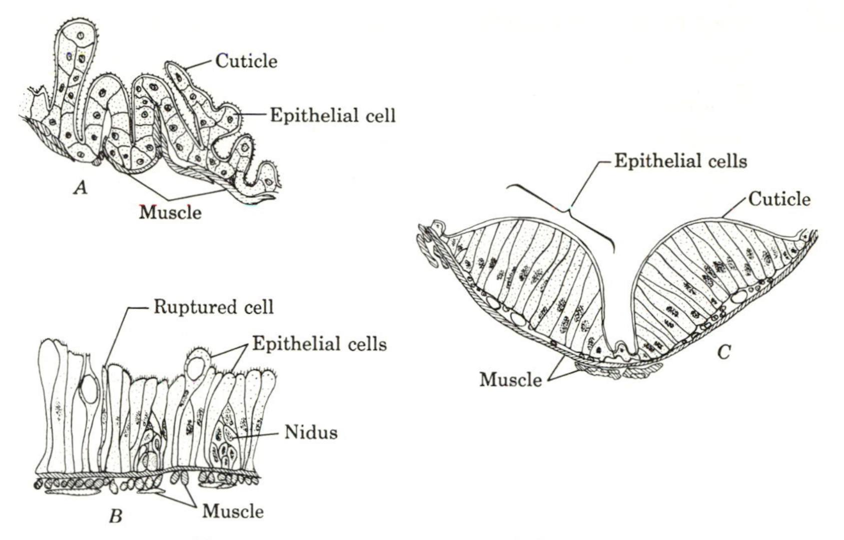

Fig. 5-10. Cell structure of portions of the digestive tract of a cockroach. (*A*) Foregut, longitudinal section; (*B*) midgut, longitudinal section; (*C*) hindgut, cross section.

Midgut or Mesenteron. In this portion of the alimentary canal, the epithelial cells (Fig. 5-10*B*) are exposed, since they do not secrete a cuticle. Some of these exposed cells perform most of the actual food absorption, and other cells carry on enzyme secretion.

The actual secretion of enzymes by the epithelial cells is accomplished by two methods: (1) *holocrine secretion*, in which the cells disintegrate in the process, emptying their contents into the lumen of the intestine; and (2) *merocrine secretion*, in which vacuoles containing the enzymes are released through the cell membrane into the lumen. The former is illustrated in Fig. 5-10*B*. This shows the clusters of regenerative cells, or nidi, which replace the cells used up during holocrine secretion.

Enzymes. In insects as a whole, the midgut produces a wide variety of enzymes that digest carbohydrates, fats, and proteins. In the main these are the same enzymes present in the mammalian system. Production of enzymes is usually correlated with diet. Omnivorous insects such as the cockroach produce the full complement of enzymes for digesting all types of food. Bloodsucking insects, however, produce chiefly proteolytic enzymes. Some insects secrete cellulase for digesting cellulose. Certain clothes moths are able to digest keratin, the proteinaceous material in mammalian hair, with the aid of a common insect proteinase combined with pecular pH conditions of the midgut. Wax moth larvae eat wax, which is probably digested chiefly by symbiotic bacteria.

Peritrophic Membrane. The epithelial cells of the midgut are exposed and delicate. If the food bolus were to be pushed over the unprotected surface of these cells, it would undoubtedly injure them severely and interfere with their functions of secretion and absorption. In vertebrates, mucous glands coat and lubricate the food boluses and hard particles to avoid abrasive injury to stomach epithelium. Insects have no mucous glands and they obtain protection for the epithelium by the formation of a *peritrophic membrane* (Fig. 5-11). This membrane forms a continuous tabular covering around the food mass. The membrane is composed of chitin; it is freely permeable to digestive enzymes and all the products of digestion. Its remarkable permeability has been demonstrated experimentally by the use of dyes.

The formation of the peritrophic membrane is a topic of considerable interest. In a great many insects, it originates from a secretion of the general surface of the midgut. This chitinous secretion is formed into a layer over the parent epithelial cells and then separated from them to form a sort of tube around the food mass. The tube usually remains attached at the anterior of the midgut where the foregut projects into it.

A peritrophic membrane is not formed in certain insect groups that take only liquid food, including Hemiptera, Anoplura, and adults of fleas, and horseflies. It is absent also in some other groups, notably the Carabidae, Dytiscidae, and Formicidae.

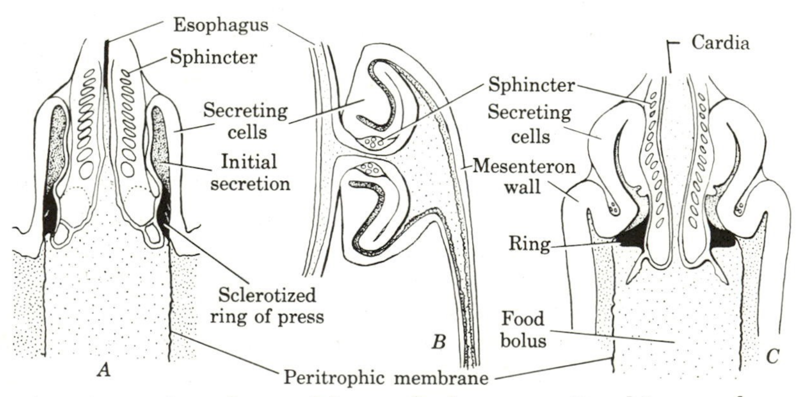

Fig. 5-11. Annular molds producing a peritrophic membrane. (*A*) Larva of mosquito *Anopheles*; (*B*) tsetse fly *Glossina*; (*C*) earwig *Forficula*. (Modified from Wigglesworth)

Cardia. In some groups of insects the peritrophic membrane is secreted by a specialized group of cells around the anterior end of the midgut. The secretion is pressed or molded into a membrane by the outward pressure of either the midgut entrance as it is distended by incoming food, or by a special organ called a *cardia*. This structure is most highly developed in the Diptera and Dermaptera (Figs. 4-4, 5-11). It consists principally of a sclerotized ring around the opening into the midgut, which presses against the walls of the gut a flow of secretion from the group of cells just anterior to it. As the membrane is formed, it is passed back through the gut as a sheath around the food.

Hindgut or Proctodeum. The function of this part of the digestive tract is primarily to eliminate the digested food bolus, although some absorption of food occurs here as well as the excretion of liquid wastes. The epithelial cells secrete a definite cuticle (Fig. 5-10C), as in the foregut, but this cuticle is readily permeable to water. The posterior part, forming the rectum, is usually heavily muscled and compresses the residue of the food after digestion to form the excrement into pellets before defecation. Two other functions are well established.

1. ***Water Absorption.*** Almost all insects must conserve water to the utmost, and to do so they rely on the hindgut to absorb water from the excrement and return it to the body. In the mealworms, the epithelial cells in the rectum may extract almost all the water from the excrement, leaving it a dry pellet. The water absorption plays an especially important role in excretion, under which it is discussed more fully.

2. ***Symbiotic Digestion.*** Termites, certain wood cockroaches, and crane fly and scarab-beetle larvae whose chief diet is wood fiber have no enzyme for digesting the cellulose they eat. They rely instead on a rich symbiotic assemblage of microorganisms in the hindgut, which digest the cellulose to form acetic acid, which is absorbed. Investigations of symbionts in other insects have brought out many apparent contradictions and questions, showing the need for more research in this field.

Adaptations to Liquid Diet. Various insects that suck blood or plant juices have evolved methods for extracting much of the water from the food before it comes into contact with the digestive enzymes. This arrangement has two advantages: (1) Some of the assimilable sugars in the food may be absorbed rapidly, and (2) the enzymes do not suffer excessive dilution. The partial dehydration is accomplished by the following methods:

1. In adults of many Diptera the midgut is divided into several sections, each with a different type of epithelium. It is thought that the first section acts as an absorption area to take most of the water out of the imbibed liquids.

2. In such bloodsucking Hemiptera as the bedbugs the first part of the midgut forms a large crop in which the blood meal is received. This "crop" absorbs most of the water and so concentrates the blood before it is passed to the region in which the enzymes are produced. Note that in the Diptera and the bedbug the water is absorbed from the midgut and passed into the insect's bloodstream. From the bloodstream it is excreted through the Malpighian tubules into the hindgut.

3. In the scale insects, cicadas, and many other Homoptera the digestive tract has a curious structure called the *filter chamber* (Fig. 5-12). The anterior part of the midgut lies beside a part of the hindgut. In some forms the two parts may be bound together by a common sheath, and in some instances the midgut may loop through an invagination of the hindgut. Because these animals live on plant juices, it has been thought that this filter chamber allowed water in newly ingested sap to pass rapidly from midgut to hindgut, thus concentrating the sap before its digestion in the posterior region of the midgut. This excess fluid is passed from the anus as honeydew. Honeydew analyses have cast considerable doubt on the correctness of these deductions, because the honeydew is found to contain large amounts of amino acids and carbohydrates. It is evident that more information is needed to obtain a better understanding of the function of the filter chamber.

Larval Adaptations. A peculiar modification of the digestive tract is found in the larvae of the higher Hymentopera and Neuroptera. The end of the midgut is closed and does not connect with the hindgut, thus during larval development the midgut becomes greatly distended with fecal matter. Before pupation the two sections of intestine become joined, and the fecal pellet for the *entire larval life* is evacuated.

Stomach Reaction. The average contents of the digestive tract in most insects are slightly acid, with a pH of 6 to 7. The saliva is usually neutral. In plant-feeding insects the intestine averages a more alkaline pH and has been recorded as high as pH 8.4 to 10.3 in the silkworm larva. Carnivorous or flesh-feeding insects usually average on the more acid side. Acidity of pH 4.8 to 5.3 has been observed in the crop of the cockroach after a carbohydrate meal; it has been suggested that this acidity is the result of fermentation by microorganisms, but it might result from acid secretion of certain cells

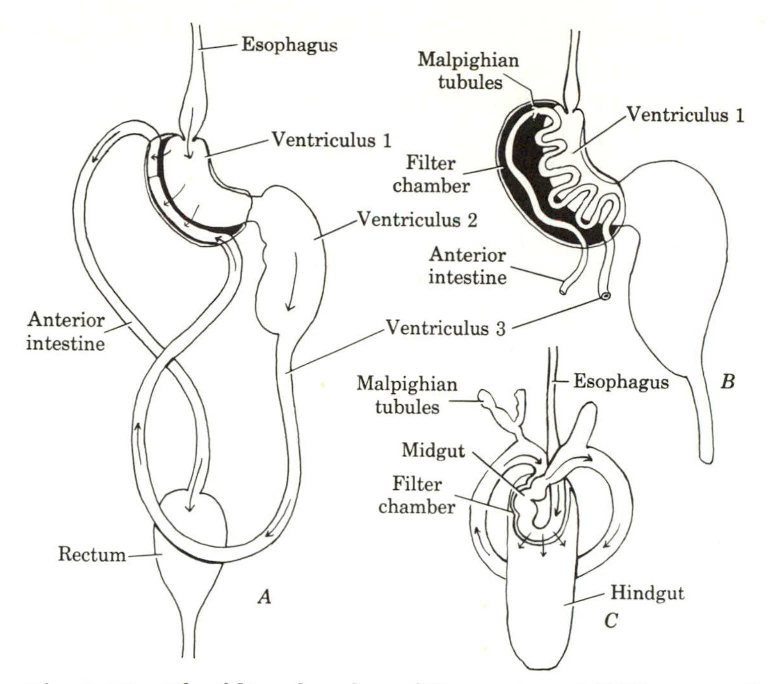

Fig. 5-12. The filter chamber of Homoptera. (*A*) Diagram of a simple type of filter chamber in which the two extremities of the ventriculus (midgut) and the anterior end of the hind intestine are bound together in a common sheath. (*B*) The ventriculus convoluted in the filter chamber and the anterior part of the hind intestine issuing from its posterior end. (*C*) The filter chamber of the scale insect *Lecanium*, diagrammatic. (Redrawn from Snodgrass, after Weber)

of the salivary glands. The greatest acidity recorded is pH 3.0 in a portion of blowfly larvae intestine.

Absorption

The products of digestion are absorbed by the epithelium of the digestive tract and passed into the blood (hemolymph). The greater part of this absorption occurs in the midgut, but various products are absorbed in the hindgut, depending on the kind of insect. Carbohydrates, such as the monosaccharide glucose, are absorbed especially in the gastric caeca. The absorbed products are passed into the blood and transported to the various tissue sites, where they are built up into the end products of growth. The rate of uptake of ions and molecules into the hemolymph depends on the movement of these substances to the absorptive surfaces of the gut and also the rate of actual transfer across the gut epithelium. Insect metabolism is so specialized that the absorptive process in the digestive tract is equally specialized to the extent that it may be unique to a single species.

Storage. The fat body plays an especially important part in the conversion process known as intermediary metabolism. Synthesis of fatty acids and proteins and conversion of carbohydrates (glucose to trehalose) occur in the fat body, which also stores energy reserves in the form of fats, glycogen, and protein. These reserves can be converted rapidly into energy according to the needs of the insect.

Parts of the midgut and other areas may also be storage sites for food reserves. In insects with complete metamorphosis these reserves are usually highest just before pupation and may be nearly depleted by the end of metamorphosis.

EXCRETION AND SALT AND WATER BALANCE

Metabolism results in the continual breakdown of organic compounds and forms a number of waste products, many of which are toxic to animal tissues and cells. Metabolic wastes are eliminated by the process called *excretion*. Nitrogenous wastes, which accumulate by the degradation of proteins and nucleic acids, may be particularly toxic and, therefore, organisms have evolved various mechanisms for their elimination. Carbon dioxide is another toxic and major waste derived from cell respiration and, therefore, although technically its removal also is excretion, it is considered under respiration.

In most insects, as in other animals, the excretory system performs a dual function: the elimination of metabolic wastes and the maintenance and regulation of dissolved salts and water balance in the internal body fluids. Both of these functions involve the production of a primary excretory fluid followed by differential resorption of some of these substances from the fluid before elimination. This major recycling process involves water, ions (particularly potassium; in aquatic insects also chloride), and inorganic and organic molecules.

In regulating the composition of internal body fluids, both intracellularly and extracellularly, the salt and water balance is critical. The fluid composition must be maintained within relatively narrow limits regardless of the external environment. Insects, because of their small size, are constantly in risk of drying up, or desiccation. Therefore, they have evolved a highly waterproof cuticle, a specialized excretory system that enables them to conserve water, and a series of spiracular valves that seal off the respiratory system so that water is not lost.

Excretory System

Malpighian Tubules and Rectum. In most insects, the excretory system is primarily composed of a number of Malpighian tubules and the rectum (Fig. 5-13). Malpighian tubules are unique to insects and are not found in other animals. They generally lie freely in the body cavity, and most open into the digestive tract at the junction of the midgut and hindgut. The urine produced by the tubules results from the exchange of diffused substances with the hemolymph. The urine, usually clear and sometimes a pasty substance, is discharged into the hindgut of the digestive tract where it passes to the rectum, which selectively reabsorbs or adds substances to the fluid before elimination.

When present, the Malpighian tubules vary in number in different species from 2 (as in hemipteran coccids) to 200 or more in some Orthoptera and honeybees. The tubules are variable in length, are

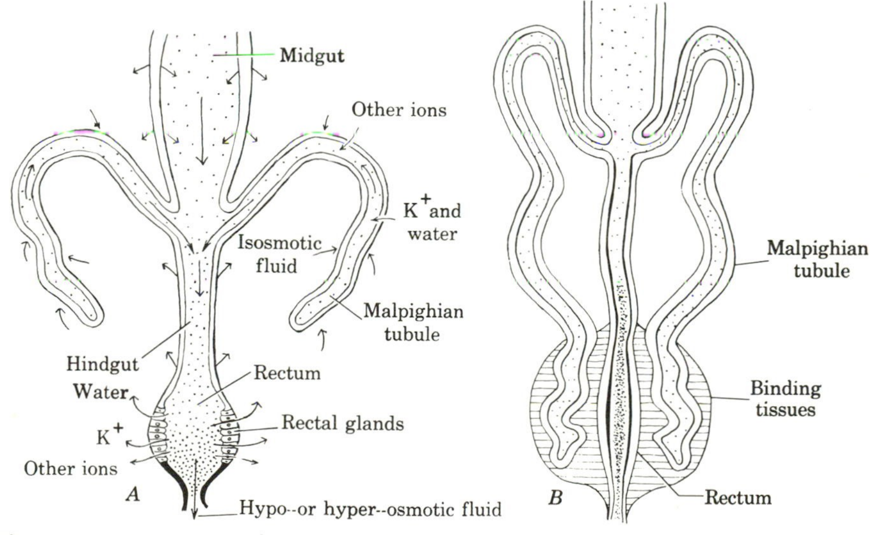

Fig. 5-13. Diagram of water and salt circulation in the excretory system of an insect. (*A*) Common type, arrows indicate direction of water and ion movements; (*B*) type in mealworm *Tenebrio*, showing close relation of rectum and tips of Malpighian tubules. (*A*, after Stobbart and Shaw, 1974; *B*, after Wigglesworth)

closed at the distal end, and are formed from a single layer of epithelial cells. The large number of tubules collectively represents a large surface area for the exchange of fluids to and from the hemolymph and the Malpighian tubule lumen. In some insects, such as most larvae of Lepidoptera and many beetles, the distal terminal ends of the Malpighian tubules open into the gut near the rectum, which increases the recycling efficiency of dissolved salts and water.

Composition of Urine. The composition of the urine depends on the life-style and life needs of the insect, as well as the food ingested. The urine generally contains water, nitrogenous compounds, pteridines, calcium carbonate (lime), calcium oxalate, and pigments. Other substances, such as carbohydrates, phosphates, salicylic acid, and ions, may be present depending on the diet of the insect. Of the nitrogenous compounds, uric acid is the most important constituent in insects. The other nitrogenous compounds—ammonia, urea, and amino acids—are generally excreted in small quantities.

In terrestrial insects, 80% of the excreted nitrogen is uric acid, which is formed in the tissues, accumulated in the hemolymph, and then filtered into the Malpighian tubules, together with amino acids, various ions, and water. Because uric acid and its ammonium salt are slightly soluble, they are eliminated as solid salts in the urine with very little loss of water by the insect. However, in many insects, insoluble substances, such as allantoin and allantoic acid, which result from the breakdown of uric acid, are the dominant nitrogenous constituents. Allantoin is dominant in the Heteroptera and also is found in various species of Orthoptera, Coleoptera, Neuroptera, Hymenoptera, and Diptera. In those Lepidoptera examined so far, uric acid dominates in the pupae, and either uric acid or allantoic acid dominates in the adults and larvae.

Blowfly larvae produce large amounts of ammonia, a highly toxic chemical, which may account for up to 90% of the nitrogenous excretion. Allantoin is the other substance excreted by the blowfly larvae. In many aquatic insects, 70 to 90% of the excreted nitrogen is in the form of ammonia, which is rapidly dispersed in the aqueous environment.

In some insects, calcium carbonate and calcium oxalate granules may be abundant and the process of formation of these granules is not known. The plant-feeding caterpillars and phasmids excrete granules and crystals of uric acid, and granules of urate, calcium carbonate, and calcium oxalate. The larvae of Diptera excrete an abundance of calcium carbonate granules.

Accessory Excretory Structures. In addition to the Malpighian tubules and rectum, other tissues and organs have a subsidiary but important role in excretion, particularly in the elimination of small molecules and water. The accessory male glands (utriculi majores) of cockroaches excrete excess uric acid. The labial glands, long convoluted tubules lying in the ventral part of the thorax, of saturnid silkmoths excrete a large volume of the alkaline fluid, potassium bicarbonate. The midgut of the larvae of the saturnid moths excretes potassium ions. The gut, particularly the midgut, plays an important role in the uptake of ions in most insects lacking specific ion-absorbing organs.

Excretion of substances of high molecular weight occurs in the *pericardial cells* along the length of the heart and aorta and in *nephrocytes*, specialized cells that are scattered around the insect body. All these cells trap the larger molecules and after destructive digestion return the smaller molecules to the hemolymph.

Storage of Excretory Material. In some insects, the excretory material, uric acid, is not eliminated but is stored in special urate cells in the fat body. This type of storage occurs particularly in cockroaches and in insects that are endoparasites of other insects, including many of the larvae of wasps. This stored uric acid may be important in insects where the Malpighian tubules carry out specialized functions, such as the secretion of silk. The stored uric acid in the fat body is available to be mobilized, particularly during metamorphosis, as a source of nitrogen that will be excreted later.

Pigments. Small quantities of excess uric acid may be deposited in the cuticle of insects as white crystalline deposits that impart a white color to the cuticle. In the butterfly family Pieridae, small quantities of uric acid and pteridines (leucopterine gives a white color; xanthopterine gives a yellow color) are deposited in the scales and wings. In the Colorado potato beetle, uric acid is deposited in the sclerites. In both groups of insects, these substances are the basis for the color pattern of the insect.

Insects Without Malpighian Tubules. In some insects, Malpighian tubules are absent and the excretory functions may be performed, in part, by the *labial glands* (*cephalic tubular glands*). In the Thysanura and the Collembola, these labial glands excrete dyes. In both groups, uric acid is stored in the urate cells of the fat body.

Regulation of Dissolved Salts and Water

The Malpighian tubules are basically secretory tubules, continuously producing excretory fluids and extracting materials from the hemolymph. The active transport of potassium ions across the tubule wall probably has a primary role in the production of the tubule fluid. A potassium-ion gradient is the cause for flow of urine in the tubule. In some aquatic and terrestrial insects, the fluid secreted by the Malpighian tubules has an extraordinarily high concentration of potassium, being 6 times to as much as 30 times higher than the concentration of potassium in the hemolymph. In contrast, the sodium concentration is lower in the tubules than in the hemolymph. In some species, such as the bloodsucking bug *Rhodinus*, sodium may replace potassium as the primary cause for flow in the tubules. The rate of fluid production in the tubules is under hormonal control. The hormones are generally produced by neurosecretory cells in the brain and in the fused ganglia of the thorax.

The excretory fluid is discharged from the Malpighian tubules, passed through the hindgut, and significantly modified in the rectum. In many terrestrial insects, some water from this fluid is resorbed in the hindgut (ileum) before it reaches the rectum. However, most water resorption occurs in the rectum and as a result of this most terrestrial insects produce relatively dry feces. The epithelium of the rectum is capable also of transferring inorganic ions from the rectum to the hemolymph.

In freshwater insects, the osmotic uptake of water through the body surface is balanced by the production of a relatively large volume of rectal fluid of low ionic content. With abundant water, and few if any salts, the rectal fluid in these insects has a low ionic concentration and is hyposmotic in relation to the hemolymph. On the other hand, in brackish-water and saltwater insects, the rectal fluid has a very high ionic concentration and it is hyperosmotic with respect to the hemolymph. In some freshwater aquatic larvae, there seems to be at least some resorption of water in the rectum. In larvae adapted to brackish conditions there is greater resorption of water in the rectum.

The Digestive Tract. In terrestrial insects, water and dissolved salts from ingested food are absorbed in the gut and are transferred to the body cavity and hemolymph. Since some water is generally lost by evaporation through the body surface of the insect, the fluid in the rectum is produced at a slower rate than the fluids that are absorbed through the gut into the body. Insects living in dry regions, such as locusts, are subject to the buildup of high concentrations of salts and water deprivation, and they have evolved a distinctive way of handling this. They pass ingested food directly to the rectum so that the water is resorbed directly by the epithelium of the rectum instead of water and dissolved salts being extracted from the breakdown of digested materials in the midgut region.

Accessory Salt- and Water-Regulating Structures. Many freshwater larvae have special regulating organs located on their external surface. The best known are the *anal papillae* of mosquito larvae (Fig. 5-14). In these insects, four thin-walled sacs arise in pairs on either side of the anus and connect with the hemocoel. These organs take up sodium, potassium, and chloride ions and are the site for the exchange of phosphate and for the osmotic uptake of water.

RESPIRATION

All animals require a supply of energy and most obtain their energy by the process of respiration. Respiration consists of the uptake, transportation, and use of oxygen by tissues and the liberation and disposal of waste products, primarily carbon dioxide. The exchange of gases between the cell and its environment is called *external* respiration, whereas the exchange and utilization of gases within the cell is called *internal* respiration or respiratory metabolism.

External Respiration

This is accomplished in most insects by the tracheal system, in essence a system of air-filled tubes, called *tracheae*, *air sacs*, and *tracheoles*, through which air is brought directly to the tissue cells. The tracheae open to the outside through a number of holes, the

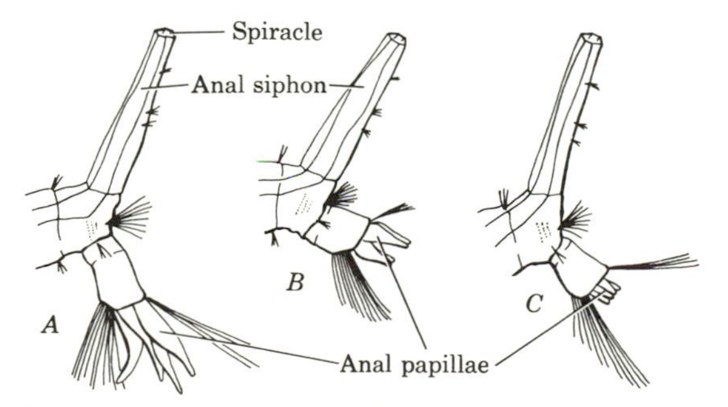

Fig. 5-14. Variation in size of the anal papillae at posterior end of larva of *Culex pipiens* when reared in different water salinities. (*A*) Distilled water; (*B*) water with 0.006% sodium chloride; (*C*) salt water, 0.65% sodium chloride. (Modified from Wigglesworth, 1965)

spiracles, which have a closing mechanism to prevent excess loss of water by evaporation.

Tracheae and Tracheoles. The tracheae are tubular invaginations of the epicuticular surface of the insect, and their general character is similar to that of the epidermis (Figs. 5-15, 5-16). A layer of flat epithelial cells secretes the cuticular lining of the tracheae and this lining is called the *intima* (Figs. 4-8, 4-9). The inner surface of the intima has spiral or annular fibrous thickenings called *taenidia* (Figs. 4-8, 4-9), which strengthen the trachea and ensure that the trachea remains round and open, even under conditions of bending and pressure. The tracheae divide and redivide, becoming smaller and smaller; finally each ends in a cluster of minute branches, the *tracheoles*. The tracheoles are less than 2 microns wide; they possess taenidia but have no regular layer of epithelial cells. The base of each cluster of tracheoles has a weblike cell, the *tracheole cell* or *tracheoblast*, with extremely thin protoplasmic extensions. The tips of the tracheoles lie alongside and between the cells of the body tissues. Most of the respiratory gas exchange between the tissues and the tracheoles apparently occurs through these tracheole tips.

The properties of the walls of the tracheae are quite different from those of the epidermis. Both tracheae and tracheoles are permeable to gases, presumably extremely permeable through the fine, delicate wall of the tracheole. The tracheae are impermeable to liquids; the spiracles at least are extremely hydrophobic, that is, the surface resists the entry of water. The tracheoles, especially the tips, are readily permeable to liquids.

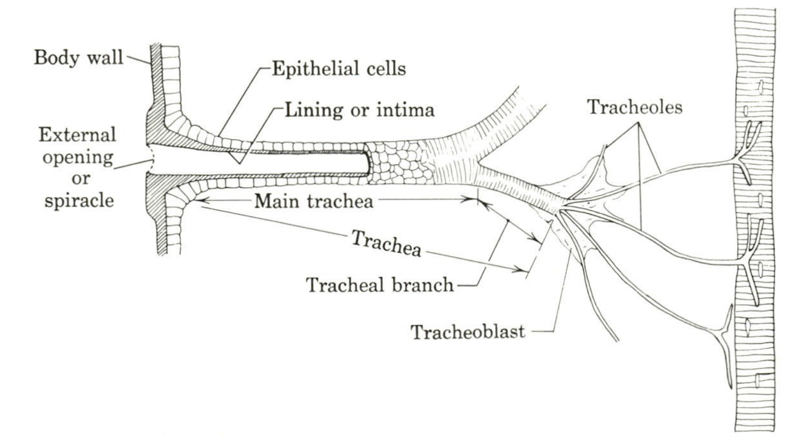

Fig. 5-15. Diagram of an open trachea of an insect.

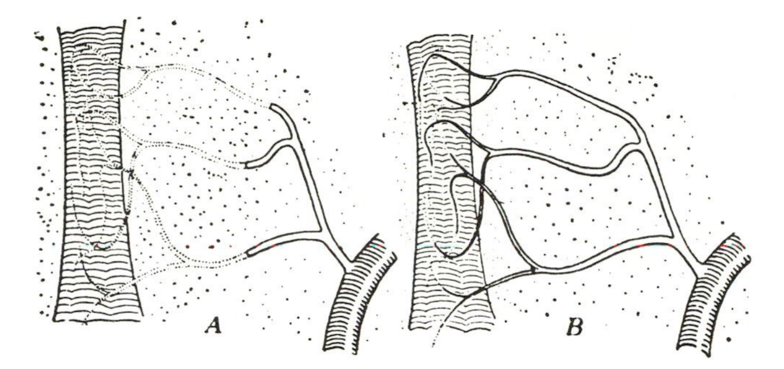

Fig. 5-16. Rise and fall of fluid in tracheoles. (*A*) High level in rested state; (*B*) low level in fatigued state. Tracheoles shown with dotted lines contain fluid, those with plain lines contain air. (From Wigglesworth, *Principles of insect physiology*, by permission of E. P. Dutton and Co.)

Air Sacs. The tracheae have deformable sections, air sacs, which permit extensive changes in volume to occur (Fig. 5-17). The air sacs are generally very thin-walled and flexible parts of tracheae that have few or no taenidia.

Tracheole Fluid. In insects, the tips of some of the tracheoles may contain a certain amount of fluid. When associated with relaxed muscle (Fig. 5-16*A*), this fluid may rise a considerable distance in the tracheoles; when the muscle is fatigued, a large amount of the fluid is withdrawn from the tracheoles into the muscle tissue (Fig. 5-16*B*). This movement of the fluid into the muscle tissue relates to changes in the osmotic pressure of the tissues that draw fluid into the tissues as a result of metabolic activity. The withdrawal of fluid into the tissue cells brings air into contact with the fatigued cells and aids in increasing the oxygen supply to those working tissues that need it the most.

Diffusion. The actual mechanics by which oxygen passes through the length of the tracheae and tracheoles, and finally into the tissue, and by which carbon dioxide is eliminated along the reverse path, has been the subject of many theories. It is now generally accepted that these gases are conveyed by diffusion (Fig. 5-18*A*), with the help of some mechanical ventilating of the abdominal tracheae and air sacs in large, active insects. Analyses have been made of dimensions of tracheae, oxygen consumption, and the diffusion coefficient of oxygen of various insects. They have shown that, even in the case of large caterpillars, diffusion alone will provide a sufficient

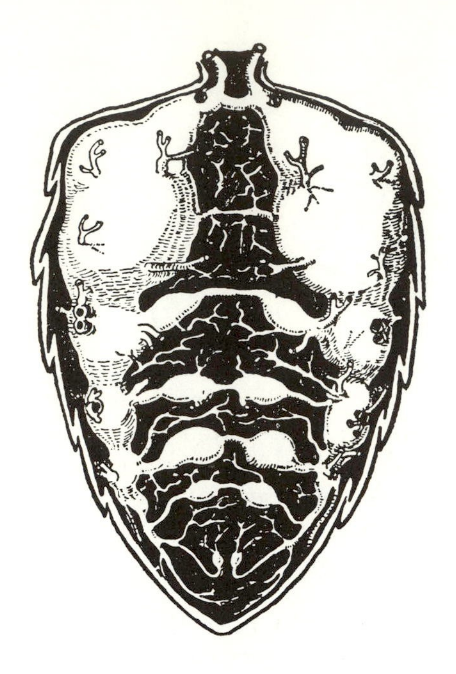

Fig. 5-17. Tracheal system in the abdomen of the honeybee worker, showing the air sacs. Dorsal tracheae and air sacs have been removed. (From Wigglesworth, after Snodgrass, *Principles of insect physiology*, by permission of E. P. Dutton and Co.)

stream of oxygen to the tracheole endings if the oxygen pressure in them is only 2 or 3% below that of the atmosphere.

This same process accounts also for the elimination of carbon dioxide. However, analysis of carbon dioxide elimination has shown that nearly a fourth of the amount produced in the insect body is eliminated over the general body surface. This is explained by the fact that carbon dioxide diffuses *through animal tissues* about 35 times faster than oxygen. Consequently, any carbon dioxide formed

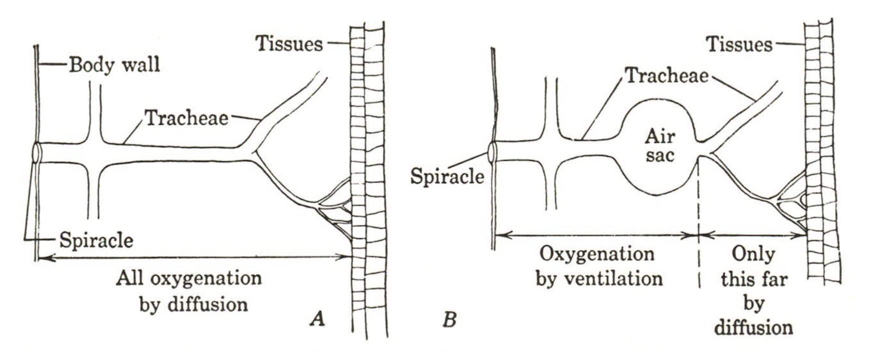

Fig. 5-18. Diagram to illustrate relation between diffusion and ventilation. (*A*) System without ventilation, relying entirely on diffusion; (*B*) system supplementing diffusion with ventilation.

in metabolism diffuses not only into the tracheoles but also into surrounding tissues in all directions and eventually to the exterior through the body wall.

Blood Respiration. In insects, the blood does not normally play an important part in transporting oxygen to the tissues. Since it is a fluid with living cells, however, it does utilize oxygen in its activities. Because the blood passes over and among many tracheae and tracheoles, it contains a ready supply of oxygen throughout the body cavity. Tissues that are widely separated from tracheae and receiving only a limited supply of oxygen from them can obtain oxygen from the blood. The blood may also supply oxygen to tissues when tracheae are incomplete.

Ventilation of the Tracheal System. For many small or sluggish insects, gaseous diffusion alone is sufficient to satisfy the needs of respiration, but it is not adequate for active, running, and flying forms with high metabolic rates and large energy consumptions. Insects in flight require immensely greater rates of oxygen supply and of carbon dioxide and water removal in comparison to the insect in its resting state. The tracheal system must be able to respond rapidly and to convey extraordinarily large volumes of oxygen to the flight muscles. Under such circumstances, diffusion is supplemented with mechanical ventilation of the tracheal system. Two principal types of structures are used for this purpose.

1. The taenidia of the tracheae give flexibility to the air-filled tubes and they expand and contract longitudinally like an accordion. Contraction may result in a reduction of as much as 30% of the expanded volume.
2. Certain portions of the tracheae may be elliptical instead of round in cross section, and have few or no taenidia. The elliptical portions can be flattened by an increase in blood pressure around them or by bending of the trachea. In many instances, sections of the tracheae form air sacs that are distinct, enlarged chambers, resembling the elliptical tracheal structures in having few or no taenidia, and that are readily compressible (Figs. 5-17, 5-18*B*). The action of these is like that of a bellows.

Both these types of structures function like lungs. The respiratory movements of the insect body cause alternate filling and emptying of these structures. When the body is contracted, the accordionlike sections are contracted, and when the blood pressure is increased, elliptical tracheae and air sacs are compressed. Both actions cause an ejection of air, including carbon dioxide, from the tracheae through the spiracles. When the body is relaxed, the air chambers expand, owing to their own elasticity, and fill up with air that is drawn in from the outside.

The effect of this ventilation (Fig. 5-18*A*) is to keep the air sacs and tracheal trunks filled with air similar in composition to that of the atmosphere. Diffusion acts along the remaining short distance to the tissues through tracheae that branch from the sacs or tracheal trunks.

Respiratory Movements. The tracheal system has no muscles of its own. Ventilation occurs by a combination of opening and closing of the spiracles and compression of the abdomen. The compression of the abdomen is usually accomplished by the dorsoventral and longitudinal muscles. The dorsoventral muscles pull the tergum and sternum toward each other and the longitudinal muscles telescope the segments. These two contracting movements usually occur simultaneously and serve to compress the air sacs.

Control of Ventilation. The ventilating movements leading to the expulsion of air (expiration) and the intake of air (inspiration) are effected by nerve impulses. There are two levels of activity.

1. The median and transverse nerves of the segmental abdominal ganglia innervate the inspiratory dorsoventral muscles. Paired lateral nerves of the abdomen innervate the dorsoventral expiratory muscles and the longitudinal expiratory and inspiratory muscles.
2. Several ganglia, primarily in the thoracic region but also abdominal ganglia of the second and third segments, act as pacemakers for ventilation. These ganglia override and coordinate the muscular rhythms of the entire insect.

Spiracular Adaptations and Evaporation. Oxygen and carbon dioxide diffuse readily through the tracheal system, and so does water in the form of water vapor. If the spiracles remain open indefinitely, the insect loses water steadily, and water is normally a precious commodity to the insect. Because of this situation, when insects successfully colonized the many terrestrial habitats, evolutionary adaptations in the spiracles occurred so as to reduce the loss of water without hindering respiration. The spiracles have structural adaptations, such as muscular closable valves and porous sieve plates, that close the spiracles for short or long periods of time and thus eliminate water loss when the insect does not need the spiracles for oxygenation.

In many insects, the opening and closing of spiracles is coordinated with the ventilating action of the abdomen, so there is a directed flow of air through the tracheal system resulting from air being drawn in through some spiracles and expelled from others.

Respiration Control. The opening and closing of the spiracles and the regulation of the respiratory movements are effected by nerve centers that are controlled by sensitive mechanisms that appear in general to be correlated with levels of carbon dioxide and oxygen in the tissues. A number of the more important and interesting generalities relating to respiration control are listed here.

1. The immediate sensory control of respiration lies in the segmental ganglia of the ventral nerve cord. Each ganglion controls only its own segment, so each segment may act as an isolated unit as far as respiration is concerned.
2. There is a modulating or coordinating center that can produce rhythmic action of all or several segments. The mode of action and identity of this center are not clearly established, although the thoracic ganglia and sometimes certain abdominal ganglia are involved.
3. During times of rest the respiratory movements may cease altogether, and the spiracles close. In some insects, an excess of oxygen will effect the same reaction.
4. Practically any external nervous stimulation (visual, tactile, etc.) will initiate or increase respiratory activities.
5. Various internal chemical stimuli will increase respiration. In most species the respiratory nerve center is stimulated by increased acidity of its receptive tissues, caused by either high concentration of carbon dioxide or acid metabolites that accumulate through lack of oxygen. The stimulation of the respiratory nerve center initiates movement of the spiracles and of the ventilating muscles. In cockroaches, it has been found that a high concentration of carbon dioxide causes respiratory activity. In mosquito larvae, carbon dioxide diffuses rapidly from the body into the aquatic environment and seldom builds up in excessive amounts. It is apparently the accumulation of acid metabolites that causes them to move to the surface for more air.
6. In many sluggish or inactive insects, the carbon dioxide is released in cyclic bursts. These range from bursts of a minute each every 2 or 3 minutes to 30-minute bursts once in 24 hours. The mechanism and significance of these bursts are not known.

Adaptations for Aquatic Life. The foregoing discussion deals with the type of respiration found in terrestrial insects. But many forms either live in water or spend a great deal of time submerged in water.

There are several types of adaptations in respiration in aquatic situations.

1. ***Diving Air Stores.*** When they dive beneath the surface, certain insects carry with them a film or bubble of air attached to some part of the body. Both adults and nymphs of water boatmen (Corixidae) and backswimmers (Notonectidae) carry a film of air in the pile on the ventral surface of the body; this film is kept in place by hydrophobic hairs that resist penetration of the air film by the water. Adults of the diving beetles (Hydrophilidae, Dytiscidae) have an air space under the wing covers or elytra, into which space the spiracles open. This air store serves not only as a supply of oxygen for the insect but also as a sort of lung and gill, obtaining added oxygen from the water and discharging carbon dioxide into it by diffusion. It cannot provide for respiratory needs indefinitely, but it enables the insect to remain under water for a considerable period before having to come to the surface for more air.

2. ***Air Tubes.*** Many insects that live submerged all the time, breathe through a tube or pair of tubes that can break through the surface of the water. Only the pair of spiracles connected with these tubes is functional; the others are either closed or not developed. The mosquito larva (Fig. 8-146) has a rigid tube; when in need of oxygen, the larva swims to the surface and thrusts the end of the tube through the surface-tension membrane and into contact with the air. The rat-tailed maggot (Fig. 8-151), a fly larva that lives in a viscous or liquid medium, does not swim to the surface but has a respiratory tube that can be extended 7 to 10 cm to the surface. Several other kinds of tubes occur in aquatic, semiaquatic, and some parasitic groups.

3. ***Cutaneous Respiration.*** Large numbers of aquatic insect larvae make no contact with the atmosphere and have no external devices or special structures for respiration. The same is true of many parasitic insect larvae that live within the tissues of their hosts. In these the gas exchange is made by diffusion through the body wall. The insect utilizes the oxygen dissolved in the water, and excess carbon dioxide diffuses into the water. There are two distinct types of cutaneous respiration. In the first (including very small or first-instar larvae) there is no tracheal system present; in these the gas exchange within the body is by diffusion through the tissues, including the blood. In the second type (including most of the larger gill-less forms such as late-stage midge larvae and many caddisfly larvae), the tracheal system is developed, but instead of spiracles there are clusters of fine tracheae in the epidermis (Fig. 5-19). Here the gas

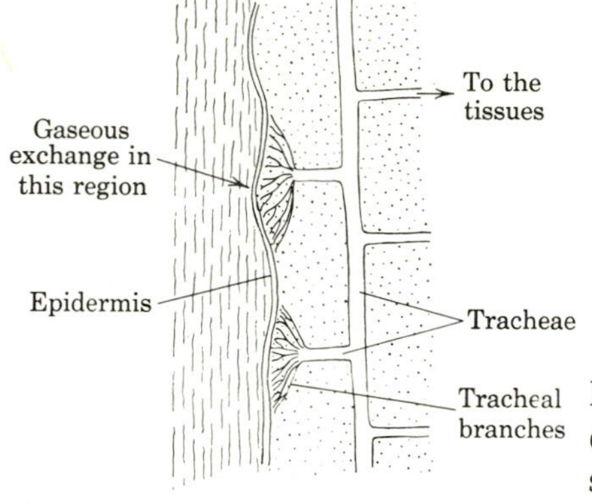

Fig. 5-19. Diagram of cutaneous respiration in aquatic insects.

exchange takes place first through the epidermis and then into the fine peripheral tracheae. From this point the diffusion pattern is the same as in a spiracular system.

Among the most conspicuous adaptations for aquatic life are the frondlike gills of damselfly nymphs (Fig. 8-13) and mayfly nymphs (Fig. 8-10). These typify many aquatic nymphs and larvae, which have developed gills for their respiratory exchange. The tracheae extend into these gills, and the diffusion of gases takes place through the epidermis between the tracheal threads and the water. An unusual structure occurs in dragonfly nymphs (Fig. 5-20). The rectum is enlarged, and gills provided with abundant fine tracheae extend into the pouch so formed. Into this rectal chamber the insect draws water and then expels it; the respiratory exchange occurs through the thin walls of the gills.

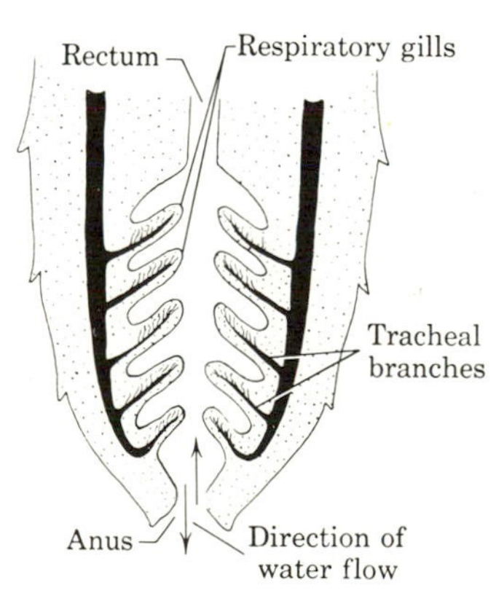

Fig. 5-20. Diagram of rectal gills from dragonfly nymphs. (Adapted from Wigglesworth)

THE BLOOD AND CIRCULATION

The blood or *hemolymph* of insects is the only extracellular liquid circulating throughout the insect body and bathing the internal organs. The hemolymph effects the chief distribution of nutrients, such as sugars and proteins, to the tissues, and carries waste products from them. It does not serve as the major system to transport oxygen and carbon dioxide; instead this is effected by the tracheae and tracheoles of the respiratory system. As a result, insects do not have the close association between respiratory and circulatory systems found in many other animals. The hemolymph is part of an open circulatory system and flows through closed ducts for only a short part of its course. Its chief progress is through the body tissues within a system of cavities, thus the blood is in direct contact with cell tissues. The body cavity with this open circulatory system is known as a *hemocoel*. In insects the blood serves as a hydraulic pressure system, which is important in the movement of soft-bodied larvae and in the expansion of the insect body after molting.

Blood Properties. Blood is composed of liquid, *plasma*, and various types of mostly colorless circulating *cells*. The plasma contains many substances, including dissolved salts, amino acids, proteins, carbohydrates, uric acid, lipids, fatty acids, organic metabolites (such as citrate, pyruvate, and succinate), and organic phsophates. The ionic composition varies in the different groups of insects. Concentrations of potassium, calcium, magnesium, and phosphate are relatively high in comparison to vertebrate blood. Magnesium ions may be present in the blood of some herbivorous insects in concentrations up to 50 times that found in vertebrates. Likewise, potassium may be present in high concentrations. Very high potassium and low sodium concentrations are found in herbivorous insects, and the concentrations of these ions reflect the foods ingested. These concentrations may be part of an adaptation of insects in relation to the evolution of angiosperm plants, which have high concentrations of potassium. The great variability of the ratio of potassium to sodium in different species of insects reflects the food ingested. A characteristic feature of insects is the wide variety of *amino acids* stored in high concentrations in the blood. The *carbohydrate trehalose*, a relatively unreactive molecule, also is stored in high concentrations in the blood. This disaccharide is the important blood sugar in insects and is the most accessible source of energy in many insects. Many of the proteins present in the blood are enzymes.

Insect blood differs greatly in its clotting properties. In many groups of insects the blood does not clot at all, and wounds are simply stopped by a plug of specialized blood cells (primarily plasmatocytes and podocytes). In other species the blood clots readily; this is probably induced by substances released by certain blood cells.

Insect blood contains no respiratory pigments—notably hemoglobin—except for *Chironomus*, which has a limited amount of hemoglobin in solution in the plasma.

Blood Cells. The blood cells or hemocytes are generally freely suspended nucleated cells, sometimes amoeboid, in the plasma. Sometimes they attach loosely to other tissues. In many insect groups, the hemocytes exhibit great diversity of size and form. Superficially there are two main categories: those with finely granular cytoplasm; and those in which the cytoplasm contains various amounts of distinctive granules. Ten types of hemocytes are recognized (Jones, 1962, 1979) (Fig. 5-21). Three are well-defined cell types in most insects (prohemocytes, plasmatocytes, and granular hemocytes); four other types may appear in many insects (cystocytes, spherule cells, adipohemocytes, and oenocytoids); and three highly specialized types in a few insects (podocytes, vermiform cells, and granulocytophagous cells). Hemocytes reproduce and multiply solely by mitosis.

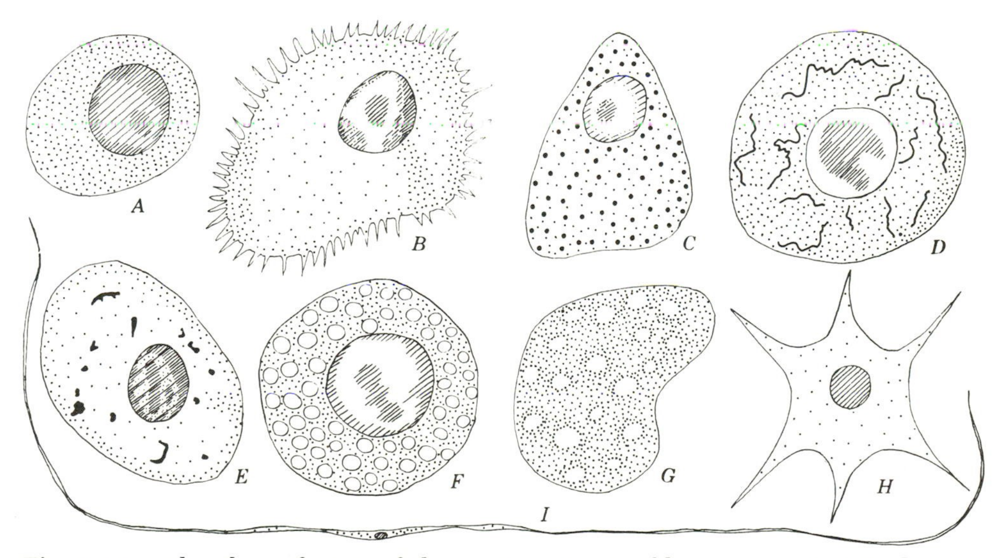

Fig. 5-21. Sketches of nine of the various types of hemocytes. (*A*) Prohemocyte; (*B*) plasmatocyte; (*C*) granular hemocyte; (*D*) oenocytoid; (*E*) cystocyte; (*F*) spherule cell; (*G*) adipohemocyte; (*H*) podocyte; (*I*) vermiform cell. (After Jones, and Chapman)

Hemocytes are as varied in function as in appearance. They ingest some living and all dead bacteria, and old or dying insect tissue, especially during metamorphosis. They extrude material by rupturing to provide nutrients for tissues, especially during metamorphosis. Certain hemocytes collect at wounds and form a plug to close breaks in the body wall; others form a partition to exclude certain parasites from the body cavity.

Functions of Blood. The blood of insects has five functions. The first four listed here are functions of the blood as a living system, whereas the fifth function is mainly mechanical.

*1. **Transportation of Nutrients and Waste Products.*** Digested food materials are absorbed by the blood from the digestive system and conveyed to the tissues, and waste products are carried from the tissues to the excretory organs. In addition, certain hormones are transported from their sites of secretion to their sites of activity.

*2. **Respiration.*** In most insects at least some of the cells are not provided with tracheoles for direct respiratory exchange. These cells presumably obtain their oxygen from the dissolved store in the blood. Much of the carbon dioxide that is generated by an insect diffuses through the tissues and finally through the cuticle; the blood aids in this process. In larvae of certain species of *Chironomus*, the blood contains dissolved hemoglobin, which aids in the passage of oxygen. This process is not nearly so effective in absorbing oxygen as mammalian hemoglobin in red blood cells.

*3. **Protection.*** The hemocytes dispose of certain bacteria and parasites. The healing of wounds is effected by the blood or its hemocytes.

*4. **Metabolism.*** The blood serves as a medium for ongoing metabolic reactions. Biochemical materials carried in the blood are converted to other substances as the blood circulates in the insect body. For example, the carbohydrate trehalose is converted to glucose.

*5. **Hydraulic Function.*** The entire volume of blood enclosed within the body wall forms a closed hydraulic system capable of transmitting pressure from one part of the body to another. This primarily mechanical process is put to many uses by an insect. The pressure of the blood is regulated by contractions of the thorax or abdomen or both. Alternate increase and decrease of blood pressure, brought about by respiratory movements and muscular activity, cause the emptying and filling of the tracheal air sacs and pouches. Localized blood pressure is responsible for stretching of the exoskeleton after molting, inflation of the wings, and frequently rupture of the eggshell at time of hatching.

Circulation. Insect blood circulation follows the path shown in Figs. 4-5 and 5-22. Blood is pumped forward by the heart from the abdomen, through the aorta, and emptied into the head where the aorta ends, commonly after passing through the brain. From the head the blood percolates back among the tissues until it reaches the abdomen, where it starts the cycle again by moving forward through the heart.

Blood is sucked into the heart through the ostia and then driven forward by peristaltic movements that flow along the entire length of the heart. The negative pressure of the heart chambers that aspirates the blood and the systolic pressure that causes the forward flow of blood are a result of the elasticity and muscular manipulation of the heart, the aliform muscles, and other muscles that may be associated with them. At times the flow is reversed, and blood pours from the heart back into the visceral cavity. In most insects the heart is unobstructed for its entire length. In a few the ostia are recessed into

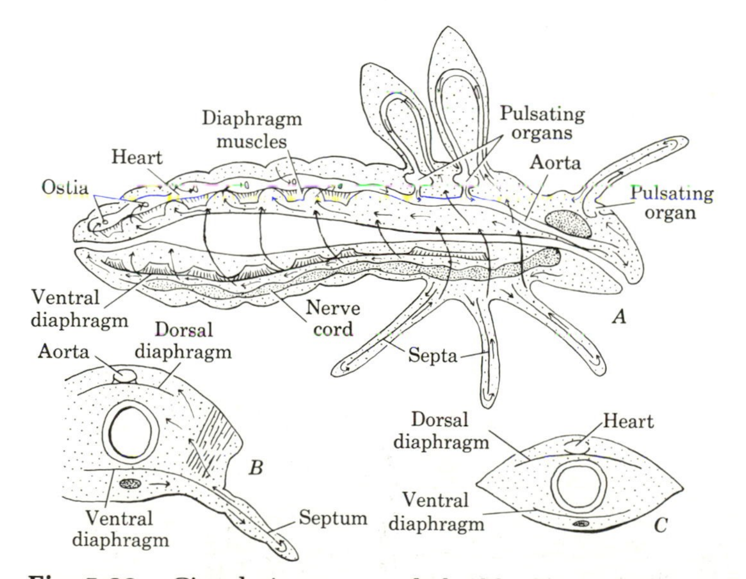

Fig. 5-22. Circulation accomplished by heart and accessory structures. (*A*) Insect with fully developed circulatory system, schematic, (*B*) Transverse section of thorax of the same, (*C*) Transverse section of abdomen. Arrows indicate course of circulation (based largely on Brocher). (Redrawn from Wigglesworth)

the heart to form valvelike flaps that divide the heart into segmental chambers.

In addition to the heart, a varied assortment of structures exist to aid the blood flow through the appendages or its distribution in the body cavity. In some insects the aorta may discharge into vessels that carry the blood in different directions. In many insects the antennae and legs are divided by longitudinal membranes or septa, so the blood enters on one side, flows the length of the appendage, and empties on the other side. Blood movements into the appendages are aided also by the respiratory movements, so that the "pulse" in the legs may synchronize with respiratory contractions and not with the heartbeats.

There are frequently supplemental blood pumps, or pulsatile organs, in the meso- and metathorax for sucking blood through the wings. In these instances the blood flows through certain veins of the wings (Fig. 5-23), and is returned either directly to the aorta or to the body cavity. When well developed, the ventral diaphragm also assists blood flow; contractions of the diaphragm muscles drive the blood both laterally and backwards.

The diagrams in Fig. 5-22 outline the direction of flow set up by the various structures and processes.

The heart in some insects is so well supplied with nerves from both the visceral nervous system and the segmental ganglia that many investigators believe all its activities are controlled by nerve impulse. It is still a moot point, however, as to whether the heartbeats are due to nerve stimulation or to muscle that possesses the ability to contract and relax periodically without nervous stimulation. In addition to a pacemaker neuron, a hormone released from the corpus cardiacum accelerates the heartbeat under certain conditions of stress.

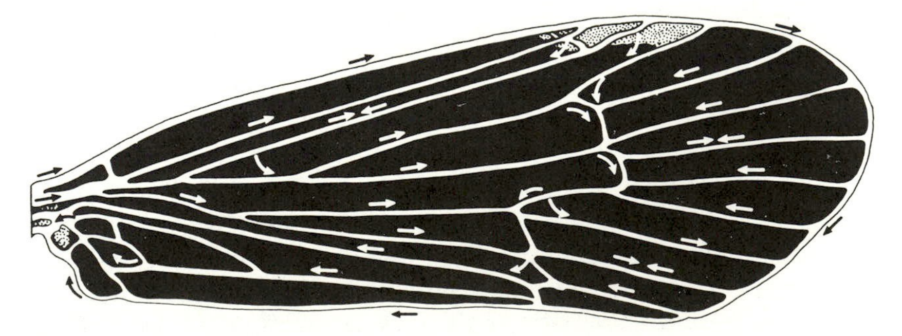

Fig. 5-23. Blood circulation in front wing of the caddisfly *Limnephilus rhombicus*. Direction of flow shown by arrows. (After Arnold, 1964)

METABOLISM

Metabolism is all the chemical processes that take place within a living organism. It is a highly integrated sequence of chemical reactions and includes both a constructive phase (*anabolism*) and a degradative phase (*catabolism*). Both phases occur simultaneously in different sequences of reactions in cells and tissues. All the metabolic reactions are accelerated (catalyzed) by specific proteins, called *enzymes*, and are accompanied by an exchange of energy.

Catabolic Phase. In the catabolic phase, complex nutrient molecules (carbohydrates, lipids, and proteins) are broken down, largely by oxidation, into simpler molecules. Catabolism is accompanied by the release of free (readily available) energy, most of which is conserved in the form of phosphate-bonded energy of adenosine triphosphate (ATP). The free energy released in the catabolic phase is used in the biosynthesis of various chemicals, ion transport, conduction of impulses in nerves, contraction in muscle, and light production (luminescence). The catabolic processes leading to the release of chemical energy are outlined in the following paragraphs.

1. Large molecules are broken down into smaller constituent units; for example, proteins are broken down first into polypetide chains and these chains are further broken down to the many different amino acids of which they are composed. Little free energy is released in these reactions.

2. Carbohydrates, such as the polysaccharide glycogen (characteristic of most animals) and the disaccharide trehalose (the characteristic blood sugar of insects), are oxidized to the monosaccharide glucose. Most free energy is obtained by the oxidation of carbohydrates. The store of carbohydrates is the ultimate source of energy for insects during flight, during metamorphosis, and during periods of starvation.

The blood sugar trehalose is converted by the enzyme trehalase to glucose as energy is needed by the insect. The adaptive advantage of insect blood carrying trehalose rather than the usual animal sugar glucose, might be that it is a relatively unreactive molecule and, therefore, can be stored in high concentrations in the hemolymph without possible side reactions with amino acids, which also are stored in high concentrations in insect blood.

3. Lipids (fats) are broken down into fatty acids and glycerol. These processes provide an energy reserve during starvation and sustained insect flight.

4. The oxidation of amino acids constitutes a limited energy reserve in some insects, such as the tsetse fly.

Anabolic Phase. In the anabolic phase, molecules representing building blocks are combined into larger macromolecules, such as proteins, nucleic acids, lipids, and polysaccharides, and are built into new tissues, or are stored in a particular form for later use. Because these processes result in increased size and complexity of structure, they require the input of free chemical energy. This free energy is furnished by the phosphate-bonded energy of ATP, which is synthesized in the catabolic phase.

Respiratory Metabolism. Respiratory metabolism, also known as internal respiration, is the source of free (readily available) energy for an organism to use in the many internal metabolic processes. Insects, like most other organisms, obtain their free energy from the burning of food, that is, the oxidation of organic substances (carbohydrates, fats, and proteins). The site of oxidation is within the cell, particularly the cytoplasm and mitochrondria (the powerhouses of a cell). The major energy-yielding reaction of all aerobic organisms is the oxidation of glucose to carbon dioxide and water. This reaction occurs in a series of complex steps involving three major stages (glycolysis, Krebs cycle, and electron-transport chain) and results in the release of free energy. The free energy becomes available because the molecules of glucose and oxygen have more energy than do the end-product molecules (carbon dioxide and water).

Intermediate Metabolism. There are many cellular reactions that are not immediately involved in the release of free energy. All these reactions are part of the *intermediate metabolism* of an organism and the fat body is an important site for many of these reactions. These reactions result in the formation of special secretions and the synthesis or breakdown of cellular constituents. Examples of these reactions are listed below.

1. ***Carbohydrates.*** Synthesis of the blood sugar trehalose, the exoskeletal material chitin (the polysaccharide *N*-acetyl-*D*- glucosamine), and glycogen. Glycogen accumulates in the fat body, halteres, flight muscle, and midgut cells.

2. ***Lipids.*** Synthesis of fatty acids, triglycerides, and waxes, which are some of the different structural components of lipids. Fatty acids and triglycerides are synthesized from amino acids; this process takes place in the fat body and elsewhere.

Lipid products form the major food reserves of insects. Much of the food eaten by insects during larval life is converted into lipids and stored in the fat body. Some of this fat is used during pupation to provide energy and chemical components for biosynthetic processes associated with metamorphosis. Some also is used in adult life as an energy reserve. Waxes have an important protective role in the cuticle and in the formation of beeswax.

3. ***Amino Acids and Proteins.*** Synthesis of some amino acids and proteins occurs in the fat body.

MUSCULAR SYSTEMS AND MOTION

Motion in insects, such as crawling, walking, leaping, jumping, and flying, also characterizes many living organisms and is produced by the action of bundles of contractile tissue called *muscles*. Muscles are grouped into coordinated systems for greater effectiveness, movement, and efficiency. In insects, muscular contractions serve to move various parts of the exoskeleton, as well as the internal organs.

Insects' muscles are similar to those of other animals, in that they are elongate and have the ability to contract when stimulated by an impulse from a nerve cell. The period of contraction is followed by a period of recovery in which the muscle cell returns to its original shape.

All insect muscles are made up only of *striated fibers*, although in some visceral muscles the striations may be difficult to detect. Each striate fiber is composed of a number of parallel threadlike *myofibrils* laid down in a matrix of cytoplasm, called *sarcoplasm*, containing many nuclei. In addition, cell organelles, called *mitochondria*, are numerous. They contain the enzymes for the oxidation process that leads to the release of free energy used in the movement of an insect. They are particularly abundant in very active muscles and comprise 40% of the muscle weight of the flight muscle of certain insects. The arrangement of myofibrils and sarcoplasm varies in different insects (Fig. 5-24) and in different muscles of the same insect.

Muscular contraction results in the movement of some part of an insect. For every movement there is a countermovement, in which the part regains its normal position. This countermovement is frequently brought about by the action of a second muscle, as in Fig. 5-25. Here the tibia articulates at each side of the femur at the point *l*; contraction of the elevator muscles *lv* raises the tibia for the initial movement; contraction of the depressor muscle *dpr* lowers the tibia for the countermovement. In other cases a single muscle is involved. In these the countermovement is brought about either by

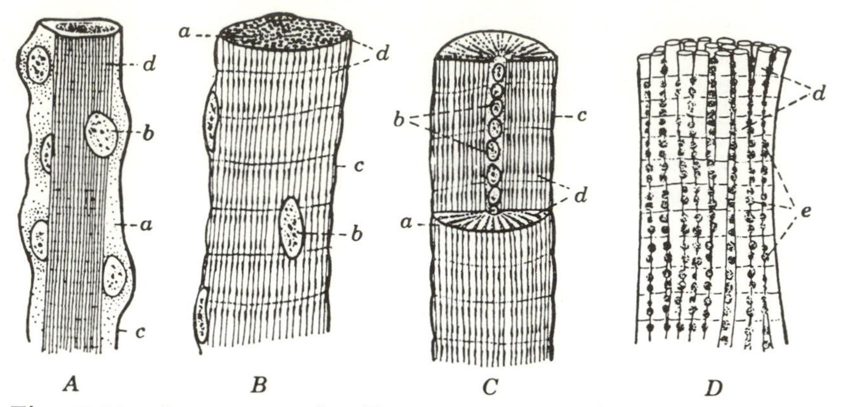

Fig. 5-24. Insect muscle fibers. (*A*) From larva of honeybee; (*B*) from leg muscles of a scarab beetle; (*C*) from leg muscles of honeybee (tubular muscles); (*D*) indirect-flight muscles of honeybee (a group of sarcostyles from fibrillar muscle). *a*, sarcoplasm; *b*, nuclei; *c*, sarcolemma; *d*, fibrils or sarcostyles; *e*, sarcosomes. (From Wigglesworth, after Snodgrass)

the pressure of the blood or by tension of flexible membranes. In countermovement by blood pressure, the hydraulic pressure of the blood simply forces back the part when muscle tension is released.

Flight. The muscular mechanism for insect flight is of unusual interest because nothing like it is found in any other animal group. Flight is attained by the stroking of the wings, which are essentially outgrowths of the lateral margin of the meso- and metathoracic tergites. In the dragonflies, damselflies, lacewing flies, and other primitive orders, the two pairs of wings move independently. In the damselflies, for example, as one pair of wings goes up, the other pair goes down. In the moths and butterflies, bees and wasps, and certain

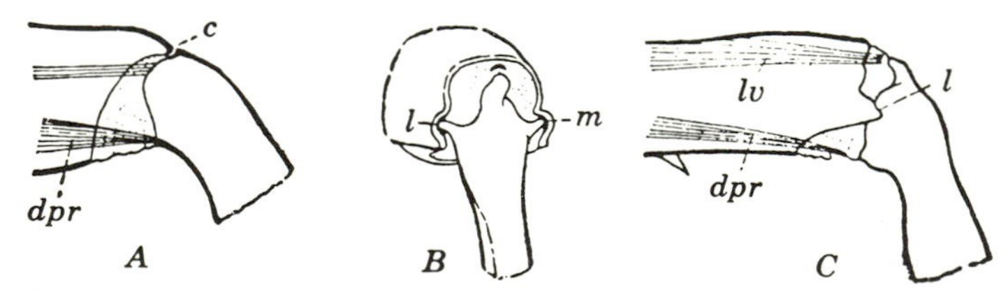

Fig. 5-25. Movement of an adult insect leg. (*A*) Monocondylic (single-socket) joint (*c*). (*B*, *C*) Dicondylic (double-socket) joint (*l* and *m*), end view and side view with levator (*lv*) and depressor (*dpr*) muscles. (After Snodgrass)

others, the two wings on the side are held together or coordinated by various types of hooks, bristles, or folds, so that both pairs work in unison. In most beetles and some of the true bugs, the front wings form hard armor plates, and only the hind wings function in flight. In the twisted-winged insects (Stylopoidea), the front pair is reduced, and in the true flies (Diptera), the hind pair is reduced; thus each of these two groups has a single pair of flight wings.

The wings of insects function as both airplane propellers and airfoils, which push a flow of air over the wings. The insect wing does not move exactly like an airplane propeller because it actually moves in a sinuous line so that if the insect is stationary, the tip of the wing describes a figure "8." The exoskeleton and musculature are the other two important components in insect flight. The tough, elastic, and flexible exoskeleton readily absorbs and releases elastic energy. These distinctive mechanical characteristics of the exoskeleton and the associated specialized muscles are vitally important in storing the kinetic energy of the upstroke of the wings for use in the propulsive downstroke. The protein *resilin*, secreted in the joints of insects and particularly in the hinges of the wings, is of the utmost importance in this economical use of energy. Resilin, because of the mobility of its amino acid chains, has almost perfect elasticity and is able to return 97% of the energy that compression and extension store in it.

Wing movements producing flight are controlled generally by 9 or 10 functionally distinct pairs of muscle groups in each of the two pterothoracic segments. The exact muscle arrangement differs in the various orders. *Direct muscles*, termed direct because they run from the pleural and sternal regions of the thorax and are inserted directly on the sclerites at the base of the wing, function in movement in synchronous flight. They include the important *basalar* and *subalar groups* that originate on the sternum, pleuron, or coxa and that insert dorsally on the basalar and subalar sclerites or the top of the pleuron. In the neopterans, the *wing-folding* muscles also link the pleuron to the third axillary sclerite. *Indirect muscles*, termed indirect because they do not run directly to insertions on the wings, are the main effectors in wing movement in asynchronous flight. They extend vertically and longitudinally across the meso- and methathoracic segments and deform the thorax by their contraction. This process moves the wings, which are attached to the thorax. The *dorsoventral indirect muscles* elevate the wings and the *dorsal longitudinal indirect muscles* depress them.

In insect flight, in different groups of insects, the wing movements are produced by two different mechanisms: synchronous and asynchronous (Fig. 5-26). These two mechanisms are effected by the

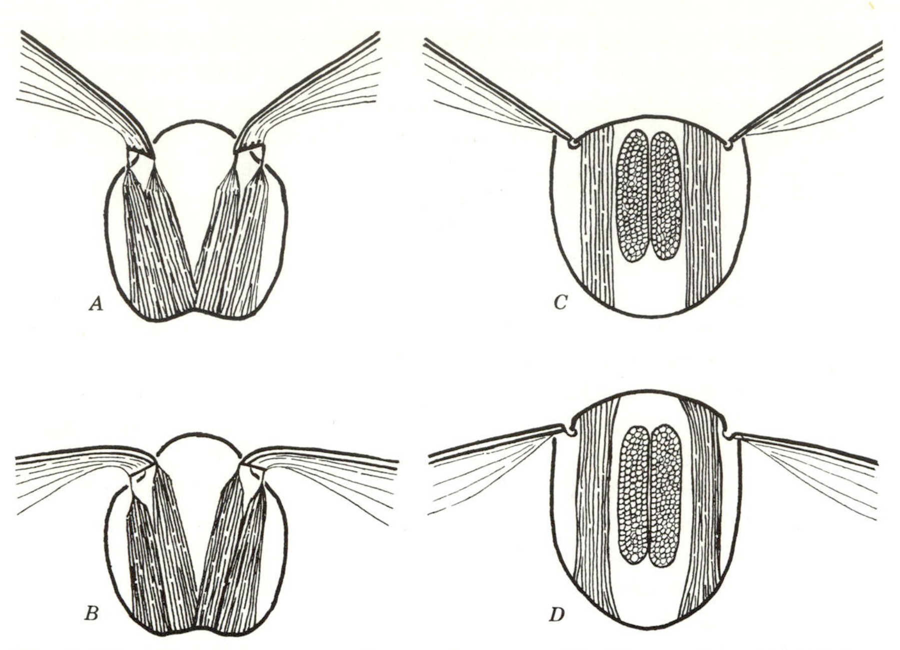

Fig. 5-26. Synchronous and asynchronous flight muscles. (*A*, *B*) Synchronous flight muscles. (*A*) contraction of inner pair of muscles in first thorax section results in a force being applied to a lever-like portion of the wing base, resulting in the wing being raised; (*B*) then contraction of a more robust pair of outer muscles creates a downward movement of the wings. (*C*, *D*) Asynchronous flight muscles. The outer bundles of muscles contract vertically and the inner ones horizontally. This is done alternately and in oscillation. The contractions deform the thorax so that the wings are driven up (*C*) and then down (*D*). (After Smith, 1965)

two kinds of flight muscle. Most winged insect orders have *synchronous direct-flight muscle* that contracts and relaxes in direct response to nerve signals (Fig. 5-26*A*, *B*). Impulses from the nervous system drive these flight muscles in an ordered cycle of contractions and provide all the information that coordinates the movements. Four orders, the beetles (Coleoptera), the wasps and bees (Hymenoptera), the flies, mosquitoes, and similar forms (Diptera), and certain true bugs and the aphids (Hemiptera) have *asynchronous indirect-flight muscle* (Fig. 5-26*C*, *D*). In these insects, the flight muscle contracts and relaxes at a much higher rate than the nerve signals it receives. The coordinated wingbeats of such insects result from the mechanical properties of the exoskeleton and the unique contractile properties of the fibrillar indirect-flight muscles. Fig. 5-26 shows diagrammatically the two flight mechanisms; in reality the distances between various points are minute and the accessory musculature is complex.

Flight Speed and Direction. In flight three types of wing movements are produced: (1) an upstroke and downstroke; (2) a deflecting or tilting; and (3) a forward and backward swing. All three types occur simultaneously but to different degrees, depending on speed and wing size.

The up-and-down stroke alone produces little more than elevation. Thus in butterflies, which rely chiefly on this stroke, flight is mostly fluttering, with very little speed in a forward direction. Addition of the deflecting movements decreases the air pressure in front of the wing and increases the air pressure behind it, in the same manner as an airplane propeller. The pressure behind pushes the insect forward at the same time that the partial vacuum in front pulls it forward. Deflection is therefore the principal agent in forward movement. Insects with highly developed deflection, such as dragonflies, are able to maintain a steady forward flight.

Rate of stroke varies exceedingly with different species. The slowest group is exemplified by the butterflies; the cabbage butterfly makes about 10 strokes per second. Faster rates are made by the honeybee, with 190 strokes per second, the bumblebee with 240, and the housefly with 330.

Greatest speed is attained by a combination of great deflection, a long narrow wing, and usually at least a fairly high rate of stroke. Fastest recorded flight of insects is developed by the sphinx moths (Sphingidae), attaining a rate of over 33 miles per hour; the horsefly (*Tabanus*) with a rate of over 31 miles per hour; and the dragonfly, which cruises at about 10 miles per hour but can speed up to about 25.

In forward flight, the path of the wing makes a figure "8" which leans forward at the top of the wing stroke. Many insects, such as the hover flies (Syrphidae), bees, and many flower-visiting moths, are able to fly backwards; this is done by reversing the path of the wings so that the figure "8" leans back at the top of the stroke.

REFERENCES

Arnold, J. W., 1964. Blood circulation in insect wings. *Mem. Ent. Soc. Can.*, **38**:1–60.

Barrington, E. J. W., 1967. *Invertebrate structure and function*, 2nd ed. London: Thomas Nelson. 549 pp.

Chapman, R. F., 1969. *The insects, structure and function.* New York: American Elsevier. 819 pp.

Eisner, T., and E. O. Wilson (Eds.), 1977. *The insects.* San Francisco: W. H. Freeman. 334 pp.

Hackman, R. H., 1974. Chemistry of the insect cuticle. In Morris Rockstein (Ed.), *The physiology of Insecta,* **6**:215–270. New York: Academic Press.

Hepburn, H. R. (Ed.), 1976. *The insect tegument.* Amsterdam, Oxford, New York: Elsevier. 571 pp.

House, H. L., 1974. Nutrition. In Morris Rockstein (Ed.), *The physiology of Insecta,* **5**:1–62. New York: Academic Press.

——— 1974. Digestion. In Morris Rockstein (Ed.), *The physiology of Insecta,* **5**:63–117. New York: Academic Press.

Jones, J. C., 1962. Current concepts concerning insect hemocytes. *Am. Zool.* **2**:*209–246.*

——— 1977. *The circulatory system of insects.* Springfield, Ill.: Thomas. 255 pp.

Locke, M., 1974. The structure and function of the integument in insects. In Morris Rockstein (Ed.), *The physiology of Insecta,* **6**:123–213. New York: Academic Press.

Maddrell, S. H. P., 1971. The mechanisms of insect excretory systems. In J. W. L. Beament, J. E. Treherne, and V. B. Wigglesworth (Eds.), *Advances in insect physiology,* **8**:200–331. London, New York: Academic Press.

Pringle, J. W. S., 1974. Locomotion: Flight. In Morris Rockstein (Ed.), *The physiology of Insecta,* **3**:433–476. New York: Academic Press.

Smith, D. S., 1965. The flight muscles of insects. *Sci. Am.* **227**:76–84.

Snodgrass, R. E., 1935. *Principles of insect morphology.* New York: McGraw-Hill. 667 pp.

Stobbart, R. H., and J. Shaw, 1974. Salt and water balance: Excretion. In Morris Rockstein (Ed.), *The physiology of Insecta,* **5**:361–446. New York: Academic Press.

Waterhouse, D. F., 1957. Digestion in insects. *Annu. Rev. Entomol.* **2**:1–18.

Wigglesworth, V. B., 1972. *The principles of insect physiology,* 7th ed. London: Chapman & Hall. 827 pp.

——— 1974. *Insect physiology,* 7th ed. London: Chapman & Hall. 166 pp.

6 Response and Behavior

A singularly important characteristic of living organisms is their ability to respond to stimuli, a characteristic commonly referred to as irritability. This provides organisms with the means to make appropriate and adaptive responses to changes in environmental conditions. These responses may be reactions to favorable or unfavorable situations, or just part of the normal range of activities of the organism. Irritability involves response from *receptors* and *sense organs* that detect changes in environmental conditions, both external and internal. It also involves the conduction, coordination, and integration, by the *nervous and endocrine systems*, of information received from the receptors of stimuli. These systems in turn provide responses and reactions to the stimuli. In addition, an organism exhibits particular *behavioral patterns* that are expressed in the ways in which it acts and adjusts in response to environmental changes.

This chapter examines receptors and sense organs, the nervous system, the endocrine system and its hormonal secretions, and behavior.

SOURCES OF INFORMATION: RECEPTORS AND SENSE ORGANS

Insects, like other multicellular organisms, possess a variety of receptor cells, *receptors* or *sensilla*, each sensitive to only one type of stimulus. These receptor cells are specialized parts of the nervous sytem that receive stimuli from the environment and transmit them to the neuromuscular system. Some receptor cells are scattered throughout tissue, such as tactile receptors in epithelial tissue. Others are aggregated, commonly in large numbers to form sense organs, such as in eyes, antennae, and many tympanal organs.

Receptors commonly are classified by the nature of the stimuli that activate them. In insects there are many kinds of receptors, including mechanoreceptors (such as sound receptors and proprioceptors), thermoreceptors, photoreceptors, and chemoreceptors. By means of these diverse receptors, insects are able to detect mechanical vibrations including a very wide range of sounds, radiant energy in the form of heat and light, mechanical pressure including the force of gravity, the amount of water and volatile chemicals in the air, and many other phenomena in the environment. They also have a precise sense of smell and taste.

Mechanoreceptors

Trichoid Sensilla. Simple sense organs, *trichoid sensilla* (Fig. 6-1), are mechanoreceptors and typically have two parts: (1) a specialized cuticular structure, such as a hair, a plate, or a thin-walled peg; and (2) a nerve ending that is in contact with this cuticular structure. A variation in the environment may cause a change in these structures, which, in turn, causes a modification in the energy potential in the associated nerve ending. This change is called a *stimulus*. The change in potential in the nerve ending starts a wave of excitation (impulse) that travels along the nerve fiber. After an impulse is generated, the nerve is briefly less sensitive and must recover before it

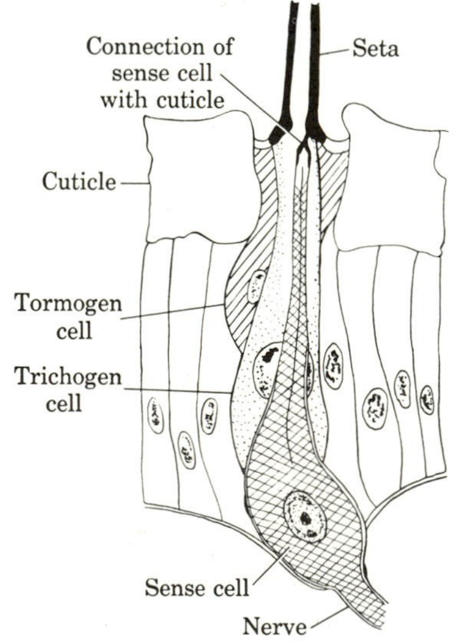

Fig. 6-1. A simple hairlike sense organ. (Redrawn from Snodgrass)

can transmit another impulse. Thus even though a nerve may be stimulated continuously, it responds in a series of discrete impulses.

Trichoid sensilla usually serve for tactile stimuli. A great variety of such hairlike sense organs receive stimuli relating to taste, smell, or humidity. They differ in that the hair has been replaced by a thin-walled peg, or seta, a bristle-like projection consisting of skeletal material, most commonly chitin, with which the tip of the nerve ending is in contact (Fig. 6-2*A*). Some of these sense organs have a group of sense cells associated with the peg or seta, allowing the accommodation of several nerve endings in the same receptor (Fig. 6-2*A*).

Most trichoid sensilla are stimulated only when being bent or straightened; this type of excitation is called *phasic reception*. Some trichoid sensilla are stimulated all the time when the setae are not in their normal position; this is called *tonic reception*.

Trichoid sensilla, distributed unevenly over the integument, are more abundant on those parts of the mouth, antennae (Fig. 6-3), and appendages that make contact with other surfaces or other parts of the insect. Some phasic sensilla are able to detect faint air currents produced by a nearby predator and serve as the initial warning system for the insect. Patches of tonic trichoid sensilla serve to orient an

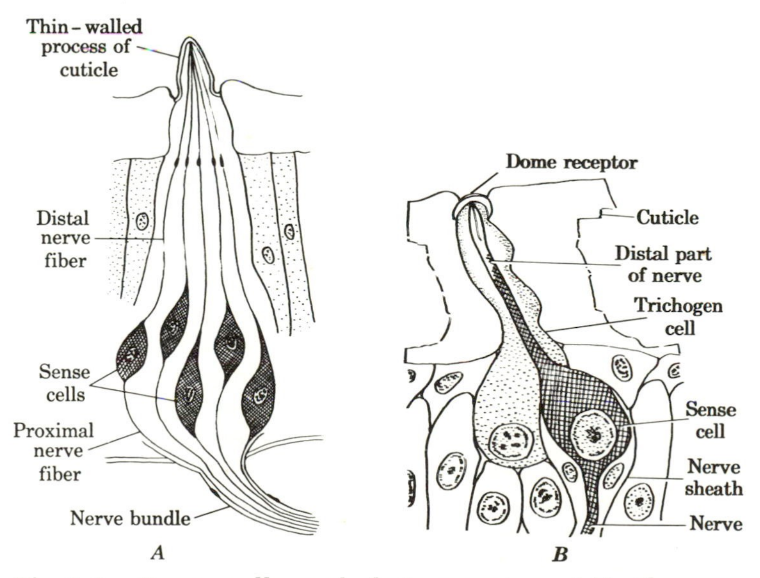

Fig. 6-2. Sense cells and their receptors. (*A*) Chemoreceptor having thin-walled peg and multiple sense cells. (*B*) Domelike sense receptor on cercus of a cockroach. (Redrawn from Snodgrass)

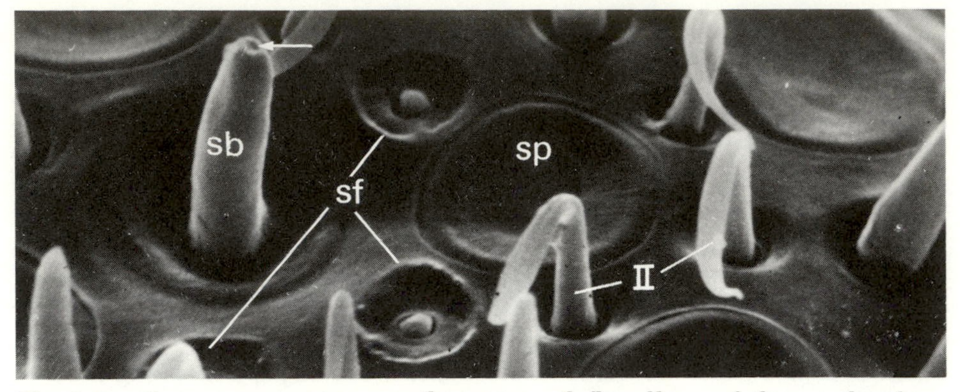

Fig. 6-3. Sensory receptors on the antennal flagellum of the worker bee, *Apis mellifera*. Four types of sensilla are shown: *II*, one of the forms of trichoid sensilla; *sf*, campaniforms; *sb*, basicones; and *sp*, placodes; ×3300. (Courtesy of A. Dietz, Univ. Georgia; from Dietz and Humphreys)

insect about the position of one part of its body relative to another part. For example, in ants, sensory setae between the head and thorax register movement of the head.

Stress Receptors (Proprioceptors). Domelike sense receptors, called *campaniform sensilla* (Fig. 6-3), occur on many areas of the integument, including the appendages. In these structures (Fig. 6-2*B*) the distal process of the sense receptor cell connects with the dome. Changes in the mechanical stresses of the integument surrounding such a sensillum cause the domelike plate to bow or flatten, and this action stimulates the nerve. These organs respond to body movements and may be important in geotropic responses and coordination of leg movements.

A large variety of sense organs, ranging from hair sensilla and stretch receptors to complex, many-nerved organs, provide information concerning the posture of the body and the position of its parts and appendages. For example, the ant *Formica* has innervated hair plates where the head and prothorax overlap and articulate (Fig. 6-4). When the head turns sideways, one set of these hair plates receives more pressure than the other. When the head is in the normal position, the pressure on both hair plates is the same. By monitoring these pressures through the nerves, the brain knows the position of the head.

Chordotonal sensilla are sensitive to changes in tension and consequently to vibrations of the substratum and of the surrounding medium. Some serve as highly evolved auditory organs. In such

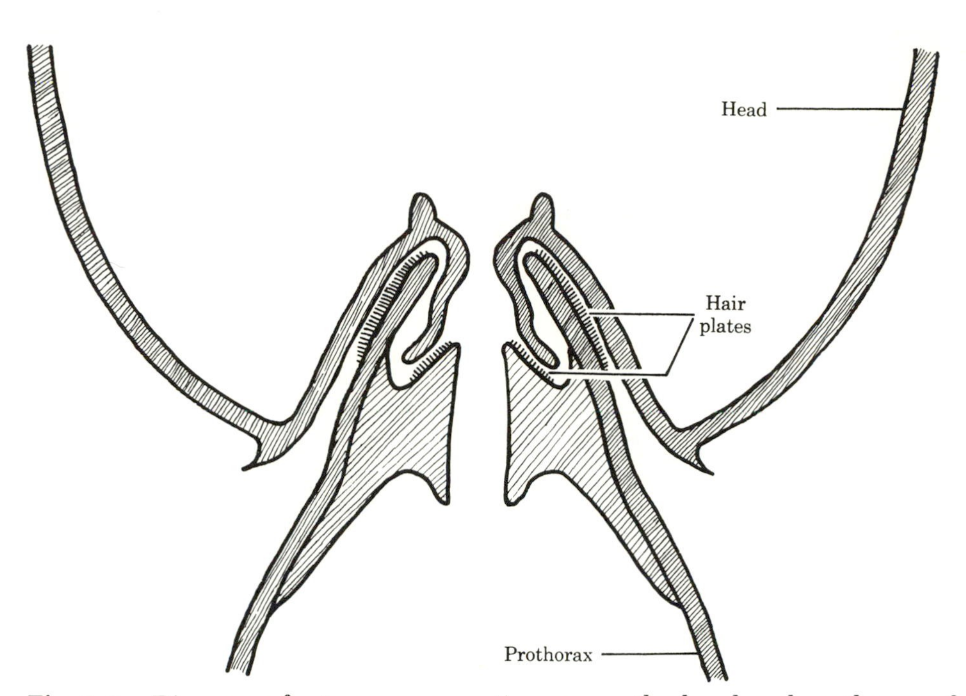

Fig. 6-4. Diagram of a transverse section across the head and prothorax of the ant *Formica polyctena,* showing hair plates on the prothorax. (After Markl, 1962)

cases, the receptors, arranged in groups that differ in their sensitivities, are stimulated by vibrations of the membranous tympanum that overlies them, allowing the insect to detect a wide range of sounds. Some chordotonal organs have an exceptional sensitivity to ultrasonic pulses generated by predaceous echo-locating bats. Chordotonal organs are subcuticular and commonly are not visible on the external surface. Each organ, or *scolopidium*, consists of three basic parts: (1) the neuron; (2) a cell (scolopale) enveloping the dendrites of the neuron; and (3) an attachment, or cap cell (Fig. 6-5*A*). Chordotonal organs occur in various parts of the integument of an insect, recording the movements of particular parts of the insect. Larval *Drosophila* have 90 such organs, each having one to five scolopidia.

Johnston's organ is a specialized chordotonal organ (Fig. 6-5*B*) that lies in the second segment of the antenna of all adult insects. It is also present in many larvae. Male mosquitoes and midges have these proprioreceptors well developed and the pedicel is enlarged to house the scolopidia. Johnston's organ serves to record move-

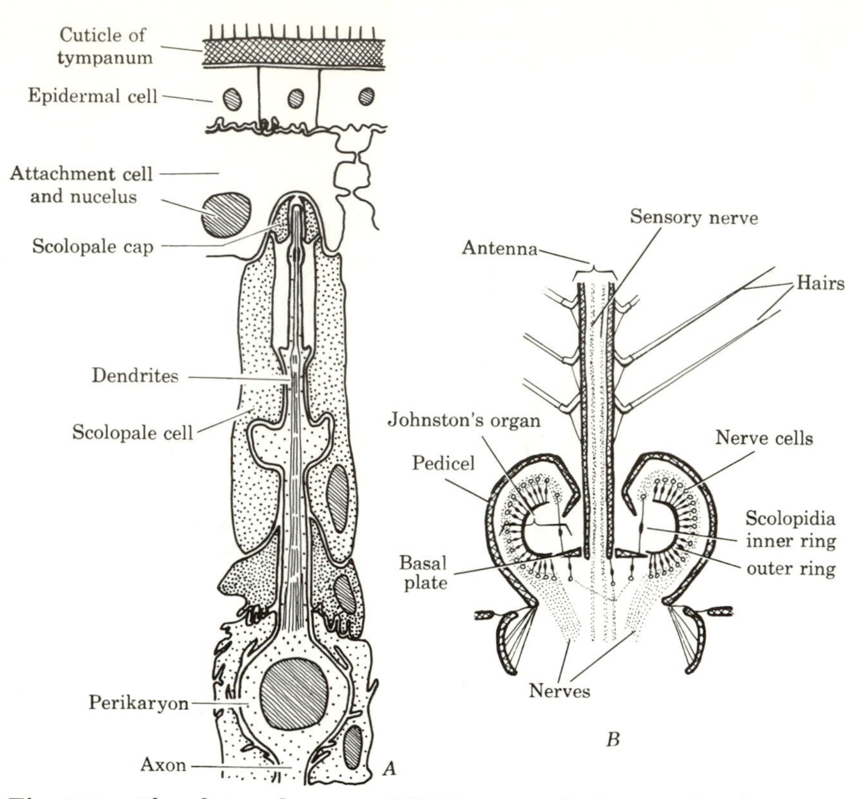

Fig. 6-5. Chordotonal organs. (*A*) Diagram of a longitudinal section through a scolopidium from the tympanum of *Locusta*. (*B*) Diagram of a section across the basal part of the antennae of a male mosquito showing Johnston's organ. (Both parts modified from Chapman, 1969)

ments of the antennal flagellum. Because this movement may result from several causes, Johnston's organ may serve a number of functions in an insect. It may be a flight-speed indicator and controller, perceive noises, or aid in orienting an insect in water or with respect to gravity.

In species of the Orthoptera, Hemiptera, and Lepidoptera, where chordotonal organs have evolved to be hearing organs, called *tympanal organs*, they occur on the thorax, abdomen, or appendages. They serve to detect predators and to receive sounds from other individuals of the same species.

Halteres. In flies, mechanoreceptors also are part of a unique balancing organ to stabilize flight. Over 400 mechanoreceptors at the base of each haltere, a modified hind wing, provide signals that serve to control the lift and to stabilize the flight of these insects. The insect in flight must correct for movement in three planes of rotation adjusting to pitch, roll, and yaw, and this is the function of the halteres.

Pulvilli. These sensory structures (Fig. 6-6) may be associated with food gathering. The pulvilli on the legs of the parasitic bee louse, *Braula coeca*, appear to stimulate the antennae of its host, the bee, which opens its mandibles slightly to allow the louse to obtain food from within its mouth.

Air and Water Current Receptors (Flow Receptors). Some structures are sensitive to air or water currents passing over the body, including patches of short hairs on the dorsal head surface in Orthoptera, Johnston's organ, and hairs on other areas of the body. Air receptors enable an insect to detect the speed of flight, and speed and direction of air currents in the immediate vicinity.

Chemoreceptors

Chemoreceptor cells are sensilla occurring as hairs, pegs, plates, and pits. They are abundant and widespread, particularly on antennae, but also are common on mouthparts, tarsi, ovipositor, and cerci. Their extreme sensitivity allows an insect to detect only a few molecules of a chemical in the environment. Olfactory receptors, those that perceive chemicals as smells, commonly have many sen-

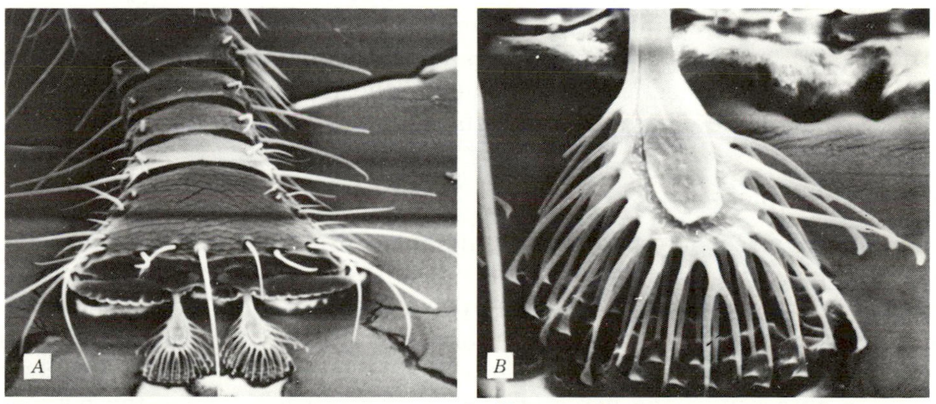

Fig. 6-6. (*A*) Sensory structures (pulvilli) on the last segment of the tarsus of the bee louse, *Braula coeca;* ×100. (*B*) Enlarged view of pulvillus (foot pad); ×500. (Courtesy of A. Dietz, Univ. of Georgia; from Dietz and Humphreys)

sory cells, each of which usually responds to a range of substances. The insect's responses to chemicals (generally gases, but also liquids) are primarily orientation and movement. An insect's abilities to smell may assist it in locating food or a mate. Its ability to taste may serve it to recognize food, an oviposition site, or a mate. In an insect, the sense of smell is considerably greater than that of taste.

Humidity Receptors. These are another type of chemoreceptor that perceives moisture in the air. The presence of water serves as a chemical stimulus to some insects. The receptor cells are sensilla, commonly occurring as hairs or pegs on the antennae. In *Drosophila* larvae, the receptors are on the ventral side of the anterior body segment. In many insects, hairs may curl or straighten depending on the humidity of the environment.

Thermoreceptors

Insects are cold-blooded and although they orient to temperature changes, few specialized temperature receptors have been positively identified. Some insects are thought to have thermoreceptors on the antennae or all over the body. Hairlike sensilla on the antennae of the bug *Rhodnius* are suspected of being thermoreceptors. In the order Orthoptera, a series of paired specialized areas on the head, thorax, and abdomen appear to be thermoreceptors. These areas differ in cuticular structure and texture from the surrounding epidermis. Different groups of Orthoptera have different arrangements of thermoreceptor areas. In the buprestid beetle *Melanophila*, sensory pits on the venter of the mesothorax are reported to be sensitive to radiant heat (infrared radiation), which enables the beetle to "home in" on recently fire-damaged trees. Some insects, parasites of warm-blooded animals, locate their hosts by detecting the temperature gradient around the hosts' body.

Sound Receptors

An insect detects sound by a number of different types of organs and mechanisms, many of which appear at various locations on the same individual. All these sound receptors have a bundle of nerves attached to a thin area of cuticle or adjacent to the cuticle. External sound waves cause this part of the integument to vibrate, which in turn causes the nerve bundle to transmit a signal to the nervous system. In the commonest type, the nerve bundle is attached to the edge of a drumlike surface or tympanum that stretches across an air space (Fig. 6-7*A*). Some of these tympanic sound receptors are extremely complex (Fig. 6-7*B*, *C*). They occur on the front legs of cric-

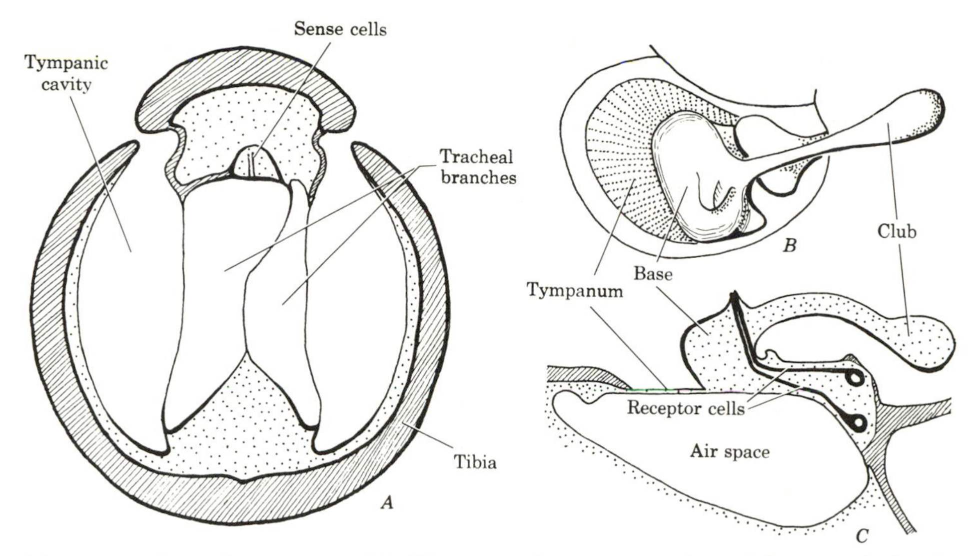

Fig. 6-7. Sound receptors. (*A*) Diagram of a cross section of the ear of a bush cricket in the region just below its "knee." (*B*) Diagram of the ear of the water boatman, from the outside; (*C*) as (*B*), but in section. The tympanum is covered largely by the base of a club-shaped cuticular mass that projects outside the insect's body. (Redrawn from Michelsen, 1979).

kets, on the mesothorax of some bugs, on the metathorax in noctuid moths, and on the abdomen in insects of several orders.

Only in vertebrates and arthropods is the sense of hearing highly developed. Hearing is an adaptation for individuals of the same species to communicate between each other and to become aware of other species, usually predators, but also competitors. Although many groups of insects produce and receive sounds, the noisy orthopterans, particularly the long-horned grasshoppers, crickets, and bushcrickets, have been the subject of most studies on insect hearing. In these, the lateral thoracic spiracles (Fig. 6-8C) lead into the prothorax, where each side branches into a large expanded tracheal sac and a long tubular trachea, that extends to a sensory organ in the foreleg tibia (Fig. 6-7*A*). The expanded tracheal sacs have a tympanal membrane on each side and are separated by other internal air sacs (Fig. 6-7*A*) that do not completely isolate the sound from the two sides of the system. The tympanal membrane, the large tracheal inflation in the prothorax, and the long tubular tracheae leading into the foreleg act together as an amplifier, or sound trumpet. In some insects, the thoracic spiracles may be closed by the individual; in others they are always open; and in some the two sides of the tracheal system may be connected by an additional tracheal tube. Thus some insects have an interconnected system with the shape of an H, having four sound-input points (two spiracles and two tympanal membranes) (Fig. 6-8C), as in quadrophonic sound. At sound waves slower than 10,000 vibrations per second, this system acts as a sensor of pressure difference; at higher frequencies it acts as a pressure receiver.

Several other structures detect sound. The simplest, and probably the least sensitive, are thin hairs on the antennae (Fig. 3-27; Fig. 6-8*A*) that vibrate as the sound waves move air particles past them. Using hair vibrations, some caterpillars are able to hear flying wasps, but only at close range where the sound level is upwards of 90 decibels, a noise level comparable to a flying wasp about 1 cm from your ear! In males of the fly families Chironomidae (midges) and Culicidae (mosquitoes), Johnston's organ is sensitive to the hum of flying females of the species.

Hawkmoths of the subfamily Choerocampinae have an ultrasonic receptor sensitive to the "radar" ultrasonic emissions of foraging bats. The receptor comprises three separate structures: the labial palp; a small lobe called a pilifer that nestles between the bases of the palp and proboscis; and the proboscis (Fig. 6-8*B*). These three structures, artificially separated in the lower part of Fig. 6-8*B*, normally are folded together as shown in the upper part. The palp is large, thin walled, and has an extensive air sac; its mesal surface is held against the smooth, bare lateral surface of the pilifer. Incoming ultrasonic signals are magnified by the palp, transmitted by direct contact to the pilifer, then in turn transmitted to the proboscis (labrum) by the long setae on the mesal side of the pilifer. These impulses activate the labial nerve, which sends a message to the brain.

Photoreceptors

Different types of photoreceptors permit various insects to perceive the form of objects, patterns, movement, distance, certain colors, light intensity, the polarization plane of light, light versus darkness, and the length of the photoperiod. These photoreceptors perceive (record) light (a form of energy) by means of a pigment that absorbs light of a particular wavelength, and this in turn stimulates associated nerve cells. Photoreceptors in insects take the form of dermal light response, lateral ocelli (stemmata), dorsal ocelli, and compound eyes.

The most primitive form of photosensitivity is a dermal light response resulting from light striking the integument. It is a diffuse sensitivity for which no light receptors have been identified. It has

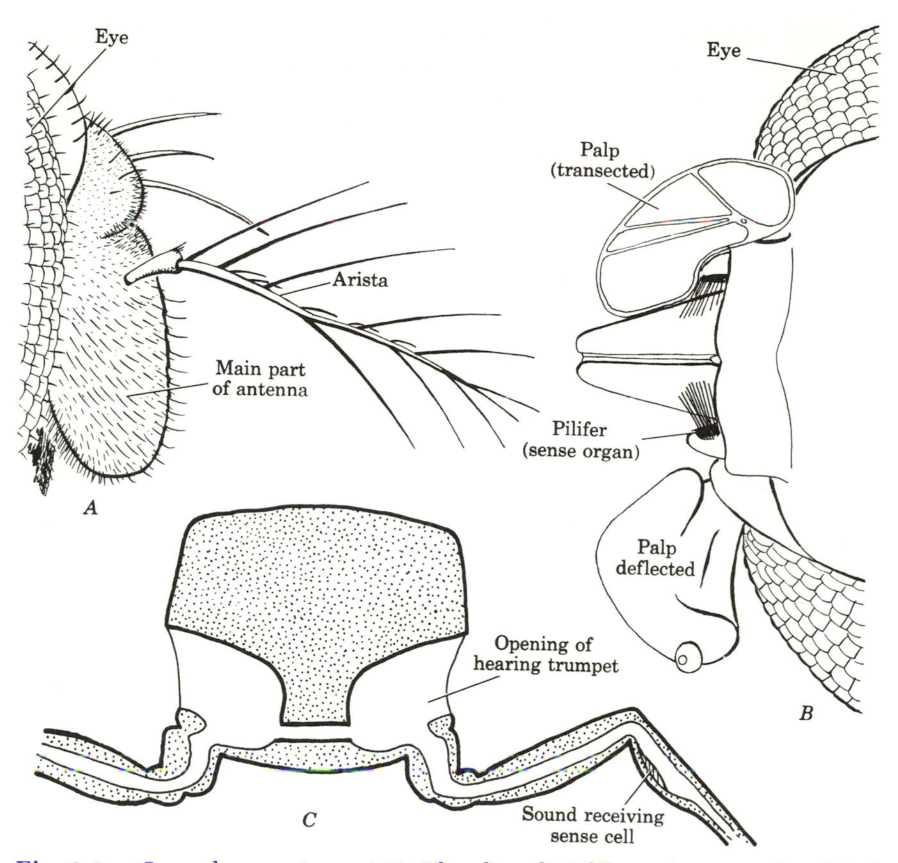

Fig. 6-8. Sound receptors. (*A*) The fine hairlike arista on the third antennal segment of the fruit fly extends laterally from an apical joint. The arista serves as a receiver for vibrations created by the calling song of other individuals of the species. (*B*) Diagram of the dorsal view of the ultrasonic receptor, the distal lobe of the pilifer, in the hawkmoth *Celerio lineata*. (*C*) Diagram of a section through the front legs and thorax of the bush cricket, showing one of the pair of ears that are acoustically separate. (Redrawn from Michelsen, 1979)

been suggested that the dermal light sense depends on the presence of small amounts of pigments similar to those in photoreceptor systems, because the maximum light sensitivity occurs at wavelengths between 470 and 580 mμ. This absorption corresponds to that of photosensitive pigments (blue-green-yellow lights).

Compound Eyes. These large visual organs, the major photoreceptor of many insects (Fig. 3-27), are aggregations of thousands of photoreceptor cells. Most adult insects have two eyes located one on each side of the head. Each compound eye is composed of many long cylindrical light sensory units, called *ommatidia*. The compound eye of a dragonfly has from 10,000 to as many as 28,000 ommatidia, and that of the common housefly about 4000. Each *ommatidium* is outlined externally as a hexagonal cuticular facet that gives the compound eye a distinctive appearance.

Each ommatidium has two major components (Fig. 6-9), as follows.

1. A light-gathering optical part consists of an external, transparent biconvex or plano-convex lens (Fig. 6-10), called the *cornea*, surrounded by cuticle and secreted by special epidermal cells. Since the cornea is continuous with the rest of the cuticle, it is an immovable lens. Beneath the cornea lies a *crystalline cone* (Figs. 6-9, 6-11) and a group of *primary pigment cells* that generally surround the crystalline cone. However, in primitive insects such as silverfish, the pigment cells lie beneath the cornea and not around the crystalline cone.

2. A sensory receptor component underlies the optical unit. It consists of sense cells grouped in clusters of usually 6 to 12 *retinular cells* whose central surfaces have many minute neurofibrillae. An axon from each of the retinular cells passes through a basement membrane to the brain. In each cluster of retinular cells, dendritic

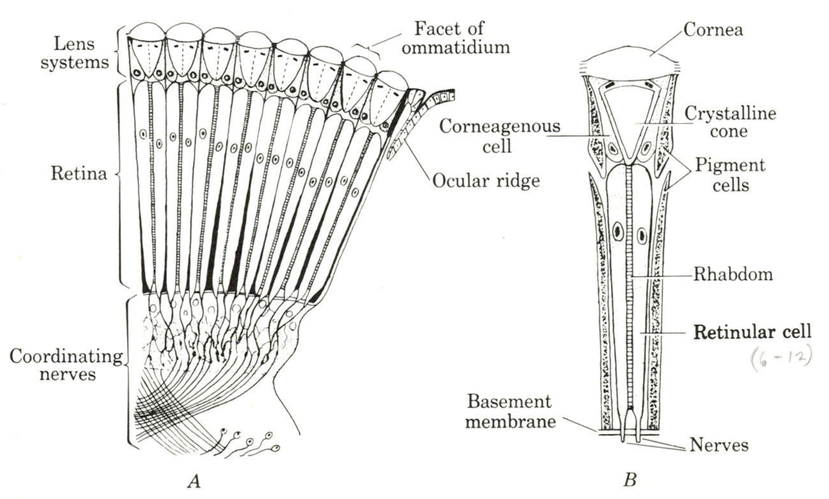

Fig. 6-9. Diagrams of a compound eye and of an ommatidium. (*A*) Vertical section of part of eye; (*B*) typical structure of an ommatidium. (Redrawn from Snodgrass)

Fig. 6-10. External surface of several lenses of the compound eye of the mosquito *Aedes aegypti*. The surface of each lens has a fine regular pattern that forms an antireflective structure; ×4500. (Courtesy Dept. Entomology, Univ. of Alberta, and D. A. Craig)

extensions from the retinular cells form a central, compound, light-sensitive rodlike structure called a *rhabdom*. The rhabdom contains visual pigments, called *rhodopsins*, which, when illuminated, undergo structural changes in molecular configuration resulting in a change of energy state. This change is relayed by the sensory neurons to the brain. The rhabdom and cluster of retinular cells are surrounded generally by secondary pigment cells that limit to varying degrees the light that may enter from adjacent ommatidia.

In the compound eye, light falling on the ommatidium is focused by the cornea, then funnelled by the crystalline cone to the rhabdom, where the change in visual pigments results in sensory information

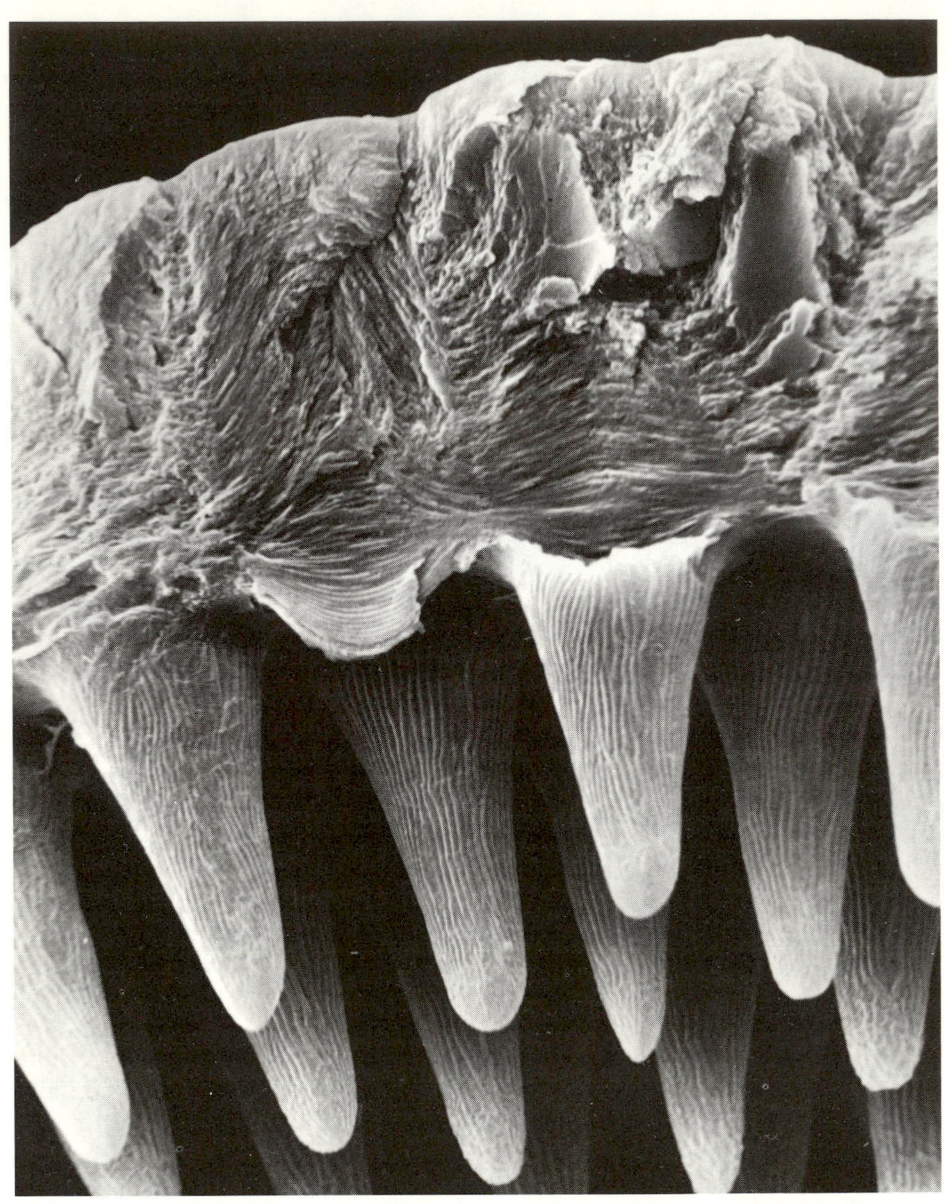

Fig. 6-11. Inside of a compound eye of an adult wireworm (Elateridae) shows crystalline cones below cornea; ×2200. (Courtesy Dept. Entomology, Univ. of Alberta, and D. A. Craig).

being relayed directly to the brain. Each ommatidium of the compound eye functions as a single photoreceptor unit that transmits a signal to the brain. The image perceived by an insect results from all the signals transmitted by the ommatidia being coordinated in the brain.

Compound eyes may function differently when the light intensity varies markedly. In many species, in bright light the secondary screening pigment may extend down between each ommatidium so that the rhabdom is stimulated only by light that enters directly above the lens of the ommatidium. The image formed is called an *apposition* (or *photopic*) image and is common to diurnal insects (Fig. 6-12*A*). This type of image may be effective in detecting movement as each ommatidium scans the moving object. In weak light, the secondary screening pigment is contracted and light entering the eye may stimulate several ommatidia. The image formed is called a *superposition* (or *scotopic*) image (Fig. 6-12*B*), and is more extensively found in nocturnal and crepuscular insects. It serves to detect changes in the intensity of light and probably does not define objects.

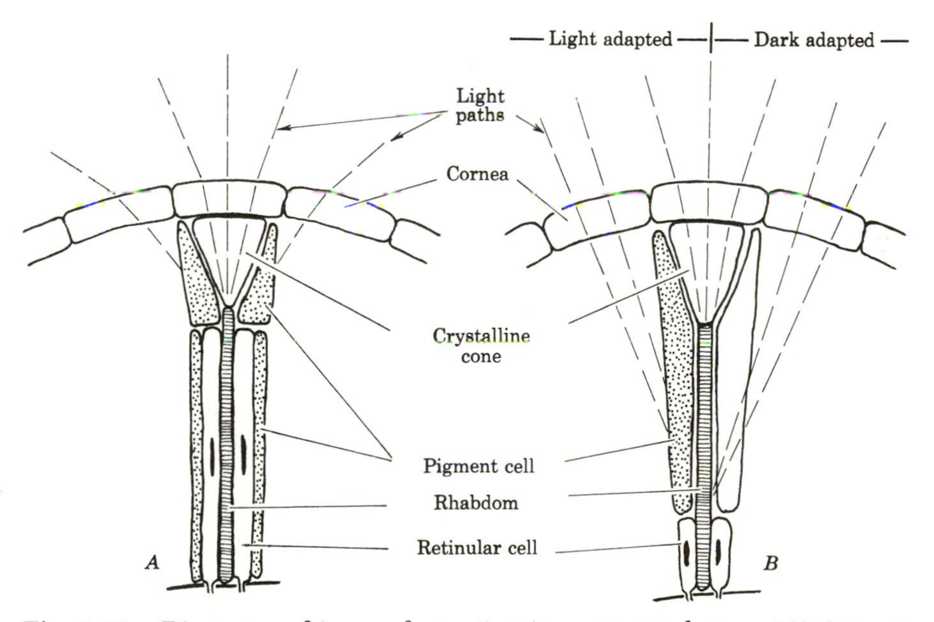

Fig. 6-12. Diagrams of image formation in compound eyes. (*A*) Apposition type of compound eye. (*B*) Superposition type: left-hand side shows light adapted type; right-hand side shows dark adapted type. (Modified from Romoser, 1973)

Although the low number of photoreceptors in a compound eye permits only a crude image to be formed in comparison to the human eye, with 130 million photoreceptors, insects have exceptional *flicker vision*. The insect eye can see successively different images at very short intervals because the rhabdoms recover rapidly from the stimulation of light. Flies detect flickers up to about 265 flickers per second, whereas the human eye detects flickers of only 45 to 53 per second and generally 20 to 30 per second. Sixty-cycle light in a fluorescent bulb is smooth, continuous light to a human, but to an insect it is a series of flickers. Flicker vision permits an insect to detect very slight movements in the environment.

Color Receptors. Many insects have green and ultraviolet receptors. Insect eyes are generally most sensitive in the ultraviolet and blue-green regions. Occasionally blue receptors are present, such as in the honeybee. The green receptors are more frequent in the ventral and frontal eye. Ultraviolet receptors present a different world to insects. As various flowers reflect ultraviolet light to different degrees, two flowers that appear identically white in color to us may be markedly different to an insect. Red receptors are generally absent in insects.

Plane of Polarization. Another distinctive character of an insect's compound eyes is its capability to perceive and to analyze polarized light from the sky. Polarized light waves vibrate only in a single plane. The pattern of polarization of the sky varies with the position of the sun, and certain insects use the pattern of polarization to determine direction. The honeybee employs these planes of polarization as navigational aids to return to the hive after a foraging flight.

Ocelli and Stemmata. In addition to compound eyes, many adult insects also have photoreceptors that consist of a single cornea (lens) and clusters of pigment cells that overlie a group of retinular cells that form several rhabdoms. These are called *ocelli*. Many adult insects have *dorsal ocelli*. Commonly three ocelli are arranged in a triangular pattern on the anterior part of the head (Fig. 3-10). In some there are only one or two. Dorsal ocelli function as pigment cups that detect changes in light intensity.

Larvae of insects that have complete metamorphosis lack compound eyes; instead they generally have *lateral ocelli*, called *stemmata*. Structurally stemmata are variable. Some, as in caterpillars, are like a single ommatidium comprising a cornea, crystalline cone, and a cluster of retinular cells that form a single rhabdom (Fig. 6-13).

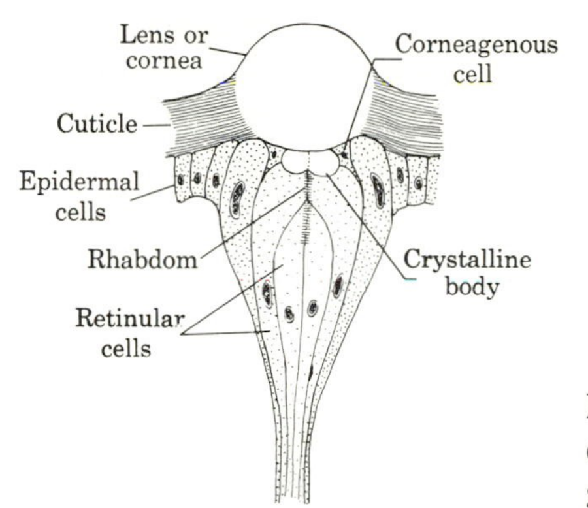

Fig. 6-13. The simple eye of a caterpillar. (Redrawn from Snodgrass)

They detect the form of objects, color, and movement. Other stemmata are structurally similar to dorsal ocelli.

NERVOUS COORDINATION AND INTEGRATION

The nervous system of an insect is the means by which it receives a wide assortment of information from both the external environment and its own internal environment. An insect coordinates the various sets of information in order to respond to particular situations and conditions. In insects, the nervous system includes a brain, a longitudinal ventral nerve cord, and many neurons (see p. 117). The system is differentiated into central and stomodeal (visceral) or sympathetic nervous systems with sensory neurons, motor neurons, and interneurons joined by synapses. A housefly has more than one million neurons and the brain and ganglia are crammed with interneurons. Sense organs, such as the compound eye and mechanical sensory receptors, have many neurons.

Neurons or Nerve Cells. A neuron, or nerve cell, is a particular kind of elongate cell capable of receiving and transmitting stimuli (impulses). In a neuron, nerve impulses travel in one direction only. Each neuron (Fig. 6-14) consists of three principal parts: (1) a *cell body* or *perikaryon*; (2) one or more *receptor fibrils* or *dendrites*; (3) a *transmitting fiber* or *axon*, which ends in a group of fibrils called *arborization*, because of a treelike pattern of branching. At least one end of each neuron is situated in the central nervous system or associated ganglia (masses of neurons grouped together). There are two chief types of neurons. In one type (Fig. 6-14*A*, *B*), one or more receptor fibers arise directly from the cell body. These are *sensory* *neurons* in which the receptor fibers are connected with sense organs and the axon runs to, and terminates in, the central nervous system. In the other type (Fig. 6-14C), the receptor fibrils are situated on what appears as a branch of the axon, called the *collateral branch*. This type includes (1) *motor neurons*, in which the receptor fibrils and cell body are situated in the nervous system and the axon forms a nerve fiber running to muscle tissue; and (2) *interneurons*, all parts of which are situated within the central nervous system.

When stimuli are received by receptor fibrils they travel as impulses to the tips of the axon. This direction of travel is not reversible. Impulses pass from one nerve cell to another through a *synapse*, an area of the nervous system in which the end fibrils of the axon arborization of one neuron and the arborizations of dendrites of another neuron almost, but do not, touch. Impulses are transmitted generally across the narrow gap between the axon of one neuron and the arborizations of another neuron by special chemical neurosecretions, commonly acetylcholine. In a generalized reaction (Fig. 6-15), an external stimulus causes some change in a sense organ, and this in turn stimulates the associated sensory neuron, the impulse from this neuron passes through a synapse to an interneuron and from this through another synapse to a motor neuron; this motor neuron then transmits an impulse through its axon to a muscle fiber, which contracts as a result of the stimulus received.

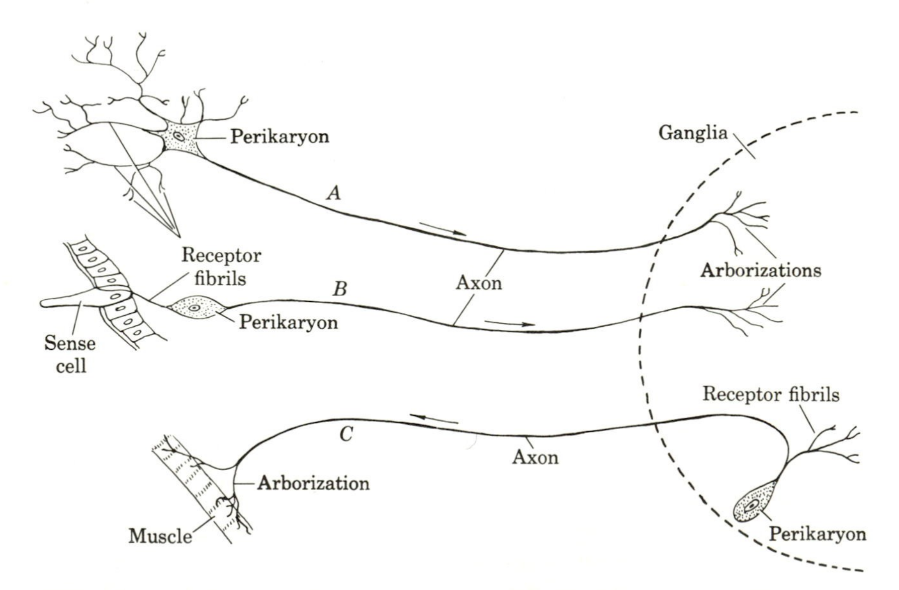

Fig. 6-14. Diagram of neurons. (*A*, *B*) Sensory types; (*C*) motor type. (Adapted from Wigglesworth)

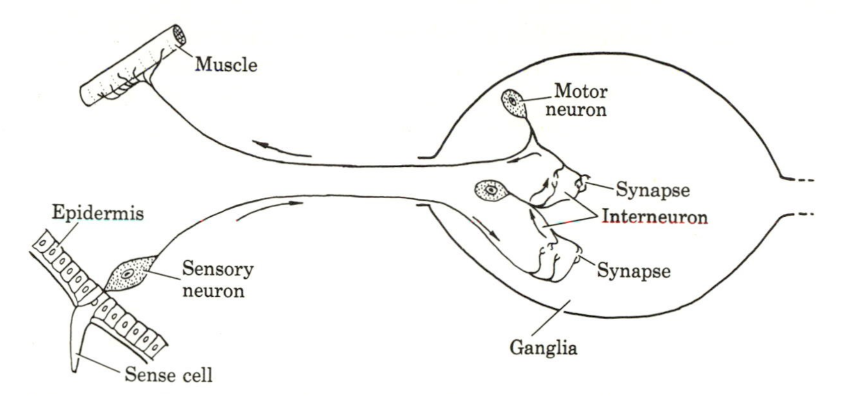

Fig. 6-15. Diagram of a simple reflex circuit of sensory, association, and motor neurons. Direction of transmission of impuse is shown by arrows. (Adapted from Wigglesworth)

Coordination. The synapses of each body segment are grouped together to form the ganglia of the central nervous system. Thus the sensory nerves all lead to these centers, and the responses go out from these to the reactive tissues. Interneurons run from ganglion to ganglion, and into the brain. They also may link the same motor cell to several sensory cells or several motor cells to one sensory cell. This whole communication system coordinates responses in different parts of the body with stimuli received at only one station. Thus a touch on a cockroach's cercus (reporting to the terminal abdominal ganglion) will cause the animal's legs (motivated by the thoracic ganglia) to respond with running movements.

ENDOCRINE SYSTEM AND HORMONES

The nervous and endocrine systems of an insect control its physiological and behavioral responses. Activities that require an insect to adapt very quickly with particular responses, such as muscular activity, are under the primary control of the nervous sytem. In contrast, long-term changes in development, growth, reproduction, and metabolism are generally under the control of the endocrine system. The endocrine system and sensory information from the environment are coordinated through the insect brain. In addition to mediating many of the physiological processes, discussed in Chap. 7, behavioral responses are also mediated by the endocrine system. This system consists of glands and special cells that secrete hormones. *Hormones* are chemical messengers that are active in minute concentrations and effect physiological and/or behavioral responses in an insect. They are very stable, and they diffuse or are transported by the body fluids to other parts of the body away from their site of origin. Hormones are better suited than neurons for the functions they perform because:

1. Millisecond timing is not essential.
2. They provide a sustained stimulus to a particular site in the body over a period of hours or days.
3. Multicellular tissues, such as the epidermis, are more easily reached via circulating body fluids than by neurons to particular cells.

Sources of the principal insect hormones are neurosecretory brain cells, corpora cardiaca, neurosecretory cells of the subesophageal ganglion, corpora allata, ventral glands, pericardial glands, prothoracic glands, oenocytes, neurosecretory cells of the ventral nerve cord, and corpus luteum (Fig. 6-16). The brain contains two paired clusters of neurosecretory cells (median and lateral). In some insects, a third small cluster of posterior neurosecretory cells is present in the protocerebrum.

Hormones, as they affect insect behavior, may be classified into three types: those having modifier effects; those showing releaser effects; and others showing both modifier and releaser effects. The modifier effect of a hormone is a slow process resulting in a change in response by the nervous system. An insect shows a specific behavioral response to a particular stimulus but, after exposure to a modifier hormone, the same stimulus produces a different response. The *releaser effect*, on the other hand, is a rapid behavioral response that is directly produced by the hormone.

Modifier effects of a hormone that induces changes in behavior are shown well by the effect of the juvenile hormone on the response of a female grasshopper *Gomphocerus rufus* to a courting male. After imaginal ecdysis, female grasshoppers are defensive when males approach. After exposure to juvenile hormone, which is secreted by the corpora allata, the defensiveness disappears and the female when approached by a courting male stridulates in response to the male's call and allows him to mount and copulate.

Releaser effects of a hormone product immediate specific behavioral response. In silkmoths and many other insects, the eclosion hormone acts as a releaser hormone because it directly produces pre-eclosion behavioral responses. Adult ecdysis occurs in silkmoths during a specific time of the day and this is synchronized with a particular photoperiod. The eclosion hormone, synthesized in

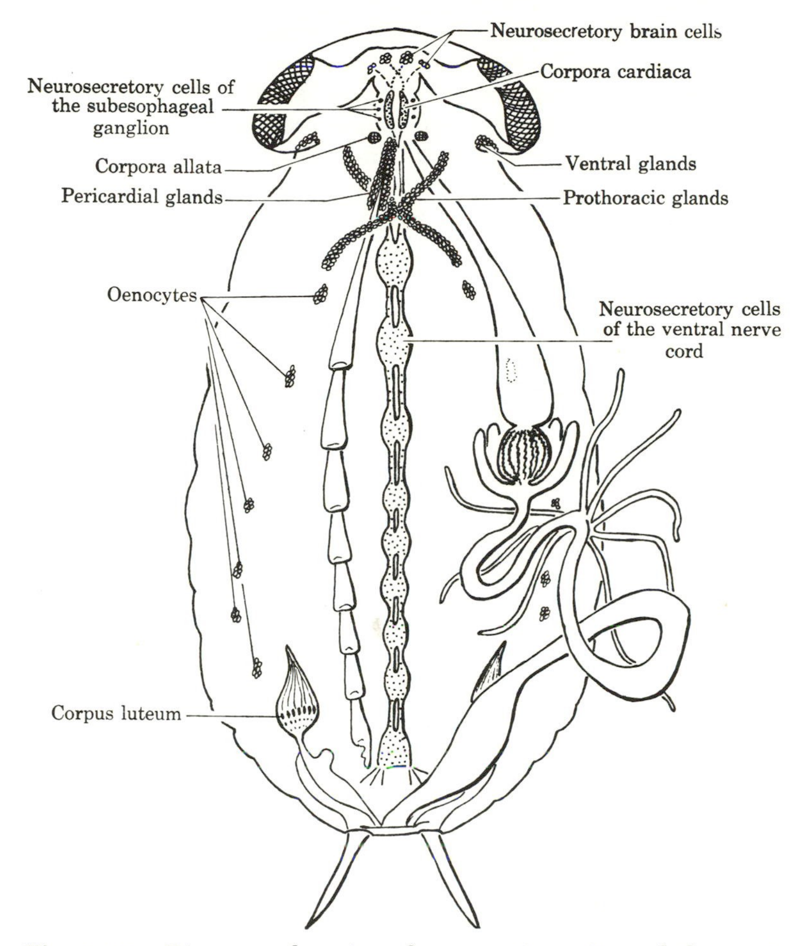

Fig. 6-16. Diagram showing the secretion sites of the main insect hormones. (Modified from Novák, 1975)

neurosecretory cells of the brain, is induced by photoperiod. The pharate silkmoth displays pre-ecdysial behavior, controlled by the eclosion hormone, through a series of stereotyped abdominal movements that initiates the sequence of emergence events.

Another example of the effects of a releaser hormone on behavioral response is found in the larvae of the locust *Schistocerca gregaria*. The secretion of the molting hormone, ecdysone, which induces molting and subsequent cuticle deposition, also causes the larvae to become inactive at a time when they are very vulnerable to prey.

COMMUNICATION

Communication involves two processes: (1) the transmission of a signal by an individual; and (2) the reception of the signal by one or more individuals. Successful transmission of the signal may result in a change in behavior in the individual or individuals receiving the signal. This change in behavior may or may not be directly observable. Three common types of insect communication are considered here: visual, auditory, and chemical.

Visual Communication

The intriguing visual displays of glowing light by plants, animals, and microbes has been observed for centuries. *Bioluminescence* in hexapods occurs in several orders including Collembola (springtails) and Coleoptera (beetles). It is most common and striking in beetles of the families Lampyridae (fireflies or lightning bugs whose larvae are glowworms) and Elateridae (click beetles). In fireflies, such as the well-studied *Photinus pyralis*, common in the eastern United States, the light-producing organ is in the abdomen and the flashing lights serve to bring the sexes together to mate. The frequency of light flashes is species specific and, even in an area where several species of fireflies are present, the distinctive flashing light patterns serve as visual communication to individuals of the same species.

The colorful and highly patterned wings in some of the Lepidoptera serve as visual signals in sexual behavior, such as courtship and territoriality. In the common orange queen butterfly *Danaus gilippus berenice* of southern Florida, courtship initially involves pursuit in flight. This occurs as a result of the males being attracted visually to the females through a combination of movement, color, and shape of the females.

Visual communication is extensively developed by insects in defense mechanisms. In many insects, the pattern of coloration, the color, or the shape serve as a warning to predators. Visual mimicry (mimicry is discussed in Chap. 10), in which an insect mimics not only objects in its surroundings but also other insects, can be an effective defense mechanism. Certain insects, including katydids and moths, imitate social wasps in shape and/or color. Among other examples, cerambycid beetles imitate ants, bees, and wasps.

Auditory Communication

Insects make sounds in various ways and most insects communicate by sound at some stage in their life. In some, the sound is produced by the insect's normal activities without aid of special sound-making structures. The most familiar example is the hum made by flying or hovering insects. The hum is produced by the extremely rapid vibrations of the wings and thoracic sclerites. In some insects, such as mosquitoes, the sound produced by the beating wings of a flying female elicits a mating response in males. In others, such sounds do not serve as a means of communication.

The features of sounds produced by insects are difficult to characterize within the framework of human hearing. Continuous sound, such as that from the vibration of wings, is less common than discrete pulses of sounds separated by intervals of silence. Insects are able to hear a greater range of sound frequencies, particularly high frequencies, than the human ear, which perceives sound of only about 60 to 1600 cycles per second (c/s). In the Lepidoptera (moths and butterflies), the very low wingbeat frequency of about 20 c/s produces a sound that is inaudible to humans. In the bee *Apis*, the wingbeat frequency is about 250 c/s. The locust *Schistocerca* has a flight noise that ranges between 60 and 6400 c/s, but mainly 3200 to 5000 c/s. The frequency of sound produced by a termite (Isoptera) banging its head against wood is about 1000 c/s. In different species of crickets (Orthoptera), sounds produced by frictional rubbing (elytral stridulation) range in frequency from a low frequency of 2000 to 10,000 c/s to very high frequencies of 5000 to 100,000 c/s. The chirp of a cricket has a remarkable constancy of frequency and amplitude and is a very pure sound. In cicadas (Hemiptera), the frequency of sound produced in the tymbal is about 4500 c/s.

In addition to sound produced by an insect during its normal activities, other mechanisms that insects use to produce sound are:

1. Pounding part of the insect body against the substrate.
2. Frictional rubbing of two parts of the body together (commonly called *stridulation*).
3. Vibrating membranes.
4. Pulsing forced airstreams.

Many variations of these mechanisms occur in thousands of species of insects.

1. Sound resulting from the impact of part of the insect body striking a substrate is produced in several ways. The grasshopper *Oedipoda* beats on the ground with its hind tibia. The male makes about 12 beats per second and beats more rapidly than the female. The deathwatch beetle produces a sequence of raps by banging its head on the substrate. Some soldier termites have oscillating movements whereby their head and the tips of the mandibles hitting the wood make a sequence of sounds that are transmitted to other ter-

mites (workers and larvae) in the colony. The other termite castes also produce sounds, but at a lower intensity, by hitting their heads on the wood.

2. Sounds produced by frictional rubbing of legs, body parts, or wings are found in many orders of insects. Such sound-making mechanisms are well developed in the Orthoptera, Heteroptera, and Coleoptera. In some grasshoppers, the front margin of the hind wing scrapes over the thickened veins of the fore wing, causing the latter to vibrate. This vibration produces a crackling sound. In other grasshoppers, the inner face of the hind femur has a file of minute teeth (Fig. 6-17). This file is rubbed over the fore wing, which then vibrates to produce sound. Various crickets have a toothed file on one elytron and a ridge forming a scraper on the opposite elytron. When the elytra close, the scraper rubs against the file causing the elytron to vibrate and to produce a sound (Fig. 6-18). In other insects, such as the beetles, the scraper and file may be on the leg and body, respectively (Fig. 6-19). Beetles use various regions of their heavily sclerotized exoskeleton, particularly the elytra, to produce sound.

3. A vibrating membrane or tymbal mechanism is found in several homopterans, particularly the cicadas, other Hemiptera, and Lepidoptera. A unique mechanism is developed in the cicadas (Fig. 6-20). They possess a set of membranes situated in ventral pouches or cavities near the base of the abdomen. One of these membranes is connected internally with a muscle fiber. The contraction of this muscle pulls the membrane inward; the relaxation of the muscle allows the membrane to snap back to its original shape. These movements are alternated with great speed to produce sound waves. The other membranes act as sound reflectors.

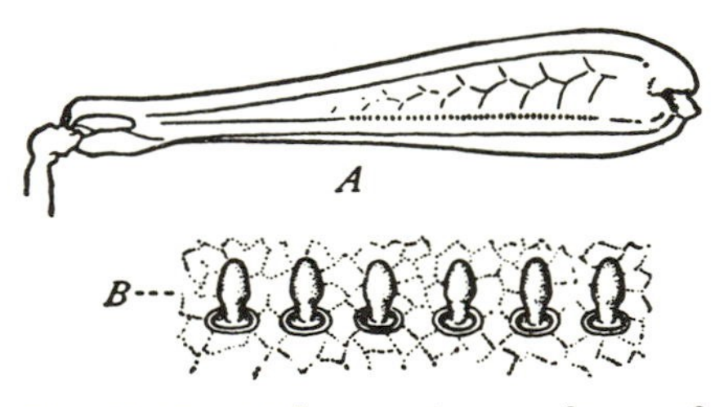

Fig. 6-17. File on inner face of hind femur of a cricket. (*A*) Hind femur of *Stenobothrus;* (*B*) file greatly enlarged. (After Comstock, *An introduction to entomology,* by permission of The Comstock Publishing Co.)

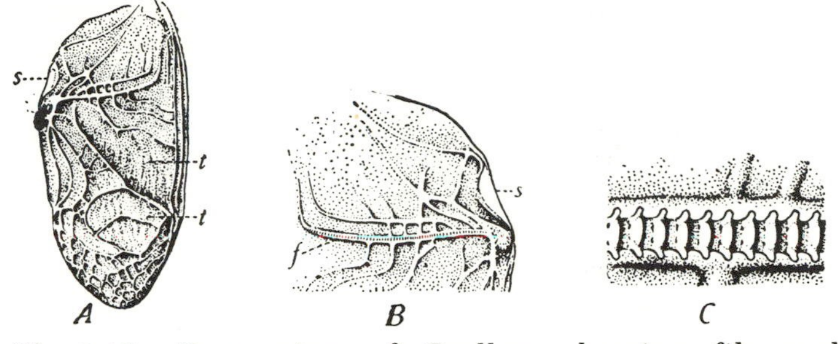

Fig. 6-18. Fore wing of *Gryllus*, showing file and scraper. (*A*) As seen from above; that part of the wing which is bent down on the side of the abdomen is not shown; *s*, scraper; *t*, tympana. (*B*) Base of wing seen from below; *s*, scraper; *f*, file. (C) File greatly enlarge. (After Comstock, 1936, *An introduction to entomology*, by permission of The Comstock Publishing Co.)

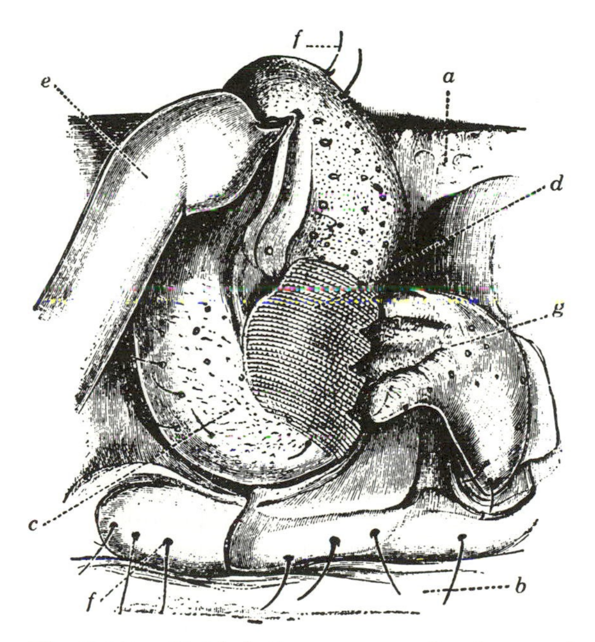

Fig. 6-19. Stridulation organ of a larva of *Passalus*. *a*, *b*, Portions of the metathorax; *c*, coxa of the second leg; *d*, file; *e*, basal part of femur of middle leg; *f*, hairs with chitinous process at base of each; g, the diminutive third leg modified for scratching the file. (From Comstock, after Sharp)

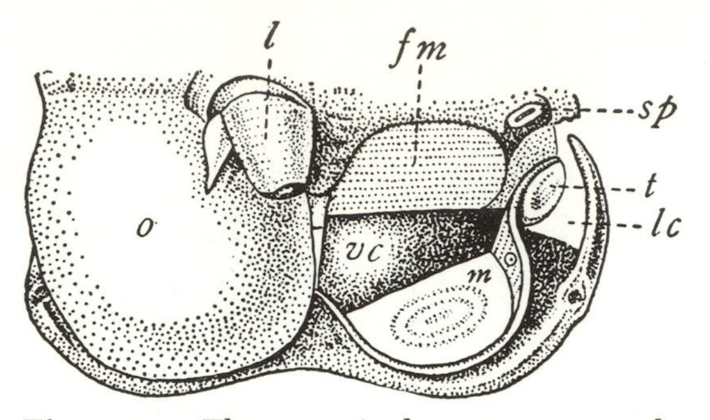

Fig. 6-20. The musical apparatus of a cicada. *fm*, Folded membrane; *l*, base of leg; *lc*, lateral cavity; *m*, mirror; *o*, operculum, that of the opposite side removed; *sp*, spiracle; *t*, timbal; *vc*, ventral cavity. (After Carlet)

4. Forced air projected through or over a small opening is still another mechanism of sound production by insects. In the hawkmoth *Acherontia atropos*, sound is emitted as air is forcibly inhaled and exhaled through the proboscis and the air streams over the epipharynx.

The sounds produced by insects may serve as signals to other species—that is, they are *extraspecific*—or as signals to other individuals of the same species—that is, they are *intraspecific*. Unorganized sounds having no regular pulse-repetition frequency apparently serve as alarm responses for defense or warning and may be either extraspecific or intraspecific. The stridulation of certain Heteroptera, beetles and lepidopteran pupae, produced by both males and females, is presumed to be sound of this type. These alarm responses are accompanied at times by other defense displays, such as a hissing noise produced when the peacock butterfly opens its wings to display eyespots.

Organized sounds with a regular pulse-repetition frequency are characteristically used in intraspecific communication. These sounds are usually associated with courtship. In the Orthoptera, five types of songs function in calling, courting, copulation, aggression, and alarm. Each type differs in pulse-repetition frequency and the form of the pulse. In some insects, such as the cicadas, sounds lead to aggregation. Males and females of a species aggregate within a particular habitat, such as certain trees. The song of a particular species of cicada serves in sexual isolation of that species from other species. In North America, three species of *Magicicada* occur in the same habitat. However, individuals of a particular species aggregate and identify each other as a result of their distinctive song. In addition, the different species sing in chorus at different times of the day.

In subsocial and social insects, sound signals identify individuals of a species and colony. Hornets produce a number of different sounds in their nests that serve as signals. In honeybees, swarming is in part initiated by the persistent buzzing sound begun by a few workers moving excitedly through the hive. In social insects, sounds also are used as directives for food finding and nest siting.

In social insects, sound is an effective medium of communication, but it is not used to the exclusion of other modes of communication, such as chemical and tactile communication. The interaction and integration of different types of communication is well illustrated in the classic social insect, the honeybee, *Apis mellifera*. The *dance of the honeybee*, once considered to be an example of visual communication, has proven to be a complex set of signals that also involves sound and proprioception. The primary mode of communication in the dance is still not established. The dance is a set of specific, stereotyped behaviors performed by a foraging bee inside the hive upon the vertical face of the honeycomb. The forager, having found a food source, brings back pollen and nectar to the hive. It then flies out again to the food source. Upon returning the second time, the forager performs the dance of the honeybee where many workers congregate. The forms of the dance differ depending on the proximity of the food. If the food is close to the hive, the *round dance* (Fig. 6-21*A*) is performed. The foraging bee moves in circles, alternating its direction of movement. The round dance is interpreted to identify that food is close by but no information about the direction or the distance of the food source is given. If the food is at intermediate distances, the *sickle dance* (an *open figure* "8") is employed and the round dance gradually changes to an open figure "8." If the food is at a distance, the *waggle dance* (Fig. 6-21*B*) is used. This is a modification of the open figure "8" dance in which the forager "waggles" her abdomen rapidly from side to side as she passes through the straight run part of the dance. It is thought that information about the distance and direction of the food source is gained from the straight (or waggle) run. A pulsed sound produced by the forager bee during the straight run and the number of pulses and the total length of sound production may be used by the other bees of the colony to establish the distance of the food source. Information from these sounds may be used in conjunction with the number of waggles in the straight run to communicate the distance of the food source to other bees of the colony. The angle between the straight (or waggle) run and a

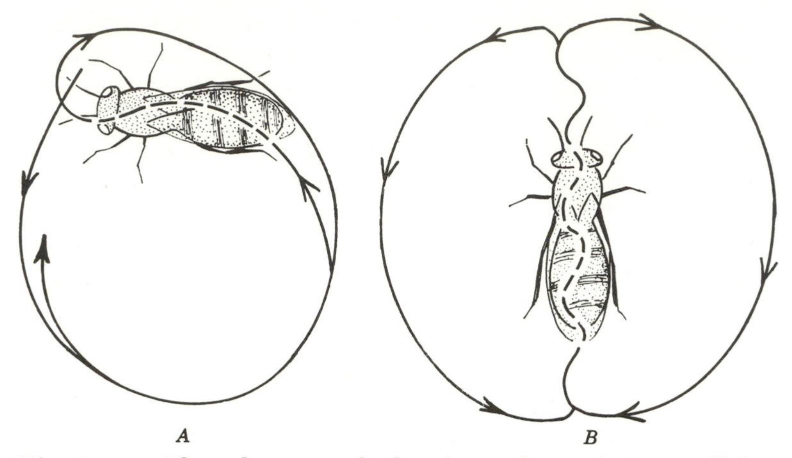

Fig. 6-21. The dances of the honeybee, *Apis mellifera*. (*A*) Round dance performed by foraging bee when the food source is close to the hive. (*B*) Waggle dance performed by foraging bee when the food source is distant from the hive. (Data from von Frisch, 1967)

vertical line is thought to give the angle between the sun and food as measured from the hive (Fig. 6-22). Straight runs vertically upwards indicate food in the direction of the sun; straight runs vertically downwards indicate the food source is in the opposite direction to the sun.

Chemical Communication

Chemical signals, including those for taste and smell, are one of the most important means of communication in insects. These signals are extraordinarily widespread in insects in all life stages and in both sexes, and they include hundreds of substances. Insects have the greatest number and probably the most refined system of chemical communication in any animal group. Insects produce and release chemicals into the external environment from specialized glands, called *exocrine glands*, such as mandibular or poison glands. These chemicals provide an effective means of chemical communication, particularly among the social insects. The chemical compounds produced by organisms that stimulate behavioral responses are collectively called *semiochemicals*. They are grouped into two categories: *pheromones*, which transmit chemical messages to individuals of the same species; and *allelochemics*, which transmit messages to individuals of a different species.

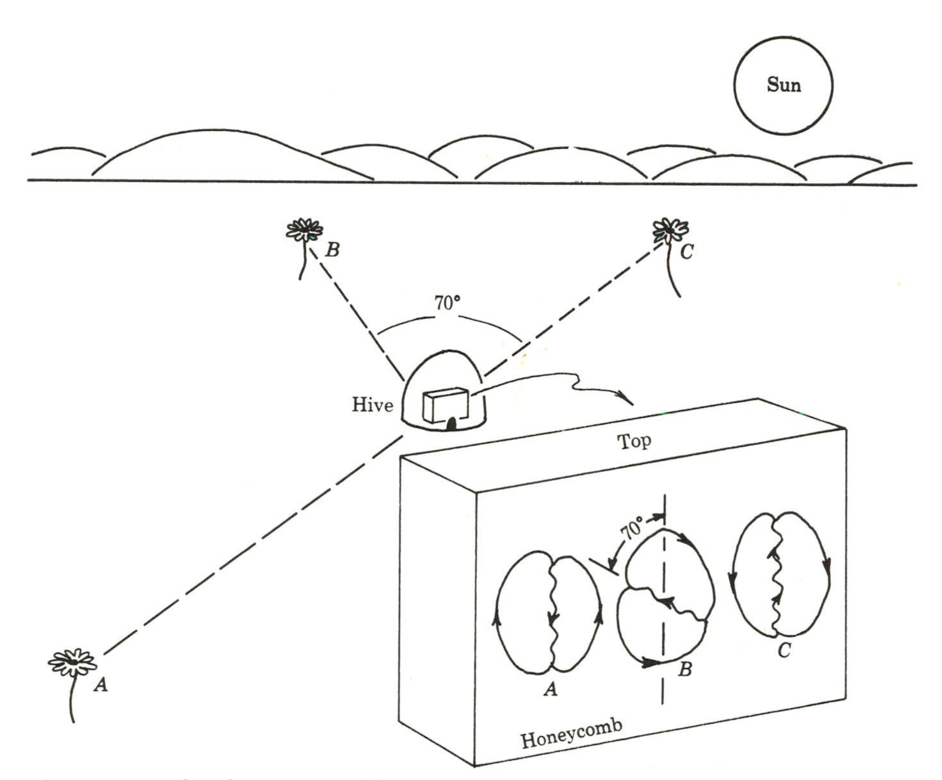

Fig. 6-22. The direction of the distant food source is given in the waggle dance of the honeybee. The angle of the waggle part of the dance from the vertical appears to indicate the angle between the food source and the sun, as projected in a horizontal plane. (Redrawn from Brown, 1975)

Pheromones

These chemical substances produce a response in an individual of the same species. They are small chemical compounds with low molecular weights ranging from 100 to 300. There are two functional groups of pheromones: releasers and primers. *Releaser pheromones* act on the recipient's nervous system to produce an immediate behavioral response. They are widespread in insects and function as sex attractants, trail identifiers, and alarm indicators. They also mediate behavioral responses involving sexual behavior, assembly and aggregation, spacing of individuals at high population densities, and identification of individuals of the same species. Generally changes in orientation and locomotion of an insect are induced by pheromones. In most species, a sexually receptive female produces a volatile pheromone that stimulates males of the same species, which then seek the female. A disturbed aphid emits a pheromone from its cornicles that produces an alarm and escape reaction in other members of the colony, who immediately disengage their probosces and drop to the ground. In certain species of sawflies, a disturbed larva emits a volatile substance that produces synchronous defensive movements in all the larvae on the leaf.

Primer pheromones act more slowly than releaser pheromones and act on the endocrine and reproductive systems of the recipient, inducing a series of physiological changes that either inhibit or enhance subsequent behavior. For example, in the migratory locust, the adult males secrete a volatile substance that increases the growth rate and causes synchronous development of young locusts. This results in large populations of locusts metamorphosing at the same time to form the immense migratory locust swarms.

Functions of Pheromone Communication. Sex pheromones, which stimulate sexual activity, comprise a wide assortment of chemical compounds. They may be produced by either sex but most commonly they are released by the female. In some instances, such as in certain beetles, both sexes release a pheromone to attract individuals of the species to aggregate on a host plant where they mate.

Assembly and Aggregation. Congregation in response to pheromones is common in many species, thus individuals of the same species participate in group activities, such as feeding, mating, or hibernation. Except in the social insects, these assemblies are temporary, such as the mating swarms of mayflies. Insect assembly is most striking in colonies of social insects tending the fertilized queen. Queens of army ant, fire ant, and termite colonies may be attended by hundreds of workers that are attracted to the queen because she is releasing pheromones.

In some eusocial insects (ants, termites, and some bees), trail pheromones are laid down by a foraging worker, presumably to facilitate its return to the nest. The worker that finds a plentiful food source returns to the colony and recruits other workers by the release of pheromones and sometimes other special stereotyped behavior. This *recruitment* is a means of communication to bring together other workers to perform specific work. Wasps also effectively use recruitment.

Control of Castes. In ants, bees, and termites, the queen produces in her mandibular glands what is termed a *queen substance* that inhibits ovarian development in the other castes and in the honeybees usually inhibits the construction of queen cells. This pheromone is transmitted orally from queen to worker. In the bees, wasps, and termites, other pheromones are important in caste determination.

Alarm and Alert. Alarm and alert pheromones are produced in response to situations presenting immediate or potential threat. The behavioral reaction elicited in many different insects generally ranges from defense to dispersal, agitation, aggregation, and recruitment. Alarm and alert chemical systems are most widespread in the social insects, particularly those species with large colonies. In pugnacious species, alarm pheromones produce an aggregative or clustering reaction in which members of the colony concentrate in the vicinity of the invader and attempt to repell it. In more timid species, alarm pheromones produce a dispersive reaction in which the colony members leave the nest, dragging the queen, the larvae and pupae, and the younger adults with them.

Spacing. At high population densities of both larval and mature stages of insects, the individuals are spaced remarkably uniformly in the limited space. This spacing generally results from the secretion of pheromones that serve as repellents or dispersers of the individual insects.

Identification. In the successful maintenance and propagation of a species, recognition of individuals of the same species is necessary for mating. In insects that are gregarious or have a social structure, a system of failproof recognition of individuals of the same species is extremely important. Generally, chemical signals, using olfactory receptors as well as tactile receptors in the antennae, serve in this identification process. A colony of honeybees with a mated laying queen will violently attack and kill a strange virgin queen that is introduced to the colony. Workers in the colony that may have come in contact with the stranger and become contaminated with her pheromones will also be attacked and killed.

Slave raiding. Many ant species raid the nests of other species and carry off the brood. When these mature in the captor's nest, they take on the food-gathering function for the colony. During the raids, the slave-making ants release a volatile pheromone that disarms the ants under attack.

Allelochemics

These are chemical substances that affect individuals of *other species* and include *allomones* and *kairomones*.

Allomones. These are substances that produce a deleterious effect on an individual of *another species* and are primarily defensive in nature. Examples include the foul-smelling emissions of both immature and adult stinkbugs and many other Hemiptera, of adult chrysopids, and of many others. The allomone of the adult bombardier beetle is in the form of a hot quinone spray delivered at the temperature of 100°C. Some allomone emissions contain cyanide.

Many kinds of insects have remarkable abilities to aim their defensive emissions in the direction of the predator. This is commonly accomplished by either twisting the body or aiming a special ejection nozzle so that the allomone spray goes in the direction from which the discharge stimulus came. Over 500 defensive chemical compounds have been identified, most of which are organic acids, alcohols, aldehydes, or paraquinones.

Kairomones. These, like allomones, produce a stimulus in an individual of *another species*. They differ from allomones in that the benefit is to the receiver and not the emitter. They include the tremendous array of odors and other substances produced by plants that aid phytophagous insects in finding their correct host plants. Insect predators and parasites of other animals (including insects) not only find their hosts but also become adjusted to the life cycles of the hosts. Thus many adult insect parasites locate their insect host by detecting odors produced by the host, and then lay an egg either in or on the host's body.

In many instances the same chemical may have a dual messenger function. For example, the sex pheromone of the bark beetle *Ips* is also a kairomone that attracts certain other beetles that prey on *Ips*. The population-regulating pheromone of the flour moth *Anagasta kueniella* acts also as a kairomone that stimulates prey searching in one of its wasp parasites.

RESPONSES AND TYPES OF BEHAVIOR

During their life cycle, insects usually change their behavioral responses because of at least three factors. Some modifications in behavior are the result of genetic factors acting on the processes of development and maturity; these are commonly called *stereotyped* or *innate behavior*. Other changes in behavior occur because of differences in environmental factors that lead to the modification of a stereotyped behavioral pattern through experience; this is called *learned behavior*. In many behavioral modifications, these two factors are not recognizable separately because the changes are the result of complex interactions of both genetic and environmental factors.

Stereotyped Behavior

Insects generally respond to a particular stimulus by one, or by a sequence, of characteristic, specific responses. The response reaction may be spontaneous the first time the insect encounters the stimulus; this is called stereotyped or innate behavior. Such behavior is inherent in the genetic composition of the species and is not learned. Reflexes, kineses, and taxes are types of stereotyped behavior.

Reflexes. The simplest response by an organism to a stimulus is a reflex, exemplified by the escape reaction of a cricket. Touching the cerci at the end of the abdomen causes the cricket to jump. In this reaction, the tactile organs of the cerci send an impulse along a nerve fiber to an interneuron in the last abdominal ganglion. The interneuron sends an impulse along the nerve cord to a thoracic ganglion that, in turn, stimulates nerves going to the legs, and thus trigger the jump reaction in the legs. In many flies and butterflies, when the front tarsi are touched with a sugar solution, the proboscis is extended. The tarsus does not move, but chemoreceptive sense organs on the tarsus send an impulse along their associated nerve cell to the thoracic ganglion, which stimulates an interneuron. This neuron sends an impulse along the nerve cord to the ganglion controlling feeding, where a motor nerve controlling proboscis extension is stimulated.

The pathway for a simple reflex involving stimulus reception and response is called a *reflex arc* and is shown in Fig. 6-15. The brain is not involved. The nerve messages proceed automatically following the original stimulus of the sense organ. Much insect behavior is of this type. Reflexes are commonly divided into two functional groups. *Phasic reflexes* are rapid, short-lived responses, involved in the rapid movements of one part of the body or the entire body. An example is the escape reaction of the cricket described above. *Tonic reflexes*, in contrast, are slow, long-lived responses that maintain the posture and the position of the insect's body in relation to its environment.

Kineses and Taxes. Directional responses to stimuli that result in movement by an insect so that it takes up a particular location or orientation are grouped as kineses and taxes. However, the interaction of environmental factors, particularly stress factors, may modify such behavioral patterns.

Kineses, exhibited by many insects, are random movements of an individual without any orientation to stimuli. A typical kinesis occurs in some cockroaches. When not otherwise stimulated, they wander around their abode in random fashion. This rate of wandering increases with the intensity of light. As a result, the cockroaches move more slowly in darker parts of their occupied area and faster in the lighter parts. This difference has the effect of causing a greater concentration of cockroaches in darker areas.

Taxes are directional movements toward or away from stimuli. A taxis differs from a kinesis in that the individual orients in a stereotyped way to a particular stimulus. Thus many moths will fly toward a light and are said to be positively phototactic. Full-grown maggots of many flies move away from the light source, being negatively phototactic. Taxes are especially important in orienting the individual in order to reach the correct spot in relation to its environmental needs, such as a food source or a place to pupate. Taxes are classified according to the type of stimulus that initiates them: for example, phototaxis—light; geotaxis—gravity; thigmotaxis—contact; chemotaxis—chemicals.

Phototaxis, Reaction to Light. Most insects have an extremely well-developed response to light, moving either toward the light source or away from it. Bees and wasps move toward light and are positively phototactic. An insect commonly reacts to light differently in various stages of its life cycle. Housefly maggots are negatively phototactic and move away from light, whereas adult houseflies are positively phototactic and move toward light. When swimming, some aquatic insects, such as mayfly nymphs, maintain their dorsoventral position (i.e., stay right side up) by orienting to light from above.

Some insects show a definite response not only to light in general but also to certain wavelengths of light. In most cases this aids in finding food or, in the case of ovipositing females, in placing their eggs on the correct type of foliage. Butterflies in search of food are guided by their perception of color in distinguishing yellow, red, and blue from green, and approach the flowers of the former colors in preference to the green foliage. But some of these same butterflies will lay eggs only on a green surface, which under natural conditions would be a healthy leaf suitable for larval food.

Geotaxis, Reaction to Gravity. Many insects when placed in a vertical tube will go steadily to the top or bottom and not wander haphazardly around the tube. Leafhoppers always go up. When the tube is inverted so that the insects are again at the bottom, they will start their upward climb again. This is a negative response to gravity, or negative geotaxis. Other insects have positive geotaxis, normally going down or toward the earth. Many soil-inhabiting larvae, such as wireworms, have this reaction. Thus if they hatch from eggs laid on or near the soil surface, they burrow down into the soil.

Thigmotaxis, Response to Contact. Many insects that normally live under bark, in soil, or in curled leaves have a well-developed touch or tactile reaction that causes them to remain in contact with some object (Fig. 6-23). This is known as positive thigmotaxis. Apparently the touch sensation inhibits activity, temporarily immobilizing the insect.

In all insects of active habits, the sense of touch serves as a detector of enemies. Frequently some area or structure at the apex of the abdomen, such as cerci, has tactile hairs of extreme sensitivity to aid in these "escape" reactions.

Chemotaxis, Response to Odors. The number of responses that insects make to various odors is legion. In relation to the environment, these are mostly correlated with searching for food, such as an insect locating food for its immediate use or a female finding a suitable place for laying eggs so that food needed by the developing immature forms is available (Fig. 6-24). In general, each insect is responsive to only the particular food odors that immediately concern the species. For example, butterfly females of the genus *Macroglossa* will oviposit only on a surface having the odor of the plant *Galium* on which the larvae feed. Other odors cause no oviposition response.

Some insects follow the trail of scent left by their prey. The braconid wasp *Microbracon* follows the scent trail of its host, the

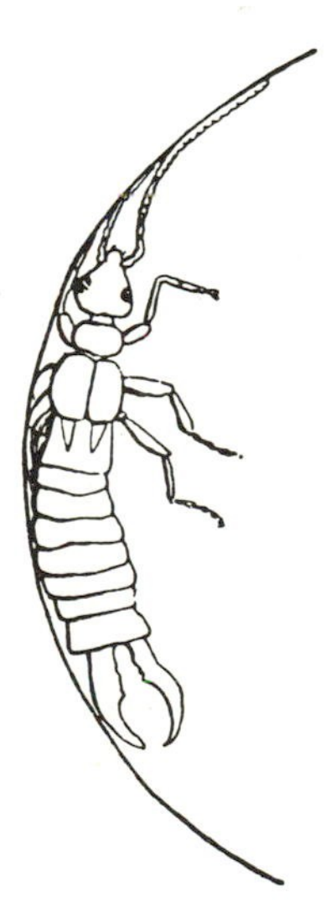

Fig. 6-23. An example of positive thigmotaxis. Position taken by the earwig *Forficula* in a circular container. (From Wigglesworth)

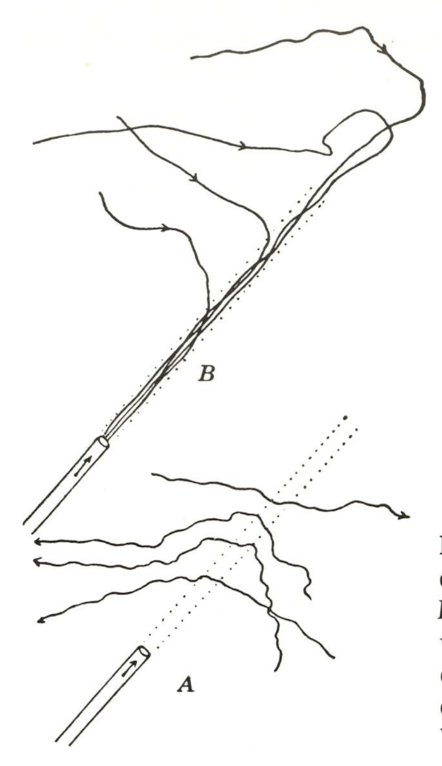

Fig. 6-24. An illustration of chemotaxis. Tracks followed by *Drosophila* flies (deprived of wings) when exposed to (*A*) an odorless stream of air, and (*B*) air carrying the odor of pears. (From Wigglesworth, after Flügge)

larva of the flour moth *Anagasta*, by running along with its antennae held close to the ground. Ants use this method to follow trails to and from the nest, which they locate by routes marked with chemical substances secreted and dropped by the ants themselves.

Thermotaxis and Hygrotaxis. Insects respond to various degrees of heat and humidity, moving toward the condition closer to their optimum. Insects that feed on warm-blooded animals use temperature as a guide to their hosts. Mosquitoes and bedbugs are positively thermotactic to temperatures near 98°F, that is, near mammal body temperature.

Coordinated Taxes. Many activities of insects are dependent on responses involving two or more taxes at the same time. For example, the ovipositing *Macroglossa* butterflies require both a green color and the odor of *Galium* to induce egg laying. Certain newly hatched caterpillars that feed in trees have both a negative geotaxis and a positive phototaxis, ensuring that the larvae travel upwards to the natural food.

Patterns of Reflexes and Taxes. Many activities of insects are essentially a correlated sequence of individual reflexes and taxes integrating sensory and motor events. The butterfly feeding pattern is an example. When hungry, the butterfly is positively phototactic to colors in the red to yellow range, orients to these colors, and eventually alights on a flower. This shuts off the locomotory response. Once on the flower, a pawing reaction begins, and a front tarsus eventually comes into contact with some nectar, which elicits proboscis extension and probing. When the chemoreceptors at the tip of the proboscis come into contact with the nectar, they trigger a reflex by which the sucking action of the food pump is started, and feeding is accomplished.

This series of events differs in an important respect from a simple reflex. Once the above sequence is started, it goes to completion without guidance from the brain. But the brain controls its starting and stopping. When the gut runs low on food, stretch receptors on the gut wall relay this information to the brain, which then *initiates* the feeding sequence. When the gut is full, the same stretch receptors convey this information to the brain, which then *inhibits* the feeding sequence. As in most organisms an insect ignores most of the sensory input from the environment most of the time and responds only to specific, selected stimuli.

Mating and egg laying are two other correlated sequences of actions. The usual mating behavior pattern involves both sexes. First, a receptive female releases a volatile sex pheromone. A male of the same species is stimulated by this scent and follows it to the female. When the female is located, the male begins an elaborate courtship ritual having a definite series of steps, such as "beating" the female with his antennae, fluttering his wings, then mounting the female. The female then extends part of the genital apparatus, the male grasps these with his genitalia, and mating begins. The sequence of actions followed by the male is under the control of the last abdominal ganglion, not of the brain. The brain, however, inhibits or initiates the sequence of actions depending on the stimulation provided by the female pheromone. In midges and some other insects the male responds initially to the sound produced by the female's wings in flight.

In some flies, moths, and butterflies, the male also produces a sex pheromone that is necessary before he is accepted by the calling female. In the boll weevil the roles are reversed: the male does the initial "calling" by pheromone release, and the receptive female tracks down the male.

There is a great variety of mating behavior patterns in insects, all of them specific for the group or species. In many flies, the male offers food or some oral secretion to the female. In many crickets, the male initiates the pattern through attraction to songs; in the fireflies he does so by a distinctive pattern of light flashes. In all of these, it appears that the action pattern is governed by the abdominal ganglion in association with the thoracic ganglia (controlling the leg or wing actions involved); the brain exercises only a release or inhibition control.

Learned Behavior

Learning may be defined as a lasting adaptive change in behavior as a result of experience or practice. Learning of several types has been demonstrated in insects, ranging from the simplest type, *habituation*, to those involving more complex learning behavior patterns. The ability to store information, called *memory*, is a necessary prerequisite that relates a stimulus and a response arising from a previous experience with that particular stimulus.

Habituation. This is considered to be the simplest form of learning. Habituation occurs when, on repeated exposure to a stimulus, the organism gradually decreases its response to the stimulus. The response may disappear completely. This lack of response may continue over long periods of time even though the stimulus is discontinued. As an example, an insect exposed to high-intensity light eventually may "tune out" the light. Habituation involves a loss of sensitivity to recurrent environmental stimuli that are unimportant in the life of the individual. Probably all insects are capable of habituation, and it is presumably an important part of behavior in most insects.

Associative Learning. Insects commonly display the ability to make learned associations between stimuli that normally produce no response and reinforced actions, such as reward or punishment. This is called associative learning. Bees can associate colors with food sources and learn to seek the "right" color rather than following the original olfactory cues. In nature, they probably rely heavily on ultraviolet light because many flowers have a central "target" ring or spot that emits ultraviolet light. Honeybees can also differentiate between certain forms of depressions and can be trained to associate these with food. Bumblebees learn the shapes of nectar-bearing plant species, and in food forays orient to the growth forms of plant species producing food at that time.

Latent Learning or Exploratory Learning. Many species of provisioning and social wasps and bees learn to locate their nests through recognizing and memorizing landmarks. In the first excursion from the nest site or colony, the foraging females or workers identify a series of landmarks that serve as cues in relocating the home site on the return trip. If a cue is removed experimentally, the homing individual has difficulty at the location of the missing cue, usually making a random or circular exploratory flight until the next cue is located. New substitute cues are soon learned. For longer distances up to at least a mile, many of these insects apparently navigate by determining the angle of the sun or the plane of polarization of the light in the sky.

Insight Learning. No insects are considered to have evolved the ability to take the next step beyond learning, that is the ability to reason. This would involve analyzing past experiences into a system of general information from which deductions could be made concerning the solution of new problems other than on a trial-and-error basis. Several examples have been proposed as reasoning in insects, but they generally appear to be behavior patterns with or without the addition of learning.

Two examples in the social insects that have been interpreted as showing reasoning involve the use of tools. Ants of the genus *Oecophylla* live in branches of tropical trees and construct tough nests of leaves sewn together. When adding to the nest, some ants hold the leaf edges together while another holds a larva that is brought in contact alternately first with one leaf edge, then the other, during which time the larva spins a silk strand that adheres to the leaf edges and sews them together. This sewing activity has been likened to the intelligent use of tools.

The use of tools has been invoked for another ant behavior pattern. Several species of the ant genus *Aphaenogaster* use leaf fragments as "tools" for transporting food that is too soft to carry in the mandibles to the nest. The ants place leaf fragments on this type of food, which adheres to the fragment; 30 to 60 minutes later they carry the fragments back to the nest. *Aphaenogaster* has a small abdomen, and it has been calculated that more liquid or gelatinous food adheres to the leaf fragment, commonly as long as the ant, than the ant could ingest at any one time.

ORGANIZATION OF SOCIAL INSECTS

Among the insects, the ultimate in behavioral complexity occurs in the social insects. These insects have a social organization that relies on reciprocal communication and cooperation among individuals living together as a colony in some type of nest. The truly social insects, called *eusocial* insects, belong to two orders. They include all the termites (Isoptera) and certain specialized members of the Hymenoptera, specifically all the ants and certain of the more highly organized wasps and bees. Eusocial behavior is characterized by three traits.

1. In a colony of a particular species, individuals cooperate in the care of the young.
2. The functions of colony activity are performed by specific groups of individuals.
3. Individuals from at least two generations overlap in their life cycles to contribute to colony activities, so the younger generation assists the parent generation during part of its life.

Presocial levels of organization are those levels where only one or two of the above traits are developed.

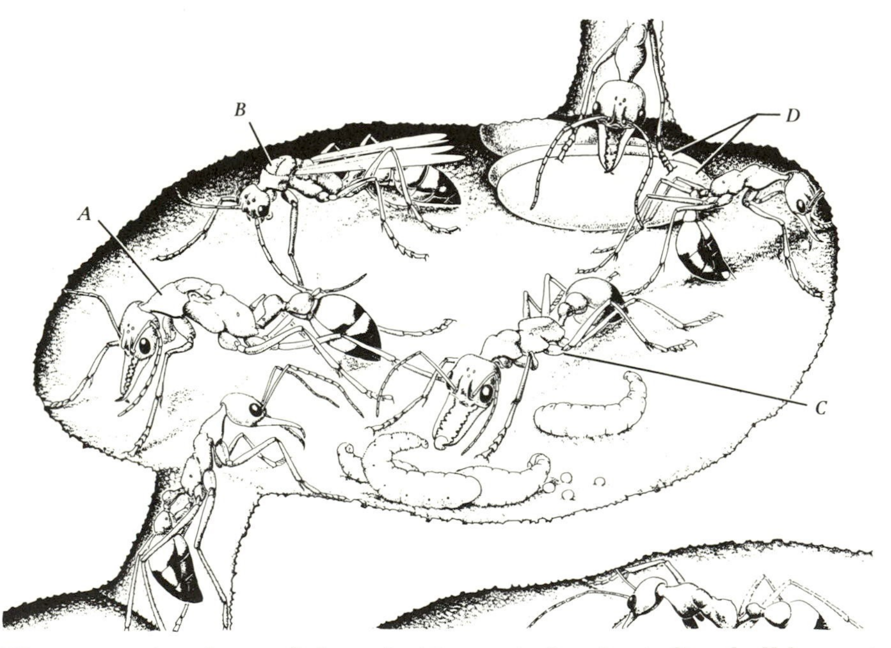

Fig. 6-25. A colony of the primitive ant, the Australian bulldog ant, *Myrmecia gulosa*, in an earthen nest. (*A*) Queen; (*B*) male; (*C*) worker offering food to larva; (*D*) cocoons with pupae. (From *The insect societies* by E. O. Wilson, 1971, original drawing by Sarah Landry; reprinted by permission of Harvard University Press. Copyright © 1971 by the President and Fellows of Harvard College.)

The eusocial insects live in colonies in which large numbers of individuals are produced. Each colony comprises from less than 100 to many millions of individuals, depending on the particular species and group of insects (Figs. 6-25, 6-26). A colony of the African driver ant may have as many as 22 million workers. In a group, such as the more highly evolved ants, for example *Myrmica rubra*, a colony contains a single queen who may live several years, who does all the egg laying for the entire colony, and who is fed by nonreproductive, sterile female individuals. The nonreproductive individuals perform many functions for the colony. Generally they have different morphological types or *castes* (see Chap. 8, termites, Figs. 8-20, 8-21; ants, Fig. 8-125; honeybee, Fig. 8-128). The smaller individuals are called *workers*, and the larger ones *soldiers* or *major workers*. The workers do the food gathering, nest excavating, and brood caretaking. The soldiers defend the nest and the vessels for storage of liquid

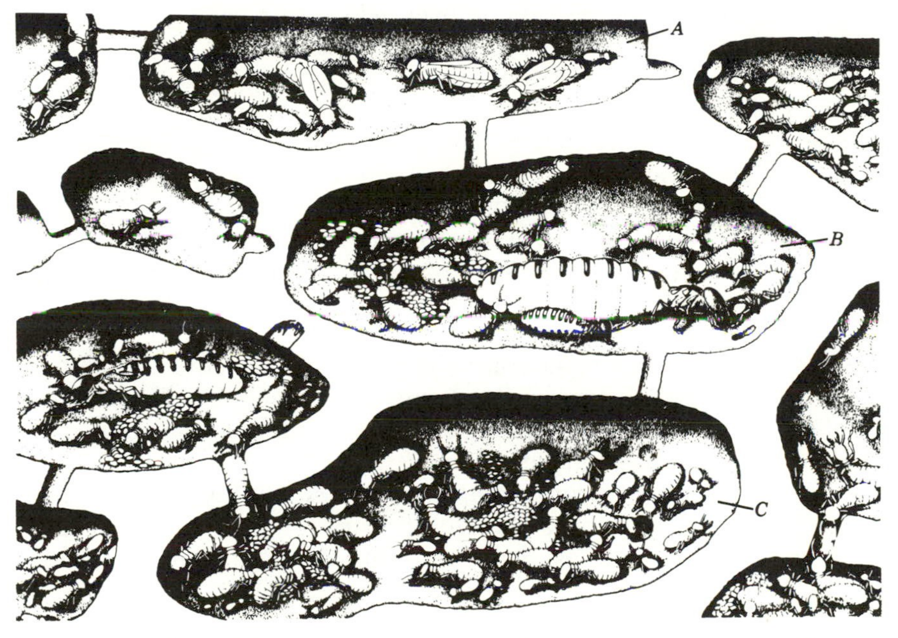

Fig. 6-26. Nest of the termite *Amitermes hastatus*. (*A*) Upper cell with reproductive nymphs. (*B*) Middle cell with primary queen and male beside her and numerous workers. (C) Lower cell with soldiers and nymphal soldiers. (From *The insect societies* by E. O. Wilson, 1971, original drawing by Sarah Landry; reprinted by permission of Harvard University Press. Copyright © 1971 by the President and Fellows of Harvard College.)

food. The nonreproductive individuals are relatively short-lived, and the queen must lay eggs relatively continuously to maintain the numbers of the colony once it has become a mature colony. In addition, *males* form another group in the colony. They perform virtually no work functions, some social functions (such as grooming), and wait for the nuptial flight to transfer sperm to the virgin queens. Virgin queens develop from broods laid by the reproductive queens. After the nuptial flight each young queen initiates her own new colony, building a nest and tending the brood after it is laid. As the colony matures, nonreproductive individuals take over the care of the brood and other colony functions.

In the evolution of social insects, two separate sequences of organization, *parasocial* and *subsocial*, are considered to have led to the eusocial level of organization (Fig. 6-27). The solitary (nonsocial) level has no cooperative brood care, no reproductive castes, and no overlap between generations. In the *parasocial sequence*, which has evolved in many of the halictid bees, adults that belong to the same generation assist one another to varying degrees. The lowest level of sociality is called *communal*. At this stage the adults cooperate in nest building but rear their brood separately. A number of species of bees of the family Halictidae are communal. Up to 50 females of a particular species share an underground cavity. Each bee apparently makes its own side tunnels or cells, in each of which an egg is laid, provisions added, and then capped. At the next level, *quasisocial*, the adults cooperate in caring for the brood but each female still lays eggs at some time in her life. At the succeeding *semisocial* stage, the quasisocial level is modified by the differentiation of a worker caste. This caste consists of nonreproductive individuals of the colony. The next level is the *eusocial* stage and is attained when individuals of semisocial colonies live long enough so that two or more generations overlap and cooperate.

The alternate sequence of *subsocial states* leading to eusocial insects is considered to be the evolutionary route taken by the ants, termites, social wasps, and a few groups of the social bees. Because

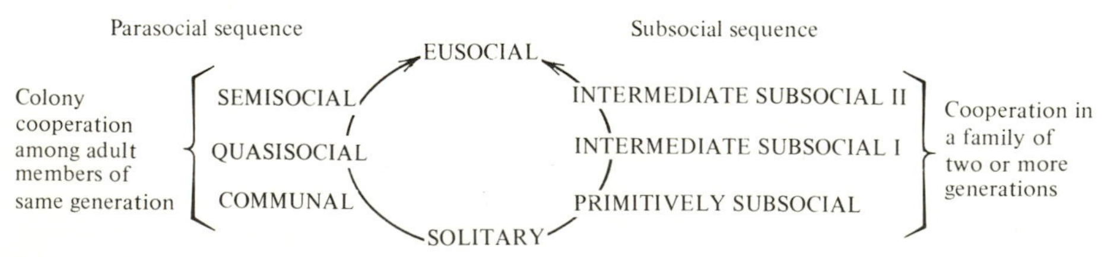

Fig. 6-27. The two sequences of organization, parasocial and subsocial, in eusocial insects.

all living ants and termites are eusocial, the gradation in the subsocial sequence has been studied in the wasps and certain bees. In this evolutionary sequence, an increasingly close association between parent and offspring develops. Following the solitary state and at the primitively *subsocial level*, a single female adult provides direct care to the brood for a time but she does not tend them through eclosion. At an *intermediate subsocial I* stage, the female is still present when the brood matures. At the *intermediate subsocial II* stage, the matured offspring help their parent in the raising of additional broods. The cooperation is between mother and offspring, not between sisters. After this stage, the group of helping offspring come to serve as permanent workers and *eusocial* organization is achieved.

Colonies of eusocial insects show two avenues of specialization. The first is an increase in numbers and degree of specialization of worker castes. The increased differentiation of worker castes may result from permanent anatomical changes between individuals so that there are a number of different morphological types. For example, in the ants, well-nourished worker ants develop large heads and mandibles and come to be part of the soldier caste. On the other hand, differentiation may arise from a series of changes in an individual so that the individual belongs to more than one caste during its lifetime. In the ant *Myrmica scabrinodis*, workers during their first season after emergence serve as nurses; in the succeeding season they become builders; when older they serve as foragers. This progressive change in function with age is best developed in honeybees.

The second avenue of specialization is communication in the colony so that the activities of the myriads of individuals are effectively coordinated. Communication, as noted earlier in this chapter, is highly developed in the social insects. Chemical communication is widely developed in the eusocial insects and includes the release of chemicals and tasting. To a lesser extent auditory communication is well developed also and includes stridulation, tapping, and many other signals. Exchange of liquid pheromones to inhibit caste development is just one of the many extraordinary features of the colonial system. The widespread communication systems elicit behavioral patterns that include alarm; attraction and assembly; recruitment for a new food source or nest site; grooming; trophallaxis (the exchange of oral and anal liquid); exchange of food particles; group interaction that either increases or inhibits a particular activity; identification and recognition of nestmates and members of particular castes; and caste determination, either inhibition or stimulation of caste differentiation.

The Presocial Insects

The eusocial insects are similar to their solitary progenitors in courtship, mating, and egg laying. But in the eusocial insects, many additional behavior patterns evolved as colony complexity increased. Presocial levels of organization include aggregation of individuals that cooperate to some extent in building nests or foraging for food and in brood care by the adults. The following are a few examples of presocial levels of organization.

Aggregation, Nest Building and Brood Care. Among the moths (Lepidoptera), there are species in which the larvae, hatching together from the same egg mass, construct a silken weblike nest, and all use it as a common abode. The nest is built around a fork or branch of a tree, all the larvae contributing to its construction. The larvae leave the nest during the day to feed on the foliage of the tree, and all return to it for resting. Larvae of the tent caterpillars (*Malacosoma*) live together in a nest for their entire larval period, leaving it finally to pupate. Larvae of fall webworms (*Hyphantria*) spin a similar nest but leave it and follow solitary existences for the last larval instar.

In the order Embioptera or web spinners, some species live in colonies that consist of interlocking silken tunnels in soil, surface cover, or the base of plants. The females exercise maternal care to a high degree, watching over the eggs and young nymphs. Some of these colonies form a solid silken mat over many square yards of ground and must contain thousands of individuals. So far, however, no closely structured relationship has been observed among individuals of such a colony and, therefore, the gregarious nature of the forms may mean little more in a social sense than the clustering aphids or scale insects on or about their parents and grandparents.

Exchange of Symbionts. From the standpoint of development toward social organization, a significant type of colony is found in the cockroaches. Wood cockroaches eat wood, which is digested by a specialized symbiotic protozoan fauna in the digestive tracts. When young cockroaches molt, they completely empty the digestive tract, so that after molting they have no symbiotic fauna, and the nymphs would soon starve if this situation were not remedied. The newly molted nymphs replenish their supply of Protozoa by eating some of the fresh excrement of another member of the colony. This necessary interchange of Protozoa requires that groups of individuals live together in a colony.

Brood Care. One prerequisite of social organization is the protection and feeding of larvae by adults. In certain Hemiptera, adults protect early young by pushing predators with forward-pointing processes on the head or pronotum. In the treehopper *Umbonia* this reaction is triggered by a pheromone released by an injured nymph. The females of certain earwig species deposit their masses of eggs in a sheltered chamber and guard them, driving away predators. After the eggs hatch, this watch is continued for a short time until the young nymphs are active enough to leave the brood chamber. At this point maternal care is discontinued, and each nymph goes its own way. Similar observations have been recorded for a few other insects, among them the mole crickets *Gryllotalpa*.

The Origin of Eusociality in Hymenoptera

In most insects, fertilized eggs develop into either females or males and each individual is diploid; that is, the individual has two sets of genes, one from each parent. Among the eusocial insects, termites have this typical diploid genetic system. However, in the other eusocial group, the Hymenoptera, all fertilized eggs develop into females, and they are diploid. All unfertilized eggs develop parthenogenetically into males and are haploid (Fig. 6-28). This reproductive pattern is called *haplodiploidy*. This different reproductive pattern leading to a different genetic composition of individuals in

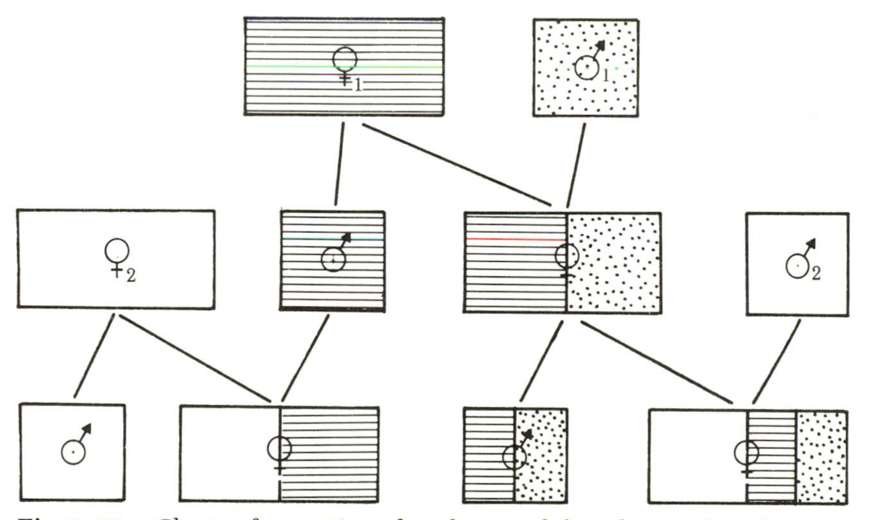

Fig. 6-28. Chart of genetic relatedness of females and males in a eusocial hymenopteran. The genetic contributions of the queen ($♀_1$) and male ($♂_1$) to their offspring are shown by hatched and stippled areas, respectively. (Redrawn from Crozier, 1979).

eusocial hymenopteran colonies is considered to explain why sociality arose at least 10 times in the Hymenoptera but only once in all other insects—that is, in the termites.

Haplodiploidy also appears to explain why hymenopteran societies have a dominance of females, in both numbers and function, whereas in the diploid termite societies, females and males share about equally in colony functions. In haplodiploid colonies of hymenopterans, sisters are more closely related genetically than a daughter and a mother. Sisters are also more closely related to each other than to their haploid brothers. The sisters, therefore, form a closely knit organization centered around assisting in the production of more sisters that are laid by a single fertile female, the queen. This may also explain why haploid hymenopteran males do not perform colony functions as do the diploid termite males.

Symbiosis in Social Insects

A great many organisms exploit the colonies of social insects. Many of these are insects and arthropods that live in the same nest as their host and obtain their nutrients from within the nest. This close and extended interaction between individuals of different species is called *symbiosis*. The individuals that live symbiotically in these nests and are the social insect symbionts are called *inquilines*. Representatives of inquilines include minute bristletails (Thysanura), a few fly larvae of the family Syrphidae, a number of butterfly larvae, a large number of beetles (notably of the families Staphylinidae and Pselaphidae), and some species of parasitic ants. Sometimes only a few inquiline species are present in a nest, but the total number of known ant and termite inquiline species runs into the thousands.

Some of the inquilines are small and extremely fast. These undoubtedly survive in the nest by agility alone. Others are fairly large and slower, often sluggish. Certain species produce various secretions that are eaten by ant hosts. Some species of staphylinid beetles have evolved large glandular areas on the abdomen that the ants lick, then in return regurgitate food that the beetle eats. Thus a definite amicable relationship exists between the host and at least some inquilines, and this is maintained by stereotyped behavior patterns. The chemical language, other than food exchange, that may exist between host and inquiline is not known. When an ant colony moves to a new nest, the inquilines commonly move with the emigrant column, and they follow ant trails. This suggests that the inquilines respond to at least some pieces of the language of the ant colony.

REFERENCES

Beroza, M. (Ed.), 1970. *Chemicals controlling insect behavior.* New York and London: Academic Press. 170 pp.

Birch, M. C. (Ed.), 1974. *Pheromones.* Amsterdam and London: North-Holland. 495 pp.

Chapman, R. F., 1969. *The insects, structure and function.* New York: American Elsevier. 819 pp.

Eisner, T., and E. O. Wilson (Eds.), 1977. *The insects.* San Francisco: W. H. Freeman. 334 pp.

Evans, H. E., and M. J. W. Eberhard, 1970. *The wasps.* Ann Arbor: Univ. of Michigan Press. 265 pp.

Fraenkel, G. S., and D. L. Gunn, 1962. *The orientation of animals,* 2nd ed. New York: Dover. 376 pp.

Heinrich, B., 1979. *Bumblebee economics.* Cambridge, Mass.: Harvard Univ. Press. 345 pp.

Hermann, H. R. (Ed.), 1979. *Social insects.* New York: Academic Press. Vol. 1, 438 pp.

Horridge, G. A. (Ed.), 1975. *The compound eye and vision of insects.* Oxford: Clarendon Press. 595 pp.

Iwata, K., 1976. *Evolution of instinct: Comparative ethology of Hymenoptera* (Trans. from Japanese), Smithsonian Inst. and National Science Foundation. New Delhi: Amerind. 535 pp.

Jacobson, M., 1972. *Insect sex pheromones.* New York and London: Academic Press. 382 pp.

Johnson, C. G., 1969. *Migration and dispersal of insects by flights.* London: Methuen. 763 pp.

Krishna, K., and F. M. Weesner (Eds.), 1969, 1970. *Biology of the termites.* New York: Academic Press. Vols. I and II.

Lipke, H., and G. Fraenkel, 1956. Insect nutrition. *Annu. Rev. Entomol.,* **1**:17–44.

Matthews, R. W., and J. R. Matthews, 1978. *Insect behavior.* New York: Wiley. 507 pp.

Michener, C. D., 1974. *The social behavior of bees: A comparative study.* Cambridge, Mass.: Belknap Press of Harvard Univ. Press. 404 pp.

Nachtigall, W., 1974. *Insects in flight* (Trans. Harold Oldroyd, R. H. Abbott, and M. Biederman-Thorson). London: Allen & Unwin. 153 pp.

Novák, V. J. A., 1975. *Insect hormones.* London: Chapman & Hall. 600 pp.

Rainey, R. C. (Ed.), 1976. Insect flight. *Symp. R. Entomol. Soc. London,* **7**:1–287.

Romoser, W. S., 1973. *The science of entomology.* New York: Macmillan. 449 pp.

Saunders, D. S., 1976. *Insect clocks.* Oxford: Pergamon. 279 pp.

Schneirla, T. C. (H. R. Topoff, Ed.), 1971. *Army ants.* San Francisco: W. H. Freeman. 349 pp.

Shorey, H. H., 1976. *Animal communication by pheromones.* New York: Academic Press. 167 pp.

Treherne, J. E., and J. W. L. Beament, (Eds.), 1965. *The physiology of the insect central nervous system*. London and New York: Academic Press. 277 pp.

Truman, J. W., and Riddiford, L. M., 1974. Hormonal mechanisms underlying insect behavior. In J. E. Treherne, J. J. Berridge, and V. B. Wigglesworth (Eds.), *Advances in insect physiology*, **10**:297–352. London, New York: Academic Press.

Wilson, E. O., 1971. *The insect societies*. Cambridge, Mass.: Belknap Press of Harvard Univ. Press. 548 pp.

See also references listed under Life Processes, The Orders of Ectognaths and Insects, and Ecological Considerations.

7
Life Cycles, Growth, and Reproduction

The start of differentiation within an egg signals the beginning of a long series of developmental changes in a growth process that leads ultimately to maturity and reproduction. This chain of events, from egg to mature adult, constitutes the life cycle of the individual. In the Insecta, there are many types of life cycles, involving different methods of development and reproduction, different relations of one generation to another, and in some species alternation of food or habitat between different divisions of a single life cycle.

DEVELOPMENT

The life cycle of the individual usually has two phases: development (from egg to adult), and maturity or adulthood. During development, the insect concentrates its energies mainly on growth, sacrificing much of its exoskeleton and much of its mobility. In contrast, the adult insect concentrates its energies on dispersal and reproduction. Because most insects start as eggs, the most universally important division point of insect development is the phenomenon of hatching from the egg. The development period within the egg is *embryonic development*; the period after hatching is *postembryonic development*. Change of form during this latter period is termed *metamorphosis*.

Embryology

The Egg, or Ovum. Insect eggs are of many shapes (Fig. 7-1); many of them are simple smooth ellipses; others may be ribbed or

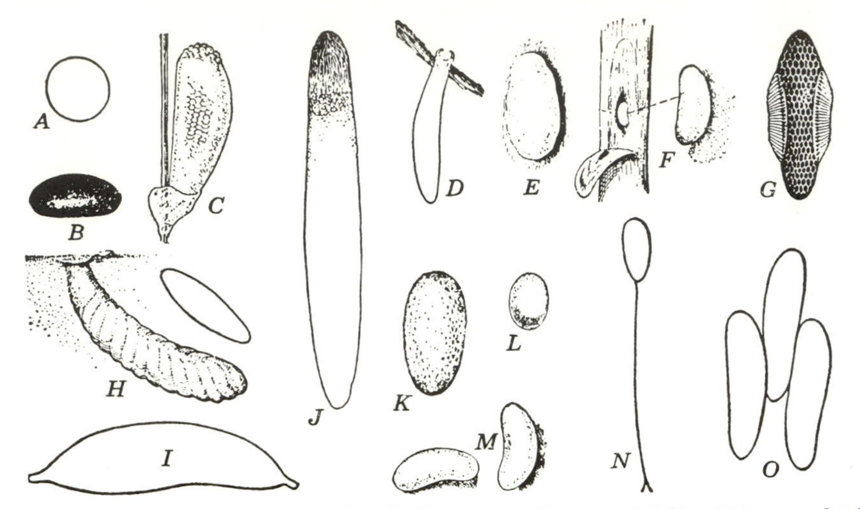

Fig. 7-1. Eggs. (*A*) A collembolan *Sminthurus viridis*; (*B*) an aphid *Toxoptera graminum*; (C) the sucking cattle louse *Solenopotes capillatus*, attached to a hair; (*D*) apple mirid *Paracalocoris colon*, in plant tissues; (*E*) the ladybird beetle *Hyperaspis binotata*; (*F*) a weevil *Sphenophorus phoeniciensis*; (G) a malarial mosquito *Anopheles maculipennis*; (*H*) grasshopper egg pod in the soil and a single egg; (*I*) an ichneumon wasp *Diachasma tryoni*; (*J*) a damselfly *Archilestes californica*, removed from water plant; (K) webbing clothes moth *Tineola bisselliella;* (L) dog flea *Ctenocephalides canis;* (*M*) pear thrips *Taeniothrips inconsequens*, removed from tissues of plant; (*N*) a lacewing *Chrysopa oculata;* (O) housefly *Musca domestica.* (After Essig from various authors)

sculptured in various ways; others are provided with processes of different kinds, such as the lateral floats of *Anopheles* eggs (Fig. 7-1G), which keep them buoyed up in water.

A typical egg (Fig. 7-2) is a bilaterally symmetrical cell encased in two coverings. The outer covering is a tough membrane, the *chorion*, and the inner is a delicate membrane, the *vitelline membrane*. Minute pores, or *micropyles*, in these membranes permit sperm entry to effect fertilization of the egg. The two membranes surround a large nucleus and a mass of cytoplasm. The cytoplasm consists of a large central area of yolk (essentially a food store) and a peripheral bounding layer, the *periplasm*, beneath the vitelline membrane. This layer is denser than the central part of the cytoplasm.

Early Cleavage. In the typical sexually reproducing insect, fertilization does not occur until the eggs are about to be laid. One to several sperm enter the egg through its micropylar openings as the egg passes down the oviduct. Following formation of the zygote, the nucleus divides, but this is not accompanied by cell division. The early divisions of the nuclei and their associated mass of cytoplasm are at first scattered throughout the yolk (Fig. 7-2*B*, *C*). This process is called *cleavage*. Each daughter nucleus gathers around it a circle of cytoplasm and the nucleus and associated cytoplasm are collectively called an *energid*. Further divisions of the energids are followed by their migration to the periplasm (Fig. 7-2*D* to *F*), where a membrane forms around each of them to form cells. Subsequent cleavage in the egg occurs only at the periplasmic layer of the egg and does not penetrate into the yolky, noncellular interior. This type of cleavage is called *superficial cleavage*. The cells make a wall

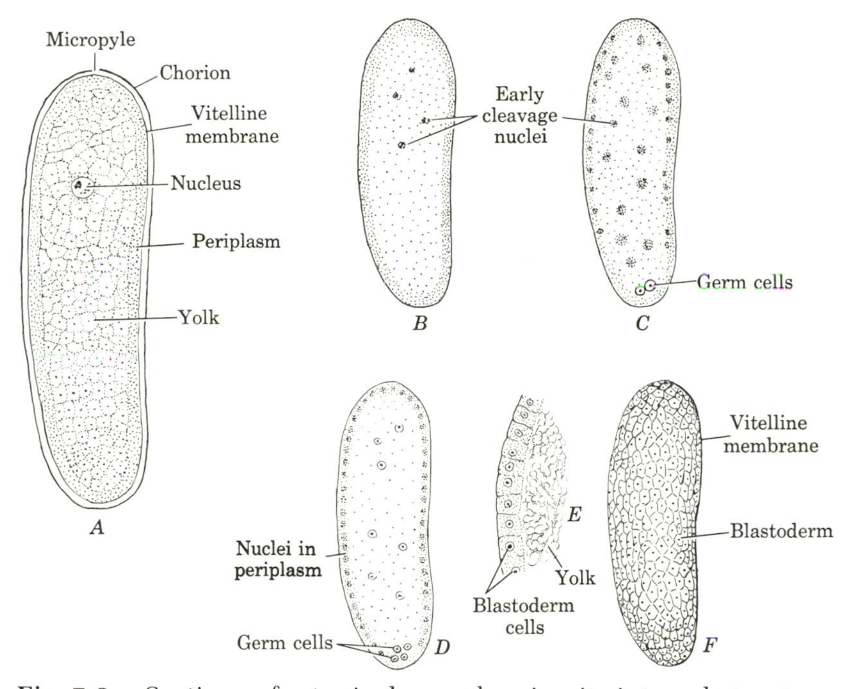

Fig. 7-2. Sections of a typical egg, showing its internal structure and early cleavage stages. (*A*) Fertilized egg; (*B*) after a few divisions of the nucleus; (*C*) after many cleavages (note migration of many nuclei to periphery); (*D*) after cleavage nuclei form a definite layer at periphery; (*E*, *F*) internal and surface views after energids have concentrated at periplasm to form a layer of cells, the blastoderm. Note the early segregation of special germ cells at the posterior pole of the egg. (Redrawn from Snodgrass, 1935)

around the egg, one cell layer thick, and this early embryonic stage is called a *blastoderm*. In the ventral region, the cells are crowded to make a thicker area, the ventral plate or *germ band* (Fig. 7-3). This is the first organized form of the embryo. In some cases it is quite extensive, lining a considerable area of the egg. In others it forms only a small platelike area that may be termed a *germ disc* rather than a germ band (Fig. 7-4, *I*).

The majority of insects have superficial cleavage. In the entognath order Collembola, the egg has little yolk and the entire egg divides during the early cleavages. This is the only order of hexapods known to exhibit *holoblastic cleavage*.

Growth of the Embryo. The germ band or disc grows by cell division and differentiation. At first, growth is largely an increase in the size of the germ disc, but this is followed rapidly by a surface partitioning whereby the body segmentation and appendages are set out. This is illustrated in Fig. 7-4, which portrays graphically the gross changes leading to the formation of the completed embryo and the first-stage nymph (Fig. 7-5).

Segmentation and Appendages. The development of these two features in the embryo parallels to a considerable extent the supposed evolutionary history of the insect group.

The body segments are first formed in the embryo by a series of transverse incisions (Fig. 7-4, IV). The segments that ultimately bear the mouthparts and fuse with the head structure appear originally as similar to the posterior segments. It is not until the appendages are well developed that the mandibular, maxillary, and labial segments become fused with the head structure (Fig. 7-4, VII).

Appendages begin to develop soon after segmentation is evident (Fig. 7-4, V). Typically each segment develops a pair of ventral appendages, but most of the abdominal appendages become only poorly defined. In many groups of insects all the abdominal appendages except the cerci are never more than small rudiments, which

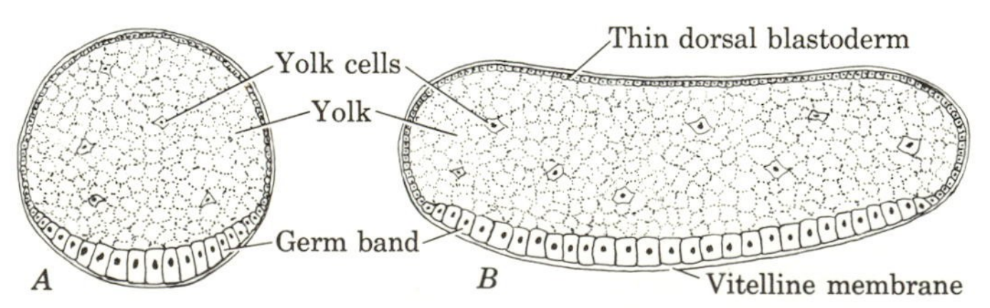

Fig. 7-3. One type of formation of the germ band on the ventral side of the blastoderm. (*A*) Cross section; (*B*) longitudinal section. (Redrawn from Snodgrass)

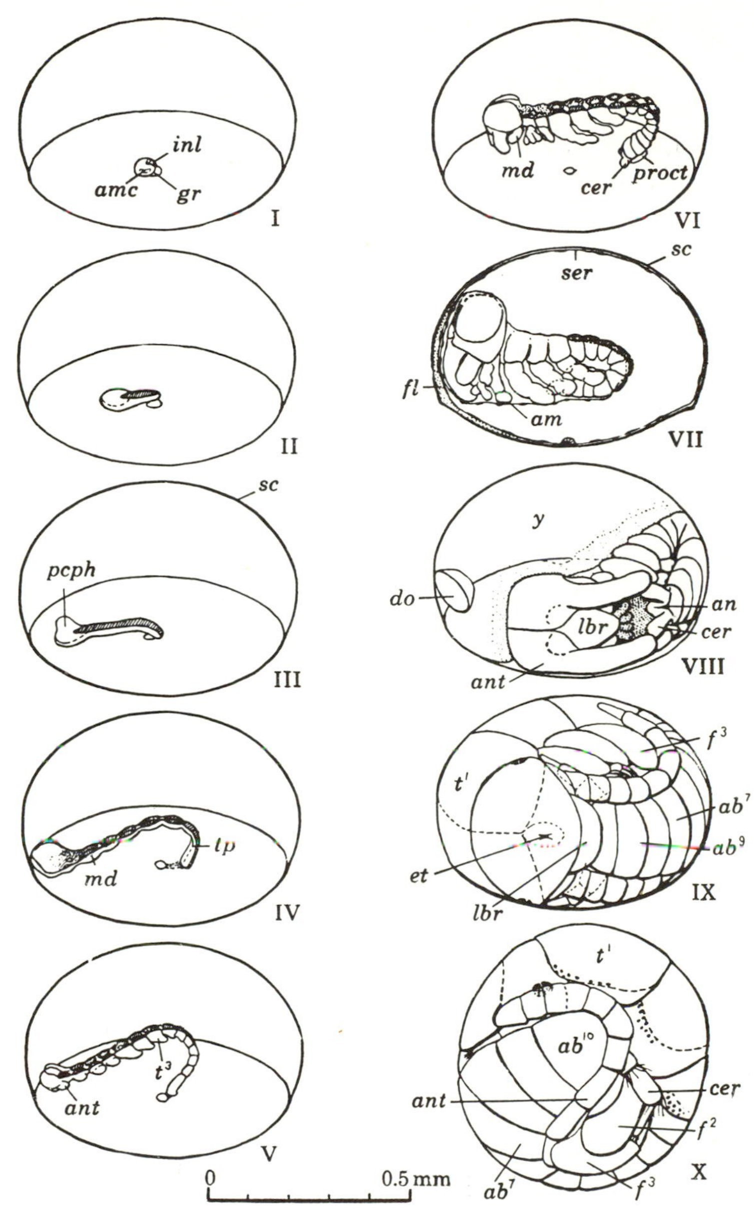

Fig. 7-4. Embryonic growth and segmentation in a stonefly. *ab*, Abdominal segment; *am*, amnion; *amc*, amniotic cavity; *an*, anus; *ant*, antenna; *cer*, cercus; *do*, dorsal organ; *et*, egg tooth; *f*, femur; *gr*, grumulus; *inl*, inner layer; *lbr*, labrum; *md*, mandible or mandibular segment; *pcph*, protocephalon (prostomium plus first post-oral segment); *proct*, proctodeum; *sc*, serosal cuticle; *ser*, serosa; *t*, thoracic segment; *tp*, tail piece; *y*, yolk. (After Miller)

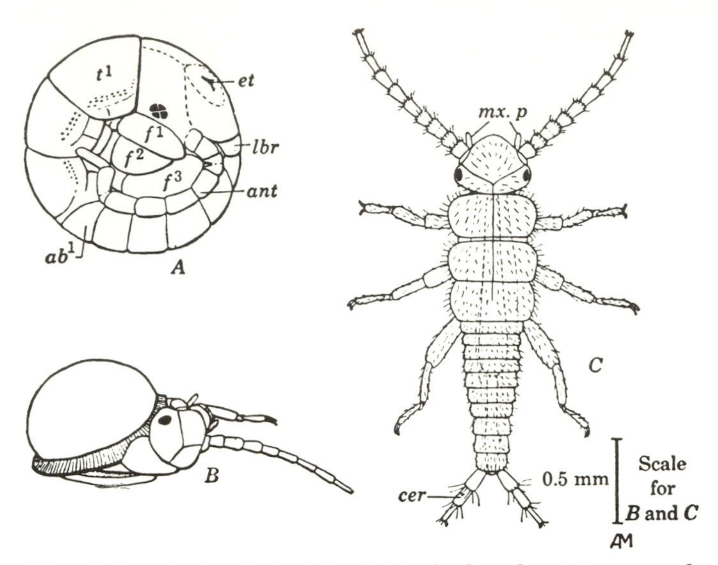

Fig. 7-5. End of embryological development in the stonefly. (*A*) Embryo in egg; (*B*) completed embryo escaping from eggshell; (*C*) newly hatched nymph. Abbreviations as for Fig. 7-4. (After Miller)

are reabsorbed at an early stage. The anterior segments and appendages develop more rapidly than the posterior ones, so that as a rule the embryo at this stage appears as in Fig. 7-6.

Body Shape. In the early stages only the appendages and the ventral portion of the body are formed. Thus in Fig. 7-4, V and VI, there are no sides or dorsum, the embryo representing the appendages and the sternal region. In essence, the body is open on top. During later growth the sides grow out and up, first over the anterior and posterior ends. Figure 7-4, VIII, shows a stage when the head is closed dorsally and also the posterior four of five segments of the abdomen; the embryo rests so that the open "top" of the intervening segments is pressed against the yolk, which fills the rest of the egg. From this stage the lateral margins of the open segments grow out and up along the sides of the egg until they meet dorsally, inclosing the yolk and completing the body closure.

Germ Layers; Tissue Determination. At first the germ band consists of only a single layer of cells, but early in embryonic life it forms a second layer. This is usually formed by *gastrulation*, that is, the infolding of a section of the germ band. In insects the common

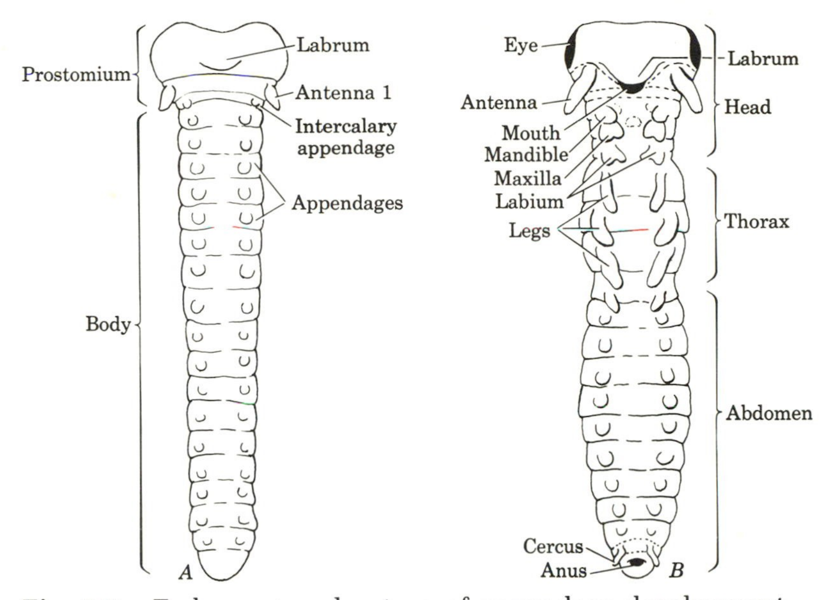

Fig. 7-6. Embryo at early stage of appendage development. (*A*) Early embryo having only small appendages on each but the last segment; (*B*) later stage in which the head and thoracic areas and their appendages are better developed. (Redrawn from Snodgrass)

methods by which this is achieved are shown diagrammatically in Fig. 7-7. Gastrulation usually begins as a longitudinal groove of the germ band (*A*); the outside edges of the groove grow toward each other, and the future second layer proliferates inward (*B*); finally the edges of the groove meet and fuse to form an outer layer or *ectoderm*. The inner layer or *mesoderm* spreads out above it (C). In Fig. 7-4, I to V, the mesoderm is shown as a darkened dorsal area. At each end of the mesoderm an invaginated cluster of cells forms, these clusters being the *endoderm*, or third germ layer.

The groups of cells that later develop into specific body segments, organs, or appendages are determined very early in embryonic life. Extensive experiments have shown that most of this determination is established by the time the germ layers are formed. A group of such apparently unspecialized cells, which are nevertheless destined to grow into a particular structure, is called an *anlage* (pl. *anlagen*).

The ectoderm gives rise to the body wall, the foregut and hindgut of the digestive tract, the nervous system, tracheal system, and many glands. The mesoderm gives rise to the muscular system, the gonads,

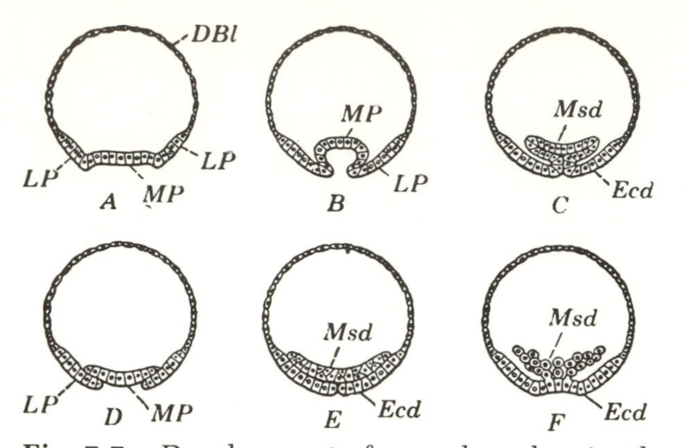

Fig. 7-7. Development of mesoderm by simple gastrulation methods. (*A*) Cross section of egg with germ band differentiated into lateral plates (*LP*) and middle plate (*MP*). (*B*) Later stage of same with middle plate curved in to form a tubular groove, edges of lateral plates coming together below it. (*C*) Still later stage, with edges of lateral plates united, forming the ectoderm (*Ecd*), and middle plate spread out above the latter as internal layer of cells, the mesoderm (*Msd*). (*D*, *E*) Second method of mesoderm formation; middle plate, separated from edges of lateral plates, become mesoderm (*Msd*) when lateral plates unite beneath it. (*F*) Third method, in which mesoderm (*Msd*) is formed of cells given off from inner ends of middle plate cells. (After Snodgrass)

the heart, and the fat body, and the endoderm gives rise to the mesenteron.

Embryonic Coverings. During much of its development, the embryo becomes partially or entirely immersed in the yolk, presumably for protection, and a pair of membranes form around it. The two principal methods followed are illustrated in Fig. 7-8. In the first, the embryo slides tail first into the yolk, pulling the membrane with it (Fig. 7-8*D*). When the embryo is completely immersed (Fig. 7-8*E*), the membrane grows over the end of the entrance cavity to form two final membranes; the outer one is the *serosa*, and the inner one (which encloses a space around the ventral aspect of the embryo) is the *amnion*. The second method by which immersion occurs in-

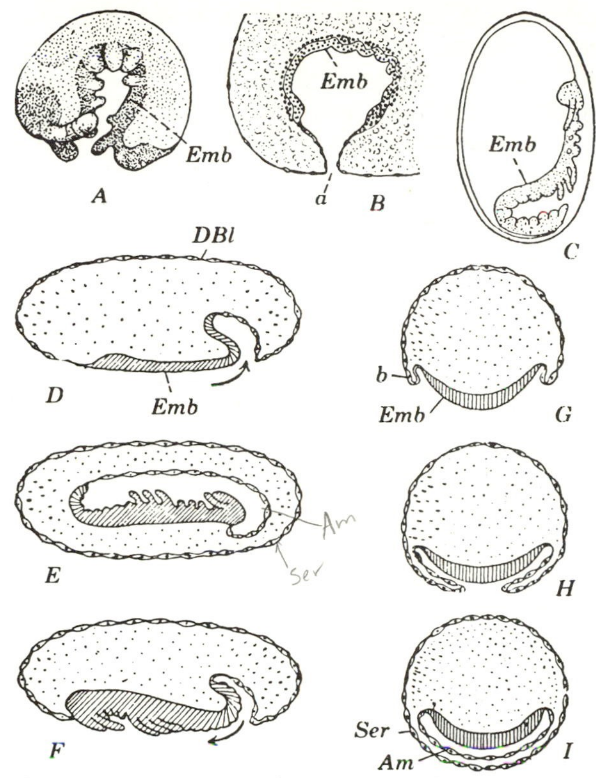

Fig. 7-8. Diagrams of position and movement of embryo within the egg, illustrating three methods. (*A*) Embryo (*Emb*) of a springtail *Isotoma cinerea*, curved into the yolk on underside of egg. (*B, C*), Embryo of a silverfish *Lepisma:* first (*B*), at early stage when deeply sunken into yolk near posterior end of egg, the opening of the cavity closed to a small pore (*a*): and second (*C*), in later stage when partially revolved to outside of egg, in which position it completes its development. (*D* to *F*) Lengthwise sections of an egg in which the embryo revolves rear end first into the yolk (*D*), becoming entirely shut in the latter (*E*) in reversed and inverted position, and then again revolves to surface (*F*) in original position before hatching. (*G* to *I*) Cross sections of an egg in which embryo becomes covered by membranes (*G*) originating in folds of the blastoderm around its edges, *b*, the folds extending beneath the embryo (*H*), and finally uniting to form two membranes (*I*), the outer, serosa (*Ser*), the inner, the amnion (*Am*). (From Snodgrass, *A* after Philiptschenko, *B* after *Heymons*)

volves only a sinking of the embryo into the yolk (Fig. 7-8G to *I*); the membrane grows over the ventral area, and the opposing membrane edges unite to form the amnion and serosa. At a later stage of development the embryo changes position again, breaking through the two membranes and assuming a position with its back to the yolk, as in Fig. 7-8*F*.

Formation of Digestive System. A detailed discussion of the origin and formation of the various insect organs is beyond the scope of this book. The formation of the digestive tract, however, is of unusual interest, and a brief outline of its early growth is illustrative of the general fashion in which organs arise. The successive stages in the formation of the digestive tract are shown diagrammatically in Fig. 7-9. In Fig. 7-9*A* there are two masses of endoderm cells, the anterior midgut rudiment, and the posterior midgut rudiment, growing inward from each end of the embryo. In Fig. 7-9*B* each of these rudiments has begun the formation of a sac, open toward the middle of the body, and beginning to inclose the central yolk mass; at the same time the ectoderm at each end has invaginated to form the beginnings of the anterior and posterior parts of the digestive tract. In Fig. 7-9*C* these developments have continued to a further stage. The completed structure is shown in Fig. 7-9*D*; the anterior and posterior sacs of the midgut have joined, completely inclosing the remains of the yolk; and openings have formed connecting the midgut with the anterior and posterior ectodermal invaginations. The digestive tract thus has three distinctive areas: (1) the anterior foregut formed of ectoderm; (2) the central midgut, of endodermal origin; and (3) the posterior hindgut, of ectodermal origin.

Hatching or Eclosion. When the embryo is full grown and ready to leave the egg, or hatch, it must force its way through the eggshell or chorion by its own efforts. Prior to hatching, the embryo may swallow air or the amniotic egg fluid to attain greater bulk or turgidity. In

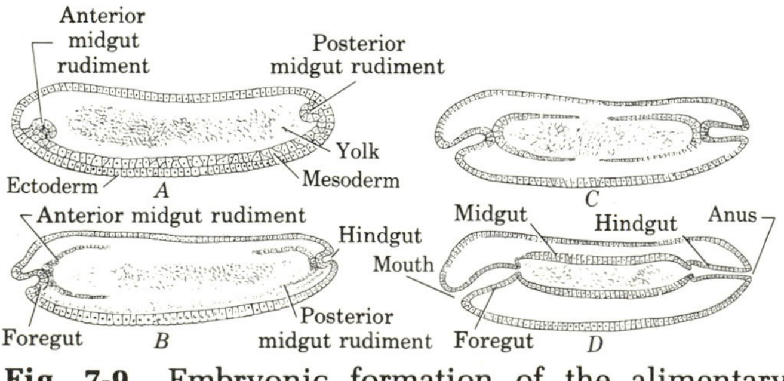

Fig. 7-9. Embryonic formation of the alimentary canal. (Redrawn from Snodgrass)

hatching, the embryo produces rhythmic muscular activity and presses against the shell or strikes it repeatedly with its head.

In some insects, such as grasshoppers, the embryo simply forces a rent in the anterior part of the eggshell. In others, such as many Hemiptera and certain stoneflies (Fig. 7-5*B*), a portion of the egg forms an easily detached cap, which the embryo pushes open like a lid. In a third group, the anterior part of the embryo is armed with an egg burster, which may be a sclerotized saw, spine, or some blades that pierce the chorion to produce the initial tear.

Once the shell is broken, the embryo works its way out of the egg. In many cases the nymph is encased in an embryonic covering or pronymphal membrane that is molted when the nymph is part way out of the egg. The cast skin remains inside or protruding from the egg. The egg burster is a thickening of this pronymphal membrane. When free from the egg and its embryonic coverings, the embryo is considered as the first-stage nymph or larva of the postembryonic period (Fig. 7-5*C*).

Polyembryony. The eggs of certain parasitic Hymenoptera frequently produce more than one embryo by a process of asexual division. In *Platygaster hiemalis*, a parasite of the Hessian fly, each fertilized egg may develop two embryos; in *Macrocentrus gifuensis*, a member of the Braconidae, each fertilized egg may develop several embryos; and in other species each fertilized egg may produce 100 to 3000 embryos. Each of these embryos develops into an active larva. The division into multiple embryos takes place before any other embryonic development occurs. The nucleus of the original zygote divides by mitosis into a number of daughter nuclei and each of these then develops into an embryo. The embryos may form either irregular groups or long chains. By means of polyembryony, a small parasite can insert a single egg into a large host and from that one egg produce enough offspring to take advantage of the great food possibilities of the host. An example of this is the minute chalcid, *Litomastix truncatellus*, which parasitizes large Lepidoptera larvae (Fig. 7-10); from only a small number of eggs over 2000 larvae usually develop in each caterpillar, and these allow no part of the host to go to waste.

Postembryonic Development; Metamorphosis

The change in form from that of a larva to the adult is called metamorphosis. These changes can dramatically alter the physical appearance of an insect from its earlier embryonic form. In some insects, the transformation of the larva into a nonfeeding stage, the *pupa*, prior to emerging as an adult, is an example of such a dramatic

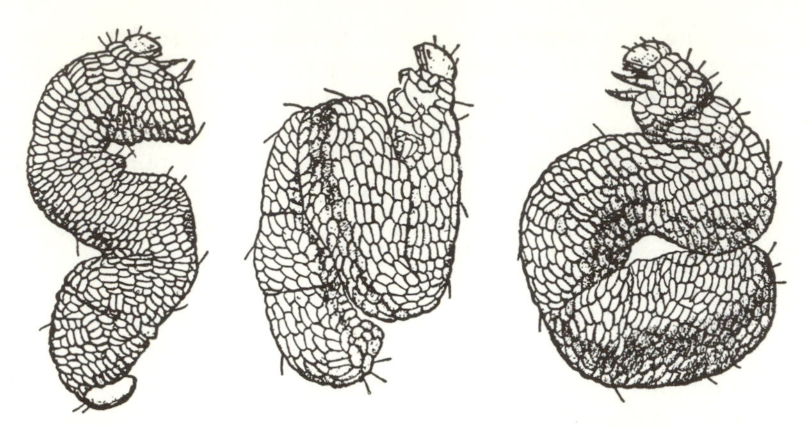

Fig. 7-10. Polyembryony in the chalcid *Litomastix*; the 2000 parasite larvae in each host caterpillar all developed from one egg. (From Clausen, 1940, after Sylvestri, *Entomophagous insects*, by permission of McGraw-Hill Book Co.)

change in form. Other insects, such as dragonflies, although they do not have a pupal stage, undergo striking changes in form during metamorphosis. In many insects, the process of growth involves periodic molting, which is discussed on page 000. Most insects molt at least three or four times, and in some there are 30 or more molts during normal development. The average is 5 or 6 molts.

The process of molting is called *ecdysis*. The old skins cast off by the insects are called *exuviae*.

Instar and Stadium. With few exceptions the molts for each species follow a definite sequence as to number, duration of time between them, and the increase in size accompanying them. The total period between any two molts is called a *stadium*. The actual insect during a stadium is termed an *instar*. Thus from time of hatching until the first molt is the first stadium. Any individual that is in this period of development is called the first instar. To put it another way, we might say of a species that the first stadium is 5 days and that the first instar is slender and yellow. In all but the primitive wingless orders, no further molts occur after the functional adult stage is reached.

Adulthood. The *adult*, or *imago*, is the stage having fully developed and functional reproductive organs and associated mating or egg-laying structures. In winged species it is the stage bearing functional wings. The only known exception to this latter is the mayfly order, Ephemeroptera, in which the stage before the winged reproductive also has wings and uses them; this curious flying preadult instar is called a *subimago*.

Metamorphosis. In the primitive wingless orders, known as the Apterygota, the newly hatched young grow into adults with little change other than in body proportions and development of the reproductive organs. These *ametabolous* forms show no distinctive external changes during metamorphosis. In the winged insects or Pterygota, the newly hatched young are still relatively simple creatures but the adults have wings and special body sclerites associated with them, so the adults are quite different in appearance from early stages of the young. This condition of exhibiting different forms in different stages of the life history is known as metamorphosis.

Gradual Metamorphosis. In the more primitive winged insects, the wings first appear in about the third instar as slight backward outgrowths of the second and third thoracic nota. These wing pads increase in size with each subsequent molt (Fig. 7-11), in the last immature instar forming pads that may extend over several segments of the abdomen. Up to this point no change occurs in the sclerites of the thorax. The complete adult structures develop inside the last

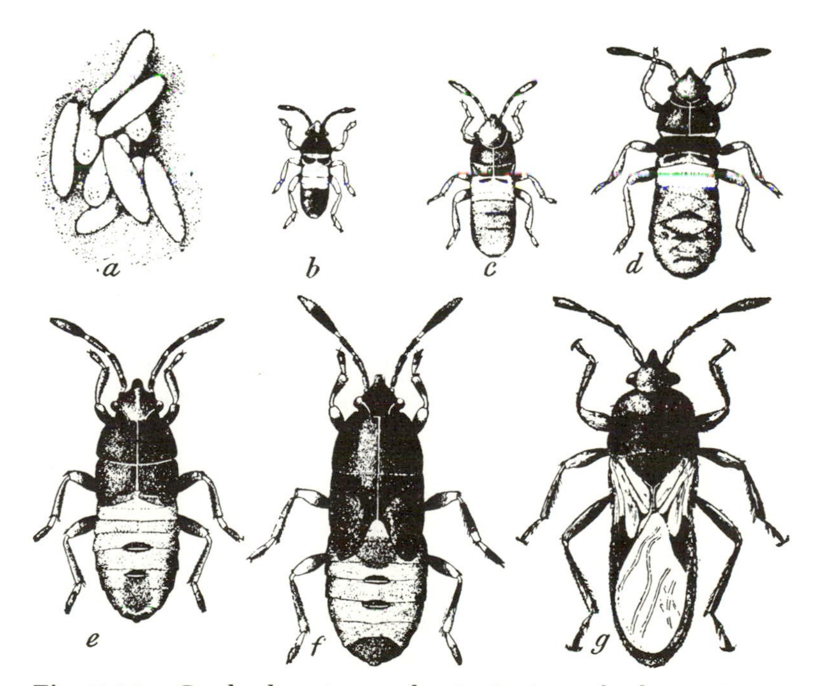

Fig. 7-11. Gradual metamorphosis; instars of a hemipteran, the chinch bug. (From U.S.D.A., E.R.B.)

immature instar, and the adult emerges fully formed. At emergence the adult wings are somewhat soft and pleated, but these soon straighten and harden into their functional condition.

This type of direct development in which the wings develop gradually as external pads is termed *gradual metamorphosis*. It is also called simple or partial metamorphosis, and *paurometabolous development*. The immature stages commonly are called *nymphs*.

The cockroaches, grasshoppers, stoneflies, leafhoppers, and bugs are among the simplest examples of gradual metamorphosis. In these forms the nymphs resemble the adults in body form except for the wings, reproductive organs, and structures associated with them. In wingless species of these groups the nymphs can be distinguished from the adults chiefly by the incompletely formed genitalia. In the two orders Ephemeroptera (mayflies) and Odonata (dragonflies), the nymphs, which are aquatic, have evolved many specializations for life in the water, such as peculiar mouthparts, well-developed lateral or anal gills, or peculiar body shape. As a result the nymphs are considerably more different from the adults than those of cockroaches and stoneflies and commonly are called *naiads*. This variant from the simpler type of gradual metamorphosis is termed *hemimetabolous development* and only occurs in aquatic insects that change from a gill-breathing naiad to a spiracle-breathing adult with wings.

Complete Metamorphosis. In many insects the wing rudiments develop internally until the preadult instar, in which stage the wings are everted as large pads. This is a nonfeeding stage in which structures of the adult are rebuilt from tissues of earlier stages. In most insects this stage is quiescent, but in the Rhaphidioptera, Megaloptera, and a few primitive Hymenoptera, it is quite active and pugnacious when disturbed. Externally in these insects toward the end of immature growth, the wings appear suddenly as well-formed structures. There are, therefore, three distinctive postembryonic stages: the early form without wing pads, called a *larva*; the usually quiescent form with wing pads, called a *pupa*; and the *adult* (Fig. 7-12). This type of development is termed *complete* or *indirect metamorphosis* and *holometabolous development*. Examples of insects having this type include moths, bees, flies, beetles, and thrips, and certain members of the Hemiptera, notably the scale insects.

One of the simplest instances of complete metamorphosis occurs in thrips, in which the larvae are very similar to the adults in body proportions and shape of antennae, legs, and mouthparts. The thrips and scale insects, incidentally, each have two pupal stages, both stages being quiescent, typical pupae. Another simple type occurs in

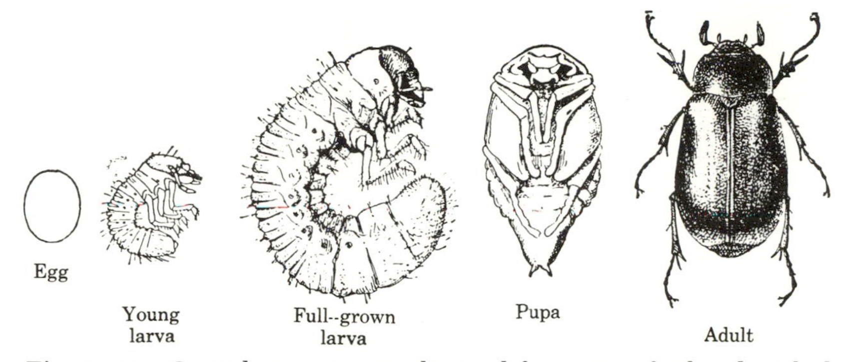

Fig. 7-12. Complete metamorphosis; life stages of a beetle, *Phyllophaga* sp. (From U.S.D.A., E.R.B.)

dobsonflies (Megaloptera). Their larvae differ from the adults in structure of antennae and mouthparts more than do those of thrips, but the legs and many other structures are remarkably like those in the adults. In both dobsonflies and thrips, as is true in insects with gradual metamorphosis, the adult thoracic sclerites differ greatly from those of the immature stages.

In other orders with complete metamorphosis the larvae have more structures unlike those of the adults. In some groups, such as moths and sawflies, the larvae have pairs of lobate abdominal legs. In larvae of all these orders the antennae, eyes, and thoracic legs are simplified or completed undeveloped. These structures, like the wings, seem to appear suddenly in the pupal stage in roughly the outlines of the adult structures. Actually the adult antennae, legs, eyes, and other structures are not formed by remodeling the reduced larval structures but, like the wings, are built up from discrete areas of dormant embryonic tissues, called anlagen. The anlagen develop in the larva to some extent as internal rudiments, called *imaginal discs*. Wing development is a good illustration of this process.

Imaginal discs begin to develop early in larval life, sometimes even in the late embryo. Typical stages in the growth of a wing imaginal disc are illustrated in Fig. 7-13 *A* to *G*; these are diagrammatic. At first an imaginal disc is only an area of thickened epithelial cells (*A*). This area enlarges (*B*) and begins to pull away from the cuticle, until it forms an internal pocket, as in *C*. One portion of the pocket wall then begins to enlarge into the pocket, as in *D*, gradually infolding to form a double-walled sac, as in *E* and *F*. Just before pupation this sac, the rudimentary wing, is usually pushed out, and the cavity flattens out, so that the sac lies directly underneath the cuticle, as in *G*. When the cuticle is shed during the

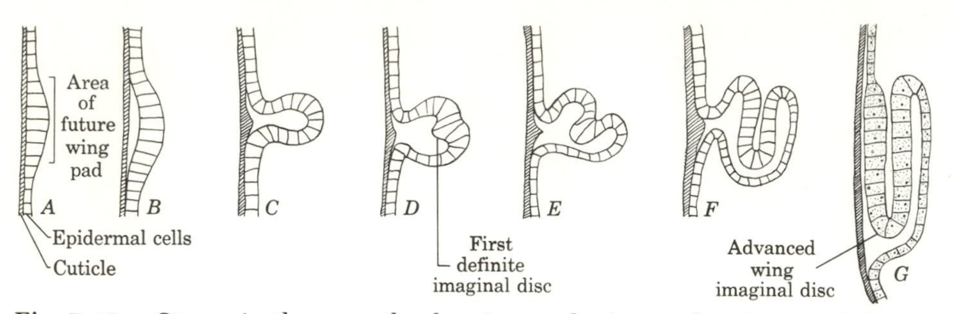

Fig. 7-13. Stages in the growth of an internal wing pad or imaginal disc.

molt to the pupal instar, the wing is finally an exposed external structure. If rudimentary wings are dissected out of the imaginal discs, they look very much like wing pads of hemimetabolous nymphs.

The development of so many adult structures from imaginal discs allows the development of specialized larval structures that are not carried over into the adult, since they are cast off with the last larval skin or dissolved during the catabolism of early pupation. This arrangement has provided a mechanism by which the larvae and adults of the same species can evolve in entirely different directions. This they have done. In general the larvae have specialized for better food gathering, the adults for better dispersal and reproduction. The culmination of this trend is found in the muscoid flies (Diptera), whose larvae (maggots) are legless and appear to lack eyes, antennae, all conventional mouthparts, and head capsule, whereas the adults are typical, fast-flying insects. Although fly larvae appear to be extremely simple creatures, their mouthparts, tracheal system, enzyme system, and musculature are specialized to a remarkable degree.

There are certain instances in which both larvae and adults have become specialized for food gathering, but in different ways. Mosquito larvae, which are aquatic, have chewing mouthparts, and most of them feed on microorganisms; the adults have piercing-sucking mouthparts, the males feeding on nectar and most of the females on avian or mammalian blood. The order Siphonaptera (fleas) offers another striking example. Flea larvae feed as scavengers on organic material in the host nest and the adults suck blood.

Hypermetamorphosis. In the majority of cases all the larval instars of a species are similar in feeding habits and general appearance, differing chiefly in size. Some groups, however, may have two or more quite distinct types of larvae in the life cycle. When this sort of development occurs, it is called hypermetamorphosis. Among the best examples are meloids and many of the parasitic Hymenoptera (Fig. 7-14). The first-instar larva is a motile form bearing bristles, tails, or other processes; this instar either penetrates the host integument or migrates through the host tissues. The later instars are sedentary and have none of the modifications of the first instar. Another striking example is the peculiar Stylopoidea (Fig. 8-107). The young larvae are active legged forms furnished with bristles and tails; the succeeding instars are grubs. The same sort of development

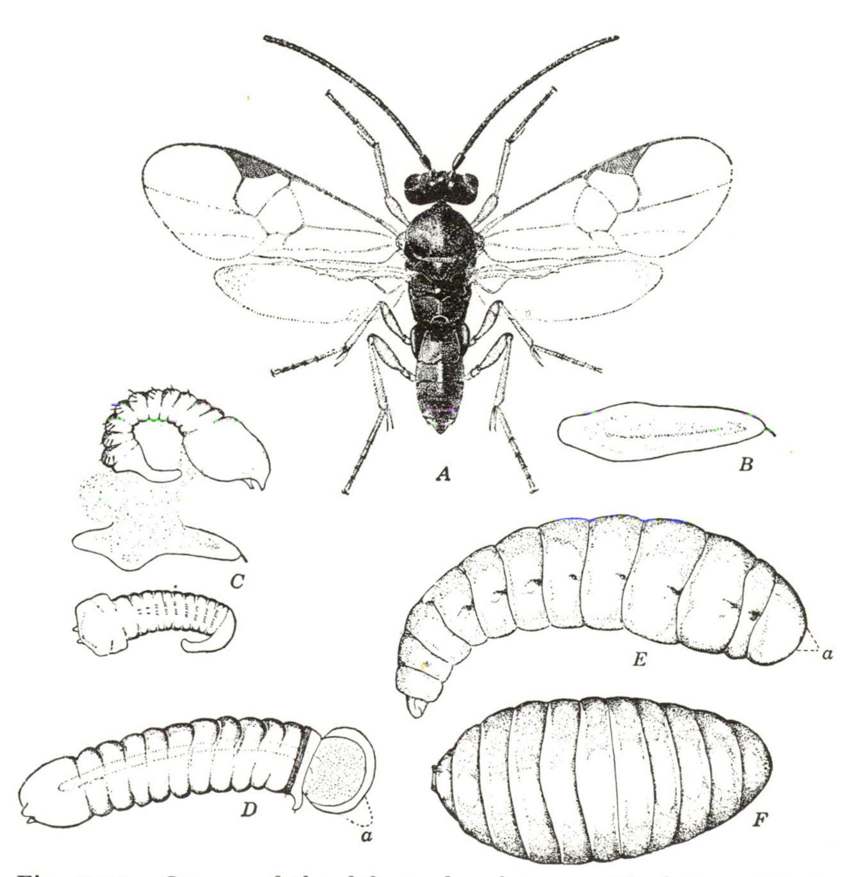

Fig. 7-14. Stages of the life cycle of a parasitic hymenopteran, *Apanteles melanoscelus*, illustrating hypermetamorphosis. (*A*) Adult; (*B*) egg; (*C*) first-instar larvae, the upper one with yolk still attached to it; (*D*) second-instar larva; (*E*) third-instar larva in feeding stage; (*F*) same, but ready to spin cocoon (prepupal stage). *a*, Anal segment. (From U.S.D.A., E.R.B.)

occurs in the blister beetles Meloidae (Fig. 8-98), although in these the difference between larval instars is chiefly general shape accompanied by a few structural changes.

Transitional cases are frequent, in which various instars differ considerably in shape and habit but have few morphological differences. For instance, certain parasitic rove beetles have slender active first-instar larvae and grublike succeeding instars. A few genera of caddisflies (Trichoptera) have free-living slender first-instar larvae and case-making stout-bodied later instars.

Histolysis and Histogenesis. During metamorphosis between larva and adult, many internal processes effect rearrangement within cells, tissues, or organs. The series of conversion processes represent two phases: *histolysis* (breakdown of body tissues) and *histogenesis* (formation and differentiation of body tissues). These processes normally begin in the last larval instar and continue through the pupal stage. During this time the larval stores of carbohydrates, fats, and glycogen in the larval fat bodies, blood sugar, and muscle provide the energy for various chemical reactions. During histolysis, phagocytes and enzymes convert the larval fat body, much of the muscle tissue, undoubtedly other tissue also—and later, phagocytes themselves—into a nutritive matrix transportable by the blood to growing tissues. Both histolysis and histogenesis occur simultaneously. Muscles are one of the tissues that are most affected by metamorphosis. Some kind of muscular degeneration succeeded by regeneration occurs in all insect groups during metamorphosis. In less advanced groups, these changes represent only cytoplasmic rearrangement within cells, whereas in higher groups a complete exchange of larval and imaginal (adult) muscles may take place through the processes of histolysis and histogenesis.

In insects with complete metamorphosis, the radical transformations occur mainly during the pupal state. At this stage there is a significant change from larval to adult characteristics. There is, however, no accompanying all-inclusive physiological change. The epidermis and tracheal system may be reconstructed, possibly at times by imaginal discs. The nervous system enlarges rapidly by growth of the constituent parts and sometimes is accompanied by fusion of certain ganglia. The heart grows without marked change. The digestive tract is changed by the growth or reduction of some parts and the remodeling of others (Fig. 4-3).

Not represented in larval structures are certain features of adults, such as the wings, antennae, compound eyes, and reproductive systems. Certain other adult features are usually radically different in size or organization from their larval counterparts, notably the appendages and the musculature, especially muscles controlling flight and reproductive activities.

Paedogenesis. This is a precocious reproductive maturity that occurs in a few insects, resulting in the production of eggs or living young by larvae or pupae. In the life cycle of the rare beetle, *Micromalthus debilis*, some larvae lay eggs or produce young. Midge larvae of the genera *Miastor* and *Oligarces* produce young but no eggs. Pupae of the midge genus *Tanytarsus* may produce either eggs or young. Paedogenesis is a freak type of metamorphosis and growth involving a maturation of the reproductive organs without a similar maturation of other adult characteristics. It often is associated with unusual generation cycles, discussed on page 255, and arises from hormonal imbalances.

MATURITY

Sexual Maturity and Mating. Adult insects are seldom sexually mature immediately on emerging from the preadult stage. In most cases the males require a few days to mature, and the females longer. Mating occurs in many forms before the females have mature eggs; the sperms are stored in the spermatheca of the female until the time for their use.

In the case of certain short-lived insects, such as many mayflies, both sexes are fully mature sexually within a matter of hours after completing the last molt. Mating takes place a day or so after emergence, and oviposition occurs shortly afterwards.

Oviposition. Most insects are oviparous; that is, they lay eggs; but the various kinds of insects differ tremendously in egg-laying habits. The Phasmidae (walkingstick insects) drop their eggs singly onto the ground; butterflies glue theirs to leaves; sawflies and some crickets (Fig. 7-15) saw out a cavity in a leaf or stem, forming a recess for each egg. Eggs may be laid separately, or they may be grouped together in large masses. Extruded eggs of stoneflies and mayflies collect as a mass at the end of the body; this mass is deposited as a unit. In the cockroaches this tendency is greatly developed. The eggs are glued together as they emerge from the body and, cemented by glandular secretions, form a compact capsule or oötheca, which is deposited. In many mayflies, oviposition is completed with the deposition of a single large mass of eggs. Oviposition in bedbugs occurs at a slower rate but continues for a period of months. These are only a few examples; additional material is incorporated in the synopsis of the orders in the following chapter.

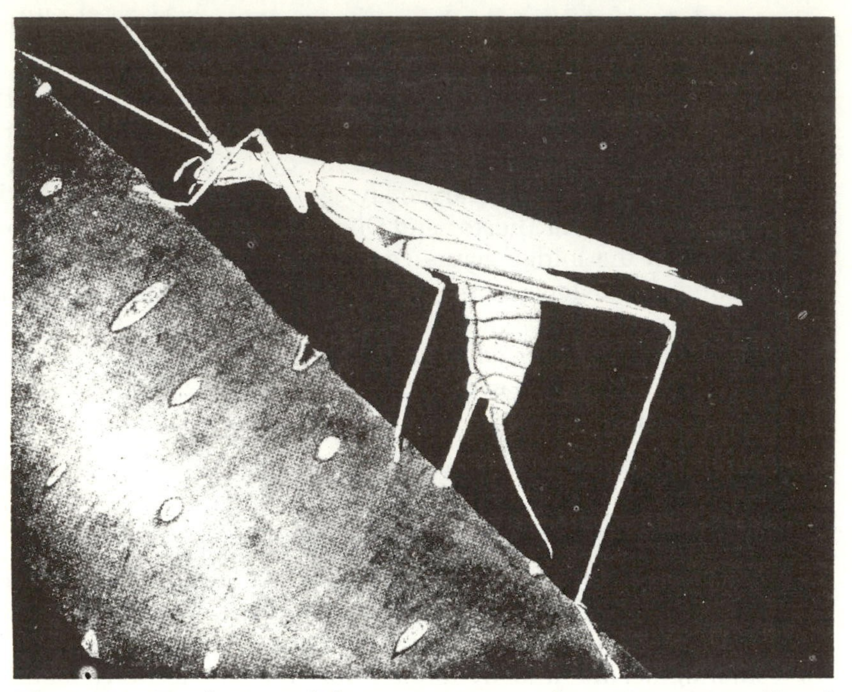

Fig. 7-15. Egg laying of the snowy tree cricket. (From Metcalf and Flint, *Destructive and useful insects*, by permission of McGraw-Hill Book Co.)

In field identification, egg form, location, and color are valuable clues to the identity of eggs of various pest species. They provide critical information as to the timing of or need for pest-management operations.

Viviparity. Not all insects lay eggs. Species of various groups are viviparous; that is, they deposit living young instead of eggs. In viviparous insects the eggs develop in the oviducts or vagina until at least the completion of embryonic growth. This phenomenon occurs in many groups scattered throughout the insect orders. There are several kinds of viviparity in adults, some cases being only slight modifications of the oviparous condition and others involving the development of special structures.

Precocious hatching of the embryo within the egg passage occurs in the flesh flies, Sarcophagidae. Eggs remain in the vagina until mature and hatch just as they are deposited. The young larvae pass through the ovipositor like eggs and at birth are in an early stage of larval development, corresponding to the point of hatching from the egg in oviparous species. In Stylopoidea (Coleoptera) the eggs hatch inside the body of the female and the minute larvae crawl out through the genital openings.

Uterine development of larvae occurs in the ked flies and the African fly genus *Glossina* (both members of the Diptera). In these the uterus is a large chamber (Fig. 7-16) provided with glands producing nourishment for the larva, which develops to maturity in the body of the female. The maggot pupates as soon as it is discharged from the vagina.

One of the more common examples of viviparity occurs in the parthenogenetic generations of aphids. The unfertilized eggs develop within the ovaries and are liberated as active nymphs equivalent to first-instar nymphs of oviparous generations.

Longevity of Adults. Adult insects have a normal life span ranging from a few days to several years, depending on the species. The length of life is correlated with fecundity, death usually occurring a short time after the completion of mating or oviposition activities. Thus the females of certain species of psocids (Psocoptera) have an adult life of about 20 days and die 5 or 6 days after final oviposition. Unmated females of these species rarely oviposit and live about 20 days longer than mated females that lay the normal number of eggs. The hibernating forms in many species are adults, and in these cases they may have a life span of nearly a year. Many adult leaf beetles, for instance, mature in July, feed for the rest of the summer, and then hibernate. They become active the following spring, mate, lay their eggs through May and June, and die soon afterwards.

Voltinism. In insects, the number of generations per year is called voltinism. A univoltine insect has one generation per year, a bivol-

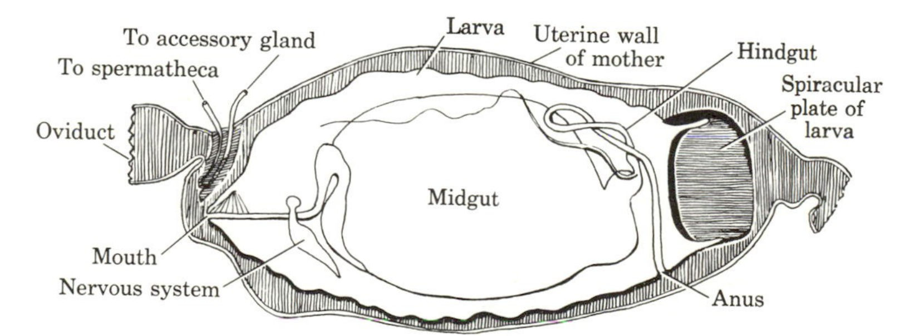

Fig. 7-16. Larva (unshaded except for its spiracular plate) of the tsetse fly *Glossina* in uterus of its parent. (Redrawn from Snodgrass)

tine insect has two, and a multivoltine insect has more than two generations per year.

Parthenogenesis. When eggs develop without being fertilized the process is called parthenogenesis. Parthenogenesis is common in certain groups of insects, particularly aphids (Hemiptera), ants, wasps, and bees (Hymenoptera), and occurs in most orders of insects except for the Odonata, Dermaptera, Neuroptera, and Siphonaptera. There are several types of parthenogenesis resulting from different genetic mechanisms. Honeybees are a well-known example of *haplodiploidy*, in which the unfertilized eggs (haploids) develop into males (drones), and the fertilized eggs (diploids) develop into females (workers), which are kept in a nonreproductive state by secretions from the reproductive female, the queen bee. This type of parthenogenesis in which only males are produced is also called *arrhenotoky*. Four insect orders—the Hymenoptera, Hemiptera, Thysanoptera, and Coleoptera—show haplodiploidy. It occurs throughout the order Hymenoptera from the primitive sawflies to the most specialized parasitic and social wasps and bees. *Thelytoky* is the type of parthenogenesis in which only females are produced from unfertilized eggs. Female individuals lay unfertilized eggs and these produce exclusively female progeny. Examples include the pear sawfly *Caliroa cerasi*, the rose slug *Endelomyia aethiops*, a number of blackflies (Simuliidae), and a large number of weevils (Curculionidae).

Suspended Activity; Quiescence and Diapause. In the life of many insects, there are more or less prolonged temporary periods of dormancy, during which visible activity and many physiological processes are suspended. These periods may occur in the egg, nymph, larva, pupa, or adult. They are characterized chiefly by a cessation of growth in immature stages and by a cessation of activity and sexual maturation in adults.

The characteristic of *quiescence* is that the state of dormancy is the *immediate consequence* of adverse conditions (for example, low temperature or summer drought), at the end of which development is immediately renewed. Thus, cinch bugs breed continuously if kept at favorable temperatures, but the adults enter a quiescent condition if kept at a low temperature.

Diapause, the temporary interruption of development in a specific stage of development, differs from quiescence in two distinct ways.

1. It is an actively induced state involving reduction of neuroendocrine activity usually at a specific point in the insect's life cycle.

2. The onset of diapause is brought about by environmental factors that, although signaling the approach of unfavorable conditions, are not in themselves adverse. The most important of these are photoperiod (day length) and temperature.

Diapause is a hereditary characteristic triggered by an internal timing mechanism that brings about a cessation of activity in advance of the unfavorable condition. Some unfavorable condition, however, is usually a necessary stimulus to break the dormancy. Such a dormant period may last weeks, months, or years. In some species, diapause occurs in each generation independently of the environmental conditions that the insects experience. In others, it occurs only in some generations, when specific environmental conditions prevail or have prevailed at an earlier stage. An example of diapause is shown in the cecropia moth (*Hyalophora cecropia*). When full grown, usually toward the end of summer, its larva constructs a cocoon and transforms into a pupa, which enters diapause. If chilled to 3 to 5°C for about 6 weeks, then returned to room temperature, the pupa will resume development. If kept constantly at room temperature, it will be 6 months to a year before diapause is broken.

Initiation of diapause also is linked with seasonal changes in day length, or photoperiodism. The exact relationship differs in various species. In many the critical period may be considerably earlier than the onset of diapause. In the moth *Polychrosis*, pupal diapause is predetermined by the photoperiodic conditions of embryonic development. In the beetle *Leptinotarsa*, on the other hand, a long photoperiod experienced by late larvae may cause them to enter diapause as soon as they dig into the ground. In the silkworm *Bombyx*, a lengthening (vernal) photoperiod triggers diapause.

The mechanism underlying the initiation and termination of diapause is under hormonal control. As a result of effects of environmental conditions (photoperiod or temperature) on the nervous system and neurosecretory cells or as a result of a genetic mechanism, a change in the hormonal system occurs. Three main types of diapause are distinguished on the basis of hormonal mechanisms (Novák, 1974).

1. All larvae and pupae undergo diapause caused by a lack of activation hormone (AH) and consequently of the molting hormone, ecdysone (MH). In the American silkmoth *Hyalophora cecropia*, pupal diapause during low winter temperatures is associated with a low concentration of ecdysone. After the pupa experiences critically low temperatures followed by rising ones, higher ecdysone concentrations follow in association with the termination of diapause and the continuation of normal development.

2. Diapause caused by a lack of AH and juvenile hormone, neotenin (JH), characterizes adult insects. In the Colorado potato beetle, adult diapause is associated with a juvenile hormone deficiency and may be terminated by increasing the JH concentration.

3. Diapause caused by the action of the neurosecretory factor produced by the subesophageal ganglia on the female, affecting the development of eggs, characterizes early embryonic diapause. In the silkworm *Bombyx*, a diapause in the egg is determined by light and temperature conditions to which the maternal insect was exposed as an embryo a full generation previously, through a release of a neurosecretory hormone.

FOOD HABITS

Food is essential to the growth of any organism and therefore is an important consideration in the life cycle of an insect. A wide range of organic substances, living and dead, are used by insects as food. According to the type of food utilized, insects may be grouped in the following manner with an example given for each category.

1. *Saprophagous*—feeding on dead organic matter.
   General scavengers—Dictyoptera (cockroaches).
   Humus feeders—Collembola (springtails).
   Dung feeders—some Scarabaeidae (dung beetles).
   Restricted to dead plant tissue—Isoptera (termites).
   Restricted to dead animal tissue—Dermestidae (larder beetles).
   Carrion feeders—Calliphoridae (flesh flies).
2. *Phytophagous*—feeding on living plants.
   Leaf feeders—Orthoptera (grasshoppers).
   Leaf miners—Agromyzidae (flies).
   Stem and root borers—Cerambycidae (beetles, round-headed borers).
   Root feeders—some Scarabaeidae (beetles, white grubs).
   Gall makers—Cynipidae (gall wasps).
   Juice suckers—Leafhoppers and aphids.
   Fruit feeders—Tephritid flies.
   Mycetophagous, fungus feeders—Mycetophagidae (fungus beetles).
3. *Zoophagous*—feeding on living animals.
   Parasites (living on another animal).
      Living on warm-blooded vertebrates—Phthiraptera (chewing and sucking lice).
      Intermittent vertebrate parasites—Blood feeders Culicidae (mosquitoes).

Entomophagous—either parasites or predators on other insects; Ichneumonidae (ichneumon wasps).

In these food-utilization categories, the gall makers and parasites involve unusual relations between insect and host.

Gall Makers. Many insects cause plants to develop abnormal outgrowths or disfigurements called galls (Fig. 7-17), and live within

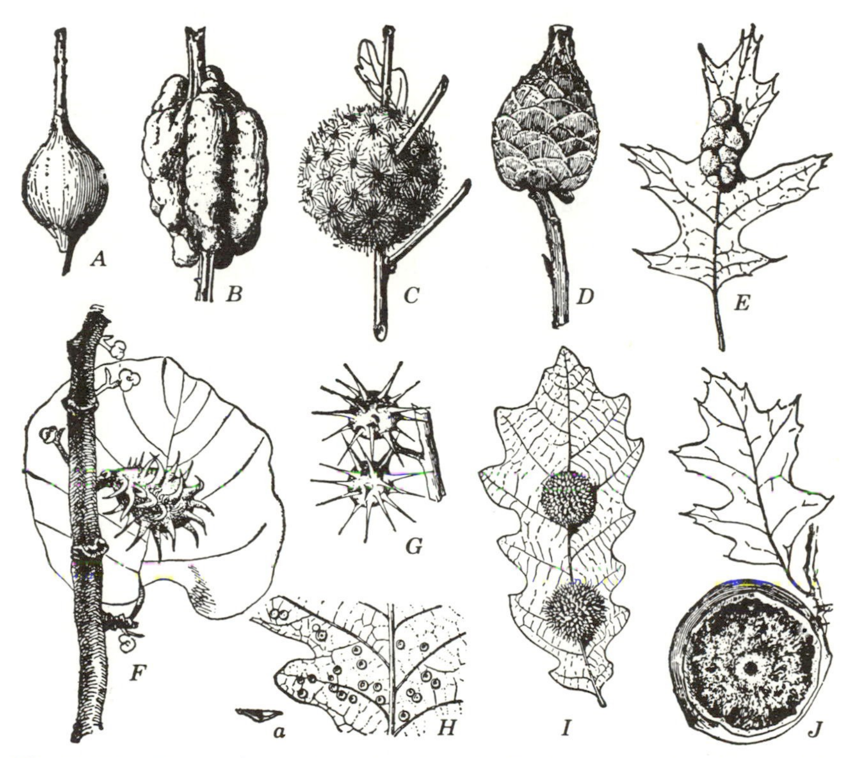

Fig. 7-17. Examples of insect galls. (*A*) Goldenrod ball gall, used by a fly *Eurosta solidaginis*; (*B*) blackberry knot gall, caused by a gall wasp *Diastrophus nebulosus*; (*C*) wool sower gall on oak twig, caused by a gall wasp *Andricus seminator*; (*D*) pinecone gall, common on willow, caused by a gall fly *Rhabdophaga strobiloides*; (*E*) oak leaf galls, caused by a gall wasp *Dryophanta lanata*; (*F*) spiny witch hazel gall, caused by an aphid *Hamamelistes spinosus*; (*G*) spiny rose gall, caused by a gall wasp *Rhodites bicolor*; (*H*) oak spangler, caused by a gall fly *Cecidomyia poculum*, one gall shown in section at *a*; (*I*) spiny oak gall, caused by a gall wasp *Philonix prinoides*; (*J*) large oak apple, caused by a gall wasp *Amphibolips confluens*. (From Metcalf and Flint, rearranged from Felt)

the shelter of these structures. Insect galls are formed by abnormal growth of particular tissues of a plant and may occur on leaves, buds, stems, or roots. Each insect produces a particular type of gall and always on the same region of the plant. The sawfly *Euura salicis-nodus* makes a gall on the stems of willow, and another sawfly *Euura hoppingi* makes a gall on the leaves of willow. In certain insects that have alternation of generations, each generation may cause the formation of a differently shaped gall.

The cause of gall formation is not exactly known, but a gall is excessive growth of plant tissue presumably brought about by a substance or substances secreted by the insect that are harmful to plant health. Aphids affect plant growth apparently by secreting a substance in their saliva that decreases plant growth hormones and increases growth inhibitors. The saliva apparently also prevents or reduces the plant's ability to seal off wounds.

Galls are of two types: open and closed (Fig. 7-18). Open galls are essentially pouchlike and have an opening to the outside. Aphids cause galls of this type. Each aphid species produces a specific type of gall. A gall begins to form around the feeding station of a single aphid. The plant tissue around this area enlarges and gradually becomes twisted or bowed to form a purselike cavity with the aphid inside. The edges of these galls are usually tightly appressed until a generation of aphids within have completed their growth; at this point the edges separate and allow the migrating aphids to escape from the gall. The galls of *Phylloxera* and mites are also of this type.

Closed galls are formed by several groups of Hymenoptera, including a few genera of sawflies, certain groups of chalcid flies, and the family Cynipidae. In these groups the females insert each egg beneath the surface into the plant tissue. The larva never leaves this haven, feeding on the inner tissues of the gall that forms around it. The insect must eat its way out. In some sawflies the larva leaves the gall and pupates elsewhere; in others the larva pupates within the gall, and the adult eats its way out.

Parasites. In the zoological sense, a parasite is an animal that lives in or on another animal, known as its *host*, from which the parasite

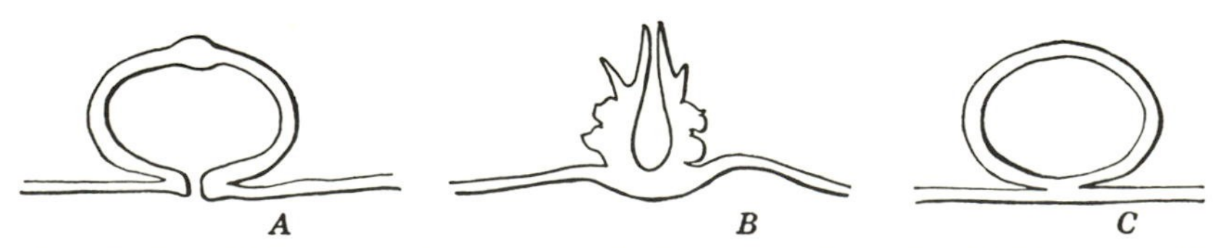

Fig. 7-18. Diagram of open and closed galls. (*A*, *B*) Open galls; (*C*) closed gall. (*A* and *B* redrawn from Wellhouse)

derives its food for at least some stage in its life history. Several insect groups are true parasites. On the basis of habits and hosts, insect parasites fall into two categories: parasites of warm-blooded vertebrates; and parasites of insects or other small invertebrates such as spiders and worms.

The insects parasitizing warm-blooded vertebrates do not kill their host, so many individuals or many generations of the parasite may live on the same host animal. Sucking lice (Anoplura) and chewing lice (Mallophaga) are examples of external parasites. They spend all their lives on a bird or mammal host, frequently occurring in large numbers on the same animal, and having continuous generations during the life span of the host. Seldom does the host die from these attacks, although its general health and resistance to other maladies may be greatly impaired. Bot flies and warble flies furnish excellent examples of internal parasites. The larvae of these flies live and mature in the nasal passages, stomach, or back of the host, and when full grown leave the host and pupate in the soil. Many individuals of the parasites may infest one host individual. Again the host is not killed, although harmed, and is attacked by successive generations of the parasites.

Insects that parasitize other insects differ from those parasitizing vertebrates in three characteristics.

1. Usually only a single parasite attacks a host individual.
2. The parasite usually kills the host.
3. Usually the adult parasite seeks out and lays eggs on or in the host.

This type of parasite is commonly called a *parasitoid*. As with the internal parasites of vertebrates, only the larvae actually live entirely on the host. This type of parasitism occurs in many families of Hymenoptera, several families of Diptera, and a few genera of Coleoptera.

Parasites of insects usually infest a host in one of three ways. The most common manner is for the female to lay an egg on the host or to insert her ovipositor through the host integument and lay an egg in the host tissue. Most of the parasitic Hymenoptera and Diptera use this method. The second manner is an indirect approach. The parasitic female deposits her eggs on leaves of the plant on which the insect host feeds. If the host eats some of the egg-infested leaves, the parasite eggs are unharmed and hatch within the digestive tract, the larvae making their way into the host tissue. The hymenopterous family Trigonalidae and many species of the Tachinidae, or tachina flies, use this method. In the third manner the eggs are laid more or less in random places, and the first larval instar finds its way to a host, as in the Meloidae (p. 395) and Stylopidae (p. 406).

Most of these parasites are internal, but in a number of parasitic Hymenoptera the larvae feed externally on the body of the host. In these instances the circumstances approach closely the condition of predatism rather than parasitism.

The habit of parasitism is an extremely specialized one. Thousands of species of Hymenoptera and Diptera are parasitic; yet each species usually parasitizes only a single host species or a group of closely related species. The parasites, called *primary parasites*, are often the host to other parasites, called *secondary parasites* or *hyperparasites*, and these may be the host to tertiary parasites. Usually secondary and tertiary parasites are much less specific in host selection than are primary parasites.

Feeding and Life-Cycle Stages. In general, the life of the adult insect is concerned primarily with reproduction. The adult feeds to maintain its metabolic losses from activity and life processes, or to furnish nutriment to the eggs or sperms developing in its body. In such groups as Orthoptera, Hemiptera, Siphonaptera, and blood-sucking members of the Diptera, the adult requires a large amount of food to supplement the reservoir carried over from the nymphal or larval stage.

In many groups sufficient stores of fats or other nutrients are carried over from the immature stages so that the adults need to do little or no feeding. This is true of caddisflies (Trichoptera), many Hymenoptera, Lepidoptera, and Diptera. In extreme cases such as mayflies (Ephemeroptera), the eggs are practically ready to deposit by the time the adult emerges, and no feeding is done in this stage. This last example is an extreme in the direction of nonfeeding on the part of adults, and there are also extremes in the opposite direction. The most outstanding North American example is the sheeptick group, Hippoboscidae (Diptera). In these the larvae develop to maturity in the body of the female, and she does all the active feeding in the life cycle.

SEASONAL CYCLES

Whereas the *life cycle* is the development of the individual from egg to egg, the *seasonal cycle* is the total successive life cycles or generations normally occurring in any one species throughout the year, from winter to winter.

The life cycle of many species consists of a single generation each year. In this case the life cycle and seasonal cycle are the same.

In cases such as the housefly there are continuous generations produced throughout the warmer months, followed by a hibernating period or diapause (see p. 244). Thus the seasonal cycle consists of several life cycles.

There are some species in which the life cycle is longer than a year in duration, as, for instance, many June beetles, whose larvae require 2 or 3 years to mature, and the 17-year locust, a cicada having a 17-year developmental period. In these insects the seasonal cycle includes only a portion of the life cycle. In most cases, however, the generations overlap so that adults of each species are produced every year; the term seasonal cycle is used here to include the activities of the combined generations of the species for the year.

Seasonal cycles made up of more than one life cycle are of two types: those having repetitious life cycles; and those having alternation of generations.

Repetitious Generations

In this category successive life cycles are fundamentally the same. One generation of houseflies, for instance, lays eggs that develop into another generation just like the first, having the same morphological characteristics, food habits, and reproductive habits.

Interruptions in the development of certain generations, as a result of estivation, hibernation, or other types of diapause, are not considered as altering fundamentally the general pattern of the life cycle. To cite the housefly again, it has successive generations during the summer, the life cycle of each varying from 4 to 5 weeks, depending on weather conditions; adults produced in autumn, however, hibernate during the winter and in spring resume normal activities. A life cycle interrupted by onset of winter differs from that of summer in no feature other than the time element.

Alternation of Generations

There are several groups of insects in which succeeding generations are quite different in method of reproduction and sometimes in habits.

Forms with Reproduction Only by Adults. Two well-known groups belong in this category: the aphids and the gall wasps. These forms are all plant feeders.

The aphids (Aphididae, Hemiptera) have varied and complicated seasonal cycles involving sexual oviparous and parthenogenetic viviparous generations, winged and wingless generations, and frequently migrations between definite and different summer and winter host plants (Table 7-1). A fairly simple seasonal cycle is exemplifed by the cabbage aphid, *Brevicoryne brassicae*. Members of this species overwinter as eggs, laid in autumn on the stems of cruciferous plants. These hatch the following spring and develop into the wingless parthenogenetic viviparous form, the stem mothers. It is interesting to note that all the eggs develop into these stem mothers. These produce parthenogenetic viviparous generations that may be winged or wingless. Similar viviparous generations are continuous throughout the summer. As a rule parthenogenetic individuals live about a month and produce 50 to 100 young. When the days become shorter in autumn, the viviparous forms produce the sexual generation, wingless females and winged males. After mating, each female lays one to several eggs, which pass through the winter.

Table 7-1. Chart of Alternation of Generations in Aphids (illustrating (A) a form with no host alternation, such as the cabbage aphid; and (B) one having an alternation of hosts, such as the plum aphid).

	(A) NONMIGRATORY SPECIES	(B) TYPICAL MIGRATORY SPECIES	
SEASON	ALL FORMS ON ONE HOST	FORMS ON PRIMARY HOST	FORMS ON SECONDARY HOST
Winter	Eggs	Eggs	
Early spring	Stem mothers Apterous viviparous females	Stem mothers Apterous viviparous females	
Late spring	Alate viviparous females (spring migrants)	Alate viviparous females (spring migrants)	Spring migrants from primary host
Summer	Alate and apterous viviparous females (these migrate from plant to plant of the same host species, or compatible related species)	A few strays	Alate and apterous viviparous females
Early Fall		Fall migrants from secondary hosts	Alate viviparous females, sometimes alate males (fall migrants)
Late fall	Sexual forms: males and oviparous females	Sexual forms: males and oviparous females	
Winter	Eggs	Eggs	

More complicated seasonal cycles have several additional specializations. Many species migrate to summer hosts. In these aphids, the winter or primary host is a perennial. The eggs are laid on such hosts in autumn, hatch in spring, and develop into wingless stem mothers. These normally produce a generation of wingless viviparous females and these in turn produce winged viviparous females that fly to summer or secondary hosts. Continuous viviparous generations, either winged or wingless, are produced on these hosts until autumn. At that time migratory forms are produced, which fly to the winter host and there produce young of the sexual generation. In certain species the winged males are produced on the summer host, migrate to the winter host, and mate with the wingless oviparous females. In other species both males and females are wingless, in which case they are both produced by winged viviparous females that have previously migrated to the winter host.

There are radical differences in habits among various generations of some aphids. In the genus *Pemphigus*, for example, the stem mother makes a gall on the leaf or petiole of poplar; her progeny migrate to the roots of Compositae and other plants and initiate a series of root-inhabiting generations that produce no galls.

The phylloxerans (Phylloxeridae, Hemiptera), closely related to the aphids, also have an alternation of parthenogenetic and sexual generations (some winged and other wingless) and complicated migration habits. They differ from aphids, however, in that all forms are oviparous. A common example is the grape phylloxera (Fig. 7-19). A series of parthenogenetic generations form leaf galls, migrating in autumn to the grape roots, where the following spring another series of parthenogenetic generations form enlargements and galls on the small grape roots. Because the insects in these generations develop and reproduce without fertilization from males, they are called *agamic*. Some of the root forms give rise to parthenogenetic agamic winged migrants in the autumn; these crawl out of the ground and fly to the grapevines, there producing wingless agamic males and females. After mating, the females each lay one egg in a crevice of the bark. The eggs overwinter and hatch the next spring, initiating a series of leaf-infesting generations.

The gall wasps (Cynipidae, Hymenoptera) contain many species having alternation of sexual and parthenogenetic generations. An example is an oak-feeding species *Andricus erinacei*, which overwinters as eggs laid in leaf or flower buds. These hatch in spring, and each larva becomes surrounded by a soft bud gall produced by the plant. The inner layer of the galls provides food for the larvae. In early summer, when those larvae mature, winged males and females

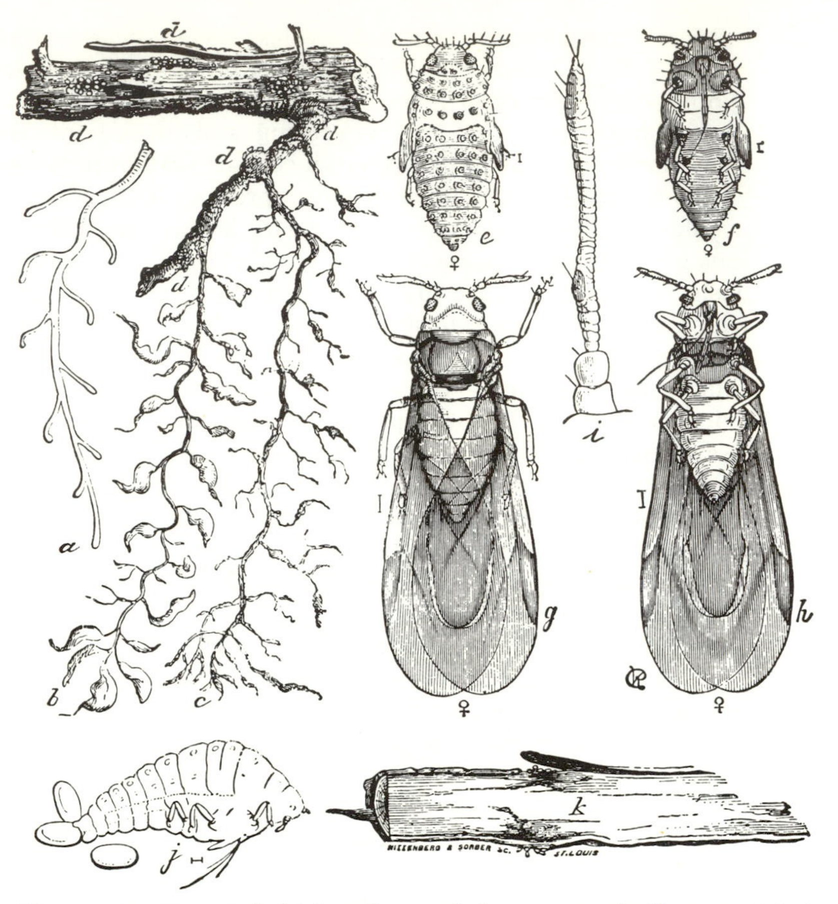

Fig. 7-19. Root-inhabiting form of the grape phylloxera. (*a*) A healthy root; (*b*) one on which the insects are working, representing the knots and swellings caused by their punctures; (*c*) a root that has been deserted by them, and where the rootlets have commenced to decay; (*d*) how the insects are found on the larger roots; (*e*) agamic female nymph, dorsal view; (*f*) the same, ventral view; (*g*) winged agamic female, dorsal view; (*h*) same, ventral view; (*i*) magnified antenna of winged insect; (*j*) side view of the wingless agamic female, laying eggs on roots; (*k*) shows how the punctures of the insects cause the large roots to rot. (From Riley)

emerge. The females lay their eggs in the veins of oak leaves. Larvae from these eggs cause pincushion-like growth on the leaf veins, called hedgehog galls. Larvae in the hedgehog galls mature in autumn, and all emerge as short-winged females, which reproduce parthenogenetically, laying the overwintering eggs in the oak buds.

Alternation of generations involving parthenogenesis provides the opportunity for rapid increase in population numbers to take advantage of short-term favorable conditions. Through parthenogenesis, isolated females are also able to establish new populations rapidly following wide dispersal (wind dispersal), or low population numbers resulting from unfavorable conditions.

Forms with Paedogenesis. Paedogenesis involving reproduction by larvae is always combined with a complex and irregular cycle of generations. The cases studied indicate that there may be successive generations of paedogenetic larvae, with irregular production of larvae that mature normally and pupate. Adults from these pupae mate normally and produce fertilized eggs. The beetle *Micromalthus debilis* has a complex cycle of paedogenetic and normal generations. The most completely studied examples are in the midge family Cecidomyiidae, especially the European *Oligarces paradoxus*. For this species a typical cycle of generations, shown diagrammatically in Fig. 7-20, is as follows: A paedogenetic larva *a* produces larvae *b*, which may grow up to be one of four types: (1) another paedogenetic larva like *a*; or (2) a female-producing larva *c*; or (3) a paedogenetic larva *d*, which gives birth only to male-producing larvae *e*; or (4) a paedogenetic larva *f*, which gives birth to both male-producing larvae *e* and other paedogenetic larvae like *a*. The male- and female-producing larvae pupate, and normal adults emerge. The females lay fertilized eggs that develop into the paedogenetic larvae *a*.

In *Oligarces paradoxus*, adults are produced more frequently as the colony becomes overcrowded or the food supply less ample. In the paedogenetic species of the midge genus *Miastor*, temperature changes have a determining effect on the type of generation produced. Thus in these midges the cycles of paedogenetic generations are correlated with day-to-day conditions as well as a more inclusive seasonal cycle.

REPRODUCTION

In insects reproduction is the function of the sexual reproductive system. Normally insect reproduction is bisexual, in that the egg produced by the female will not develop unless fertilized by spermatozoa produced by the male. Except in a few species, only one sex is represented in any one individual. In most insect species, therefore, the physiology of reproduction deals with the development and maturation of spermatozoa in the male and of eggs or ova in the female and the manner in which they are brought together to effect fertilization. Mating is discussed in Chap. 6, Response and Behavior.

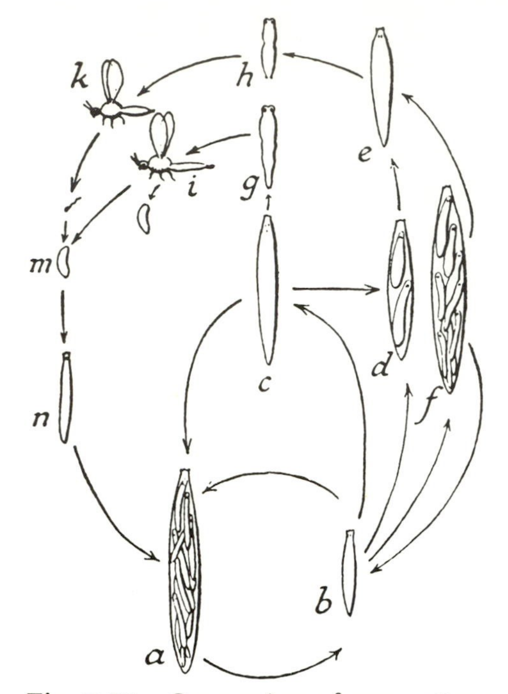

Fig. 7-20. Succession of generations in the midge *Oligarces paradoxus*. (*a*) Paedogenetic larva giving rise to (*b*) undetermined daughter larva, which may develop into (*a*) again or into (*c*) female-imago-producing larva, or into (*d*), which gives rise to male-imago-producing larvae (*e*) or into (*f*), which gives rise both to undetermined daughter larvae and to male-imago-producing larvae; (g) female pupa; (*h*) male pupa; (*i*) female imago; (*k*) male imago; (*l*) sperm; (*m*) egg; (*n*) young larva from egg. (From Wigglesworth, after Ulrich)

Development of Spermatozoa. Spermatozoa are produced in the follicles of the testis (Fig. 7-21). The upper portion of the follicles contain primary germ cells called spermatogonia. These divide repeatedly to form cysts, which move to the base of the follicle because of the pressure of their own increase in size. At the base of the follicle each cell in a cyst undergoes repeated division and may

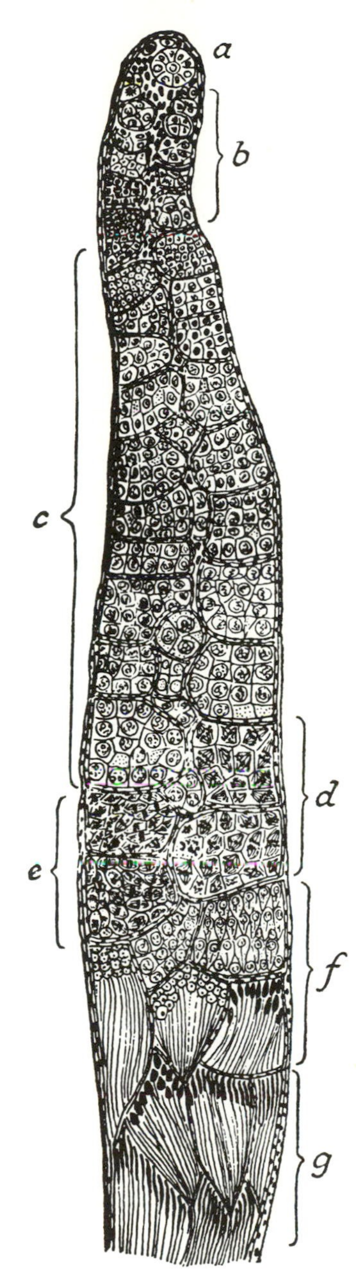

Fig. 7-21 Longitudinal section of testis follicle of a grasshopper, semischematic, (*a*) Apical cells surrounded by spermatogonia; (*b*) zone of spermatogonia) (*c*) zone of spermatocytes; (*d*) cysts with mitoses of second maturation division; (*e*, *f*) zone of spermatids; (*g*) zone of spermatoza. (From Weber, after Depdolla)

increase in number 5 to 250 times. In the next cell division following this mulitplication stage there occurs the reduction division of the chromosomes. This is followed by a transformation period in which the round cells develop into slender flagellate spermatozoa. These mature spermatozoa escape from the duct of the follicle (*vas efferens*) into the genital ducts (*vas deferens*); they are stored in an enlarged or coiled portion of this duct, the seminal vesicle, until mating. At the time of mating, the spermatozoa are transferred to the female, where they are stored until needed for fertilization.

Maturation of Eggs. The eggs are developed in the ovarioles of the ovary. The tip of the ovariole, the *germarium*, contains primary germ cells (*oogonia*), which differentiate to produce developing eggs (*oocytes*) as well as *nutritive cells* (trophocytes or nurse cells) and *follicular epithelial cells*. The oocytes usually appear in successive stages of growth down the length of the ovariole, deriving nourishment from the follicular epithelium and also from nutritive cells if these are present in the ovariole (Fig. 7-22). The oocyte enlarges from

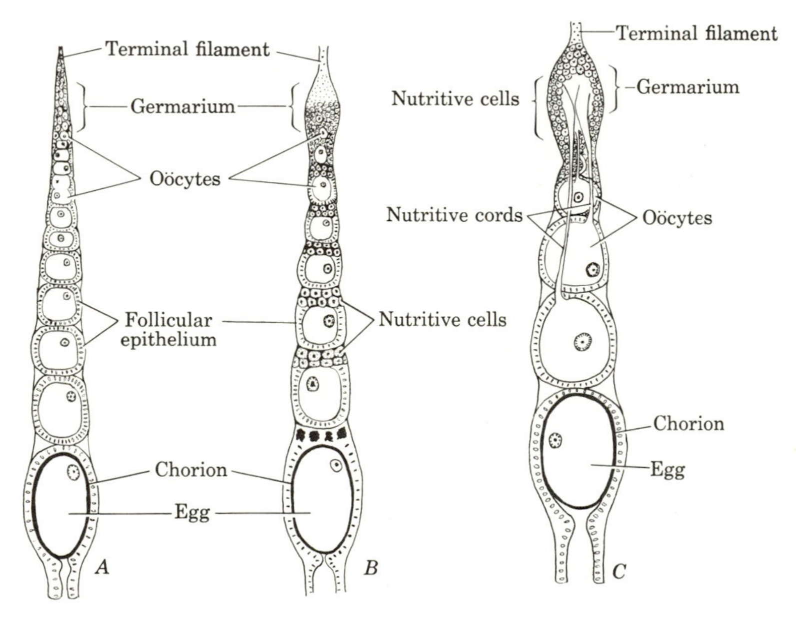

Fig. 7-22. Longitudinal section of ovarioles. (*A*) Simple panoistic type having only oocytes and follicular epithelium; (*B*) polytrophic type having oocytes, follicular epithelium, and nutritive cells; (*C*) telotrophic type having nutritive cells connected to oocytes by nutritive cords. (Redrawn from Weber)

the accumulation of yolk, composed of proteins, lipids, glycogen, and many other compounds. When the oocyte stops accumulating yolk, the follicular epithelium secretes the chorion or eggshell and the vitelline membrane, and then degenerates, as do the nutritive cells.

Below the oocyte at the end of the ovariole is a plug of epithelial cells that seals the ducts leading from ovariole to oviduct. When the oocyte is fully developed, this plug breaks down and the oocyte or egg is released into the oviduct. The portion of the ovariole that contained the released egg shrinks, and a new plug forms below the next oocyte.

The eggs at time of discharge into the oviduct are surrounded by the eggshell or chorion. The eggshell is perforated in one or more places by minute pores or micropyles. It is through these minute openings that the spermatozoa gain entrance to the interior of the egg.

Fertilization. As the eggs pass down the oviduct (by peristaltic action of the oviduct muscles) into the vagina, they come to lie at the opening of the spermathecal duct. From this duct spermatozoa emerge and enter the egg micropyle. After the spermatozoa enter the egg, the egg nucleus undergoes two divisions, one a reduction division, with the production of the female pronucleus and polar bodies. The spermatozoan loses its tail and changes to the male pronucleus. The male and female pronuclei unite for form the zygote.

From this usual sequence of events there are many deviations. The following are interesting examples. In the bedbug *Cimex* the spermatozoa migrate from the spermatheca, into the follicular structure of the ovarioles and from there into early-stage oocytes. Fertilization is thus accomplished before the eggshell is formed. In parthenogenetic species such as the European spruce sawfly *Diprion hercyniae*, fertilization does not occur, but the diploid chromosome count is restored by fusion of a polar body with the female pronucleus.

Sperm Transfer and Mating. At each mating a large number of spermatozoa are transferred from the male to the female. The female stores and controls the spermatozoa so that only a small number are liberated at a time as successive eggs pass down the oviduct. In this way a separate mating is not necessary for the fertilization of each egg. As a consequence a large number of insects mate only once in their lifetime, and most of the remainder mate only a few times.

Immediately after the adult ecdysis, insects will not mate, but in a few hours or at most a day or two most of them become receptive to mating stimuli. These are of many kinds: by peculiar movements, such as the dancing of swarms of male mayflies; by sound, as the chirping of crickets and grasshoppers; by color reactions, as in some butterflies; and chiefly by a wide variety of scents. These phenomena are discussed in Chap. 6. The gonads seem to have little influence on mating behavior, since many species mate before the female ovaries are well developed, and castrated males will mate normally but without transfer of spermatozoa.

The mechanics of spermatozoa transfer may be divided into several distinctive types. In many forms, such as some of the true bugs, the penis is inserted into the female spermatheca and the spermatozoa are placed directly in this storage chamber. In many moths, grasshopppers, and beetles, the penis discharges the spermatozoa into the female bursa copulatrix; after mating, the spermatozoa migrate from this structure to the spermatheca. The mechanism of this transfer is not well known, but it may be initiated or caused by *prostaglandins*, which are synthesized in the male reproductive tissue and introduced into the female with the spermatozoa.

In many insects of this group having a bursa copulatrix, the spermatozoa are transferred in a membranous sac or *spermatophore*, formed by the secretion of the male accessory glands. This sac of spermatozoa is deposited in the bursa or vagina, and its contents are transferred to the spermatheca. After this transfer the empty spermatophore is ejected by the female.

Longevity of Spermatozoa. Apparently the secretions of the female spermatheca or its associated glands can keep spermatozoa viable for a considerable period. The honeybee can sustain its spermatozoan store for several years, and ants for more than 16 years. In moths the spermatozoa remain alive in the spermatheca for several months. In females of a few insects, such as the bedbug *Cimex*, the spermatozoa not utilized in a few weeks are digested and absorbed by the body tissues, and mating occurs from time to time, replenishing the supply.

THE ROLE OF HORMONES

In insects, processes of growth, development, and reproduction are directly affected and mediated by the secretion of hormones of the endocrine system (see also Chap. 6). Two divisions of hormones have been established. Hormones that are synthesized within the body of the insect that affect that individual are called *endogeneous hormones*. Hormones that are produced in one individual and affect another individual are called *exogenous hormones*. The best-known example of exogenous hormones occurs in fleas. Eggs of the rabbit flea *Spilopsyllus cuniculi* do not mature until the flea has fed on a pregnant rabbit, at which time the rabbit's blood is laden with the hormones controlling its own reproductive cycle. These hormones stimulate maturation in the flea. This relationship probably evolved as a mechanism that increased the probability of the next generation of fleas having abundant food.

Hormones commonly are categorized according to where they originated in the body and the way in which they reached the circulating body fluids. There are three types.

1. *Glandular hormones* originate in special ductless glands that secrete solely or primarily hormones, commonly a specific hormone.
2. *Tissue hormones* are produced by various nonglandular tissues, and the secretion of the hormone is solely secondary in function; for example, histamine is produced by damaged tissue.
3. *Neurohormones* originate in special cells of the central nervous system called the neurosecretory cells, and they travel in the form of neurosecretory granules along the axons of these cells to special glandular organs (for example, the corpora cardiaca in insects). They are stored in these organs and later pass into the body fluids.

The neurosecretions commonly are regarded as a third system of physiological and behavioral control that lies intermediate between the nervous and endocrine systems. This third system induces long-term endocrine responses by short, usually repeated, nervous stimuli. For example, certain hormonal secretions are induced in the prothoracic glands and corpora allata by the *activation hormone*, which is a neurohormone formed in specific neurosecretory cells in the protocerebrum of the brain, as a result of changes in photoperiodism.

Neurohormones originate in many different parts of the central nervous system, including the cerebral ganglion, subesophageal ganglion, ventral ganglia including thoracic and abdominal ganglia, stomodeal ganglia, and neurohaemal organs. Their diverse effects include the following.

1. Influence on the frequency and amplitude of spontaneous rhythmic activity of muscle (for example, muscle in the heart, Malpighian tubules, oviducts, gut, etc.).
2. Influence on movements of pigments, including dispersion and concentration of pigments in the epidermis and chromatophores, and development of various pigments in the epidermis.

3. Influence on cyclical, seasonal, and diurnal activity.
4. Influence on water balance within the insect.
5. Activation of enzymes (for example, midgut proteinase activity).
6. Activation of secretion of other endocrine glands.
7. Induction of diapause.
8. Influence on the development of gonads.
9. Influence on metabolic processes, such as the level of trehalose in the hemolymph.
10. Influence on adult emergence (eclosion).
11. Control of tanning in the cuticle of newly molted insects and postmolt deposition of the endocuticle. This neurosecretion is called *bursicon*.

Physiological Effects on Hormonal Control

Glandular hormones and neurohormones have widespread effects on an insect's life cycle, as illustrated by their effects on metamorphosis. The metamorphosis hormones influence postembryonic development. Presently three are recognized and all are essential for the normal course of postembryonic development.

1. Activation Hormone (AH). A neurohormone produced by specific neurosecretory cells in the protocerebrum of the insect brain, this steroid hormone conditions the reactivation of the body after each molt and the production of two other hormones, the molting hormone and the juvenile hormone. Activation hormone also has an indirect effect on ovarian development and the male accessory glands through its activation of the corpora allata. Interruption of the production of this hormone plays an important role in the control of diapause (Chap. 7, p. 245).

2. Molting Hormone (MH) (Ecdysone). A hormone produced by the prothoracic glands or corresponding tissues and by the ovaries of some female insects, MH is a steroid that is synthesized by the insect from cholesterol. It has two forms (α and β) that are generally similar in activity. This hormone initiates the molting process and thereby indirectly regulates growth and morphogenesis. In fly larvae, it initiates puparium formation. It stimulates development in general, including DNA, RNA, and protein synthesis, and enzyme activity.

3. Juvenile Hormone (JH) (Neotenin). This hormone is produced by the corpora allata and is released by nervous or hormonal stimulation from the brain. It has several functions: It conditions the growth of larvae in all metabolous insects, stimulates the ovarian follicles of adult females in most insects, and controls development and activity of several other structures and functions of the insect body, such as the fat body, which do not develop or function in its absence. The role of JH is shown in the following example. During the development of an insect, when the molting hormone is circulating in the young larva, the JH stimulates the epidermis to reconstruct a larval cuticle with larval characters. When levels of the JH are reduced, the epidermis develops pupal cuticle with pupal characteristics. When the JH is absent, the epidermis produces adult characteristics. When present in adults, the JH stimulates egg development. Juvenile hormone has some general effects on metabolism. For example, in the male locust *Schistocerca*, it controls color change as the insect sexually matures. In several species of locusts, JH is involved in the change from a solitary to gregarious migratory individual. In termites and aphids, JH plays a role in differentiation of various castes or types of individuals. In termites, soldier differentiation requires a greater amount of JH than that needed for normal termites.

REFERENCES

Agrell, I. P. S., and A. M. Lundquist, 1973. Physiological and biochemical changes during insect development. In Morris Rockstein (Ed.), *The physiology of Insects*, 2nd ed., **1**:159–247. New York: Academic Press.

Anderson, D. T., 1973. *Embryology and phylogeny in annelids and arthropods*. Oxford: Pergamon. 495 pp.

Askew, R. R., 1971. *Parasitic insects*. New York: American Elsevier. 316 pp.

Chapman, R. F., 1969. *The insects: Structure and function*. New York: American Elsevier. 819 pp.

Clausen, C. P., 1940. *Entomophagous insects*. New York: McGraw-Hill. (Reprinted 1962, New York: Hafner). 688 pp.

Counce, S. J., and C. H. Waddington (Eds.), 1971, 1972. *Developmental systems: Insects*. London and New York: Academic Press. Vol. 1, 304 pp.; Vol. 2, 615 pp.

de Wilde, J., and A. de Loof, 1973. Reproduction. In Morris Rockstein (Ed.), *The physiology of Insecta*, 2nd ed., **1**:11–95. New York: Academic Press.

Dixon, A. F. G., 1973. *Biology of aphids*. London: Edward Arnold. 58 pp.

Engelmann, F., 1970. *The physiology of insect reproduction*. Oxford: Pergamon. 307 pp.

Felt, E. P., 1940. *Plant galls and gall makers*. Ithaca, N.Y.: Comstock. 364 pp.

Hagan, H. R., 1951. *Embryology of the viviparous insects*. New York: Ronald Press. 472 pp.

Johannsen, O. A., and F. H. Butt, 1941. *Embryology of insects and myriapods*. New York: McGraw-Hill. 462 pp.

Lawrence, P. A. (Ed.), 1976. Insect development. *Symp. R. Entomol. Soc. London*, **8**:1–240.

Matsuda, R., 1976. *Morphology and evolution of the insect abdomen*. Oxford: Pergamon. 534 pp.

Novák, V. J. A., 1975. *Insect Hormones*. London: Chapman & Hall. 600 pp.

Snodgrass, R. E., 1935. *Principles of insect morphology*. New York: McGraw-Hill. 667 pp.

Tauber, M. J., and C. A. Tauber, 1976. Insect seasonality: Diapause maintenance, termination, and postdiapause development. *Annu. Rev. Entomol.*, **21**:81–107.

White, M. J. D., 1973. *Animal cytology and evolution*, 3rd ed. Cambridge: Cambridge Univ. Press. 961 pp.

8
The Orders of Entognaths and Insects

One of the greatest marvels of the living world is that entognaths and insects, starting from a single simple form, evolved into the multitude and diversity of species found in the world today. The study of this evolution is the drama of the fortuitous development of structures that endowed their possessors with added advantages in the struggle for existence. A sufficient number of early, primitive insect types have persisted to the present time, and enough fossils are known to allow us to reconstruct the main trends in the evolution of these fascinating creatures (Fig. 8-1).

In general, the evolutionary relationships shown in Fig. 8-1 follow the hypotheses proposed by many authors, including Hennig (1969), Hamilton (1972) and Kristensen (1975). The similarities shown among the hexapod arthropods indicate that they undoubtedly had a common origin. Although many agree in general with the Hennig–Kristensen ideas on insect phylogeny, a considerable difference of opinion exists as to how the various orders should be grouped into a classification. Here, we recognize two classes of hexapods (Fig. 8-1), the class Entognatha and the class Insecta, modifying the usage of Hennig (1969), Tuxen (1970), and Lauterbach (1972).

CLASS ENTOGNATHA

The earliest hexapods differed from their centipede-like ancestors chiefly in having a three-segmented thorax with its three pairs of legs, and the abdominal legs greatly reduced or absent. In these early

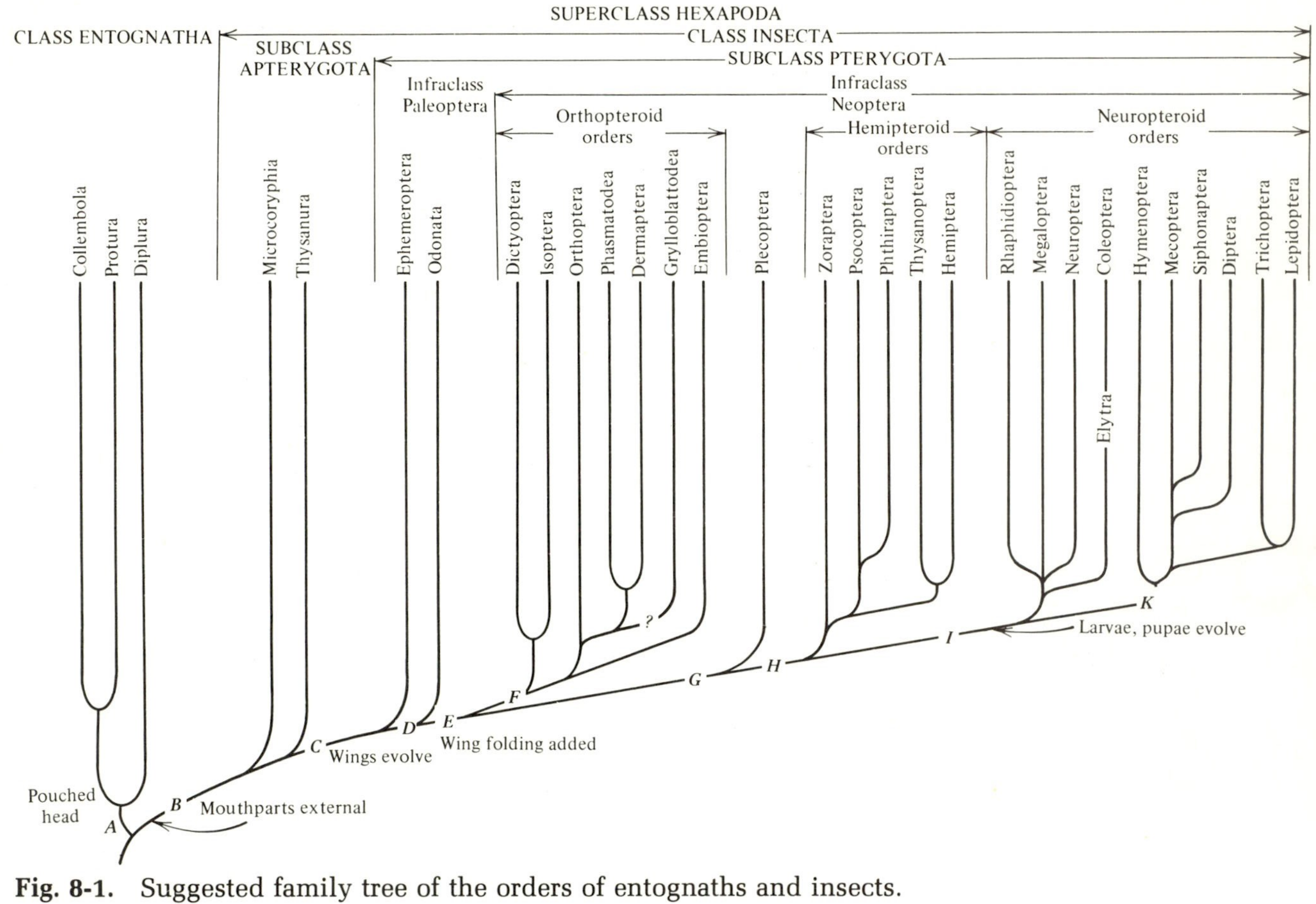

Fig. 8-1. Suggested family tree of the orders of entognaths and insects.

forms the young differed little in appearance from the adults and wings had not yet evolved. Six of these early wingless types are known. In three of them—the Diplura, Collembola, and Protura (Fig. 8-1A)—the internal head structure is little changed from that in the centipede-millipede groups, but the sides of the mouth cavity have fused with the sides of the labium and grown outward. This compound structure forms a cavity surrounding the functional mouthparts, the mandibles and maxillae. In addition, these three orders lack compound eyes. These features are the basis for grouping these three orders into a primitive branch called the class Entognatha, which means "mouthparts inside." The most primitive entognathans comprise the order Diplura (Fig. 8-1). Considerable evidence suggests that two other orders, the Collembola and Protura, arose from a Diplura-like ancestor in which the abdominal spiracles and cerci atrophied. In spite of their many features in common, the Collembola and Protura have evolved into remarkably dissimilar animals. In the Protura the eyes are absent, the antennae are nearly atrophied, and the front legs have become somewhat antenna-like; in the Collembola the abdominal segments are reduced in number and the vestigial legs of the fourth abdominal segment have fused and developed into a forked leaping organ or spring.

CLASS INSECTA

Subclass Apterygota

The primitively wingless orders of insects (Fig. 8-1*B*) are represented by the extant Microcoryphia (machilids) and Thysanura (silverfish), which form the subclass Apterygota. They have mouthparts that are external to the head (ectognathous) and are similar in this feature to other insects. These orders developed several body and leg proportions that led to the evolution of insect flight. Most important of these were the longer and stronger thoracic legs, providing speed and agility in running and jumping, and an increase in the relative size of the thorax compared with the abdomen, which moved the body's center of gravity forward. Of the living orders of Apterygota, the Microcoryphia have styli in abdominal segments 2 to 9 and are more primitive than the Thysanura, which lack styli on segments 1 to 6. The fossil order Monura resembled the Microcoryphia, but still had styli on segment 1 and was, therefore, more primitive than the other two. In the Thysanura line, the body became wider and flatter, suggesting the possibility of at least some aerial planing.

Subclass Pterygota

Infraclass Paleoptera

It is probable that some Thysanura-like insect developed more efficient planing habits and structures that eventually led to wings and flight, giving rise to the winged insects of Pterygota (Fig. 8-1C). The first successful wings were sharply and deeply pleated like a fan, but could not be folded and laid over the back at rest. Insects with wings of this type belong to the infraclass Paleoptera (ancient wings) and include several extinct orders and two living orders, Ephemeroptera (mayflies) and Odonata (dragonflies and damselflies).

At this point in insect evolution a drastic change occurred in the life history. In the Apterygota the individuals continued to molt regularly after becoming functional adults. In the Pterygota molting ceased as wings and functional reproductive structures were developed. The only exception is the Ephemeroptera, in which functional wings develop in the next to last instar. This preadult instar (called *subimago*) is typically not reproductively functional; reproductivity occurs in the true adult instar, the *imago*. From this it is inferred that primeval winged insects (of which we have no record) continued to molt after maturity as in the Apterygota, but that strong selection pressures led to the evolution of forms in which postadult molting did not occur. The Ephemeroptera may represent the last stage of such an evolutionary change. Further support for this inference comes from mayfly wings. They are the only ones in which radius and sector are not fused, marking them as the most primitive of living Pterygota. The Odonata and their fossil relatives represent the next step (Fig. 8-1*D*) in pterygote evolution, in which radius and sector became fused at their bases.

Infraclass Neoptera

In some paleopteran lineage (Fig. 8-1*E*), a set of mechanisms developed by which the wings could be folded and laid over the back at rest. Concurrently much of the wing fluting was reduced, resulting in larger flat areas on parts of the wing. Insects with wings of this type are called Neoptera (modern wings).

The first abundant members of the Neoptera may have been the extinct suborder Protoblattoidea and order Protorthoptera, diverse groups of fairly large running insects having completely veined wings, a typical sawlike ovipositor, and long cerci. From the Protorthoptera, or similar early forms, arose the lineages represented by the living neopteran orders.

The most primitive branch arising from the Protorthoptera evolved into the orthopteroid branch, represented by seven extant orders (Fig. 8-1*F*). The Embioptera (web spinners) have a greatly reduced wing venation, no distinct ovipositor, and peculiar silk-producing front legs. In both front and hind wings, radius and sector are fused for only a short distance, suggesting that this order is the most primitive in the orthopteroid branch. A second lineage contains the Dictyoptera (cockroaches and mantids) and the Isoptera (termites), both of which have a primitive venation and, in early fossil forms, a well-developed sawlike ovipositor that now is greatly reduced in all living forms.

Another orthopteroid lineage, characterized by a modified wing venation, embraces the Dermaptera (earwigs), Phasmatodea (walking sticks), and Orthoptera (grasshoppers and crickets). The Orthoptera have a well-developed sawlike ovipostior; this has been lost in the other two orders, which have evolved other distinctive characteristics.

Also in the orthopteroid branch is the order Grylloblattodea (grylloblattids), in which all known species lack wings. The placement of this order is open to question. The females have a primitive sawlike ovipositor and the male genitalia have several primitive features. Some authors consider the group as a subgroup of the Orthoptera, others as possibly an extremely ancient member of the orthopteroids, which is the view followed here.

In these seven orders (and presumably in related fossil orders also), the wings develop as external wing pads, and except for the lack of wings and difference in size, the nymphs look much like the adults and usually live and feed the same way. These living orders are collectively termed the orthopteroid orders.

In another line, arising from the same neopteran base (Fig. 8-1*G*), one of the mesopleural sclerites, the trochantin, became reduced to a thin strap forming a movable leg articulation, and certain adult characters became suppressed in the immature stages. It is possible that the order Plecoptera (stoneflies) is an early offshoot of this line, for it has the narrow trochantin but no appreciable suppression of adult characters in the nymph. Beyond the possible point of origin of the stonefly line, however, a form arose in which the ocelli became suppressed in the immature stages. This ancestor gave rise to two vigorous groups, the hemipteroid orders and the neuropteroid orders (Fig. 8-1*H*).

In the hemipteroids, the adult tarsi were reduced to three segments. The first members of the line were probably general feeders much like the order Psocoptera (psocids). The most primitive branch seems to be represented by the Zoraptera (zorapterans), a group of colonial insects resembling psocids but preserving distinct cerci (see page 100). In the Psocoptera, the lacinia of the maxilla evolved into a slender, chisel-like piece, which forms the basis of sucking-type mouthparts in two other orders. The first step in this development is exemplified by the order Thysanoptera (thrips), which have evolved a rasping-sucking set of mouthparts. The end stage is seen in the Hemiptera (bugs), in which the mouthparts form a slender piercing-sucking organ. Another branch of the hemipteroid line evidently arose directly from a Psocoptera-like ancestor, and became established as skin scavengers of animals that evolved into the ectoparasitic order Phthiraptera (chewing and sucking lice). In most hemipteroid nymphs the wings develop gradually as external wing pads (paurometabolous). In the Thysanoptera and two groups of Hemiptera (Psyllidae and Coccoidea), the wings at first grow internally and are everted only at a late (often preadult) instar that is a resting stage and thus a typical pupa. In these groups the prepupal instars commonly are called larvae.

In the neuropteroid line (Fig. 8-1*I*), the trend in juvenile suppression of adult characters became accelerated, the immature stages (or larvae) having greatly reduced eyes, antennae, and body sclerotization. The last immature instar developed into a typical pupa in which the various parts are re-formed into those of the adult. The adults of primitive members retained five-segmented tarsi.

The primeval neuropteroid insect can be reconstructed with considerable reliability by adding together primitive characters still persisting in living orders. It was probably of medium size for an insect. In the adults, the two pairs of wings were about the same size and each had a fairly large number of veins and crossveins; the abdomen was well sclerotized, and it possessed a sawlike ovipositor much like that of a sawfly. The larva was elongate, predaceous, fast running, and resembled present-day snakefly larvae and some beetle larvae.

The relationships between the neuropteroid orders have been a real puzzle, but in recent years additional morphological information has permitted the reconstruction of what appears a highly probable solution to the main branchings. The primeval neuropteroid line first gave rise to two branches, the Neuroptera–Coleoptera and the Hymenoptera–Mecoptera. In the Neuroptera–Coleoptera branch, the ovipositor became highly modifed by a reduction of valvula 1 and the loss of valvula 2; the larval tarsi are primitive in retaining two tarsal claws, at least in the primitive families of each order. In the ancestor of the Hymenoptera–Mecoptera branch the ovipositor was of the primitive three-bladed type; the larval tarsi, however, had only one claw.

The Neuroptera–Coleoptera branch gave rise to the Coleoptera, in which the fore wings became hardened and formed a protective cover over the abdomen; and to a complex of three orders, the Megaloptera, Neuroptera, and Rhaphidioptera. The latter three share several small morphological features, but between themselves each order is highly distinctive. The Megaloptera have the most primitive venation, are aquatic, and the larvae have long, segmentally arranged abdominal filaments and gills. The Rhaphidioptera have generalized terrestrial larvae; the adults have a long ovipositor and modified venation. The Neuroptera are extremely diverse; their larval mandibles and maxillae fit together to form a pair of puncturing-sucking structures used in extracting food from their prey.

The Hymenoptera–Mecoptera branch (Fig. 8-1K) gave rise to the Hymenoptera (bees and wasps), in which the venation became peculiarly modified but the ovipositor retained the primitive sawlike condition, and a group of five orders that lost the saw and sheath of the ovipositor, the latter becoming a simple extensile tube. The ancestral line of this latter group appears to have divided into the Trichoptera–Lepidoptera stem, in which the longitudinal veins and especially the crossveins became reduced in number; and the Mecoptera–Diptera stem, in which the lower edge of the pronotum fused with the anterior edge of the mesopleuron below the spiracle. In the Trichoptera (caddisfly) line arising from the Trichoptera-Lepidoptera stem, the larvae became aquatic and their spiracles atrophied; in the Lepidoptera (moth and butterfly) line arising from the Mecoptera–Diptera stem, little change seems to have occurred; in the Diptera (true fly) line, the venation became greatly reduced, the second pair of wings evolved into small knoblike balancing organs called halteres, and the pronotum fused very solidly with the adjacent mesothoracic parts.

The order Siphonaptera (fleas) has no winged species and has become so modified for existence as an ectoparasite that only a few clues to its ancestory remain. Fleas share two derived characters with the Mecoptera and Diptera. In primitive members of all three orders, the mandibles are flat and bladelike, fitted for cutting and piercing the tissues of their hosts, and the ovipositor is a simple tube. These features suggest that the fleas belong in this complex (Fig. 8-1K). Certain internal structures of the proventriculus are remarkably like those of the Mecoptera. Unlike the Mecoptera, the flea pronotum is not fused with the mesopleuron; however, this is a connection that could have been lost without difficulty in the evolution of the wingless conditions. Because of the uniqueness of the proventricular structure, it seems probable that the fleas are a highly modified offshoot of primeval Mecoptera.

During their evolutionary history insects have multiplied into more species than any other group of organisms. The 27 living orders of insects comprise nearly 1000 families and many thousand genera, and at the present, about 900,000 described species. These do not include a large number of fossil orders, genera, and species. All 27 orders, over 500 families, several thousand genera, and over 100,000 species, are known from North America. Many more species than this occur. It has been estimated that nearly 25,000 species remain to be discovered from North America and at least a million more from the entire world.

NAMES OF ORDERS AND FAMILIES

In many cases more than one name may have been applied to the same order or family. Certain of these different usages are based on contentions regarding either the descriptive propriety or the date of priority of the names in question. In this book the attempt is made in all such cases to employ the name in most common use. Notations are given concerning synonymous names that the beginning student is likely to encounter when using reference material.

ZOOGEOGRAPHIC REGIONS

Various large regions of the world have many distinctive taxonomic elements in their biotas. On the basis of relative similarities and differences between their animal components, six major realms or zoogeographic regions of the world are recognized by zoologists, as shown in Fig. 8-2. These regions are mentioned frequently in taxonomic literature.

General References on the Orders of Insects

After each order are listed references that will be helpful in obtaining a more thorough knowledge of the respective order. The following list of general references is concerned with classification, evolution, the rarer or exotic families of insects, terminology, or in collecting or rearing insects.

References

Borror, D. J., D. M. DeLong, and C. A. Triplehorn, 1976. *An introduction to the study of insects*, 4th ed. New York: Rinehart. 852 pp.

Brues, C. T., A. L. Melander, and F. M. Carpenter, 1954. Classification of insects. *Mus. Comp. Zool. Harvard Coll. Bull.*, **108:**1–917.

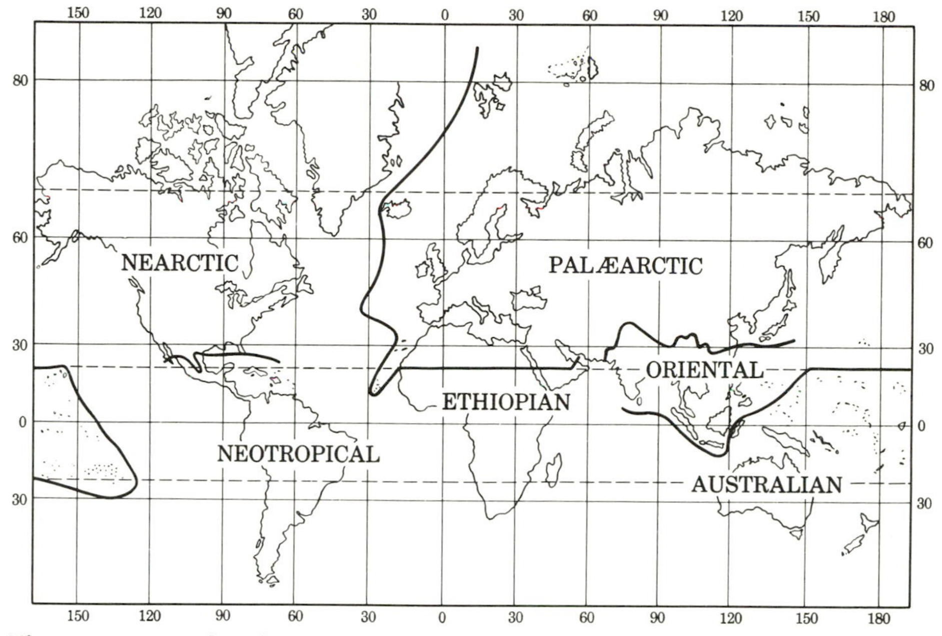

Fig. 8-2. Faunal realms of zoogeographic regions of the world. (After Sclater and Wallace)

de la Torre-Bueno, J. R., 1950. *A glossary of entomology*. Brooklyn, N.Y.: Brooklyn Entomology Society. 336 pp.

Foote, R. H., 1977. *Thesaurus of entomology*. College Park, Md.: Entomological Society of America. 188 pp.

Hamilton, K. G. A., 1972. The insect wing, Part IV. Venational trends and the phylogeny of the winged orders. *J. Kans. Entomol. Soc.*, **45**(3):295–308.

Hennig, W., 1969. *Die Stammesgeschichte der Insekten*. Frankfurt am Main: W. Kramer. 436 pp.

Horn, D. C., 1976. *Biology of Insects*. Philadelphia: Saunders. 439 pp.

Insects—The Yearbook of Agriculture, 1952. Washington, D.C.: USDA.

Insects of Hawaii, 1948–1981. Vols. 1–14 to date. Honolulu: Univ. of Hawaii Press. Authors to volumes are: E. C. Zimmerman, Vols. 1–9; D. E. Hardy, Vols. 10–14; J. M. Tenorio, Suppl. to Vol. 11.

Kristensen, N. P., 1975. The phylogeny of hexapod "orders": A critical review of recent accounts. *Z. Zool. Syst. Evolutionsforsch.*, **13**:1–44.

Lauterback, K.-E., 1972. Beschreibung zweier neuer europäischer Inocelliiden (Insecta, Raphidioptera), zugleich ein Beitrag zur vergleichenden Morphologie und Phylogenie der Kamelhalsfliegen. *Bonn. Zool. Beitr.* **23**:219–252.

Linsenmaier, W., 1972. *Insects of the world*. New York: McGraw-Hill. 392 pp.

Mackerras, I. M., 1970 *Evolution and classification of the insects*. In

Waterhouse, D. F., et al. (Eds.), *The insects of Australia*. Carlton, Victoria: Melbourne Univ. Press. pp. 152–167; Suppl. 1974, pp. 26–27.

Merritt, R. W., and K. W. Cummins (Eds.), 1978. *An introduction to the aquatic insects of North America*. Dubuque, Iowa: Kendall/Hunt. 442 pp.

Peterson, A., 1948, 1951. *Larvae of insects*. Ann Arbor: J. W. Edwards. Pt. 1, 315 pp.; Pt. 2, 416 pp.

1953. *A manual of entomological techniques*, 7th ed. Ann Arbor: J. W. Edwards. 367 pp.

Richards, O. W., and R. G. Davies, 1977. *Imms' general textbook of entomology*, 10th ed. London: Chapman & Hall. Vols. 1 and 2, 1354 pp.

Romoser, W. S., 1973. *The science of entomology*. New York: Macmillan. 449 pp.

Ross, H. H., 1953. How to collect and preserve insects, 3rd ed. Ill. Nat. Hist. Surv. Circ., **39**:1–59.

——— 1974. *Biological systematics*. Reading, Mass.: Addison-Wesley. 345 pp.

Smith, R. C., et al., 1944. *Insects in Kansas*. Topeka: Kansas State Board of Agriculture. 440 pp.

Tietz, H. M., 1963. *Illustrated keys to the families of North American insects*. Minneapolis: Burgess.

Tuxen, S. L. (Ed.), 1970. *Taxonomist's glossary of genitalia in insects*, 2nd ed., rev. Copenhagen: Munksgaard. 359 pp.

Usinger, R. L. (Ed.), 1965. *Aquatic Insects of California, with keys to North American genera and California species*. Berkeley: Univ. of California Press. 508 pp.

Waterhouse, D. F., et al., 1970. *The insects of Australia*. Carlton, Victoria: Melbourne Univ. Press, 1029 pp.; 1974, *Suppl*. 146 pp.

KEY TO THE ORDERS OF COMMON ENTOGNATHA AND ADULT INSECTA

1. Wings well developed (Fig. 8-9), the front pair sometimes forming short, hard wing covers (Fig. 8-32) 2
   Wings reduced to small pads, or no wings developed 32
2. Front wings hard, horny, opaque, and veinless, in repose lying over the body and forming wing covers (Figs. 8-32, 8-79) 3
   Front wings either transparent, membranous, or with definite veins or one-half horny, one-half membranous 6
3. Front and hind wings both long, narrow, and the same shape; ventral region of head capsule prolonged into a beaklike structure at the end of which are situated typical chewing mouthparts, much as in Fig. 8-130 (male *Boreus*) .... **Mecoptera**, p. 434
   Front and hind wings either broad or quite dissimilar in shape; head capsule not prolongated 4
4. Mouthparts forming a needle-like piercing beak (Fig. 8-49); venation usually indicated but indistinct .... **Hemiptera**, p. 335
   Mouthparts with mandibles fitted for chewing (Fig. 3-8) 5

5. Abdomen terminating in a pair of external forceps-like appendages (Fig. 8-32) .... **Dermaptera**, p. 310
Abdomen either without terminal appendages, or these pointed and stylelike (Fig. 8-87) .... **Coleoptera**, p. 373
6. Having only 1 pair of wings, the hind pair at most forming small, clubbed balancing organs, or halteres (Fig. 8-135) 7
Having 2 pairs of wings, although the hind pair may be small 12
7. Wings leathery or parchment-like 8
Wings membranous, sometimes dark in color 10
8. Mouthparts forming a piercing-sucking beak (Fig. 8-49) .... **Hemiptera**, p. 335
Mouthparts fitted for chewing, with generalized parts as in Fig. 3-8 9
9. Hind legs greatly enlarged for leaping (Figs. 8-23 to 8-32) .... **Orthoptera**, p. 304
Hind legs not greatly enlarged, fitted for running (Figs. 8-16, 8-18) .... **Dictyoptera**, p. 295
10. Abdomen without terminal filaments; hind wings represented by halteres .... **Diptera**, p. 440
Abdomen with 1, 2, or 3 terminal filaments, as in Figs. 8-9, 8-61 11
11. Halteres present; terminal appendages short, antennae long (Fig. 8-61, males of Coccidae) .... **Hemiptera**, p. 335
Halteres not developed; terminal appendages very long, antennae short, as in Fig. 8-9 .... **Ephemeroptera**, p. 286
12. Front wings, and usually also hind wings, clothed with overlapping scales (Fig. 8-16), except for window-like areas (Fig. 8-177) .... **Lepidoptera**, p. 479
Wings bearing only scattered scales, having chiefly fine hair, bristles, or no vestiture 13
13. Tarsus ending in a round bladder-like structure, without evident claws (Fig. 8-47); wings long and narrow, at most with 2 longitudinal veins .... **Thysanoptera**, p. 331
Tarsus without a terminal bladder, sometimes having large pads and/or distinct claws (Fig. 8-52 *C* to *E*); wings not as above 14
14. Mouthparts forming slender stylets fitted together for piercing and sucking, and housed in a beak that is either triangular or rodlike (Figs. 8-49*A*, 8-52*G*) .... **Hemiptera**, p. 335
Mouthparts not forming a beak, either vestigial or of the chewing or lapping type; mandibles not forming slender stylets 15
15. Abdomen having 2 or 3 terminal filaments as long as the body, and head having short hairlike antennae (Fig. 8-9) .... **Ephemeroptera**, p. 286
Either abdomen having short terminal filaments or none, or antennae long and slender (Figs. 8-18, 8-20) 16
16. Front wings leathery, hind wings membranous, the front ones when folded forming a protective covering over the hind pair (Figs. 8-16,

8-23) 17
Both pairs of wings of about the same texture 18

17. Hind legs greatly enlarged for leaping (Fig. 8-23) .... **Orthoptera**, p. 304
Hind legs not greatly enlarged, fitted for running (Figs. 8-16, 8-18) .... **Dictyoptera**, p. 295

18. Both front and hind wings having many veins and many crossveins, forming a close network over part of the wing surface (Figs. 8-20, 8-36, 8-76) 19
Wings with crossveins few in number (Fig. 3-43), or longitudinal veins reduced (Fig. 8-109), or both 28

19. Front and hind wing each with radius and branches of sector sclerotized and forming a heavy anterior band, remainder of venation semimembranous or subatrophied (Fig. 8-20) .... **Isoptera**, p. 300
Venation sclerotized throughout 20

20. Antennae short and setalike, 5- to 8-segmented; wings long and exceedingly reticulate (Figs. 8-12, 8-14) .... **Odonata**, p. 289
Antennae elongate, usually with more than 20 segments; wings various (Fig. 8-36) 21

21. Tarsus 2- or 3-segmented .... **Plecoptera**, p. 316
Tarsus 5-segmented 22

22. Head prolonged into a trunklike beak, and having typical chewing mouthparts (Fig. 8-130) .... **Mecoptera**, p. 434
Head not prolonged into a beak 23

23. Prothorax greatly lengthened (Figs. 8-74, 8-78) 24
Prothorax little, if at all, longer than wide (Figs. 8-3, 8-76) 25

24. Front legs arising from posterior margin of prothorax and shaped like the other legs (Fig. 8-74) .... **Raphidioptera**, p. 365
Front legs arising from anterior margin of prothorax and fitted for seizing prey (Fig. 8-78, Mantispidae) .... **Neuroptera**, p. 368

25. Mesoscutellum wide and short, its anterior margin truncate (Fig. 8-3*B*); posterior angle of lateral margin of pronotum forming a point that is fused to the upper anterior edge of the mesopleuron, the mesothoracic spiracle situated above this fusion (Fig. 8-3C) .... **Mecoptera**, p. 434
Mesoscutellum longer, its anterior margin pointed (Fig. 8-3*A*); posterolateral angle of pronotum not fused with mesopleuron, the mesothoracic spiracle therefore not above a sclerotized bridge 26

26. Ocelli present (Corydalidae) (Fig. 8-75) .... **Megaloptera**, p. 366
Ocelli absent 27

27. Pronotum as long as mesonotum and head (Fig. 8-75) (Sialidae) .... **Megaloptera**, p. 366
Pronotum shorter than either mesonotum or width of head through eyes (Fig. 8-3*A*) .... **Neuroptera**, p. 368

28. Tarsus 2- or 3-segmented 29
Tarsus 5-segmented 31

29. Pronotum minute: either short wide species, less than 4 mm long; or stout species up to 8 mm, with the mesonotum produced high above pronotum (Fig. 8-38) .... **Psocoptera**, p. 321

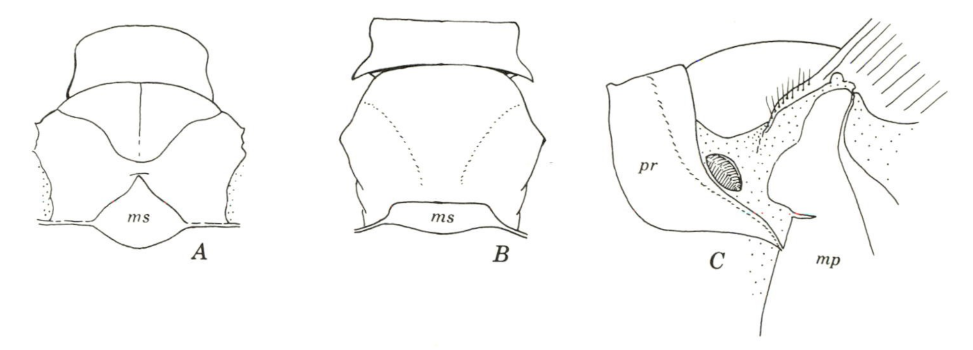

Fig. 8-3. Thoracic structures of insects. (*A*), Pro- and mesonotum of *Hemerobius* (Hemerobiidae); (*B*) same of *Brachypanorpa* (Panorpodidae); (C) anterior pleural region of *Brachypanorpa* (Panorpodidae). *ms*, Mesoscutellum; *mp*, mesopleuron; *pr*, pronotum.

Pronotum large and flat 30

30. Size less than 3 mm; front wings with only 3 principal veins (Fig. 8-37) .... **Zoraptera**, p. 319
    Size 5 mm or more; front wings with a much more extensive venation (Fig. 8-36) .... **Plecoptera**, p. 316
31. Mandibles sclerotized and large; wings clothed with minute setae, the venation either greatly reduced or forming a series of irregular cells (Fig. 8-109) .... **Hymenoptera**, p. 408
    Mandibles difficult to detect, subatrophied; either small species (less than 6 mm) with extremely hairy wings, or venation composed of regularly branching veins (Fig. 8-160) .... **Trichoptera**, p. 473
32. Abdomen ending in 2 or 3 long "tails" (Figs. 8-4*A*, 8-8) 33
    Abdomen without long terminal tails 35
33. Having 3 terminal tails (Fig. 8-8) 34
    Having 2 terminal tails (Fig. 8-4) 55
34. Abdominal segments 2 to 9 having styliform appendages (Fig. 8-7) .... **Microcoryphia**, p. 284
    Abdominal segments 1 to 6 without styliform appendages (Fig. 8-8) .... **Thysanura**, p. 284
35. Abdomen ending in a pair of strong sclerotized pincer-like jaws or forceps (Fig. 8-4*B*) 36
    Abdomen without stout terminal forceps 37
36. Tarsus 1-segmented; head without eyes (Fig. 8-4*B*) .. **Diplura**, p. 279
    Tarsus 3-segmented; head having conspicuous eyes, as in Fig. 8-32 .... **Dermaptera**, p. 310
37. Tarsus 4- or 5-segmented 38
    Tarsus 1- to 3-segmented 48
38. Base of abdomen constricted to a narrow joint hinged to forward part of body (Figs. 8-111*D*, 8-123, 8-126) .... **Hymenoptera**, p. 408
    Abdomen not constricted and hinged at its base 39

39. Antenna minute, flattened, and indistinctly segmented (Fig. 8-132) or round, sometimes with a terminal hair much as in Fig. 8-140*J* 40
Antenna slender and long, many-segmented (Figs. 8-27, 8-167) ... 41
40. Antenna with many indistinct segments; body bilaterally compressed, with distinct segmentation .... **Siphonaptera**, p. 436
Antenna globular, appearing as one segment, sometimes with a terminal hair; body not greatly compressed from side to side, and often without distinct segmentation on abdomen. .... **Diptera**, p. 440
41. Head prolonged into a beaklike projection at the end of which are located a set of chewing mouthparts, as in Fig. 8-130 .... **Mecoptera**, p. 434
Head not prolonged into a beak 42
42. Mouthparts vestigial or composed chiefly of a short coiled tube, mandibles indistinct; body densely hairy or scaly (Fig. 8-167) .... **Lepidoptera**, p. 479
Mouthparts having sclerotized and massive mandibles, mouthparts of simple chewing type (Fig. 3-8) body never hairy 43
43. Pronotum enlarged, forming a long neck (Fig. 8-18); a large saddle (Fig. 8-25); or a shield partly or entirely hiding the head (Fig. 8-16) 44
Pronotum small, at most only slightly larger than mesonotum (Fig. 8-34) 46
44. Hind legs with femora greatly enlarged for leaping (Fig. 8-27) .... **Orthoptera**, p. 304
Hind legs not so enlarged 45
45. Either front legs fitted for grasping prey (Fig. 8-18), or body fairly flat (Fig. 8-16) .... **Dictyoptera**, p. 295
Front legs simple, not of a grasping type, and body round and robust .... **Orthoptera**, p. 304
46. Elongate sticklike insects, thorax as wide as or wider than abdomen (Fig. 8-22) .... **Phasmatodea**, p. 303
Body not sticklike (Fig. 8-34) 47
47. Thorax with all segments wide (Fig. 8-34) ... **Grylloblattodea**, p. 312
Stocky insects, prothorax constricted and markedly narrower than either head or abdomen (Fig. 8-20) .... **Isoptera**, p. 300
48. Head indistinct, antennae and legs short; insect body often covered by waxy filaments or plates, or by a detachable scale (Fig. 8-61) .... **Hemiptera**, p. 335
Head distinct; other characteristics diverse but body never covered by a scale 49
49. Head with antennae forming only a low, small blister (Fig. 8-6) .... **Protura**, p. 280
Head with antennae projecting, finger-like 50
50. Tarsus ending in a bladder-like pad; mouthparts together forming a conical structure (Fig. 8-47) .... **Thysanoptera**, p. 331
Tarsus ending in claws that may be small and sharp or large and hooked 51
51. Mouthparts forming a distinct external tubular beak (Fig. 8-49) .... **Hemiptera**, p. 335

Mouthparts not forming an external beak 52

52. Pronotum forming a large sclerite, often saddle-shaped and hind legs enlarged for leaping (Fig. 8-25) .... **Orthoptera**, p. 304
    Either pronotum reduced to a narrow sclerite or hind legs not particularly enlarged 53
53. Antenna 13- to 50-segmented (Fig 8-39) .... **Psocoptera**, p. 321
    Antenna 3- to 6- segmented 54
54. Leg having tibia and tarsus united, abdomen of many species having a ventral spring (Fig. 8-6). Free-living species ..... **Collembola**, p. 282
    Leg having tibia and tarsus separate, articulating with a joint (Figs. 8-40, 8-41, 8-45); abdomen never having a spring. Ectoparasites of warm-blooded animals and birds .... **Phthiraptera**, p. 324
55. Tarsi 1-segmented; mouthparts slender and inconspicuous, withdrawn inside the head .... **Diplura**, p. 279
    Tarsi 2- or 3-segmented; mouthparts conspicuous, not withdrawn into the head .... **Plecoptera**, p. 316

CLASS ENTOGNATHA

Order DIPLURA*: Campodeids and Japygids

These are wingless, blind, slender entognaths of small size, with long, many-segmented antennae, well-developed legs, and a pair of conspicuous cerci that are either segmented or forceps-like. Mouthparts of chewing type hidden within the ventral pouch of the head. Metamorphosis is not marked.

The young and adults differ chiefly in size and sexual maturity. The genae of the head and the labium form a ventral pouch in which the other mouthparts are situated. The legs are not so well developed as in the Thysanura, and the abdomen has neither vestigial paired appendages nor a caudal filament. In the Campodeidae (Fig. 8-4*A*) and allied families, the abdomen has a pair of many-segmented cerci; in the Japygidae (Fig. 8-4*B*), the cerci are forceps-like.

The species of the order occur under leaves, stones, logs, or debris, or in the soil. Their movements are at most moderately rapid, and they seldom if ever come out into the light. Little is known about details of their life history; none of the species is of economic importance.

References

Smith, L. M., 1960. The family Projapygidae and Anajapygidae (Diplura) in North America. *Ann. Entomol. Soc. Am.*, **53**:575–583.

*Some authors use the name Entotrophi for this order.

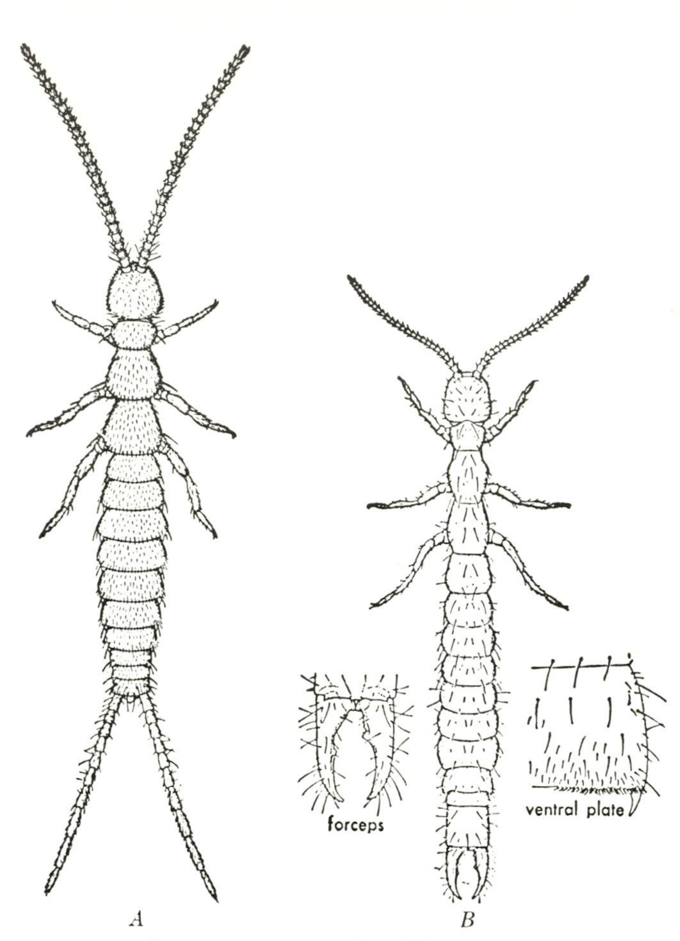

Fig. 8-4. Diplura. (A) *Campodea folsomi*; (B) *Japyx diversiunguis*. (From Essig, 1942, *College entomology*, by permission of Macmillan Co.)

Order PROTURA: Proturans

The adults are small and slender (Fig. 8-5), ranging from 0.5 to 2 mm in length. The head is cone shaped; it has no eyes; the antennae are reduced to minute, ocellus-like structures; and the mouthparts consist of stylet-like mandibles, small and generalized maxillae, and a poorly developed membranous labium that is fused at the sides with the cheeks. The three pairs of thoracic legs are similar in general appearance; the first pair serve as tactile organs. The first three abdominal segments have a pair of small ventral styli.

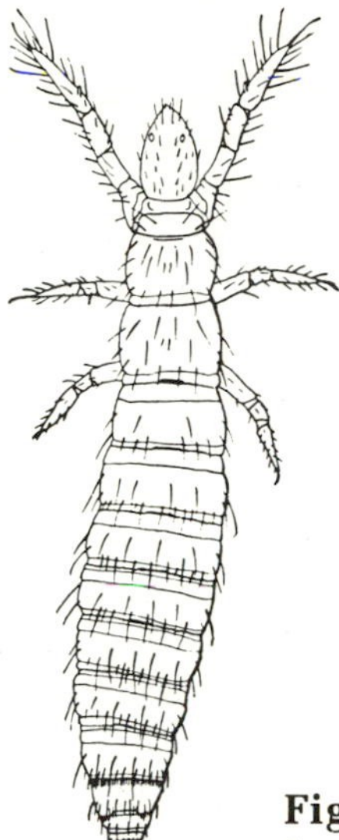

Fig. 8-5. A proturan, *Acerentulus barberi.* (Redrawn from Ewing, 1940)

The nymphs are similar to the adults in general appearance. In development they exhibit *anamorphosis*, that is, adding segments to the body at each molt. The abdomen of the first-stage nymph, the protonymph, has 9 segments; the abdomen of the deutonymph has 10 segments; that of the tritonymph has 11 segments; and, finally, the abdomen of the adult has 12 segments. The head and thorax are not affected in this manner. Proturans are moderately rare. They live in humus and soil, preferring damp situations. An ideal habitat for many species is old leaf mold along the edge of woods. Both adults and nymphs feed on decayed organic matter, and both may be found together during most of the year. Specimens may be collected either by examining leaf mold or by drying it in a Berlese funnel. Study specimens should be preserved in 70% ethyl alcohol, then slide mounted.

The order Protura has at times been considered as constituting a separate class, the Myrientomata.

References

Ewing, H. E., 1940. The Protura of North America. *Ann. Entomol. Soc. Am.*, **33**:495–551.

Tuxen, S. L., 1964. *The Protura. A revision of the species of the world with keys for determination.* Paris: Hermann and Cie. 360 pp.

Order COLLEMBOLA: Springtails

Minute to medium-small entognaths comprise this order; antennae and legs are well developed; mouthparts are of chewing type, but in some forms the maxillae and mandibles are long, sharp, and stylet-like; in the entire order the genae or cheeks have grown down and fused with the sides of labium, forming a hollow cone into which the other mouthparts appear to be retracted. The abdomen with six segments, frequently has a ventral jumping organ or furcula and a button-like structure, the tenaculum. Metamorphosis is absent; both sexes are usually similar.

The adults range in length from $^1/_5$ mm in the minute genus *Megalothorax* to over 10 mm in the larger species of the Entomobryidae and Poduridae. In the suborder Anthropleona the body is long and cylindrical; in the suborder Symphypleona the abdomen is round and more or less globular (Fig. 8-6).

The antennae are four- to six-segmented, the last segment sometimes with many fine annulations. The eyes are either lacking or represented by a series of isolated ommatidia. Also they lack Malpighian tubes and a tracheal system. Most members of the Collembola have a ventral springing organ or *furcula*; this is coupled with a ventral button or *tenaculum* when not in use. By means of the fur-

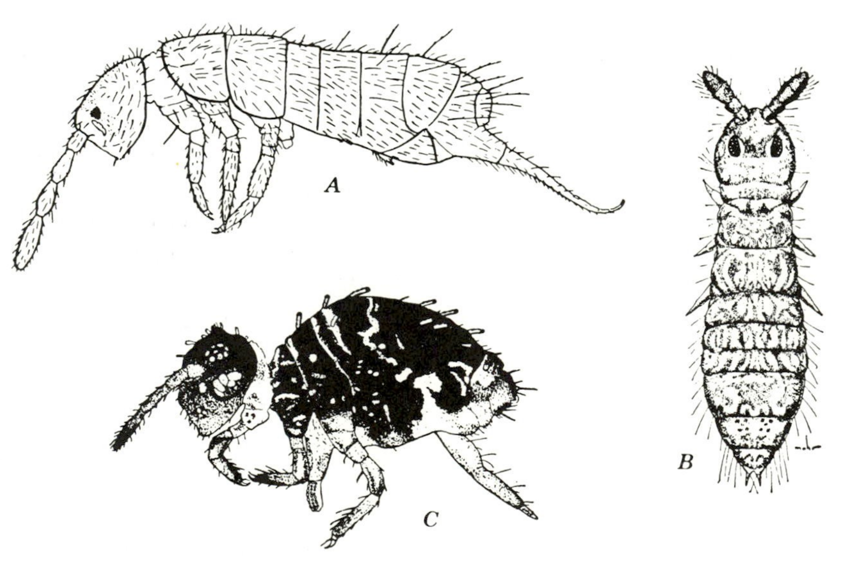

Fig. 8-6. Collembola. (*A*) *Isotoma andrei*; (*B*) *Achorutes armatus*, suborder Arthropleona; (*C*) *Neosminthurus clavatus*, suborder Symphypleona. (*A*, *C*, after Mills, 1934; *B* from Illinois Natural History Survey)

cula these little animals can execute a leap of some distance, which has earned them their name of springtails. The young are similar to the adults in both appearance and habits, differing chiefly in size and sexual maturity. Many of the species are white or straw-colored, others are blue, gray, yellow, mottled, or marked with distinctive patterns.

Springtails are found abundantly in many types of moist situations, including deep leaf mold, damp soil, rotten wood, the edges of ponds or streams, snow, and fleshy fungi. A few species, especially members of the family Sminthuridae, attack plants, and may be of local economic importance. A small gray species, *Achorutes armatus* (Fig. 8-6*B*), is sometimes destructive to mushrooms in commercial production. Egg-laying habits are known for only a few species, which lay their eggs singly or in clusters in humus or soil.

One of the most interesting features of the Collembola is their wide distribution.

KEY TO COMMON FAMILIES

1. Thorax and abdomen together comprising an almost globular or ovoid mass; all but anal segments of abdomen coalesced so that little external sign of segmentation remains (Fig. 8-6C) (suborder *Symphypleona*) .... **Sminthuridae**
   Thorax and abdomen tubular and more elongate, the segments of the abdomen indicated by external sutures (Fig. 8-6*A*, *B*) (suborder *Arthropleona*) 2
2. Dorsum of prothorax forming at least a semisclerotized plate similar in texture to that of mesothorax (Fig. 8-6*B*) .... **Hypogastruridae**
   Dorsum of prothorax completely membranous, in contrast to sclerotized or semisclerotized dorsal plate of mesothorax (Fig. 8-6*A*) .... **Entomobryidae**

References

Christiansen, K., 1964. Bionomics of Collembola. *Annu. Rev. Entomol.*, 9:147–178.

Joose, E. N. G., 1976. Littoral apterygotes (Collembola and Thysanura). In L. Cheng (Ed.), *Marine insects*. Amsterdam: North-Holland. pp. 151–186.

Maynard, E. A., 1951. *The Collembola of New York*. Ithaca, N.Y.: Comstock. 339 pp.

Mills, H. B., 1934. *A monograph of the Collembola of Iowa*. Ames, Iowa: Collegiate Press. 143 pp.

Richards, W. R., 1968. Generic classification, evolution, and biogeography of the Sminthuridae of the world (Collembola). *Mem. Entomol. Soc. Can.*, **53**:1–54.

Salmon, J. T., 1964–1965. An index of the Collembola. *Bull. R. Soc. N.Z.*, **7**:1-644, 3 vols.
Scott, H. G., 1961. Collembola: pictorial keys to neartic genera. *Ann. Entomol. Soc. Am.* **54**(1):104–113.

CLASS INSECTA

SUBCLASS APTERYGOTA

Order MICROCORYPHIA: Bristletails

These small- to medium-sized wingless insects have large eyes, long multisegmented antennae, long cerci, and a long mesal caudal filament (Fig. 8-7). The mouthparts are of chewing type, the abdomen with paired stylets on segments 2 to 9 (Fig. 3-48). They make up the family Machilidae.

The young and adults are extremely similar in shape and habits, differing chiefly in size and sexual maturity. The body is deep and tapers posteriorly. The abdomen has 10 complete segments; the eleventh segment forms the caudal filament. The legs are moderately stout. All species are very swift runners and agile dodgers. A common American genus is *Mesomachilis*, which feeds on humus and is found among leaves and around stones. It is thought that they lay eggs singly in cracks and crevices. These insects molt continuously throughout their life.

References

See under Thysanura.

Order THYSANURA: Silverfish, Firebrats

Similar in general to members of the previous order, differing in having a flatter, wider body, styliform appendages on only abdominal segments 7 to 9, and small or no eyes (Fig. 8-8).

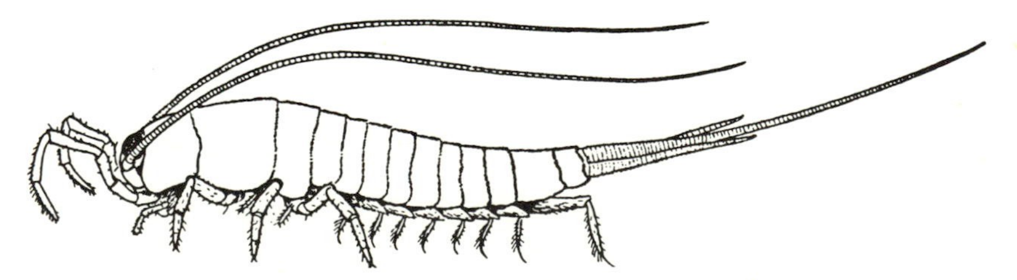

Fig. 8-7. Microcoryphia. *Machilis* sp. (Adapted from Snodgrass)

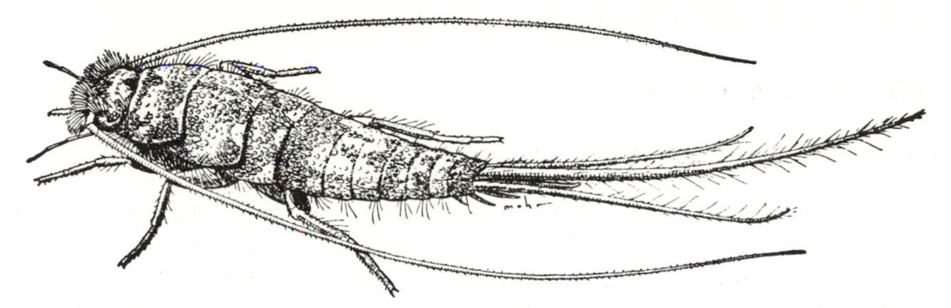

Fig. 8-8. Thysanura. *Thermobia domestica,* a common silverfish. (From Illinois Natural History Survey)

Most of the American species are domestic and belong to the family Lepismatidae. Like the Microcoryphia, they are extremely fast runners. They lay eggs singly in cracks, crevices, and secluded places. The young grow slowly, maturing in 3 to 24 months, and have a large and indefinite number of molts. Molting continues after adulthood is reached. The rare family Nicoletiidae contains a few small, ovate forms that live in ant nests and a few elongate subterranean species.

Economic Status. Silverfish, *Lepisma saccharina*, firebrats, *Thermobia domestica*, and other domestic species feed commonly on starch. They cause considerable damage to books and clothing by chewing off the starch sizing; other articles containing glue or sizing are attacked.

References

Adams, J. A., 1933. Biological notes upon the firebrat, *Thermobia domestica* Packard. *J. N.Y. Entomol. Soc.*, **41**:557–562.

Delany, M. J., 1957. Life histories of the Thysanura. *Acta Zool. Cracov.*, **2**:61–90.

Remington, C. L., 1954. The suprageneric classification of the order Thysanura. *Ann. Entomol. Soc. Am.*, **47**:277–286.

Slabaugh, R. E., 1940. A new thysanuran, and a key to the domestic species of Lepismatidae . . . in the United States. *Entomol. News*, **51**:95–98.

Smith, E. L., 1970. Biology and structure of some California bristletails and silver fish (Apterygota: Microcoryphia, Thysanura). *Pan-Pacific Entomol.*, **46**:211–225.

Wygodzinsky, P., 1972. A revision of the silverfish (Lepismatidae, Thysanura) of the United States and the Caribbean area. *Am. Mus. Novit.* No. 2481. 26 pp.

SUBCLASS PTERYGOTA

Infraclass Paleoptera

Order EPHEMEROPTERA*: Mayflies

These small to large, soft-bodied, slender insects have gradual metamorphosis. Adults have two pairs of net-veined wings, the metathoracic pair small and completely atrophied in a few genera; legs are usually well developed; antennae inconspicuous and hair-like; mouthparts vestigial; eyes large; and abdomen with a pair of cerci and in many species with a median terminal filament, all very long and tail-like (Fig. 8-9). Nymphs are aquatic, varied in shape, and similar in general structures to adults, but with well-developed chewing mouthparts; and usually with series of tracheal gills on the abdomen (Fig. 8-10).

Metamorphosis in the mayflies is characterized by a feature unique among living insects. The nymphs follow the usual type of gradual development with wings developing in external pads. When full grown, they swim to the surface of the water or crawl up on some support, and the winged form escapes from the nymphal skin. This winged form is capable of flight and looks like an adult but in most species is not yet sexually mature. The term *subimago* is applied to this stage. In a few genera, the subimago is the final form, the sexes mating and laying eggs in this stage. In most mayflies, however, the subimago molts again and produces the mature adult or *imago*. The adults apparently take no solid food, probably imbibing only water during their short life. In certain genera, especially *Hexagenia* and *Ephemera*, mass emergence of adults may take place, resulting in the

*The name Ephemerida is sometimes used for this order.

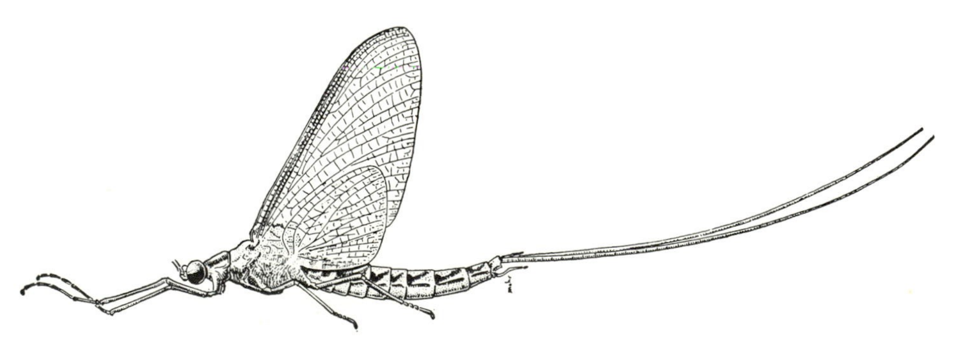

Fig. 8-9. Ephemeroptera. *Hexagenia limbata.* (From Illinois Natural History Survey)

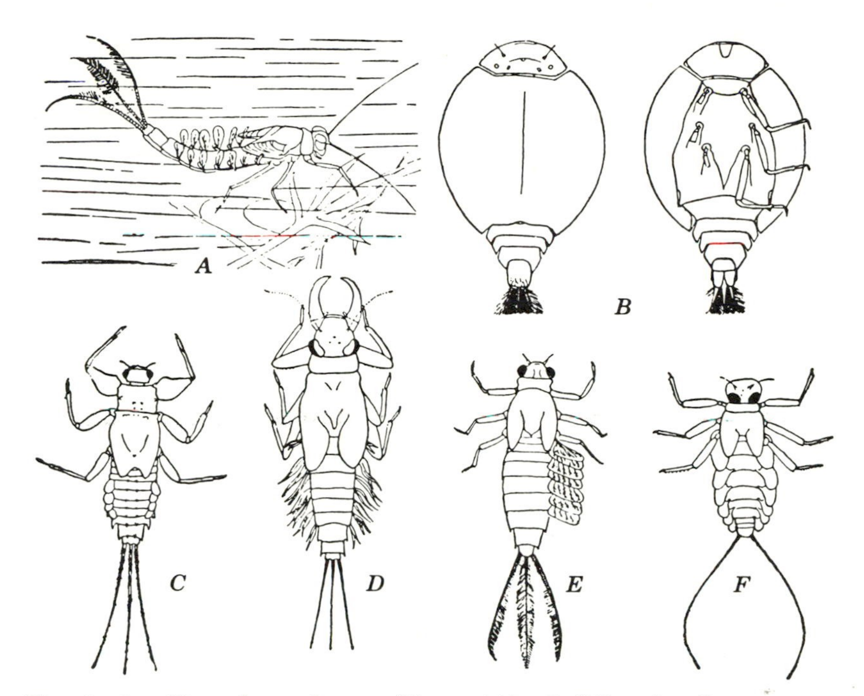

Fig. 8-10. Nymphs of mayflies. (*A*) *Callibaetis flunctuans*; (*B*) *Prosopistoma foliaceum*, dorsal and ventral aspects; (*C*) *Ephemerella grandis*; (*D*) *Paraleptophlebia packii*; (*E*) *Siphlonurus occidentalis*; (*F*) *Iron longimanus*. (From Essig, 1942, after various sources, *College entomology*, by permission of Macmillan Co.)

appearance of clouds of these insects over lakes and along streams. The adults mate in dancing swarms. In most genera, the female extrudes masses of eggs from the abdomen, swoops down to the water and releases the eggs into it. In the genus *Baetis*, the female crawls into the water and lays her eggs under stones. Each female may lay several hundred to several thousand eggs.

The complete winged life of many species of mayflies is extremely short, ranging from an hour and a half to a few days; mating normally occurs the same day adulthood is achieved, and the eggs are laid immediately. These eggs hatch in a few weeks or a month. In certain genera, such as *Callibaetis* and *Cloeon*, the adult females live much longer, from 2 to 3 weeks. In these longer-lived forms the eggs are fertilized and held in the body until the embryos are mature. When laid, the eggs hatch almost immediately on touching the water.

Nymphs (Fig. 8-10) live in a great variety of lake, pond, and stream situations. The nymphs of some species mature in 6 weeks; others may require 1, 2, or 3 years to attain their full growth. Their food consists of microorganisms and fragments of plant tissue.

Nymphs of the Ephemeridae and its allies live in mud, burrowing through it by means of their large shovel-like front legs. Many nymphs of other families occur under stones and logs. Those which live in rapid mountain streams may have the entire venter of the body developed into a disclike suction cup that enables them to attach firmly to smooth surfaces. Nymphs of a few genera live in small pools or ponds and are free swimmers along the bottom or in the shallows.

Mayflies play an extremely important role in the fish-food economy of most North American waters. They are the most abundant insect group in many types of fishing waters. Studies of fish-stomach contents indicate that, by and large, mayflies and chironomids (midges) are undoubtedly the two most important insect groups from the standpoint of fish food.

Well-illustrated keys for both the adults and nymphs of the families of Ephemeroptera are given by Edmunds in Merritt and Cummins (1978).

References

Berner, L., 1950. The mayflies of Florida. *Univ. Fla. Publ. Biol. Sci. Ser.*, **4**(4):1–267.

Burks, B. D., 1953. The mayflies or Ephemeroptera of Illinois. *Bull. Ill. Nat. Hist. Surv.*, **26:**1–216.

Day, W. C., 1956. *Ephemeroptera.* In R. L. Usinger (Ed.), *Aquatic insects of California.* Berkeley: Univ. of California Press. pp. 79–105.

Edmunds, G. F., 1972. Biogeography and evolution of the Ephemeroptera. *Annu. Rev. Entomol.*, **17:**21–42.

1978. *Ephemeroptera.* In R. W. Merritt and K. W. Cummins (Eds.), *Introduction to the aquatic insects of North America.* Dubuque, Iowa: Kendall/Hunt. pp. 57–80.

Edmunds, G. F., Jr., and R. K. Allen, 1957. A check list of Ephemeroptera of North America north of Mexico. *Ann. Entomol. Soc. Am.*, **50:**317–324.

Edmunds, G. F., Jr., R. K. Allen, and W. L. Peters, 1963. An annotated key to the nymphs of the families and subfamilies of mayflies (Ephemeroptera). *Univ. Utah Biol. Ser.*, **13:**1–49.

Edmunds, G. F., Jr., S. L. Jensen, and L. Berner, 1976. *The mayflies of North and Central America.* Minneapolis: Univ. of Minnesota Press. 330 pp.

Edmunds, G. F., Jr., and J. R. Traver, 1959. The classification of the Ephemeroptera. I. Ephemeroidea: Behningiidae. *Ann. Entomol. Soc. Am.*, **52:**43–51.

Flannagan, J. F., and K. E. Marshall (Eds.), 1980. *Advances in Ephemeroptera Biology.* New York: Plenum. 566 pp.

Koss, R. W., 1968. Morphology and taxonomic use of Ephemeroptera eggs. *Ann. Entomol. Soc. Am.*, **61**:696–721.

Needham, J. G., J. R. Traver, and Yin-chi Hsu, 1935. *The biology of mayflies.* Ithaca, N.Y.: Comstock. 759 pp.

Westfall, M. J., Jr., 1978. *Odonata.* In R. W. Merritt and K. W. Cummins (Eds.), *Introduction to the aquatic insects of North America.* Dubuque, Iowa: Kendall/Hunt. pp. 81–98.

Order ODONATA: Dragonflies and Damselflies

These medium-sized to large predaceous insects have gradual metamorphosis. Adults are slender or stout-bodied, with two pairs of nearly similar net-veined wings; legs well developed; antennae hair-like; mouthparts mandibulate, of the chewing type; eyes large; abdomen without long "tails." Nymphs are aquatic; mouthparts of chewing type, with labium elongate and hinged to form a stout grasping organ for seizing prey; legs stout; three leaflike terminal gills are present in the suborder Zygoptera.

Adult Odonata feed on insect prey captured on the wing. They devour mosquitoes, midges, horseflies—in fact, almost any insect that the odonate can tackle and catch successfully. The nymphs are aquatic, living chiefly in ponds, lakes, and backwaters of streams. They do not swim, but instead walk along the bottom or among debris or vegetation. The nymphs also are predaceous, catching aquatic insects, crustaceans, and the like, trapping them with the extensile spined labium.

The eggs are laid in or near the water in a variety of ways. Some are thrust into aquatic vegetation or rotten wood; others may be deposited in masses on some object just beneath the water surface, or laid in ribbons or rings in the water, or thrust into wet mud near the water's edge. Females of many species dip down to the surface and wash the eggs off the end of the abdomen. Others crawl beneath the water to deposit eggs.

Nymphs of the smaller species mature in a year. In the case of larger species, development may take 2 to 4 years. Hibernation is passed in the nymphal stage. When full grown, the nymphs crawl out of the water and attach to a stick, stem, or other object for the last molt. The newly emerged adults harden and color relatively slowly, many of them requiring 1 or 2 days for the process.

A peculiar characteristic of the order is the method of mating (Fig. 8-11). Before mating, the male bends the tip of the abdomen forward and transfers the spermatozoa to a bladder-like receptacle situated in the second abdominal sternite. In mating, the male, using its terminal claspers, grasps the female around the neck; the female then bends her abdomen forward to the second sternite of the male, at

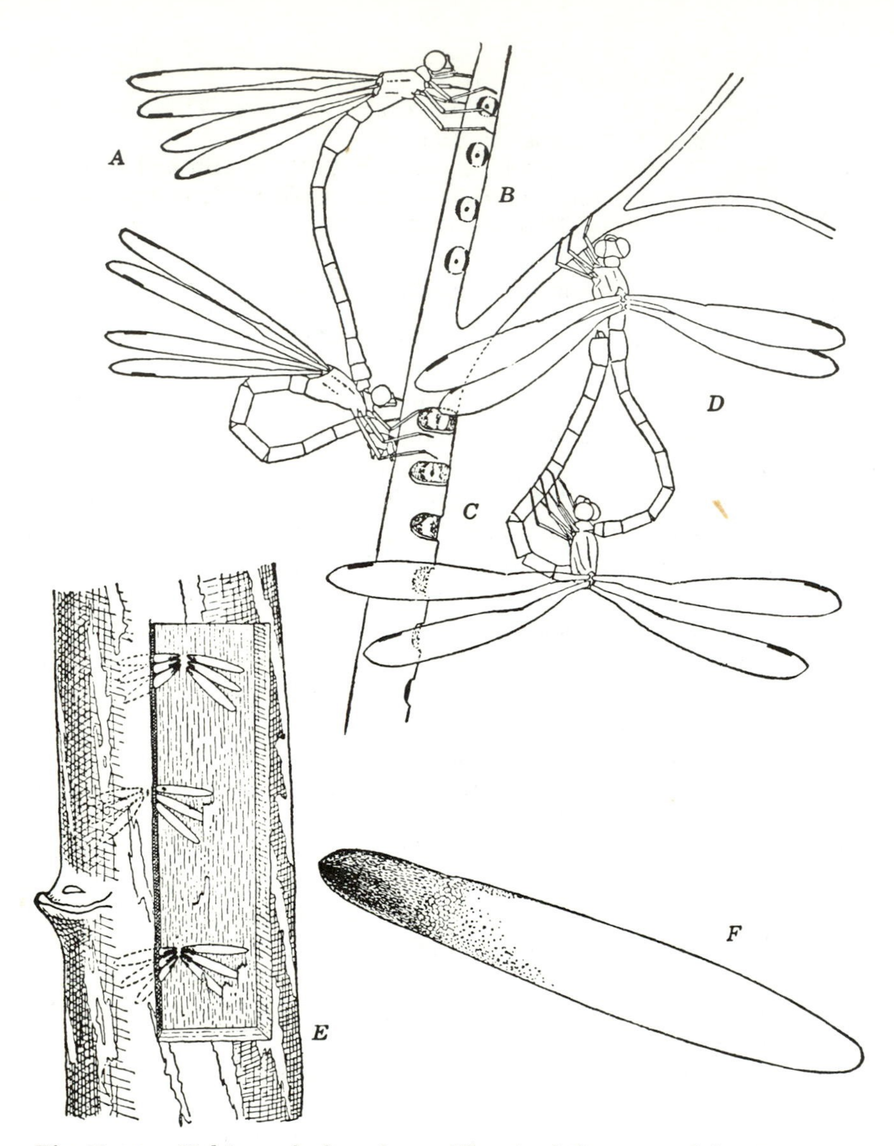

Fig. 8-11. Habits of the damselfly *Archilestes californica*. (*A*) Ovipositing; (*B*) scars from oviposition, 1 year old; (*C*) scars 2 years old; (*D*) in copulation; (*E*) bark cut away showing eggs in cambium; (*F*) egg. (After Kennedy)

which place the actual transfer of spermatozoa is effected. This unusual procedure is known in no other order of insects.

The Odonata includes three different types of insects that look and act strikingly different but that are separated by only a limited number of diagnostic characters. Present-day forms of one suborder, Anisozygoptera, are known to occur only as rarities in the oriental region. The two suborders that occur in North America may be separated by the following key.

DIAGNOSIS OF SUBORDERS

Nymphs provided with terminal leaflike tracheal gills (Fig. 8-13). Adults having fore and hind wings of similar shape and venation, when at rest held together and extending parallel to the abdomen (Fig. 8-12) .... **Zygoptera**, damselflies

Nymphs without external gills (Fig. 8-15). Adults with hind wings much wider than fore wings, especially at the base, extended outwards when at rest (Fig. 8-14) .... **Anisoptera**, dragonflies

Well-illustrated keys for both the adults and nymphs of the families of Odonata are given by Westfall in Merritt and Cummins (1978).

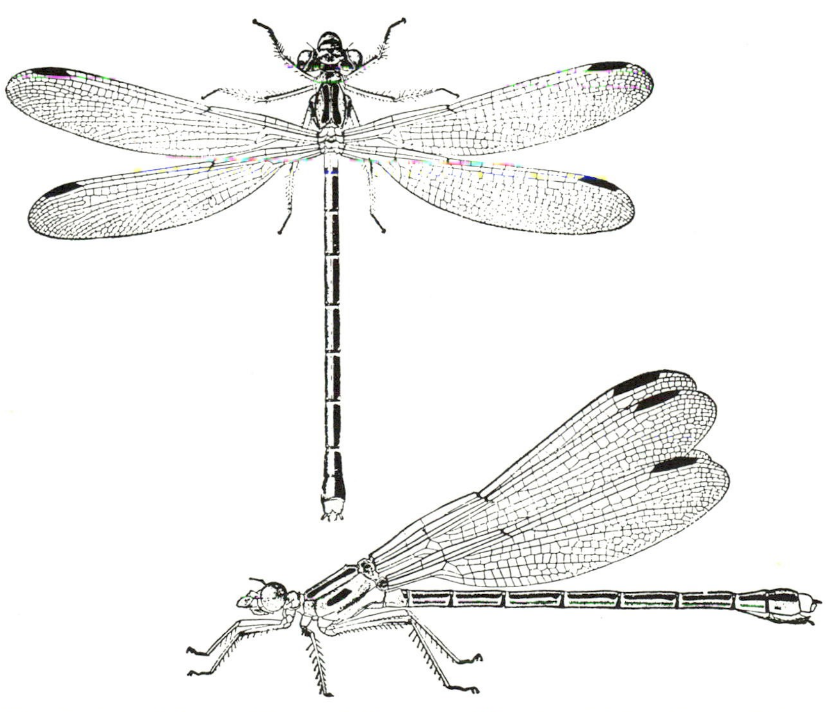

Fig. 8-12. A damselfly *Archilestes californica*. (After Kennedy)

SUBORDER ZYGOPTERA

DAMSELFLIES. Damselflies are slender and delicate, with a fluttering flight quite in contrast to the rapid and positive movements of the dragonflies. The damselfly adults have a thorax of very peculiar shape (Fig. 8-12); the meso- and metathorax together are somewhat rectangular and tilted backward 70 to 80 degrees in relation to the linear axis of the entire body. The wings at rest are held together above the back at right angles to the upper margin of the meso- and metathorax. Because they are tilted to such a degree, the folded wings are nearly parallel to and held just above the abdomen.

Most of the adults are somber-hued, but a few have red or black banding on the wings or metallic green or bronze body and wings.

The nymphs (Fig. 8-13) also are slender and possess three large caudal tracheal gills. They frequent the stems of aquatic vegetation more than the actual bottoms of ponds or streams.

SUBORDER ANISOPTERA

DRAGONFLIES. The adults of this suborder (Fig. 8-14) are stout bodied, with strong, graceful, and superbly controlled flight. The thorax is not tilted as in the damselflies, and the wings at rest are extended to

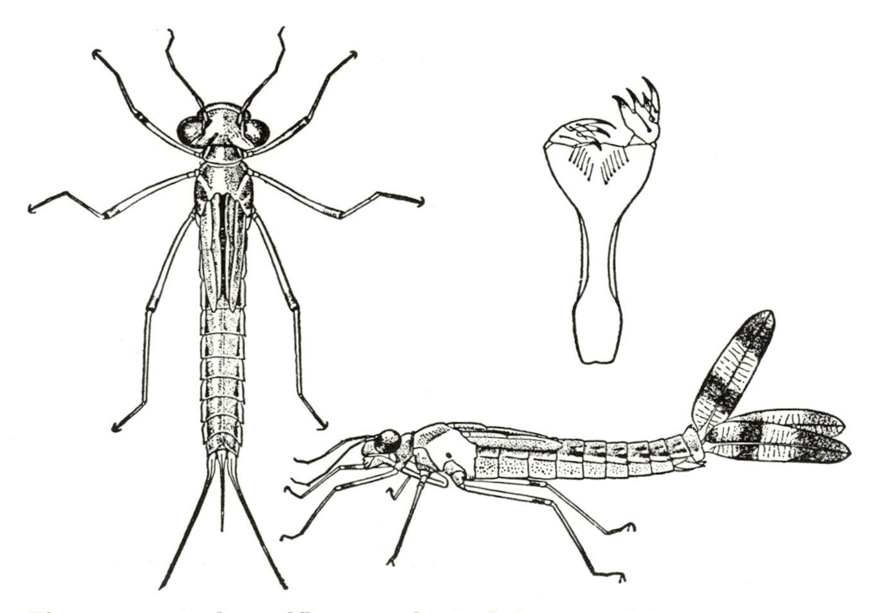

Fig. 8-13. A damselfly nymph *Archilestes californica*. Insert is labium showing grasping teeth. (After Kennedy)

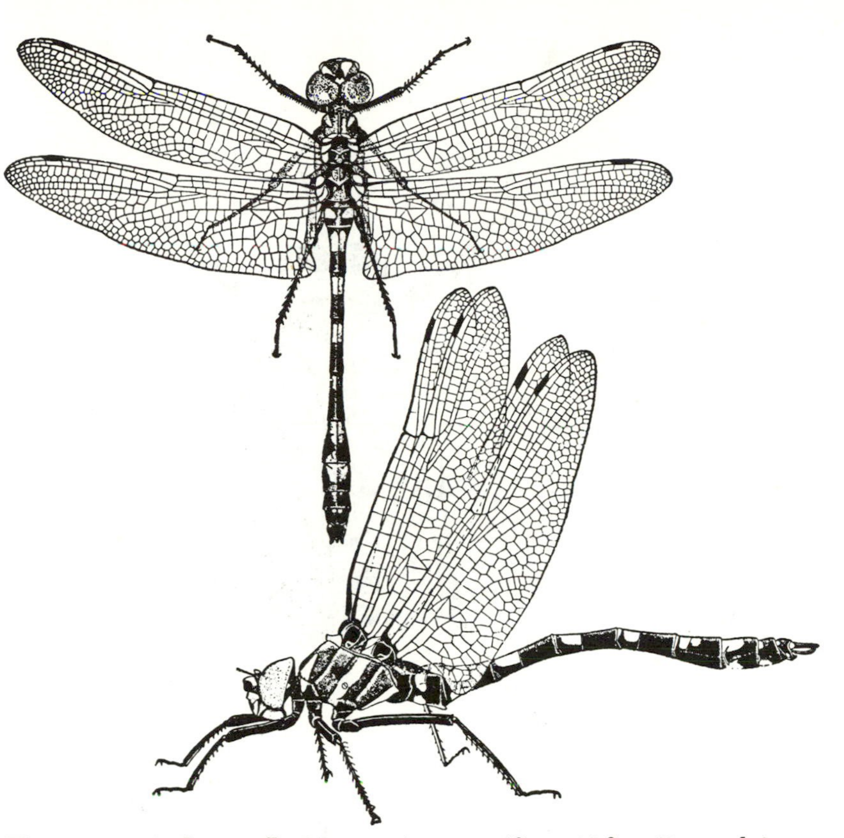

Fig. 8-14. A dragonfly *Macromia magnifica*. (After Kennedy)

the side. Many species are gaudily colored and have conspicuous mottling or spotting on the wings. Older specimens frequently develop a pale-blue waxy bloom over the body and wings that may obscure the original colors and markings.

The adults, especially the larger ones, are showy, flying with great speed and poise, although not gaudily colored. Many dragonflies, usually males, may establish a regular beat, their territory. These fly back and forth, patrolling their beat at regular intervals, and looking for flying insects as prey. When one of these is sighted, the dragonfly wheels from its course in pursuit of the prey; when the prey is captured, the dragonfly wheels back to its regular beat. Sometimes rivals clash, and there is displayed a real show of aerial acrobatics to the accompaniment of clicking of mandibles and rustle of wings.

The nymphs (Fig. 8-15) are also stout, many of them frequenting the ooze or mud in the bottoms of ponds and lakes. They have no

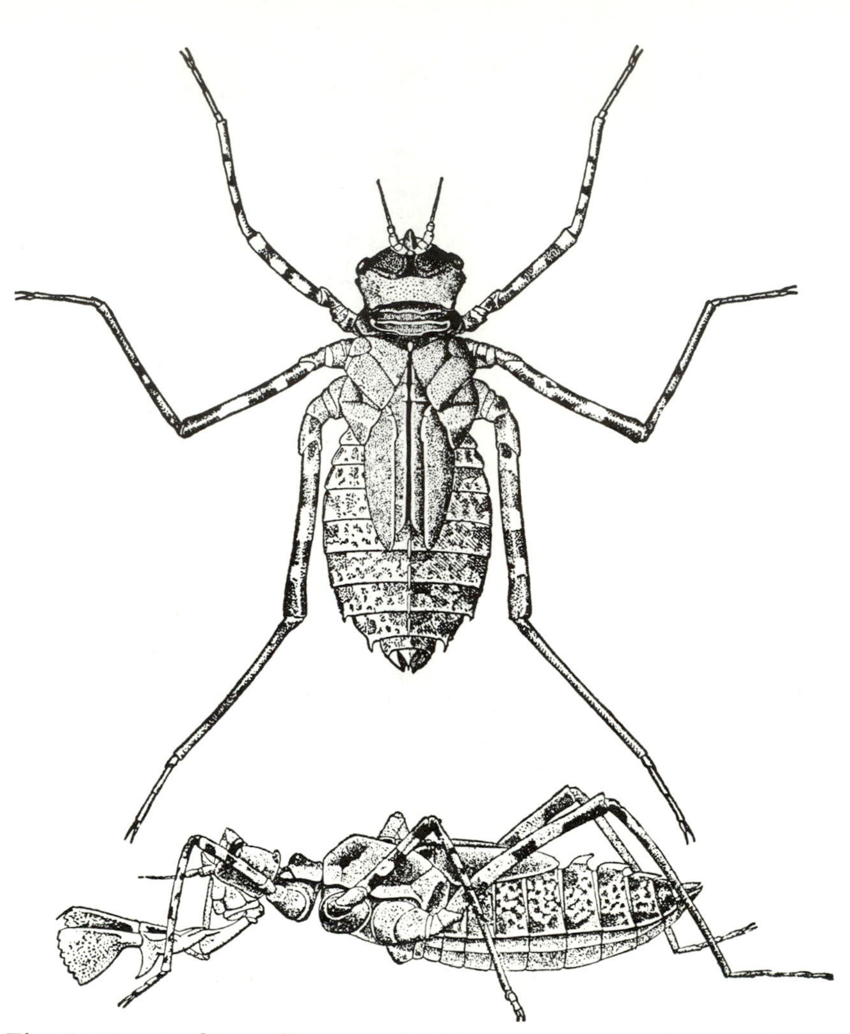

Fig. 8-15. A dragonfly nymph *Macromia magnifica*. (After Kennedy)

external gills but have a rectal respiratory chamber (see p. 161) in which the gaseous exchange takes place. Such a respiratory chamber is found in no other group of insects.

References

Borror, D. J., 1945. A key to the New World genera of Libellulidae. *Ann. Entomol. Soc. Am.*, **38**:168–194.

Corbet, P. S., 1963. *A biology of dragonflies*. Chicago: Quadrangle Books. 247 pp.

1980. Biology of Odonata. *Annu. Rev. Entomol.*, **25**:189–218.
Corbet, P. S., C. Longfield, and N. W. Moore, 1960. *Dragonflies.* London: William Collins. 260 pp.
Kormondy, E. J., 1961. Territoriality and dispersion in dragonflies (Odonata). *J. N.Y. Entomol. Soc.*, **69**:42–52.
Needham, J. G., and M. J. Westfall, Jr., 1955. *A manual of the dragonflies of North America (Anisoptera).* Berkeley: Univ. of California Press. 615 pp.
Smith, R. F., and A. E. Pritchard, 1956. *Odonata.* in R. L. Usinger (Ed.), *Aquatic insects of California.* Berkeley: Univ. of California Press. pp. 106–153.
Walker, E. M., 1953. *The Odonata of Canada and Alaska. Part I, General, Part II, The Zygoptera—damselflies.* Toronto: Univ. of Toronto Press. Vol. I, 292 pp.
1958. *The Odonata of Canada and Alaska. Anisoptera.* Toronto: Univ. of Toronto Press. Vol. 2, 318 pp.
Walker, E. M., and P. S. Corbet, 1975. *The Odonata of Canada and Alaska. Anisoptera, Libellulidae.* Toronto: Univ. of Toronto Press. Vol. 3, 307 pp.
Westfall, M. J., Jr., 1978. Odonata. In R. W. Merritt and K. W. Cummins (Eds.), *Introduction to the aquatic insects of North America.* Dubuque, Iowa: Kendall/Hunt. pp. 81–98.
Wright, M., and A. Petersen, 1944. A key to the genera of anisopterous dragonfly nymphs of the United States and Canada (Odonata, suborder Anisoptera). *Ohio J. Sci.*, **44**:151–166.

Infraclass Neoptera

The great majority of living winged insects and their allies belong to this group, as do a considerable number of fossil orders. For convenience, the neopterous orders may be discussed as a series of four evolutionary clusters: the orthopteroid orders, the plecopteroid orders, the hemipteroid orders, and the neuropteroid orders.

The Orthopteroid Orders

Here belong several orders that appear to be the most primitive elements of the Neoptera. They are all paurometabolous and have many characters in common with the Paleoptera.

Order DICTYOPTERA*: Cockroaches and Mantids

In this order belong a varied assemblage of insects, including wingless, short-winged, and long-winged forms. The long-winged forms have two pairs of net-veined wings: the fore wings, called *tegmina*,

* The name Cursoria, Oothecaria, Blattiformia, Blattopteriformia, and Blattopteroidea are sometimes used for this order.

are leathery or parchment-like; the hind pair are membranous, larger, and folded beneath the tegmina in repose. The head bears antennae, eyes, and mouthparts of a generalized chewing type. Metamorphosis is gradual. Adults and nymphs are terrestrial. The hind legs are in about the same proportions as the middle legs, fitted for running. Classified in the Orthoptera by some authors.

KEY TO SUBORDERS

Front legs similar to middle legs in general shape; pronotum wide (Fig. 8-16); broad, flat insects (Fig. 8-16) .... **Blattaria**

Front legs large, with series of strong teeth on opposing tibia and femur, fitted for grasping prey (Fig. 8-18) .... **Mantodea**

SUBORDER BLATTARIA

COCKROACHES. Cockroaches are rapid-running flattened insects, with long slender antennae, well-developed eyes, and chewing mouthparts having mandibles, maxillae, and labium very similar in type to Figs. 3-16, 3-17, and 3-18, respectively. In species with well-developed wings (Fig. 8-16), both pairs have many veins and a very large number of crossveins; the fore pair are narrower, thickened, and leathery or parchment-like, called *tegmina*, serving chiefly as a cover for the hind pair when not flying; the hind pair are thin, much larger, chief agent in flight, and are folded fanlike beneath the tegmina when not in use. Many species have only padlike wings or no wings at all. The prothorax is large and conceals much of the head. The abdomen is large, many-segmented, and bears a pair of apical cerci.

Cockroaches frequent dark, humid situations. Typically they be-

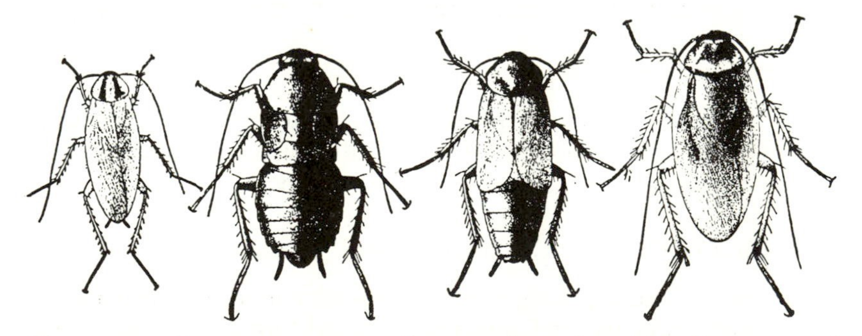

Fig. 8-16. Common cosmopolitan cockroaches. The German cockroach *Blattella germanica*; the Oriental cockroach *Blatta orientalis*, female and male; and the American cockroach *Periplaneta americana*. (From Connecticut Agricultural Experiment Station)

long to the tropics, where occurs a great variety of species, large and small. An extensive native fauna occurs in the southern portion of the United States, especially in humid regions. A limited number of outdoor species are found in the areas to the north, where they occur chiefly under bark of dead trees or fallen logs. The more conspicuous elements of the cockroach fauna of the northern states are not native to this continent but are a group of cosmopolitan species that are household pests. They find in human dwellings and heated buildings the semitropical conditions that enable them to thrive and multiply throughout the entire year. They are almost omnivorous in habit, eating a wide variety of animal and vegetable foods. The nymphs are similar to the adults in general structure and usually occur and feed along with the adults. In certain species in which wings are never developed, it is sometimes necessary to examine the genitalia to differentiate adults and nymphs.

The egg-laying habits of cockroaches are unusual. As the successive individual eggs are extruded from the oviduct they are grouped in an egg chamber and "glued" together by a secretion into a capsule or *ootheca*. These are definite in shape and sculpture for the species. The eggs in each usually number 15 to 40, arranged in symmetrical double rows (Fig. 8-17). The ootheca is formed over a period of several days. In different species the hatching process is quite different. In some species the ootheca is deposited from a few to several days after it is formed, but long before hatching (e.g., *Periplaneta*, *Blatta*); in others, it is extruded but carried with its base firmly embedded in the brood chamber of the female until about hatching time (e.g., *Blattella*); and in others, the ootheca is carried within the brood chamber and the eggs may hatch there (e.g., *Pycnoscelus*, *Diploptera*). In the latter two categories the eggs obtain needed substances from the mother's body. Several species, such as, the greenhouse roach *Pycnoscelus surinamensis*, are parthenogenetic.

The nymphs are extremely active but grow relatively slowly. The smaller species may attain maturity in a few months, but the larger

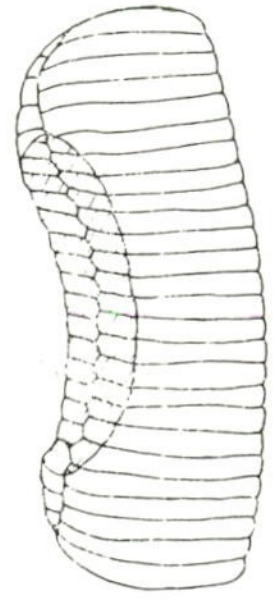

Fig. 8-17. Ootheca of *Blatella germanica*, with lateral portion cut away to show arrangement of individual eggs.

species may require a year or more. Many species are gregarious in habit, the adults and nymphs running together.

Of unusual interest is the wood roach *Cryptocercus punctulatus* found in the Pacific Northwest and the southeastern states. The species lives in colonies in decaying logs, feeds on rotten wood, and has developed a close approach to true social life. The family unit forms the colony, several generations living together. The first steps in the digestion of cellulose material are accomplished by certain Protozoa, which are always abundant in the intestinal fauna of *Cryptocercus*, not unlike those found in termites.

Economic Importance. Cockroaches are one of the most disagreeable pests of human habitations. They get into many kinds of food, eat part of it, discolor and spot it with fecal material, and leave behind a disagreeable odor. In addition to the actual spoilage they cause, these scurrying insects are regarded as a general nuisance and a sign of unclean conditions. As a consequence the nation foots a large bill for the control of these insects in warehouses, eating places, and homes.

North of the frost line, four cosmopolitan domestic species are most abundant: the small German cockroach *Blattella germanica*; the larger Oriental cockroach or "water bug" *Blatta orientalis*; the brown-banded cockroach *Supella longipalpa*; and the American roach *Periplaneta americana*, the largest of the four and sometimes nearly 3 or 4 cm in length. In local areas the Australian cockroach *Periplaneta australasiae* is abundant; this is another cosmopolitan species as large as the American roach. To the south of the frost line, other more tropical species invade buildings; some attain the size of a mouse.

SUBORDER MANTODEA

PRAYING MANTIDS. Predaceous insects of medium to large size, they have an elongate prothorax and large spined grasping front legs (Fig. 8-18). The middle and hind legs are usually slender. The pronotum does not cover most of the head. Otherwise the mantids are similar to the cockroaches in general features, the structure of their mouthparts, internal organs, and genitalia. The mantids in North America consist of a single family, the Mantidae, represented by only a few dozen native species. The tropics support a larger fauna. Two species in the North American mantid fauna were introduced : *Mantis religiosa* from Europe and *Paratenodera sinensis* from the Orient. Both species probably came into the United States as oothecae (egg masses) on nursery stock or packing.

The mantids may be long-winged, short-winged, or completely

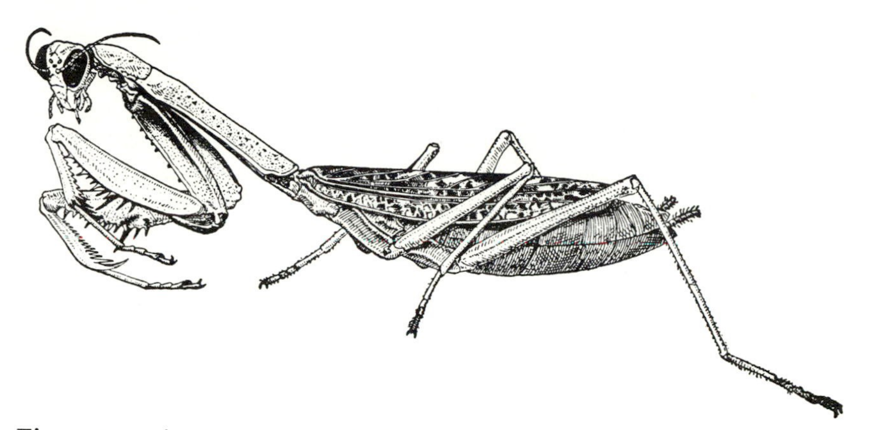

Fig. 8-18. A praying mantis *Stagmomantis carolina*. (From Illinois Natural History Survey)

wingless. Many are green, brown, or mottled; a few species have brighter colors, and some have definite patterns.

All the species are predaceous in habit, feeding on other insects, which they capture by means of the prehensile front legs. Cannibalism is not unusual; in fact, in certain species it is customary for the female to seize and devour the male after mating is completed.

Mantid eggs are deposited in large masses of definite pattern. In these masses, or oothecae, the eggs are arranged in a series of rows, glued together with secretion, and the whole mass glued to a branch or other object (Fig. 8-19). In the northern areas there is a single

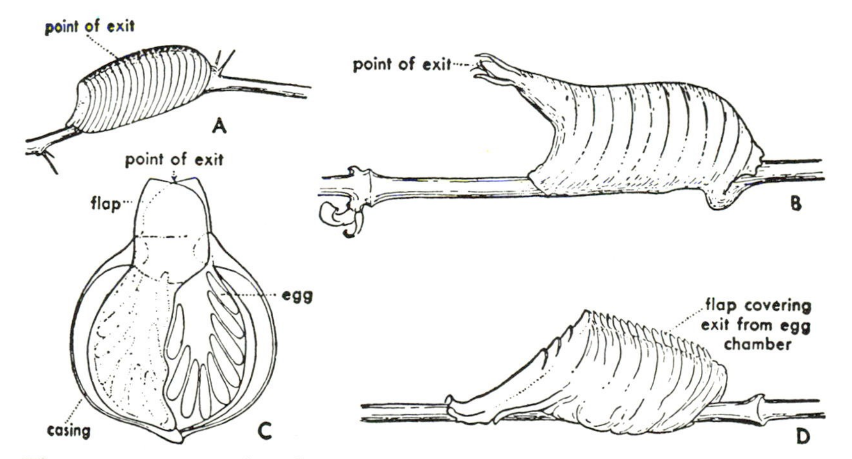

Fig. 8-19. Mantid oothecae or egg capsules. (*A*) Generalized type; (*B*) *Oligonyx mexicanus*; (*C*) sectional and (*D*) exterior aspects of *Paratenodera sinensis*. (From Essig, 1942, *College entomology*, by permission of Macmillan Co.)

generation per year, and the winter is passed in the egg stage. It is interesting to gather these oothecae in late winter or early spring and bring them into the laboratory, and see the young mantids emerge sometime later. The eggs are frequently parasitized by some Hymenoptera that are as odd looking as the young mantids; the parasites normally emerge from the ootheca some time after the hatching date of the mantids.

References

Gurney, A. B., 1951. Praying mantids of the United States. *Smithsonian Inst. Ann. Rep., 1950.* pp. 339–362.

Guthrie, D. M., and A. R. Tindall, 1968. *The biology of the cockroach*. New York: St. Martin's Press. 408 pp.

Hebard, M., 1917. The Blattidae of North America north of the Mexican border. *Mem. Am. Entomol. Soc.*, **2**:1–284.

McKittrick, F. A., 1964. Evolutionary studies of cockroaches. *Cornell Univ. Agric. Exp. Stn. Mem.*, **389**:1–197.

Roth, L. M., 1970. Evolution and taxonomic significance of reproduction in Blattaria. *Annu. Rev. Entomol.*, **15**:75–96.

Roth, L. M., and E. R. Willis, 1960. The biotic associations of cockroaches. *Smithsonian Misc. Collections*, **141**:1–470.

Order ISOPTERA: Termites, White Ants.

Termites are medium-sized insects having gradual metamorphosis, living in large colonies much like those of ants, and having several different social castes. A typical colony, for example, has three castes: sterile workers, sterile soldiers, and sexual forms (reproductives) (Fig. 8-20). The workers are white and sometimes appear to be translucent. They are wingless, and have round heads, long antennae, chewing mouthparts, and small eyes or none at all. The legs are well developed and all about equal in size. The soldiers have bodies similar to those of the workers, but their heads are enlarged and have massive mandibles. The reproductives are of two types: One type is white, wingless or with only short wing pads; the other type includes fully formed sclerotized winged males and females. These have round heads, long antennae, chewing mouthparts, well developed eyes, and two pairs of similar transparent wings. After dispersal flights and mating, the wings fall off, each one leaving only a short stub or scale that persists for the life of the individual. Termite workers superficially resemble ants, but the termites do not have the constricted "waist" of the ants.

The termites in North America feed on cellulose, in almost all cases obtained from dead wood. The colonies, which may number several thousand individuals, are located in dead trees or logs or in

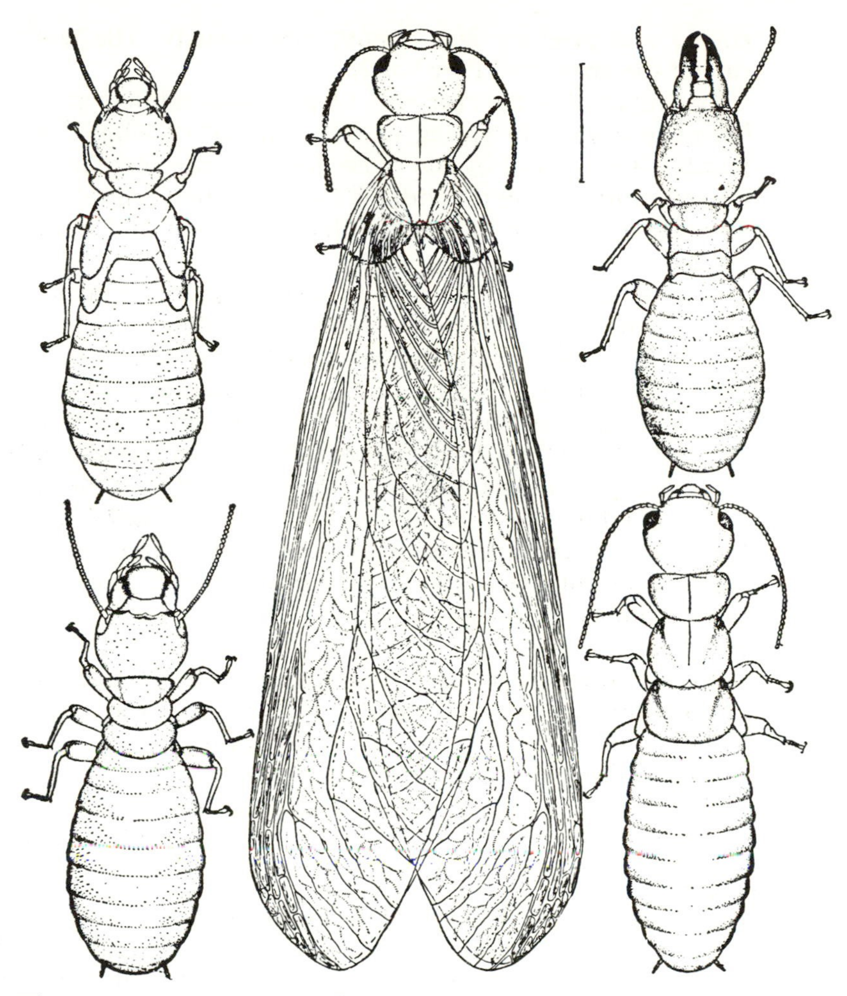

Fig. 8-20. Castes in a colony of termites, order Isoptera. Center, first form of winged reproductive; upper left, second form reproductive; upper right, soldier; lower left, worker; lower right, first-stage reproductive after the wings have broken off. (From Duncan and Pickwell, *A world of insects*, by permission of McGraw-Hill Book Co.)

the ground with covered runways connecting the nest to a log or stump that provides a food supply. The workers and, in some groups, the nymphs do the foraging for the colony, feeding both the soldiers and reproductives. The soldiers afford protection from enemies from the outside, taking up strategic stations near the exits of the colony. The reproductives are the only fertile members of the colony and produce eggs almost continuously. The workers take care of the eggs until they hatch.

During most of the year only workers and soldiers are produced, but once a year, in spring or fall, a brood of winged males and females is produced by the more northern species. These are fully formed reproductive individuals called the *first reproductive caste*. They leave the nest in swarms, disperse, mate, and form new colonies.

A new colony is established by a single pair of winged individuals. The male and female lose their wings after the dispersal flight and, in North American species, together eat out a small nest in a dead stump or log. They feed as normal individuals, and the female produces eggs that develop into workers and soldiers. When a sufficient number of these have matured, these sterile castes take over the activities of nest expansion and the feeding of both the female and male, called the *queen* and *king* of the colony. If either of these die, their place is taken by the worker-like fertile forms known as the *second reproductive caste*. These are produced in small numbers in most colonies and appear to be held in reserve for substitution purposes.

In certain genera of North American termites there are no soldiers, but instead a caste called *nasutes* (Fig. 8-21). These have a curious snoutlike head; they produce a droplet of liquid with high deterrent quality and use it to repel enemies of the colony.

The nearctic fauna contains about 50 species distributed in four families. Many more species occur in other parts of the world, especially the tropics.

Economic Status. Every year termites cause a large loss to buildings and other structures. In their search for cellulose, several kinds of termites invade foundation woodwork of buildings and may spread from that point through the woodwork into upper parts of buildings. They may cross masonry or metal in their progress. Over these nonwood areas they build covered runways out of excrement,

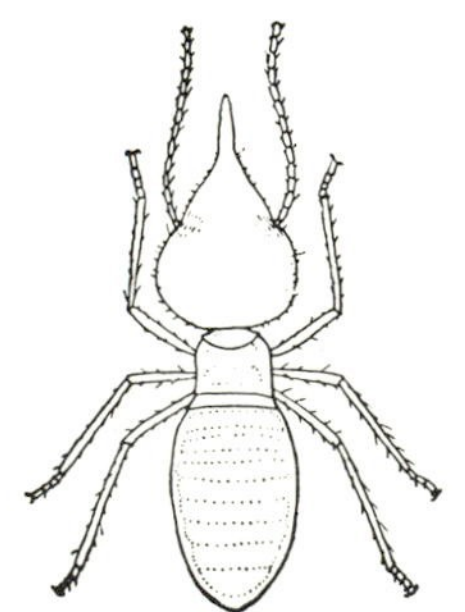

Fig. 8-21. Nasute of a termite. (Adapted from Banks and Snyder, 1920)

soil, and chewed wood and by this means always keep a contact with the soil from which they derive needed moisture. Books or wooden furniture may be attacked if these are stationary for long periods and in contact with wood. In the wild, termites eat dead trees and in this fashion are an important link in the processes leading to a recycling of nutrients in dead plant tissue.

References

Banks, N., and T. E. Snyder, 1920. A revision of Nearctic termites. *U.S. Natl. Mus. Bull.*, **108**:1–226.

Ebling, W., 1968. Termites: Identification, biology, and control of termites attacking buildings. *Calif. Agricultural Experiment Station Extension Service Manual*. Vol. 38, 68 pp.

Emerson, A. E., 1933. A revision of the genera of fossil and recent Termopsinae (Isoptera). *Calif. Publ. Entomol.*, **6**:165–196.

Krishna, K., and F. M. Weesner (Eds.), 1969, 1970. *Biology of termites*. New York: Academic Press. Vol. 1, 600 pp., Vol. 2, 643 pp.

Lee, K. E., and T. G. Wood, 1971. *Termites and soil*. New York: Academic Press. 251 pp.

Snyder, T. E., 1949. Catalogue of the termites (Isoptera) of the world. *Smithsonian Misc. Collections*, **112**(3953):1–490.

1954. *Order Isoptera of the United States and Canada*. New York: National Pest Control Assn. 64 pp.

Weesner, F. M., 1965. *The termites of the United States. A handbook*. Elizabeth, N.J.: National Pest Control Assn. 70 pp.

Order PHASMATODEA: Walking Sticks

These large, sluggish insects are either leaf mimics or stick mimics (Fig. 8-22). The North American species are wingless and sticklike, except for a single winged Florida species, *Aplopus mayeri*.

All but one of the North American species have five-segmented tarsi and belong to the worldwide family Phasmatidae. The apterous local Californian genus *Timema* has three-segmented tarsi and is the sole known member of the family Timemidae.

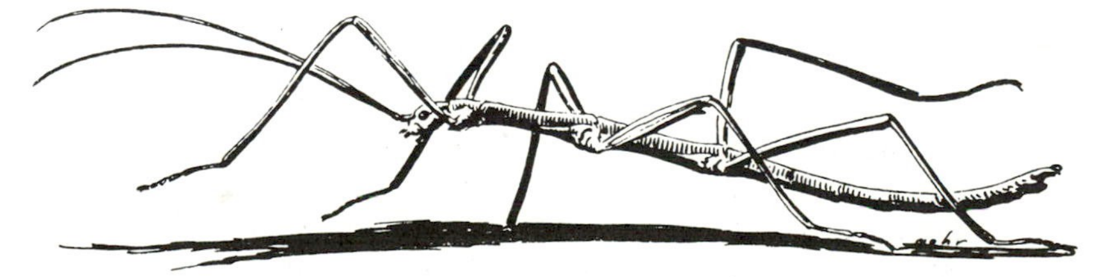

Fig. 8-22. A walking stick, *Diapheromera femorata*. (Original by C. O. Mohr)

The resemblance of so many species to sticks has given the suborder the name walking sticks. The head is round and has long slender antennae, small eyes, and simple chewing mouthparts. The body and legs are long, sometimes thorny or extremely slender. The smaller species may be only 1/2 in. long (over 12 mm); the largest, the southeastern *Megaphasma dentricus*, attains a length of 6 in. (125 to 150 mm). Some of the tropical species are broad and leaflike, with leaflike expansions on the leg segments.

All members of the phasmids are leaf feeders, most of them frequenting trees. They sometimes are sufficiently abundant to defoliate large areas of woodland. The insects themselves are never conspicuous. Their sticklike appearance and green or brown coloring gives them almost perfect protection from observation without close scrutiny. They move very slowly and feign death if disturbed.

The eggs are laid singly and simply dropped, falling to the ground. The winter is passed in this stage, the adults dying with the advent of cold weather. There is only one generation a year.

References

Bedford, G. O., 1978. Biology and ecology of the Phasmatodea. *Annu. Rev. Entomol.*, **23**:125–149.

Beier, M., 1957. Ordnung Cheleutoptera Crampton 1915 (Phasmida Leach 1815). *Bronn's Kl. Ordn. Tierreichs*, (5) (3)**6**:305–454.

1968. Phasmida (Staborder Gespenstheuschrecken). *Handb. Zool.* **4**(2)2/10:1–56.

Key, K. H. L., 1970. Phasmatodea. In D. F. Waterhouse, et al. (Eds.) *The insects of Australia*. Carlton, Victoria: Melbourne Univ. Press. pp. 348–359; Suppl. 1974, pp. 48–49.

See also under Orthoptera.

Order ORTHOPTERA: Grasshoppers and Their Allies

Medium-sized to large insects, they usually have the hind legs elongate, their femora enlarged for leaping (Fig. 8-23). In almost all forms the pronotum is large and produced downward at the sides to form a large collar back of the head. The head is large, with long antennae, well-developed eyes, and chewing mouthparts of a simple type. In many species the wings are large and functional; the tegmina (thickened forewings) are invariably leathery, and the hind wings membranous, pleated fanwise in repose. Other species may be short-winged or completely wingless. To this order belong the grasshoppers, crickets, katydids, mole crickets, and pygmy locusts, altogether making up an array of forms varied in size, shape, color, and habits. Seven or eight families usually are recognized in the North American fauna, including several hundred species.

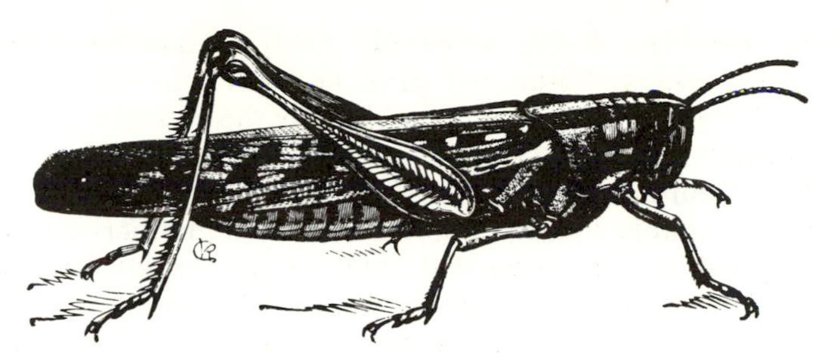

Fig. 8-23. The American grasshopper *Schistocerca americana americana*. (From Illinois Natural History Survey)

KEY TO COMMON FAMILIES

1. Front tibiae and tarsi enlarged for digging, the former having a group of large heavy sharp processes, the latter forming 2 or more heavy flanged knifelike processes (Fig. 8-30) .... **Gryllotalpidae**, p. 309
   Front tibiae and tarsi lacking heavy black processes 2
2. Hind tarsi minute or absent, but the tibial spurs forming large flat structures (used for jumping on mud) (Fig. 8-31) .... **Tridactylidae**, p. 309
   Hind tarsi well developed, projecting beyond tibial spurs (Figs. 8-23, 8-28) 3
3. Antennae much shorter than body and relatively heavy (Fig. 8-23) .. 4
   Antennae much longer than body, and slender (Fig. 8-26) 5
4. Pronotum extending backward into a long shield covering all or nearly all of abdomen; tegmina short and ovate (Fig. 8-24) .... **Tetrigidae**, p. 306
   Pronotum extending only over thorax, tegmina various but often extending beyond apex of abdomen (Fig. 8-23) .... **Acrididae**, p. 305
5. All tarsi 3-segmented; ovipositor needle-shaped .... **Gryllidae**, p. 308
   Middle tarsi and usually all tarsi 4-segmented; ovipostior sword-shaped .... **Tettigoniidae**, p. 307

Acrididae. This family contains the grasshoppers (Fig. 8-23). The antennae are short, seldom half the length of the body, and, because of this characteristic, the family is often called the short-horned grasshoppers. Most of the group are grass or herb feeders, but a few feed on the foliage of trees. The eggs are deposited in masses in the soil. The female works the end of the abdomen down into the soil to form a chamber; into the bottom of this chamber she starts depositing eggs, and, as she gradually withdraws the abdomen from the chamber, more eggs are laid. When the chamber is filled with eggs, she secretes a weatherproof cap covering the opening and protecting the eggs from enemies and the elements.

Many members of the subfamily Oedipodinae have brightly banded hind wings of blue, red, pink, and black. In the field, males of many of these species attract attention by the crackling noise they make in flight.

To the Acrididae belongs the interesting and important group known as migratory locusts. These are species that periodically develop populations of a size that staggers the imagination. Under these conditions the locusts soon completely denude the area in which they develop and after maturity migrate in huge swarms to other areas. These swarms may travel many hundred miles, eating all the foliage and visiting complete destruction on farm crops in their path. Every continent has its particular migratory species. In North America the most important are species of the *Melanoplus mexicanus* complex. In 1873 one of these, *Melanoplus spretus*, the Rocky Mountain locust, swarmed from the Rocky Mountains eastward to about the Mississippi River.

Several other species cause serious but less spectacular damage year after year. The most persistent are other species of the genus *Melanoplus*, including *femur rubrum*, *bivittatus*, and *differentialis* and *Camnula pellucida*. All these species eat a wide variety of crops and the species occurring in the western states are extremely destructive to rangeland following overgrazing.

Tetrigidae. This family contains the grouse locusts or pygmy grasshoppers. At first glance these appear similar to short-horned grasshoppers, but they differ in having the tegmina greatly reduced and the pronotum produced posteriorly into a long, narrow shield extending over the entire length of the body (Fig. 8-24). The North American species are few in number and all small, seldom more than 15 mm long. They occur in a variety of situations, especially moist places near water. Certain species of the family display extraordinary variations in color pattern and were employed in genetic research some years ago.

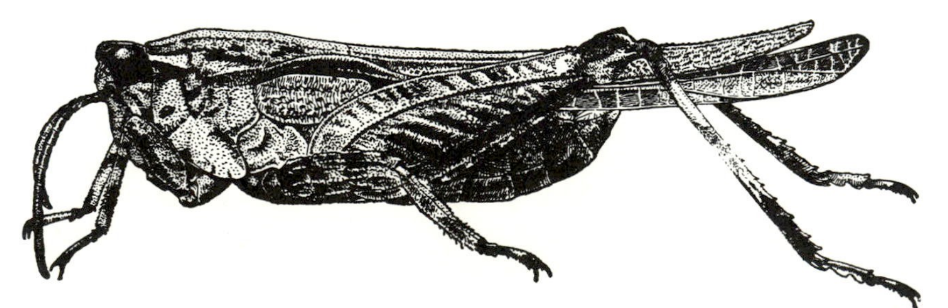

Fig. 8-24. A pygmy grasshopper *Acridium ornatum*. (From Illinois Natural History Survey)

Tettigoniidae. Here belong the long-horned grasshoppers in which the antennae are long and slender, as long as or longer than the body, and the tarsi are four-segmented. The family is a large one, embracing the meadow grasshoppers (Fig. 8-25); the cone-headed grasshoppers; the various types of katydids (Figure 8-26). Best known are the katydids, which are large, usually green or pinkish insects with wide wings. The katydids produce a musical series of chirps and as insect musicians are as renowned as the crickets.

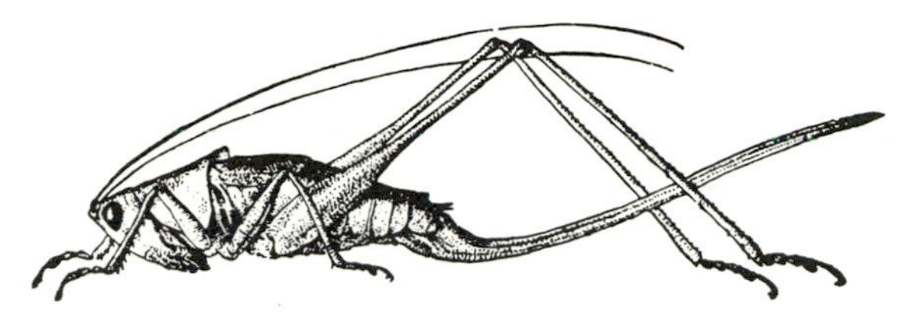

Fig. 8-25. A female meadow grasshopper *Conocephalus strictus*. (From Illinois Natural History Survey)

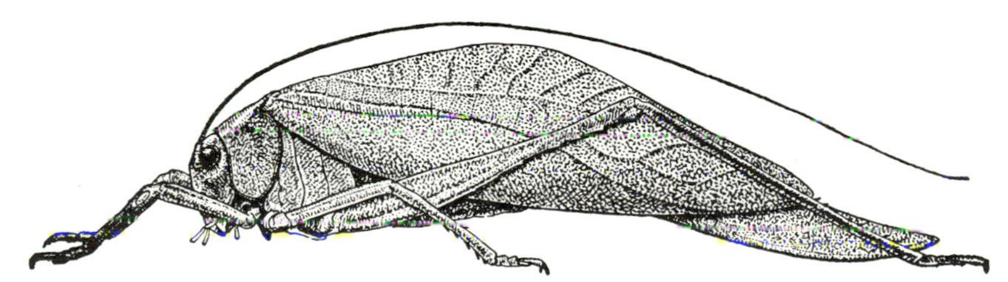

Fig. 8-26. The bush katydid *Microcentrum rhombifolium*. (From Illinois Natural History Survey)

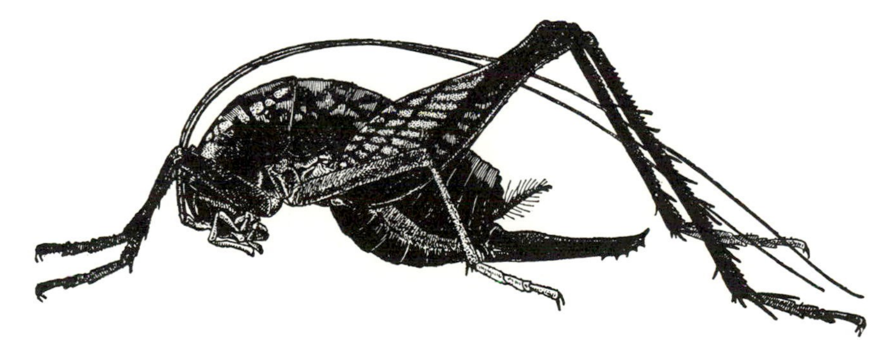

Fig. 8-27. A camel or cave cricket *Ceuthophilus maculatus*. (From Illinois Natural History Survey)

The most destructive member of the Tettigoniidae is *Anabrus simplex*, the Mormon cricket. This is a large wingless western species that often occurs in outbreak numbers in the Great Basin region of the Rocky Mountains, inflicting great damage on natural range and cultivated grain and grass crops.

Gryllacrididae. To this minor family belong several wingless longhorns of unusual interest. These include the cave or camel crickets belonging to *Ceuthophilous* (Fig. 8-27) and its allies, the western Jerusalem or sand crickets, *Stenopelmatus*.

Gryllidae. A varied assemblage of crickets comprise this family; these have long antennae, as have the Tettigoniidae, but all the tarsi

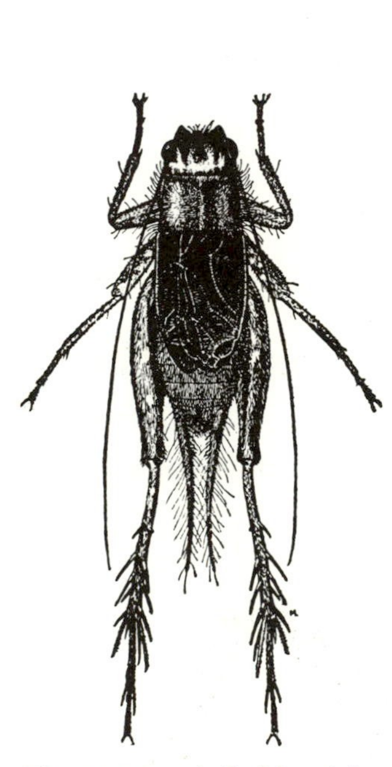

Fig. 8-28. A field cricket *Nemobius fasciatus*. (From Illinois Natural History Survey)

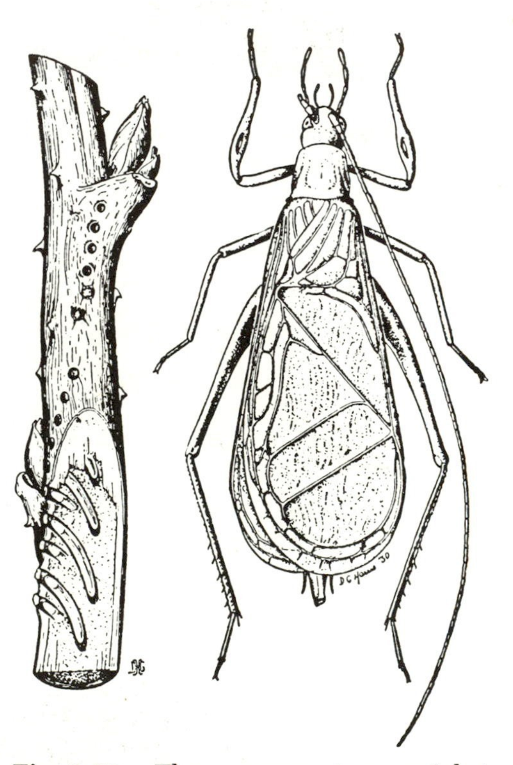

Fig. 8-29. The snowy tree cricket *Oecanthus niveus*. Egg punctures and eggs exposed to view in a raspberry cane, and adult male. The males are among the most fascinating insect musicians. (From Essig, 1942, after Smith, *College entomology*, by permission of Macmillan Co.)

are three-segmented. A number of genera such as *Nemobius* (Fig. 8-28) live in open fields or in woodland grasses. Other genera frequent shrubs or trees. One of these, *Oecanthus*, containing the tree crickets, has an awl-shaped ovipositor with which it drills holes into pithy stems and deposits its eggs in these holes (Fig. 8-29). In local areas female tree crickets may seriously injure raspberry canes in this manner.

The mole crickets represent two other families, the Gryllotalpidae and the Tridactylidae. The **Gryllotalpidae** (Fig. 8-30) are about an inch (25 mm) long and have long scooplike front legs used in digging. These crickets make burrows in fairly light soil and feed on small roots and insects that they encounter underground. The adults rarely emerge from their burrows and are seen only occasionally. **Tridactylidae**, or pygmy mole crickets, are much smaller, at the most 5 mm long (Figure. 8-31). They occur at the edge of lakes and streams, where they may be found either burrowing in the sand or leaping about near the shore line.

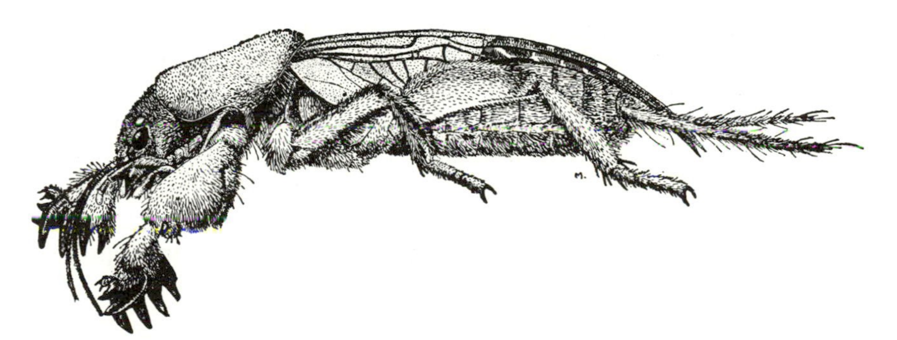

Fig. 8-30. Mole cricket *Gryllotalpa hexadactyla*. (From Illinois Natural History Survey)

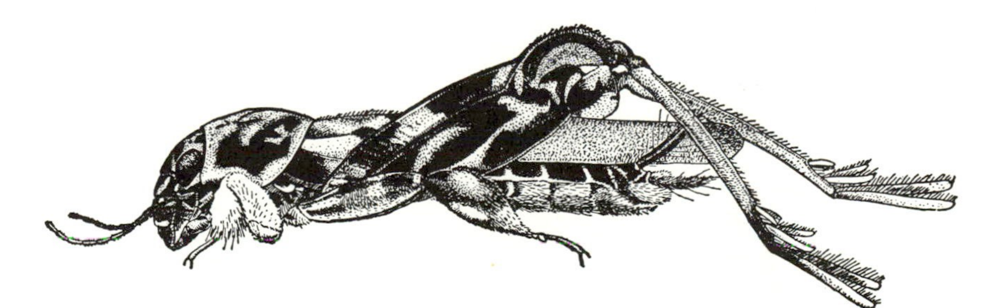

Fig. 8-31. Pygmy mole cricket *Tridactylus minutus*. (From Illinois Natural History Survey)

References

Ball, E. D., E. R. Tinkham, Robert Flock, and C. T. Vorhies, 1942. The grasshoppers and other Orthoptera of Arizona. *Univ. Ariz. Tech. Bull.*, **93**:257–373.

Blatchley, W. S., 1920. *Orthoptera of northeastern America*. Indianapolis, Ind.: Nature Publishing Co. 784 pp.

Helfer, J. R., 1963. *How to know the grasshoppers, cockroaches and their allies*. Dubuque, Iowa: W. C. Brown. 353 pp.

Key, K. H. L., 1970. Orthoptera. In D. F. Waterhouse, et al. (Eds.), *Insects of Australia*. Carlton, Victoria: Melbourne Univ. Press. pp. 323–347; Suppl. 1974, pp. 45–47.

Rehn, J. A. G., and H. J. Grant, Jr., 1961. A monograph of the Orthoptera of North America (north of Mexico). *Monogr. Acad. Nat. Sci. Philadelphia*, **1(12)**:1–225.

Uvarov, B. P., 1966. *Grasshoppers and locusts. A handbook of general acridology.* Vol. 1. *Anatomy, physiology and development, phase polymorphism, introduction to taxonomy.* Cambridge: Cambridge Univ. Press. 481 pp.

Order DERMAPTERA*: Earwigs

These medium-sized elongate heavily sclerotized insects have on their abdomen a pair of stout forceps in American species, the modified cerci (Fig. 8-32). The mouthparts are a simple chewing type; the compound eyes are large, the ocelli usually indistinct or lacking; the antennae are long, multisegmented, and slender. Wings are sometimes lacking; if present, the first pair forms short usually truncate veinless hard wing covers, and the second pair is fan-shaped, with a peculiar radial venation (Fig. 8-33). When not in flight, the second

* Sometimes called the Euplexoptera.

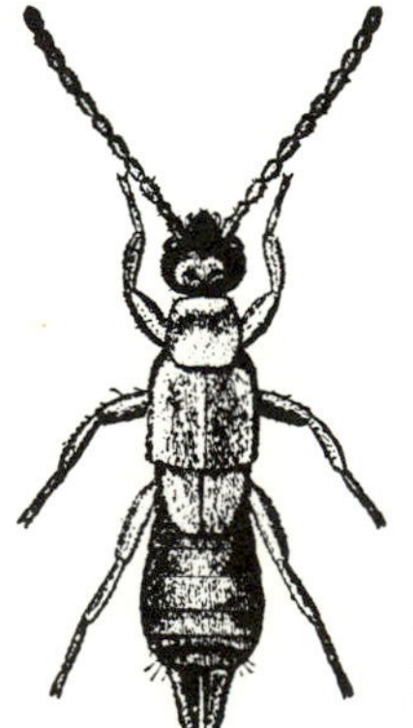

Fig. 8-32. An earwig *Labia minor*. (From Illinois Natural History Survey)

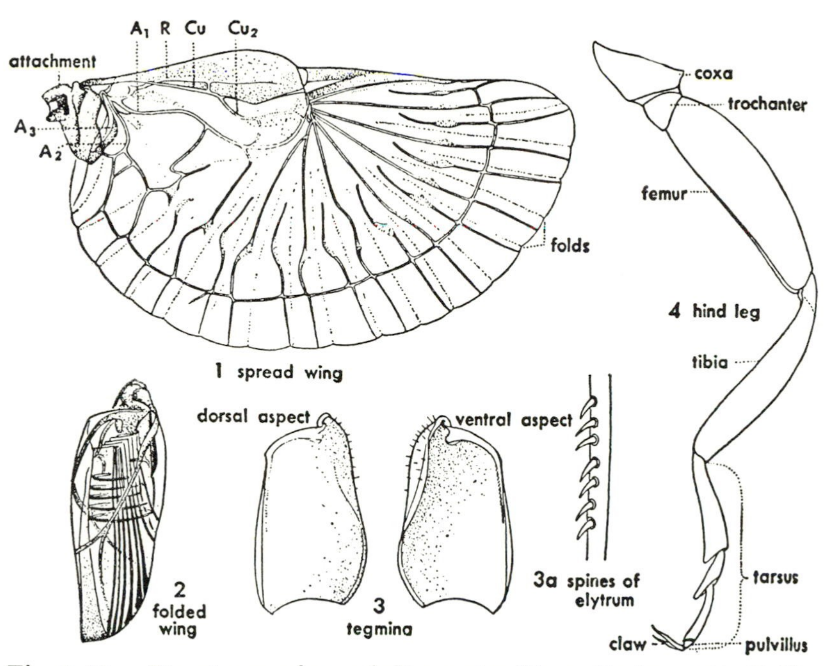

Fig. 8-33. Structures of an adult earwig. (From Essig, 1942, *College entomology*, by permission of Macmillan Co.)

pair folds into a complicated compact mass almost entirely covered by the wing covers or elytra. Metamorphosis is gradual.

The earwigs in North America vary from about 5 to 15 mm in length but are otherwise relatively uniform in shape and habits. They are nocturnal, roaming actively at night, and are omnivorous in food habits. Some species are apparently predaceous; others feed chiefly on decayed vegetation, or occasionally on living plant tissue. During the day they hide in a wide variety of tight places—under bark and boards, in the soil, and in cracks and crevices of every sort.

In temperate regions there is only one generation a year. The female lays a large cluster of white ovate eggs in a chamber in the ground in some protected spot. She watches over these for the period required for hatching, and then feeds and guards the young until they are ready to fend for themselves. The young pass through four to six molts, maturing fairly rapidly.

Earwigs are chiefly tropical, with a few representatives extending north into temperate areas. Less than 20 species occur in America north of Mexico, representing four families and several genera. The most widely distributed is the small *Labia minor*, which is an introduced species, as are most of our other earwigs.

Two curious families usually placed in this order occur only in the Old World: the viviparous Arixeniidae from southeastern Asia, associated with bats; and the Hemimeridae from Africa, ectoparasites on banana rats. This last family has one-segmented cerci that are not forceps-like; it has often been considered a separate order, the Diploglossata.

Economic Status. In certain areas of North America, the cosmopolitan European earwig *Forficula auricularia* has become a pest of great importance. It is especially abundant on the West Coast, where it is destructive to roses, dahlias, and other flowers, eating off the petals at the base and causing them to drop. Aside from this habit, it is chiefly a general feeder around the garden and home.

References

Cantrell, I. J., 1968. An annotated list of the Dermaptera, Dictyoptera, Phasmotoptera and Orthoptera of Michigan. *Mich. Entomol.*, **1**(9):299–346.

Giles, E. T., 1963. The comparative external morphology and affinities of the Dermaptera. *Trans. R. Entomol. Soc. London*, **115**:95–164.

1970. Dermaptera. In D. F. Waterhouse, et al. (Eds.), *Insects of Australia.* Carlton, Victoria: Melbourne Univ. Press. pp. 306–313.

Hebard, M., 1917. Notes on earwigs of North America, north of the Mexican boundary. *Entomol. News*, **28**:311–323.

1934. The Dermaptera and Orthoptera of Illinois. *Bull. Ill. Nat. Hist. Surv.*, **20**:125–279.

Hinks, W. D., 1955, 1959. *A systematic monograph of the Dermaptera of the world based on the material in the British Museum (Natural History).* Pt. I. *Pygidicranidae, Subfamily Diplatyinae.* 132 pp. Pt. II. *Pygidicranidae excluding Diplatyinae.* 218 pp. London: British Museum, Natural History.

Popham, E. J., 1965. A key to the Dermaptera subfamilies. *Entomologist*, **98**:126–136.

Order GRYLLOBLATTODEA: Grylloblattids

This order is composed of one family, the Grylloblattidae, with a small number of species distributed in three genera. These are medium-sized wingless elongate insects (Fig. 8-34), the head bearing long antennae, small eyes, and chewing mouthparts of generalized shape. The legs are slender, but well developed, and have five-segmented tarsi. The abdomen of the female bears at its apex a stout but primitive type of ovipositor. Both sexes have a pair of eight- or nine-segmented cerci.

In North America grylloblattids have been found near the snow line of mountains in western Canada, California, Montana, and Washington. They are known also from Japan and Siberia. They live

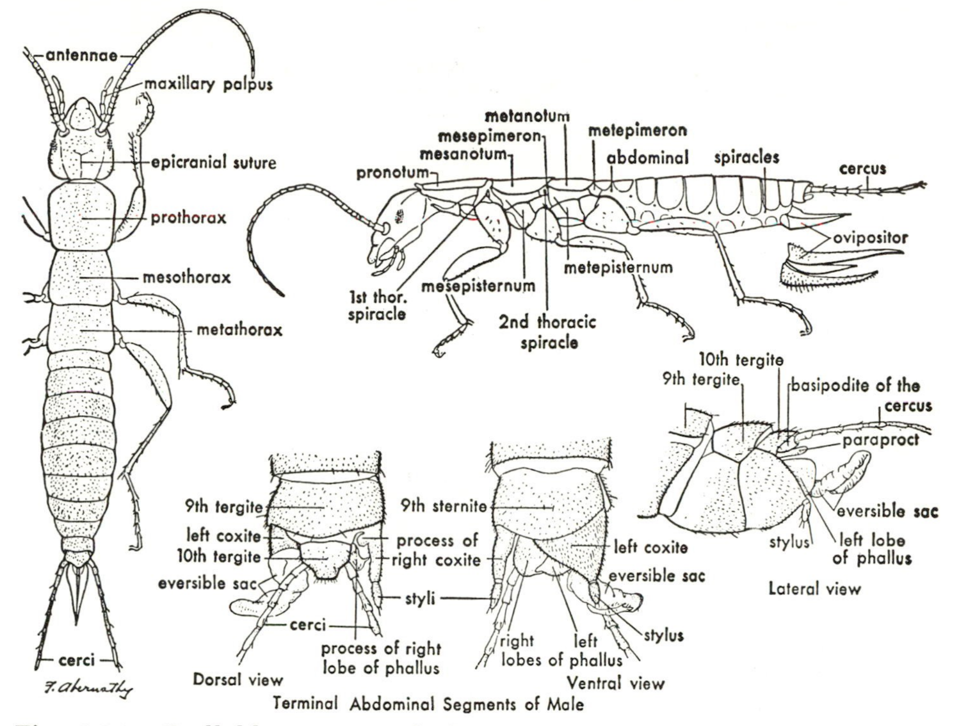

Fig. 8-34. *Grylloblatta campodeiformis*. (From Essig, 1942, *College entomology*, by permission of Macmillan Co.)

in soil or rotten wood, or under logs or stones, always in places that are covered with snow for much of the year. They are most active in the cooler periods of the year, including winter. They feed on dead or dying insects and other dead organic matter. The females deposit black eggs in moss or soil.

References

Gurney, A. B., 1948. The taxonomy and distribution of the Grylloblattidae. *Proc. Entomol. Soc. Wash.*, **50**:86–102.

1961. Further advances in the taxonomy and distribution of the Grylloblattidae (Orthoptera). *Proc. Biol. Soc. Wash.*, **74**:67–76.

Key, K. H. L., 1970. Grylloblattodea. In D. F. Waterhouse, et al. (Eds.), *Insects of Australia*. Carlton, Victoria: Melbourne Univ. Press. pp. 304–305.

Order EMBIOPTERA: Embiids, Web Spinners

These elongate flattened insects (Fig. 8-35) have curious enlarged front tarsi, used for spinning silken webs in which the insects live. Mouthparts are the chewing type, primitive in structure; eyes well

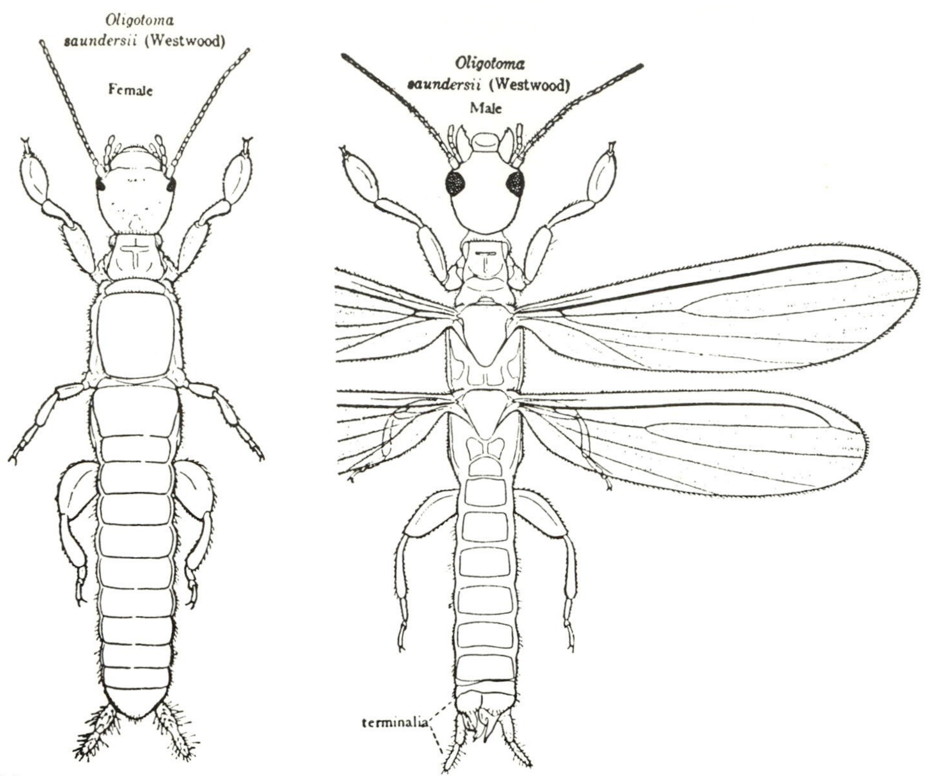

Fig. 8-35. A web spinner, *Oligotoma saundersii*. (From Essig, 1942, *College entomology*, by permission of Macmillan Co.)

developed, ocelli absent; antennae many-segmented and elongate; legs short but stout, the tarsi three-segmented; cerci one- or two-segmented. Females are always wingless; males usually with two pairs of long membranous similar wings with reduced venation. Metamorphosis is gradual.

To this order belong a small number of peculiar tropical and semitropical insects, living in silken tunnels spun on their food supply. They feed on a wide variety of plant materials, especially dried grass leaves. Their tunnels may be found under loose bark, among lichens, or on the ground. The ground nets are often among matted leaves, or under dry cattle droppings or stones. Sometimes these nets are found around the bases of plants. In arid regions the insects may be active at the ground surface during the wet seasons and retire into the soil during the dry season. The embiids themselves are active and rapid in movement. The winged males fly readily and are frequently attracted to lights.

The web spinners live in large colonies, with numerous interlocking tunnels, and are gregarious. Most species have both males and females, but a few are parthenogenetic, and of these only females are known. The eggs are elongate and relatively large. They are laid in clusters attached to the walls of the tunnels. The female exhibits considerable maternal interest in both eggs and newly hatched nymphs, remaining near them and attempting to drive away enemies.

The nymphs are remarkably similar to the adults, many of which are wingless. In those species with winged males, there is a noteworthy phenomenon. In the male nymphs the wing pads develop internally as imaginal buds until the penultimate molt, and appear as typical wing pads only in the last nymphal instar. This is what happens in holometabolous insects, so that this last embiid nymphal stage might well be called a pupa.

About 70 species of the order have been found in the Americas, representing 17 genera and 6 families. Most of these occur in the tropical areas, but 5 species extend north into the southern portion of California, Arizona, Texas, and Florida. Two of these, *Oligotoma saundersii* and *onigra*, are tropicopolitan, and have been transported by commerce to most of the equatorial world. A few additional species are occasionally found by quarantine inspectors in shipments of material to the United States from other countries.

All species of the order are remarkably uniform in general appearance. In fact, to date few characters have been discovered to use for the identification of the females, and almost the entire classification of families, genera, and species is based on characteristics of the males.

References

Davis, C., 1940. Family classification of the order Embioptera. *Ann. Entomol. Soc. Am.*, **33**:677–682.

Ross, E. S., 1940. A revision of Embioptera of North America. *Ann. Entomol. Soc. Am.*, **33**:629–676.

1944. A revision of the Embioptera, or webspinners, of the New World. *Proc. U.S. Natl. Mus.*, **94**(3175):401–504.

1970a. Biosystematics of Embioptera. *Annu. Rev. Entomol.*, **15**:157–172.

1970b. Embioptera. In D. F. Waterhouse, et al. (Eds.) *The insects of Australia*. Carlton, Victoria: Melbourne Univ. Press. pp. 360–366.

The Plecopteroid Orders

The only living member of this group is the order Plecoptera, which has a coxal articulation (Fig. 3-37) that is more advanced than in the orthopteroid orders. Several fossil orders also are placed in this group.

Order PLECOPTERA: Stoneflies

Moderate-sized to large insects, the Plecoptera have aquatic nymphs and gradual metamorphosis. The adults (Fig. 8-36) have chewing mouthparts, frequently reduced in size and sclerotization; long, many-segmented antennae; distinct eyes and ocelli; cerci ranging from short and one-segmented in some families to long and multisegmented in others. Typically the adults have two pairs of long wings of similar texture, each having a moderate number of veins and often a large number of crossveins. Many species have abbreviated wings or have lost the wings in one or both sexes.

All stoneflies' nymphs (Fig. 8-36) are aquatic. They have long antennae and a pair of long multisegmented cerci, chewing-type mouthparts, well-developed eyes and ocelli, and body proportions similar to the adults.

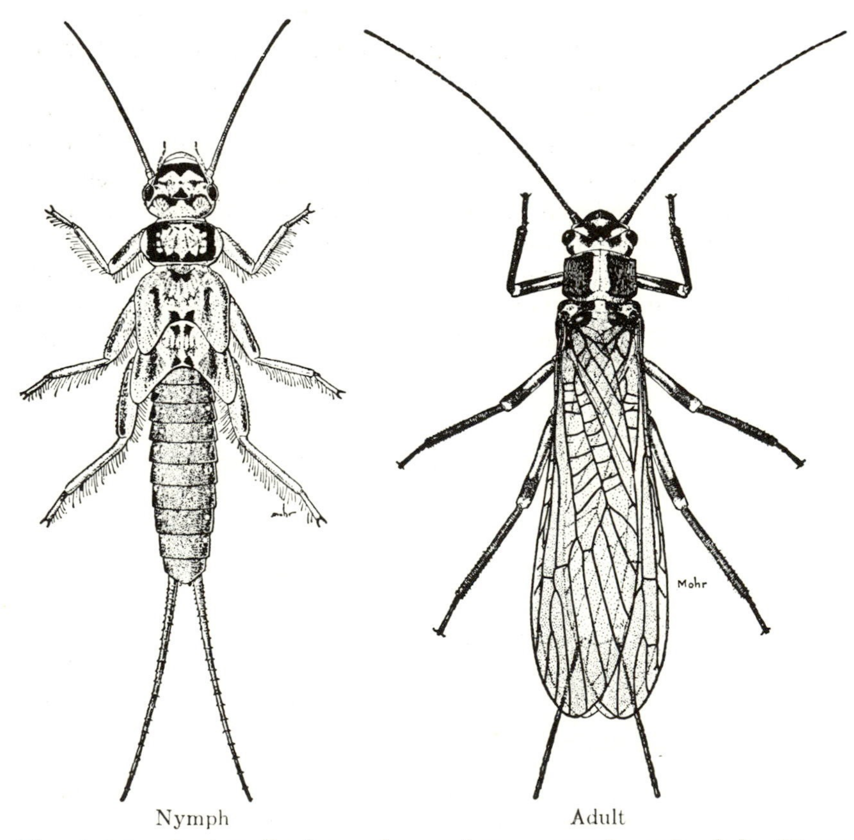

Fig. 8-36. A stonefly *Isoperla confusa*, nymph and adult. (From Illinois Natural History Survey)

The nearctic stoneflies include about 300 species comprising six families. Their generalized mouthparts, antennae, and wings, together with their simple type of metamorphosis, indicate that the order is a primitive offshoot of an ancestral orthopteroid stock.

The nymphs of this order are one of the abundant components of stream life. They range in body length from about 5 to over 20 mm and present a varied appearance, including drab plain forms, spotted patterns, and forms striped with yellow, brown, or black. Many of them breathe by means of external finger-like gills. The gills are single in some and arranged in tufts in others. Some nymphs have no external gills and use the cuticle or wing pads for respiration. As a rule, the nymphs are found in cool unpolluted streams; a few species occur also along the wave-washed shore area of some of the colder lakes. The nymphs live in a variety of situations, frequently specific for the species. They are found under stones, in cracks of submerged logs, in masses of leaves that accumulate against stones or around branches trailing in the water, and in mats of debris. The majority of the nymphs are vegetarian, feeding on dead organic matter presumably incrusted with algae and diatoms. Others are predaceous, feeding on small insects and other aquatic invertebrates.

The females lay several hundred to several thousand eggs, discharging them in masses into the water. The eggs soon hatch. The smaller species and some large ones mature in 1 year, but other large species require 2 years to complete their development. When full grown, the nymphs crawl out of the water and take a firm hold on a stone, stick, tree trunk, or other object preparatory to the final molt. At molting a dorsal split occurs in the nymphal skin; then the adult emerges in about a minute or less. After another few minutes the wings have expanded and hardened enough for flight. The adults live for several weeks.

One wingless species of stonefly, *Capnia lacustra*, is unique among insects in that all stages are completely aquatic. The adults apparently never surface, living, mating, and reproducing at depths of 100 to 400 feet in Lake Tahoe, California–Nevada. Both nymphs and adults have cutaneous respiration.

There is a peculiarity about certain groups of stoneflies that is only rarely encountered among insects. Winter signals the end of the active season and the beginning of the quiescent period for most insects. With many of the stoneflies the opposite is the case. Apparently the first-instar nymphs do not develop further during the warmer months of the year. With the approach of winter, nymphal development becomes accelerated, and the adults emerge during the coldest months of the year, beginning in late November or early December, and continuing through March. The adults are active on the warmer winter days and may be found crawling over stones and tree trunks, mating, and feeding on green algae. They show a decided preference for concrete bridges and may be collected in great numbers there. This group is called the fall and winter stoneflies and includes roughly the families Capniidae, Leuctridae, Nemouridae, and Taeniopterygidae. The latter three have members that appear later in the year, and their emergence overlaps that of the spring and summer species.

This peculiar growth behavior of the fall and winter stoneflies indicates a physiological adjustment to the warm and cold seasons quite different from that in most insects. When discovered, the controls and mechanisms for this adjustment will make an interesting story.

Well-illustrated keys for both nymphs and adults for the families of Plecoptera are given by Harper in Merritt and Cummins (1978).

References

Bauman, R. W., A. R. Gaufin, and R. F. Surdick, 1977. The stoneflies (Plecoptera) of the Rocky Mountains. *Mem. Am. Entomol. Soc.*, **31**:1–208.

Claassen, P. W., 1931. Plecoptera nymphs of America north of Mexico. *Thomas Say Foundation, Entolmol. Soc. Am.*, **3**:1–199, 35 plates.

Frison, T. H., 1935. The stoneflies, or Plecoptera, of Illinois. *Bull. Ill. Nat. Hist. Surv.*, **20**(4):281–471.

1942. Studies of North American Plecoptera. *Bull. Ill. Nat. Hist. Surv.*, **22**(2):235–355, 126 figs.

Harper, P. P., 1978. *Plecoptera.* In R. W. Merritt and K. W. Cummins (Eds.), *Introduction to the aquatic insects of North America.* Dubuque, Iowa: Kendall/Hunt. pp. 105–118.

Hitchcock, S. W., 1974. The Plecoptera or stoneflies of Connecticut. *Bull. Conn. State Geol. Nat. Hist. Surv.*, **107**:1–262.

Hynes, H. B., 1976. Biology of Plecoptera. *Annu. Rev. Entomol.*, **21**:135–153.

Illies, J., 1965. Phylogeny and zoogeography of the Plecoptera. *Annu. Rev. Entomol.*, **10**:117–140.

Jewett, S. G., Jr., 1956. Plecoptera. In R. L. Usinger (Ed.), *Aquatic insects of California.* Berkeley: Univ. of California Press. pp. 155–181.

Needham, J. H., and P. W. Claassen, 1925. A monograph of the Plecoptera or stoneflies of America north of Mexico. *Thomas Say Foundation, Entomol. Soc. Am.*, **2**:1–397.

Ricker, W. E., 1943. Stoneflies of southwestern British Columbia. *Ind. Univ. Studies, Sci. Ser.*, **12**:1–145.

1952. Systematic studies in Plecoptera. *Ind. Univ. Studies., Sci. Ser.*, **18**:1–200.

Riek, E. F., 1970. Plecoptera. In D. F. Waterhouse, et al. (Eds.), *Insects of Australia.* Carlton, Victoria: Melbourne Univ. Press. pp. 314–322.

Ross, H. H., and W. E. Ricker, 1971. The classification, evolution, and dispersal of the winter stonefly genus *Allocapnia*. *Ill. Biol. Monogr.*, **43**:1–240.

Zwick, P., 1973. Insecta: Plecoptera. *Phylogenetisches System und Katalog. Tierreich*, **94**:1–465.

The Hemipteroid Orders

The five orders comprising this group are characterized chiefly by the evolution of a beak and sucking mouthparts, beginning as a simple elongation of certain parts of the feeding mechanism. In addition, certain adult structures such as ocelli were suppressed in the immature stages.

Order ZORAPTERA: Zorapterans

This order of minute insects, 1.5 to 2.5 mm long, have both winged and wingless adult forms (Fig. 8-37). In both, the head is distinct and oval and has chewing mouthparts, long nine-segmented antennae, and one-segmented cerci. The wingless forms are blind, with only occasional vestiges of eyes or ocelli; the winged forms, or alates, have compound eyes and distinct ocelli. The alates have two pairs of delicate membranous wings, each with only one or two veins, which may be branched. These wings are shed by the adults much as in the termites, leaving only small stubs attached to the body. Metamorphosis is gradual.

The order contains one family, the Zorotypidae, with only a single genus, *Zorotypus*. From the entire world less than 20 species are known, most of them found in the tropics. Two occur in the more southern areas of North America, *Z. snyderi* described from the West Indies and southeastern North America, and *Z. hubbardi*, which has been collected in many localities in the southern states and as far north as Washington, D.C., and central Illinois.

Zorapterans live in rotten wood or under dead bark and are usually found in colonies of a few to a hundred individuals. Their food, as far as is known, consists mainly of small arthropods, especially mites and small insects. Whether they are scavengers or predators has not been established, but observations on culture specimens indicate the former.

The wingless and winged adults have similar genitalia and reproductive habits. Eggs of only *Z. hubbardi* have been observed, laid without definite anchor lines or matrix in the runways of the colony. The creamy-colored oval eggs hatched in about 3 weeks. Collection

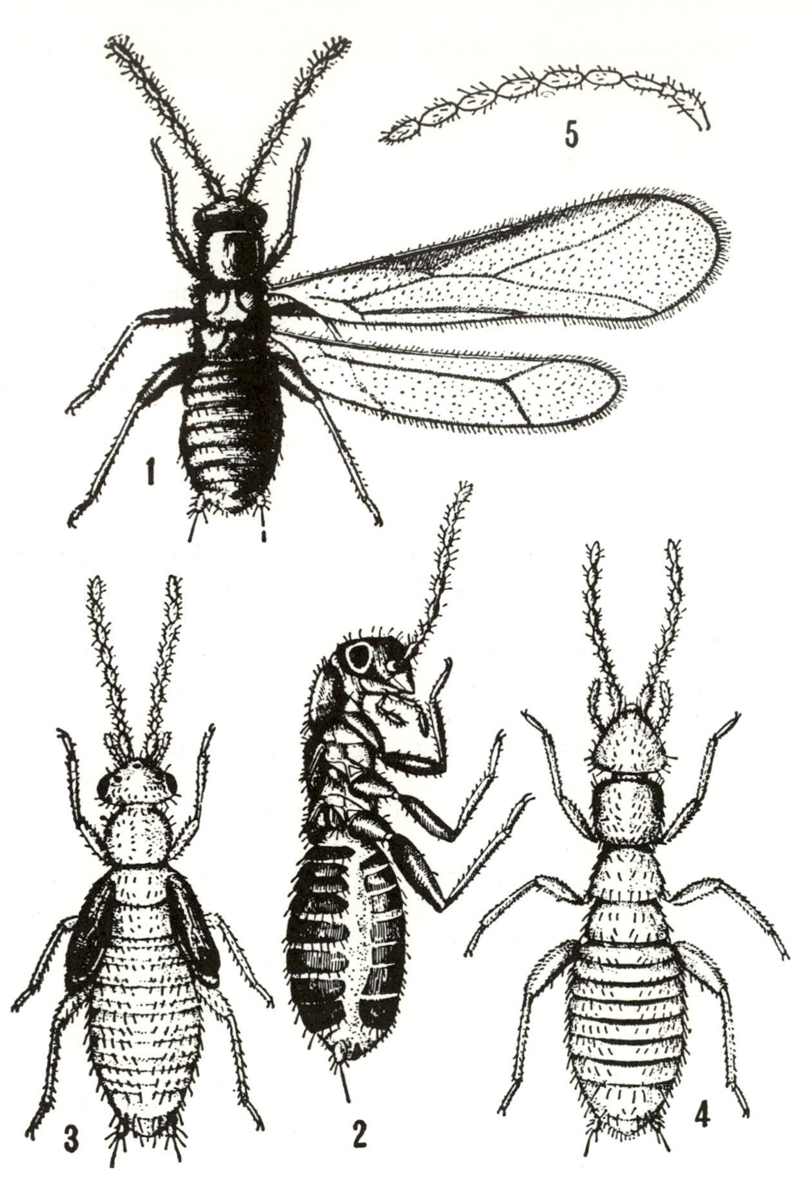

Fig. 8-37. Forms of *Zorotypus hubbardi*. (1) Winged adult female; (2) adult female that has shed her wings; (3) nymph of winged form; (4) wingless adult female; (5) antenna of adult wingless *Zorotypus snyderi*. (After Caudell)

observations over several years suggest that nymphs require several months to become adults.

Although the development of winged and wingless forms might indicate a forerunner of a caste system, no evidence of social life has been observed in the Zoraptera. There is apparently no division of labor, care of young, or social interrelationship between individuals. The gregarious nature of the colonies is very similar to the conditions found in many species of Psocoptera.

References

Gurney, A. B., 1938. A synopsis of the order Zoraptera, with notes on the biology of *Zorotypus hubbardi* Caudell. *Proc. Entomol. Soc. Wash.*, **40**(3):57–87.

Smithers, C. N., 1970. Zoraptera. In D. F. Waterhouse et al. (Eds.), *Insects of Australia*. Carlton, Victoria: Melbourne Univ. Press. pp. 302–303.

Order PSOCOPTERA*: Psocids, Booklice

Pscopterans are small insects, ranging in length from 1.5 to about 5 mm, with chewing mouthparts, long 13- to 50-segmented antennae, small prothorax, and no projecting cerci (Fig. 8-38). Two pairs of wings are well developed in some forms, the front pair much larger than the hind pair, both of similar texture and with a reduced and simple venation. In other forms the wings may be small and scalelike or absent. Metamorphosis is gradual.

Most of the members of this order are either inconspicuously colored or exhibit marked protective coloration. For this reason they are seldom collected by the beginning student, although they occur abundantly in many habitats. Their food consists chiefly of fungus mycelium, lichens, dead plant tissue, and dead insects, even of their own species. They live in a wide variety of situations out-of-doors: on clumps of dead leaves, dried standing grass, dead or dying leaves of corn plants, bark of tree trunks, in the leaf cover on top of the ground, on shaded rock outcrops, under fence posts, and in bird and rodent nests. Several species live on moldy or partially moldy foods, bookbindings, and almost anything with available starch or fungus mycelium.

Some of the species are stocky and move slowly, even when disturbed. Many of them are more slender, and a few are quite flat. These usually move with considerable speed, and a few are among the most rapid dodgers to be found among the insects. Studies to date indicate that the entire life span from egg to death of the adult is

* The names Corrodentia and Copeognatha are sometimes used for this order.

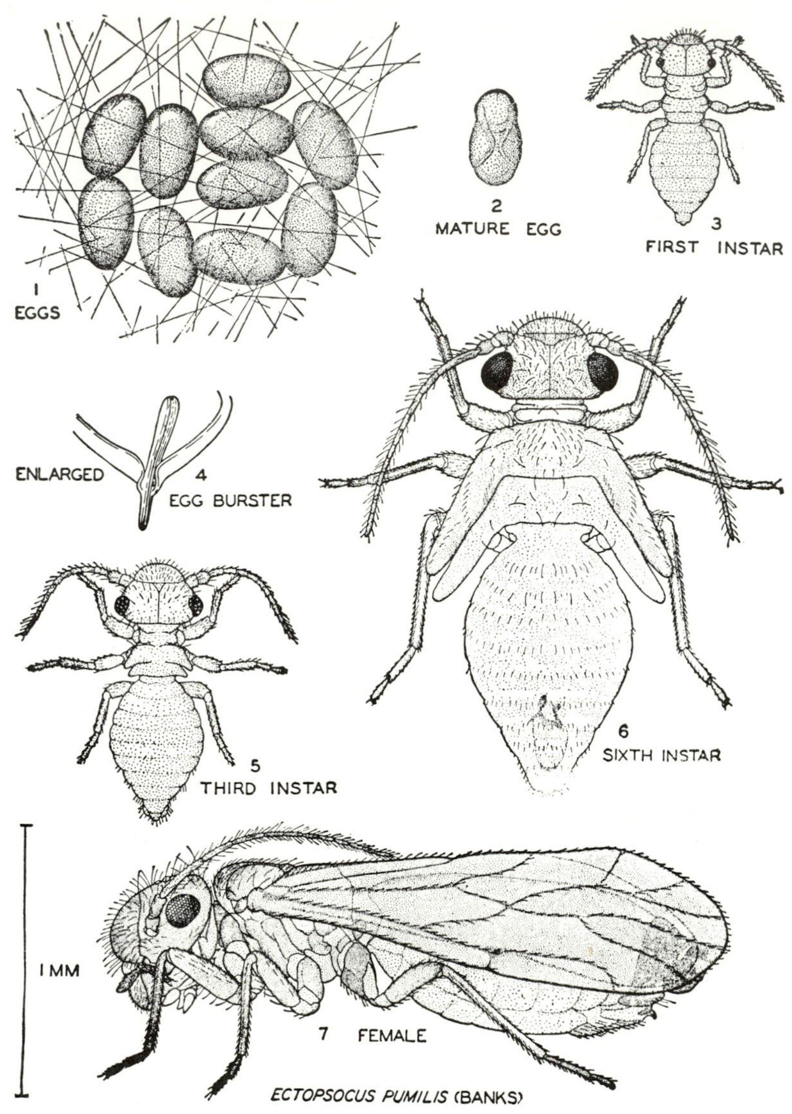

Fig. 8-38. A winged psocid *Ectopsocus pumilis*, and its life history stages. (After Sommerman)

between 30 and 60 days, of which about half is spent in the adult stage. The eggs are laid on the leaf surface or other spot that the adult frequents. Depending on the species, eggs are deposited singly or in groups up to about 10. After oviposition the female spins strands of silk over the eggs and anchors them to the surface of the support. In some species only a few strands are spun over the eggs; in others a dense web may be spun over each group of eggs. The eggs hatch in a few days, and the nymphs pass through six nymphal stages and become adults in 3 or 4 weeks.

In the more northern states the winter is passed in the egg stage by some species and as nymphs or adults by others. Species inhabiting warm buildings continue to breed throughout the year.

About 150 species are known from North America, representing about 12 families and many genera. The group is worldwide in distribution, with an estimated number of species nearing 900.

Economic Status. Several species of psocids cause considerable waste of food and damage to libraries. They consume only small quantities of foodstuffs, because they feed chiefly on mold. At times, however, they become extremely abundant, spread through an entire building, and get into every possible hiding place. In this way they may contaminate otherwise marketable goods to such an extent that quantities of the material must be discarded. Their damage to libraries is more direct. They eat the starch sizing in the bindings of books and along the edges of the pages, defacing titles and necessitating rebinding and repairs. The two most common species are the common booklouse, *Liposcelis divinatorius*, a minute wingless species (Fig. 8-39); and *Trogium pulsatorium*, another small species having the wings reduced to small scales.

References

Chapman, P. J., 1930. Corrodentia of the United States of America: I. Suborder Isotecnomera. *J. N.Y. Entomol. Soc.*, **38**:219–290.

Gurney, A. B., 1950. Corrodentia. In *Pest Control Technology, Entomological Section.* New York: National Pest Control Assn. pp. 129–163.

Mockford, E. L., 1951. The Psocoptera of Indiana. *Proc. Ind. Acad. Sci.*, **60**:192–204.

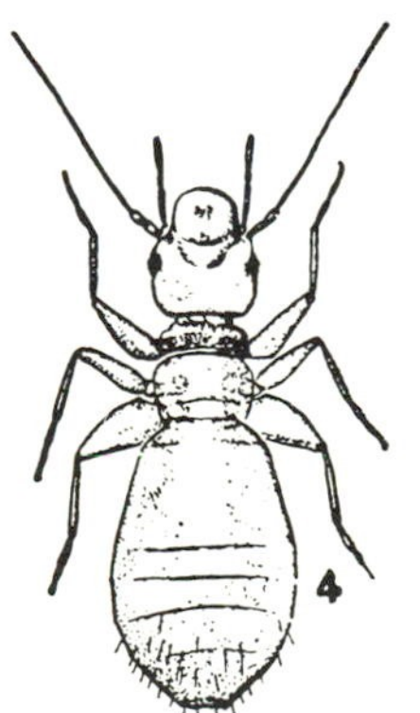

Fig. 8-39. A wingless booklouse *Liposcelis divinatorius*. (From U.S.D.A., E.R.B.)

1963. The species of Embiodopsocinae of the United States (Psocoptera: Liposcelidae). *Ann. Entomol. Soc. Am.*, **56**:25–37.

Mockford, E. L., and A. B. Gurney, 1956. A review of the psocids, or book-lice and bark-lice of Texas (Psocoptera). *J. Wash. Acad. Sci.*, **46**:353–368.

Smithers, C. N., 1970. Psocoptera. In D. F. Waterhouse, et al. (Eds.), *Insects of Australia.* Carlton, Victoria: Melbourne Univ. Press. pp. 367–375.

Sommerman, K. M., 1944. Bionomics of *Amapsocus amabilis* (Walsh). *Ann. Entomol. Soc. Am.*, **37**:359–364.

Order PHTHIRAPTERA*: Chewing and Sucking Lice

These small to medium-sized wingless insects are usually somewhat flattened (Figs. 8-40 to 8-45) and live as ectoparasites on birds and mammals. They have various types of mouthparts; only short three- to five-segmented antennae, sometimes hidden in a recess of the head; reduced or no compound eyes; and no ocelli. The thorax is small, the segments sometimes indistinct; the legs short but stout; and the abdomen has from five to eight distinct segments. Wings are absent. Metamorphosis is gradual with only three instars.

The Phthiraptera contain three distinctive suborders with an interesting evolutionary history. Members of the primitive suborder Mallophaga live on sloughed skin, dried blood at wounds, and other organic material on the body of the host. This suborder has a simple, reduced type of chewing mouthparts. One line of the Mallophaga apparently began to break the skin of the host and feed on exuding blood. Concurrently with this development the mouthparts became reduced to only the labium and mandibles, and an esophageal pump evolved, used for sucking up this food. The only known representative of this stage is the elephant louse *Haematomyzus*. From a primitive member of this line arose a branch in which a wholly new set of piercing-sucking stylets (Fig. 8-42) evolved in company with the blood-sucking habit. This branch developed into the suborder Anoplura, the sucking lice.

KEY TO SUBORDERS

1. Mandibles sclerotized, toothed, and functional, situated at end of a beak-like projection of the head or on ventral side of head (Fig. 8-41); mouthparts not styliform 2

   Mandibles apparently absent; mouthparts composed of long stylets retractable into a cavity in the head (Fig. 8-42) .... **Anoplura**

* Some authors divide this into two orders: Mallophaga (chewing lice) and Anoplura (sucking lice).

2. Front of head produced into a narrow beak longer than the rest of the head. Contains only *Haematomyzus elephantis*, occurring on Indian elephant .... **Rhyncophthirina**
Front of head not produced into such a beak (Figs. 8-43 to 8-45). On many kinds of mammals and birds .... **Mallophaga**

SUBORDER MALLOPHAGA*

CHEWING LICE. The chewing lice average about 3 mm in length, a few species attaining 10 mm. They vary considerably in shape and habits, some being long and slender, others short and wide; some are active and rapid of movement, others sedentary and sluggish. There is little correlation between speed and shape. Their mouthparts are of the chewing type, but greatly reduced and difficult to interpret without careful study.

There are several hundred species of Mallophaga in North America, comprising about six families and many genera. Each species occurs on only one species of host, or on a group of closely related species. The turkey louse, for instance, occurs only on turkeys, but the large poultry louse occurs on many kinds of domestic fowl, such as chickens, turkeys, peacocks, guinea hens, and pigeons. The small family Trichodectidae occurs only on mammals, and the large family Menoponidae occurs only on birds.

All the Mallophaga live entirely on the host body and have continuous and overlapping generations throughout the year. They feed on scaly skin, bits of feather, hair, clotted blood, and surface debris. The eggs are glued to the hair or feathers of the host and thus kept under incubator conditions. The eggs of various species differ in shape; some are long and simple, as in Fig. 8-40*B*; others are ornamented with tufts of barbs or hair, as in Fig. 8-41*B*.

KEY TO FAMILIES

1. Maxillary palps present; antennae arising from ventral portion of head and usually situated in grooves or cavities (Fig. 8-41) (series **Amblycera**) 2
Maxillary palps absent; antennae arising from or near lateral margin of head and not situated in grooves (Fig. 8-40) (series **Ischnocera**) 5
2. Tarsus with 1 claw or none. On guinea pigs .... **Gyropidae**
Tarsus with 2 claws. On birds 3
3. Entire head triangular in outline (Fig. 8-41), the posterolateral areas posterior to the eyes considerably expanded laterally; antennae in

* Some authors divide this suborder into two suborders: the Amblycera, which have maxillary palpi; and the Ischnocera, which lack maxillary palpi.

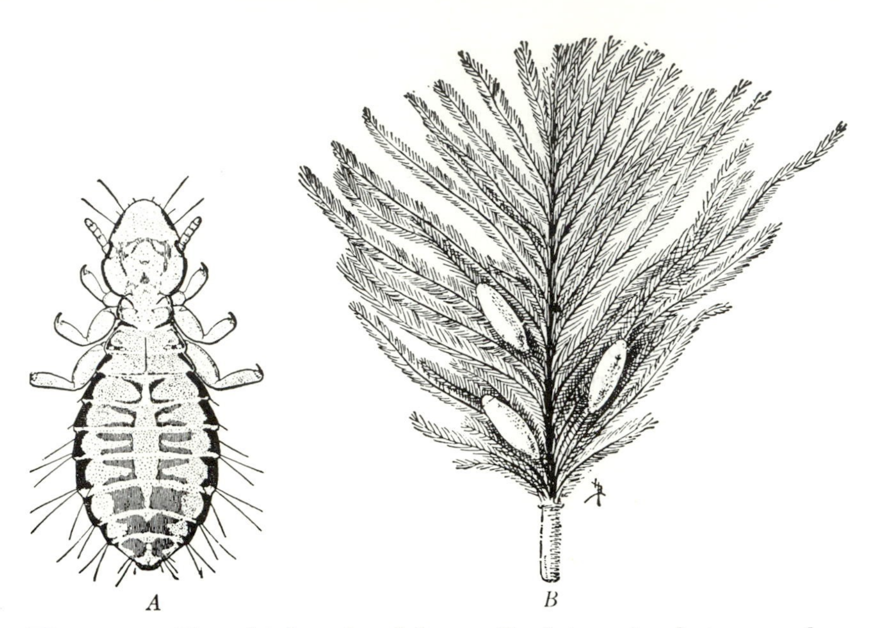

Fig. 8-40. The chicken head louse *Cuclotogaster heterographus*. (*A*) Adult; (*B*) eggs on feather. (*A*, from Illinois Natural History Survey; *B*, from U.S.D.A., E.R.B.)

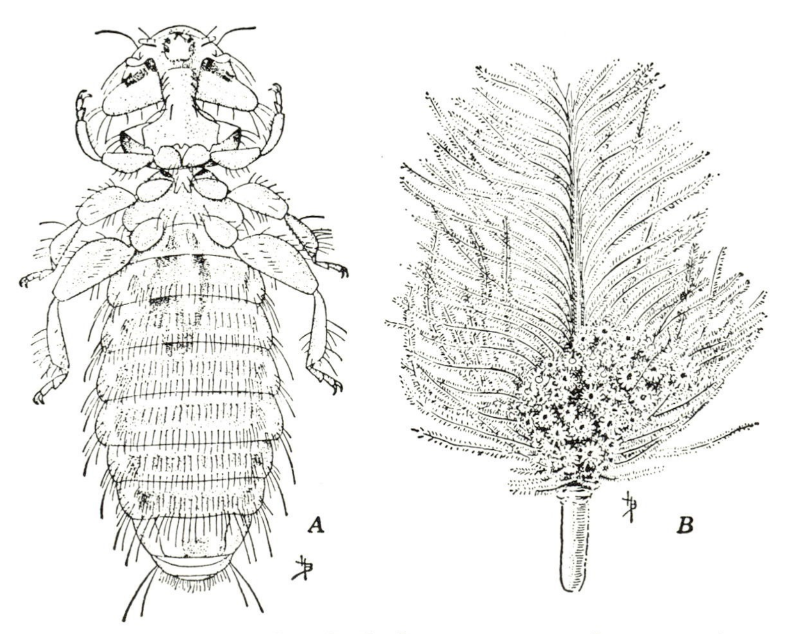

Fig. 8-41. The chicken body louse *Menacanthus stramineus*. (*A*) Adult; (*B*) eggs on feather. (From U.S.D.A., E.R.B.)

grooves that are completely open laterally .... **Menoponidae**
Head more elongate and the anterior portion comparatively wide; antennae in nearly circular cavities opening ventrally 4

4. Sides of head with conspicuous swellings anterior to eyes .... **Laemobothriidae**
Sides of head straight or nearly so, without such swellings .... **Ricinidae**
5. Tarsus with a single claw. On mammals .... **Trichodectidae**
Tarsus with 2 claws. On birds .... **Philopteridae**

Economic Status. Many species of Mallophaga infest domestic birds and animals and cause a considerable loss. Poultry are the most important group attacked. Shaft lice, *Menopon gallinae* and chicken body lice, *Menacanthus stramineus* (Fig. 8-41) cause loss of weight and reduction of egg laying in chickens, turkeys, and other fowl. The chicken head louse *Cuclotogaster heterographus* (Fig. 8-40) occasionally occurs in outbreak form and causes the death of broods of young chicks. Several other species infest fowl, but the aforementioned, because of their reproductive capacity, are the most common and destructive.

Domestic animals are attacked by various species of the genus *Trichodectes*. Dogs, cats, horses, cattle, sheep, and goats may suffer considerable loss of condition if badly infested with these lice. There is evidence that the biting sheep louse *T. ovis* injures the base of the wool and causes commercial depreciation by lowering the staple length of the sheared product.

SUBORDER ANOPLURA

SUCKING LICE. The North American species represent about 20 genera and 100 species ranging in length from 2 to 5 mm. All of them occur normally on mammalian hosts and feed on blood, which is sucked through a tube formed by an eversible set of fine stylets (Fig. 8-42). Occasionally poultry may have a small infestation of sucking lice, but all cases on record have been accidental colonizations by a common mammalian species. The entire life cycle is spent on the host. The eggs are glued to a hair and soon hatch into nymphs that are very similar to the adults in both appearance and habits. Breeding occurs continuously throughout the year.

KEY TO FAMILIES

1. Body with a dense covering of short, stout spines or of spines and scales; parasitic on marine mammals such as seals, sea lions, and walrus .... **Echinophthiriidae**

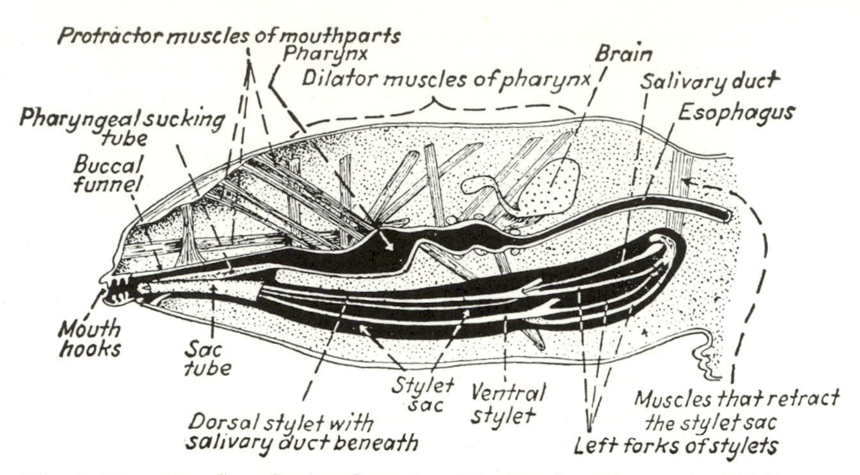

Fig. 8-42. Head and mouthparts of *Pediculus humanus*. (After Metcalf and Flint, *Destructive and useful insects*, by permission of McGraw-Hill Book Co.)

Body chiefly with discrete rows of spines or hairs (Figs. 8-43 to 8-45), never with scales; on land animals 2

2. Head with no trace of eyes (Fig. 8-43) .... **Haematopinidae**
   Head with eyes or eye tubercles present (Figs. 8-44, 8-45) 3
3. Three pairs of spiracles, forming an oblique row on each side, on what appears to be the first abdominal segment (really the fused first three) (Fig. 8-44) *(Pthirius)* .... **Pthirudae**
   Only one pair of spiracles on first abdominal segment (apparent as well as real) (Fig. 8-45) *(Pediculus)* .... **Pediculidae**

Economic Status. Sucking lice are a real concern on two counts: (1) losses inflicted on livestock, and (2) their menace to humans.

LOSSES TO LIVESTOCK. Horses, cattle, sheep, goats, dogs, and cats are attacked by several species of lice. The loss is partly a result of irritation and partly loss of blood, with resultant poor condition of the animal and failure to gain weight normally. Frequently lice will cause sheep and goats to rub against fences or trees, with heavy damage to the wool. In the main, poorly kept animals are the principal individuals badly attacked, but this is not always the case. An outbreak allowed to go unchecked will usually spread through an entire herd. Most of the lice attacking domestic animals belong to the family Haematopinidae, of which *Haematopinus asini*, the horse sucking louse (Fig. 8-43), is a common example.

MENACE TO HUMANS. Two species of Anoplura are external parasites of humans. They are both widespread in distribution and are most

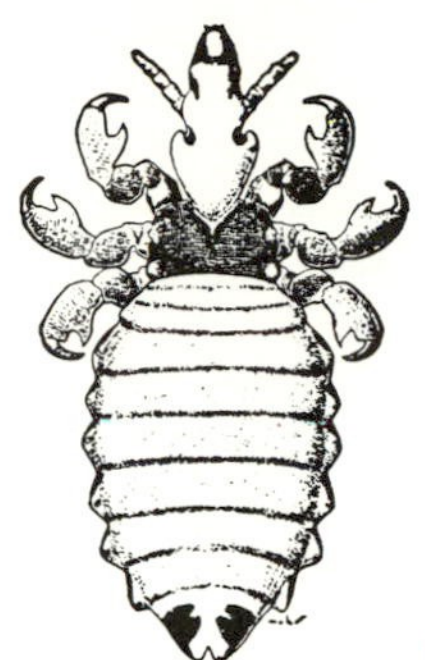

Fig. 8-43. The horse sucking louse *Haematopinus asini.* (From Illinois Natural History Survey)

abundant under crowded insanitary conditions. They spread from person to person in crowded situations or by clothing and bedding.

Crab Louse. *Pthirius pubis* (Fig. 8-44) is a very small crablike species infesting hairy portions of the body, especially the pubic region. It is seldom found on the head. This species inflicts painful bites and causes severe irritation. It has not been incriminated in the dispersal of any disease.

Body Louse or Cootie. *Pediculus humanus* (Fig. 8-45) is a larger louse about 4 or 5 mm long, which occurs on the hairy parts of the body. There are two forms of this species: the head louse, which occurs chiefly on the head and glues its eggs to the head hairs; and the body louse, which occurs chiefly on the clothes and reaches to the adjacent body areas to feed. The body louse glues its eggs to strands of the clothing. Under condition of regular head washing and clothes change cooties are seldom a nuisance. Under insanitary conditions they may develop in tremendous numbers and produce constant irritation.

The cootie has vied with the mosquito in shaping the destiny of history. Cooties transmit typhus fever and trench fever, which until

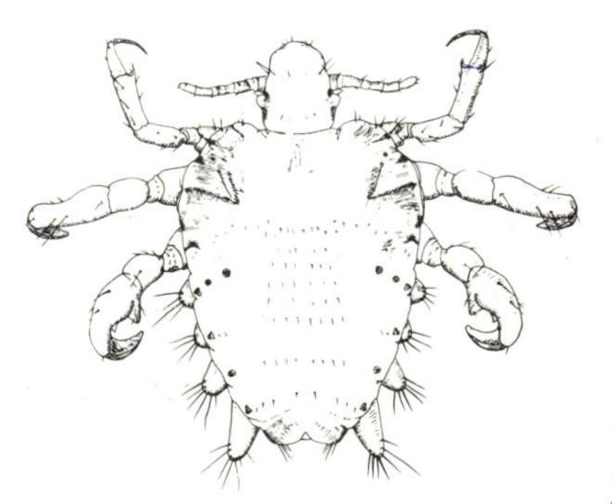

Fig. 8-44. The crab louse *Pthirius pubis.* (Redrawn from Ferris)

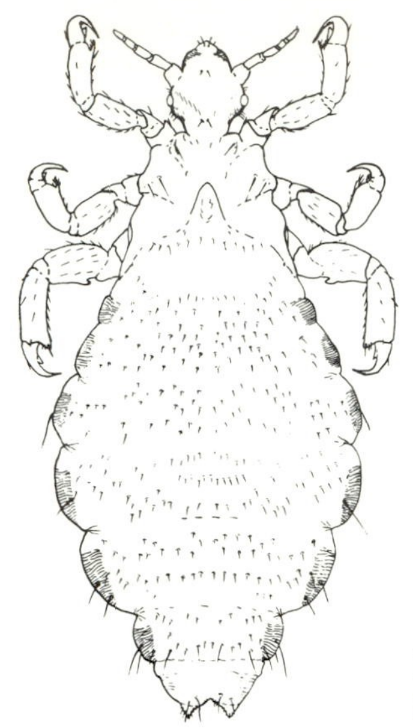

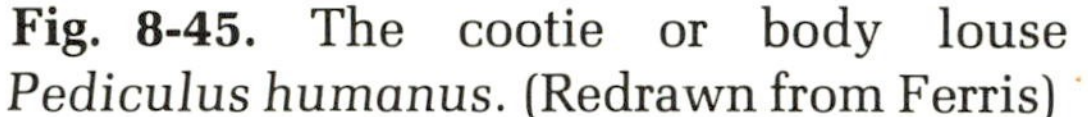

Fig. 8-45. The cootie or body louse *Pediculus humanus*. (Redrawn from Ferris)

recent years have been the scourges of northern armies, especially in winter. Under insanitary crowded camp or trench conditions, soldiers with heavy clothing provided ideal hosts for cooties. Typhus and trench fever have occurred in outbreak form and with disastrous results throughout many European armies. Napoleon's army in Russia was decimated as much by louse-borne disease as by hunger and exposure. The opposing Russian army suffered fully as much from typhus as the French army. Many claim that the cooties won the campaign, defeating both armies.

In World War II the control of lice by treatment of entire city populations chiefly with DDT, stopped outbreaks of typhus that had reached epidemic proportions. Especially effective results achieved in Naples in 1944 represent one of the most significant modern advances in the annals of preventive medicine.

References

Clay, T., 1970. The Amblycera (Phthiraptera: Insecta). *Bull. Brit. Mus. Nat. Hist. Entomol.*, **25**:73–98.

Emerson, K. C., 1964. *Checklist of the Mallophaga of North America (North of Mexico)*. Dugway, Utah: Proving Ground. 275 pp.

Ewing, H. E., 1929. *A manual of external parasites*. Baltimore: Thomas. pp. 127–157.

Ferris, G. F., 1934. A summary of the sucking lice (Anoplura). *Entomol. News*, **45**:70–74; 85–88.

1951. The sucking lice. *Pacific Coast Entomol. Soc., Mem.* **1**:1–320.
Hopkins, G. H. E., and T. Clay, 1952. A list of Mallophaga taken from birds and mammals of North America. *Proc. U.S. Natl. Mus.*, **22**(1183):39–100.
Kim, K. C., and H. W. Ludwig, 1978. The family classification of the Anoplura. *Syst. Entomol.*, **3**:249–284.
Riley, W. A., and O. A. Johannsen, 1938. *Medical entomology, 2nd ed.* New York: McGraw-Hill. 483 pp.

Order THYSANOPTERA*: Thrips

Small elongate insects (Fig. 8-47), most thysanopterans are between 2 and 3 mm long, with six- to nine-segmented antennae, large compound eyes, and compact mouthparts that form a lacerating-sucking cone. Many forms are wingless; others may have short wings or well-developed wings. In the latter case there are two pairs, both very long and narrow, with only one or two veins or none; the front pair are often larger, and both have a long fringe of fine hair along at least the hind margin. The legs are stout, the tarsi ending in a blunt tip containing an eversible pad or bladder. Metamorphosis is homometabolous, there being a definite pupa. Thrips occur commonly in flowers, and they may be found by breaking open almost any blossom and looking around the bases of the stamens or pistils. A large number of diverse forms can be taken by sweeping grasses or sedges in bloom. Many species are destructive to various plants and are found on the leaves of infested hosts. A large number of species, predaceous on mites and small insects, occur under bark of dead trees and in ground cover. The name thrips is both singular and plural.

Thrips' mouthparts are of an unusual type (Fig. 8-46). The various parts fit together to form a cone; some of the parts are needle-like stylets, which pierce and lacerate the food tissues; the juices thus released are sucked up into the stomach by a pump in the head capsule that pulls the liquid food through the cone formed by the mouthparts.

The metamorphosis of thrips is as unusual as their morphological features. The early stages are similar to the adults in structure of legs and mouthparts and in general shape; their feeding habits are also the same as those of the adults. These points are characteristic of insects having gradual metamorphosis. The first two instars have no wing pads (Fig. 8-47); the pads appear suddenly in the third instar as

* In the European literature the name Physopoda is commonly used for this order.

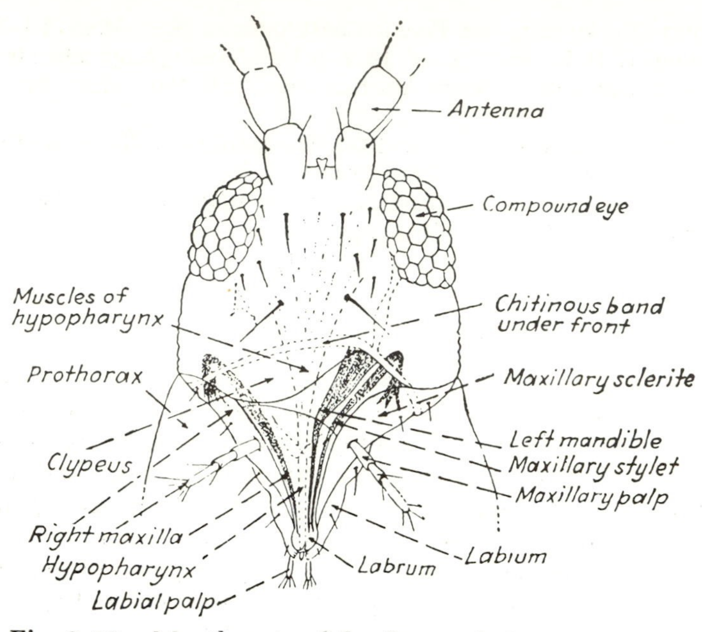

Fig. 8-46. Mouthparts of the flower thrips. (From Metcalf and Flint, *Destructive and useful insects*, by permission of McGraw-Hill Book Co.)

fairly large structures (Fig. 8-47); in the fourth (last) instar the wing pads are greatly enlarged (Fig. 8-47, 4). This fourth instar is quite unlike the others in habits. It does no feeding and is completely quiescent, with the antennae held back over the top of the head and pronotum. Certain thrips have an additional fifth instar. In some species having this, the fourth instar enters the soil and forms a cocoon, in which the quiescent fifth stage is passed. This feature is similar in so many respects to holometabolous development that the quiescent stage is called a *pupa*, and the first two instars are called *larvae*. The third-stage form, the active stage with the wing pads, is called a *propupa*. In some groups this form is not developed, the larvae transforming directly to the quiescent pupae.

The order contains several families represented in North America by about 500 species, some of them measuring 5 or 6 mm in length. A large number of species occur in only one species of plant, but a few common species, such as the flower thrips, *Frankliniella tritici*, feed on a great variety of plants and frequent blossoms of almost any species of plant. A few species are predaceous, feeding on red spiders and other mites, and minute insects.

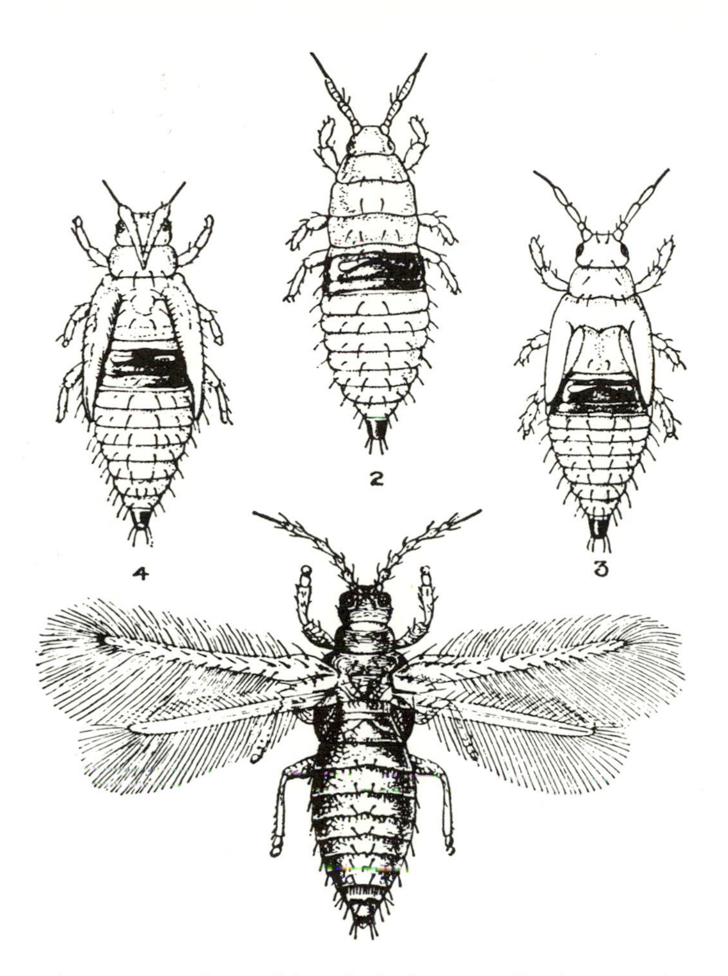

Fig. 8-47. The red-banded thrips *Heliothrips rubrocinctus*. (1) Adult; (2) nymph or larva; (3) propupa; (4) pupa. (After U.S.D.A., E.R.B.)

KEY TO COMMON FAMILIES

1. Major anal setae of last segment arising from a ring at apex of segment, which forms an undivided tube in both sexes (Fig. 8-48*F*) (suborder Tubulifera) .... **Phlaeothripidae**
   Major anal setae of last segment arising from body of segment (Fig. 8-48*E*), in female, last segment divided ventrally by saw slit, (Fig. 8-48*A*), in male, last segment sometimes tubular (suborder Terebrantia) 2
2. Sensory areas on antennal segments 3 and 4 small and circular or blister-like, never forming conical or finger-like protuberances

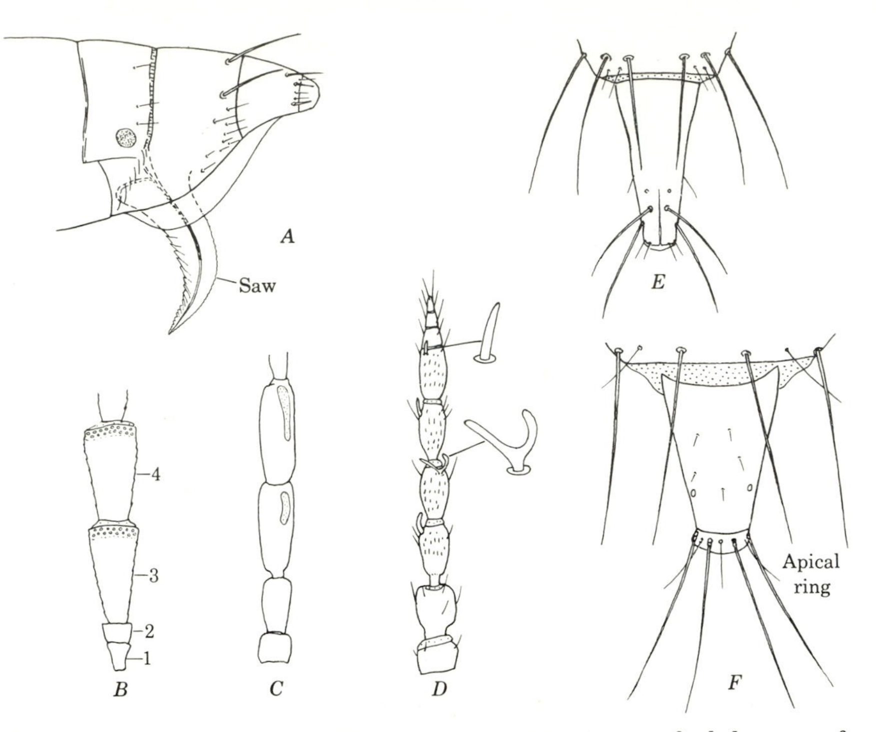

Fig. 8-48. Structures of Thysanoptera. (*A*) Apex of abdomen of *Anaphothrips*, Thripidae; (*B*) base of antenna of *Heterothrips*, Heterothripidae; (*C*) same of *Aeolothrips*, Aeolothripidae; (*D*) antenna of *Anaphothrips*, Thripidae; (*E*) dorsum of apex of abdomen of *Oxythrips*, Thripidae; (*F*) same of *Allothrips*, Phlaeothripidae. (Redrawn from various sources)

(Fig. 8-48*B*, *C*); antennae always 9-segmented 3

Sensory areas on segments 3 and 4 conelike or finger-like, projecting from segments (Fig. 8-48*D*); antennae 8- or 9-segmented; female saw curved down at apex (Fig. 8-48*A*) .... **Thripidae**

3. Apex of third antennal segment having a complete band of small, circular sensoria (Fig. 8-48*B*); female saw curved down at apex, as in Fig. 8-48*A* .... **Heterothripidae**

Apex of third antennal segment with one or two ovoid or elongate sensoria (Fig. 8-48C); female saw curved up at apex ...... **Aeolothripidae**

Economic Status. Several species of this order inflict considerable damage on commercial crops. The following, all belonging to the family Thripidae, are among the most injurious thrips over the United States as a whole.

The onion or tobacco thrips *Thrips tabaci* is a widespread species varying from lemon yellow to dark brown, which is especially injurious to onions, beans, and tobacco.

The greenhouse thrips *Heliothrips haemorrhoidalis* is a dark species with the body ridged to give it a checked or reticulate surface. The species is cosmopolitan. In the temperate region it is chiefly a greenhouse pest, attacking many kinds of hothouse plants.

The pear thrips *Taeniothrips inconsequens* is brown with gray wings. It attacks pears, plums, and related plants and produces a curious silvery blistered appearance on the injured leaves.

References

Ananthakrishnan, T. N., 1979. Biosystematics of Thysanoptera. *Annu. Rev. Entomol.*, **24:**159–184.

Lewis, T., 1973. *Thrips, their biology, ecology and economic importance.* New York: Academic Press. 349 pp.

Mound, L. A., 1977. Species diversity and the systematics of some New World leaf-litter Thysanoptera. *Syst. Entomol.*, **2:**225–244.

Preisner, H., 1949. Genera Thysanopterorum. *Soc. Found 1er Entomol. Bull.*, **33:**31–157.

Stannard, L. J., Jr., 1957. The phylogeny and classification of the North American genera of the suborder Tubulifera. *Ill. Biol. Monogr.*, **25:**1–200.

1968. The thrips or Thysanoptera of Illinois. *Bull. Ill. Nat. Hist. Surv.*, **29**(4):215–552.

Watson, J. R., 1923. Synopsis and catalogue of the Thysanoptera of North America. *Univ. Fla. Agr. Exp. Stn. Bull.*, **168:**1–100.

Order HEMIPTERA: Bugs and Their Allies

A large assemblage of diverse insects, the hemipterans are characterized chiefly by (1) piercing-sucking mouthparts that form a beak, (2) gradual metamorphosis, and (3) usually the possession of wings (Fig. 8-49). With few exceptions the compound eyes are large, the antennae 4- to 10-segmented, the individual segments frequently long, two pairs of wings are present and have relatively simple or reduced venation, and the abdomen has no cerci. In many families the abdomen has a well-developed sheath and sawlike ovipositor much as in Fig. 3-49.

The order derives its name, half-wing, from the structure of the front wing in many families (Fig. 8-49), in which the basal portion (the *corium*) is hard and thick and the apex (the *membrane*) is thinner and transparent. The corium approaches in texture the hard

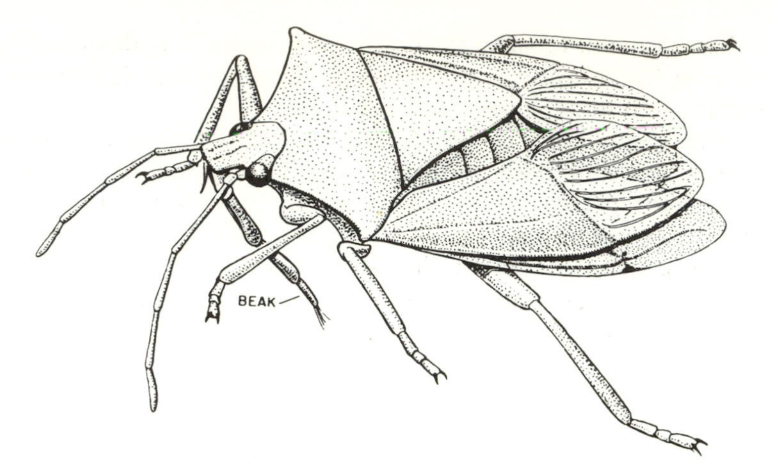

Fig. 8-49. A typical stinkbug, illustrating beak and wings. (From Illinois Natural History Survey)

wing cover or elytron of beetles; hence the name "hemelytron" is often applied to this half-hard, half-soft type of wing.

All but a few of the species comprising the Hemiptera can be placed in two large groups (Fig. 8-50). In one, the suborder Homoptera, the membranes of the front wings are typically entirely translucent (coreaceous in a few groups), the beak arises from the posterior portion of the head, and the head has a typical tentorium. In the other, the suborder Heteroptera, the bases of the front wings are thickened and the apices membranous, the beak typically arises from the front end of the head, the cheeks fuse behind the beak to form a gula, and the head lacks a tentorium. A third group is known only from the discontinuous circum-Antarctic genus *Peloridium*, containing small forms that occur on the bark of *Nothofagus*. *Peloridium* is in many ways intermediate between the Homoptera and Heteroptera, lacking a gula as in the Homoptera and possessing other structures like those of the Heteroptera. Some authors place it as an aberrant member of the family Peloridiidae; others consider it as a separate intermediate suborder, the Coleorhyncha.

Comparing these suborders with other related orders of insects, it seems probable that the ancestral form of the entire order was a plant feeder and morphologically was in general like some of the existing Homoptera, especially small members of the family Cicadidae with their abundant wing venation and primitive type of sawlike ovipositor.

The early homopterous lineage arising from this primeval hemipteran appears to have evolved a slender flagellum and

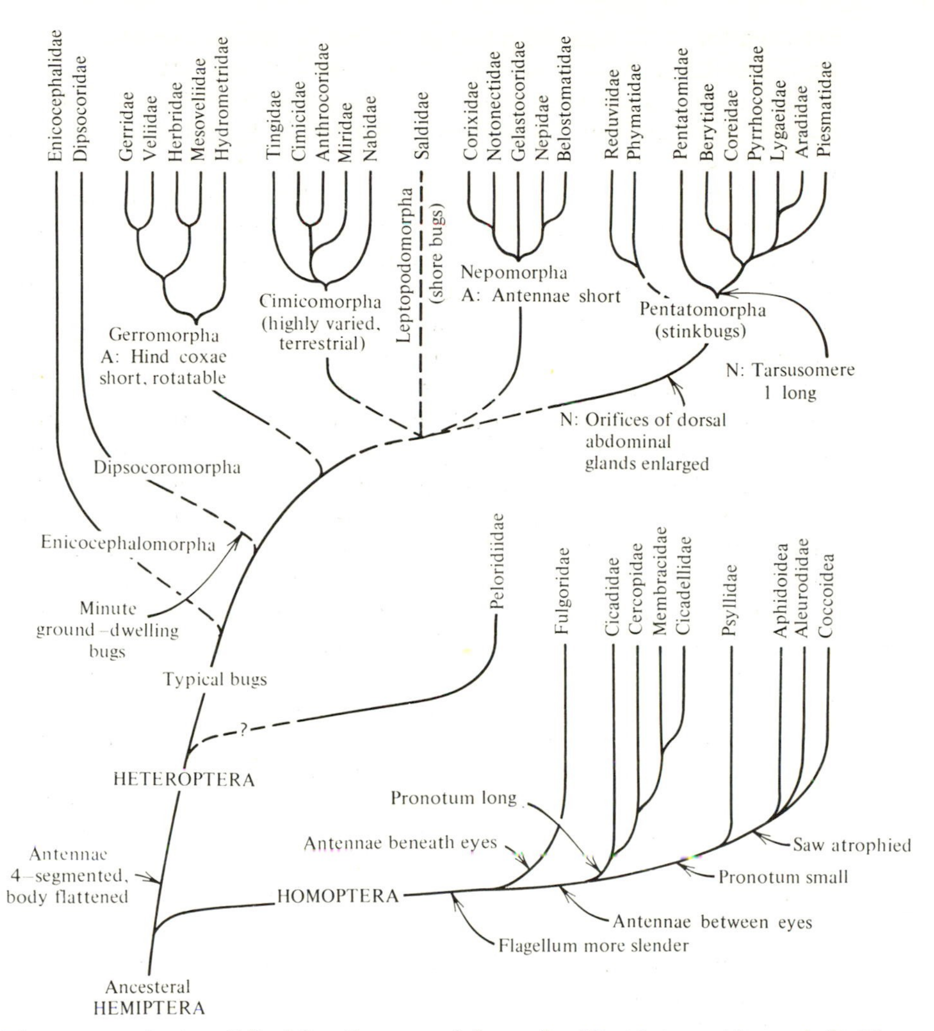

Fig. 8-50. A simplified family tree of the order Hemiptera. (A = adult; N = nymph)

sound-producing organs. In an early homopterous offshoot, represented by the family Fulgoridae, the antennae became situated beneath the eyes instead of between them. The continuing homopteran line gave rise to two main branches. In one the pronotum became large and considerably overlapped the mesonotum; this branch includes the cicadas (Cicadidae), spittle bugs (Cercopidae), treehoppers (Membracidae), and leafhoppers (Cicadellidae). In the other branch the pronotum remained small but the flagellum thickened, and the beak became angled so that it appears to arise from between the front legs; the most primitive family of this branch, the Psyllidae, has a typical sawlike ovipositor, but this structure is lost in the more specialized portion of the branch that includes the aphids (Aphidoidea), whiteflies (Aleurodidae), and scale insects (Coccoidea).

All of the suborder Homoptera are a plant-feeding group. Because of this uniformity of feeding habits and the lack of morphological specialization, the suborder Homoptera as a whole might be considered little specialized. In the aphids, whiteflies, and scale insects, however, great biological specializations evolved, including holometabolous development, complex life histories involving as many as seven different stages or morphs in the annual cycle of a single species, and seasonal alternation of hosts. In the other large homopterous branch, the leafhoppers evolved into such a multitude of species that their number is equaled by very few other families.

In the heteropterous line, the beak became anterior and the cheeks fused behind it, forming an extremely strong bridge or gula across the back of the head below the occipital foramen. This suggests evolutionary changes associated with a dietary change to a predaceous habit. The base of the fore wings became thickened and the antennae became reduced to four segments. This ancestral heteropteran gave rise to a large number of families; several have become highly modified for aquatic or amphibious life and at present these can be placed in the family tree only tentatively.

The evolutionary changes that occurred within the Heteroptera are difficult to unravel. The various families have been grouped into a number of apparent evolutionary lines, but how these lines are joined together to make an overall family tree for the suborder is not yet understood in spite of a wealth of characters on their embryology and morphology uncovered by detailed studies by Cobben (1968, 1978) and others. Seven of the major groups are known from North America, the name of each being the stem of a family name ending -omorpha. For example, the water strider group, including the family Gerridae and its allies, is termed Gerromorpha.

The most primitive existing bugs seem to be some minute forms, the Enicocephalomorpha and Dipsocoromorpha. These, the jumping ground bugs, live in woodland leaf litter. They are treated no further here.

The water striders (Gerromorpha) have specialized tarsal hairs and rotatable hind coxae, both of which allow them to skim rapidly over the surface of the water on which they live. Five families occur in North America; some members are marine. The Gerridae are the most abundant.

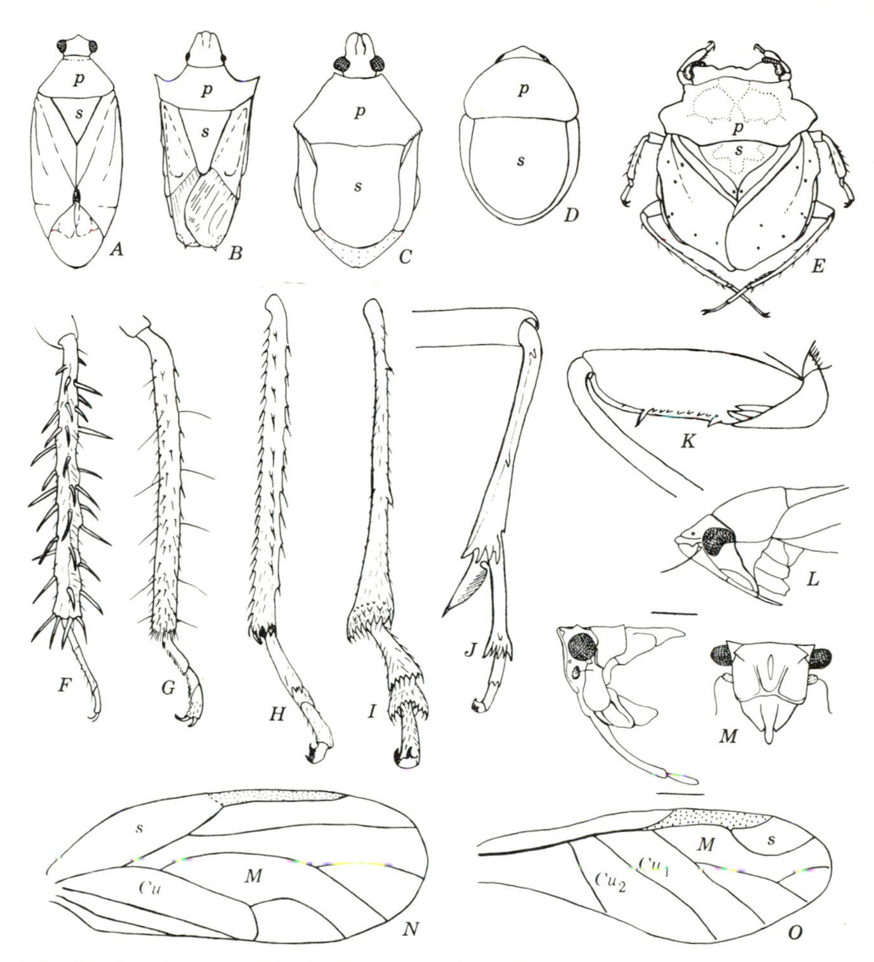

Fig. 8-51. Diagnostic characters of Hemiptera. (*A*) Outline of *Lygus*, Miridae; (*B*) outline of *Solubia*, Pentatomidae; (*C*) outline of *Stiretrus*, Pentatomidae; (*D*) outline of *Corimelaena*, Pentatomidae; (*E*) *Gelastocoris*, Gelastocoridae; (*F*) hind tibia and tarsus of *Pangaeus*, Pentatomidae; (*G*) hind tibia and tarsus of *Thyanta*, Pentatomidae; (*H*) hind tibia and tarsus of *Paraulacizes*, Cicadellidae; (*I*) hind tibia and tarsus of *Aphrophora*, Cercopidae; (*J*) hind tibia and tarsus of *Stenocranus*, Fulgoridae; (*K*) front femur of *Pacarina*, Cicadidae; (*L*) head of *Gypona*, Cicadellidae; (*M*) head, lateral and anterior views, of *Poblicia*, Fulgoridae; (*N*) wing of *Psylla*, Psyllidae; (*O*) wing of *Aphis*, Aphididae.

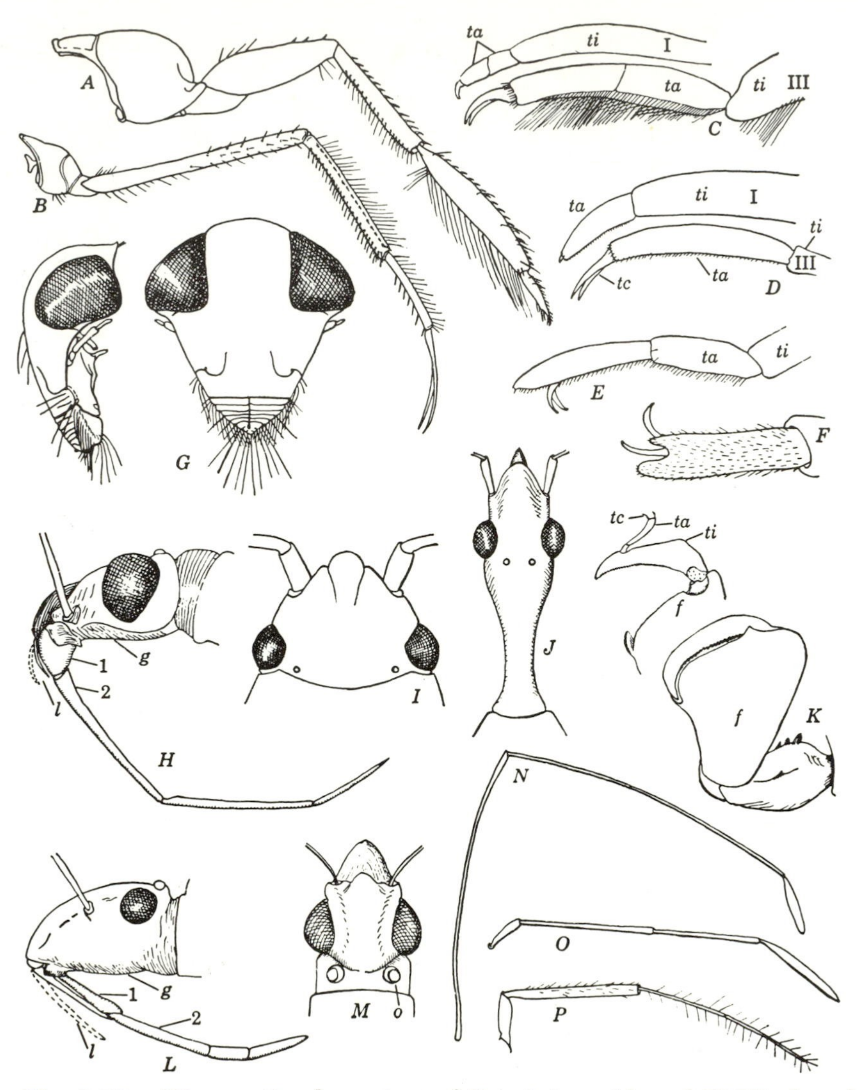

Fig. 8-52. Diagnostic characters of Hemiptera. (*A* and *B*) Hind and middle legs of *Corixa*, Corixidae; (*C*) front and hind tarsi of *Belostoma*, Belostomatidae; (*D*) front and hind tarsi of *Nepa*, Nepidae; (*E* and *F*) front tarsus, lateral and dorsal aspect of *Gerris*, Gerridae; (*G*) head of *Corixa*, Corixidae; (*H*) head of *Nabis*, Nabidae; (*I*) head of *Lygaeus*, Lygaeidae; (*J*) head of *Myodocha*, Lygaeidae; (*K*) front leg, inset showing structure of tarsus of *Phymata*, Phymatidae; (*L*) head of *Alydus*, Coreidae; (*M* and *N*) head and antenna of *Jalysus*, Neididae; (*O*) antenna of *Myodocha*, Lygaeidae; (*P*) antenna of *Lyctocoris*, Anthocoridae. *f*, femur; g, gula; *l*, labrum; *o*, ocellus; *ta*, tarsus, *tc*, tarsal claws; *ti*, tibia.

The remaining four major groups share such a mixture of ancestral states that they must be related in some fashion, but exactly how is not yet known nor is it clear that each group is monophyletic. The Cimicomorpha are a varied group of terrestrial and ectoparasitic forms having a reduced fore-wing venation. The Leptopodomorpha are shore inhabiting and have a unique basic venation (Fig. 8-53). The short-horned bugs (Nepomorpha) have extremely short antennae that in repose are almost hidden in lateral grooves of the head. All but the shore-inhabiting Gelastocoridae are aquatic. The typical Pentatomorpha have several diagnostic characteristics in the nymph. The first tarsomere is long. Laterally the abdomen has a row of peculiar small setae called *tricobothria*, and the orifices of the dorsal abdominal stink glands are heavily sclerotized. In almost all our species the adult antenna is four-segmented rather than five. The Reduvidae and its allies possibly belong to this group, having massive orifices of the abdominal dorsal stink glands.

More details of typical families of the four more common of these groups follow.

KEY TO COMMON FAMILIES

1. Hind leg without tarsal claws and having both tarsus and pretarsus flattened and bearing a dense fringe of long hair down each side (Fig. 8-52*A*); middle tarsus having normal tarsal claws (part of **Heteroptera**) 2
   Hind tarsus having tarsal claws similar to those on middle tarsus; hind leg usually without a long fringe but occasionally having one (Fig. 8-52*C*) 3
2. Beak forming a triangular striated piece that appears as a ventral sclerite of the head (Fig. 8-52*G*); middle tarsus having extremely long claws (Fig. 8-52*B*); front tarsus comblike . . . . **Corixidae**, p. 358
   Beak cylindrical and rodlike, curving back from the ventral portion of the head, as in (Fig. 8-52*L*); front and middle legs usual in shape . . . . **Notonectidae**, p. 359
3. Beak arising from posterior margin of head (Fig. 8-51*L*); no gula present behind it **(Homoptera)** 4
   Beak arising from front or venter of head (Fig. 8-52*L*); the venter of the head posterior to beak forming a sclerotized bridge or gula **(Heteroptera)** 15
4. Having wings, which are sometimes reduced to short scales 5
   Completely wingless species 13
5. Front femur greatly enlarged in comparison with middle femur (Fig. 8-51K); three ocelli present . . . . **Cicadidae**, p. 347
   Front femur no larger than middle femur; three, two, or no ocelli present 6
6. Antennae arising from sides of head, situated beneath or behind eyes (Fig. 8-51M) . . . . **Fulgoridae**, p. 346

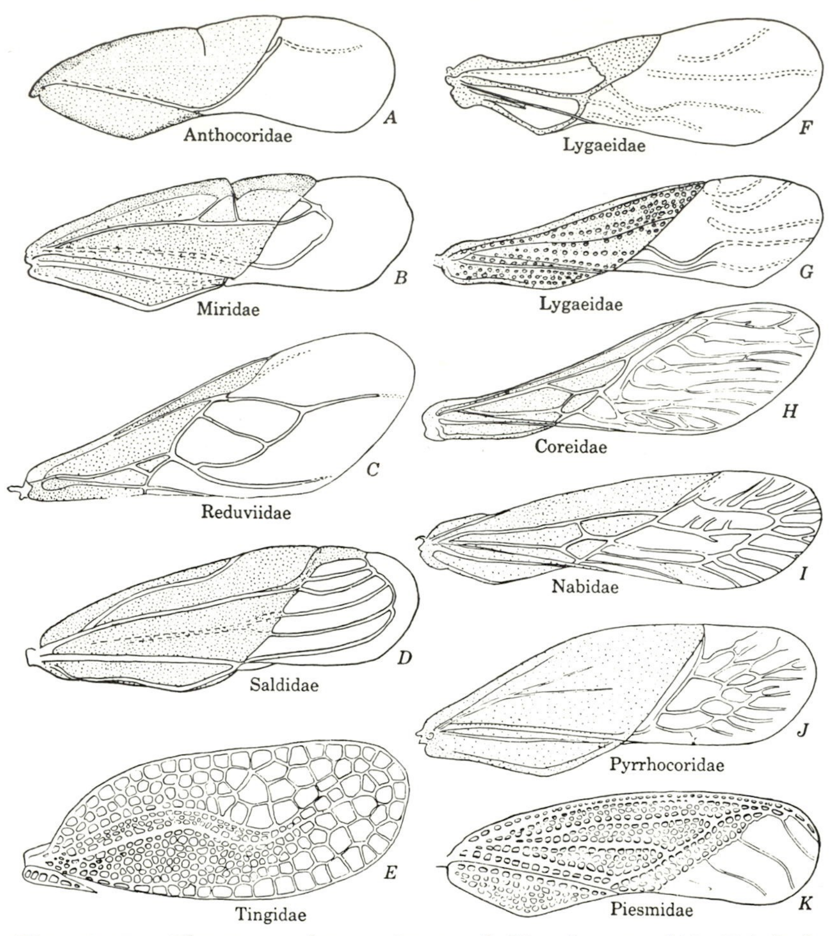

Fig. 8-53. Elytra or fore wings of Hemiptera. (*A*) *Triphelps*; (*B*) *Lygus*; (*C*) *Pselliopus*; (*D*) *Salda*; (*E*) *Gargaphia*; (*F*) *Blissus*; (*G*) *Myodocha*; (*H*) *Alydus*; (*I*) *Nabis*; (*J*) *Euryophthalmus*; (*K*) *Piesma*.

Antennae arising from front of head between eyes (Fig. 8-51*L*) 7

7. Pronotum enlarged dorsally into a large structure that covers most of head and body and may be highly ornamented with spines and processes (Fig. 8-54) .... **Membracidae**, p. 346
   Pronotum much smaller, without dorsal enlargement 8
8. Pronotum forming a broad shield that covers the greater part over the mesonotum (Fig. 8-57), tarsus 3-segmented (Fig. 8-51*H*, *I*) 9
   Pronotum forming a narrow collar that does not extend back over the mesonotum (Fig. 8-60); tarsus 1- or 2-segmented 10
9. Hind tibia bearing a double row of spines down its entire length, its apex usually not enlarged (Fig. 8-51*H*) ....**Cicadellidae**, p. 347

Hind tibia with only scattered spines except at apex, which is enlarged and armed with prominent crown of spines (Fig. 8-51*I*) .... **Cercopidae**, p. 347

10. Having only 1 pair of wings .... male **Coccoidea**, p. 350
    Having 2 pairs of wings 11
11. Wings milky-opaque, covered with a fine powdery white wax .... **Aleurodidae**, p. 348
    Wings transparent or patterned, not covered with a waxy secretion 12
12. Front wing with *S* very long, arising before stigma, and *Cu* branched (Fig. 8-51*N*); abdomen never with cornicles .... **Psyllidae**, p. 348
    Front wing with *S* short, arising from some part of the stigma, branched in some species (Fig. 8-51*O*); abdomen in many species having a pair of lateral tubes or cornicles (Fig. 8-60). .... **Aphidoidea**, p. 350
13. Eyes large, antennae situated at sides of head below or behind eyes (Fig. 8-51*M*) .... **Fulgoridae**, p. 346
    Either eyes rudimentary or absent, or antennae situated on front of head between eyes (Fig. 8-60) 14
14. Tarsus 1-segmented; body covered with a hard shell, waxy secretions, or a detachable scale (Fig. 8-61); abdomen never having cornicles ... . **Coccoidea**, p. 350
    Tarsus 2-segmented; body at most with waxy secretions; abdomen often having a pair of conspicuous cornicles or tubes (Fig. 8-60) .... **Aphidoidea**, p. 350
15. Antennae shorter than head, usually recessed in a concavity beneath the eyes or under the lateral margin of the head, as in Fig. 8-52*G* 16
    Antennae at least as long as the head, usually extending free from it (Fig. 8-64), sometimes fitting into a pronotal groove when at rest 18
16. Ocelli present. Small toadlike bugs (Fig. 8-51*E*) found along the margins of lakes and streams .... **Gelastocoridae**
    Ocelli absent. Forms living in water, sometimes flying and attracted to lights 17
17. Tarsi 1-segmented, front tarsus with only a minute claw or none (Fig. 8-52*D*); apex of abdomen with a long or short respiratory tube (Fig. 8-68*B*), each blade of which is concave mesally, the two fitting together to make a hollow tube; hind legs slender and without fringes of long hair .... **Nepidae**
    Tarsi 2-segmented, front tarsus with a stout curved claw (Fig. 8-52*C*); apex of abdomen at most with a pair of short flat respiratory filaments; hind tibia and tarsus often flattened, always having fringes of long hair for swimming .... **Belostomatidae**, p. 359
18. Head extremely long and slender, slightly bulbous at apex, where beak arises, the eyes situated at the middle of what appears to be a long neck; rest of body also very slender (Fig. 8-68*C*) .... **Hydrometridae**
    Head much stouter (Fig. 8-52*I*) or eyes not situated on the neck (Fig. 8-52*J*) 19

19. Front leg having femur and tibia chelate, forming a large grasping device, femur swollen and triangular, tibia curved and closing against the end of femur (Fig. 8-52*K*)........................ **Phymatidae**
    Front leg not chelate 20
20. Claws of front tarsus inserted before apex (Fig. 8-52*E*, *F*) 21
    Claws of front tarsus attached at apex, as in Fig. 8-52*C* 22
21. Middle pair of legs attached far from front legs, close to hind legs; hind femur very long (Fig. 8-64), beak 4-segmented ..... **Gerridae**, p. 355
    Middle pair of legs attached about midway between front and hind legs; hind femur only moderately long; beak 3-segmented .. **Veliidae**
22. Scutellum very large, reaching about one-half or more distance from posterior margin of pronotum to end of folded wings (Fig. 8-51*B* to *D*); antennae usually 5-segmented 23
    Scutellum much smaller, reaching about a quarter of the distance from pronotum to tip of body (Fig. 8-51*A*); antennae usually 4-segmented 26
23. Tibia armed with rows of thick thornlike spines (Fig. 8-51*F*) (Cydnidae) 24
    Tibia having series of short even spines, occasionally with a few scattered, very slender hairs (Fig. 8-51*G*) (Pentatomidae) 25
24. Scutellum triangular and not very large, as in Fig. 8-51*B* .. **Cyeninae**
    Scutellum large and U-shaped, covering most of abdomen (Fig. 8-51*D*) .... **Thyreocorinae**
25. Scutellum U-shaped and very wide, covering almost all of abdomen, the sides of scutellum curved mesad at extreme base, as in Fig. 8-51*D* .... **Scutellerinae**
    Scutellum V-shaped (Fig. 8-51*B*) or, if U-shaped, then never larger than in Fig. 8-51*C*, and slightly contracted just beyond base .... **Pentatominae**
26. Front wing abbreviated, with no membrane, and reaching at most to middle of abdomen (Fig. 8-67) 27
    Front wing normal, with a large apical membrane, or reaching well beyond middle of abdomen (Fig. 8-53) 28
27. Body flat and wide; front wing short, broad, and scalelike, only barely reaching over base of abdomen; sides of pronotum large, round, and flangelike (Fig. 8-67); beak 3-segmented; antennae long and slender .... **Cimicidae**, p. 357
    Body narrower, or otherwise different from foregoing, having either a 4-segmented beak, different-shaped wing, or short antennae. A few genera, most of them rare, difficult to key to family, belonging to **Anthocoridae**, **Miridae**, **Aradidae**, **Lygaeidae**, or **Nabidae**; and all nymphs of Heteroptera families listed beyond this point. The wingless species of these groups are keyed no farther here.
28. Hemelytra large, covering entire abdomen and reticulate over their entire surface with a netlike pattern, with little or no distinction between corium and membrane (Fig. 8-53*E*) .... **Tingidae**, p. 355
    Hemelytra with a definite apical membrane (Fig. 8-53, all except *E*) 29

29. Membrane of hemelytron having 1 or 2 large basal cells, and none, 1, or 2 short spurlike veins extending distally from these (Fig. 8-53*B*, *C*) 30
Membrane of hemelytron having either no closed cells (Fig. 8-53*A*) or at least 5 or 6 veins (including the costa) running through the membrane (Fig. 8-53*D*, *H*, *I*, *J*) 31
30. Ocelli prominent, two in number; membrane of hemelytron having a long vein proceeding from top of upper closed cell (Fig. 8-53*C*) .... **Reduviidae**, p. 359
Ocelli absent; membrane with a vein proceeding only from bottom to lower closed cell, or such a vein lacking (Fig. 8-53*B*) .... **Miridae**, p. 356
31. Antenna having first 2 segments stout, last 2 threadlike, forming a slender terminal filament (Fig. 8-52*P*); ocelli present but small; hemelytral membrane with only 1 or 2 weak veins (Fig. 8-53*A*) .... **Anthocoridae**
Antenna having 1 or both of the 2 apical segments as thick as the first or second (Fig. 8-52*N*, *O*) 32
32. Hemelytral corium extending markedly beyond a ridgelike oblique vein near apex of corium (Fig. 8-53*K*); corium entirely reticulate. .... **Piesmatidae**
Hemelytral corium not extending beyond an apical oblique vein (Fig. 8-53*H* to *J*), or not having such a vein (Fig. 8-53*G*) 33
33. No ocelli present 34
Two ocelli present 35
34. Flat wide warty bugs (Fig. 8-71*A*, *B*); tarsus 2-segmented, the first segment short; hemelytra often small, the periphery of the abdomen extending considerably beyond them .... **Aradidae**, p. 360
Stout insects, the body deep; tarsus 3-segmented, the first segment long; hemelytra larger (Fig. 8-53*J*); covering all abdomen except tip and sides near apex .... **Pyrrhocoridae**
35. Hemelytral membrane having 4 or 5 large and fairly regular closed cells and no other venation (Fig. 8-53*D*); oval, fairly flat bugs found on stream and lake shores .... **Saldidae**
Hemelytral membrane having either an irregular network of cells or only 1 or 2 small, well-sclerotized ones (Fig. 8-53*F* to *I*) 36
36. Membrane having a series of about 15 irregular veins, at least on apical portion (Fig. 8-53*H*, *I*) 37
Membrane having only 5 or 6 veins across it (Fig. 8-53*F*, *G*) 38
37. First segment of beak short and conelike, thicker than the second (Fig 8-52*H*); front femur thickened, front tibia armed inside with a double row of short black teeth .... **Nabidae**
First segment of beak cylindrical and long, similar in general shape to second segment (Fig. 8-52*L*); front femur usually more slender, front tibia never with inner rows of black teeth .... **Coreidae**, p. 360
38. Each ocellus situated behind an eye, at the base of a distinct swelling (Fig. 8-52*M*); extremely slender and elongate bugs, with long and slender legs and antennae; last segment of antenna short and oval, forming

a small club (Fig. 8-52*N*) . . . . **Berytidae**
Ocelli situated closer to or between eyes, and not at the base of a swelling (Fig. 8-52*I*, *J*); chiefly robust short insects or having short legs; antennae either short (Fig. 8-52*O*) or not clubbed . . . . **Lygaeidae**, p. 361

SUBORDER HOMOPTERA

CICADAS, LEAFHOPPERS, APHIDS, SCALE INSECTS AND OTHERS. This suborder contains the cicadas, leafhoppers, aphids, scale insects, and their allies, all plant feeders. The North American fauna is composed of about a dozen families or superfamilies.

Fulgorid–Cicada. In these homopterans (Fig. 8-50), the flagellum of the antenna is needle-like and the tarsus is almost invariably three-segmented. They include many forms of bizarre appearance. Some of the Membracidae, or treehoppers (Fig. 8-54), have the pronotum greatly enlarged and ornamented with ridges, horns, or prongs. The Fulgoridae are a large family, and many resemble leafhoppers (Fig. 8-55). Some of the fulgorids have large foliaceous

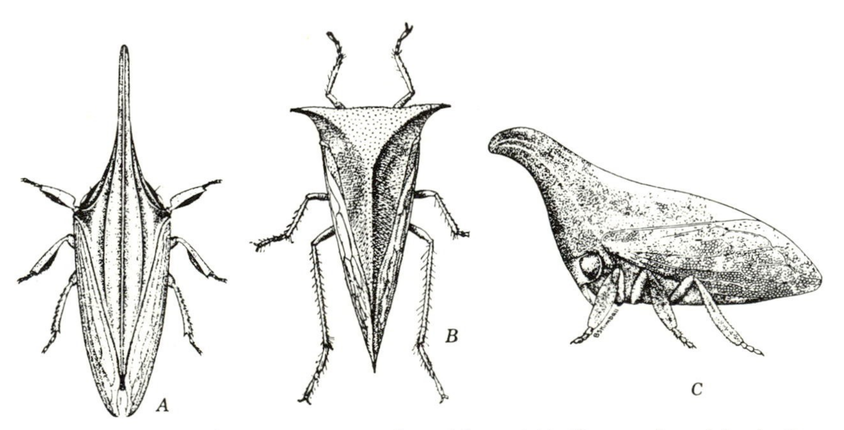

Fig. 8-54. Treehoppers, Membracidae. (*A*) *Campylenchia latipes;* (*B*) *Ceresa bubalis;* (*C*) *Enchenopa binotata.* (*A*, *C*, from Kansas State University; *B*, from U.S.D.A.)

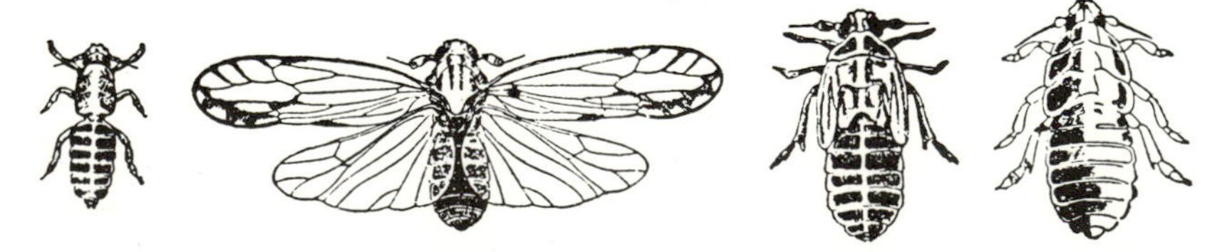

Fig. 8-55. A fulgorid *Peregrinus maidis*. (After Thomas)

wings, and others, such as our native *Scolops* and the South American lantern fly, or peanut bug, *Lanternaria phosphorea*, have bizarre projections of the head. Another oddity is the spittle bug family, Cercopidae. The nymphs of this family produce masses of white froth or spittle-like substance and live hidden beneath it. Two well-known and abundant families are the Cicadidae (cicadas) and the Cicadellidae (leafhoppers).

CICADIDAE, CICADAS. These are large insects, many North American species measuring 5 cm or more. They are distinguished structurally from related families by having three distinct ocelli on the dorsum of the head. The males have highly developed musical organs, and during warm days and summer evenings they make a shrill noise. The nymphs have enlarged front legs, presumably for digging, and are subterranean, feeding on sap from the roots of deciduous trees.

The nymphal period is long, 2 to 5 years for most species. The periodic cicadas, *Magicicada septendecim* and its immediate relatives (Fig. 8-56), constitute a group of southern species with a nymphal life of 13 years and a group of more northern species with a nymphal life of 17 years. These species have attracted widespread attention because of the periodic nature of their cycles. In some areas only a single brood occurs, and there the adults appear only every 13 or 17 years. On these occasions they usually appear in huge swarms, and the ovipositing females may cause serious damage to the twigs and branches of fruit and hardwood trees.

CICADELLIDAE, LEAFHOPPERS. This family is the largest in the entire order Hemiptera, represented in North America by over 2500

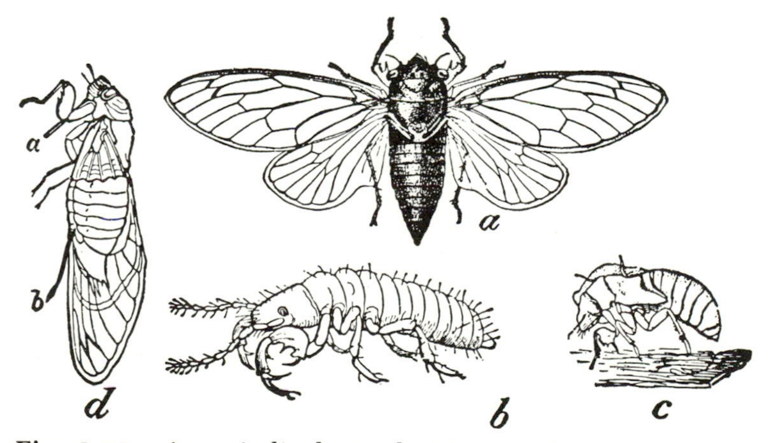

Fig. 8-56. A periodical cicada *Magicicada septendecim*. (*a* and *d*) Adults; (*b*) nymphs; (*c*) shed nymphal skin. (From U.S.D.A., E.R.B.)

species. Leafhoppers are not only numerous in species but also extremely abundant in numbers of individuals. They are probably collected in general sweeping more commonly than any other insect group. Most of these are less than 10 mm long and have long hind tibiae bearing longitudinal rows of spines, but with neither large spurs nor a crown of spines at the tip. Although a few species are broad or angular, most are slender and nearly parallel-sided (Fig. 8-57). Female leafhoppers have strong ovipositors that they use to cut slits for eggs in plant stalks (usually herbs) or leaves. Leafhoppers are often destructive to certain crops, not only by direct damage caused by feeding, but also because they transmit many plant diseases. The beet leafhopper *Circulifer tenellus* transmits the virus that causes curly top of beets, a most destructive disease to the sugar-beet crop; and the plum leafhopper *Macropsis trimaculata* transmits another destructive virus that causes peach yellows.

Psyllid–Aphid. The antennae are either short and stout or long and threadlike (Fig. 8-60) in at least some stage of the life cycle. The wing venation is greatly reduced, and the tarsi have only one or two segments. Because of the occurrence of diverse body forms within the life cycle of a single species, it is difficult to characterize the families with a brief description.

Two families, the jumping plant lice or Psyllidae (Fig. 8-58) and the whiteflies or Aleurodidae (Fig. 8-59) have a simple life cycle in which adults of both sexes are winged and similar in general appearance. In many Psyllidae and all Aleurodidae the later nymphal

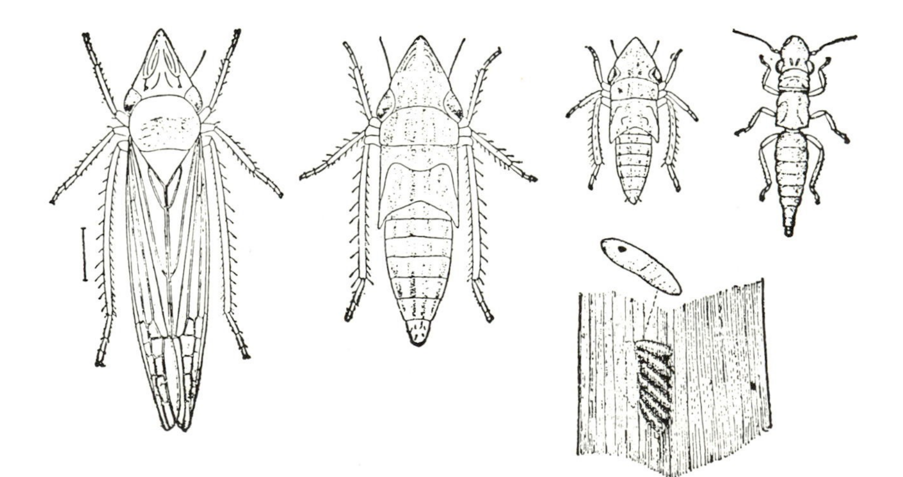

Fig. 8-57. A leafhopper *Draeculacephala mollipes*, adult, nymphs, and eggs. (From U.S.D.A., E.R.B.)

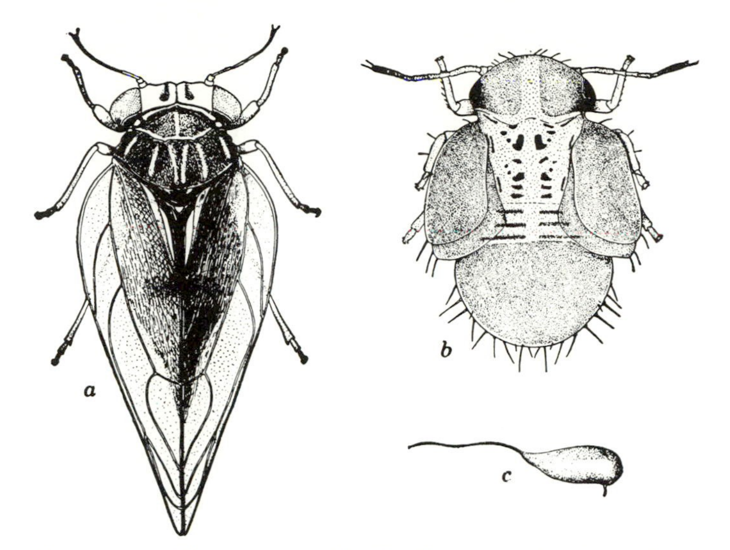

Fig. 8-58. The pear psylla *Psylla pyricola*. (*a*) Adult; (*b*) nymph; (*c*) egg. (From Connecticut Agricultural Experiment Station)

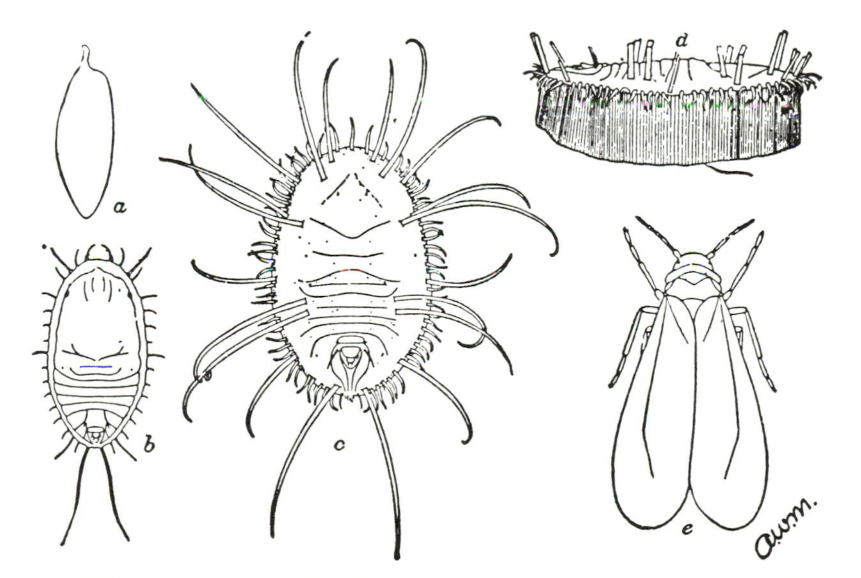

Fig. 8-59. A whitefly *Trialeurodes vaporariorum*. (*a*) egg; (*b*) larva, first instar; (*c*) puparium, dorsal view; (*d*) puparium, lateral view; (*e*) adult. (After Morrill)

instars are flat, inactive or sluggish, and scalelike in appearance. The members of both families are small.

All the other families of the psyllid–aphid branch are segregated into two major groups: (1) the aphids and their allies, the superfamily Aphidoidea, and (2) the mealybugs and scale insects, the superfamily Coccoidea. Each group contains several families differentiated chiefly by biological characteristics and including many species of great economic importance.

THE APHIDOIDEA. (Fig. 8-60) are characterized by (1) the presence of several veins and a stigmal area in the fore wings of the winged forms; (2) the existence of two-segmented tarsi in most species; and (3) the existence of a complex system of alternating generations including wingless, winged, parthenogenetic, and sexual forms in the life cycle of a single species. This phase is discussed more fully in Chap. 7. The Aphididae, or plant lice, is the most important family in the group. Many species of great economic concern are members of this family, for example, the melon aphid *Aphis gossypii*, a pest of cucurbits and cotton; and the green peach aphid *Myzus persicae*, a pest of many crops and the disseminator of many plant diseases.

THE COCCOIDEA (Fig. 8-61) differ in several important respects from the aphids; (1) The females are always wingless, extremely sluggish or completely fixed in position, and are covered by a waxy secretion

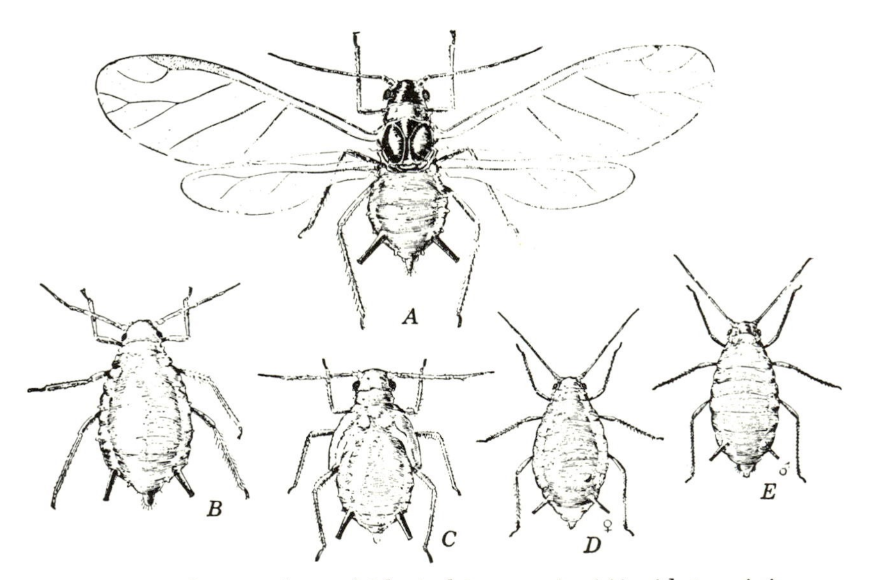

Fig. 8-60. The apple aphid *Aphis pomi*. (*A*) Alate viviparous female; (*B*) apterous viviparous female; (*C*) nymph of alate; (*D*) oviparous female; (*E*) male. (From U.S.D.A., E.R.B.)

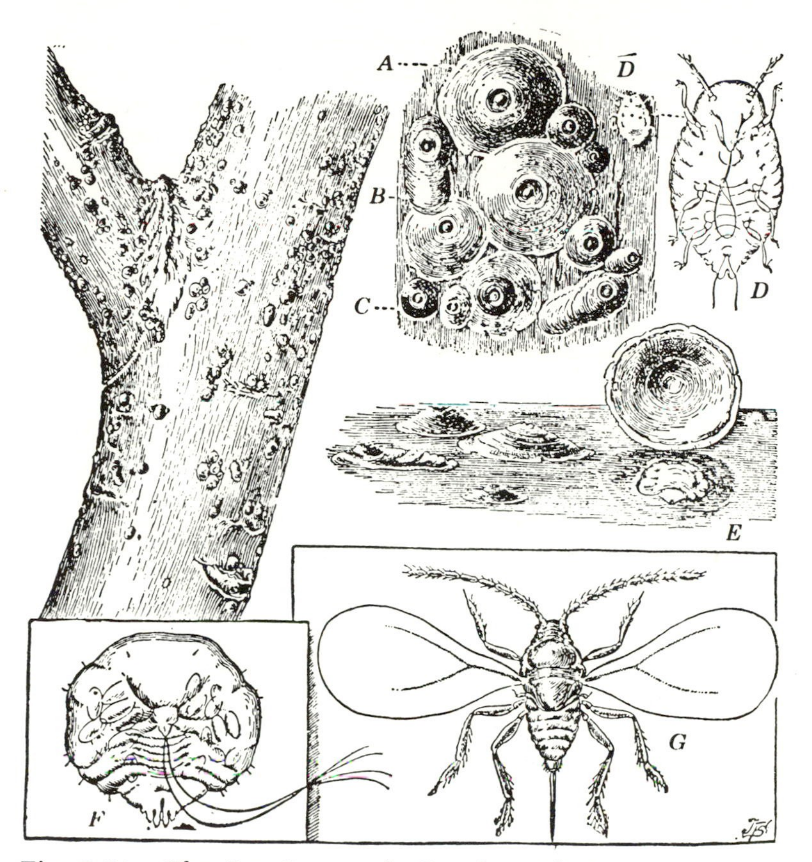

Fig. 8-61. The San Jose scale *Quadraspidiotus perniciosus*, infesting apple. (*A*) Scale of adult female; (*B*) scale of male; (*C*) first-instar young; (*D*) same, more enlarged; (*E*) scale lifted to expose the female body beneath; (*F*) body of the female; (*G*) adult male. (From U.S.D.A., E.R.B.)

or a tough scale, or have a hard integument as in the family Coccidae; (2) the males are small and delicate and have a single pair of wings with only one or two simple veins; and (3) the life cycle is relatively simple.

The family Diaspididae (Fig. 8-61) is one of the most important in the scale insect group. The females are the sedentary, small, scalelike or cushion-like insects found on many species of trees. The actual insect is a delicate oval body hidden beneath the scale, which is a protective covering. The appendages are extremely reduced, the body at maturity being little more than an egg sac. As the eggs are gradually discharged, the body shrinks, so that the entire egg mass is laid within the protective covering of the scale. The first-instar nymphs are minute and extremely active. They crawl with rapidity in all directions and thus effect the widespread distribution of these scale insect species. After the first molt, the nymphs become sedentary, and each forms a scale. Several species of the family are among the most destructive insects known to commercial agriculture. The San Jose scale *Quadraspidiotus perniciosus* (Fig. 8-61) is a persistent pest of deciduous fruit trees and many ornamentals; before advent of oil sprays it threatened to wipe out several of the fruit crops in many areas in the United States. The cosmopolitan oystershell scale *Lepidosaphes ulmi* (Fig. 8-62) is a common pest of almost all deciduous trees and shrubs in the United States.

The family Eriococcidae or mealybugs (Fig. 8-63) are another important group, attacking many hosts, especially greenhouse and household plants in the more northern areas. The mealybugs make no scale but secrete waxy filaments that are especially noticeable along the periphery of the body.

SUBORDER HETEROPTERA

BUGS. This suborder contains a wide variety of forms ranging from a few millimeters to a few centimeters in length, and including terrestrial, semiaquatic and aquatic types. The antennae are four- or five-segmented, and the eyes are well developed except in the ectoparasitic family Polyctenidae.

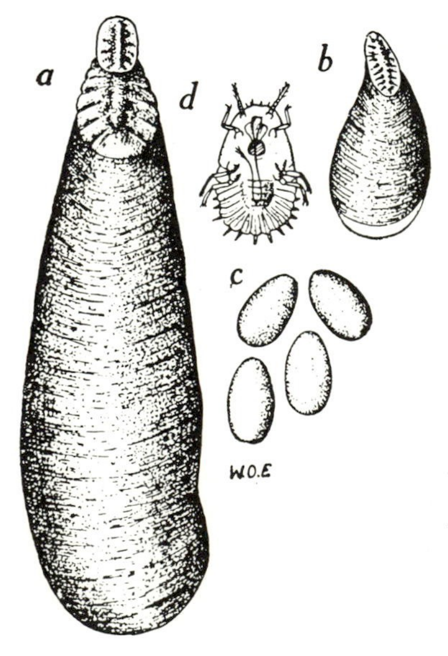

Fig. 8-62. Oystershell scale *Lepidosaphes ulmi*. (*a*) Female scale; (*b*) male scale; (*c*) eggs; (*d*) first-instar nymph. (After Blackman and Ellis)

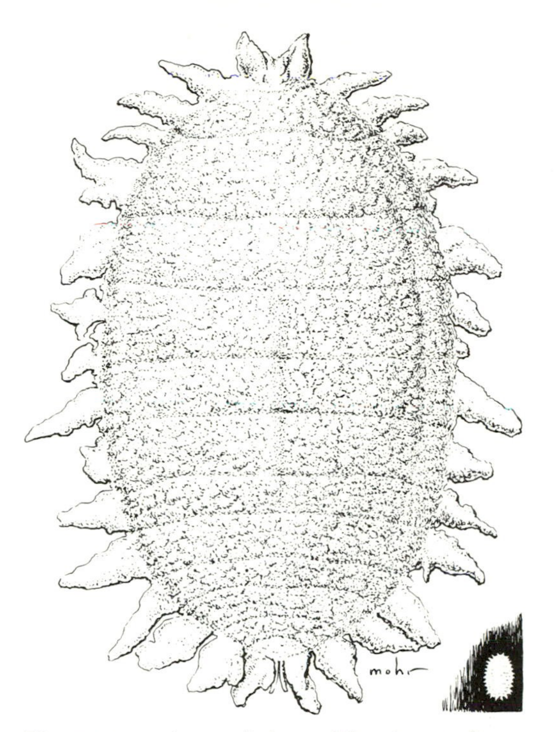

Fig. 8-63. A mealybug. The inset shows natural size. (Original by C. O. Mohr)

The nymphs of all forms resemble the adult in general outline, but typically differ in having dorsal stink glands on the abdomen. In this respect they differ also from the nymphs of the Homoptera.

The eggs are laid singly or in groups glued to stems or leaves, inserted into plant stems or, rarely, into damp sand.

The suborder contains predaceous species and plant-feeding species; often both types occur in the same family. The predaceous species feed chiefly on smaller insects. Some species of the plant bugs, Miridae, are wholly predaceous; in others the predaceous habit is only partially developed, and insect blood merely serves to supplement the principal diet of plant juices. The mixture of food habits in the same families has resulted in some queer anomalies. In the stinkbug family Pentatomidae, the harlequin bug *Murgantia histrionica* is a serious pest of cabbages; a predatory species *Perillus bioculatus* is one of the most effective natural enemies of the Colorado potato beetle.

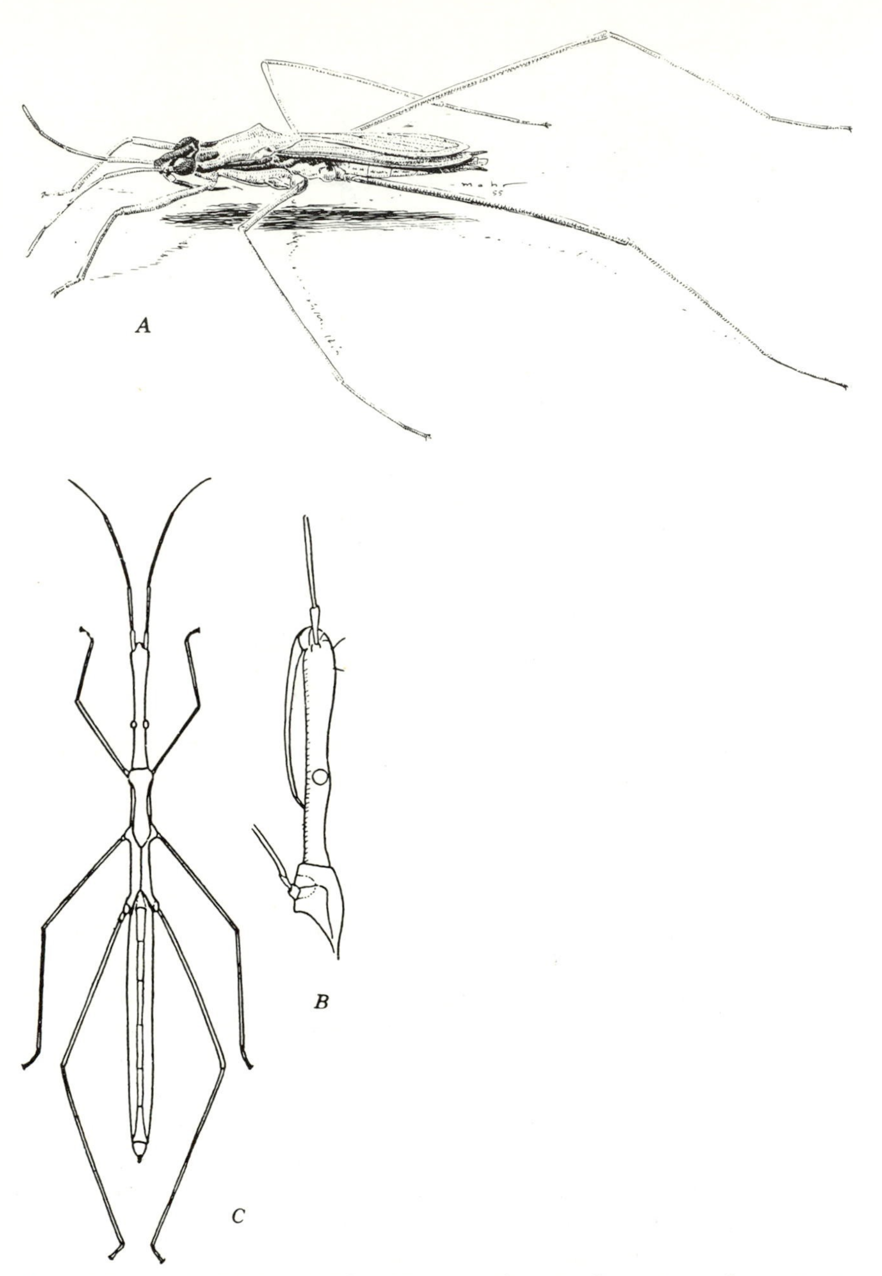

Fig. 8-64. (*A*) Water strider *Gerris rufomaculata*, Gerridae. (*B*, *C*) *Hydrometra*, Hydrometridae. (From *Hemiptera of Connecticut*)

Water Striders (Gerromorpha)

GERRIDAE, WATER STRIDERS. These are slender bugs with long legs (Fig. 8-64). They live on the surface of the water, inhabiting chiefly ponds, the margins of lakes, and the more sluggish backwaters and edges of rivers and streams. The tarsi are fitted with sets of nonwetting hair that allow the bugs to run with amazing speed and stand on the water surface with ease. All species are predators or scavengers, feeding chiefly on other insects that occur on the water surface. They lay their eggs in masses attached to aquatic plants or thrust them into submerged stalks. Several other closely related families live as striders or skaters on the water surface; from these the Gerridae differ in having very long hind femora, which extend considerably beyond the apex of the abdomen.

Bedbugs and Their Allies (Cimicomorpha)

TINGIDAE, LACE BUGS. These are small delicate plant-feeding insects (Fig. 8-65*A*, *B*), usually occurring in large colonies. The pronotum and hemelytron are wide, reticulate, and lacelike, extending well beyond the sides of the body; in certain genera the pronotum has a large bulbous mesal lobe that extends forward above the head (Fig. 8-65*A*). The antennae and beak are four-segmented, ocelli are lacking, and the tarsi are two-segmented. The nymphs differ considerably from the adults in general appearance; some are comparatively smooth and scalelike; others are armed with large numbers of long spines. The eggs are laid in or on the leaves of the host plant. The

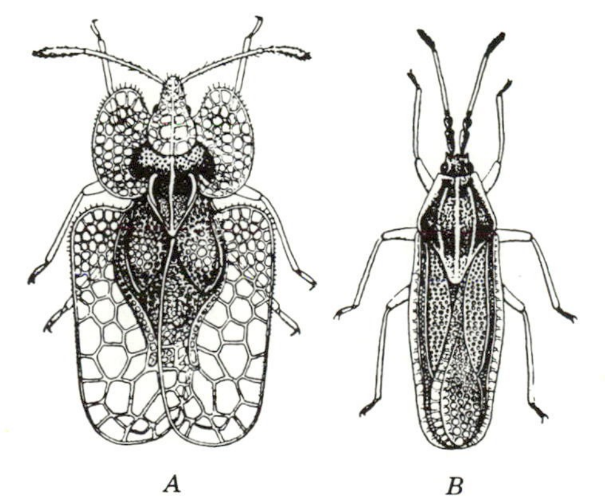

Fig. 8-65. Lace bugs. (*A*) *Atheas exiguus*; (*B*) *Corythuca floridanus*. (After Heidemann)

Tingidae are represented in North America by over 200 species, most of them specific to a single host genus or species. A colony of lace bugs produces a characteristic white-spotted appearance of the leaves that readily betrays the presence of the colony. Examination of alder, oaks, sycamores, hawthorns, apples, birches, and other trees will net many species of lace bugs. Shrubs and herbs also support a considerable fauna.

MIRIDAE, PLANT BUGS. This family is a large one and, in North America, it contains nearly 2000 species, over a third of all known North American Heteroptera. The plant bugs (Fig. 8-66) belong to the series of families having a four-segmented beak and no ocelli. With few exceptions they possess fully developed wings; the hemelytron usually has a distinctive sclerite or *cuneus* in the sclerotized portion and one or two simple cells in the membrane (Fig. 8-53).

Most of the species are plant feeders, many attacking only one or a very limited number of host species. Plant bug feeding causes a bleaching *(etiolation)* or blossom blight of the host and frequently results in marked commercial loss of certain crops. Some of the most destructive economic species are the cotton fleahopper *Pseudatomoscelis seriatus*; the garden fleahopper *Halticus bracteatus*, which damages alfalfa, clover, and garden crops such as beans; and the tarnished plant bug *Lygus lineolaris*, a general feeder and a local

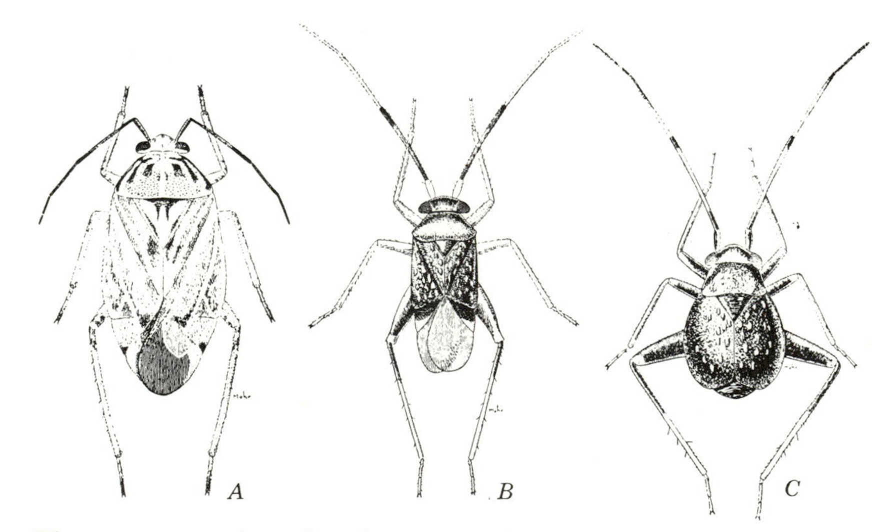

Fig. 8-66. Miridae, plant bugs. (*A*) The tarnished plant bug *Lygus lineolaris*; (*B* and *C*) the garden fleahopper *Halticus bracteatus*, male and female. (From Illinois Natural History Survey)

pest of many crops. Certain genera, including a few striking ant mimics, are predaceous on aphids and other insects.

Plant bug females insert their eggs into herbaceous material. Most of the species have only a single generation a year; winter usually is passed in the egg stage. A few species, including the tarnished plant bug, hibernate as adults and deposit their eggs the following spring.

CIMICIDAE, BEDBUGS. This family includes only a few species of wide flat insects that feed on the blood of birds and mammals. The fore wings or hemelytra are represented only by short scalelike pads; the hind wings are completely atrophied. Cimicidae live in bird or mammal nests and in dwellings. Humans are attacked by the bedbugs *Cimex lectularius* (Fig. 8-67) and *C. hemipterus*, which may become important pests in living quarters of all kinds. During the day the bedbugs hide in cracks and crevices of woodwork, furniture, and debris, emerging at night to seek a blood meal. The female lays up to 200 cylindrical whitish eggs, depositing them in crevices.

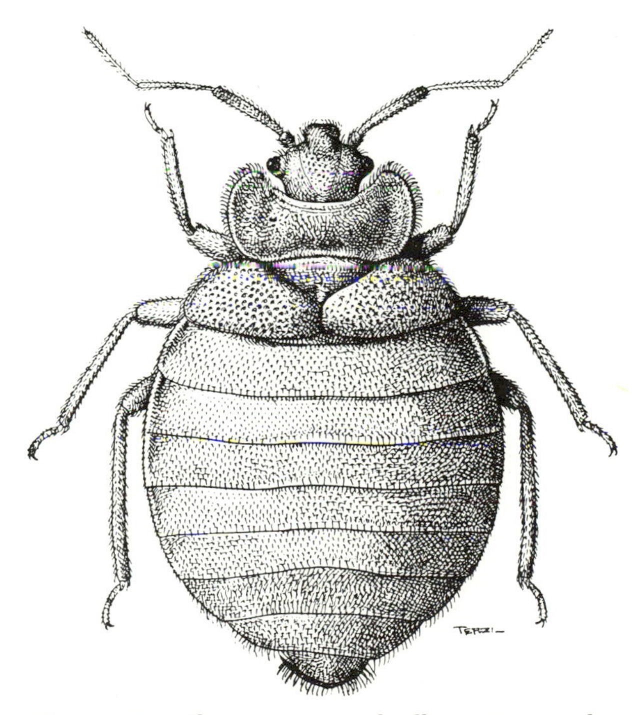

Fig. 8-67. The common bedbug *Cimex lectularius*. (By permission of British Museum of Natural History)

Short-Horned Bugs (Nepomorpha)

The members of this group have short antennae, usually recessed under the head and not visible from the dorsal aspect, and are either aquatic or shore inhabiting. Of the nine families recognized in the North American fauna, the Corixidae, Notonectidae, and Belastomatidae are the most common.

CORIXIDAE, WATER BOATMEN. These bugs (Fig. 8-68) are characterized by the short stout labium that looks more like the lower sclerite of the head than like a beak (Fig. 8-52*G*). The front legs are short, flattened, or scoop-shaped; the hind legs are long, flattened, and fringed with combs of bristles. Both nymphs and adults are truly aquatic, swimming in the water and incapable of more than clumsy flopping on land. The fringed hind legs are used for swimming; they swim dorsal side up. The adults leave the water for dispersal flights and may be observed in swarms over bodies of water. Sometimes these swarms are attracted to lights. The eggs are attached to solid supports such as stones, sticks, and shells in the water. Certain forms, such as *Ramphocorixa*, more often lay their eggs on the body or appendages of the crayfish *Cambarus*, which in some localities may be literally plastered with corixid eggs.

The water boatmen differ from all other Hemiptera in their feeding habits. They feed in the ooze at the bottom of the water, the stylets of the mouthparts darting in and out in unison like a snake's

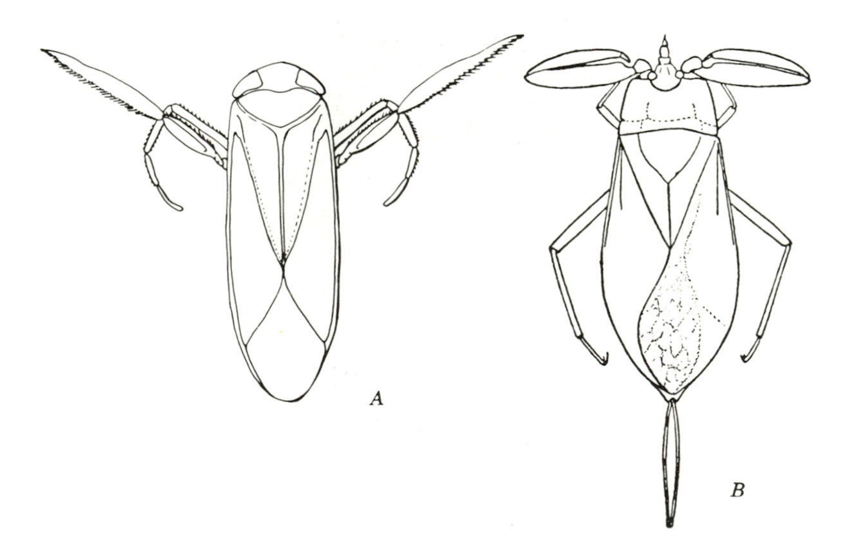

Fig. 8-68. (*A*) Dorsal aspect of *Arctocorixa*, Corixidae; (*B*) dorsal aspect of *Nepa*, Nepidae. (*A* adapted from Hungerford, 1959; *B* from *Hemiptera of Connecticut*)

tongue. These stylets draw into the pharynx an assortment of diatoms, algae, and minute animal organisms that constitute their food.

NOTONECTIDAE, BACKSWIMMERS. In form these aquatic bugs superficially resemble the water boatmen, particularly in the long-fringed oarlike hind legs used for swimming. They are very different, however, in many ways. Most conspicuous is their habit of always swimming on their backs. The coloration of the backswimmers is modified to match this change in swimming position. The ventral side, which is uppermost, is dull brown to match the stream or pond bottom. The dorsal side, which is hidden from above when the insect is swimming, is usually whitish, creamy, or lightly mottled. The beak in the Notonectidae is stout and sharp, used to suck the body contents from small aquatic animals such as Crustacea and small insects. Many species of backswimmers deposit their eggs on the surface of objects in the water; others insert their eggs into the stems of aquatic plants.

BELOSTOMATIDAE, GIANT WATER BUGS. Members of this family are wide and stout, with grasping front legs and crawling and swimming middle and hind legs (Fig. 8-69). They include some of the largest North American Hemiptera, for example, *Lethocerus americanus*, which attains a length of 8 to 10 cm. The giant water bugs have strong beaks. They are predaceous, feeding on insects, snails, small frogs, and small fish. Commonly they are attracted to lights where they draw considerable attention, owing to their ungainly movements and large size.

Stinkbugs and Their Allies (Pentatomorpha)

REDUVIIDAE, ASSASSIN BUGS. The family Reduviidae is placed only tentatively in the Pentatomorpha for reasons given on pg. 341. These are sluggish predaceous insects, usually medium-sized to large, that feed on other insects. Most North American species have fully de-

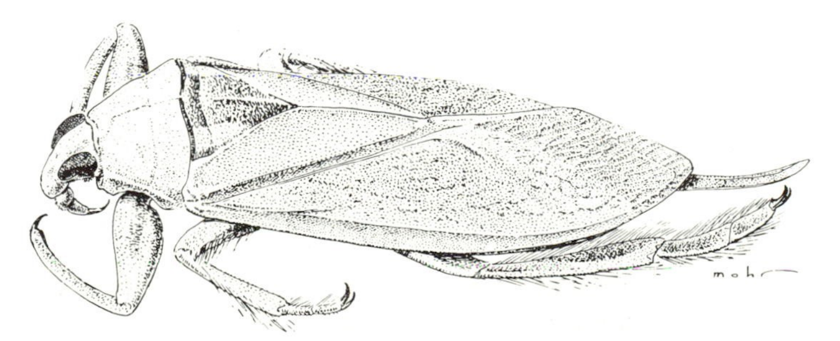

Fig. 8-69. A giant water bug *Lethocerus*, Belostomatidae. (Original by C. O. Mohr)

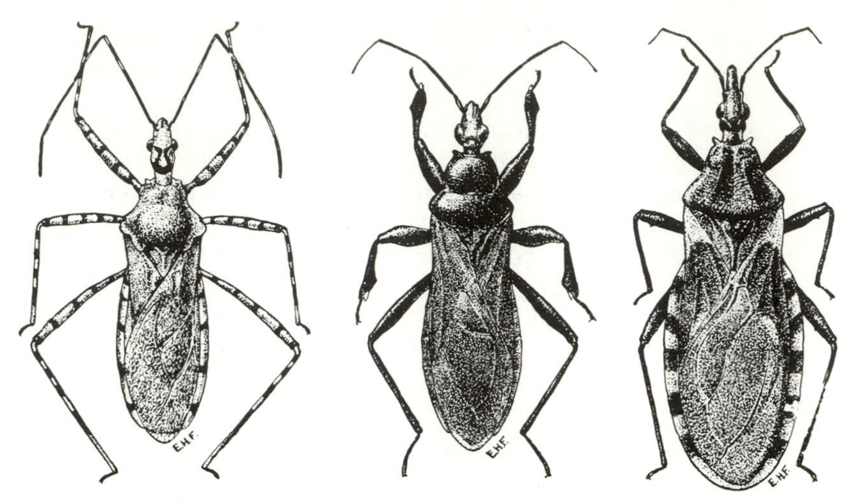

Fig. 8-70. Three Reduviidae. Left to right, *Pselliopus barberi*, *Melanolestes picipes*, and *Triatoma sanguisuga*. (Drawings loaned by R. C. Froeschner)

veloped wings (Fig. 8-70) and several have foliaceous hairs on the side of their legs. The nymphs of certain species secrete a sticky substance over the dorsum, on which are carried bits of leaves and debris, providing the animal with very good camouflage. The eggs are laid singly or in clusters, glued to plants or other supports. Assassin bugs, including *Reduvius personatus*, sometimes attack humans, inflicting a painful burning wound. All the Reduviidae are terrestrial.

COREIDAE, SQUASH BUGS, COREID BUGS. In general characters the Coreidae are like the Lygaeidae, differing chiefly in having many veins in the hemelytron membrane. Many of these bugs resemble the lygaeid bugs in general shape. Others, such as *Acanthocephala*, bear a striking resemblance to the Reduviidae. Coreid bugs feed on plants. The most widely known is the squash bug *Anasa tristis* (Fig. 8-71C), which attacks squash, cucumbers, and other cucurbit crops. Its eggs are laid in patches on the leaves and stem of its hosts. There are several generations each year, and winter is passed in the adult stage.

ARADIDAE, FLAT BUGS. This family includes a group of moderate-sized species that are the flattest members of the Heteroptera. They live under the bark of dead trees and are thought to feed on fungi. The tarsi are two-segmented, the antennae and beak four-segmented, and ocelli are lacking. The wings are greatly reduced in size and when folded occupy only a small area of the dorsum (Fig. 8-72*A*, *B*).

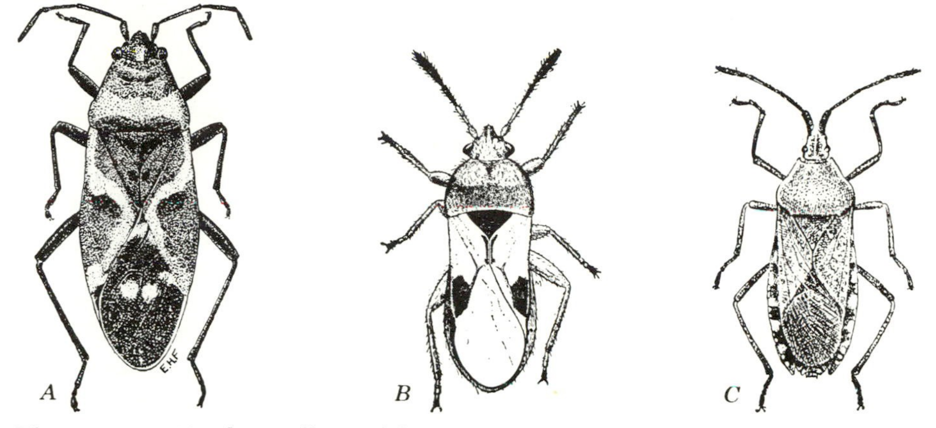

Fig. 8-71. (*A*) The milkweed bug *Oncopeltus fasciatus*; (*B*) the chinch bug *Blissus leucopterus leucopterus*, Lygaeidae; (*C*) the squash bug *Anasa tristis*, Coreidae. (*A* after Froeschner, *B* from Illinois Natural History Survey; *C* from U.S.D.A.)

LYGAEIDAE. CHINCH BUGS, LYGAEID BUGS. Most North American members of this family are fairly small somber-colored or pale forms. A few genera, such as the milkweed bug *Oncopeltus fasciatus* (Fig. 8-71*A*), are strikingly marked with red and black. The diagnostic family characters include the four-segmented beak and a hemelytron with a few irregular veins crossing the membrane (Fig. 8-53*B*). Most species have distinct ocelli. In North America the most important member of this family is the chinch bug *Blissus leucopterus leucopterus* (Fig. 8-71*B*), which is one of the major insect pests of corn and small grains in the corn-belt states. The chinch bugs hibernate in ground cover as adults. In early spring they feed on grasses and small grains

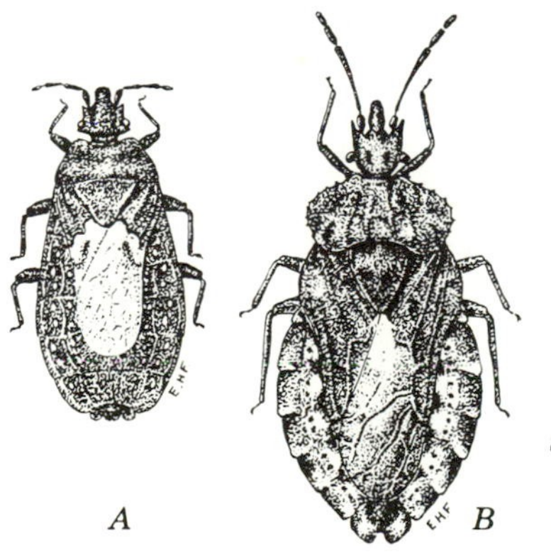

Fig. 8-72. Aradidae. Flat bugs (*A*) *Aradus acutus*; (*B*) *Neuroctenus simplex*, two common species in eastern and central North America. (After Froeschner)

and lay their eggs on the roots and crown of the food plants. The eggs hatch in about 2 weeks, the nymphs feeding on the same plants and maturing in 6 weeks. By the time this brood matures, the original food crop is almost invariably either mature and becoming dry, or it has been overpopulated and offers little in the way of nourishment. When this happens, the entire brood moves out in search of more succulent food. The exodus takes the form of a mass migration, not by flight but by foot. Mature nymphs and newly emerged adults make up the hurrying mass of insects on the march. The search for a better food supply usually ends in a field of corn, now well established and in prime growth. The individuals that reach the corn plants establish themselves and produce the second generation. Both the migrating first generation and the second generation feeding to maturity do extensive damage to the corn crop. When mature, the second generation goes into hibernation until the following spring.

PENTATOMIDAE, STINKBUGS. To this family (Figs. 8-51*B*, *C*, *D*, and 8-73) belong numerous large or medium-sized bugs, most of them broad, many of them mottled with shades of green, gray, or brown. A few

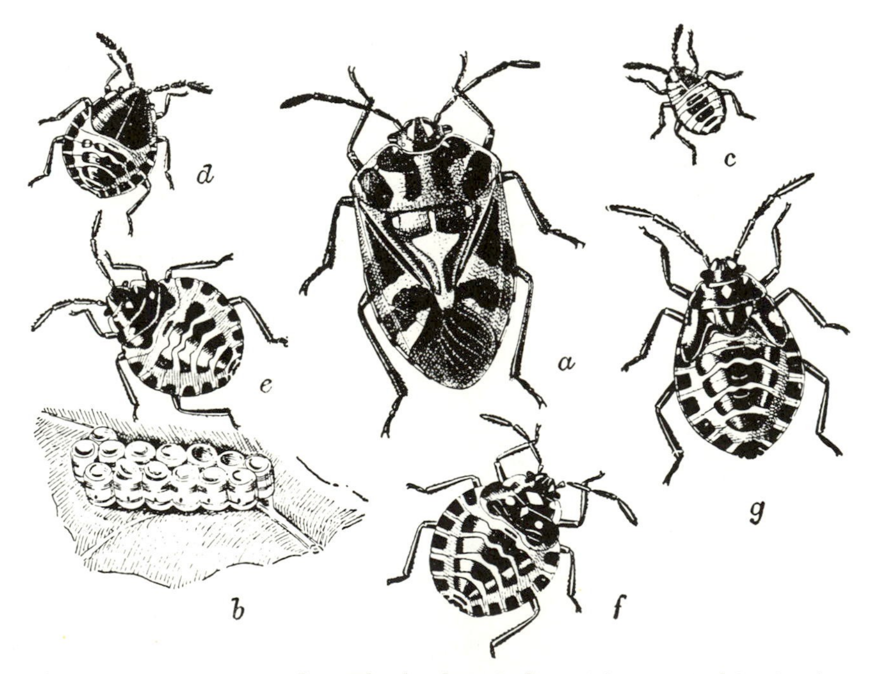

Fig. 8-73. Pentatomidae. The harlequin bug, *Murgantia histrionica*. (*a*) Adult; (*b*) egg mass; (*c*) first stage of nymph; (*d*) second stage; (*e*) third stage; (*f*) fourth stage; (g) fifth stage. (From U.S.D.A., E.R.B.)

are brightly patterned. Some species are predaceous, feeding on a wide variety of other insects. Others are entirely phytophagous, of which the harlequin bug *Murgantia histrionica* (Fig. 8-73) is a familiar example. This bug feeds on cruciferous plants and often does serious damage to cabbage. Three subfamilies, the Scutellerinae, Cydninae, and Thyreocorinae, are classed as separate families in some works. The identifying characteristics of these subfamilies, given in the key to families, hold fairly well for the nearctic fauna but break down when considered for the world fauna as a whole.

References

Blatchley, W. S., 1926. *Heteroptera or true bugs of eastern North America.* Indianapolis, Ind.: Nature Publishing Co. 1116 pp.

Britton, W. E., 1923. The Hemiptera or sucking insects of Connecticut. *Bull. Conn. State Geol. Nat. Hist. Surv.*, **34**:1–807.

Brooks, A. F., and L. A. Kelton, 1967. Aquatic and semi-aquatic Heteroptera of Alberta, Saskatchewan and Manitoba (Hemiptera). *Mem. Entomol. Soc. Can.*, **51**:1–92.

China, W. E., and N. C. E. Miller, 1959. Checklist and keys to the families and subfamilies of the Hemiptera-Heteroptera. *Bull. Brit. Mus. Nat. Hist. Entomol.*, **8**(1):1–45.

Cobben, R. H., 1968. *Evolutionary trends in Heteroptera.* Part 1: *Eggs, architecture of the shell, gross embryology and eclosion.* Wageningen, The Netherlands: Centre for Agricultural Publishing and Documentation. 475 pp.

1978. Evolutionary trends in Heteroptera, Part II. Mouthpart structures and feeding strategies. *Entomologie, Landbouwhogesch. Wageningen Meded.*, **289**:1–407.

DeCoursey, R. M., 1971. Keys to the families and subfamilies of the nymphs of North American Hemiptera-Heteroptera. *Proc. Entomol. Soc. Wash.*, **73**:413–428.

DeLong, D. M., 1971. The bionomics of leafhoppers. *Annu. Rev. Entomol.*, **16**:179–210.

Dixon, A. F. G., 1973. *Biology of aphids.* London: Edward Arnold. 58 pp.

Evans, J. W., 1963. The phylogeny of the Homoptera. *Annu. Rev. Entomol.*, **8**:77–94.

Froeschner, R. C., 1941–1944. Contributions to a synopsis of the Hemiptera of Missouri, Pts. 1–4. *Am. Midl. Naturalist*, **26**:122–146; **27**:591–609; **31**:638–683; **42**:123–188.

Herring, J. L., and P. D. Ashlock, 1971. The key to the nymphs of the families of Hemiptera (Heteroptera) of America north of Mexico. *Fla. Entomol.*, **54**:207–213.

Hungerford, H. D., 1959. *Hemiptera.* In W. T. Edmundson (Ed.), *Freshwater biology.* New York: Wiley. pp. 958–972.

Kennedy, J. S., and H. L. G. Stroyan, 1959. Biology of aphids. *Annu. Rev. Entomol.*, **4**:139–160.

McKenzie, H. L., 1967. *Mealybugs of California*. Berkeley and Los Angeles: Univ. of California Press. 526 pp.

Menke, A. S., 1979. The semiaquatic and aquatic Hemiptera of California (Heteroptera: Hemiptera). *Bull. Calif. Insect Surv.*, **21**:1–166.

Miller, N. C. E., 1956. *Biology of the Heteroptera*. London: Leonard Hill. 162 pp.

Milne, L. J., and M. Milne, 1978. Insects of the water surface. *Sci. Am.*, **238**:134–143.

Mound, L. A., and S. H. Halsey, 1978. *Whitefly of the world: A systematic catalogue of the Aleyrodidas (Homoptera) with host plant and natural enemy data*. New York: Wiley. 336 pp.

Slater, J. A., and R. M. Baranowski, 1978. *How to know the true bugs (Hemiptera-Heteroptera)*. Dubuque, Iowa: W. C. Brown. 256 pp.

Smith, C. F., 1972. Bibliography of the Aphididae of the world. *N.C. Agric. Exp. Stn. Tech. Bull.*, **216**:1–717.

Torre-Bueno, J. R. de la, 1939. A synopsis of the Hemiptera-Heteroptera. *Entomol. Am.*, **19**:141–310.

Usinger, R. L., 1966. Monograph of Cimicidae (Hemiptera-Heteroptera). *Thomas Say Foundation, Entomol. Soc. Am.*, **7**:1–585.

Usinger, R. L., and R. Matsuda, 1959. *Classification of the Aradidae (Hemiptera-Heteroptera)*. London: British Museum, Natural History. 410 pp.

Van Duzee, E. P., 1917. Catalogue of the Hemiptera of America north of Mexico. *Univ. Calif. Tech. Bull.*, **2**:1–902.

Woodward, T. E., J. W. Evans, and V. F. Eastop, 1970. Hemiptera. In D. F. Waterhouse, et al. (Eds.), *The insects of Australia*. Carlton, Victoria: Melbourne Univ. Press. pp. 387–464; *Suppl.* 1974, pp. 52–57.

Wygodzinsky, P., 1966. A monograph of the Emesinae (Reduviidae, Hemiptera). *Bull. Am. Mus. Nat. Hist.*, **133**:1–614.

The Neuropteroid Orders

The 10 neuropteroid orders all have complete metamorphosis and because of this are frequently termed the Holometabola. Living species form two major branches: the Neuroptera-Coleoptera branch, having several primitive larval characteristics and some peculiar developments of the female saw; and the Hymenoptera-Mecoptera branch, having a larval tarsus with only one claw.

The Neuroptera-Coleoptera Branch

The four orders comprising this branch—Raphidioptera, Megaloptera, Neuroptera, and Coleoptera—have larvae with two claws on each tarsus (a primitive character) and a peculiar reduction or fusion on the valvulae that form the ovipositor.

Order RAPHIDIOPTERA*: Snakeflies

Large insects similar in many features to the Megaloptera but distinguished by the long serpentine neck (Fig. 8-74). Snakeflies have long antennae, chewing mouthparts, large eyes, and two pairs of similar transparent net-veined wings. The female has a conspicuous terminal ovipositor. The larvae are terrestrial. They have segmented antennae, faceted eyes, well-developed thoracic legs, but no processes or appendages on the abdomen. The entire known world fauna of five genera and about a hundred species is divided into two families, the Raphidiidae (having ocelli) and the Inocelliidae (lacking ocelli). Two genera occur in North America, both confined to the West: *Agulla* (Raphidiidae) and *Inocellia* (Inocelliidae). The larvae occur under loose bark of conifers and are predaceous on other insects. When mature, the larvae do not spin cocoons but form an oval retreat in a sheltered position, and here the pupal stage is passed. The adults are also predaceous. They are occasionally swept from foliage and are indeed strange-appearing creatures to find in the net.

References

Carpenter, F. M., 1936. Revision of the nearctic Raphidiodea (recent and fossil). *Am. Acad. Arts. Sci. Proc.*, **71**:89–157.

Woglum, R. S., and F. A. McGregor, 1958. Observations on the life history

*Sometimes called the Raphidiodea.

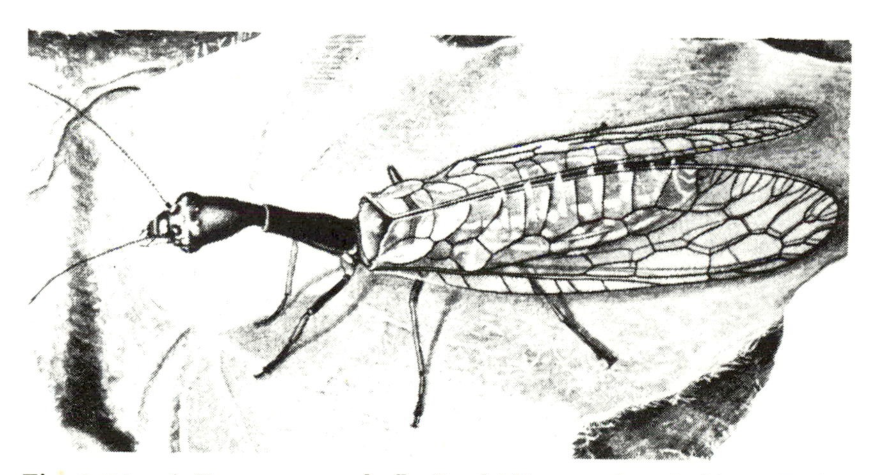

Fig. 8-74. A European snakefly *Raphidia ratzeburgi*. (From Essig, 1942, *College entomology* by permission of Macmillan Co.)

and morphology of *Agulla bractea* Carpenter. *Ann. Entomol. Soc. Am.*, **51**:129–141.

1959. Observations on the life history and morphology of *Agulla astuta* (Banks) (Neuroptera: Raphidiodea: Raphidiidae). *Ann. Entomol. Soc. Am.*, **52**:489–502.

Order MEGALOPTERA*: Dobsonflies and Alderflies

These large insects have aquatic larvae and terrestrial adults and pupae. The adults (Fig. 8-75) have long antennae, chewing-type mouthparts, large eyes, and two pairs of wings. The wings are similar in texture and venation and have all the major veins plus some additional terminal branches and a large number of crossveins. The pronotum is large and wide, the abdomen without projecting cerci.

*Sometimes called the Sialoidea.

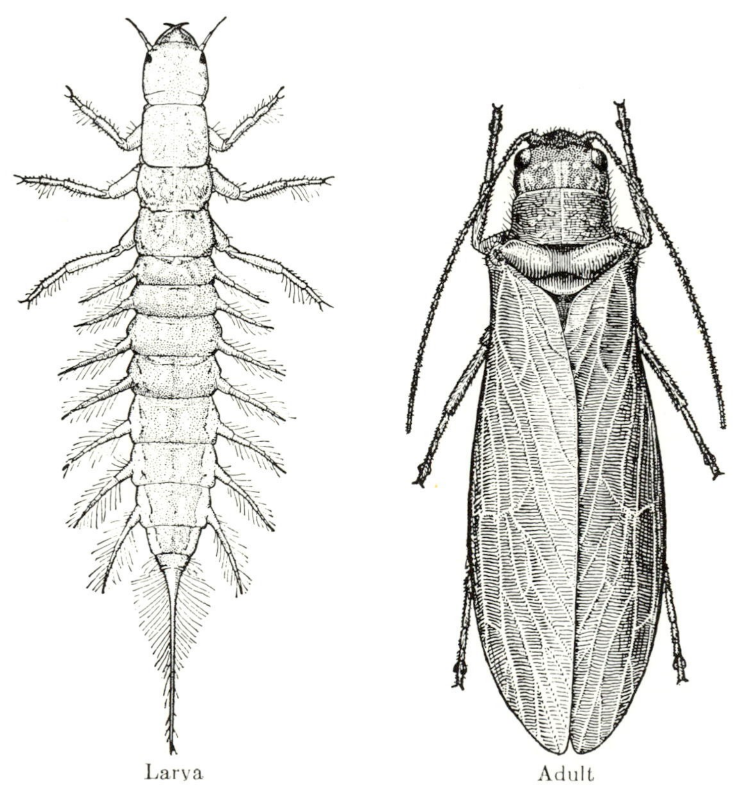

Fig. 8-75. An alderfly *Sialis* sp., larva and adult. (From Illinois Natural History Survey)

The larvae (Fig. 8-75) have strong biting mouthparts; distinct, segmented antennae; large eyespots each composed of a group of about six facets *(stemmata)*; well-developed thoracic legs, and paired abdominal processes or gills. The apex of the abdomen has a long mesal process in Sialidae, and a pair of stout hooked larvapods in Corydalidae.

KEY TO FAMILIES

Head with no ocelli; tarsi with fourth segment dilated and bilobed (Fig. 8-75) Alderflies . . . . **Sialidae**
Head with 3 ocelli; tarsi with all segments cylindrical. Dobsonflies . . . . **Corydalidae**

Both families together are represented in North America by only a few genera and less than 50 species. The adults range in color from black to mottled or yellow; in one genus *(Nigronia)* the wings are banded with black and white.

The larvae are aquatic, occurring in both lakes and streams. They are predaceous on small aquatic animals. The mature larvae of *Corydalus* may attain a length of 80 mm. They are ferocious larvae, highly prized for bait by fishermen, and called hellgrammites. The smaller species of the order mature in a year and have an annual life cycle. Some hellgrammites require 2 or 3 years to reach full growth.

When mature, the larvae leave the water and make a pupal cell in damp earth or rotten wood nearby. Here the larvae transform to pupae. Megalopteran pupae are active if irritated and capable of considerable locomotion. The pupal stage usually lasts about 2 weeks.

The adults are good fliers, but not agile compared to some of the flies and moths. Some *Corydalus* adults may have a wing span of 13 cm. and are among our largest North American insects. The females lay their eggs in large clusters of several hundred each on stones and other objects overhanging the water. These hatch soon after deposition, and the minute larvae fall or twist their way into the water.

References

Parfin, S., 1952. The Megaloptera and Neuroptera of Minnesota. *Am. Midl. Naturalist*, **47**:421–434.

Ross, H. H., 1937. Nearctic alderflies of the genus *Sialis*. *Bull. Ill. Nat. Hist. Surv.*, **21**(3):57–78.

Weele, W. H. van der, 1910. Megaloptera. *Coll. Zool. Selys Longchamps, Brussels*, fasc., **5**(115):1–93.

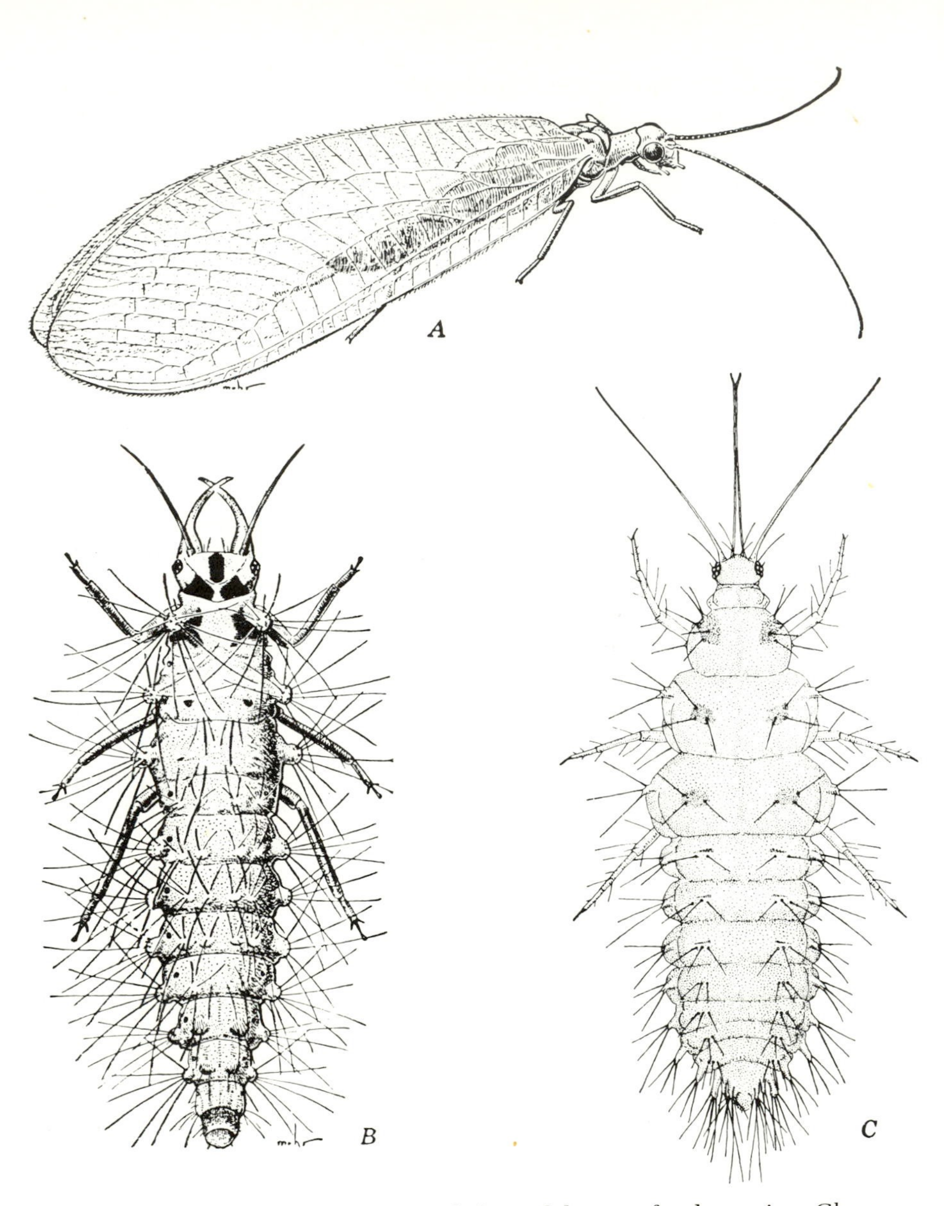

Fig. 8-76. Neuroptera. (*A*, *B*) Adult and larva of a lacewing *Chrysopa* sp.; (C) larva of a sponge feeder *Sisyra* sp. (*A*, *B from Illinois Natural History Survey*; *C* after Townsend)

Order NEUROPTERA: Lacewings, Mantispids

The adults are minute to large insects, usually with two pairs of clear wings having many veins and crossveins, with chewing-type mouthparts, long and multisegmented antennae, and large eyes (Fig. 8-76*A*). The larvae are varied: Most of them are terrestrial and predaceous; one family (Sisyridae) is aquatic, and the larvae feed in freshwater sponges. All the larvae have thoracic legs, but no abdominal ones, well-developed heads, and mandibulate mouthparts.

KEY TO FAMILIES

1. Front legs with apical segments enlarged for grasping (Fig. 8-78) . . . . **Mantispidae**
   Front legs with apical segments slender, same as other legs (Fig. 8-76) 2
2. Wings with very few veins or crossveins (Fig. 8-77*B*). Minute insects covered with a waxy bloom and gray in appearance . . . . **Coniopterygidae**
   Wings with veins and crossveins numerous (Fig. 8-77*C* to *G*). Larger insects never covered with waxy bloom 3
3. Front wings with a regular, fencelike series of 12 or more crossveins (gradate veins) between *R* and *S* (Fig. 8-77*E*) 4
   Front wings either with only 1 to 5 well-separated crossveins between *R* and *S* (Fig. 8-77*C*, *G*), or *R* and stem of *S* fused (Fig. 8-77*D*) 6
4. Antennae long and slender (Fig. 8-76), tapering to apex . . . . **Chrysopidae**
   Antennae either short and clavate or knobbed at apex 5
5. Antennae short, gradually thickened toward apex . . . **Myrmeleontidae**
   Antennae long, knobbed at apex . . . . **Ascalaphidae**
6. Front wings with 2 or more branches of S arising from fused *R* and *S* (Fig. 8-77*D*) . . . . **Hemerobiidae**
   Front wings with all branches of S arising from a separate *S* stem (Fig. 8-77*C*, *F*, *G*) 7
7. Front wings with almost all costal crossveins forked (Fig. 8-77*F*, *G*) 8
   Front wings with few or no costal crossveins forked and with apical margin evenly rounded, as in Fig. 8-77*A*, *C* 9
8. Front wings only slightly incised and with recurrent costal vein (Fig. 8-77*G*) . . . . **Polystoechotidae**
   Front wings markedly incised and with no recurrent costal vein (Fig. 8-77*F*) . . . . **Berothidae**
9. Front wings with *Sc* and *R* not fused before apex; basal vein *(b)* present (Fig. 8-77*A*) . . . . **Dilaridae**
   Front wings with *Sc* and *R* fused some distance before apex; basal vein absent (Fig. 8-77*C*) . . . . **Sisyridae**

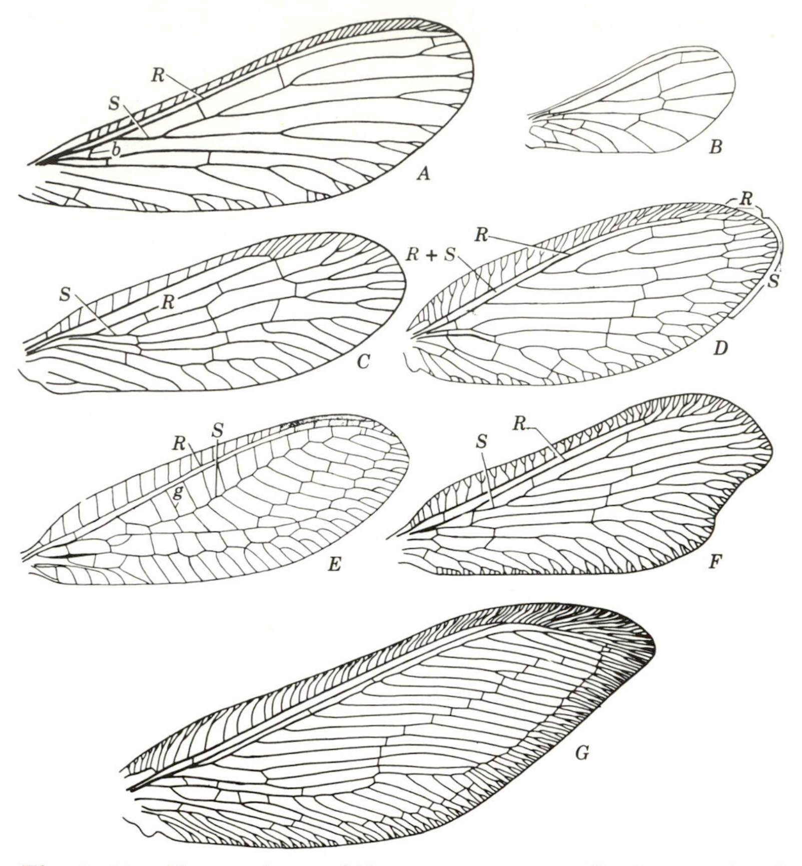

Fig. 8-77. Front wings of Neuroptera. (*A*) *Nallachius*, Dilaridae; (*B*) Semidalis, Coniopterygidae; (*C*) *Climacia*, Sisyridae; (*D*) *Hemerobius*, Hemerobiidae; (*E*) *Chrysopa*, Chrysopidae; (*F*) *Lomamyia*, Berothidae; (*G*) *Polystoechotes*, Polystoechotidae. *b*, Basal vein; g, gradate vein. (From various sources)

SPONGE FEEDERS. The spongeflies, or spongillaflies, are a small family comprising the Sisyridae. The adults look like small, typical lacewings, but the larvae are robust creatures that live in and eat freshwater sponges. Their mouthparts form a long beak that sticks out in front of the larva (Fig. 8-76C). Only a few species are known to occur in the nearctic region.

SEDENTARY PREDATORS. The mantispids (Fig. 8-78) are another small family, Mantispidae. The adults have a striking resemblance to praying mantids. The front legs are greatly enlarged, fitted for grasping insect prey and are attached at the anterior end of the very long pronotum. The larvae feed on egg sacs of spiders or contents of wasp nests. The first-instar larvae are slender and active, and hunt for a suitable food reservoir. Once this is found, the larvae enter a parasitoid stage, and succeeding instars are grublike and have degenerate legs.

ACTIVE PREDATORS. The larvae of all the other families are active predators. The adults have transparent and abundantly veined wings, which give them the name "lacewings" (Fig. 8-76*A*). Most of these insects are relatively slow fliers. The eggs are laid either attached directly to foliage or at the end of a long hairlike stalk, which is attached to a leaf (Fig. 7-1*N*). This latter method is used only by the Chrysopidae. The larvae are active, but sluggish and soft-bodied, and frequently bear warts, tubercles, and long hair. The mouthparts are modified for sucking body juices from the prey. The mandibles and ends of the maxillae are long, bladelike, and sickle-shaped, and a maxillary blade fits beneath each mandible; each of these opposing pieces has a groove, the two fitting together to form a canal from near the tip of the mandible into the mouth opening. The two man-

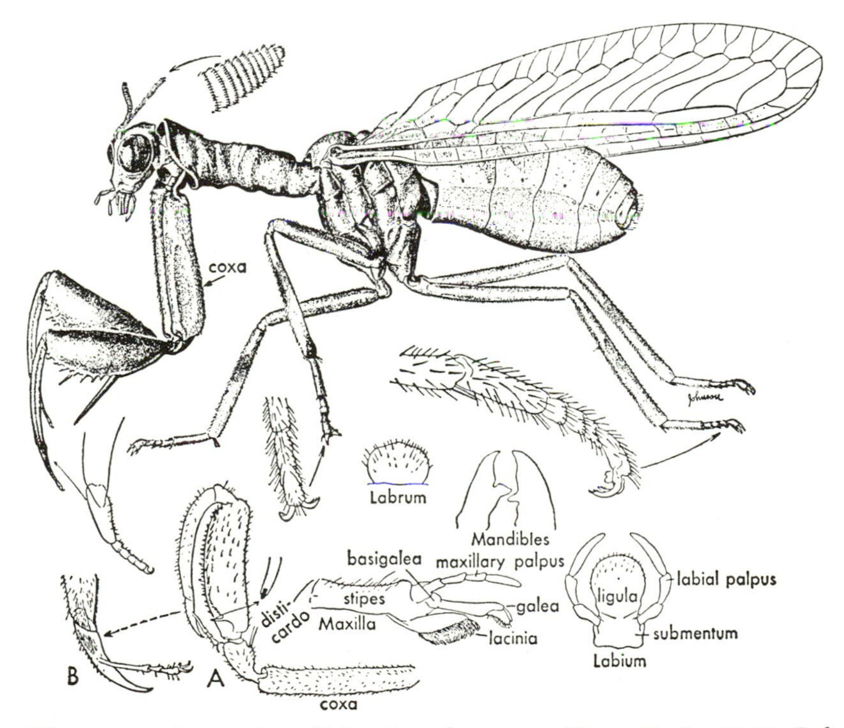

Fig. 8-78. A mantispid *Mantispa brunnea*. (From Essig, 1942, *College entomology*, by permission of Macmillan Co.)

dibular-maxillary blades are thrust into the body of the prey from opposite sides, and its body juices are sucked out through the canals.

The larvae of Chrysopidae (Fig. 8-76*B*) and Hemerobiidae crawl freely about on plants and feed on aphids, other small insects, and insect eggs. Their frequent attacks on aphids have earned them the name "aphidlions." When full grown, the larvae spin a woolly ovoid cocoon under a leaf or in some sheltered spot, and pupation ensues. The larvae of Myrmeleontidae live in sandy soil and dig cone-shaped pits that trap ants and other prey that fall into them. These larvae, called "antlions," differ from the aphidlions only in being more robust. The antlion digs the pit by throwing out sand from the center by upward jerks of the head, using the long mandibles as shovels. The pitfalls may be 2 to 3 cm. deep, with sides sloping as much as the texture of the loose sand will allow. The antlion stays in the soil with its head just below the bottom of the crater, constantly in wait for unwary prey. These curious larvae are known to most people by the name "doodlebug." When mature, the larva forms a cocoon in the soil and pupates in it.

References

Carpenter, F. M., 1940. A revision of nearctic Hemerobiidae, Berothidae, Sisyridae, Polystoechotidae, and Dilaridae. *Proc. Am. Acad. Arts Sci.*, **74**(7):193–280.

Froeschner, R. C., 1947. Notes and keys to the Neuroptera of Missouri. *Ann. Entomol. Soc. Am*, **40**(1):123–136.

Gurney, A. B., and S. Parfin, 1959. *Neuroptera.* In W. T. Edmundson (Ed.), *Freshwater biology.* New York: Wiley. pp. 973–980.

MacLeod, E. G., and P. A. Adams, 1967. A review of the taxonomy and morphology of the Berothidae with a description of a new subfamily from Chile (Neuroptera). *Psyche*, **74:**237–265.

New, T. R., 1975. The biology of Chrysopidae and Hemerobiidae (Neuroptera), with reference to their usage as biocontrol agents: A review. *Trans. R. Entomol. Soc. London*, **127:**115–140.

Parfin, S., 1952. The Megaloptera and Neuroptera of Minnesota. *Am. Midl. Naturalist*, **47:**421–434.

Riek, E. F., 1970. *Neuroptera.* In D. F. Waterhouse, et al. (Eds.), *The insects of Australia.* Carlton, Victoria: Melbourne Univ. Press. pp. 472–494.

Smith, C., 1934. Notes on the Neuroptera and Mecoptera of Kansas, with keys for the identification of species. *J. Kans. Entomol. Soc.*, **7**(4):120–145.

Tauber, C. A., 1969. Taxonomy and biology of the lacewing genus *Meleoma* (Neuroptera: Chrysopidae). *Univ. Calif. Publ. Entomol.*, **58:**1–94.

Toschi, G. A., 1965. Taxonomy, life history and mating behavior of green lacewings of Strawberry Canyon (Neuroptera: Chrysopidae). *Hilgardia*, **36:**391–431.

Withycombe, C. L., 1925. Some aspects of the biology and morphology of the Neuroptera with special reference to the immature stages and their possible phylogenetic significance. *Trans R. Entomol. Soc. London, 1924*. pp. 303–411.

Order COLEOPTERA: Beetles and Weevils

The adults usually have two pairs of wings: The first pair is veinless, hard, and shell-like, and folds together over the back to make a stout wing cover; the second pair, used for flight, is membranous, usually veined, and in repose folds up under the wing covers or elytra (Fig. 8-79). The body is normally hard and compact. The mouthparts are of the chewing type; the antennae are well developed, usually 10- to 14-segmented; the compound eyes are usually conspicuous; and the legs are heavily sclerotized. The larvae normally have distinct head capsules, chewing mouthparts, antennae, and thoracic legs, but no abdominal legs. The pupae (Fig. 8-96) have the adult appendages folded against but not fused with the body. The adults vary in size from less than a millimeter long to several inches (up to 100 mm). The shape and coloring varies just as much, but the most brilliantly colored, weird-shaped, and gigantic species occur in the tropical regions of the world.

Most kinds of beetles are either plant feeders or are predaceous on other insects. Usually both adults and larvae of the same species have similar food habits; that is, both forms will be phytophagous, or both will be predaceous. The June beetles, which are phytophagous, utilize different parts of plants for food during their development. The adults feed on the foliage of forest trees, but the larvae, known as white grubs, feed on the roots of trees, shrubs, herbs, or grasses. The predaceous beetles are active hunters, stalking their prey. Certain

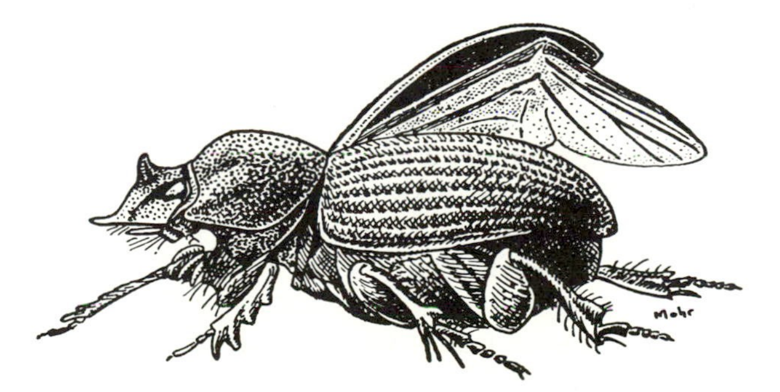

Fig. 8-79. A scarab beetle *Copris minutus*. (From Illinois Natural History Survey)

groups have more specialized food habits. Some are endoparasites of other insects or feed on insect egg masses.

The order as a whole is terrestrial. Certain families, however, are aquatic, both larvae and adults living in water. The larvae usually leave the water to pupate, making an earthen cell in nearby soil. As is the case with the land forms, the aquatic group includes both herbivorous and predaceous species, the latter predominating.

The great majority of beetles have a single generation per year and a simple life cycle similar to the following. The oval or round eggs are laid in spring or early summer and hatch in 1 or 2 weeks. The larvae have three larval instars and are voracious feeders, usually attaining full growth during the summer and pupating in the soil. The adults emerge in a few weeks, feeding and maturing throughout the remainder of the summer and autumn. With the advent of cold weather they hibernate. The following spring these adults emerge and lay eggs, and the cycle begins again. The old adults usually die soon after egg laying is completed.

There are many deviations from this biological pattern. For example, some ladybird beetles have continuous and overlapping generations throughout the warmer seasons. In other groups, notably many species whose larvae live in soil or rotten wood, the winter is passed in the larval stage, and the adults occur for only a limited span during spring or summer.

Since they first differentiated from their neuropteroid ancestors over 300 million years ago, the beetles have evolved into more species than any other group of living things. Probably about 200,000 species have persisted to the present; of these about 25,000 occur in North America, representing about 150 families. By adding together primitive characters of living beetles, it appears (Fig. 8-80) that all of them arose from an ancestral beetle in which the abdominal sterna of the adults were free and distinct and the larval legs possessed a tarsal segment. From this ancestor two lines arose. In one line the larval tarsi persisted but the first three abdominal sterna of the adult fused and the hind coxae became immovably attached to and divided the first abdominal sternum (Fig. 8-81). This line gave rise to the suborder Adephaga, including a primitive family Gyrinidae (which has a sawlike ovipositor); the small family Rhysodidae, found in rotten logs; the terrestrial predators related to the ground beetles (Carabidae); and the predaceous water beetles (Dytiscidae and others).

In the other line arising from the primeval coleopteran, the abdominal segments remained separate but the larval tarsi were lost. The suborder Archostemata seems to be the most primitive existing member of this line, because the tarsal segment of the larval legs is

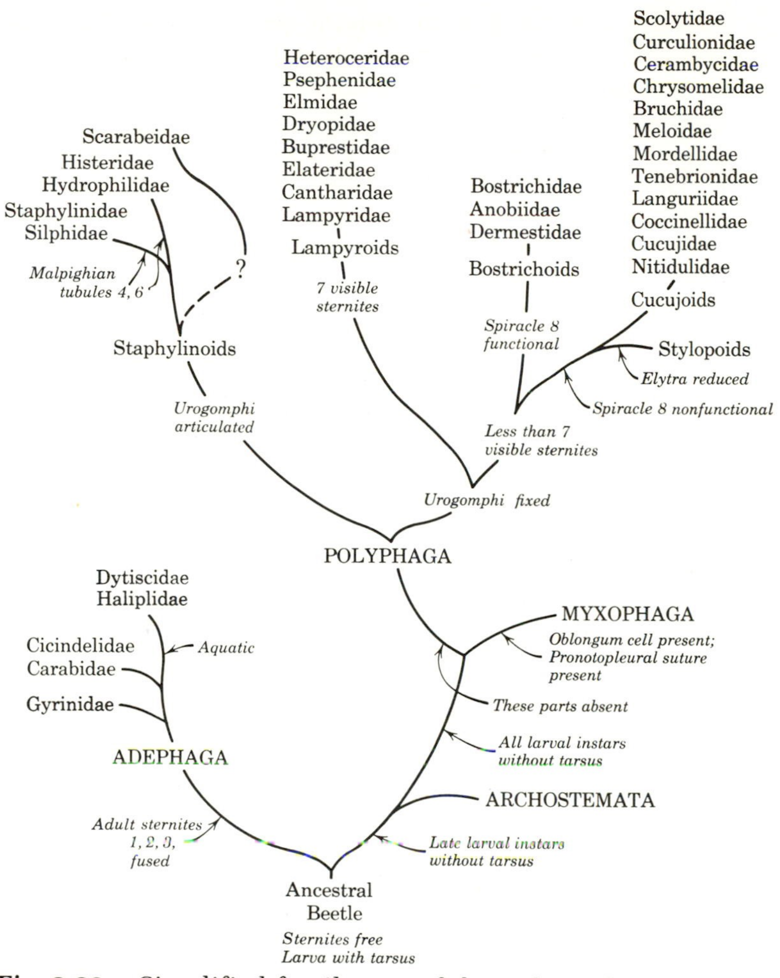

Fig. 8-80. Simplified family tree of the order Coleoptera. (Compiled chiefly from Crowson, 1955)

still present in early instars and lost only in later instars. This suborder contains only a few small, rare families including the Cupedidae and Micromalthidae. In a line arising from an ancestor having this condition the larval tarsus was lost completely. This line evolved into the suborder Polyphaga, containing the great bulk of living beetles. In the Polyphaga, certain families have become extremely specialized and quite different morphologically from more primitive members of their own suborder. Adults of certain minute fungus beetles, for example, have lost all their hind-wing venation, most of their tarsal segments, and have undergone peculiar changes of body sclerites associated with minute size. In describing the various branches it is therefore impossible at this state of our knowledge to do more than indicate the characters of primitive members of each subdivision.

The most primitive members of the line that gave rise to the Polyphaga are a few rare, small beetles (Hydroscaphidae and Sphaeriidae) possessing the crossvein setting off the oblongum cell in the hind wing and the suture between the pronotum and propleuron. These are considered a separate suborder, the Myxophaga. The oblongum cell and pronotopleural suture were lost in the line that evolved into the suborder Polyphaga. The ancestor of this suborder gave rise to two branches. In the more primitive (the staphylinoids), the urogomphi (cercus-like organs) of the larvae remained articulated; in the other the urogomphi fused solidly with their parent segment.

The staphylinoids contained two major lineages: one had four malpighian tubules, including the rove beetles (Staphylinidae), carrion beetles (Silphidae) and their allies; the other had six malpighian tubules, including the Histeridae and Hydrophilidae. The affinities of the scarab beetles (Scarabaeidae) are difficult to determine; it has been suggested that they are an unusual offshoot of the staphylinoids.

The other branch of the typical Polyphaga gave rise in turn to two subsequent branches. In the lampyroids the adult body of the more primitive families retained at least seven visible abdominal sternites. The lampyroids evolved into an amazing assortment of families, including the soldier beetles (Cantharidae) and fireflies (Lampyridae), the click beetles (Elateridae) and flat-headed borers (Buprestidae), many water beetles (Psephenidae and allies), and many others.

In the sister branch of the lampyroids, the abdomen has at most six visible sternites. This line gave rise to the bostrichoids, in which all the spiracles in the adult remained functional, and the cucujoids, in which the spiracle on the adult eighth abdominal segment became nonfunctional. The bostrichoids evolved into the larder beetles and powder-post beetles (Dermestidae, Anobiidae, and Bostrichidae) and their allies. The cucujoids evolved into many families diverse in both habits and appearance. More common forms include the ladybird beetles (Coccinellidae), the darkling beetles (Tenebrionidae), and the principal group of plant-feeding beetles (Chrysomelidae, Scolytidae, Curculionidae, and allies).

The structures of the stylopids (superfamily Stylopoidea) are so highly modified that the exact affinities of the group are obscure. In the adult the spiracles of the eighth abdominal segment are nonfunctional. On this and other tenuous evidence it is probable that the group arose either as a sister group of the cucujoids or from some ancestor among the cucujoids.

KEY TO COMMON FAMILIES

1. Front of head produced into a definite beak (Fig. 8-104) that may be long or short, the antennae arising from side of beak; palpi vestigial .... **Curculionidae**, p. 402
   Front of head not produced into a beak; if slightly so, antennae arising between the eyes, or maxillary palpi prominent (Fig. 8-81) 2
2. Middle and hind legs very wide and flat, almost paper thin, fitted for swimming, the basitarsus large and triangular, the next two produced laterally to form long swimming "fingers"; front legs tubular, fitted for grasping (Fig. 8-82*L*); each eye completely divided, one part on dorsum of head, the other on ventral aspect of head (Fig. 8-82*K*) .... **Gyrinidae**, p. 386
   Middle and hind legs having some or most segments robust and not flattened, occasionally furnished with rows of long hairs and fitted for swimming in this fashion; eye seldom divided, and then only by a continuation of a head flange or by an antennal base (Fig. 8-84*K*) 3
3. Maxillary palpi longer than antennae and slender, resembling antennae (Fig. 8-82*Q*); chiefly aquatic .... **Hydrophilidae**
   Maxillary palpi shorter than antennae, and not antenna-like; antennae various 4
4. Antenna with each of last 3 to 7 segments enlarged on one side to form an eccentric plate or lamella (Fig. 8-82*A*, *C*), each lamella situated nearly at a right angle to long axis of antenna 5
   Antenna not lamellate, but frequently having the end segments fairly evenly enlarged to form a club (Fig. 8-82*E*) or most of the segments with anterior projections giving a saw tooth (Fig. 8-82*G*, *J*) or pectinate outline 6
5. Elytra short and squarely truncate, exposing full tergites of abdomen, these tergites heavily sclerotized and hard (Fig. 8-85*B*) .... **Silphidae**
   Elytra longer, usually rounded at apex usually covering entire dorsum of abdomen but in a few species exposing 1 or 2 tergites (Fig. 8-88) .... **Scarabaeidae**, p. 389
6. Elytra short, exposing 5 or more sclerotized abdominal tergites (Fig. 8-87) 7
   Elytra covering all or most of abdomen, never more than 2 or 3 sclerotized tergites visible from above; occasionally the abdomen of a female of this group may be extremely distended with eggs, and a third or fourth segment may project from beneath the elytra, but all except the apical two are soft and semimembranous 8
7. Elytra truncate, parallel-sided, and abutting evenly down the meson (Fig. 8-87), abdomen hard and regular in outline .... **Staphylinidae**, p. 388

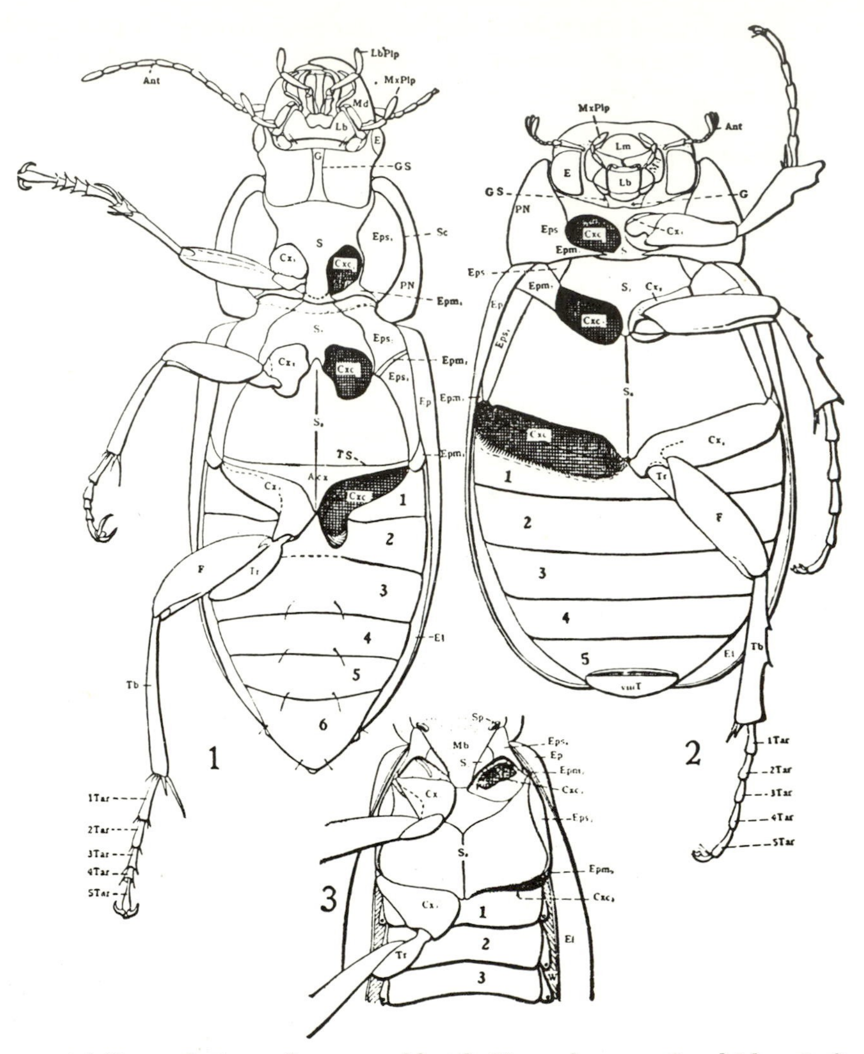

Fig. 8–81. (1) Ventral view of a ground bettle *Harpalus* sp., Carabidae. Left legs are removed. (2) Ventral view of a May beetle *Phyllophaga* sp., Scarabaeidae. Right legs are moved. (3) Ventral view of part of thorax and abdomen of a soldier beetle *Chauliognathus pennsylvanicus*, Cantharidae. Acx, Antecoxal piece; Ant, antennae; Cx_{1-3}, first, second, and third coxal cavities; E, eye; El, elytron; Ep, epipleuron; Epm_{1-3}, epimera of the pro-, meso-, and metathorax; Eps_{1-3}, episterna of pro-, meso-, and metathorax; F, femus; G, gula; GS, gular suture; Lb, labium, LbPlp, labial palpus; Lm, labrum; Mb, membrane; Md, mandible; MxPlp, maxillary palpus; PN, pronotum; S_1, S_2, S_3, pro-, meso-, and metasterna; Sc, suture separating pronotum from episternum; Sp, spiracle; 1 Tar to 5 Tar, the five tarsal segments; Tb, tibia; Tr, trochanter; TS, transverse suture; 1 to 6, abdominal sternites; viiiT, eighth tergite. (After Matheson, *Entomology for introductory courses*, by permission of Comstock Publishing Co.)

Elytra ovate, overlapping considerably at base; abdomen flabby and shrinking irregularly when the specimen dries ..... **Meloidae**, p. 395

8. Hind tarsus 4-segmented, front and usually middle tarsi 5-segmented 9

 Either hind tarsus having 3 or 5 segments, or all tarsi having the same number of segments 11

9. Body narrow and deep, bilaterally compressed, and with the hind coxa forming a large plate that appears as a major sclerite in the side of the thorax (Fig. 8-82*M*); small beetles often found abundantly in flowers .... **Mordellidae**

 Body wider than deep, often flattened, hind coxae no larger than in Fig. 8-81, 2 10

10. Each front coxal cavity closed posteriorly by a projection of the pleuron that meets the apex of the sternum (Fig. 8-84*D*); lateral edge of pronotum forming a sharp flange or delineated by a ridge or carina (compare Fig. 8-84*H*) .... **Tenebrionidae**, p. 398

 Front coxal cavities open posteriorly, the posteromesal corner of propleuron not extending mesad of outer portion of coxa (Fig. 8-84*C*); lateral edge of pronotum rounding inconspicuously into pleural region .... **Meloidae**, p. 395

11. Head not retracted within prothorax 12

 Head retracted within prothorax, so that only anterior portion protrudes (Fig. 8-84*H*) 14

12. Ventral portion of head having a large convex gular region with a single gular suture down the middle; and the palpi very short or indistinct (Fig. 8-82*N*), antennae always elbowed, with a long first segment, and sometimes ending in a flat club (Fig. 8-82*B*) 13

 Ventral portion of head having either a small gular area or two gular sutures, and the maxillary palpi usually much longer (Fig. 8-81, 1); antennae not elbowed or the first segment shorter in proportion to the remainder 14

13. Antenna long (Fig. 8-82*P*), usually ending in a small cylindrical, elliptical club; side of head having a deep groove for reception of antenna .... **Curculioniade**, p. 402

 Antenna short (Fig. 8-82*N*, *O*), ending in a large flat club or comb (Fig. 8-82*B*); side of head without antennal groove .... **Scolytidae**, p. 403

14. Hind and middle tarsi either 3- or 4-segmented, or fourth segment very small in comparison with the third (Fig. 8-83*b*, *c*) 15

 Hind and middle tarsi 5-segmented, the fourth as large as or as thick as the third (Fig. 8-83*a*) 27

15. All tarsi having the third segment enlarged (Fig. 8-83*b*, *c*), deeply bilobed or channeled dorsally; the fourth segment extremely small, either sunken into the cleft of the third or arising from dorsum of the base of the third (Fig. 8-83*b*); fifth segment large and normal; frequently the fourth segment cannot be seen, or it appears as a minute subdivision of the base of the fifth segment, in which case each tarsus appears 4-segmented 16

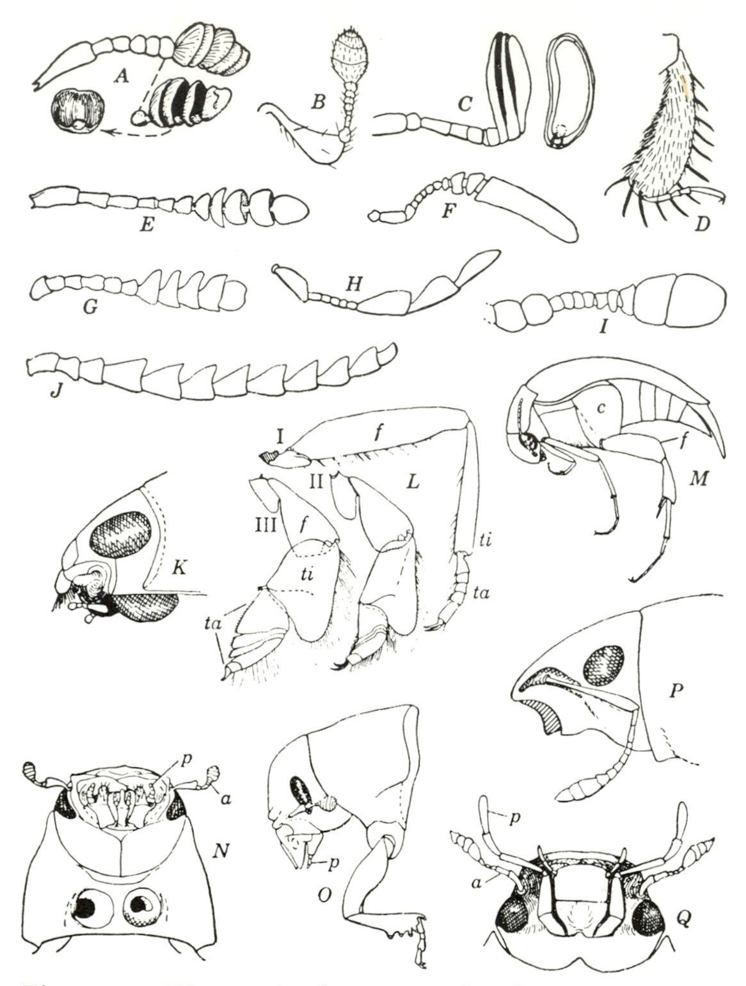

Fig. 8-82. Diagnostic characters of Coleoptera. (*A*) Antenna of *Nicrophorus*, Silphidae, inserts showing concave end segments and an end view of one; (*B*) antenna of *Scolytus*, Scolytidae; (*C*) antenna of *Thyce*, Scarabaeidae; (*D*) tibia and tarsus of *Heterocerus*, Heteroceridae; (*E*) antenna of *Silpha*, Silphidae; (*F*) antenna of *Attagenus*, Dermestidae; (*G*) antenna of *Languria*, Languriidae; (*H*) antenna of *Stegobium*, Anobiidae; (*I*) antenna of *Anthrenus*, Dermestidae; (*J*) antenna of *Melanotus*, Elateridae; (*K*) head of *Dineutes*, Gyrinidae; (*L*) legs of *Dineutes*, Gyrinidae; (*M*) profile of *Mordellistena*, Mordellidae; (*N* and *O*) head and prothorax of *Dendroctonus*, Scolytidae; (*P*) head of a short-snouted weevil, Curculionidae; (*Q*) head of *Tropisternus*, Hydrophilidae. *a*, Antenna; *c*, coxal plate; *f*, femur; *p*, maxillary palpus; *ta*, tarsus; *ti*, tibia.

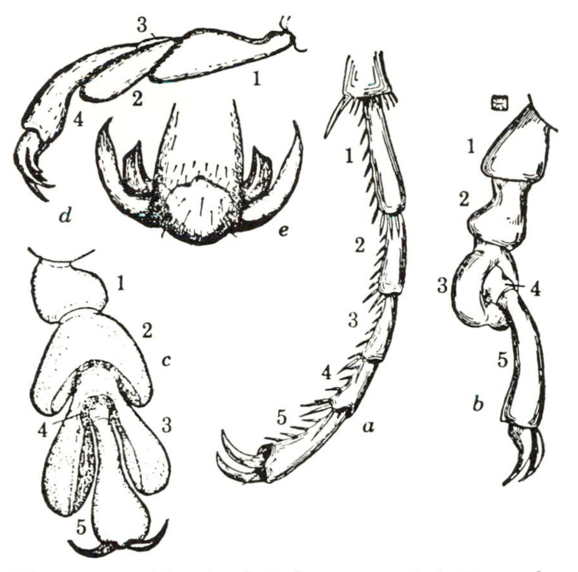

Fig. 8-83. Tarsi of Coleoptera. (*a*) *Harpalus* Carabidae; (*b*) *Leptinotarsa*, Chrysomelidae; (*c*) *Chelymorpha*, Chrysomelidae; (*d*) *Epilachna*, Coccinellidae; (*e*) toothed tarsal claws. (From Matheson, *Entomology for introductory courses*, by permission of the Comstock Publishing Co.)

Either tarsi only 3-segmented, or third segment not enlarged, channeled, or bi-lobed 42

16. Antennae about as long as or longer than the body (Fig. 8-103) . . . . Many **Cerambycidae**, p. 400
    Antennae distinctly shorter than body 17
17. Last tergite (pygidium) exposed, almost completely visible beyond or below end of elytra (Fig. 8-84*E*, *F*) 18
    Last tergite almost entirely or completely covered by elytra 21
18. Hind tibia little, if any, longer than basitarsus, and having a pair of long apical spurs (Fig. 8-84*P*) . . . . **Bruchidae**, p. 400
    Hind tibia much longer than basitarsus, frequently with only short spurs or none 19
19. Elytra short, each only about twice as long as wide; stocky short species, the pygidium oblique, and at a definite angle to dorsal contour of the body (Fig. 8-84*F*) 20
    Elytra long, each four times or more as long as wide; elongate species, the pygidium nearly horizontal, following dorsal contour of the body (Fig. 8-84*E*) . . . . some **Cerambycidae**, p. 400
20. Anterior part prothorax forming a cylinder against which the flat head

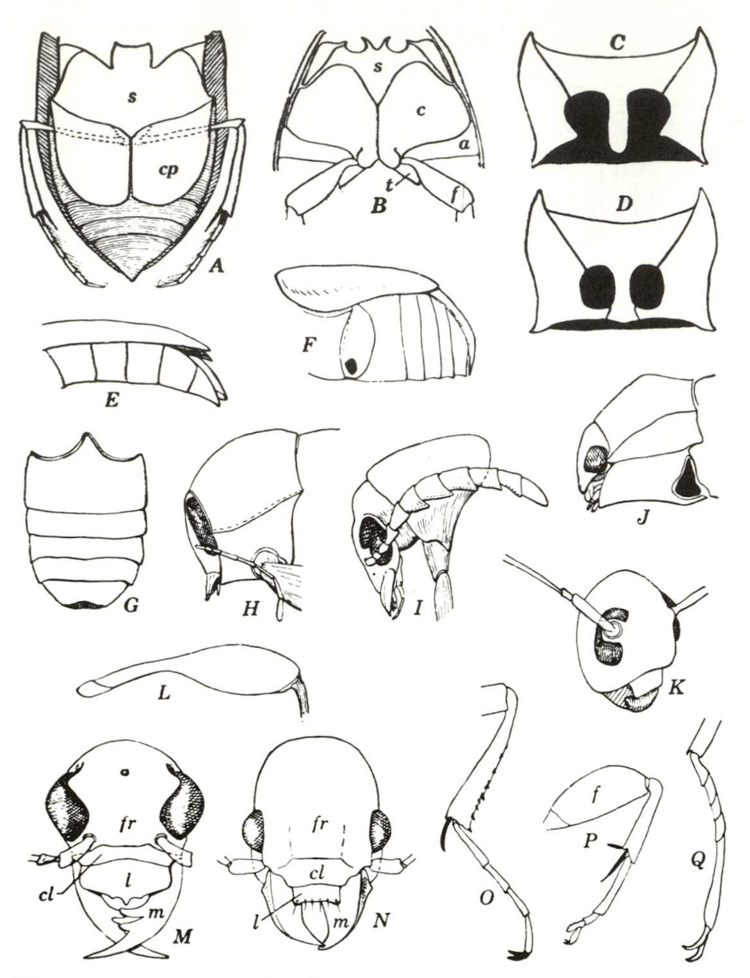

Fig. 8-84. Diagnostic characters of Coleoptera. (*A*) Venter of *Haliplus*, Haliplidae; (*B*) venter of *Dytiscus*, Dytiscidae; (*C*) prosternum showing open coxal cavities, diagrammatic; (*D*) prosternum showing closed coxal cavities, diagrammatic; (*E*) abdomen of *Leptura*, Cerambycidae; (*F*) abdomen of *Bruchus*, Bruchidae; (*G*) venter of *Criocerus*, Chrysomelidae; (*H*) head and prothorax of *Cryptocephalus*, Chrysomelidae; (*I*) head and prothorax of *Acanthoscelides*, Bruchidae; (*J*) head and prothorax of *Helichus*, Dryopidae; (*K*) head of *Saperda*, Cerambycidae; (*L*) hind femur of *Neoclytus*, Cerambycidae; (*M*) head of *Cicindela*, Cicindelidae; (*N*) head of a Carabidae; (*O*) hind leg of *Amphicerus*, Bostrichidae; (*P*) hind leg of *Amblycerus*, Bruchidae; (*Q*) hind tarsus of *Psephenus*, Psephenidae. *a*, Abdomen; *c*, coxa; *cl*, clypeus; *cp*, coxal process; *f*, femur; *fr*, frons; *l*, labrum; *m*, mandible; *s*, sternum, *t*, trochanter.

fits like a lid; eyes oval, fitting against margin of prothorax (Fig. 8-84*H*) .... some **Chrysomelidae**, p. 399

Anterior part of prothorax narrow, head projecting freely beyond it; eyes incised and V-shaped, head constricted behind them (Fig. 8-84*I*) .... most **Bruchidae**, p. 399

21. Last 3 to 5 antennal segments enlarged to form a large loose club (Fig. 8-82*G*), elongate, smooth, and highly polished species .... **Languriidae**

Antenna of uniform thickness throughout, or widening gradually to form a club, or forming a round compact club; form various 22

22. Hind femur greatly enlarged, short, and oval in outline; flea beetles and their relatives .... **Chrysomelidae**, p. 399

Hind femur elongate or more parallel-sided, or constricted at base and enlarged only at apex (Fig. 8-84*L*) 23

23. First abdominal sternite very long, its mesal length nearly equal to that of the following four sternites, combined (Fig. 8-84*G*), several genera of varied form (e.g., **Donacia**, **Crioceris**) .... **Chrysomelidae**, p. 399

First abdominal sternite considerably shorter in proportion to following segments 24

24. Either antenna as long as or longer than body, or hind femur having basal portion slender and apical portion enlarged and clavate .... (Fig. 8-84*L*) .... **Cerambycidae**, p. 400

Antenna shorter than body, and hind femur without a basal stalklike portion 25

25. Tibia having a pair of well-developed tibial spurs at apex .... **Cerambycidae**, p. 400

Tibia having either no tibial spurs or very minute ones 26

26. Mesal margin of eye deeply incised, with the antenna situated in the incision (Fig. 8-84*K*), or the eye completely divided, the antenna situated between the two parts .... **Cerambycidae**, p. 400

Either eye not incised, or antenna not at all in incision of eye .... **Chrysomelidae**, p. 399

27. Hind coxa having a wide long ventral plate that covers coxa, trochanter, and most of the femur (Fig. 8-84*A*), small to medium-sized stout aquatic beetles .... **Haliplidae**

Hind coxa at most having only a small outer plate, which does not cover trochanter or femur 28

28. Apices of hind coxae forming a double-knobbed process; base of each coxa extremely large and platelike, appearing as a dominant sclerite of the sternum (Fig. 8-84*B*); hind legs fringed for swimming .... **Dytiscidae**

Hind coxa neither with such a knobbed apex nor with such a platelike base (Fig. 8-81, 1, 2) 29

29. First abdominal sternite completely divided by the hind coxal cavities, the sternite appearing as a pair of triangular sclerites, one on each side of the coxae; first three abdominal sternites immovably united, the separating sutures appearing partly as extremely fine lines (Fig. 8-81, 1) 30

Either first abdominal sternite not completely or not at all cut into by the coxal cavities, or the first three sternites with separating sutures extremely well developed across the entire width of the segment, as are those between the more apical segments (Fig. 8-81, 2) 31

30. Clypeus fairly narrow, the antennal sockets wider apart than the width of the clypeus (Fig. 8-84*N*) .... **Carabidae**, p. 385
Clypeus much wider (Fig. 8-84*M*), the antennal sockets situated closer together than the width of the clypeus .... **Cicindelidae**

31. Abdomen having 6 or more exposed sternites 32
Abdomen having not more than 5 exposed sternites (Fig. 8-81, 2) 35

32. Last 3 to 5 antennal segments enlarged to form a club (Fig. 8-82*E*) .... **Silphidae**
Antenna the same width throughout, often beadlike or serrate 33

33. Tarsus slender and smooth, segments 1 to 4 very short and ringlike, segment 5 long, about equal in length to first 4 combined (Fig. 8-84*Q*); aquatic forms .... **Psephenidae**
Tarsus stout and densely setose, segment 5 no longer than segment 1, some of the basal segments elongate, as in Fig. 8-83*a* 34

34. Prothorax with broad anterior and lateral margins forming a hood that covers most or all the head from above (Fig. 8-90); head partially retracted, viewed from side .... **Lampyridae**, p. 390
Prothorax not having wide margins, head not retracted in pronotum, hanging down or projecting forward freely .... **Cantharidae**

35. Antenna elbowed and capitate, the club appearing as one round segment, some having faint cross sutures; very hard shining black beetles having short stout legs, the tibiae expanded and spurred for digging, much as in Fig. 8-81, *2* .... **Histeridae**
Antenna not elbowed, and either not thickened toward tip, or the enlarged portion composed of 2 or 3 well-separated segments or a single elongate segment; legs or body shape different from above 36

36. Antennae elongate and serrate throughout (Fig. 8-82*J*) 37
Antennae short, filiform, or clavate 38

37. Pronotum having a sharp projection at each posterolateral corner (Fig. 8-91) .... **Elateridae**, p. 391
Pronotum having posterolateral corners rounded and not pointed (Fig. 8-92) .... **Buprestidae**, p. 392

38. Prosternum produced anteriorly to form a long concave shelf under the head, but the latter largely retracted into the opening of the prothorax (Fig. 8-84) 39
Prosternum not so produced, much shorter than pronotum and proportioned more as in Fig. 8-84*I*; head usually held downward against chest 40

39. Front coxae round, situated some distance from lateral edge of pronotum; antennae elongate or definitely clavate .... **Elmidae**
Base of front coxae triangular, the point of the triangle extending nearly to lateral edge of prosternum (Fig. 8-84*J*); antennae short, the

flagellum of equal thickness throughout, with wide flat segments ... . **Dryopidae**, p. 392

40. Antennae ending in a distinct club composed of 1 to 3 segments (Fig. 8-82*F*, *I*) .... **Dermestidae**, p. 393
    Last 3 segments of antenna greatly enlarged but well separated to form a chain (Fig. 8-82*H*) 41
41. Tibiae without spurs; small convex forms including drugstore and cigarette beetles, infesting dried food products .... **Anobiidae**
    Tibiae with distinct spurs (Fig. 8-84*O*); includes many subcylindrical forms ornamented with horny processes and ridged areas; the powder post beetles .... **Bostrichidae**
42. Front and middle tarsi 4-segmented, hind tarsus 3-segmented .... **Cucujidae**
    All tarsi having the same number of segments 43
43. Tarsi 4-segmented, all segments well marked, and second and third of about equal width 44
    Tarsi 3-segmented, or third segment minute and hidden at the base of an enlarged second segment (Fig. 8-83*d*) 45
44. Tibiae dilated and armed with a series of stout spurs, fitted for digging (Fig. 8-82*D*), molelike beetles covered with dense pile, inhabiting wet mud or sand banks .... **Heteroceridae**
    Tibiae either not dilated or not armed with a series of spurs; flat beetles found under bark or in stored grain and feed .... **Cucujidae**
45. Second segment of tarsus dilated, with a large ventral pad (Fig. 8-83*d*); almost hemispherical beetles, usually polished and strikingly patterned .... **Coccinellidae**, p. 394
    Second segment of tarsus not much if at all wider than first segment; beetles often somewhat angular in outline, and flat; frequenting flowers, and tree sap .... **Nitidulidae**

SUBORDER ADEPHAGA

This suborder contains eight families, most of them predaceous, feeding on other insects. Two common families—the Cicindelidae (tiger beetles) and Carabidae—are terrestrial. Three other common families—the Gyrinidae, Haliplidae, and Dytiscidae—are aquatic, both adults and larvae living in the water.

CARABIDAE, GROUND BEETLES. These are active terrestrial species with long slender antennae of even thickness, long elytra, and long legs suitable for running (Fig. 8-85). They vary in size from 1 mm in length to large metallic-colored species 35 mm long. The mouthparts are well developed, and the mandibles are long, strong, and sharp. The adults are either diurnal or nocturnal, feeding on a wide variety of animals, including snails, worms, and adult and immature insects of many kinds. The larvae are slender with strong mouthparts,

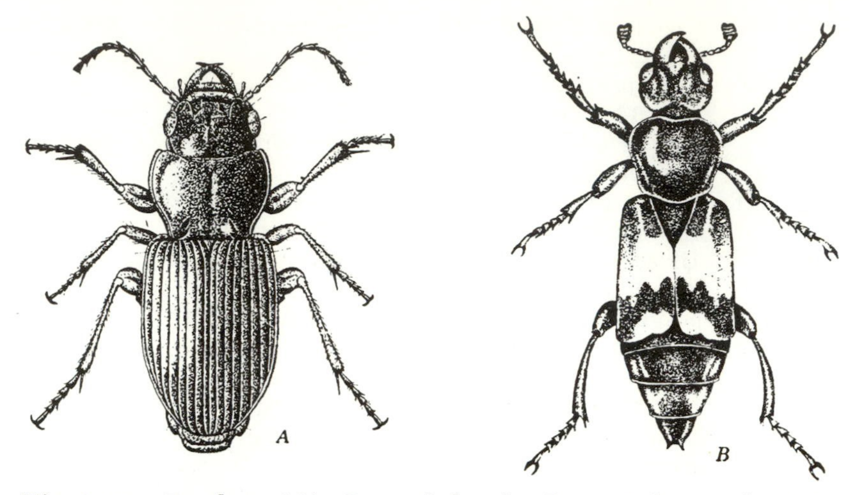

Fig. 8-85. Beetles. (*A*) Ground beetle *Pterostichus substriatus*; (*B*) carrion beetle *Nicrophorus marginatus*. (From Kansas State University)

well-developed legs, and a pair of terminal *urogomphi* (cercus-like organs). They are predaceous, feeding on other insects. They are usually found in the soil or under ground cover, where they hunt their prey. The family is a large one, embracing over 2000 nearctic species.

GYRINIDAE, WHIRLIGIG BEETLES. Aquatic beetles, the adults hard, convex, shining, and dark. The middle and hind legs are broad, fringed with hair, and used for swimming. The head has two peculiarities: (1) short antennae with an earlike expansion of the third segment; and (2) eyes that are completely divided into an upper and a lower half, so that the beetle appears to have two pairs of eyes (Fig. 8-82*K*). The adults are common in lakes and streams. They congregate in swarms of thousands, and zigzag along the surface, moving at high speed and leaving silvery crisscrossing wakes that are a familiar sight to every naturalist. During these surface gyrations, the divided eyes give the beetles the opportunity of seeing into the air with the dorsal eyes and into the water with the ventral pair. The larvae (Fig. 8-86) are slender, white-bodied, and elongate. Each segment of the abdomen has a pair of long tracheal gills, and at the end of the abdomen are a pair of hooks. Both adults and larvae are predaceous, the adults feeding chiefly on small organisms falling or alighting on the water surface, the larvae on small organisms they find in sheltered places on the bottom of the pond or stream.

SUBORDERS ARCHOSTEMATA AND MYXOPHAGA

In North America, the Archostemata are represented by two small families inhabiting old logs, Cupedidae and Micromalthidae. The latter includes only a single rare, small species (adults 1.8 to 2.5 mm), which is remarkable for its paedogenetic larvae (p. 241).

Of the three families of Myxophaga, two occur in North America. The minute (1.3 mm) Hydroscaphidae are aquatic and the larger (2 to 7 mm) Scaphidiidae live in ground cover and old logs.

SUBORDER POLYPHAGA

This includes the great bulk of the beetles, containing over 130 families, and embracing forms diverse in appearance and life history. Some families, such as the Chrysomelidae and Coccinellidae, are abundant and include species of considerable economic importance. Many families, however, are rare or seldom seen except as a result of specialized collecting and have no known economic importance. The families mentioned below are only a few of the many beetle families in the suborder Polyphaga. For families not discussed here, students will find a great deal of information on life history and identification in the books listed at the end of the section on Coleoptera.

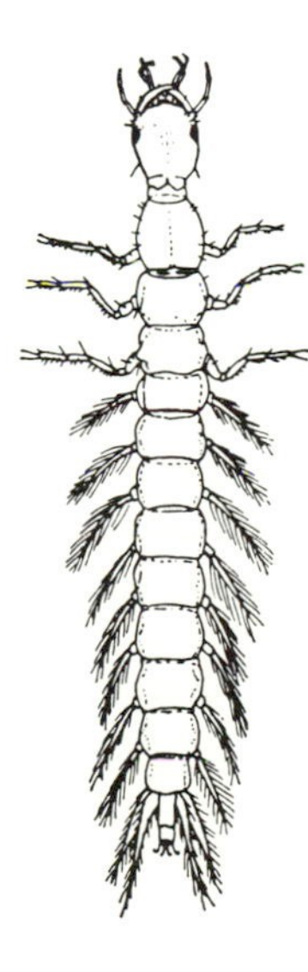

Fig. 8-86. Larva of whirligig beetle. (Redrawn from Boving and Craighead)

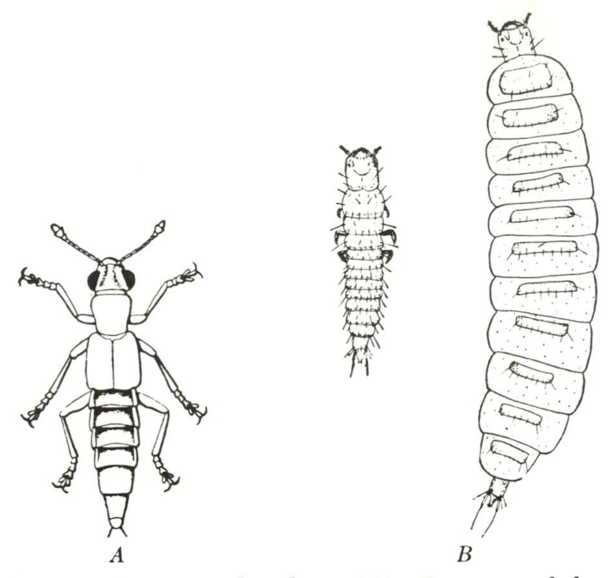

Fig. 8-87. Rove beetles. (*A*) *Stenus* adult (*B*) First-instar larva of *Aleochara curtula*: left, before feeding; right, after feeding. (*A* after Sanderson; *B* modified from Kemner)

STAPHYLINOIDS

STAPHYLINIDAE, ROVE BEETLES. Slender elongate beetles (Fig. 8-87*A*) with short truncate elytra beneath which the pair of flying wings are folded. The antennae are fairly long, either filiform or slightly enlarged at the tip. The larvae are also elongate and look much like small carabid larvae. Both adults and larvae are scavengers or predators, with the exception of a few species that have parasitic larvae. Adult rove beetles are found on flowers, in ground cover, under bark, in rotting organic material, in ant and termite nests, and in many other situations. The larvae are more secretive and occur chiefly in humid places. The species that are predaceous feed on mites, small insects, insect and mite eggs, and especially on small dipterous larvae. Certain species are valuable factors in the natural control of pests. An interesting example is the genus *Baryodma*, whose larva is a parasite of the cabbage-maggot pupa. During the first instar the larva is a slender free-running form with relatively large tergal plates. It searches in the soil for a fly puparium and enters it by boring an opening. In the puparium the beetle larva feeds on the fly pupa and grows amazingly, becoming greatly distended and grublike before molting. Later instars are always grublike. The same phenomenon occurs in some species of *Aleochara* (Fig. 8-87*B*).

SCARABAEIDAE, THE LAMELLICORN BEETLES, OR SCARABS. This is one of the largest families of beetles, characterized most conspicuously by the lamellicorn antennae, in which the apical segments are leaflike and appressed in repose. The scarabs vary greatly in size and shape; most of them are stout and very hard shelled. The larvae are sluggish, stout, usually white, and with a characteristic curved outline (Fig. 8-88). A number of groups of scarabs are scavengers, and feed on dung, rotting hides, or fungi. Of unusual interest are *Canthon* and certain other genera; the adult fashions a ball of dung, rolls it away, and buries it. Eggs are laid on this ball, and the developing larvae utilize it as food. The remainder of the scarabs are phytophagous, many species of great economic importance. The most publicized member of the family is the Japanese beetle *Popillia japonica* (Fig. 8-88); the larvae feed on grass roots and are especially destructive to lawns and golf courses, and the adults defoliate fruit and shade trees. Another group of destructive species are the June beetles (Fig. 7-12), members of the genus *Phyllophaga*; the adults defoliate deciduous trees, and the larvae, known as white grubs, eat the roots of various grass crops, including corn, small grains, and pasture plants.

In the tropics there are many species of large scarabs, often brilliantly colored or ornamented with spines or projections on the head or pronotum. In the United States the largest species is *Dynastes tityus*, the hercules beetle, which is larger than a shrew (Fig. 8-89). The larva lives in rotten wood.

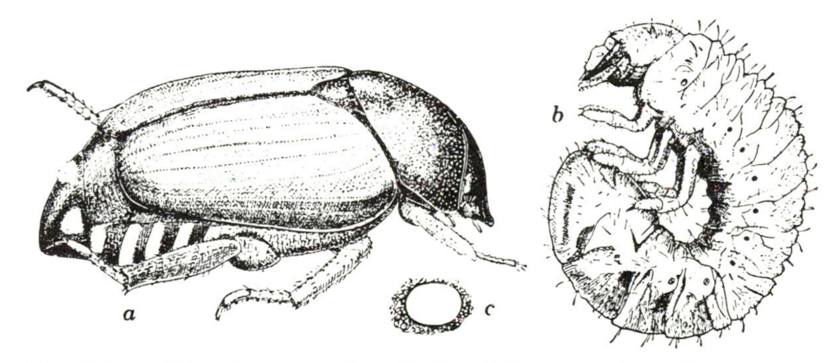

Fig. 8-88. The Japanese beetle *Popillia japonica*. (From Connecticut Agricultural Experiment Station)

Fig. 8-89. *Dynastes tityus*, Illinois' largest beetle, and a shrew *Sorex longirostris*, Illinois' smallest mammal. (Drawing loaned by C. O. Mohr)

LAMPYROIDS

LAMPYRIDAE, FIREFLIES. These are moderate-sized soft-bodied beetles, having serrate antennae and having the margins of the pronotum projecting like a flange or shelf, which partially covers the head. The elytra are relatively soft. A common eastern species is *Photurus pennsylvanicus* (Fig. 8-90). The adults occur in summer and fly actively on warm nights, almost invariably following a dipping up-and-down course only a few feet above the ground. As they start their "upstroke," each individual flashes a bright light, a mating signal. When swarms of adults are on the wing, the entire countryside is lighted up with these tiny dots of light. From this comes

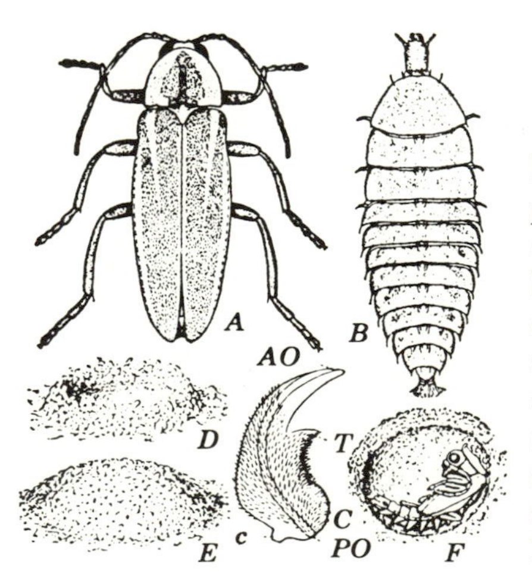

Fig. 8-90. A firefly *Photurus pennsylvanicus*. (*A*) Adult male; (*B*) mature larva; (*C*) left mandible of larva to show mandibular canal. *AO*, Opening of canal; *PO*, opening of canal to mouth; *c*, condyle; *T*, tooth on mandible. (*D*) Beginning of pupal chamber; (*E*) completed pupal chamber; (*F*) pupa in chamber. (From Matheson, after Hess)

the name firefly or "lightning bug." In certain genera the females are wingless and grublike; these and the larvae are also luminous and are called glowworms. Both larvae and adults are predaceous, although some adults may feed partly on plant material or not at all.

ELATERIDAE, CLICK BEETLES OR WIREWORMS. The adults are trim and hard bodied (Fig. 8-91), having five-segmented tarsi, a large pronotum with sharp posterior corners, serrate antennae, and a long stout sharp process projecting backward from the prosternum. If placed on their backs, these beetles can spring several inches into the air, at the same time making a loud click, usually alighting right side up. This leap is engineered by using the ventral process of the prosternum as a sort of spring release when the body is tensed. The adults are frequently encountered during spring and summer. These beetles and their acrobatics are well known to youngsters in rural districts.

The larvae, called wireworms (Fig. 8-91), are wormlike and hard-bodied and live in soil or rotten wood. The soil-inhabiting larvae feed on roots of grasses and related plants and are extremely destructive to many of the grain crops. Especially injurious are species of

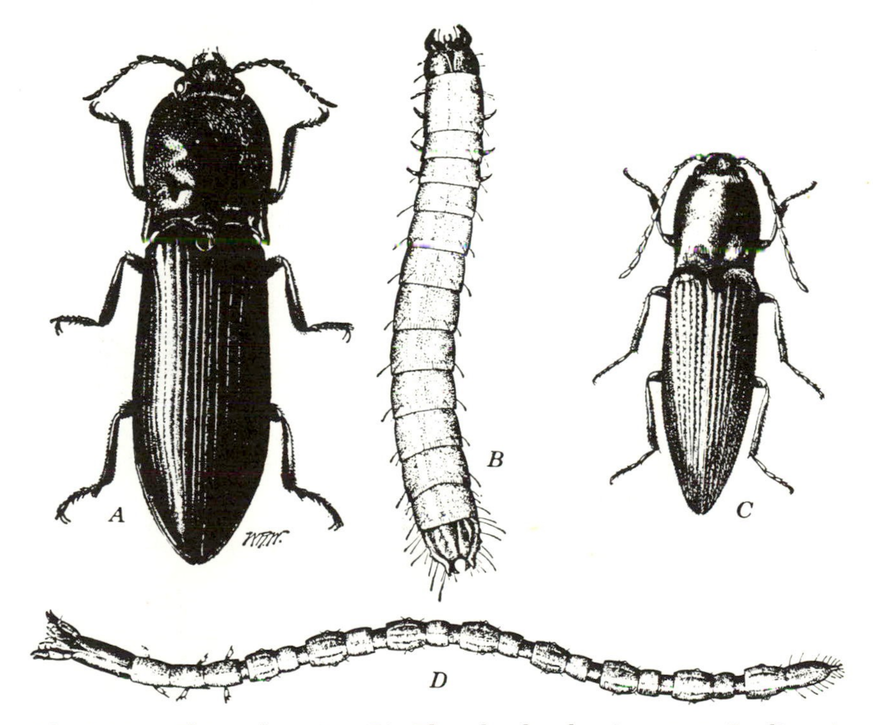

Fig. 8-91. Elateridae. (*A*, *B*) The dry-land wireworm *Ludius inflatus*, adult and larva; (*C*, *D*) the sand wireworm *Horistonotus uhlerii*, adult and larva. (From U.S.D.A., E.R.B.)

the genera *Melanotus*, *Agriotes*, and *Monocrepidius*. Several hundred species occur in North America. Of unusual interest to the collector of immature insects are the odd wormlike larvae of the group to which *Horistonotus* belongs (Fig. 8-91*D*).

BUPRESTIDAE, METALLIC OR FLATHEADED WOOD BORERS. These beetles (Fig. 8-92) resemble the click beetles in that the adults have serrate antennae, five-segmented tarsi, and hard bodies, but they differ in having the first two sternites of the abdomen fused and in the coppery or bright metallic coloring of the body and elytra. They are usually more robust and have a shorter pronotum. The larvae of the larger species (Fig. 8-92) bore in wood, attacking live trees or newly felled or killed trees, and feed either beneath the bark or into the solid wood. These larvae are legless and elongate, having the thorax expanded and flattened. They attack a wide variety of trees, including deciduous and coniferous species. The flatheaded apple tree borer *Chrysobothris femorata* is often an orchard pest of importance. A few genera of small buprestids have leaf-mining larvae, which are more cylindrical than the wood borers and have minute legs.

DRYOPIDAE AND THEIR ALLIES, THE DRYOPID BEETLES. In this group of aquatic beetles both the adults and larvae live in the water. Both are sluggish, crawling over stones or submerged wood, and feeding on surface encrustments. The adult is clothed with fine hair that holds a film of air when under water. The insect uses this film as a means of gas exchange with the surrounding water. The larvae respire by means of tracheal gills situated on various parts of the body, many retractile within ventral pouches. As with other aquatic Coleoptera, the mature larvae leave the water and pupate in damp soil. Most of

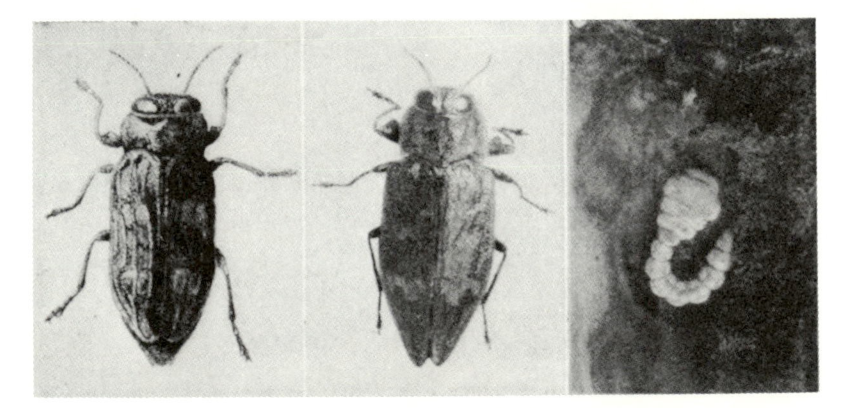

Fig. 8-92. Buprestidae. Left, the flatheaded apple tree borer, *Chrysobothris femorata*; center, the Pacific flatheaded borer, *C. mali*; right, larva in burrow. (From U.S.D.A., E.RB.)

the dryopids live in cold rapid streams and are frequently found in large numbers in siftings from gravel bars. The adults leave the water periodically for mating or dispersal flights.

BOSTRICHOIDS

DERMESTIDAE, CARPET BEETLES. Convex oval beetles (Figs. 8-93, 8-94) having short clubbed antennae, five-segmented tarsi, and abdomens with only five sclerotized sternites. The larvae are elongate or oval, clothed distinctively with large tufts or bands of long barbed hair. They feed on dried animal products, including fur, skins, and dried meat. The last is attacked readily by species of the genus *Dermestes*, especially *D. lardarius*, the larder beetle. Some of the smaller species, particularly the carpet beetle *Anthrenus scrophulariae* (Fig. 8-94) and the black carpet beetle *Attagenus megatoma* (Fig.

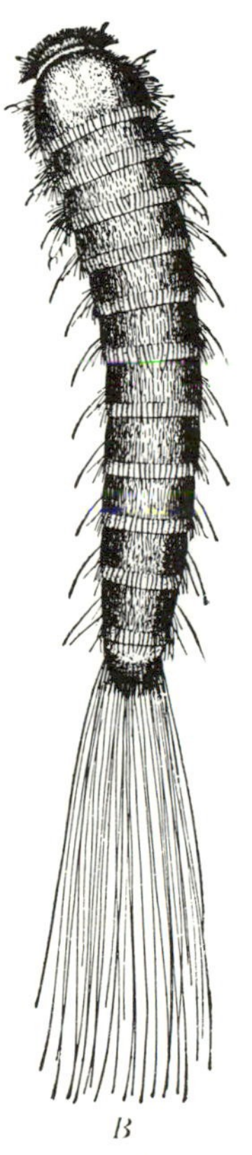

Fig. 8-93. The black carpet beetle *Attagenus megatoma*. (*A*) Adult; (*B*) larva. (From Connecticut Agricultural Experiment Station)

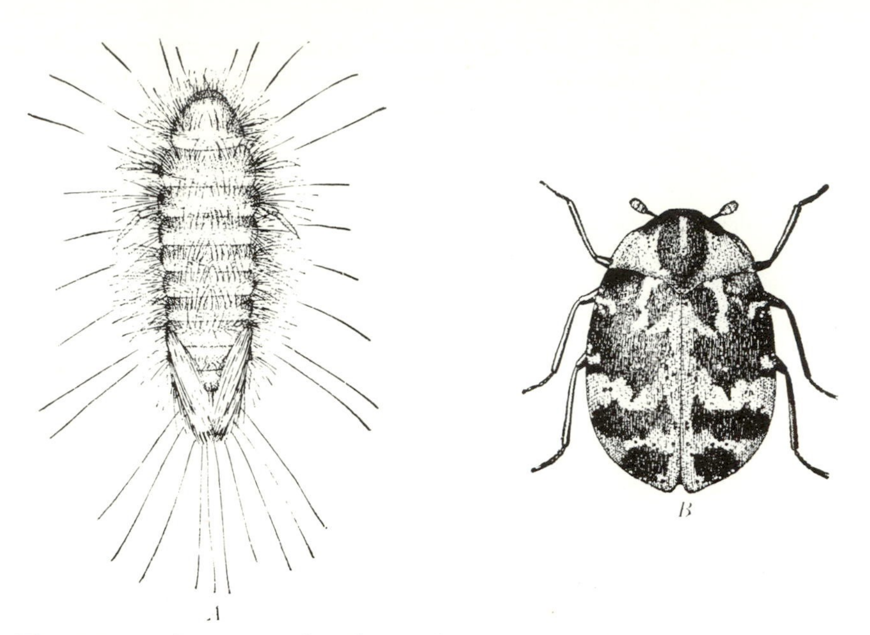

Fig. 8-94. The carpet beetle *Anthrenus scrophulariae.* (*A*) Larva; (*B*) adult. (From Connecticut Agricultural Experiment Station)

8-93) attack furs, carpets, and upholstery, in fact, anything made from animal hair. These pests are so widespread that they are a constant menace to household goods and many stored materials. In nature the species feed on dead insects or animal carcasses. The adults feed on the same material as the larvae, but during dispersal flights they feed on pollen and at this time are often found on garden flowers.

CUCUJOIDS

COCCINELLIDAE, LADYBIRD BEETLES. Moderately small round convex shining beetles (Fig. 8-95), sometimes patterned with red, yellow, black, or blue markings, are also called ladybugs or lady beetles. The antennae are short and clavate. The tarsi are four-segmented but appear to be three-segmented; the third segment is extremely minute, situated between the padlike second segment and the large end segment bearing the claws. The larvae either are warty creatures or are covered with a waxy secretion and have extremely short antennae but long legs. Two categories of ladybird beetles are of economic importance. Species of the first feed on aphids and scale insects and function as effective means of natural control against some of these

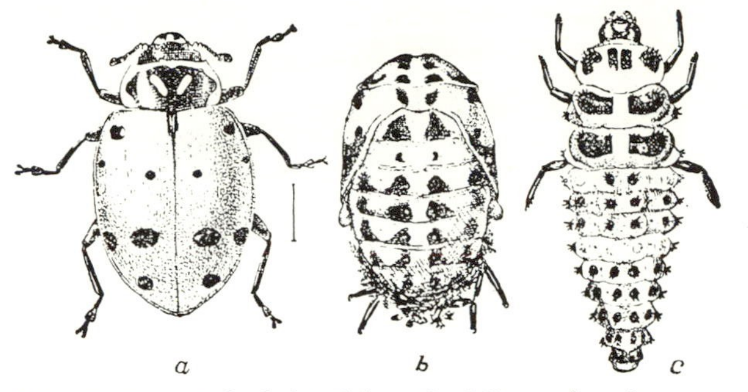

Fig. 8-95. A ladybird beetle *Hippodamia convergens*. (*a*) Adult; (*b*) pupa; (*c*) larva. (From U.S.D.A., E.R.B.)

pests. The best known of these predators is the vedalia beetle *Rodolia cardinalis*, a native of Australia, which was introduced into California for the control of the cottony cushion scale. The vedalia has been very effective in this capacity. A common and widespread native species is *Hippodamia convergens* (Fig. 8-95). The second important category of ladybird beetles includes plant-feeding species. In North America the Mexican bean beetle, *Epilachna varivestis* (Fig. 8-96) is one of the most destructive defoliators of beans in many central and southern areas. Both adults and larvae feed on the foliage.

MELOIDAE, BLISTER BEETLES. The adults (Fig. 8-97) are moderately large beetles with relatively soft bodies and elytra, long simple antennae, and a prominent round or oval head that is well set off from the thorax. The tarsi of the front and middle legs are five-segmented, but the tarsi of the hind legs are four-segmented. This characteristic marks off a group of some 20 beetle families sometimes referred to as

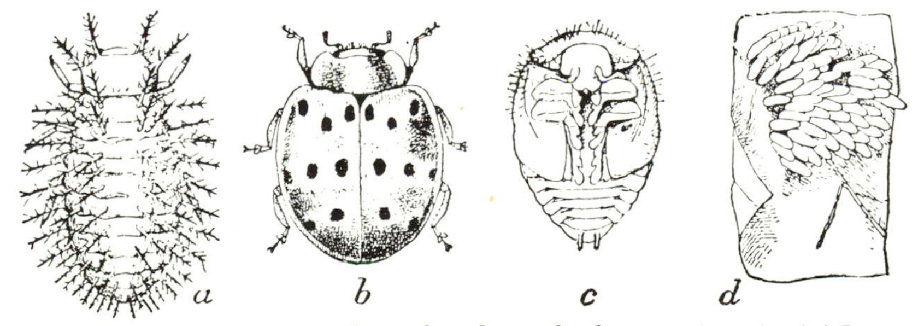

Fig. 8-96. The Mexican bean beetle *Epilachna varivestis*. (*a*) Larva; (*b*) adult; (*c*) pupa; (*d*) eggs. (From U.S.D.A., E.R.B.)

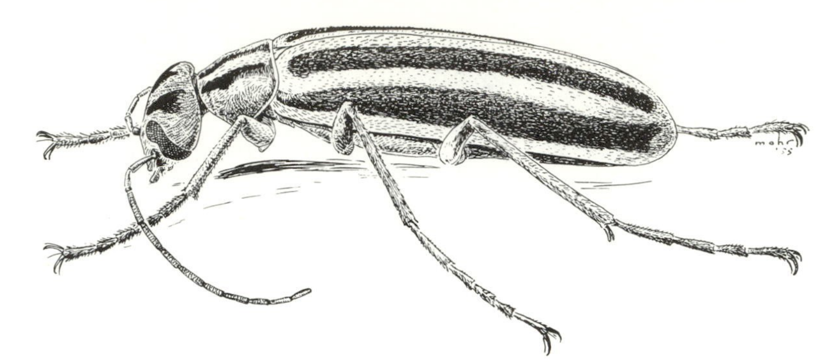

Fig. 8-97. A blister beetle, *Epicauta vittata*. (Original by C. O. Mohr)

the Heteromera. The blister beetles contain an oil, cantharidin, which is a powerful skin irritant and causes the formation of large water blisters on human skin. A sufficient amount of cantharidin to cause irritation is picked up by just handling the live adults.

In most subfamilies the larvae feed on bee larvae or as inquilines on provisions of bee nests. In one subfamily, containing the common and widespread genera *Epicauta* and *Henous*, the larvae feed on grasshopper egg pods. In all of them the life history is complex. The eggs are laid in masses of 50 to 300 about 2 to 3 cm. deep in the soil. The first instar is an active, slender, well-sclerotized form called a *triungulin*. In the bee inquiline groups the triungulins crawl into flowers, attach to bees, and are thus carried to bee nests, where the triungulins dismount and begin feeding. In the grasshopper egg predators the triungulins simply search out the grasshopper egg pods.

The development of the larvae is peculiar in many ways. This is well illustrated by *Epicauta pennsylvanica* (Fig. 8-98). The triungulin seeks out a grasshopper egg mass, digs down to it, punctures an egg, and starts feeding on its contents. In a few days, when the triungulin is swollen and fully fed, the first molt occurs. The grub-like thin-skinned second instar is relatively inactive and continues feeding on the grasshopper egg mass.

The third, fourth, and fifth instars follow in rapid succession and are similar to the second. The fifth instar, when full grown, leaves the food mass, burrows a few inches farther into the soil, and makes an earthen cell in which it molts to form the sixth instar larva. This sixth instar, called the *coarctate* form, is unique among beetles. It is nonfeeding, heavily sclerotized, oval in shape, and rigid. Only humplike legs are present. Usually the winter is passed in this stage.

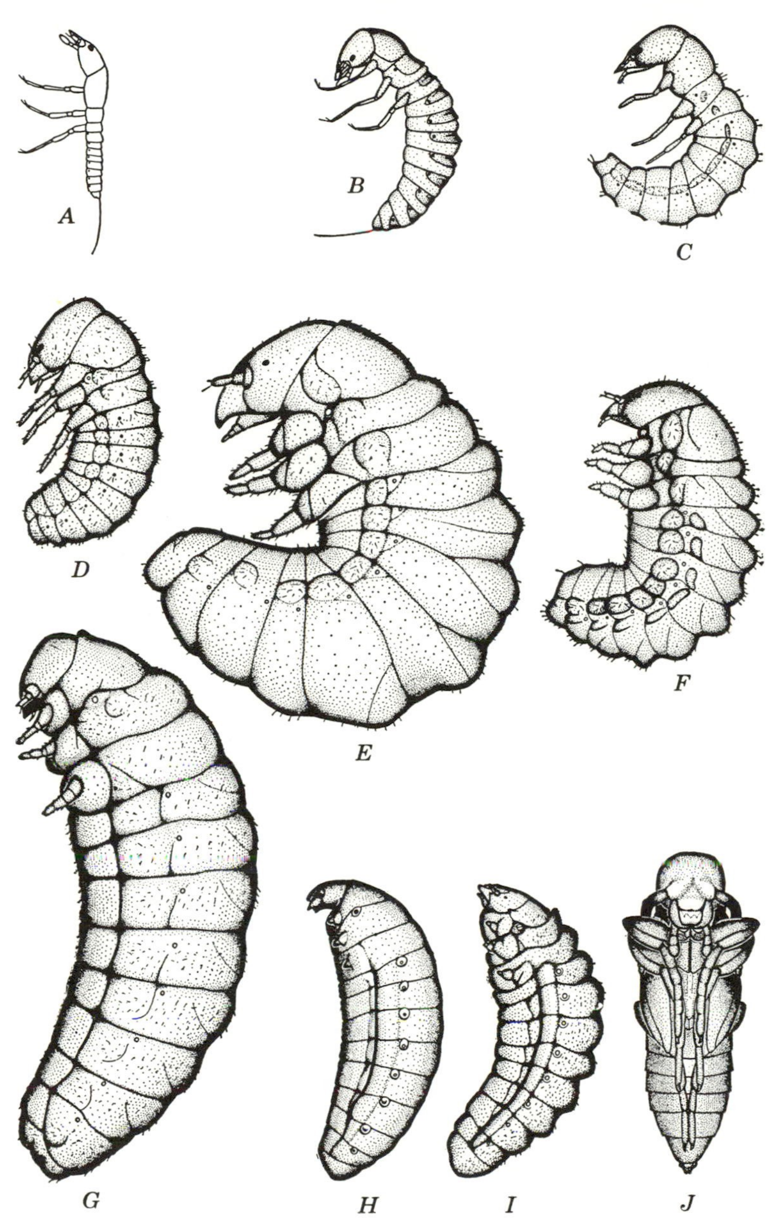

Fig. 8-98. Immature stages of the black blister beetle *Epicauta pennsylvanica*. (*A*) Unfed first instar, (*B*) fully fed first instar; (*C*, *D*, *E*) second, third, and fourth instars; (*F*) newly molted fifth instar; (*G*) gorged fifth instar; (*H*) sixth instar; (*I*) seventh instar; (*J*) pupa. *A* to *E*, ×17; *F*, *G*, ×9; *H* to *J*, ×5. (After Horsfall)

The next summer the coarctate larva molts to form a seventh instar much like the fifth; this larva gives rise to the pupa, which transforms in a few weeks to the adult. The coarctate larva is extremely resistant to desiccation, which provides a margin of safety to the species in drought years. For, if conditions are too dry during the summer after hibernation, the coarctate larva will not molt, but will "lay over" an additional year or even 2 years, if necessary, when less arid conditions prevail and the normal life cycle can be resumed.

Meloid adults are leaf feeders and cause appreciable damage to potatoes, tomatoes, squash, certain legumes, and other crops. The three-striped blister beetle *Epicauta vittata* (Fig. 8-97) is a colorful representative often injurious to these plants. One of the species frequently feeding on potatoes, *Meloe angusticollis*, is unusual among beetles in that the elytra overlap at the base. Another short-winged meloid is the squash blister beetle *Henous confertus*.

TENEBRIONIDAE, THE DARKLING BEETLES. These are hard-shelled beetles, normally dark in color, and oval or parallel-sided in outline (Fig. 8-99); tarsi of the front and middle pairs of legs are five-segmented, those of the posterior legs four-segmented; antennae moderately long, usually filiform or clavate. The larvae feed chiefly on dead plant material and fungi, especially bracket types, or mycelium in rotten wood; a few are predaceous, and a few attack stored grain. The larvae are elongate and cylindrical, with stout legs. Those of the western genus *Eleodes* feed on plant roots and are called false wireworms because of their resemblance to true wireworms (Elateridae larvae). Another western false wireworm is *Embaphion muricatum*, whose larvae are destructive to wheat. Some of the species attacking stored grains and prepared foods are widespread and cause large commercial loss. Among this group are the

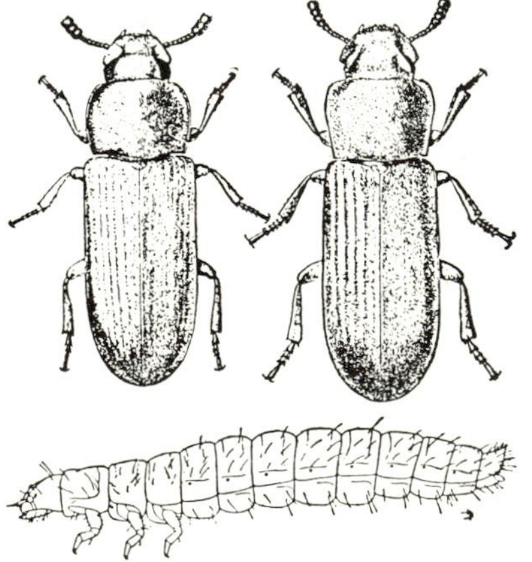

Fig. 8-99. The red flour beetle *Tribolium castaneum*, and the confused flour beetle *Tribolium confusum*. Below, larva of *T. castaneum*. (From U.S.D.A., E.R.B.)

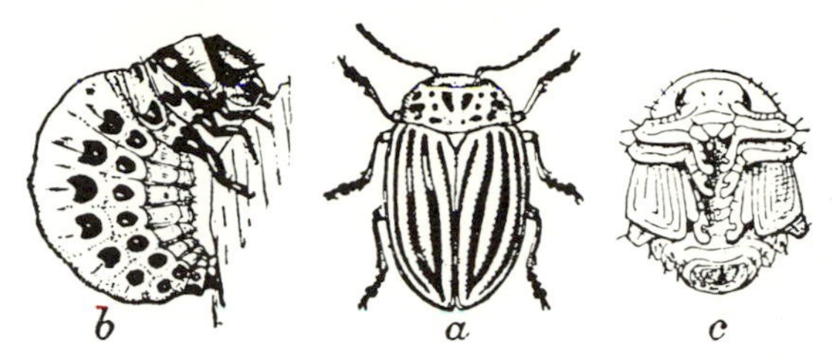

Fig. 8-100. The Colorado potato beetle *Leptinotarsa decemlineata*. (*a*) Adult; (*b*) larva; (*c*) pupa. (From U.S.D.A., E.R.B.)

mealworms *Tenebrio* sp., which are relatively large, the adults reaching a length of 15 mm and the larvae 25 or 30 mm; and the confused flour beetle *Tribolium confusum*, and several related species of *Tribolium*, which are very small, the adults being only 3 or 4 mm in length.

CHRYSOMELIDAE, THE LEAF BEETLES. These comprise a large family, the species small to moderate in size, usually oval, stout, or wide bodied, and having filiform fairly long antennae (Fig. 8-100). The most outstanding characteristics are found in the tarsi (Fig. 8-83*b*, *c*). These appear four-segmented; the third segment is enlarged to form a large kidney-shaped pad; the last segment, really the fifth, is long and slender and appears to be attached within the median incision of the third; actually the fourth segment is an extremely reduced ring at the base of the fifth, but it is so small that is is seldom seen without one's first making a special preparation of the leg. The larvae of the leaf beetles are varied, but most of them are stout grubs with short legs and antennae; a number bear spines and processes. Some of the leaf-mining species are long and flat. The eggs are laid in soil or under bark or deposited on stems or leaves.

With few exceptions adult Chrysomelidae feed on plant foliage, and their larvae on roots or leaves. Many attack commercial crops, and the family includes a large number of important economic species. The Colorado potato beetle, *Leptinotarsa decemlineata* (Fig. 8-100), is one of the most destructive insects attacking potato; both larvae and adults feed on the foliage. The asparagus beetle, *Crioceris asparagi*, is a common showy species wherever asparagus is grown; both larvae and adults feed on the foliage; the eggs are black and stuck by one end into the emerging stalks of the plants. The larvae of many species, such as *Acalymma vittata* (Fig. 8-101), are known as rootworms. The adults of many small species jump like fleas and for this reason are called flea beetles. Of these, the genera *Phyllotreta* and *Epitrix* contain several species whose adults eat

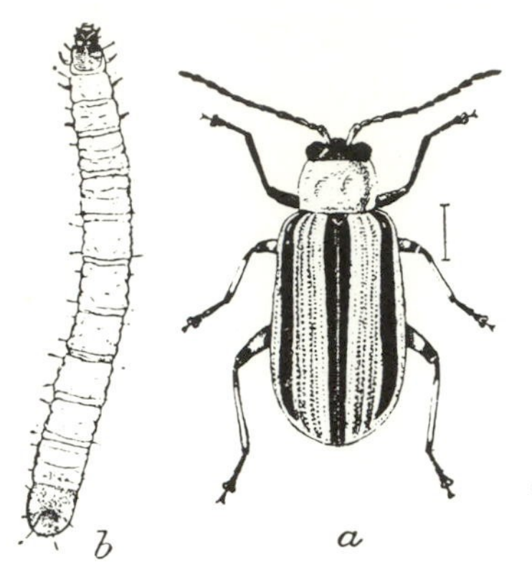

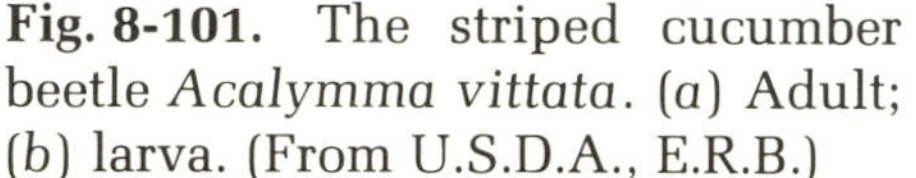

Fig. 8-101. The striped cucumber beetle *Acalymma vittata*. (*a*) Adult; (*b*) larva. (From U.S.D.A., E.R.B.)

holes in leaves, and the elongate larvae eat roots of cabbage, turnips, potatoes, cucumbers, and other plants. An important species is the tobacco flea beetle *Epitrix hirtipennis*. Not all flea beetles have root-feeding larvae; those of the genus *Haltica*, for instance, are leaf feeders like the adults.

BRUCHIDAE, BEAN AND PEA WEEVILS. This is a most interesting family closely related to the Chrysomelidae. The adults are short and stout; the larvae are grublike and almost legless, living inside legume seeds. Several species are pests of considerable importance in various kinds of stored peas and beans; a common example is the pea weevil *Bruchus pisorum* (Fig. 8-102).

CERAMBYCIDAE, LONGHORN BEETLES. Elongate beetles, many of them attractively colored, having long legs and unusually long antennae (Fig. 8-103). In other characters the longhorns are almost identical with the Chrysomelidae, including the curious tarsi with the en-

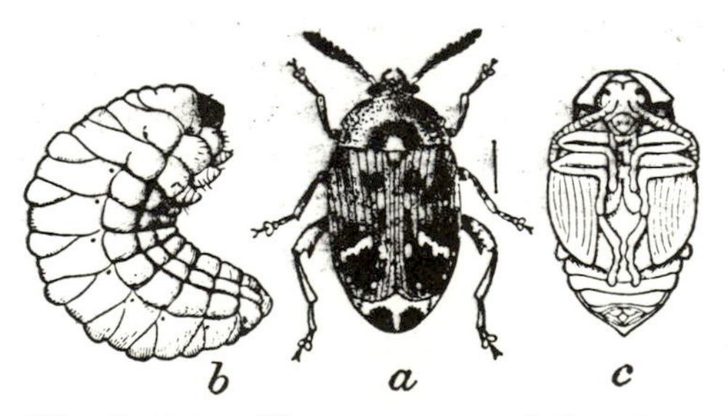

Fig. 8-102. The pea weevil *Bruchus pisorum*. (*a*) Adult; (*b*) larva; (*c*) pupa. (From U.S.D.A., E.R.B.)

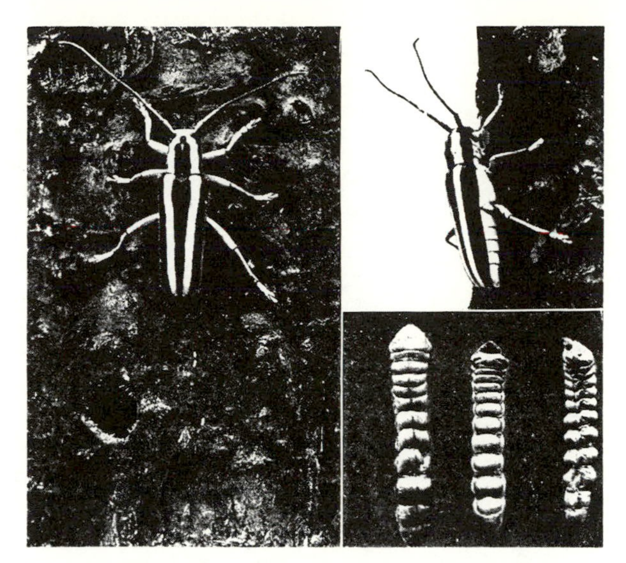

Fig. 8-103. The roundheaded apple tree borer *Saperda candida*; larvae, adults, and exit holes, natural size. (After Rumsey and Brooks)

larged third segment and the reduced fourth. The larvae of the longhorns are cylindrical and elongate, with a round head, and either no legs or minute ones; they are known as roundheaded borers. Most of them bore either in the cambium layer or through the heartwood of trees; a few bore in the roots and lower stems of succulent herbs such as milkweeds and ragweeds, or in the stems of shrubs such as willow and raspberries. Certain species are of considerable economic importance.

The roundheaded apple tree borer *Saperda candida* (Fig. 8-103) is a brown and white striped species that is locally a serious pest of apple, the larvae boring through the trunk and making extensive tunnels. The locust borer *Megacyllene robinae* has a handsome adult with a geometrical yellow pattern on black; its larvae bore in young black locust trees and weaken them so that wind breaks them easily; many black locust plantings, established for soil-erosion purposes, have been entirely destroyed in this manner.

Most of the longhorn species are pests of forest trees, attacking both deciduous and coniferous species. Under improper or careless lumbering conditions or unusual weather, various longhorn species may become abundant enough to cause considerable loss to commercial stands of trees.

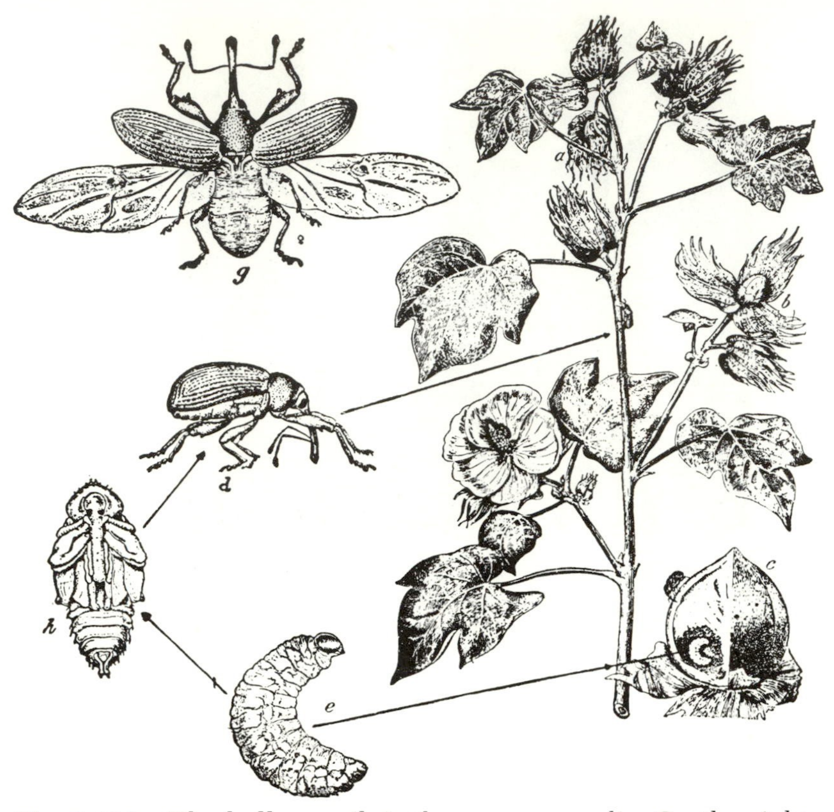

Fig. 8-104. The boll weevil *Anthonomus grandis*. On the right a cotton plant attacked by the boll weevil, showing (*a*) a hanging dry infested square; (*b*) a flared square with weevil punctures; (*c*) a cotton boll sectioned to show attacking weevil and larva in its cell (g) adult female with wings spread as in flight; (*d*) adult from side; (*h*) pupa ventral view; (*e*) larva.

CURCULIONIDAE, TYPICAL WEEVILS. The head has a definite beak, sometimes elongate and curved (Fig. 8-104), antennae usually elbowed and clubbed; body, elytra, and legs very hard, forming a solid well-armored exterior. The larvae are legless grubs, usually having dark head capsules and white bodies.

The Curculionidae is one of the largest insect families, containing over 75,000 species, many of which are extremely important pests. The larvae feed on plant material in a variety of ways; some feed on roots, leaves, or fruits, such as hazelnuts, acorns, cherries, and plums; others bore stems; still others feed on rotten wood or stored grain. The adults usually feed on the leaves or fruiting bodies of the plant species that serves as host for the larvae.

The boll weevil *Anthonomus grandis* (Fig. 8-104), is one of the most serious cotton pests in the United States and is high on the list of insects causing excessive commercial damage. The adults attack the plants early in the season, feeding on the leaves and in the flowers and young bolls. The females deposit their eggs in feeding holes in the bolls, one egg to each hole. The larvae feed inside the boll, thus making a cell in which they pupate. Adults emerge in a few weeks, cause additional damage by their feeding, and then go into hibernation at the onset of winter. The boll weevil is a native of tropical America and entered Texas from Mexico about 1890. Since that time it has spread gradually throughout most of the cotton-growing region of the United States.

The plum curculio *Conotrachelus nenuphar* attacks plums, cherries, and related fruits; the larvae feed in the body of the fruit. This species is frequently a serious pest.

Of especial importance to stored grains are the granary weevil *Sitophilus granarius* and the rice weevil *Sitophilus oryza*; the larvae of both species live and feed inside grain kernels, and the adults feed either in the old larval burrows or promiscuously on the grain.

SCOLYTIDAE, THE BARK BEETLES. The species of this family feed chiefly in trees, either alive or dead. The beetles have only an indistinct snout and are almost cylindrical in body shape (Fig. 8-105). The larvae are legless grubs with typical weevil characteristics. Both adults and larvae feed in galleries (Fig. 8-106). In some species these galleries are in the cambium layer of the tree and if sufficiently numerous result in girdling the tree. Many species of this family

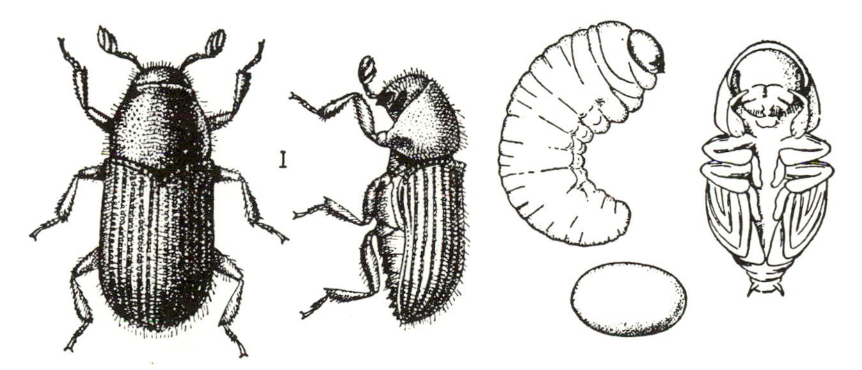

Fig. 8-105. The peach bark beetle *Phloeotribus liminaris*. (From U.S.D.A., E.R.B.)

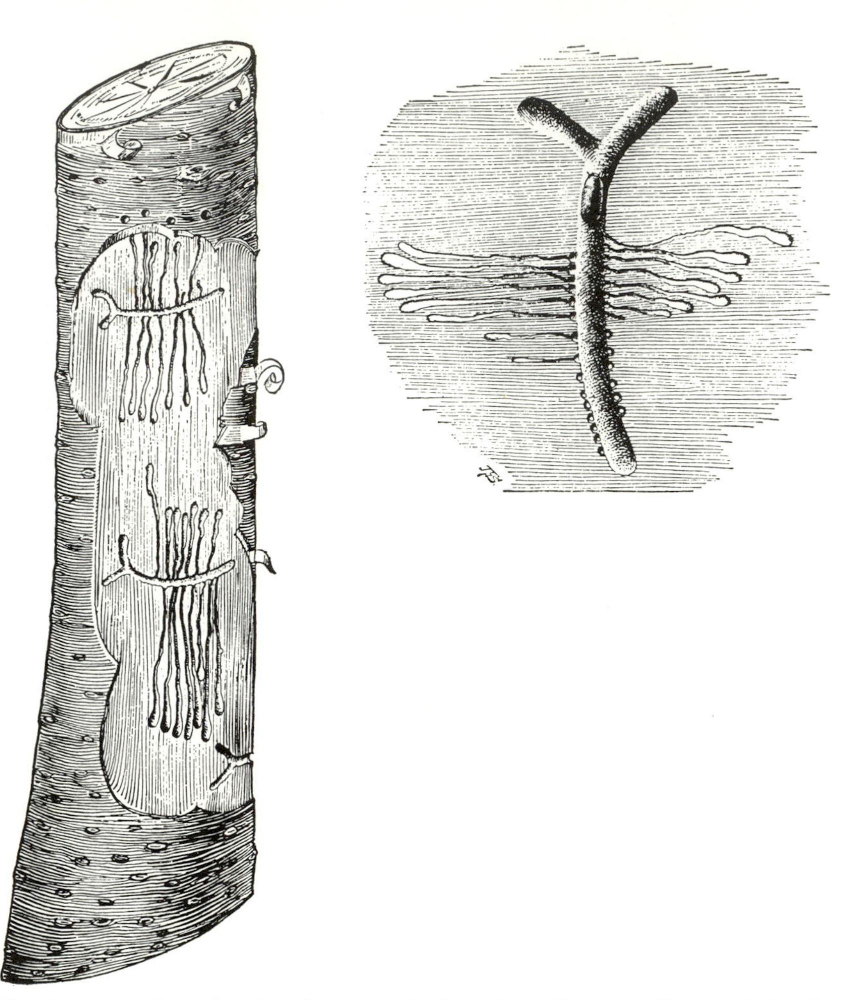

Fig. 8-106. Brood chambers and larval galleries of peach bark beetle. At right is shown a brood chamber with egg pockets and early larval galleries. (From U.S.D.A., E.R.B.)

attack commercial timber, especially pine, and sometimes cause the premature death of large stands of it.

STYLOPOIDS

The stylopoids or twisted-wing insects (Fig. 8-107), comprise the superfamily Stylopoidea (sometimes called the order or suborder Strepsiptera) and are a remarkable group of Coleoptera. They include two families, the Stylopidae and Mengeidae, both seen only rarely by collectors.

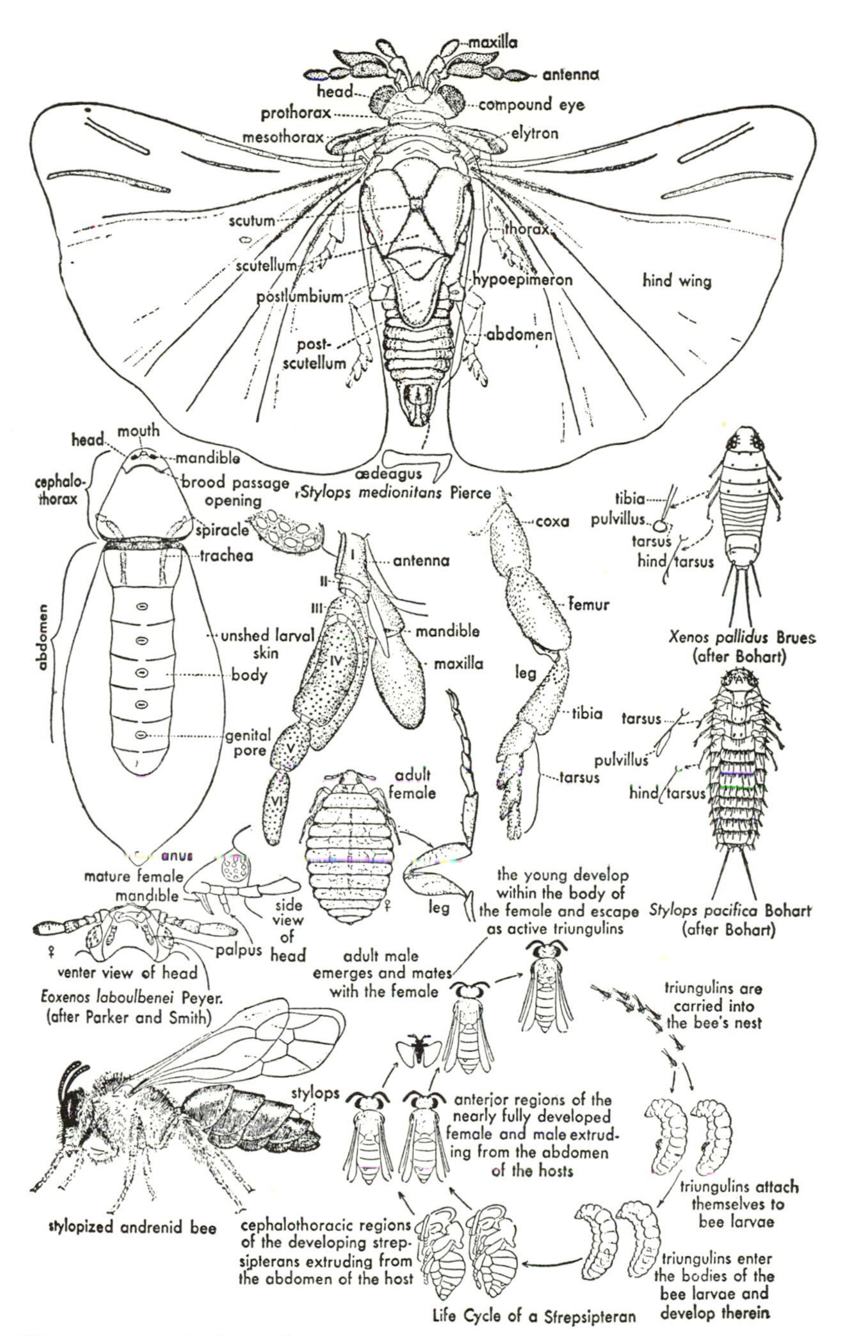

Fig. 8-107. Stylopoidea, important characteristics and life cycle. (From Essig, 1942, *College entomology*, by permission of Macmillan Co.)

STYLOPIDAE. The male is winged and free flying, more like a fly than a beetle, its elytra reduced to twisted finger-like organs, the hind wings large and fan-shaped, and the metathorax greatly enlarged. The antennae are short and have at least some leaflike segments; the eyes are berry-shaped, with each ommatidium protruding and distinct. The female is extremely degenerate, being only a sac enveloped by the last larval skin, and remains embedded in the host, with only an anterior portion, called a cephalothorax, projecting on the outside of the host integument.

The female is viviparous; the first-instar larvae develop inside her body and escape to the outside by crawling through a slit in the exposed cephalothorax. These larvae, called *triungulinids* because of their similarity to meloid triungulins, have three pairs of legs, well-developed sclerites, reduced mouthparts and ocelli, several well-developed eye facets, and one or two pairs of long terminal filaments. The larvae can run and jump with great agility. Each attaches and burrows into an individual host and soon molts into a legless grub. The grub lies quiescent in the host body and absorbs its food by diffusion from the host bloodstream. Thus there is no actual destruction of tissues like that caused by the usual type of insect parasitic larva that macerates and ingests its food by mouth. When full grown, the stylopid larva pushes its anterior end between the abdominal sclerites of the host so that the head and thoracic region is exposed, forming a round or flattened structure. If the larva is a male, it transforms within the larval skin to a typical beetle pupa; when mature, the adult breaks through the exposed larval skin and escapes to the outside. If the larva is a female, it has no definite pupal stage, molting directly into the saclike adult female that remains within the larval skin. The eggs develop and hatch within the body of the mother, until she is merely a sac of eggs or young. The number of progeny of one of these females is enormous; counts of young range from 2500 to 7000 per female.

The North American species of stylopids attack chiefly adults of various bees and wasps. In other parts of the world different groups of stylopids attack Hemiptera (Pentatomidae, Cicadellidae, Fulgoridae, and certain allied families) and Cursoria (Mantidae).

The closely allied family Mengeidae is similar in most respects to the Stylopidae, but has a free-living larviform female. The host of the only nearctic genus is unknown, but European species parasitize silverfish.

The Stylopidae and their close allies are regarded by some authors as comprising a separate order, the Strepsiptera. Most workers agree, however, that this group originated with the beetle complex. The triungulinid larva, elytroid front wings, reduced hind-wing venation, and parasitic habit suggest a relationship near the Meloidae.

References

Arnett, R. H., Jr., 1960. *The beetles of the United States. (A manual for identification).* Washington, D.C.: Catholic Univ. Press. 1112 pp. (Reprinted 1968. Am. Entomol. Institute, Ann Arbor, Mich.)

——— 1967, Present and future systematics of the Coleoptera in North America.

Annu. Rev. Entomol., **60**:162–170.

Blatchley, W. S., 1910. *The Coleoptera or beetles known to occur in Indiana.* Indianapolis, Ind.: Nature Publishing Co. 1386 pp.

Blatchley, W. S., and C. W. Leng, 1916. *Rhyncophora or weevils of northeastern America.* Indianapolis, Ind.: Nature Publishing Co. 682 pp.

Bohart, R. M., 1941. A revision of the Strepsiptera. *Univ. Calif. Publ. Entomol.*, **7**:91–160.

Bradley, J. C., 1930. *Manual of the genera of beetles of America, north of Mexico.* Ithaca, N.Y.: Daw, Illston & Co. 360 pp.

Britton, E. B., 1970. Coleoptera (beetles). In D. F. Waterhouse, et al. (Eds.), *The insects of Australia.* Carlton, Victoria: Melbourne Univ. Press. pp. 495–621; Suppl. 1974, pp. 62–89.

Clausen, C. P., 1940. *Entomophagous insects: Coleoptera.* New York: McGraw-Hill. pp. 522–584.

Crowson, R. A., 1955. *The natural classification of the families of Coleoptera.* London: Nathaniel Lloyd. 187 pp.

1960. The phylogeny of the Coleoptera. *Annu. Rev. Entomol.*, **5**:111–134.

Dillon, E. S., and L. S. Dillon, 1961. *A manual of common beetles of Eastern North America.* Evanston, Ill.: Row, Peterson. 884 pp.

Edwards, J. G., 1949. *Coleoptera or beetles east of the Great Plains.* Ann Arbor, Mich.: Edwards Bros. 181 pp.

Hatch, M. H., 1957–1973. *The beetles of the Pacific Northwest.* Seattle: Univ. Wash. Publ. Biol. and Univ. of Washington Press. 5 vols.

Hinton, H. E., 1955. On respiratory adaptation, biology and taxonomy of the Psephenidae with notes on related families (Coleoptera). *Proc. Zool. Soc. London*, **125**:543–568.

Leech, H. B., and H. D. Chandler, 1956. Aquatic Coleoptera. In R. L. Usinger (Ed.), *Aquatic insects of California.* Berkeley: Univ. of California Press. pp. 293–371.

Leech, H. B., and H. W. Sanderson, 1959. Coleoptera. In W. T. Edmondson (Ed.), *Freshwater biology.* New York: Wiley. pp. 981–1023.

Thiele, H.-U., 1977. *Carabid beetles in their environments.* Berlin: Springer-Verlag. 369 pp.

The Hymenoptera-Mecoptera Branch

This branch may be divided into two subbranches: (1) the Hymenoptera, which have retained an almost archaic type of bladelike or sawlike ovipositor, but have developed a highly specialized wing in which many longitudinal veins fused but the crossveins did not; and (2) the Mecoptera and its allies, in which the sawlike ovipositor was completely lost but in which the venation of the primitive members changed little from the neuropteroid archetype. The Mecoptera and its four related orders are further subdivided into two additional groups: the Mecoptera–Siphonaptera–Diptera, in which the mandibles become sharp and swordlike; and the Trichoptera–Lepidoptera, in which the wing venation lost most of its crossveins.

Order HYMENOPTERA: Sawflies, Ants, Bees, and Wasps

A major order, this includes many different body shapes with a size range from 0.1 mm in minute parasitic forms to at least 50 mm in some of the wasps (Fig. 8-108). Integument is heavily sclerotized, and the pleural sclerites are considerably coalesced. Mouthparts are of the chewing type, in many forms modified for lapping or sucking. Wings are well developed, reduced, or absent; if well developed, they are transparent, the two pairs similar in texture, and without scales; they have a great range in venation. Legs of primitive forms have the base of the femur set off as an extra trochanter-like segment. Generalized forms (Fig. 8-109) have a considerable reduction and coalescence of veins, but there are a moderate number of crossveins. Antennae range from 3- to about 60-segmented, and are of many

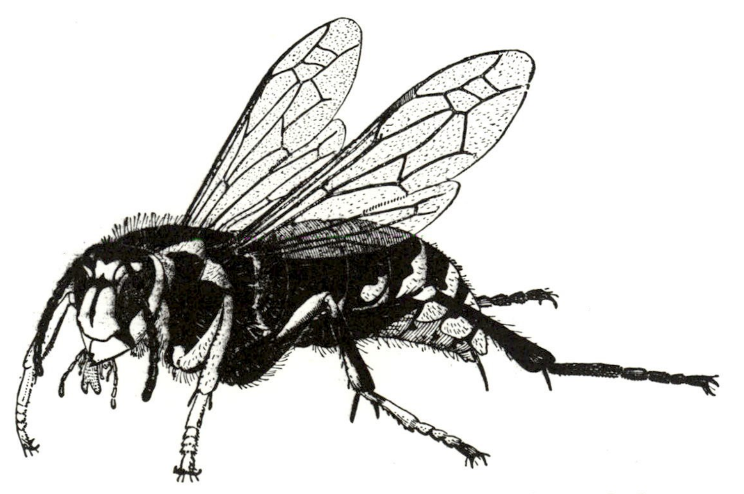

Fig. 8-108. The bald-faced hornet *Vespula (Dolichovespula) maculata*, a representative hymenopteran. (From Illinois Natural History Survey)

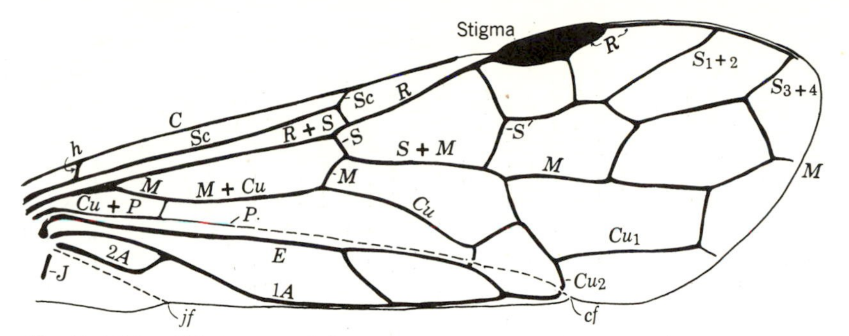

Fig. 8-109. Diagram of hymenopterous wing, combining primitive veins of several archaic families.

shapes. Larvae are caterpillar-like or grublike, all having a distinct head and chewing mouthparts, some with thoracic or abdominal legs or both, and others without any legs.

In number of known species, the Hymenoptera is one of the largest orders of insects. It also contains groups such as the ants, wasps, and bees that are extremely abundant in numbers of individuals and therefore conspicuous and common elements of the fauna. These forms, however, are advanced members of the order.

Of living Hymenoptera, the most primitive are plant-feeding sawflies of the relatively rare families Xyelidae and Pamphiliidae (Fig. 8-110). In these, the larvae have antennae possessing up to seven segments and well-developed legs on each thoracic segment. An early offshoot of this line evolved into the typical sawflies, all leaf feeders, including the common families Argidae, Tenthredinidae, Cimbicidae, and Diprionidae. In all of these and the Xyelidae, the larvae have a pair of legs on each of several abdominal segments; because of this they are often mistaken for caterpillars of Lepidoptera. In these groups, the female cuts slits in leaves with her saw, and lays an egg in each slit. The slit heals over and the egg is then encased in plant tissue.

Another early line arising from presumably a pre-xyelid ancestor evolved into stem borers, comprising the group called the horntails. In these, the saw became narrow, fitted for drilling into harder plant tissue, and the larval legs became greatly reduced. The family Cephidae appears to be the most primitive member of this line; other members include the hard-bodied Siricidae and Xiphydriidae, and the peculiar rare family Orussidae whose larvae are parasitic on wood-boring beetle larvae.

The sawflies and horntails constitute the suborder Symphyta, in which the abdomen is joined broadly to the thorax (Fig. 8-111*A*, *B*).

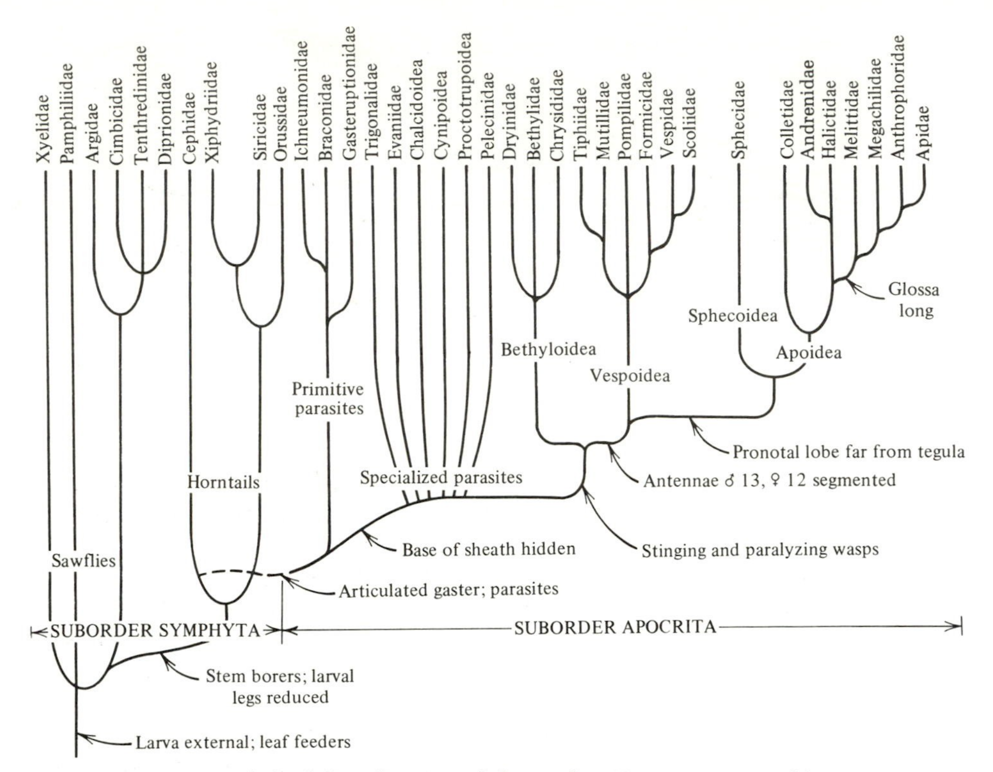

Fig. 8-110. Simplified family tree of the order Hymenoptera. (Arrangement of stinging wasps modified from Brothers, 1975)

In the other and much larger portion of the Hymenoptera, the suborder Apocrita, the first abdominal segment is fused solidly with the metathorax and there is a hingelike articulation between the first and second abdominal segments (Fig. 8-111*D*), the rest of the abdomen becoming a single flexible unit called the *gaster*. The larvae of the Apocrita are primarily parasites or predators on immature stages of other insects and are legless. The ovipositor of the female adult is round and slender, adapted to pierce substrates in which the host larvae live, to penetrate the host itself, or to sting.

From which group of Symphyta the Apocrita arose is far from settled. The Apocrita may well have arisen from the base of the horntail line. It seems plausible that in its progress through host plant tissue, the stem-boring larva of cephid or horntail might have overtaken and eaten larvae of related species; that this relationship became habitual; and that finally it became established as the normal food relationship of the species. The parasitic habit probably originated in this way. That some such change occurred within the horntails, resulting in the parasitic Orussidae, strengthens this supposition regarding the Apocrita. The hinged gaster of the Apocrita is only a slight modification of that found in the Cephidae, suggesting that an ancestral branch of the latter may have given rise to the Apocrita.

The earliest known Apocrita are today represented by primitive parasites including the rare Gasteruptionidae and the abundant Braconidae and Ichneumonidae. In these the base of the saw and sheath are exposed, much as in the sawflies. In a line from some early ancestor of these groups arose a form in which the seventh

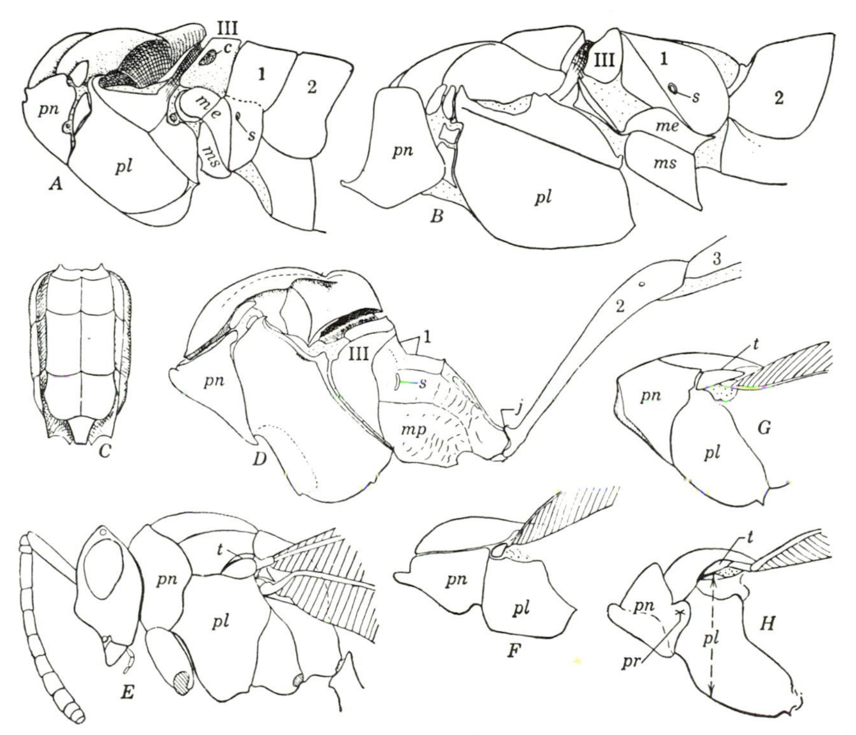

Fig. 8-111. Diagnostic characters of Hymenoptera. (*A*) Thorax of *Arge*, Argidae; (*B*) thorax of *Janus*, Cephidae; (*C*) venter of abdomen of *Chrysis*, Chrysididae; (*D*) thorax of *Eremotylus*, Ichneumonidae; (*E*) head and thorax of *Chalcis*, Chalcididae; (*F*) thorax of *Proctotrupes*, Proctotrupidae; (*G*) thorax of *Ancistrocerus*, Vespidae; (*H*) thorax of *Sceliphron*, Sphecidae. *c*, Cenchrus; *j*, basal articulation of gaster; *me*, metaepimeron; *mp*, metapleuron; *ms*, metaepisternum; *pl*, mesopleuron; *pn*, pronotum; *pr*, pronotal lobe; *s*, first abdominal spiracle; *t*, tegula. *III*, Metanotum. 1, 2, 3, Segments of abdomen.

sternite grew forward, concealed the base of the saw or ovipositor, and became a ventral trough or guide for it. This form gave rise to a large group of specialized parasites, including the Cynipoidea, Chalcidoidea, Proctotrupoidea, and several distinctive but smaller families such as the Pelecinidae, Evaniidae, and Trigonalidae. It is a peculiar circumstance that independently in several lines the basal segment of the femur re-fused with the apical portion and the number of antennal segments decreased. In both the Cynipoidea and Chalcidoidea, plant-feeding forms arose; the other members are typical parasites whose larvae kill their hosts by the simple mechanism of gradually eating their tissues. In all of these parasites, the female does not paralyze the prey. These are often called the Terebrantia.

From some specialized parasite, a new kind of parasitic habit evolved. The female uses her ovipositor to sting and immobilize or paralyze the host larvae, then lays an egg on it. The prey is thus kept alive while the egg hatches and the larva commences feeding; in addition, the immobilized prey cannot move into a new situation detrimental to the wasp larva. Most of these wasps prey on insect larvae that live in protected situations, such as soil-inhabiting beetle larvae.

The most primitive of these stinging forms (the Aculeata) are a group of diverse families comprising the superfamily Bethyloidea, and including the Bethylidae and the cuckoo wasps (Chrysididae). From the bethyloid ancestor there arose a lineage in which the antennal segments became stabilized at the number of 13 in the male, 12 in the female. This lineage gave rise to the remainder of the Hymenoptera, the superfamilies Vespoidea, Sphecoidea, and Apoidea.

The Vespoidea evolved into a most diverse assemblage of families, including the velvet ants (Mutillidae), the more typical solitary wasps Tiphiidae and Scoliidae, the ants (Formicidae), and the vespids (Vespidae), including a number of subfamilies that are social insects.

From the ancestor of the Vespoidea arose a line in which the pronotum evolved a posterior lobe that was widely separated from the tegula (Fig. 8-111*H*). This line gave rise to the solitary wasps (Sphecoidea) and the bees (Apoidea). In the primitive Sphecoidea, as in some Vespidae, the females first construct a nest (probably excavated in the ground), then hunt and sting the prey, bring it to the nest, and lay an egg on it. In the more highly evolved Sphecidae, the provisions are normally procured before the egg is laid.

A branch of the typical sphecoid line gave rise to the bees. The major changes in this evolution were a switch in food from insects or spiders to pollen, and the evolution of branched hairs on the body and appendages associated with gathering pollen. Although most of the bees are solitary, social or partially social life has evolved in many groups, including the bumblebees and honeybees (Apidae).

The evolution of the higher Hymenoptera (the wasps and bees) therefore appears to have occurred chiefly through changes in behavior patterns without drastic accompanying changes in external morphology.

KEY TO SUBORDERS AND COMMON FAMILIES

1. First abdominal segment solidly joined with second, at most a shallow constriction between them, first tergite forming a distinct plate or pair of plates (Fig. 8-111*A*, *B*) (suborder **Symphyta**, p. 418) 2
   Juncture of first and second abdominal segments constricted to form a ball-and-socket joint (Fig. 8-111*D*); the first tergite is fused solidly to the thorax, and the remainder of the abdomen forms an articulating unit called the gaster (suborder **Apocrita**, p. 420) 9
2. Antenna 3-segmented (Fig. 8-112*A*), the third sometimes split longitudinally to form a lyre-shaped prong (Fig. 8-111*B*) .... **Argidae**
   Antenna at least 6-segmented, the end segment never cleft (Fig. 8-112*C* to *I*) 3
3. Third antennal segment at least as long as combined length of the succeeding 9 segments, the segments beyond the third forming a slender terminal filament (Fig. 8-112*E*) .... **Xyelidae**

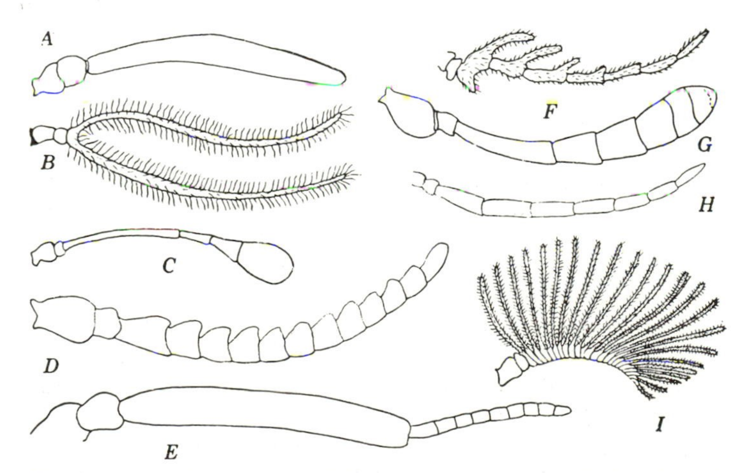

Fig. 8-112. Antennae of Hymenoptera. (*A*) *Sofus* female, Argidae; (*B*) *Sphacophilus* male, Argidae; (*C*) *Trichiosoma*, Cimbicidae; (*D*) *Augomonoctenus* female, Diprionidae; (*E*) *Pleroneura*, Xyelidae; (*F*) *Cladius* male, Tenthredinidae; (*G*) *Tenthredo*, Tenthredinidae; (*H*) *Pseudodineura*, Tenthredinidae; (*I*) *Monoctenus* male, Diprionidae.

Third antennal segment not longer than the combined length of the next 3 or 4 segments, or antenna clavate (Fig. 8-112*C*) 4

4. Antenna capitate (Fig. 8-112*C*); lateral edge of abdomen sharp and angular; large robust species (Fig. 8-115) .... **Cimbicidae**
   Antenna pectinate, serrate, filiform, or in a few species as clavate as Fig. 8-112*G*; lateral edge of abdomen round 5
5. A shallow but distinct constriction between first and second abdominal tergites, and cenchri absent (Fig. 8-111*B*) .... **Cephidae**
   No constriction between first and second abdominal tergites, and cenchri (*c*) well developed, forming a pair of velvety pads on the metanotum (Fig. 8-111*A*) 6
6. Front tibia having only one apical spur .... **Siricidae**, p. 420
   Front tibia having two apical spurs 7
7. Antenna 7- to 9-segmented (Fig. 8-112*F* to *H*) ....
   .... **Tenthredinidae**, p. 418
   Antenna having 10 or more segments (Figs. 8-112*D*, *I*) 8
8. Antenna narrow and filiform, proportioned as in (Fig. 8-112*H*)
   .... **Tenthredinidae**, p. 418
   Antenna serrate in females (Fig. 8-112*D*), pectinate in males (Fig. 8-112*I*) .... **Diprionidae**, p. 420
9. Petiole composed of two segments, usually one or both bearing a dorsal hump or node (Fig. 8-125) .... **Formicidae**, p. 428
   Petiole consisting of only one segment (Figs. 8-126, 8-127) 10
10. First segment of gaster forming an isolated petiole bearing a dorsal node or projection (Fig. 8-126); includes winged and wingless forms .... **Formicidae**, p. 428
    First segment of gaster either expanded posteriorly or not bearing a dorsal node 11
11. Wings completely atrophied or reduced to small pads 12
    Wings well developed, reaching to or beyond middle of abdomen ...
    13
12. Body fuzzy with dense hair (Fig. 8-123) .... **Mutillidae**, p. 426
    Body smooth or with only inconspicuous hair. A few species in each of several families of parasitic habit, keyed no further here.
13. Front wing without a stigma (a thickened area along the anterior margin of the wing), and with sclerotized venation reduced to a single anterior vein, sometimes with a "tail" at its tip (Fig. 8-113*D*), sometimes completely atrophied 14
    Front wing either having a definite stigma or having a more extensive venation (Fig. 8-113*F*) 15
14. Lateral corner of pronotum extending to the tegula (Fig. 8-111*F*). Several families of small parasitic wasps, chiefly .... **Proctotrupoidea**
    Lateral corner of pronotum not extending to the tegula (Fig. 8-111*E*)
    .... **Chalcidoidea**, p. 423
15. Pronotum having each posterolateral corner forming a round earlike or epaulet-like lobe ending below level of tegula (Fig. 8-111*H*) 16
    Pronotum having posterolateral corner truncate or angulate, often abut-

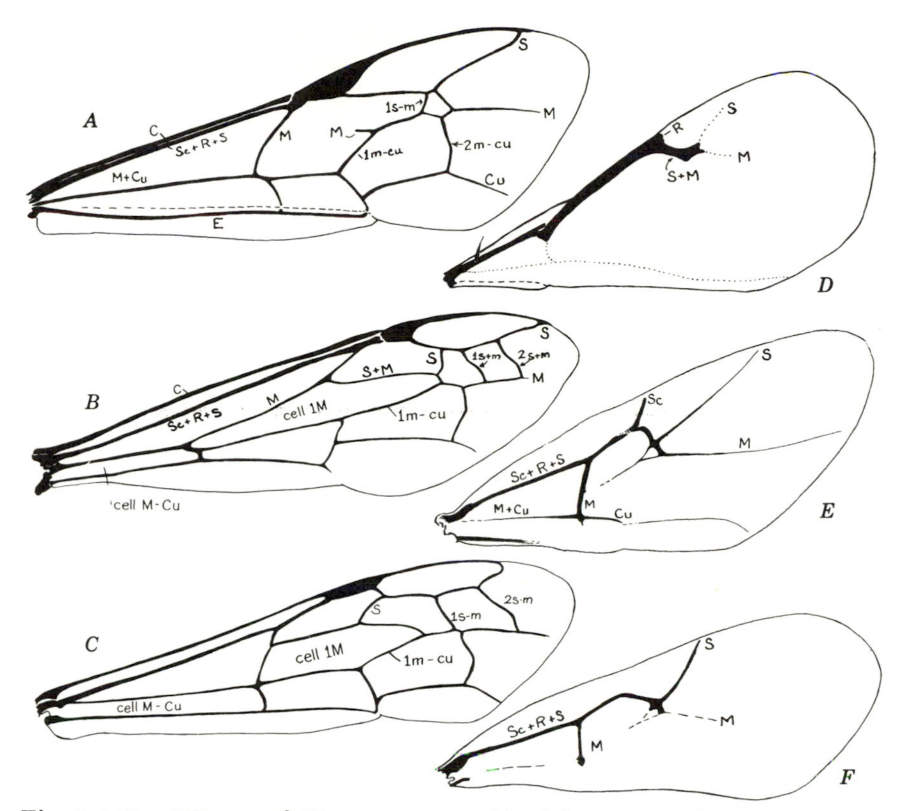

Fig. 8-113. Wings of Hymenoptera. (*A*) Ichneumonidae; (*B*) *Vespa*, Vespidae; (*C*) *Myzine*, Tiphiidae; (*D*) *Tetrastichus*, Chalcididae; (*E* and *F*) two types of Cynipidae.

ting tegula (Fig. 8-111G), or rounded but not earlike (Fig. 8-114*A*) 23

16. Hind basitarsus no wider than succeeding segments, the plantar surface clothed only with dense, short pile; body and appendages without branched hairs, each hair simple, neither branched nor fringed (Sphecoidea) .... **Sphecidae**, p. 430
    Hind basitarsus slightly wider (in numerous forms many times wider) than succeeding segments, the plantar surface with moderately long and abundant hair; body and appendages having branched or spiral hairs; each hair has many branches or whorls and may appear fringed (Fig. 8-129) (*Apoidea*, p. 432) 17
17. Head having two sutures below each antenna, delimiting a large subantennal sclerite (Fig. 8-114*H*) .... **Andrenidae**
    Head having only one suture below each antenna (Fig. 8-114G) ... 18

18. Labium with mentum and submentum virtually absent; front wing with free basal part of *M* (basal vein) usually strongly curved .... **Halictidae**
Labium with mentum and submentum present (Fig. 8-114*E*, *F*); front wing with basal vein straight 19
19. Labium with glossa short and its apex either rounded, truncate, or incised (Fig. 8-114*E*) .... **Colletidae**
Labium with glossa elongate and narrow or pointed at apex (Fig. 8-114*F*) 20
20. Labial palp with segments similar and cylindrical, as in Fig. 8-114*E* .... **Melittidae**
Labial palp with first two segments elongate and sheathlike (Fig. 8-114*F*) 21
21. Front wing with 3 submarginal cells (crossveins 1*s*–*m* and 2*s*–*m* both present as in Fig. 8-113*C*) .... **Apidae**
Front wing with 1 or 2 submarginal cells (one or both of these crossveins lacking) 22
22. Labrum longer than wide and widened at extreme base .... **Megachilidae**
Labrum wider than long or narrowed at extreme base .... **Apidae**
23. Front wing having first 2 longitudinal veins running very close together, obliterating the costal cell (Fig. 8-113*A*), and antenna with more than 16 segments 24
Either front wing with an open costal cell between first 2 longitudinal veins (Fig. 8-113*B*, *C*) or antenna with no more than 14 segments 25
24. Front wing having crossvein 2m–*cu* (Fig. 8-113*A*) .... **Ichneumonidae**, p. 421
Front wing lacking crossvein 2*m*–*cu* .... **Braconidae**, p. 422
25. Gaster with only 3 apparent dorsal segments, the tergites heavily sclerotized; the 3 large sternites each divided longitudinally into a pair of concave armored plates (Fig. 8-111*C*). Robust, hard, shining metallic wasps capable of curling up into a ball. Cuckoo wasps .. **Chrysididae**
Gaster with at least 5 apparent dorsal segments 26
26. Front wings having no definite stigma, but instead a clear triangular area bounded posteriorly by a vein (Fig. 8-113*E*, *F*) but without an anterior vein .... **Cynipoidea**, p. 424
Front wings having a thickened stigma, or an anterior vein (Fig. 8-113*B*, *C*) 27
27. Gaster extremely elongate and slender (Fig. 8-113*C*); head and thorax black and shining. Females only (males are rare) of *Pelecinus polyturator* .... **Pelecinidae**
Gaster much shorter and thicker; texture and color of various sorts 28
28. Corner of pronotum ending below level of wing and not in vicinity of tegula; mesopleuron almost always having a straight, fine transverse suture at about its midpoint (Fig. 8-114*A*); although in some species it is difficult to detect .... **Pompilidae**

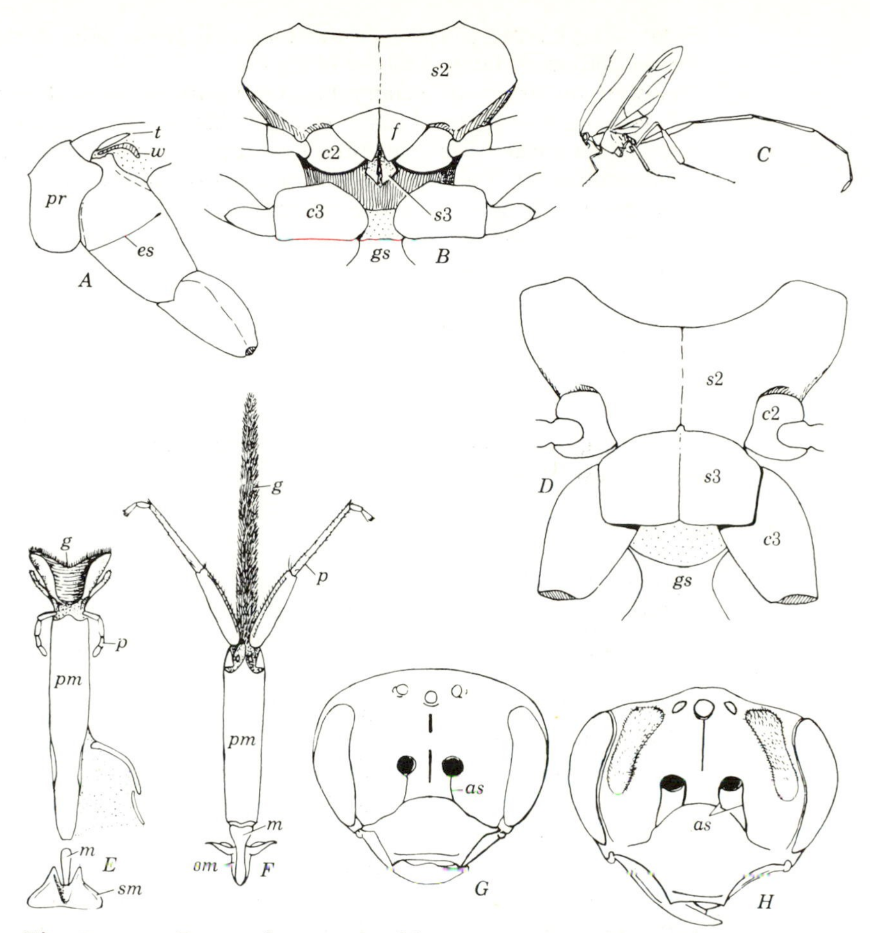

Fig. 8-114. Parts of wasps and bees. (*A*) Mesopleuron of *Ceropales*, Pompilidae; (*B*) mesosternum of *Myzine*, Tiphiidae; (*C*) outline of *Pelecinus*, Pelecinidae; (*D*) thoracic sterna of *Scolia*, Scoliidae; (*E*) labium of *Colletes*, Colletidae; (*F*) labium for *Anthidium*, Megachilidae; (*G*) head of *Halictus*, Halictidae; (*H*) head of *Andrena*, Andrenidae. *as*, Antennal sutures; *c*, coxa; *es*, mesopleural suture; *f*, flaplike processes of mesosternum; *g*, glossa; *gs*, base of gaster; *m*, mentum; *p*, palpus; *pm*, prementum; *pr*, pronotum; *s*, sternum; *sm*, submentum; *t*, tegula; *w*, wing base. (*E* to *H* after Michener)

Corner of pronotum ending at or above level of wing and usually abutting against the tegula (Fig. 8-111G); mesopleuron with a crooked suture or none 29

29. Front wing having cell 1*M* longer than cell *M–Cu* (Fig. 8-113*B*); either wings pleated lengthwise when folded or antennae clublike .... **Vespidae**, p. 426

Front wing having cell 1*M* shorter than cell *M–Cu* (Fig. 8-113*C*), or former cell open due to atrophy of 1*m–cu* 30

30. Metasternum large and rectangular, fused with the mesosternum, the two forming a large, level plate overlying the bases of the 4 posterior coxae; hind coxae wide apart (Fig. 8-114*D*) .... **Scoliidae**, p. 425

 Metasternum small and triangular, or inconspicuous, not fused with mesosternum; hind coxae close together (Fig. 8-114*B*) 31

31. Hind margin of mesosternum produced into a pair of triangular plates partially overlapping the bases of the middle coxae (Fig. 8-114*B*) .... **Tiphiidae**, p. 426

 Hind margin of mesosternum without such lobes 32

32. Second segment of gaster large and bulbous compared with third segment (Fig. 8-123); body often conspicuously hairy .... **Mutillidae**, p. 426

 Second segment of gaster not markedly wider than third segment; body never conspicuously hairy .... **Tiphiidae**, p. 426

SUBORDER SYMPHYTA

SAWFLIES AND HORNTAILS. The Symphyta, with the exception of the small parasitic family Orussidae, are a plant-feeding group. The larvae either feed externally on foliage or mine in leaves, leaf petioles, or stems. The adults of many groups feed on the pubescence of the host plant, cropping it by means of their sickle-shaped mandibles as a cow does grass; in other groups they may be predaceous on smaller insects or feed on nectar and pollen. The group is a large one; the North American forms represent 12 families and include in their host selection a great diversity of plant groups.

Distinguishing features of sawfly larvae are: a distinct head, with simple chewing mouthparts; antennae slender or platelike with one to seven segments; eyes with only a single lens; and abdominal legs (when present) without hooks or crochets.

The adults of most sawflies are compact and fairly robust. Of the leaf-feeding families, the largest common species is *Cimbex americana* (Fig. 8-115), in which the antennae are capitate; the males and females are differently colored. The females of all but a few species have a well-developed saw used to cut egg slits in leaves or petioles.

The Tenthredinidae is the largest family, characterized chiefly by the simple 9- to 16-segmented antennae. Most of the species are external leaf feeders, and among them are several of economic importance, such as the rose slug *Endelomyia aethiops*; the imported currant worm *Nematus ribesi* (Fig. 8-116) and the larch sawfly *Pristiphora erichsoni*. In certain species the larvae mine in the leaf tissue, for example, *Heterarthrus nemorata*, one of the birch leaf miners. Species of other genera, including *Euura*, produce true galls.

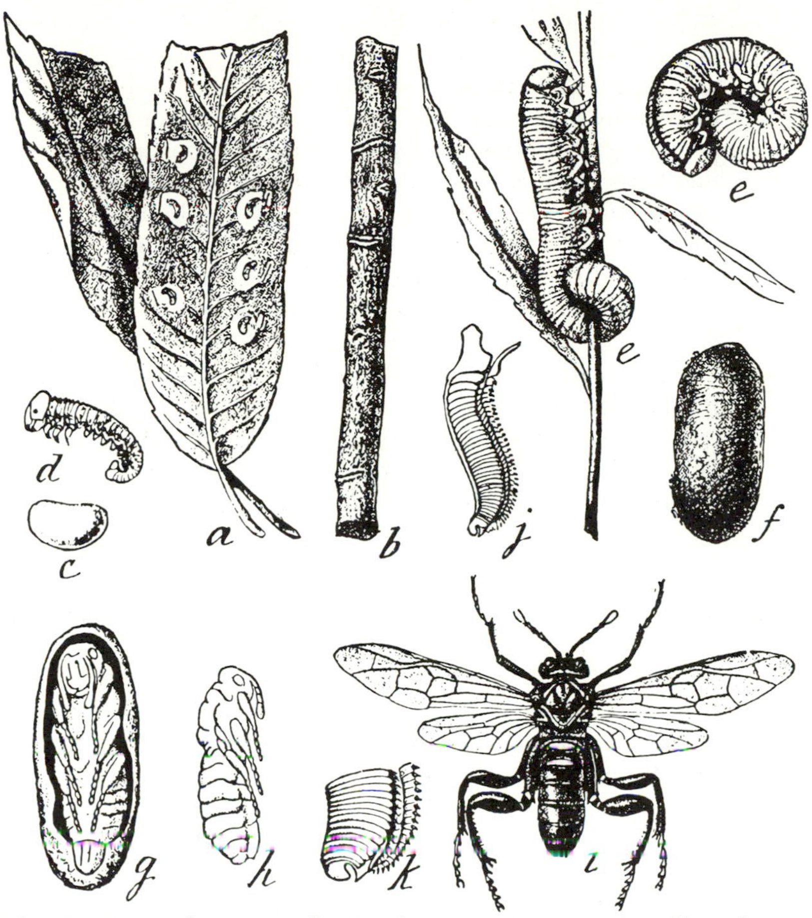

Fig. 8-115. A large sawfly *Cimbex americana*. (*a*) Willow leaves showing location of eggs; (*b*) twig showing incisions made by adult; (*c*) egg; (*d*) newly hatched larva; (*e*) mature larvae; (*f*) cocoon; (g) open cocoon showing pupa; (*h*) pupa, side view; (*i*) mature sawfly; (*j*, *k*) saw of female. (After Riley)

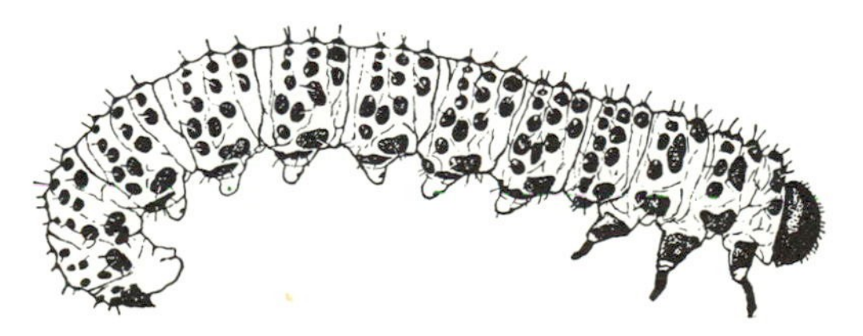

Fig. 8-116. The feeding-stage larva of the imported currant worm *Nematus ribesi*. (From Connecticut Agricultural Experiment Station)

The Diprionidae is another economically important family. In this family the antennae are at least 13-segmented, serrate in the female and pectinate in the male (Fig. 8-117). All the species are stout and more or less drab in color. The larvae are caterpillar-like and external feeders on coniferous needles. Many species are among the major defoliators of spruce and pine forests. Of special note are the ravages caused to spruce in northeastern America by the European spruce sawfly *Diprion hercyniae*. Especially injurious to young pines is the redheaded pine sawfly *Neodiprion lecontei* (Fig. 8-117).

The Siricidae contain some of the largest members of the suborder. They are elongate, sometimes attaining a body length of 40 mm. The larvae bore in tree trunks and are cylindrical and almost legless. Both adults and larvae have a horny spikelike projection at the posterior end of the body, the character that gives them the name "horntails." *Tremex columba* (Fig. 8-118) is a common species attacking maple, elm, beech, oak, and some other deciduous trees.

SUBORDER APOCRITA

ANTS, BEES, AND WASPS. In general the Apocrita are more graceful, active, and more rapid of movement than the Symphyta. The larvae are chiefly internal or external parasites, or are fed by the adults, or make plant galls. They are legless, have a distinct exposed head capsule bearing greatly reduced mouthparts and antennae, and frequently exhibit hypermetamorphosis (p. 238).

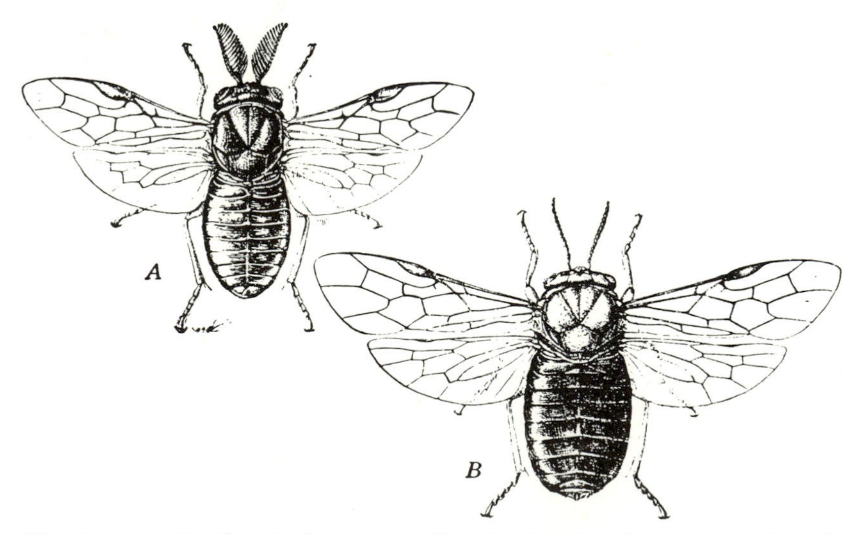

Fig. 8-117. Redheaded pine sawfly *Neodiprion lecontei*. (*A*) Male; (*B*) female. (From U.S.D.A., E.R.B.)

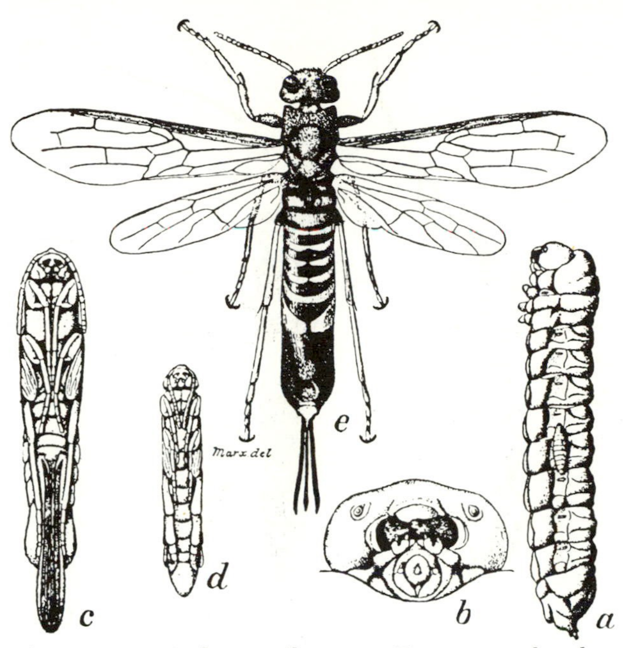

Fig. 8-118. A horntail wasp *Tremex columba.* (*a*) Larva; (*b*) larval head, ventral aspect; (*c*, *d*) female and male pupa; (*e*) adult. Note small parasite larva attached to horntail larva. (After Riley)

In the Apocrita the first segment belonging to the abdomen is fused solidly with the thorax. Hence what appears to be the thorax includes a segment of the abdomen, and what appears to be the abdomen really does not have its anterior segment. These two body regions are called the *mesosoma* and *metasoma*, respectively. In older literature the latter is called the gaster.

Primitive Parasites

Ichneumonidae, The Ichneumon Wasps. These usually slender wasps have long and many-segmented antennae (Fig. 8-119) and have subcosta fused with the stem of radius on the front wings. All members of this family are parasites on insects or spiders. Their usual hosts are the larvae of Lepidoptera, for example, *Glypta rufiscutellaris*, a parasite of the oriental peach moth. In addition, a number of ichneumon wasps parasitize the larvae of Coleoptera, Hymenoptera, and Diptera, and a few other insects. The adult ichneumon wasp female deposits its eggs on or inside the body of the host. If the eggs are laid on the epidermis of the host, the newly

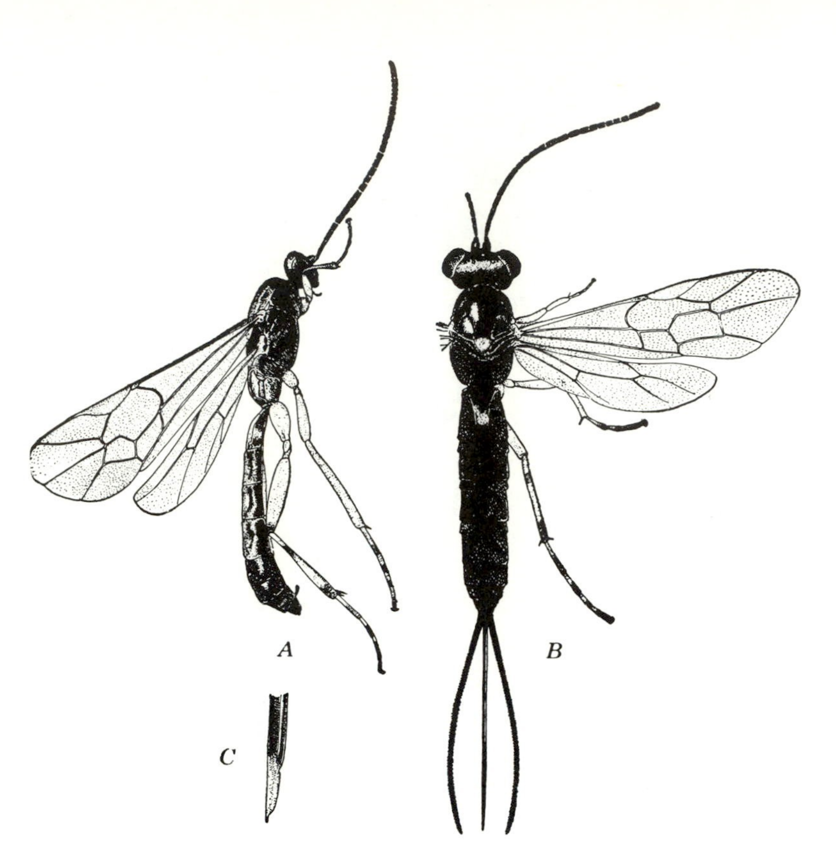

Fig. 8-119. An ichneumon wasp *Glypta rufiscutellaria*. (*A*) Male; (*B*) female; (*C*) tip of ovipositor. (From Connecticut Agricultural Experiment Station)

hatched larvae may bore into the body. The larvae develop into legless grubs that either attach to the outside of the host or develop within the body of the host. When mature, the larvae may pupate within the host or leave it to spin cocoons. Ichneumon wasps have a wide range in size. Many of the small forms only a few millimeters long parasitize small moth larvae.

Braconidae. This is a large family closely related to the ichneumon wasps. The species average smaller than the ichneumon wasps, and many braconids have reduced wing venation. A number of species are important as parasites of economic pests. One of these, *Apanteles melanoscelus* (Fig. 7-14) has been imported for biological control of gypsy moth larvae. This small parasite exhibits the interesting hypermetamorphosis prevalent among most of the parasitic families of Hymenoptera. In Fig. 7-14 are illustrated the different shapes of the larva in various stages of development; the anal vesicle *(a)* may be used to identify the posterior end of the larva.

Specialized Parasites

Superfamily Chalcidoidea, The Chalcid Wasps. These small wasps are sometimes less than a millimeter in length, with a greatly reduced wing venation (Fig. 8-120), and usually with elbowed antennae. These wasps are largely internal parasites, especially of larval Lepidoptera and of the larvae of other parasitic Hymenoptera, which they attack within the body of the primary host. These parasites of parasites are called *hyperparasites*. The few chalcids that are not parasitic develop in various seeds, or in plant stems, especially grasses. To this nonparasitic group belongs the cloverseed chalcid *Bruchophagus platyptera*, whose larva develops in the seeds of clover and alfalfa; the wheat jointworm *Harmolita tritici*, whose larva bores in the stems of wheat; and the wheat strawworm *Harmolita grandis* (Fig. 8-120). Locally and sporadically the wheat jointworm causes serious damage to the crop.

Of unique interest is the specialized life history of certain tiny chalcid wasps belonging to the family Agaontidae. These develop in the seeds of figs. The males are wingless and live only within the fig fruit in which they develop, fertilizing the females even before the latter emerge from the fig seed. The females are winged and fly from flower to flower in search of suitable seeds for oviposition, carrying the pollen on their bodies and pollinating each flower visited. This is the only method by which figs are pollinated. Many commercial varieties of figs do not require pollination to develop their fruits, but the fruit of the choice Smyrna fig will not develop without pollina-

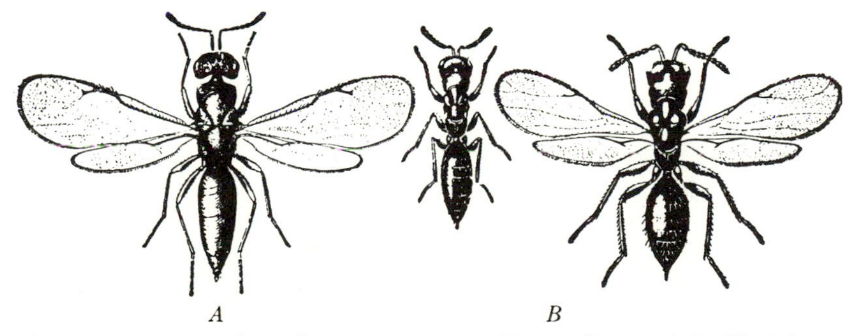

Fig. 8-120. (*A*) The wheat jointworm *Harmolita tritici*. (*B*) Wingless and winged form of the wheat strawworm *Harmolita grandis*. *(From U.S.D.A., E.R.B.)*

tion. In order to grow these in North America it was necessary to introduce the fig wasp *Blastophaga psenes* to effect pollination.

Superfamily Cynipoidea, the Gall Wasps. These are small wasps, most of them characterized by the large triangular cell near the apex of the front wing (Fig. 8-113*E*) and by the deep but bilaterally compressed abdomen. Many groups of the superfamily are parasitic on dipterous larvae, aphids, and other insects, but the best-known group produces galls on plants. As a matter of fact, the gall wasps themselves are seldom seen, but every naturalist is familiar with some of the many different types of galls that are produced on the leaves, stems, or roots of oak, roses, and other plants by the larvae of these insects. One of these is shown in Fig. 8-121. There are hundreds of species of gall wasps, nearly every species producing a different type of gall. Some species that live on oaks have an alternation of generations, with one generation producing a gall on the roots and the alternate generation making a gall on the leaves or twigs.

Stinging Wasps

In the terebrantian line that evolved into the stinging wasps, the ovipositor became a stinging organ, and its associated glands secreted a venom discharged in the act of stinging. Against enemies such as mammals, this venom produces a painful sensation, and at times is lethal. When injected into the prey (usually insects or spiders),

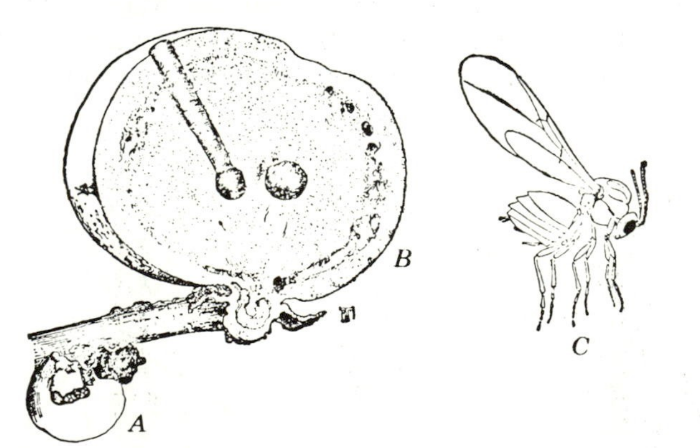

Fig. 8-121. A cynipid (*C*) and its gall; (*A*) immature gall and (*B*) a section of mature gall showing cells and exits. (From Essig, 1942, *College entomology*, by permission of Macmillan Co.)

this secretion typically causes complete motor paralysis without death resulting. Such a paralysis is used by wasps that lay an egg on the prey or provision a nest with prey. The induced paralysis has a triple advantage: It keeps the prey edible until the wasp larva hatches and begins feeding, it ensures that the prey will not move away from the legless wasp larva, and it prevents the attraction of scavenger insects to the odor of dead insects.

The stinging wasps, usually called the Aculeata, evolved into a diverse array of forms that can be designated as the following four superfamilies.

Superfamily Bethyloidea. Here belong a group of primitive forms including the Bethylidae, which superficially resemble ants and which parasitize a variety of insects, and the Chrysididae (cuckoo wasps), metallic green or blue in color, most of which are external parasites of wasp or bee larvae.

Superfamily Vespoidea. In this and the remaining Hymenoptera the antennal segments have become stabilized at 12 in the female, 13 in the male. Sociality has twice evolved independently. The family Scoliidae is typical of a generalized member of the Vespoidea.

SCOLIIDAE, THE SCOLIID WASPS. These fairly large insects have wings in both sexes, and most of the species are black or are banded or spotted with black and yellow, such as *Scolia dubia* (Fig. 8-122). The female wasps dig through the soil in search of their prey, white grubs (larvae of the beetle family Scarabaeidae). When the female encounters a suitable host larva, she stings it, thereby paralyzing it; digs a crude

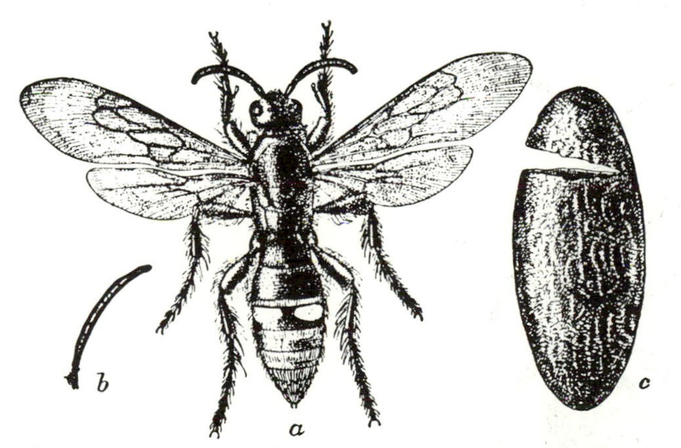

Fig. 8-122. A scoliid wasp *Scolia dubia*. (*a*) Female wasp; (*b*) antenna of male; (*c*) cocoon showing escape opening. (From U.S.D.A., E.R.B.)

cell around it; lays an egg on the doomed larva; and then moves on in search of another victim. The egg soon hatches into a legless grub that attaches to the paralyzed beetle larva and begins eating it. Within a period of about 2 weeks, the wasp larva has consumed the host and is full grown. It then spins a cocoon in the earthen cell and usually passes the winter in this stage. The next spring or summer the larva pupates, and later the adult chews its way out of the cocoon and digs to the surface.

MUTILLIDAE AND TIPHIIDAE. Closely related to the Scoliidae are many families of somewhat similar habits. Species of the family Mutillidae are parasites of wasps and bees. In many mutillids the females are wingless and have a close rsemblance to ants. The Mutillidae females, however, lack the "node" on the petiole of the gaster (Fig. 8-123); and in addition are covered with dense velvety or silky pile. From this latter character the family has received the name velvet ants. These velvet ants have a powerful sting and use it freely if interfered with.

Many oriental species of *Tiphia*, of the related family Tiphiidae, have been brought to the United States and propagated for parasitizing grubs of the Japanese beetle, and a few have shown definite promise of assisting in the control of the beetle.

VESPIDAE, YELLOW JACKETS AND HORNETS, WASPS. This family contains species varying from 10 to about 30 mm in length, many of them having elaborate yellow and black or white and black markings (Fig. 8-108). In all subfamilies except the Masarinae the wings in repose are folded longitudinally like a fan. In the Masarinae the antennae are capitate. Most of the Vespidae are solitary in habit. The adults make a burrow in wood or soil or construct a pottery container

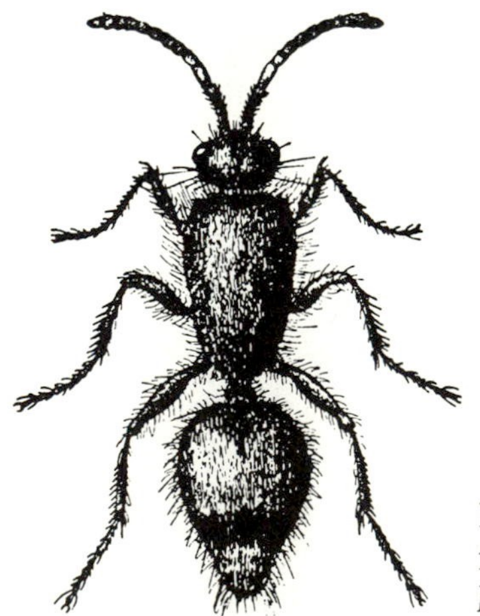

Fig. 8-123. A mutillid wasp female *Dasymutilla bioculata*. (After Washburn)

for the abode of the grub. The nest is usually stocked by the adult with paralyzed caterpillars or with pollen and honey. Certain of the Vespidae are social in habit. By masticating wood fibers with an oral secretion, they produce a paper that they fashion into a platelike or baglike nest. The most familiar of these are the platelike nests of *Polistes* (Fig. 8-124), which are made up of a single horizontal comb of larval cells. These are commonly found hanging from eaves of buildings and in similar sheltered places. Colonies of *Polistes* rarely have over a few dozen members. The largest colonies of vespids found in North America are made by the bald-faced hornet *Vespula (Dolichovespula) maculata*. These colonial nests, oval in shape and with an opening at the bottom, are most often attached to tree branches. Each contains several layers of larval cells, or combs, arranged one above the other. The workers in a colony forage for insect prey, such as flies and caterpillars. These are crushed and mangled by the wasps and fed to the maggot-like larvae in the cells of the nest.

In temperate regions of North America the colonies die out at the end of autumn. In late summer a brood of males and females is produced. With cold weather, the workers and males die; the au-

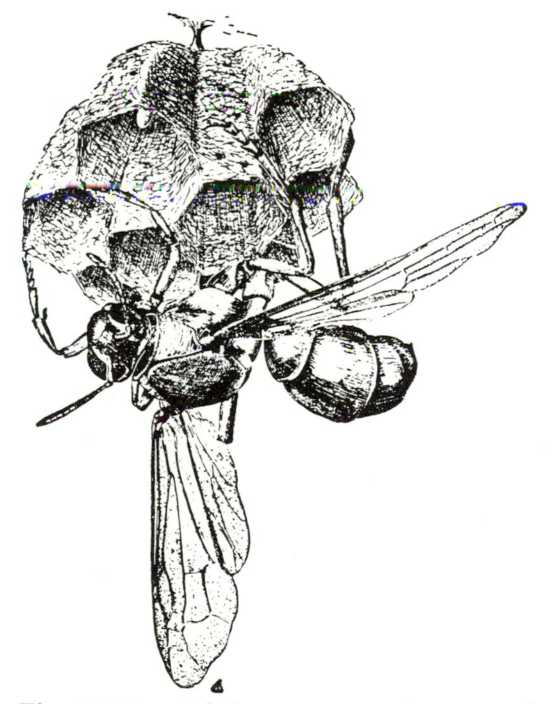

Fig. 8-124. Adult queen and nest of *Polistes*. *(From U.S.D.A., E.R.B.)*

tumn brood of females, by this time fertilized, hibernate in rotten logs or stumps. These females emerge the following spring and begin new colonies.

FORMICIDAE, ANTS. In these insects the first segment of the gaster forms a petiole or stalk and bears a dorsal projection or node (Figs. 8-125, 8-126). This structure differentiates ants from other antlike wasps. In addition to the normal males and females, ant species usually have a third form, the nonreproductive workers, which are always wingless. These workers are the ants we usually see scurrying about. They perform most of the work of the colony, such as building the nest, excavating the subterranean chambers, and gathering food for the colony.

A typical colony starts with the swarming flights of the winged reproductive males and females. At periodic intervals (frequently once a year) large numbers of winged males and females are produced in an established colony. When weather conditions are favorable, the sexual forms leave the nest as a swarm, embark on their nuptial flight, and mate in flight. The male dies soon after mating.

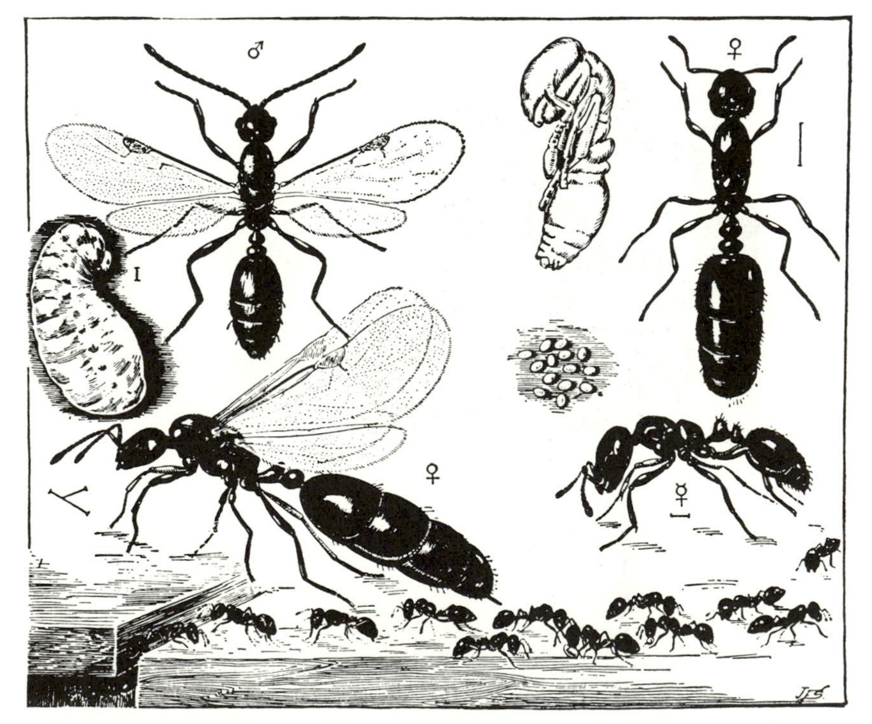

Fig. 8-125. The little black ant *Monomorium minimum*, showing several stages and activities. (After Marlatt, U.S.D.A., E.R.B.)

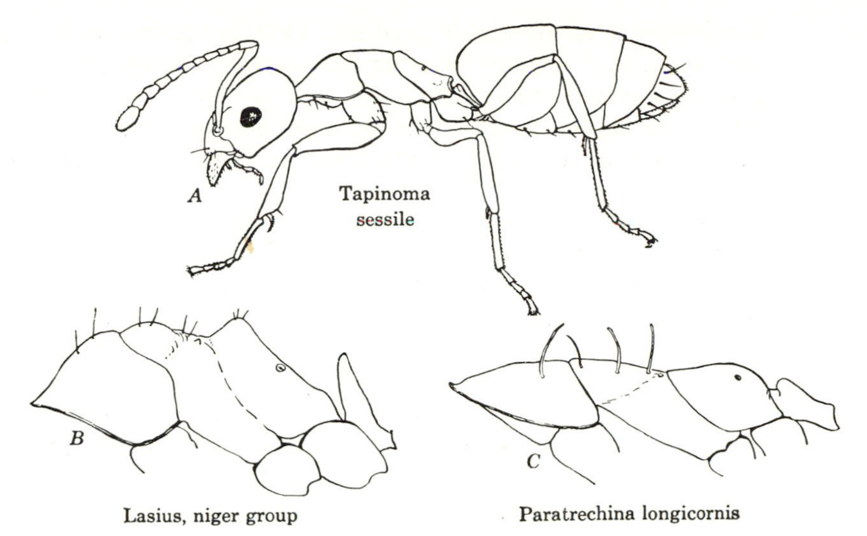

Fig. 8-126. Lateral aspect of three ant genera to illustrate the dorsal node on the petiole. Note that in *A* and *C* this node is small but distinct.

The fertilized female seeks a suitable nest site in the ground, an old log, or other situation, bites off her wings, and seals up a small hollow, which becomes the first nest chamber. The female remains in this chamber for several weeks, during which time the eggs are laid and the grublike larvae are fed to maturity by the female. The food is produced apparently from the histolysis of the wing muscles and the fat body of the female. It is extruded from her mouth as a secretion. The mature larvae pupate and soon emerge as small workers. These break out of the nest chamber, seek food, and henceforth keep the female, or queen, and the next brood of workers provided with food. Subsequent broods help with the task of keeping the colony provisioned. The female continues to lay eggs, without further fertilization, for several years.

Colonies of many species contain only a few dozen or a few hundred individuals, whereas those of other species may attain a population of many thousands. The small colonies are usually situated under stones, in stumps, logs, or in galleries in the soil. Many of the large colonies build large mounds of earth, sticks, and debris, interspersed with a complex system of galleries and chambers.

In the main, ants are omnivorous, feeding on living or dead animal matter (especially other insects), vegetable substances such as fungi, and exudates or secretions of plants, such as nectar, wound discharges, and glandular products. Certain insects such as aphids and some scale insects produce honeydew or other secretions; the ants tend these insects with great care and "harvest" the secretions produced. Insect honeydew is an excellent food source, containing sugars, amino acids, and many other nutrient dietary requirements of ants.

Several ant species invade houses or stores and are among the most persistent domestic pests. In the northern states the thief ant *Solenopsis molesta*, Pharaoh ant *Monomorium pharaonis*, and the odorous house ant *Tapinoma sessile* are common household species. In the southern states the introduced Argentine ant *Iridomyrmex humilis* is an exceedingly common household pest. The introduced fire ant *Solenopsis invicta*, which inflicts a painful bite and sting, is a pest in pastures, but is a valuable predator on boll weevil. It preys on other ant species and locally has almost replaced the native ant populations. Other southern ants of great interest include the leaf-cutting ants (e.g., *Atta texana*), which cut leaves of trees into small sections, carry these to the nest where they form fungus gardens. The ants eat the fungi.

Some of the ants feed primarily on plant seeds. Of these ants, various species of the genus *Pogonomyrmex*, known as "agricultural ants," have become abundant and are destructive in the grain and grass areas of the Great Plains and westward.

OTHER VESPOID WASPS. Several small families of wasps are closely related to the Scoliidae and Vespidae. Many of these families are rare or their members are small and secretive, and seldom collected. One is frequently seen: the green, metallic cuckoo wasps or Chrysididae that lay their eggs in the burrows of other kinds of wasps, and whose larvae feed on the provisions stored by the host wasp.

Solitary Wasps and Bees

In these wasps and the bees, each corner of the pronotum ends in a round lobe that is situated below the level of, and does not touch, the tegulae.

Superfamily Sphecoidea, the Solitary and Digger Wasps. In these insects the hairs are simple and undivided, in contrast to those of the bees, and the hind tibiae and tarsi are not modified for pollen gathering. The group is a large one and includes a great diversity of sizes, shapes, and colors. As currently defined it comprises the heterogeneous and large family Sphecidae. The habits of nearly all Sphecidae are essentially the same. The female wasp makes a mud nest or a nest excavated in pith, wood, or soil and provisions it with a particular kind of paralyzed prey. An egg is deposited in each stocked compartment of the nest, this egg hatching into a legless grub that feeds on the provender stored for it by the parent. In temperate climates the larva overwinters in its cocoon, pupating the following year in early summer.

The various wasp groups are usually specific in the prey they choose for provisioning the larval cells. The Pemphredoninae (often only 2 or 3 mm long) capture aphids, the Sphecinae usually use caterpillars, the Trypoxylinae use spiders, and so on. One of the most interesting and showy species in the central and eastern states is the cicada killer *Sphecius speciosus*. This is a large black and yellow species that attains a length of 40 or 50 mm. It commonly captures and paralyzes the common cicada *Tibicen linnei* and carries it to a burrow in the ground, provisioning each burrow with one cicada. When attacked by the wasp, the cicada makes a loud piercing noise, but this subsides in a moment as the cicada is stung and becomes paralyzed.

The most familiar of the solitary wasps are the thread-waisted mud daubers, especially species of *Sceliphron* (Fig. 8-127). These

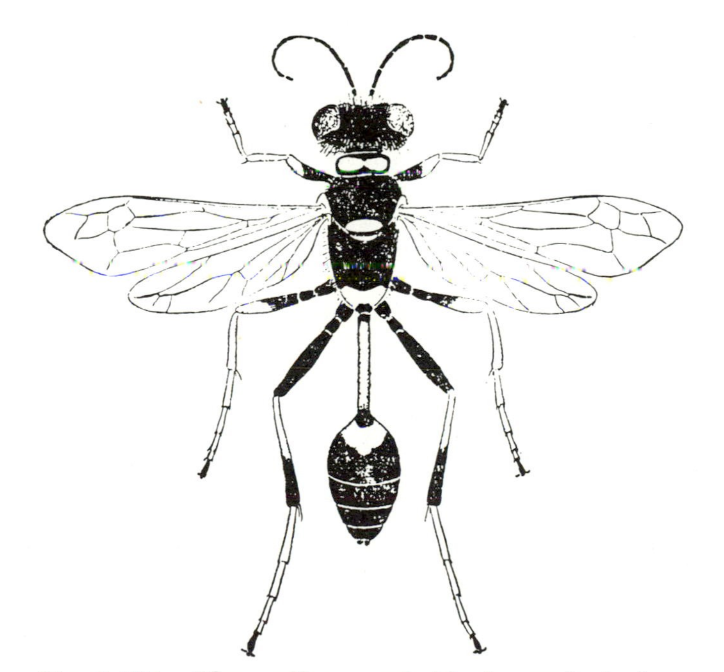

Fig. 8-127. The yellow and black mud dauber wasp, *Sceliphron servillei*, which builds series of mud cells on stones and walls and provisions them with spiders. (From Essig, *Insects of Western North America*, by permission of Macmillan Co.)

mud daubers build a mud nest of several cells. The nests are common under bridges, eaves of houses, or other sheltered places. The wasps gather the mud at the edge of nearby pools or puddles, flying back and forth from water's edge to nest with mouthfuls of "plaster." The cells are provisioned with spiders.

Superfamily Apoidea, The Bees. This group (Fig. 8-128) includes all the native and domestic bees, comprising the six families keyed on p. 415. They are morphologically very similar to the Sphecidae, differing in having branched body hairs (Fig. 8-129), which give them a fuzzy or velvety appearance. Typically they have the hind legs or the venter of the abdomen provided with areas of fairly long hairs used in collecting pollen; in the males of some genera and in both sexes of parasitic genera these structures are lacking. Most of our 3000 species are solitary in habit, making cells in burrows or cavities as do many solitary wasps. The bees provision the nest cells with honey and pollen, which constitutes the food of the larvae. Certain genera of bees are "parasitic," laying their eggs in the cells or nests of other bees. The intruder larva matures faster than the host larva and eats up the stored food. These "parasitic" bees are "cuckoo bees" or "cowbird bees."

Several genera of the North American fauna, including the bumblebees (the genus *Bombus*, family Apidae), have developed social living. In the bumblebees, fertilized females overwinter in log cavities or ground cover. They emerge in spring, find a protected site in a hole in the ground or in a deserted mouse nest, and begin a colony. A brood of eggs and young is raised, the mother, or queen, feeding the larvae on honey. This brood is composed of sterile females or workers that take over the task of gathering food for suc-

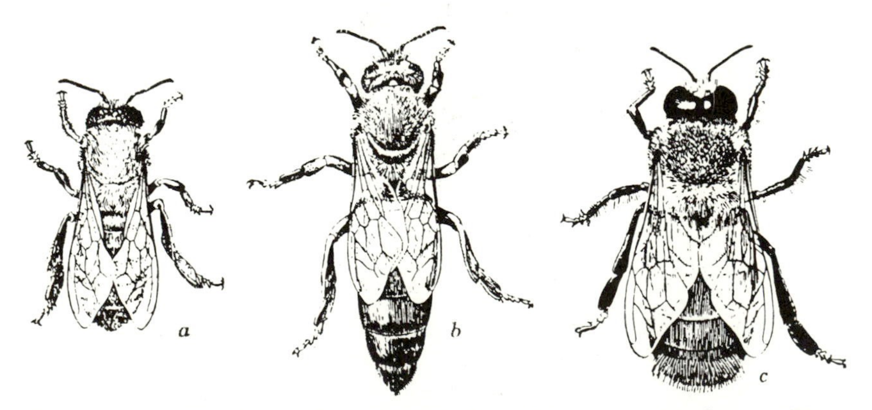

Fig. 8-128. The honeybee. (*a*) Worker; (*b*) queen; (*c*) drone. (From U.S.D.A., E.R.B.)

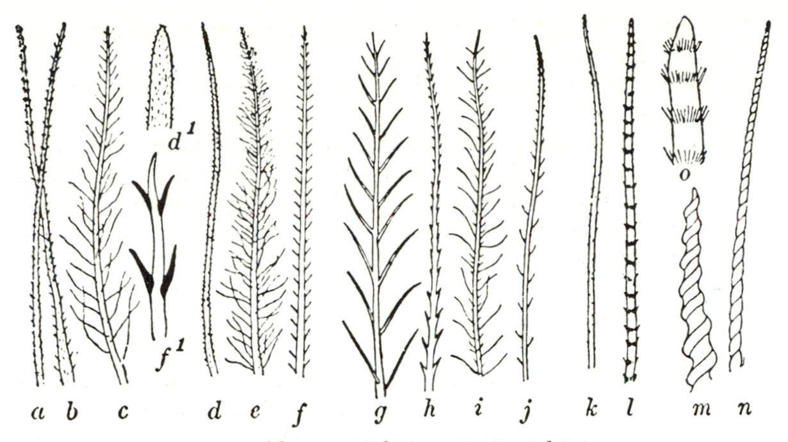

Fig. 8-129. Hairs of bees. (After J. B. Smith)

ceeding larvae. Toward fall no more workers are produced, but instead a swarm of males and functional females. The males die soon after mating, and the workers also die with the approach of winter. The new brood of fertilized queens disperses for hibernation, and the entire colony is disbanded.

The habits of the domestic honeybee *Apis mellifera* are much more specialized than those of the bumblebee. In the first place the colonies do not die out during the winter; their members live during this period on honey stored up throughout the summer. Individual queens have lost the ability to forage for themselves; hence they cannot start a new colony alone but must be accompanied by some workers from the parent colony. The phenomenon of "swarming" is the colonization flight in which the old queen sets out to form a new nest.

The honeybee is of considerable importance in that it affords a large cash return from the sale of honey. The greatest role of bees, including native species and the honeybee, is the pollination of a great variety of wild and domestic plants, including most commercial fruits and legume crops.

References

Crozier, R. H., 1977. Evolutionary genetics of the Hymenoptera. *Annu. Rev. Entomol.*, **22**:263–288.

Evans, H. E., and M. J. W. Eberhard, 1970. *The wasps*. Ann Arbor: Univ. of Michigan Press. 265 pp.

Krombein, K. V., P. D. Hurd, Jr., D. R. Smith, and B. D. Burks, 1979. *Catalog of Hymenoptera in America north of Mexico*. Washington, D.C.: Smithsonian Inst. Press. 3 vols., 2736 pp.

Malyshev, S. I., 1968. *Genesis of the Hymenoptera and phases of their evolution*. London: Methuen. 319 pp.
Matthews, R. W., 1974. Biology of Braconidae. *Annu. Rev. Entomol.*, **19**:15–32.
Michener, C. D., 1944. Comparative external morphology, phylogeny, and a classification of the bees. *Bull. Am. Mus. Nat. Hist.*, **82**:151–326.
——— 1953. Comparative morphological and systematic studies of bee larvae with a key to the families of hymentoperous larvae. *Univ. Kans. Sci. Bull.*, **35**:987–1102.
——— 1974. *The social behavior of the bees. A comparative study*. Cambridge, Mass.: Harvard Univ. Press. 404 pp.
Michener, C. D., and A. Fraser, 1978. A comparative anatomical study of mandibular structure in bees. *Univ. Kans. Sci. Bull.*, **51**:463–482.
Mitchell, T. B., 1960–1962. Bees of eastern United States. *N.C. Agric. Exp. Stn. Tech. Bull.*, **141**:1–538; **152**:1–557.
Muesebeck, C. F. W., et al., 1951. *Hymenoptera of America north of Mexico, synoptic catalogue*. USDA, Agric. Monogr. Vol. 2, 1420 pp.
Richards, O. W., 1977. *Hymenoptera. Introduction and key to families*, 2nd ed. *Handbook for the identification of British insects*. London: Royal Entomological Society. Vol. 6, No. 1, 100 pp.
Ross, H. H., 1937. A generic classification of the Nearctic sawflies. *Ill. Biol. Monogr.*, **15**(2):1–173.
Spradberry, J. P., 1973. *Wasps. An account of the biology and natural history of social and solitary wasps*. Seattle: Univ. of Washington Press. 416 pp.
Townes, H., 1969. The genera of Ichneumonidae. *Mem. Am. Entomol. Inst.*, **11**:1–300; **12**:1–537; **13**:1–307.
Viereck, et al., 1916. The Hymenoptera of Connecticut. *Bull. Conn. State Geol. Nat. Hist. Surv.*, **22**:1–824. 10 plates, 15 figs.

Order MECOPTERA: Scorpionflies

The adults, ranging in size from small to medium, either have two pairs of large net-veined wings (Fig. 8-130), or have the wings short or aborted. The antennae are long; eyes large; and legs slender, in some families long and spindly. The mouthparts are of the chewing type, often situated at the end of a snoutlike elongation of the head. The larvae are grublike or caterpillar-like, always with thoracic legs and in some groups having abdominal larvapods also (Fig. 8-131).

The adults are omnivorous, feeding chiefly on small insects, but supplementing their diet with nectar, pollen, petals, fruits, and mosses. The winged forms are active fliers. The males of the Panorpidae (Fig. 8-130) have a large bulbous genital capsule that resembles to some extent the abdomen of a scorpion, and from this the order derives its name "scorpionflies."

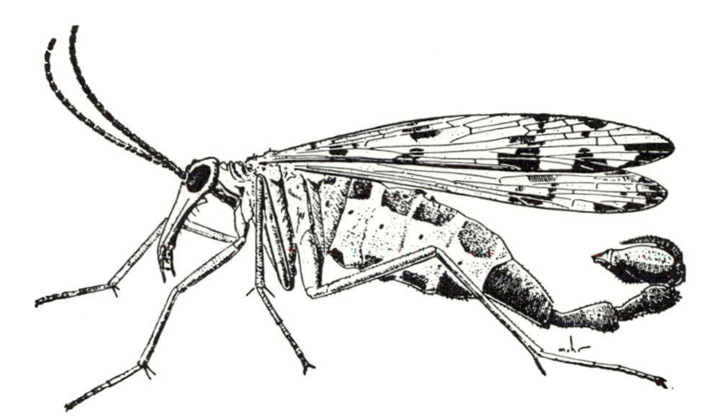

Fig. 8-130. A male scorpionfly *Panorpa chelata*. (From Illinois Natural History Survey)

The eggs are ovoid and are laid in or on the ground, either singly or in clusters of 100 or more. The larvae live in moss, rotten wood, or the rich mud and humus around seepage areas in densely wooded situations. Their food consists of various types of organic matter. Pupation occurs in the soil. There is only one generation per year.

In the small species found in the genus *Boreus* (family Boreidae), the adults have small short wings and, since they mature in winter or early spring, are often found running about on the snow. The robust, grublike larvae live in the rotting bases of moss mats. Because they differ from other Mecoptera in many details of both adults and larvae, the Boreidae are considered a separate order Neomecoptera by some authors. The structure of the mouthparts, however, indicates that the Boreidae are a specialized offshoot related to the families Panorpodidae and Panorpidae, and are not a distinctive order.

KEY TO FAMILIES

1. Ocelli lacking; wings always well developed, oval, less than 3 times as long as wide, with very reticulate venation. Contains only the rare eastern *Merope tuber* .... **Meropeidae**

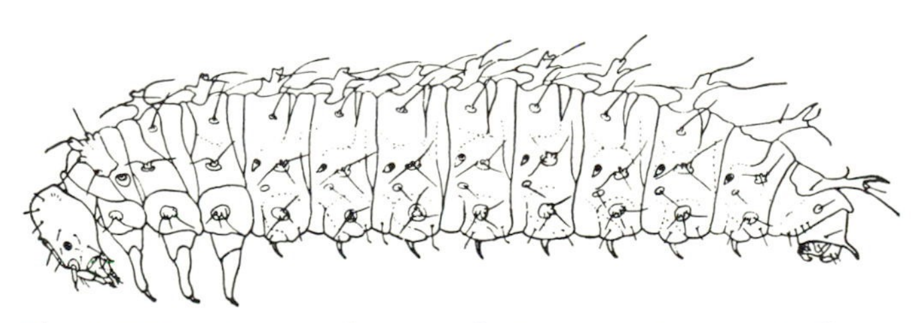

Fig. 8-131. Larva of *Apterobittacus apterus*. (Redrawn from Applegarth)

Ocelli present; if well developed, wings more than 3 times as long as wide, with only moderately reticulate venation (Fig. 8-130) 2

2. Tarsus raptorial, with a single claw . . . . **Bittacidae**

   Tarsus with 2 claws 3

3. Wings, whether well developed or short, with a well-defined and complete set of veins 4

   Wings reduced to small oval pads or short, hard, tapered structures, devoid of venation. Contains *Boreus* and *Hesperoboreus* . . . . **Boreidae**

4. Head produced into a ventral beak as long as in Fig. 8-130. Sole N. Am. genus, *Panorpa* . . . . **Panorpidae**

   Head short, without a ventral beaklike prolongation. Sole N. Am. genus, *Brachypanorpa* . . . . **Panorpodidae**

References

Byers, G. W., 1954. Notes on North American Mecoptera. *Ann. Entomol. Soc. Am.*, **47**:481–510.

——— 1965. Families and genera of Mecoptera. *Proc. 12th Int. Congr. Entomol.*, 1964. 123.

——— 1971. Ecological distribution and structural adaptation in the classification of Mecoptera. *Proc. 13th Int. Congr. Entomol., 1968.* 486.

Carpenter, F. M., 1931. A revision of the nearctic Mecoptera. *Mus. Comp. Zool. Harvard Coll. Bull.*, **76**:206–277.

Hinton, H. E., 1958. The phylogeny of the panorpoid orders. *Annu. Rev. Entomol.*, **3**:181–206.

Penny, N. D., 1975. Evolution of extant Mecoptera. *J. Kans. Entomol. Soc.*, **48**:331–350.

Order SIPHONAPTERA: Fleas

All adults are wingless (Fig. 8-132), have distinctive laterally compressed bodies, and have long stout spiny legs and short clubbed

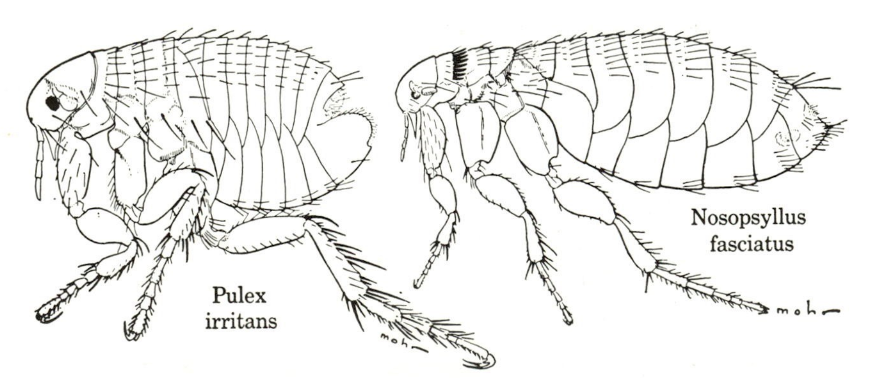

Fig. 8-132. Two common fleas: the human flea and the rat flea. (From Illinois Natural History Survey)

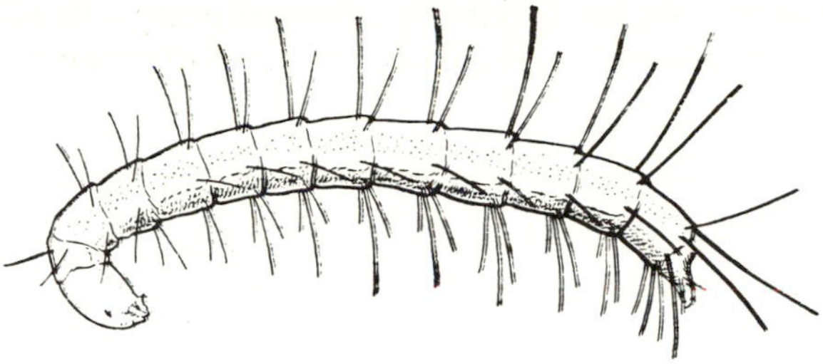

Fig. 8-133. Larva of a flea. (From Illinois Natural History Survey)

antennae that in repose fit into a depression along the side of the head. The mouthparts are fitted for piercing skin and sucking blood and consist of a beak, a pair of palps, and a pair of short bladelike maxillae. These insects are small, most of them 2 to 4 mm, but a few species attain a length of 6 to 8 mm. The larvae are slender and wormlike (Fig. 8-133). They have round heads, no legs, and long hairs on each body segment. The segments of thorax and abdomen are similar in appearance. The mouthparts are minute and inconspicuous, of the chewing type. The pupae are formed in a cakelike cocoon made of earth or grass; they have rudimentary wings.

Fleas feed on the blood of mammals or birds and are found on the body of the host or in the nest or runways of the host. The eggs are minute, whitish, and oval. They are dropped by the female either on the host or in the nest. The eggs have no adhesive, so that, if laid on the host animal, they slip through the hair and fall into the litter of the nest or retreat, where they soon hatch. The larvae live in the soil or debris in the nest and feed on anal secretions of the adults. When full grown, the larvae spin an irregular cocoon in the nest of the host and pupate. The eggs, larvae, and pupae are seldom seen except by diligent search.

The adult fleas of most species are extremely active. They slip through hair or feathers with great ease; and in many species the body has combs of spines that further aid this progress. In some species the adults stay on the body of the host almost all the time; in others the adults stay in the nest and get on the host only for feeding periods. For this reason it is necessary to collect both from the bodies and in the nests of the hosts to be sure of finding all the species connected with them.

The order Siphonaptera is not a large one. The North American fauna includes about 60 genera and over 200 species. The great percentage of these occur with native mammals, especially various kinds of mice, shrews, ground and tree squirrels, gophers, and rabbits. Bears, beavers, coyotes, and many others support a flea fauna. A few fleas prey on birds and occasionally are a pest of domestic fowl.

KEY TO COMMON FAMILIES

1. Thorax extremely short, all three segments combined shorter than the first abdominal tergum (Fig. 8-134) .... **Tungidae** *(Hectopsyllidae)*
   Thorax considerably longer than first abdominal tergum (Fig. 8-132) 2
2. Abdominal terga with only a single row of setae (Fig. 8-132), *Pulex*; when telescoped, abdomen therefore appears to have a series of similar rows of setae .... **Pulicidae**
   Abdominal terga with a double row of setae, the basal row composed of short setae, the apical row of long setae (Fig. 8-132) *Nosopsyllus*; when telescoped the abdomen appears to have an alternation of short-spined rows and long-spined rows of setae 3
3. Head without black teeth, armed only with setae and the genal process (Fig. 8-132), *Nosopsyllus* .... **Ceratophyllidae** *(Dolichopsyllidae)*
   Head with at least one, and usually with a comb of long black teeth (actually flat, black, toothlike setae) along ventral margin or on genae, like those on prothorax of *Nosopsyllus* 4
4. Head with two black teeth situated at tip. Bat fleas ... **Ischnopsyllidae**
   Head with either more than two teeth, forming a comb, or with teeth situated some distance from tip. Not on bats ...... **Hystrichopsyllidae**

PULICIDAE. Many species of this family are of especial importance to humans. The dog and cat fleas, *Ctenocephalides canis* and *C. felis* and the human flea *Pulex irritans* attack humans and invade dwellings, causing great discomfort and inconvenience. The bites inflicted by the fleas are painful, usually cause hard itching swellings, and may be the source of secondary infections through scratching. Most important of the fleas are certain rat species, particularly *Xenopsylla cheopis* (and *Nosopsyllus fasciatus* of the Ceratophyllidae), which transmit the dread bubonic plague from rats to humans. In their role as disseminators of this disease, which many times has spread like wildfire through crowded cities in many parts of the world, fleas have been responsible for millions of human deaths. At present, public health organizations all over the globe keep close watch on the rat-flea-plague focal points, in an effort to break up new incipient outbreaks of the disease, through control of the rats and fleas.

TUNGIDAE. Two species of Tungidae, the sticktight and chigoe fleas, have unusual habits in contrast with other fleas. The females of both species are minute and attach themselves firmly to the host, feeding more or less continuously. The sticktight flea *Echidnophaga gallinacea* (Fig. 8-134) attacks domestic fowl, attaching to the face and

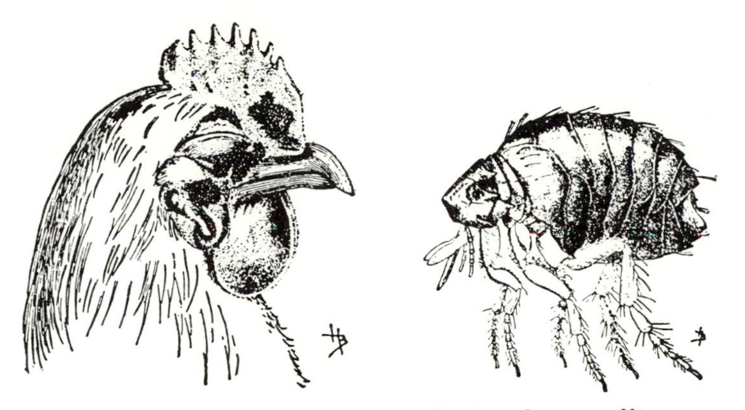

Fig. 8-134. The sticktight flea *Echidnophaga gallinacea*. Left, infested head of rooster (dark patches are clusters of fleas); right, adult female. (From U.S.D.A., E.R.B.)

wattles. The chigoe *Tunga penetrans* attacks humans, especially on the feet. It burrows beneath the skin and forms a painful swelling out of which protrudes only the end of the flea's abdomen. Both species drop eggs to the ground and have typical flea metamorphosis.

References

Dunnet, G. M., 1970. Siphonaptera. In D. F. Waterhouse, et al. (Eds.), *The insects of Australia*. Carlton, Victoria: Melbourne Univ. Press. pp. 647–655.

Dunnet, G. M., and D. K. Mardon, 1974. A monograph of Australian fleas (Siphonaptera). *Aust. J. Zool. Suppl. Ser.*, **30:**1–273.

Ewing, H. E., and I. Fox, 1943. *The fleas of North America*. USDA Misc. Publ. 500. 128 pp.

Fox, I., 1940. *Fleas of eastern United States*. Ames, Iowa: Collegiate Press. 191 pp.

George, R. S. (Ed.), 1974. *Provisional atlas of insects of the British Isles, Pt. 4. Siphonaptera*. Abbots Ripton, U.K.: Inst. Terrestrial Ecol., Monks Wood Exp. Stn.

Haddow, J., R. Traub, and M. Rothschild, 1975. *The distribution of the Ceratophyllidae with an illustrated key to the genera*. Cambridge: Cambridge Univ. Press.

Holland, G. P., 1949. The Siphonaptera of Canada. *Can. Dept. Agric. Tech. Bull.*, **70:**1–306.

——— 1964. Evolution, classification, and host relationships of Siphonaptera. *Annu. Rev. Entomol.*, **9:**123–146.

Hopkins, G. H. E., and M. Rothschild (Eds.), 1953–1971. *An illustrated catalogue of the Rothschild Collection of fleas in the British Museum (Natural History)*. London: British Museum, Natural History. Vols. I–V.

Hubbard, C. A., 1947. *Fleas of western North America.* Ames, Iowa: Iowa State College Press. 532 pp.
Layne, J. N., 1971. Fleas (Siphonaptera) of Florida. *Entomologist*, **54**:35–51.
Rothschild, M., 1965. Fleas. *Sci. Am.*, **213**(6):44–53.
1975. Recent advances in our knowledge of the order Siphonaptera. *Annu. Rev. Entomol.*, **20**:241–259.
Rothschild, M., and R. Traub, 1971. A revised glossary of terms used in the taxonomy and morphology of fleas. In G. H. E. Hopkins and M. Rothschild (Eds.), *An illustrated catalogue of the Rothschild Collection of fleas in the British Museum (Natural History).* London: British Museum, Natural History. Vol. 5, pp. 8–85.

Order DIPTERA: Flies

Typical adults (Fig. 8-135) have a single (front) pair of membranous wings, rarely scaled. The wings have few crossveins and a moderate number of veins. The hind wings are represented only by a pair of slender knobbed balancing organs, called *halteres*. Mouthparts are of various types; in some groups they are modified for piercing and sucking, in other groups for rasping and lapping. The body form is diverse. In a few groups the adults are completely apterous. The eyes are usually large; the antennae vary from 3- to 40-segmented. These are holometabolous insects with legless larvae, usually either with a distinct mandibulate head (Fig. 8-146) or with an internal sclerotized skeleton attached to a pair of hooklike mandibles. The pupa is either free or formed within the skin of the third instar larva.

The order Diptera is a large one, including over 15,000 North American species. The food and habitat of the adults are usually very different from those of the larvae.

Adults of many families feed chiefly on nectar and plant sap or on free liquids associated with rotting organic matter. Certain groups,

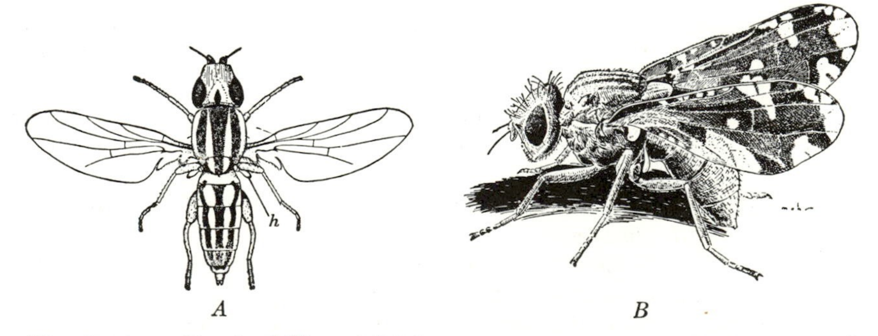

Fig. 8-135. Typical flies. (*A*) *Meromyza americana;* (*B*) *Eurosta solidaginis. h,* Haltere. (From Illinois Natural History Survey)

such as mosquitoes and horseflies, feed on animal blood; these have mouthparts highly modified for piercing and sucking. In a few groups—for example, the botflies—the mouthparts are so vestigial that it is doubtful if the adults take any nourishment.

As a group, fly larvae are moisture loving, the great majority living in water, in rotting flesh, inside the bodies of other animals, in decaying fruit or other moist organic material, or inside living plant tissue. A few live in relatively dry soil or move about exposed to the air, but these are the exceptions rather than the rule.

For the most part, fly eggs are simple, ovoid or elongate, and are normally laid singly, in, on, or near the larval food. Some, such as those of *Drosophila*, have lateral or polar floats that prevent them from sinking into semiliquid food and drowning. Eggs of certain mosquitoes are sufficiently well protected against the elements to withstand months of alternate drying, wetting, and freezing. In some groups, such as the flesh flies and some parasitic flies, the eggs may hatch just before leaving the body of the female and are deposited as minute larvae. This habit is carried to extreme development in the sheep-tick groups (Glossinoidea) where the larvae hatch and grow to their full size in the body of the female. Of unusual interest in the order is the paedogenesis exhibited by some species of midges (see p. 255).

Phylogeny. Living Diptera exhibit an extraordinary range of differences in the larval head and its parts and in the adult antennae, wing venation, mouthparts, and other structures (Fig. 8-136). In the group of families embracing the fungus gnats and craneflies, the condition of these structures approaches the condition found in more primitive insects such as the Mecoptera. These fly families are therefore considered the primitive end of the evolutionary scale in the Diptera and designated the suborder Nematocera.

By adding together these primitive conditions we obtain the following partial reconstruction of the ancestral dipteran. In the larva the head formed a distinct, typical, sclerotized capsule, the mandibles moved sidewise as in a grasshopper, and the abdomen bore small lateral functional spiracles on the first eight segments. In the adult, the antennae were long and composed of many freely articulating segments, and vein *E* extended as a free vein to the wing margin.

All the Nematocera possess the capsule-like larval head, sidewise moving larval mandibles, and primitive adult antennae. Only in the Cecidomyiidae (gall midges) and Mycetophilidae (fungus gnats) are all the larval abdominal spiracles about equal, indicating that these two families represent two of the most archaic lines in the order. In

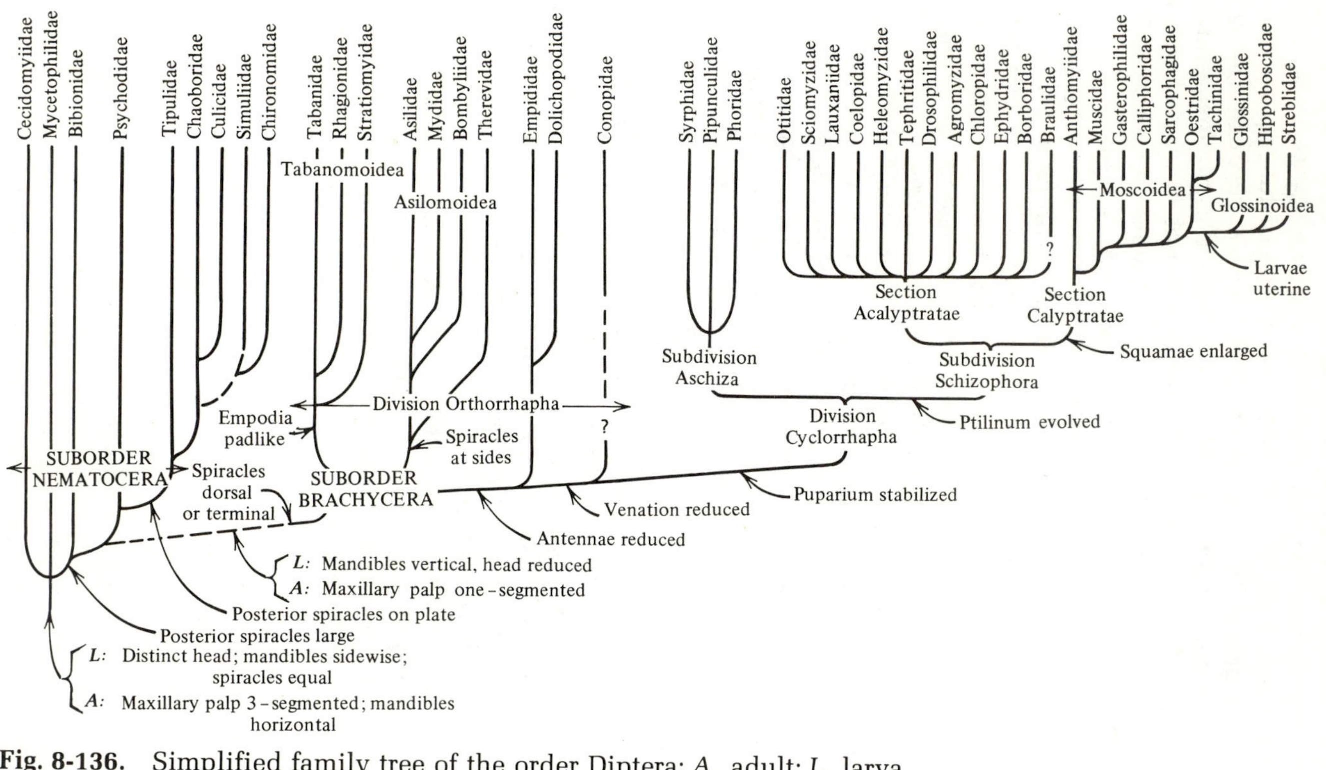

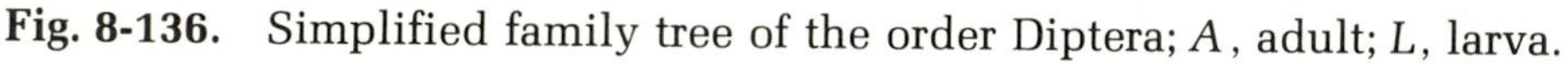

Fig. 8-136. Simplified family tree of the order Diptera; *A*, adult; *L*, larva.

the larvae of the Bibionidae (March fly) and the Psychodidae (moth fly) lineages, the posterior abdominal spiracles became much larger, the others smaller, and respiratory gaseous exchange in the larva was effected chiefly or entirely by the posterior pair of spiracles. Probably from the base of the psychodid line arose the Tipulidae (crane fly) and Culicidae (mosquito) branch, in which anterior larval abdominal spiracles were either extremely minute and functionless or atrophied, and the entire respiratory exchange occurred through the posterior spiracles.

These spiracular adaptations were associated first with larval life in shallow water, mud, or extremely wet organic media. The larva fed on submerged organic material, but kept its posterior spiracles above the water surface–air interface. This larval feeding pattern is the basis for practically all later evolutionary developments of the Diptera. In mosquitoes the larvae evolved into swimmers able to live in water of some depth. In the higher Diptera, the larvae of various families exploited other essentially aquatic media, including the internal tissues of other insects, dung, and semiliquid organic material such as decaying mushrooms and rotting flesh.

In two common families of Nematocera, the Simuliidae (blackflies) and Chironomidae (midges), the larvae became completely aquatic and lost their posterior spiracles. The exact relationship of these families is not known. In the accompanying tree, these families are placed tentatively primarily on similarities to other families with respect to wing venation.

From the base of either the psychodid or tipulid line evolved a dipteran that changed in many respects. The larval mandibles became hooklike and moved up and down, in and out, in a shredding motion; the larval head lost much of its sclerotization; the adult maxillary palps were reduced from three to one segment; and in the wings, veins E and Cu_2 typically met at the wing margin. This marked the beginning of the suborder Brachycera. In one of the primitive lineages including the Asilidae (robber flies) and their allies, the posterior larval spiracles are at the sides of the segment, much as in the nematoceran Bibionidae. The larvae of the asiloid branch are predaceous or parasitic on other insects. In a related lineage, the posterior larval spiracles were close together and situated in a dorsal or terminal depression or cavity, and the segments posterior to them became reduced to small ventral lobes situated under the eighth segment. This ancestor gave rise to two branches: the Tabanidae (horsefly) branch in which the adult tarsal empodia became pulvillus-like; and the Empididae (smoke fly) branch in which the empodia remained slender but the wing venation became reduced. During the evolution of the Brachycera the adult antennae became progressively solidified and reduced. By the time the empidid line arose, the antennae consisted of only three segments, the last one bearing a long style or arista.

In the families discussed up to this point, the pupa is typically free, although in a few isolated families (e.g., Cecidomyiidae and Stratiomyidae) the pupa forms inside the old larval skin, which hardens to form a protective covering or *puparium*. The emerging adult makes a longitudinal, T-shaped dorsal slit in the puparium, through which it emerges. These families form the division Orthorrhapha.

From the base of the empidid branch arose a form in which the formation of the puparium became fixed and the adult emerges by pushing off the front end of the puparium. This line evolved into the division Cyclorrhapha. Its most primitive members constitute the subdivision Aschiza, which includes the Syrphidae (flower flies) and its allies.

In a second line of the Cyclorrhapha an eversible sac or *ptilinum* evolved, which is extruded through flexible sutures above and beside the antennae, and which assists in pushing off the front end of the puparium. This line, the subdivision Schizophora, is the most diverse of the dipteran lineages.

The Schizophora include two large sections. The Acalyptratae is the more primitive line and is exemplified by the Otitidae (picture-winged flies) and the Drosophilidae (vinegar gnats). In the other line, Calyptratae, the connecting membranes at the bases of the wing became enlarged, forming distinct lobes (squamae). Early stages in this development are seen in the Anthomyiidae; more advanced stages occur in the Tachinidae and Sarcophagidae.

In one line of Cyclorrhapha, comprising the superfamily Glossinoidea, the larvae develop in a type of uterine pouch inside the mother fly. The exact origin of the Glossinoidea is not clear but they probably arose as an early branch of the Calyptratae.

One pair of structures of great evolutionary interest are the adult mandibles. The Simuliidae, Psychodidae, Ceratopogonidae, Culicidae, Tabanidae, and Rhagionidae all have adult mandibles. They are bladelike, have typical mandibular attachments and muscles, and in shape are suggestive of the adult mandibles of Siphonaptera and some Mecoptera. The number of families in which mandibles still occur in the adults indicates strongly that the primeval dipteran had these structures, and that they have become atrophied in many branches of the order. They are not present in the lineage that arose from the base of the Tabanidae line and later evolved into the Cyclorrhapha.

1. Abdomen only indistinctly segmented; the coxae of the 2 legs of each segment far apart. Adults living as parasites on birds, mammals, or bees 2
   Abdomen having distinct segments; the 2 legs of each segment held fairly closely together, sometimes the coxae almost contiguous. Not living as ectoparasites in the adult stage 4
2. Mesonotum short, resembling the abdominal segments; minute (1.5 mm long), wingless, parasitic on honeybees. Includes only *Braula caeca* . . . . **Braulidae**, p. 466
   Mesonotum different in appearance from abdominal segments (Fig. 8-159) 3
3. Palpi long and slender, forming a sheath for the mouthparts. Living on birds and mammals with the exception of bats . . . . **Hippoboscidae**, p. 471
   Palpi broader than long, not encasing the mouthparts. Parasitic on bats . . . . **Streblidae**
4. Antenna having more than 3 segments (Fig. 8-140*F* to *H*), not counting a style or arista, borne by the third 5
   Antenna having 3 segments or less; usually the third bears a style (Fig. 8-140*I*) or arista (Fig. 8-140*J*) 21
5. Small mothlike flies, never longer than 5 mm, having body and wings densely clothed with hair or scales; wings having about 10 longitudinal veins, and having crossveins only at extreme base (Fig. 8-137*A*); aquatic or semiaquatic moth flies . . . . **Psychodidae**
   Appearance not mothlike, or venation of a different type 6
6. Mesonotum having a distinct V-shaped suture (Fig. 8-140*A*); elongate species having long legs (Fig. 8-144) . . . . **Tipulidae**, p. 456
   Mesonotum with suture transverse, indistinct, or not present (Fig. 8-140*B*, *C*) 7
7. Antenna having 6 or more well-marked ringlike or beadlike segments (Fig. 8-140*F*) 8
   Antenna having only 4 or 5 segments (Fig. 8-140*G*) or the terminal segments sometimes indistinctly subdivided (Fig. 8-140*H*) 17
8. Wing having cell Cu_2 either open at apex (Fig. 8-137*F*) or lost due to atrophy of veins 9
   Wing having cell Cu_2 entirely at apex by fusion of veins Cu_2 and 1E (Fig. 8-137*H*) 17
9. Wing having both S_{1+2} and M_{1+2} unbranched, the venation fairly parallel (Fig. 8-137*B*) 10
   Wing having either S_{1+2} or M_{1+2} unbranched, the venation frequently markedly divergent (Fig. 8-137*E*) 11
10. Mouthparts forming a long slender beak (Fig. 8-146) . . . . **Culicidae**, p. 456
    Mouthparts not forming a long beak. Mosquitolike midges . . . . **Chaoboridae**

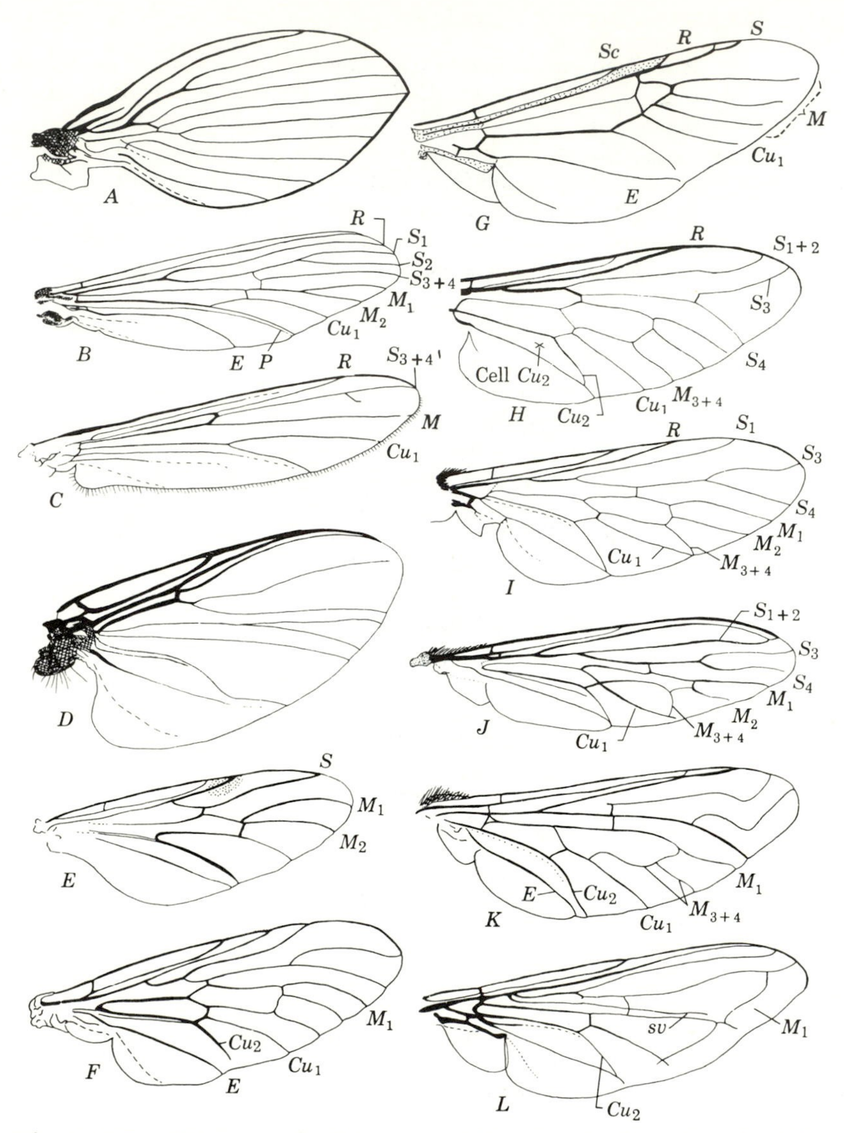

Fig. 8-137. Wings of Diptera. (*A*) *Psychoda*, Psychodidae; (*B*) *Aedes*, Culicidae; (*C*) *Chironomus*, Chironomidae; (*D*) *Simulium*, Simuliidae; (*E*) *Bibio*, Bibionidae; (*F*) *Rhagio*, Rhagionidae; (*G*) *Stratiomys*, Stratiomyidae; (*H*) *Tabanus*, Tabanidae; (*I*) *Thereva*, Therevidae; (*J*) *Asilus*, Asilidae; (*K*) *Anthrax*, Bombyliidae; (*L*) *Scaeva*, Syrphidae. *sv*, Spurious vein. See Chap. 3 for explanation of abbreviations.

11. Ocelli present 12
Ocelli absent 14
12. Anterior margin of wing only slightly more heavily sclerotized than apical and posterior margins .... **Cecidomyiidae**, p. 453
Anterior margin of wing having a sclerotized thickening that stops abruptly at or just beyond juncture with S_{3+4} (Fig. 8-137*G*) 13
13. Antennae inserted below level of eyes (Fig. 8-140*L*); front femur often enlarged; the March flies .... **Bibionidae**
Antennae inserted on a level with middle of eyes; coxae often greatly elongated; the fungus gnats .... **Mycetophilidae**
14. Wing having 2 or 3 strong parallel veins near anterior margin and a group of 6 or 8 oblique very weak veins running from anterior region to or near posterior margin of wing (Fig. 8-137*D*); antennae short, 12-segmented, the last 10 annular and closely knit (Fig. 8-142) .... **Simuliidae**, p. 453
Wing having a different venation, either more veins equally sclerotized, or most of them longitudinal in general course rather than oblique (Fig. 8-137*C*); antennae usually elongate, with well-separated segments 15
15. Anterior margin of wing only slightly more heavily sclerotized than apical and posterior margins .... **Cecidomyiidae**, p. 453
Anterior margin of wing having a sclerotized thickening that stops abruptly at or just beyond juncture with S_{3+4} (Fig. 8-137*C*) 16
16. Postnotum very large, projecting some distance to scutellum (Fig. 8-140 *C*); slender elongate flies .... **Chironomidae**, p. 458
Postnotum smaller, scarcely projecting at all from beneath the scutellum (Fig. 8-140*B*); small stouter flies, the punkies or "no-see-ums" .. .. **Ceratopogonidae**
17. Tarsus having 3 whitish pulvillar pads (Fig. 8-140*D*); the middle one (the empodium) is sometimes dorsal of the lateral pulvilli, which are sometimes small 18
Tarsus at most having 2 pulvillar pads, the empodium reduced to a seta (Fig. 8-140*E*); the 2 pulvilli may be reduced, in which case the tarsus lacks pads 19
18. Wing with branches of S or *R* +S close to front margin, forming a group of narrow cells along it (Fig. 8-137*G*); tibia without apical spurs; soldier flies .... **Stratiomyidae**, p. 462
Wing with branches of S not crowded to front margin, S_3 and S_4 diverging to form a triangular cell embracing apex of wing (Fig. 8-137*H*) .... **Tabanidae**, p. 461
19. Antenna having third segment elongate fourth clavate, without arista or style (Fig. 8-140*G*); very large species .... **Mydidae**, p. 461
Antenna having apical segment not clavate 20
20. Top of head sunken to form a deep excavation between eyes (Fig. 8-140*M*); wing and M_2 present and M_{3+4} having its base almost or entirely free from Cu_1, the latter 2 veins sometimes fused for a short distance at base and/or apex (Fig. 8-137*J*) .... **Asilidae**, p. 460

Top of head without a deep excavation; wing having M_2 atrophied, and M_{3+4} fused near base for a considerable distance with Cu_1 (Fig. 8-137*K*); hover flies .... **Bombyliidae**

21. Wing at most having S_c and stem of $R+S$ sclerotized, remaining venation consisting of 3 or 4 weak veins arranged as in Fig. 8-138*A*; antenna composed of one segment and its arista .... **Phoridae**, p. 464
 Wing having a more extensive venation (Figs. 8-138*B* to *D*); antenna may have 1 to 3 segments 22
22. Tarsus having 3 pulvillar pads (Fig. 8-140*D*), 2 lateral and 1 mesal 23
 Tarsus having only 2 (lateral) pulvillar pads (Fig. 8-140*E*) or none 24
23. Antenna having distinct but faint annular lines on third segment (Fig. 8-140*I*); wing having *S* and its branches forming a series of narrow cells along anterior margin (Fig. 8-137*G*) .... **Stratiomyidae**
 Antenna without annulations on third segment; wing having some branches of *S* ending at or below apex of wing (Fig. 8-137*F*); predaceous snipe flies .... **Rhagionidae**
24. Wing having *S* with 3 branches (Fig. 8-137*I* to *K*) 25
 Wing having *S* with only 1 or 2 branches (Fig. 8-138*B*, *D*) 28
25. Top of head sunken to form a deep excavation between eyes (Fig. 8-140*M*) .... **Asilidae**, p. 460
 Top of head flat or convex, confluent in outline with top of eyes 26
26. Wing having M_{3+4} not fused at base with Cu_1 but frequently fusing with Cu_1 near margin: M_2 present (Fig. 8-137*I*). Moderate-sized species similar to Asilidae in habits .... **Therevidae**
 Wing having M_2 atrophied and M_{3+4} fused at base with Cu_{1a} (Fig. 8-137*K*) 27
27. Wing having Cu_2 reaching wing margin or fusing with *E* near wing margin (Fig. 8-137*K*) .... **Bombyliidae**
 Wing having Cu_2 fused with *E* considerably before wing margin, as in Fig. 8-137*H* .... **Empididae**
28. Wing with Cu_2 straight and long, angled only slightly from stem of *Cu* (Fig. 8-137*L*) 29
 Wing with Cu_2 markedly angled from its parent vein, appearing like a crossvein (Fig. 8-138*B* to *D*); spurious vein never developed 30
29. Wing with M_1 sinuate; a linear veinlike thickening (the spurious vein) usually present between *S* and *M* (Fig. 8-137*L*) .... **Syrphidae**, p. 462
 Wing with M_1 short and straight, appearing to be a crossvein; no spurious vein present .... **Conopidae**, p. 460
30. Lower squama large and platelike, nearly twice as long as upper squama and projecting considerably beyond it (Figs. 8-138*H*, 8-139*E*) 31
 Lower squama at most slightly longer than upper squama (Fig. 8-139*D*), frequently appearing as only a cordlike band (Fig. 8-139*A*, *B*) 35

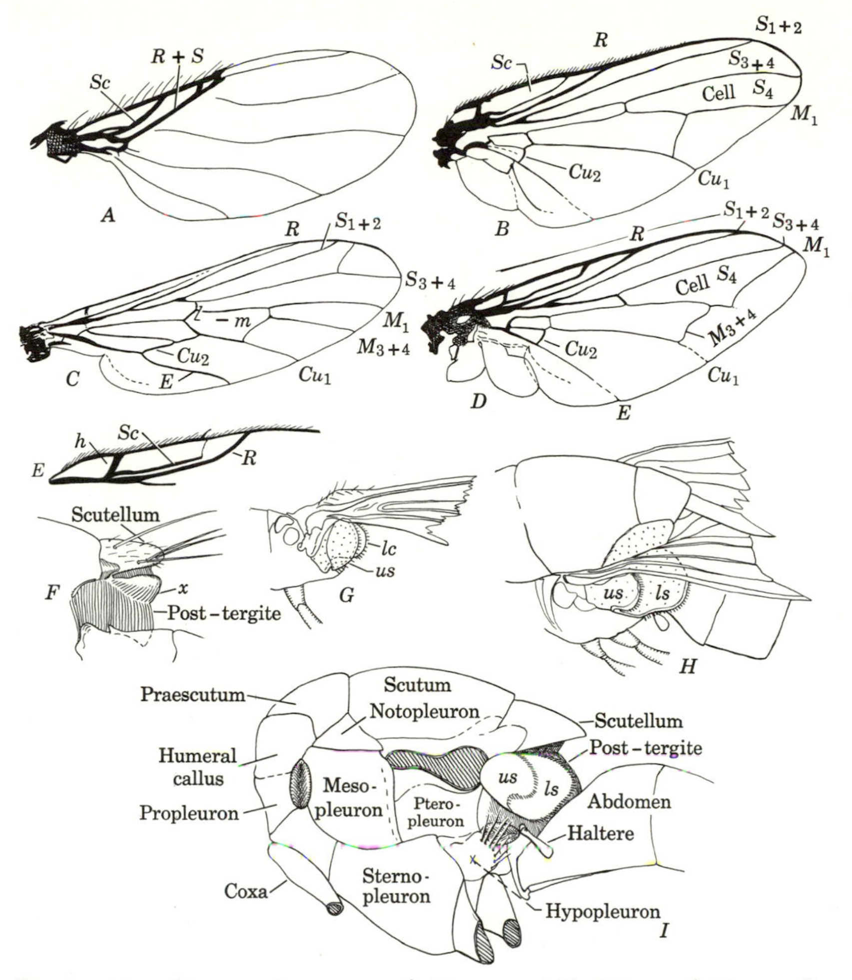

Fig. 8-138. Diagnostic parts of Diptera. (*A*) Wing of *Megaselia*, Phoridae; (*B*) wing of *Hylemya*, Anthomyiidae; (*C*) wing of *Empis*, Empididae; (*D*) wing of *Sarcophaga*, Sarcophagidae; (*E*) costal region of *Rhagoletis*, Tephritidae; (*F*) scutellum of *Zenillia*, Tachinidae; (*G*) squamae of *Anthomyia*, Anthomyiidae; (*H*) squamae of *Musca*, Muscidae; (*I*) thorax, diagrammatic, of higher Diptera. *ls*, Lower squama; *us*, upper squama; *x*, secondary convexity on scutellum.

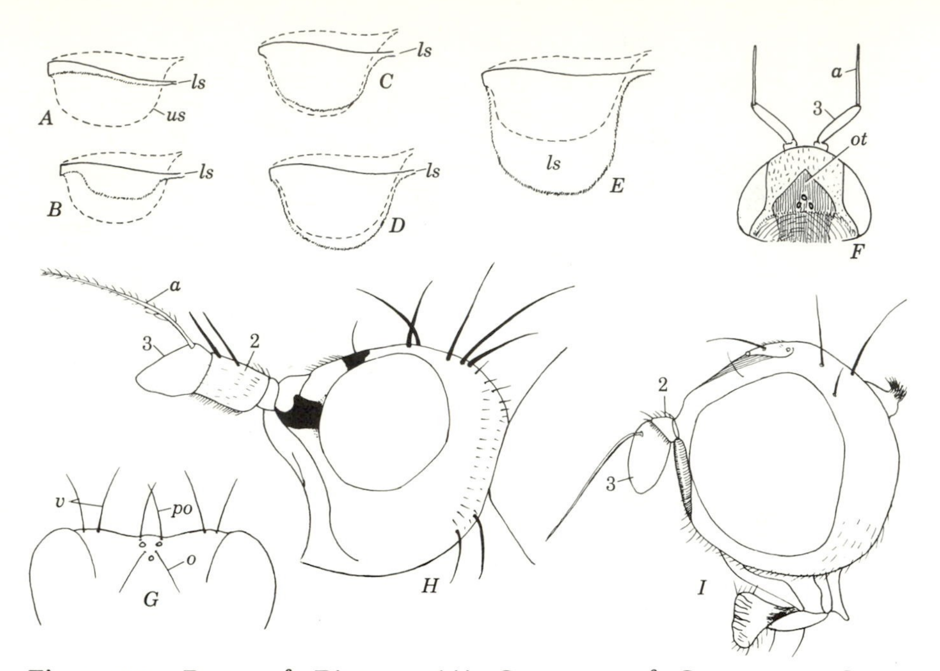

Fig. 8-139. Parts of Diptera. (*A*) Squamae of *Scopeuma*, Scopeumatidae; (*B*) of *Pegomyia*, Anthomyiidae; (*C*, *D*) of *Fannia*, Fanniidae; (*E*) of *Musca*, Muscidae; (*F*) head of *Agromyza*, Agromyzidae; (G) diagram of upper part of head showing ocelli and associated bristles; (*H*) head of *Euthycera*, Sciomyzidae; (*I*) of *Nemopoda*, Sepsidae. *a*, Arista; *ls*, lower squama; *o*, ocellar bristle; *ot*, ocellar triangle; *po*, postocellar bristles; *us*, upper squama; *v*, vertical bristle. In *A* to *E* the upper squama is shown in broken outline. (*H*, *I* after Curran)

31. Oral opening either a minute circle or a small triangular cleft, as in Fig. 8-140*P*; body fuzzy with long dense hair .... **Oestridae**, p. 469
    Oral opening large (Fig. 8-140*Q*); body usually not fuzzy but often spiny in appearance 32
32. Hypopleura with only weak, scattered hairs or none .... **Muscidae**, p. 466
    Hypopleura with a row of bristles (Fig. 8-139) 33
33. Mesopostnotum having a convex bump below the scutellum (Fig. 8-138*F*) .... **Tachinidae**, p. 469
    Mesopostnotum having no well-developed extra bump, at most a gentle convexity below scutellum; the flesh flies 34
34. Arista of antenna with feathering not extending much beyond middle; body dull black or striped with gray and black .... **Sarcophagidae**, p. 469
    Arista feathered to tip (Fig. 8-140J) or body entirely metallic blue or green .... **Calliphoridae**, p. 469

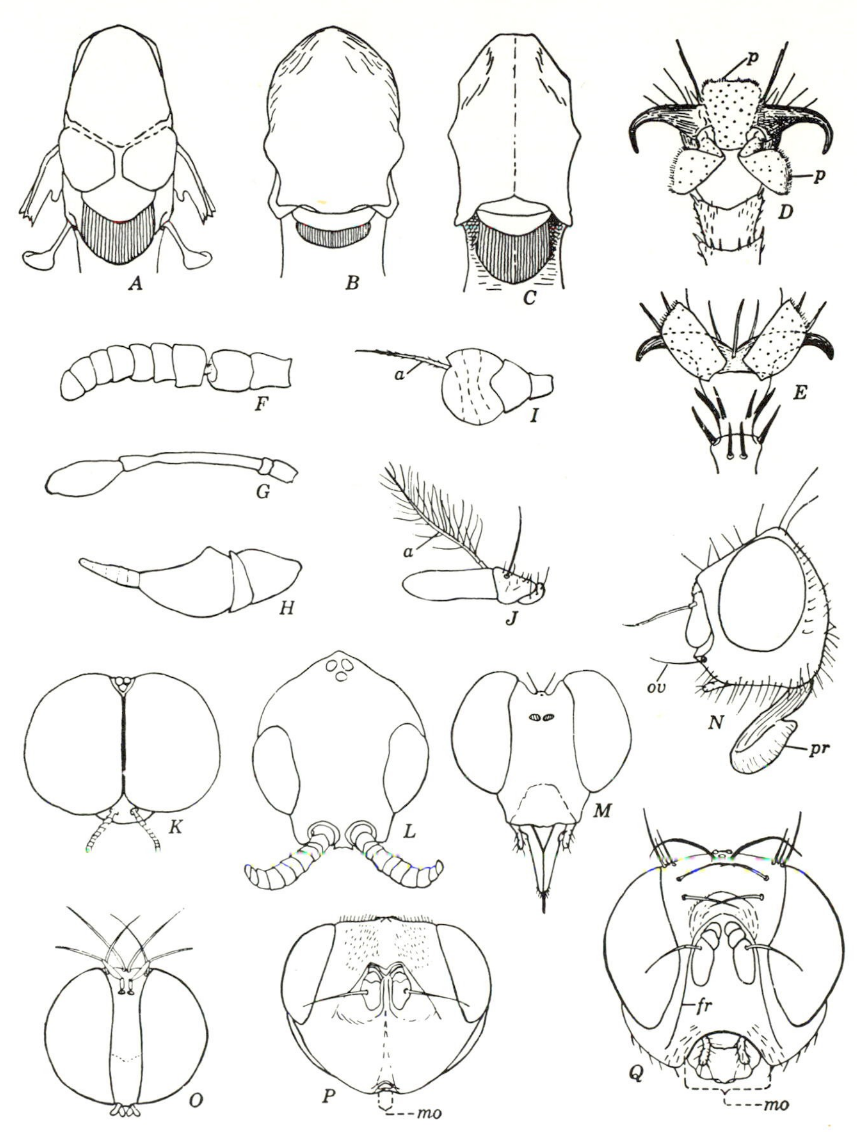

Fig. 8-140. Diagnostic parts of Diptera. (*A*) Mesonotum of *Helobia*, Tipulidae; (*B*) mesonotum of *Palpomyia*, Ceratopogonidae; (*C*) mesonotum of *Chironomus*, Chironomidae; (*D*) tarsal claws and pads of *Tabanus*, Tabanidae; (*E*) tarsal claws and pads of *Zenillia*, Tachinidae; (*F*) antenna of *Bibio*, Bibionidae; (*G*) antenna of *Mydas*, Mydaidae; (*H*) antenna of *Tabanus*, Tabanidae; (*I*) antenna of *Geosargus*, Stratiomyidae; (*J*) antenna of *Pollenia*, Calliphoridae; (*K* and *L*) face of *Bibio*, Bibionidae, male and female; (*M*) face of *Asilus*, Asilidae; (*N*) head of *Hylemya*, Anthomyiidae; (*O*) face of *Dolichopus*, Dolichopodidae; (*P*) face of *Gasterophilus*, Gasterophilidae; (*Q*) *Tephrita*, Tephritidae. *a*, Arista; *fr*, frontal ridge or suture; *mo*, mouth opening, *ov*, oral vibrissa; *p*, pulvillar pads; *pr*, proboscis.

35. Oral opening round and minute, mouthparts vestigial, cheeks inflated (Fig. 8-140*P*) .... **Gasterophilidae**, p. 467
Oral opening large, mouthparts well developed 36
36. Front of head lacking sutures or ridges running ventrolaterally from base of antennae (Fig. 8-140*O*) 37
Front of head having a pair of frontal sutures or ridges *(fr)* each running from near the antenna toward the oral excavation (Fig. 8-140*Q*) . . . 38
37. Crossvein *s–m* situated at the base of the wing, and inconspicuous, sometimes difficult to find; free end of M_{3+4} lacking; predaceous species in moist habitats .... **Dolichopodidae**
Either crossvein *s–m* situated at least one-third distance from the base to apex of wing, or M_{3+4} present (Fig. 8-138*C*) .... **Empididae**
38. *Sc* distinctly sclerotized for its entire length, curving gently to join *C* before *R* does (Fig. 8-138*B*, *D*) (this character best seen from anterodorsal view of wing) 39
Sc either partially atrophied, or fused at tip with *R*, or abruptly angled (Fig. 8-138*E*) 47
39. Head nearly spherical, eyes occupying most of lateral aspect, and second antennal segment small (Fig. 8-139*I*) .... **Sepsidae**, p. 464
Either head not spherical, or eyes situated a considerable distance from ventral margin, or second antennal segment massive (Fig. 8-139*H*) 40
40. Dorsum of head and mesonotum flat, all but the lateral areas clothed only with abundant, short, stiff setae; seashore species . . . **Coelopidae**
Dorsum of mesonotum convex and arched, often with long bristles scattered over much of its area 41
41. Oral vibrissae present (compare Fig. 8-140*N*) 42
Oral vibrissae absent 45
42. Lower squama differentiated as a flap of the axillary cord (Fig. 8-139*B* to *D*) 43
Lower squama not differentiated, being simply a thin, fuzzy edge of the axillary cord (Fig. 8-139*A*) 44
43. Empusal *(E)* vein reaching wing margin, although the end of the vein may be faint (Fig. 8-138*B*) .... **Anthomyiidae**, p. 466
Empusal *(E)* vein reaching only about halfway to wing margin .... **Fanniidae**
44. Postocellar bristles long and convergent (Fig. 8-139*G*); body never hoary .... **Heleomyzidae**, p. 464
Postocellar bristles either short, absent, or divergent; body often hoary with long, thick hair .... **Scopeumatidae**
45. Some or all of the tibiae with a preapical dorsal bristle 46
Tibiae without preapical bristles .... **Otitidae**, p. 464
46. Postocellar bristles convergent; second antennal segment always small; small flies, rarely over 6 mm long .... **Lauxaniidae**, p. 464
Postocellar bristles parallel, divergent, or absent; second antennal segment often massive, as large as third (Fig. 8-139*H*); moderate-sized flies .... **Sciomyzidae**, p. 464

47. Wing having costal cell wide and having *Sc* ending abruptly or angled abruptly much before apex of cell, either *Sc* beyond this point weak or atrophied, or *C* with a distinct break, at which point there are several stout bristles (Fig. 8-138*E*) .... **Tephritidae**, p. 465
Wing either having costal cell narrow, or *Sc* gradually fading out toward its apex, or fusing with *R*, or absent 48
48. Posterior basitarsus short and enlarged .... **Borboridae**, p. 465
Posterior basitarsus little if at all thicker than succeeding segments and usually longer than second tarsal segment 49
49. Wing with *E* and Cu_2 almost entirely atrophied, cell Cu_2 therefore open or absent 50
Wing with *E* and Cu_2 present and enclosing cell Cu_2 51
50. Dorsum of head with a large, triangular, sclerotized area (the ocellar triangle, Fig. 8-139*F*, *ot*) flanked by wide membranous areas; wing having vein Cu_{1a} slightly sinuate .... **Chloropidae**, p. 465
Dorsum of head with sclerotized areas either very small, or quadrangular, or occupying all of dorsal aspect; wing with vein Cu_{1a} not sinuate .... **Ephydridae**
51. Postocellar bristles convergent, as in Fig. 8-139G ....
.... **Drosophilidae**, p. 465
Postocellar bristles parallel or divergent .... **Agromyzidae**, p. 465

SUBORDER NEMATOCERA

The North American fauna contains representatives of about 20 families of Nematocera. Midges, crane flies, mosquitoes, and blackflies are examples of the more abundant and conspicuous families. The larvae of all families have a well-defined sclerotized head, which is retracted into the thorax only in the Tipulidae.

CECIDOMYIIDAE (ITONIDIDAE), THE GALL GNATS. The adults (Fig. 8-141) are inconspicuous fragile flies, having greatly reduced wing venation, and elongate beadlike antennae. The adults feed only on aqueous materials such as sap; most of the food is consumed during the larval stage. Many genera of the family are gall makers, and it is by these galls, especially on willows, deciduous trees, and many herbs, that the family is known to most observers. The family exhibits a wide range of other habits. Some larvae are predaceous, feeding on mites and small insects; others feed on developing plant seeds; some feed on decomposing organic matter; and still others feed in or on the tissues of leaves or stems of plants. To the latter category belongs the most notable economic species of the family, the Hessian fly *Mayetiola destructor* (Fig. 8-141); the larvae feed in the lower stems of grasses and are especially injurious to wheat and barley.

SIMULIIDAE, BLACKFLIES. This family, often referred to as buffalo gnats, has aquatic larvae and pupae and bloodsucking adults, like the

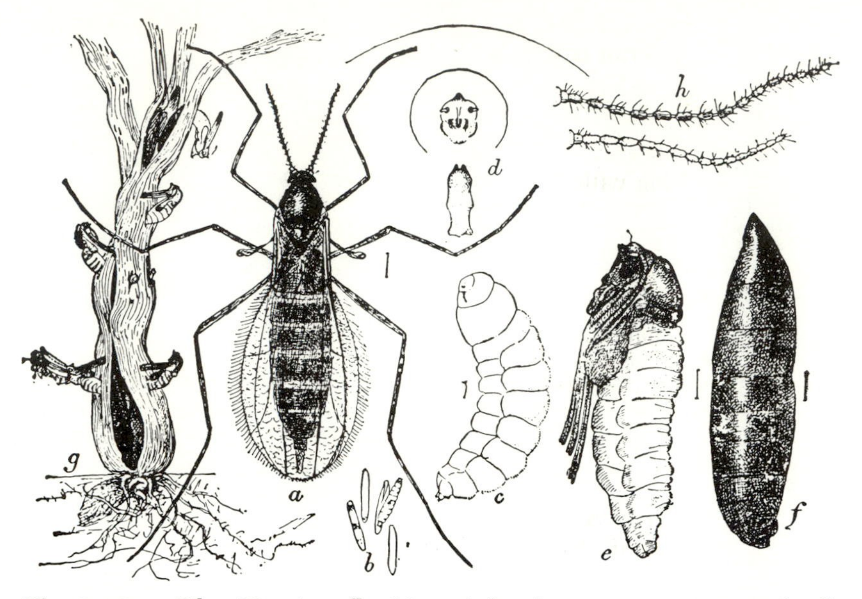

Fig. 8-141. The Hessian fly *Mayetiola destructor.* (*a*) Female fly; (*b*) eggs, one hatching; (*c*) larva; (*d*) head and breastbone of same; (*e*) pupa; (*f*) puparium; (*g*) infested wheat stem showing emergence of pupae and adults; (*h*) antennae, male and female. (From U.S.D.A., E.R.B.)

Culicidae. The blackflies are short, stubby, and humpbacked, with short legs and compact, usually 11-segmented antennae (Fig. 8-142). Like the mosquitoes, only the females bite. Horses, cattle, ducks, and many wild animals are attacked. Certain species attack humans also, gathering around the ears, eyes, and exposed areas of the face, hands, and ankles. They draw a great amount of blood and produce burning welts on the victim's skin. These welts may itch and burn for a week or more.

The eggs are laid in clusters near the water edge or in the water. The larvae are sedentary, elongate, and slightly vasiform (Fig. 8-143), and occur only in running water; the posterior end is anchored by a hooked sucker-disc to some support such as a rock, log, or trailing leaves; the head has a pair of feathery branched rakes that are used as strainers to obtain food from the running water, primarily microorganisms and organic material. The pupae are also aquatic, formed in a slipper-shaped cocoon (Fig. 8-143). On the prothorax of the pupa are a pair of long respiratory processes, usually branching into many slender filaments.

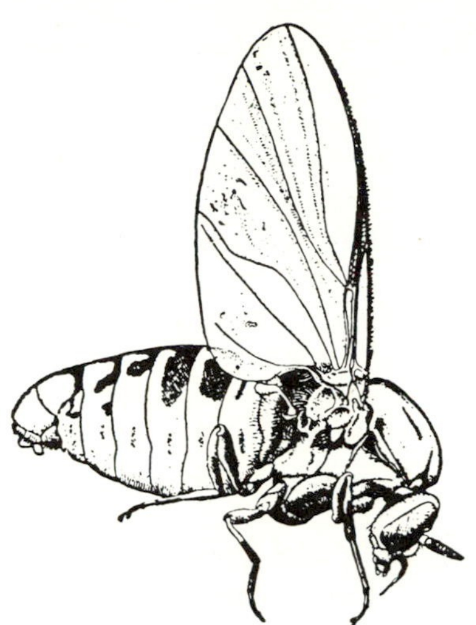

Fig. 8-142. Adult blackfly *Simulium vittatum.* (After Knowlton and Rowe)

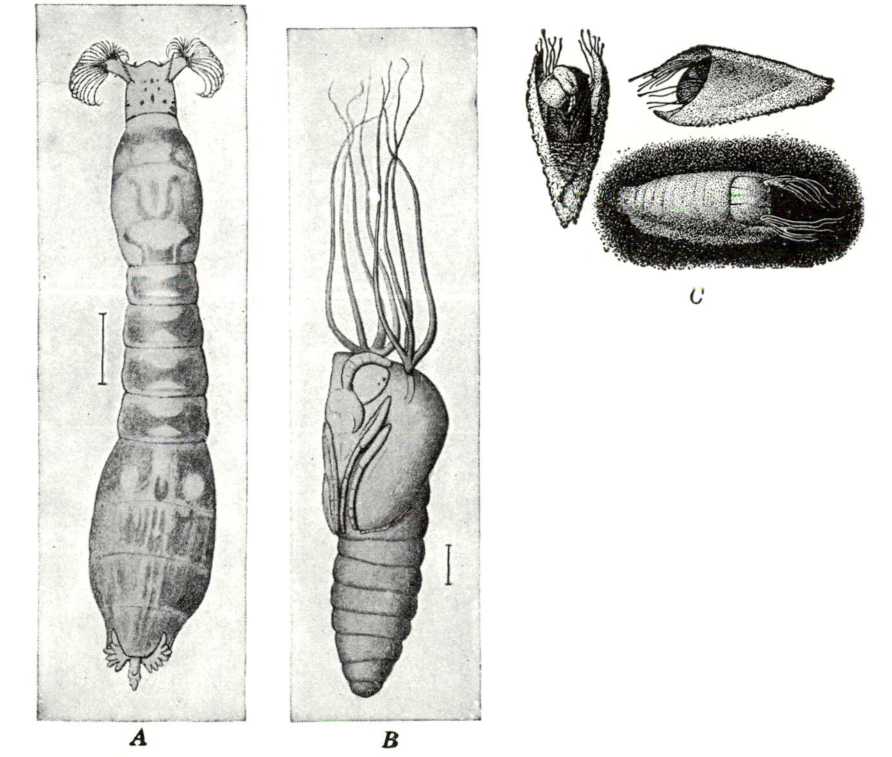

Fig. 8-143. (*A*) Larva, and (*B*) pupa of a blackfly *Simulium johannseni* and (C) pupal cases of *S. venustum*. (From Illinois Natural History Survey)

TIPULIDAE, CRANE FLIES. The adults are long-legged slender-winged flies (Fig. 8-144), the antennae are threadlike, with many distinct segments, and the mesonotum has a V-shaped transverse furrow or suture. The adults are extremely abundant in moist woods and sheltered ravines, and along wooded stream banks. The larvae are elongate and wormlike (Fig. 8-145); many are aquatic, living especially in submerged clusters of rotting leaves; many feed in leaf mold; and a few feed on living plant roots or mine in leaves. They vary greatly in size and appearance but have in common a stout head that is partly retracted within the thorax (in certain species it is further retractile and may be completely hidden in the thorax when the insect is disturbed) and strong toothed mandibles that work from side to side.

CULICIDAE, MOSQUITOES. Long-legged slender insects (Fig. 8-146), having beadlike antennae (in the males they are plumose as in

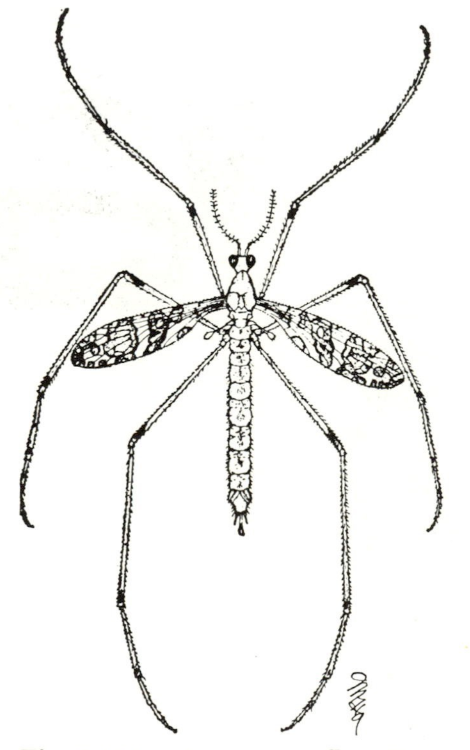

Fig. 8-144. A crane fly *Epiphragma fascipennis*, adult female.

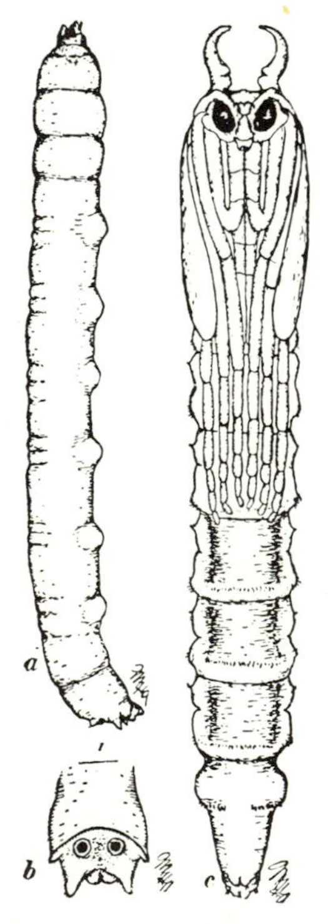

Fig. 8-145. *Epiphragma.* (*a*) Larva; (*b*) end of larva from above; (*c*) pupa. (After Needham)

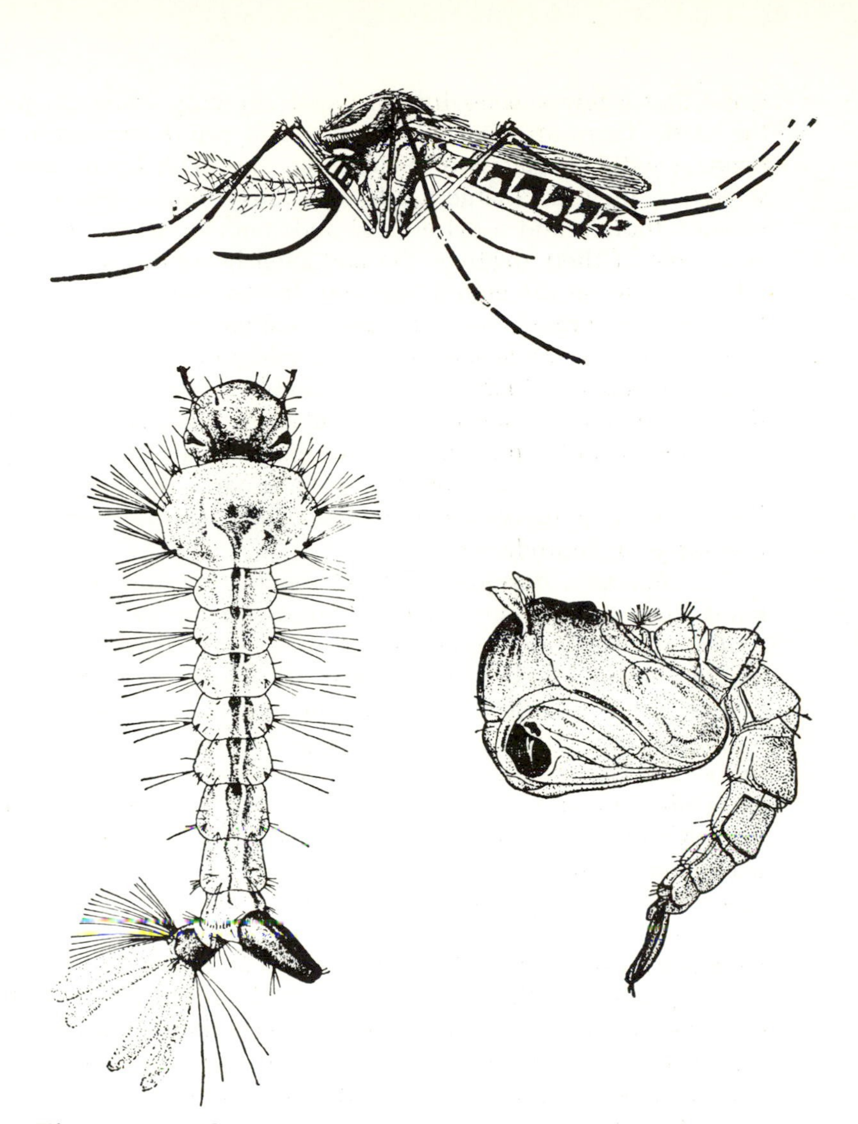

Fig. 8-146. The yellow fever mosquito *Aedes aegypti.* Adult above; larva at left; pupa at right. (From U.S.D.A., E.R.B.)

the Chironomidae) and many-veined wings. The mouthparts form a beak, composed of a highly modified assemblage of piercing-sucking parts (see Figs. 3-30, 3-31).

In certain genera eggs are laid on water, either singly or glued together to form rafts; they hatch in a few days. In other genera eggs are laid in humus or just above the water line on the sides of containers; these eggs hatch at some later time when the water rises and inundates them. The larvae are aquatic, most of them living in still water, but a few species live in slowly moving water. In parts of the world there are species that breed in rapid streams, quite in contrast with the habits of nearctic species. The larvae are called wrigglers; they have large heads with fairly long antennae, a large swollen thorax, and a cylindrical abdomen. In all the mosquitoes and some of their relatives the abdomen bears a dorsal breathing tube or plate on the eighth segment. In the main, mosquito larvae feed on microorganisms and organic matter in or on the water. A few groups are predaceous and feed solely on other mosquito larvae. The pupae, which are called tumblers, are also aquatic, free living, and active. Their breathing tubes are situated on the thorax. Adult females of a few species and males of all mosquito species feed only on nectar or water. But unfortunately in most species the females seek a blood meal, which under natural conditions is necessary for reproduction.

Economically, mosquitoes are of tremendous importance to humans. Some species, principally those of the genera *Anopheles* and *Aedes*, transmit an imposing list of human diseases, including malaria, dengue, yellow fever, and filariasis. For this reason mosquitoes are of utmost importance in a consideration of medical entomology. As a direct nuisance mosquitoes also have an economic influence. The severity of their attacks decreases property values, especially in resort areas, and has undoubtedly had an influence in the settlement of extreme northern areas where mosquitoes are unusually abundant.

CHIRONOMIDAE (TENDIPEDIDAE), **MIDGES.** These are frail insects (Fig. 8-147), frequently mistaken for mosquitoes, but they do not bite and have several structural differences that set them off. The male antennae are plumose with whorls of long silky hair. The larvae of all but a few species are aquatic, some free living, others spinning a loose web of bottom particles and silk and in certain genera making a definite case. They are slender and wormlike, and have a small but distinct sclerotized head, and a 12-segmented body. In some groups the prothorax, or the last body segment, or both, may have a pair of nonjointed leglike protuberances or *pseudopods*. The larvae feed on organic matter on the bottom of bodies of water and are found in rivers, lakes, ditches, and stagnant ponds. The pupae are also aquatic, some of them free living, but most of them staying in the web or case made by the larva. In many bodies of water these midge larvae are tremendously abundant and form one of the principle items of fish food. When the adults emerge, they appear at night in clouds and blanket near-by lights in a humming mass. The eggs are laid in water and hatch in a few days.

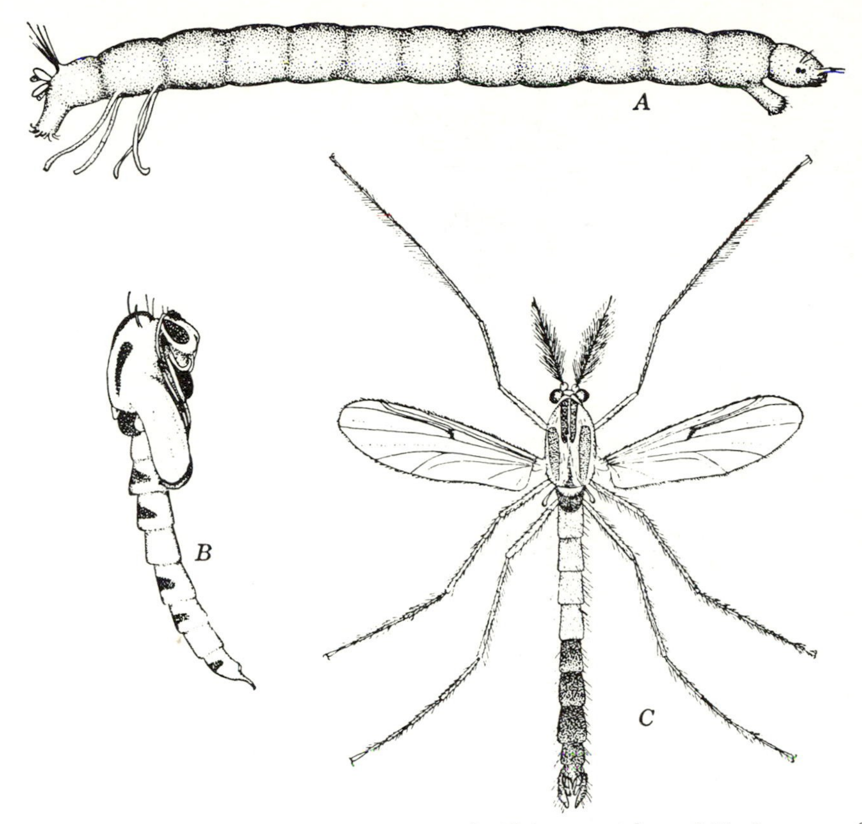

Fig. 8-147. Life history stages of Chironomidae. (*A*) Larva of *Chironomus tentans*; (*B*) pupa of *Cricotopus trifasciatus*; (*C*) adult male of *Chironomus ferrugineovittatus*. (From Illinois Natural History Survey)

SUBORDER BRACHYCERA

The adults of this suborder are mostly larger and stouter bodied than the Nematocera and are stronger on the wing.

DIVISION ORTHORRHAPHA

The larvae have hooked or blade-shaped mandibles that work up and down, rather than sideways; the head is frequently much retracted within the thorax (Fig. 8-148), and may have stout internal supports extending far into the body. About 15 families are represented in North America; some of them are abundant, and in one, the Tabanidae or horseflies, the adults are voracious bloodsuckers.

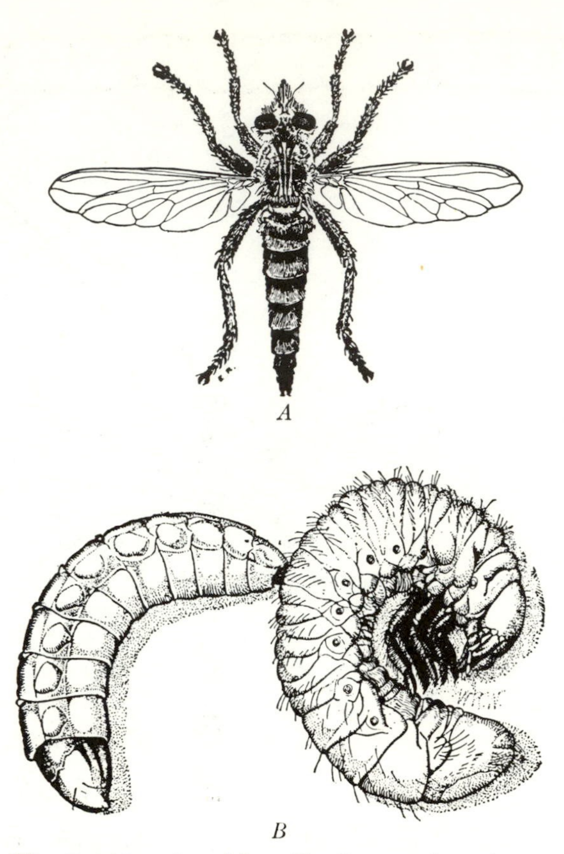

Fig. 8-148. A robber fly *Promachus vertebratus.* (*A*) Adult; (*B*) legless larva attacking white grub. (From Illinois Natural History Survey)

Other families of the Orthorrhapha contain species of great diversity as regards structural peculiarities and habits. Most of them, for example the families Dolichopodidae and Empididae, are predaceous on other insects in both adult and larval stages and have free pupae; larvae of most species live in rotten logs or in soil and pupate there. Larvae of the Conopidae are internal parasites of other insects.

ASILIDAE, ROBBER FLIES. These are also large flies (Fig. 8-148), usually with a humped thorax and elongate abdomen, but in some genera the body is stout, hairy, and brightly colored, resembling a bumblebee. In many genera also the lower part of the face has a beardlike brush of hair called a *mystax*. These flies are predators on other insects, capturing and eating bumblebees and dragonflies in addition to smaller forms. The adults are not easily frightened and are both conspicuous and easily taken in a sweep net. The larvae are found chiefly in soil and rotten wood and are predaceous on insects found there. Some of them are important natural enemies of white grubs and other soil-inhabiting species attacking cultivated crops (Fig. 8-148).

MYDIDAE, MYDAS FLIES. This is a small family, but it contains some of the largest and showiest species. The antennae are clubbed, and the wings have a modified venation, the ends of many veins bending forward towards the anterior margin of the wing. The common eastern species *Mydas clavatus* is black with an orange band and has a wing spread of 55 mm. The larvae, resembling large asilid larvae, are predaceous and found in rotten logs.

TABANIDAE, HORSEFLIES AND DEERFLIES. The adults (Fig. 8-149) are large-headed stout-bodied insects, often attaining a length of 25 to 30 mm, having strong wings with fairly complete venation and frequently a striking color pattern. The venation features a wide V-shaped cell S_3, which embraces the apex of the wing. The females are bloodsuckers; their mouthparts are developed for cutting skin and sucking the blood that oozes from the wound. The males feed on nectar. Eggs are laid in masses on stems or other objects growing over water; the newly hatched young crawl or drop into their breeding place. The larvae live in swamps or sluggish streams, staying in the bottom mire, and are often abundant in damp pasture soils. They are predaceous, feeding on snails, insects, and other aquatic organisms. The larval body is tough and leathery, usually white or banded; the head is completely retractile within the thorax. The

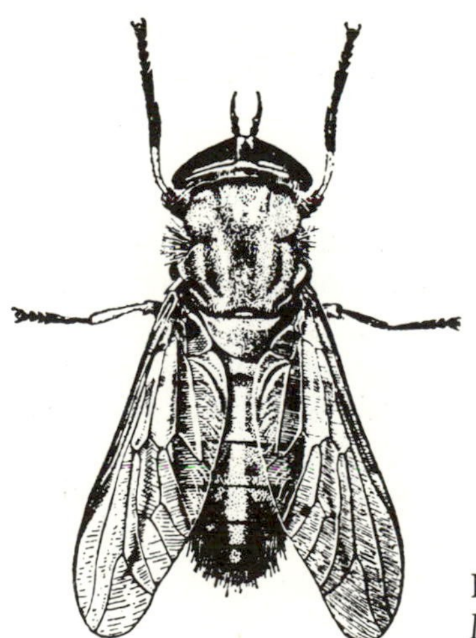

Fig. 8-149. A horsefly *Tabanus lineolus*. *(From U.S.D.A., E.R.B...)*

pupae are cylindrical and brown, normally formed in a mass of peat or dead vegetation that is damp but above the free-water level.

Farm livestock, especially cattle and horses, are often bothered by adult horseflies. Locally these pests may assume major importance in reducing condition of stock. Many species of horseflies and deerflies abound in marshy areas of the northern and humid montane areas of the continent and discourage vacationers and settlers.

Closest relatives of the Tabanidae are the Rhagionidae (snipe flies) that have predaceous larvae and adults. Some species of the genus *Symphoromyia* (Fig. 3-26) have bloodsucking adults that attack humans and other vertebrates. More distant relatives of the Tabanidae include the Stratiomyidae (soldier flies) whose larvae are predators or scavengers and whose adults are predaceous.

DIVISION CYCLORRHAPHA

In these flies the pupa is always contained in a puparium from which the adult emerges through a round opening by pushing off the head end. In the more specialized families, the Schizophora, this is done by expanding the ptilinum, which is pushed out between the frontal sutures at the time of emergence, then completely retracted. In a few primitive families, the Aschiza, the frontal sutures are not complete dorsally and there is no ptilinum.

Subdivision Aschiza

SYRPHIDAE, FLOWER FLIES, SYRPHID FLIES. They are small to large flies (Fig. 8-150) characterized by the upturned ends of some of the wing veins, and usually the presence of a veinlike thickening or *spurious vein* in front of *M*. These flies feed almost exclusively on flowers and have a remarkable ability for hovering apparently motionless in the air. Many are brilliantly striped or marked with yellow, red, white, and black; within the family are mimics of various wasps, bumblebees, and other bees.

The larvae are extremely diverse. Many feed on aphids, crawling from colony to colony and devouring huge numbers of individuals. Other species feed in decaying liquid organic matter, for example the rat-tailed maggots of the genus *Eristalis* (Fig. 8-151).In these maggots the posterior spiracles are situated at the end of extensile "tails" that can be extended to the surface of the food and so provide a contact with air.

Other larval foods include debris in burrows of wood-boring insects, leaves and sheaths of grasses, and debris in nests of bumblebees, ants, and wasps. Larvae of two species are serious pests of bulbs, eating out the centers of planted tulips and related species.

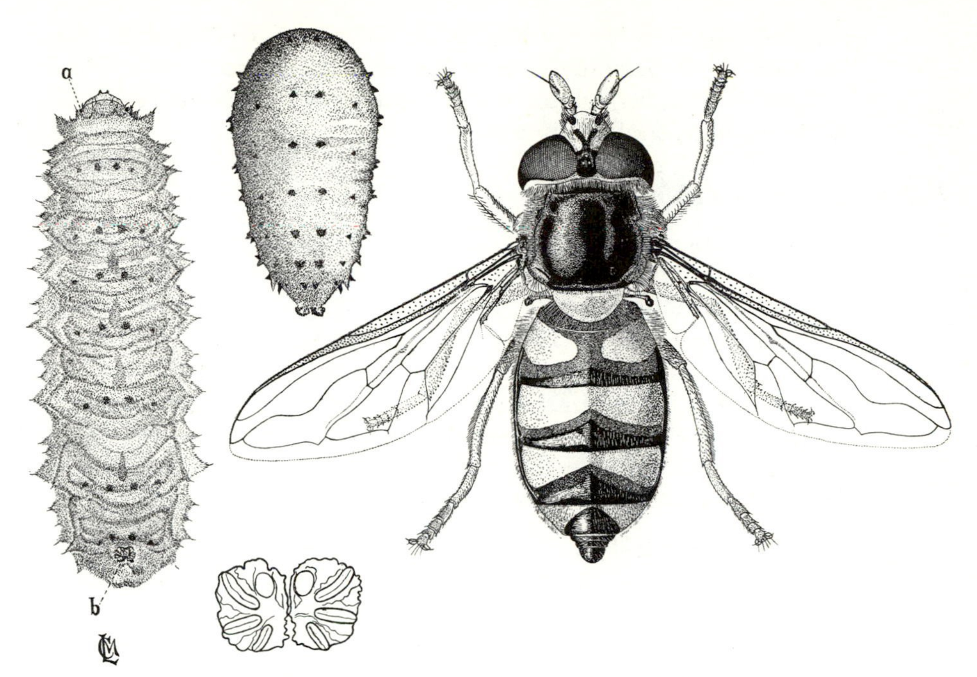

Fig. 8-150. A flower fly *Didea fasciata*. Right, adult; left, larva; center, puparium; *a*, Anterior spiracle; *b*, caudal spiracles. (After Metcalf)

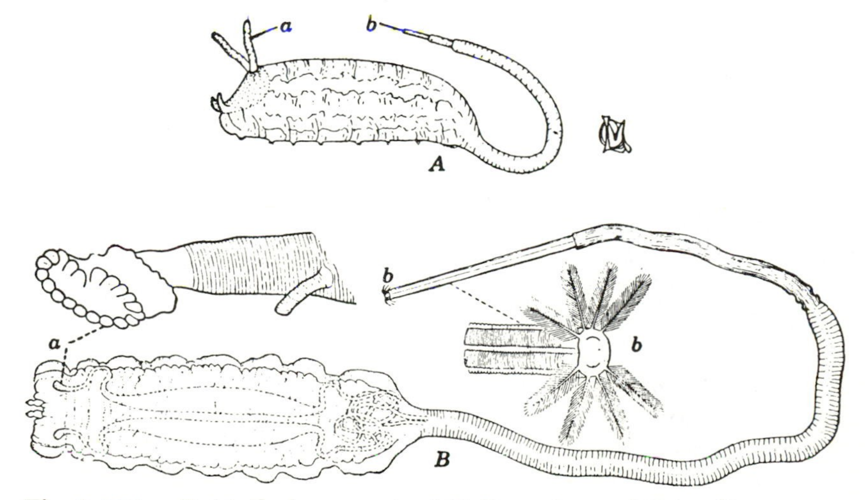

Fig. 8-151. Rat-tailed maggots. (*A*) Puparium of *Eristalis aeneus*; (*B*) larva of *Eristalis tenax*. *a*, Anterior spiracle; *b*, posterior spiracles on long "rat tail." (After Metcalf)

Allied families include the Phoridae having small species of varied habits, but principally with saprophytic larvae, and the Pipunculidae (Dorilaidae) having small species with larvae that are parasitic in leafhoppers (Hemiptera).

Subdivision Schizophora

The Schizophora comprises one of the largest branches of the Diptera and is divided into two major sections: the Acalyptratae, in which the membranes at the base of the wings are not produced into folds or lobes; and the Calyptratae, in which those membranes form three lobes. Two of these three lobes lie one above the other at the base of the wing and are called the *upper* and *lower squamae* or *calypters*. The third lobe lies outboard of these along the rear edge of the wing and is called the *alula*. In most calyptrates, such as the muscids, the squamae are large enough to cover the halteres.

Section Acalyptratae. This section contains a large number of families. Primitive families include the Otitidae (picture-winged flies), Lauxaniidae, Coleopidae, and Heleomyzidae, whose larvae feed on various decaying organic material; Sepsidae, whose larvae feed in dung; and Sciomyzidae, whose larvae are aquatic and feed chiefly on aquatic or amphibious snails. More specialized families include the Borboridae, the little dung flies; Agromyzidae, the leaf miners; Chloropidae, chiefly scavengers; and Ephydridae, the shore flies and brine maggots. The following are typical families.

SCIOMYZIDAE. This cosmopolitan family includes medium-sized flies that are parasites on terrestrial and freshwater gastropods and are capable of attacking and killing snails many times their size. Commonly their free-living larvae and floating puparia are seen in ponds and reservoirs in which snails are abundant. Because the snails frequently are the vectors of diseases of humans and domestic animals, particularly in subtropical and tropical areas, the Sciomyzidae are considered a potentially effective biological control for snails.

LAUXANIIDAE. This is another widely distributed and common family. Adults of the Lauxaniidae occur in a broad range of environmental conditions, from coastal swamps, grasslands, forests of all types, to mountain tops. The larvae are not well known; however, most that have been studied are saprophagous.

HELEOMYZIDAE. This family is largely restricted to temperate forests where it is common and occurs in a great range of habitats. In some species the larvae live in fungi, but in most the larvae are associated with debris in the burrows of other insects, particularly boring beetle larvae. The closely related family Coelopidae has larvae specialized to living on decaying kelp and sea grass stranded on the beach.

TEPHRITIDAE (TRYPETIDAE, TURPANEIDAE), FRUIT FLIES. Here belong many brightly colored species frequently with mottled or banded wings. The larvae of several species live in the seed heads or fruits of plants. Important commercially are the cherry fruit fly *Rhagoletis cingulata* (Fig. 8-152), whose maggots tunnel in cherry fruits, and the apple maggot *Rhagoletis pomonella*, which tunnels through apples. The Mediterranean fruit fly *Ceratitis capitata*, discovered in Florida in 1929, is a widespread tropical species as is the Oriental fruit fly; the larvae feed in fleshy fruits.

DROSOPHILIDAE, VINEGAR GNATS. These include small flies (Fig. 8-153), which are common at times in most houses and stores. The larvae breed in decaying fruits and other organic materials. *Drosophila melanogaster* is used as an experimental subject by geneticists and from the standpoint of chromosome and gene mapping is undoubtedly the best-known animal in the world.

AGROMYZIDAE, LEAF MINERS. This is a family of small or minute flies. The larvae include leaf and stem miners, several gall makers, and some fruit borers. They are important pests of cultivated plants in many areas.

OTHER ACALYPTRATAE FAMILIES. The Borboridae, small dung flies, are widely distributed small to minute flies that wave their wings when at rest. They are usually associated with animal dung or decaying organic matter in which their larvae live. The Ephydridae, shore flies, usually live near water and their larvae are mostly either aquatic or live in stems of hydrophilic plants, although some occur away from water habitats. The Chloropidae, stem flies, are adapted to

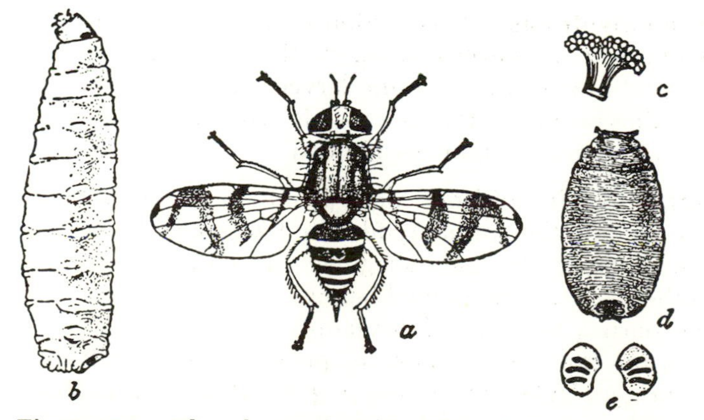

Fig. 8-152. The cherry fruit fly *Rhagoletis cingulata*. (*a*) Fly; (*b*) maggot; (*c*) anterior spiracles of same; (*d*) puparium; (*e*) posterior spiracular plates of pupa. (From U.S.D.A., E.R.B.)

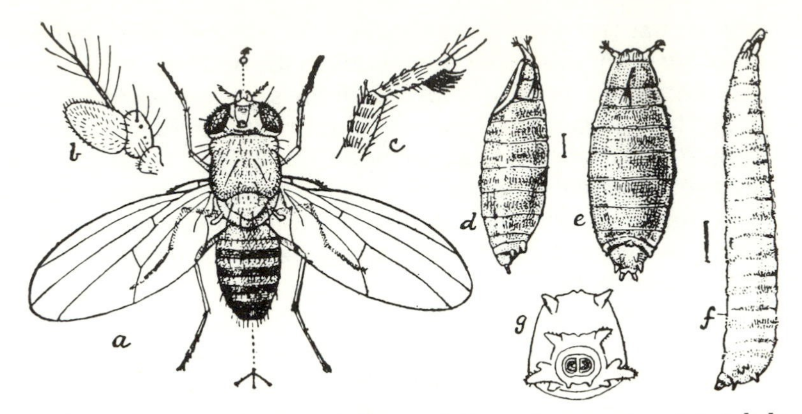

Fig. 8-153. Vinegar gnats *Drosophila melanogaster.* (*a*) Adult; (*b*) antenna of same; (*c*) base of tibia and first tarsal joint of same; (*d*) puparium, side view; (*e*) puparium from above; (*f*) full-grown larva; (g) anal spiracles of same. (From U.S.D.A., E.R.B.)

widely varied habitats. In some species the larvae feed on young plant shoots and, in at least one species, the larvae live in the subcutaneous layers of the skin of frogs and are blood feeders. Several are predators in the egg cocoons of other insects and spiders.

BRAULIDAE, BEE LOUSE. The minute parasite of the honeybee, *Braula caeca*, is the sole member of the Braulidae. It is able to distract the bee and to remove regurgitated honey directly from the bee's mouthparts.

Section Calyptratae. This section is characterized by the development of upper and lower *squamae* in the adults and the enlargement of the posterior spiracles in the larvae. The section contains about ten families occurring through a wide ecological range in the superfamilies Muscoidea and Glossinoidea.

Superfamily Muscoidea. More primitive members of this superfamily include the Anthomyiidae and Muscidae; more specialized families include the Sarcophagidae and Calliphoridae (flesh flies) and other families mentioned below.

ANTHOMYIIDAE. Members of this family are similar in general appearance to the housefly. Economic species in the family include the onion maggot *Hylemya antiqua* (Fig. 8-154) and the cabbage maggot *Hylemya brassicae*, which feed on the roots of their respective hosts.

MUSCIDAE, THE HOUSEFLY AND ITS ALLIES. This family contains probably the world's commonest and most ubiquitous insect, the housefly

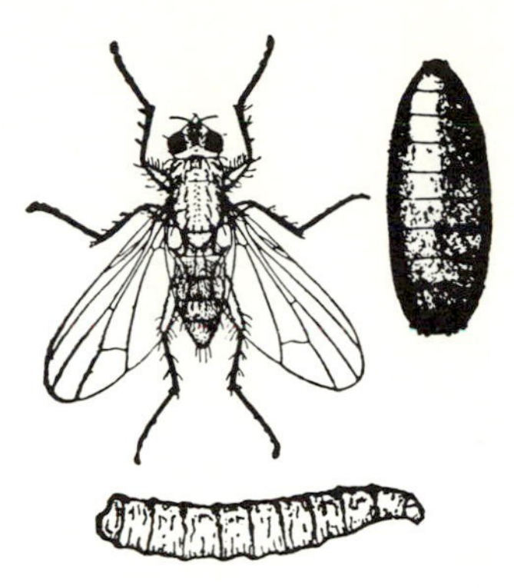

Fig. 8-154. Life history stages of the onion maggot *Hylemya antiqua*: adult, larva, and puparium. (From Illinois Natural History Survey)

Musca domestica (Fig. 8-155). Housefly larvae are white maggots that breed in many types of decaying matter. The adult flies transmit several dangerous and widespread diseases, including typhoid fever, several kinds of dysentery, cholera, and trachoma. Another widespread member of this family is the stable fly *Stomoxys calcitrans*, the adult of which inflicts a painful bite and attacks humans and domestic animals. Its larvae breed in decaying organic matter; rotting piles of new grass and lawn clippings and manure are high on the list of favorites.

GASTEROPHILIDAE, BOTFLIES. These are moderate-sized flies about the size of a honeybee, somewhat hairy in appearance, and banded with black, yellow, or red (Fig. 8-156). The larvae are internal parasites of horses, mules, and some of the larger wild mammals. In North America only the genus *Gasterophilus* is represented by four species that infest horses. The horse botfly *G. intestinalis* lays its eggs on the

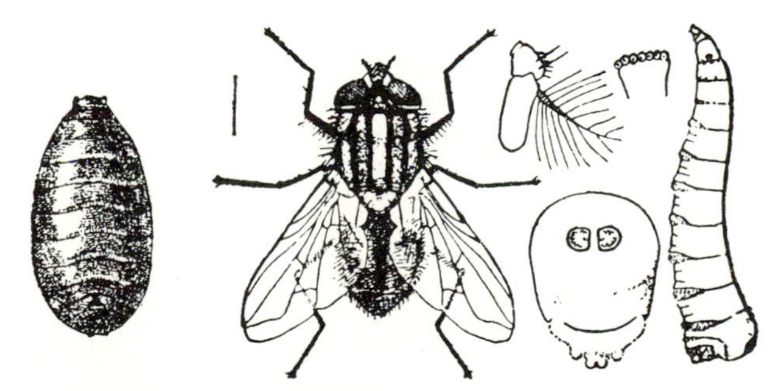

Fig. 8-155. Stages of the housefly *Musca domestica*. Puparium at left; adult next; larva and enlarged parts at right. (From U.S.D.A., E.R.B.)

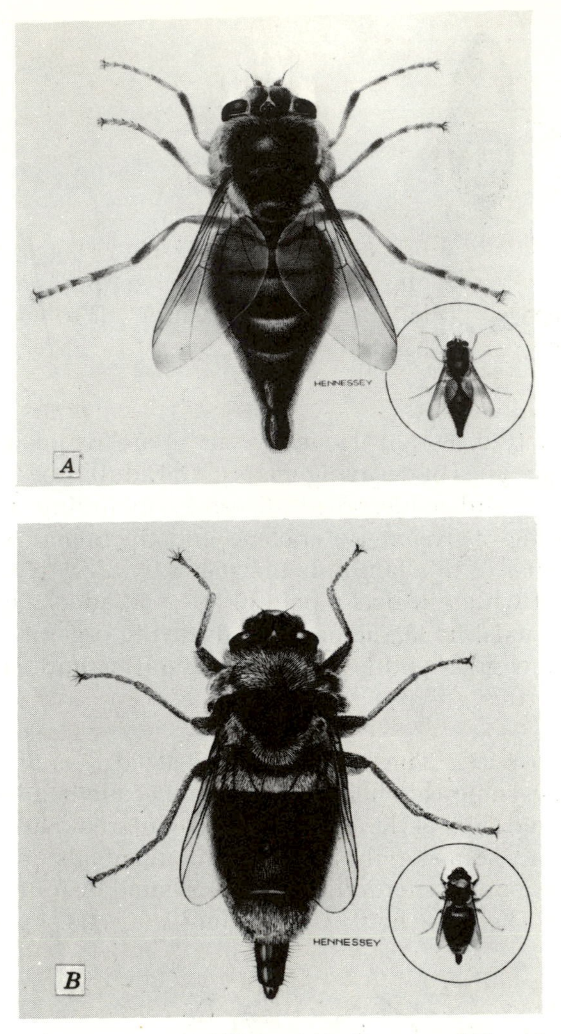

Fig. 8-156. Botflies. (*A*) *Gasterophilus intestinalis*; (*B*) *G. haemorrhoidalis*. (From Canadian Department of Agriculture)

hairs of the horse's legs and forequarters. When the horse licks these eggs, they hatch, and the young larvae work their way through the mouth and throat tissues into the horse's stomach. Here the young bots attach to the lining and feed, growing into stout spiny grubs 15 mm or more in length. They pass the summer and winter in the horse; in spring they loosen their hold, are passed by the horse, and pupate in the ground. The adults emerge in a few weeks.

CALLIPHORIDAE, BLOWFLIES, BLUEBOTTLES. These large, solidly built flies are cosmopolitan in distribution and are noisy daytime fliers. Adults are attracted to moisture apparently because they feed mainly on liquified decomposition or on nectar and other sweet liquids. The larvae of some live in animal flesh, several parasitize earthworms, and several live in termite or ant colonies. One group is parasitic on terrestrial snails. Although numerous, blowflies transmit fewer diseases and are economically less important than other related fly families. A few species of blowflies oviposit on raw, already infected skin of sheep and the larvae commence carrion feeding causing cutaneous myiasis on the sheep. This condition leads to further infestations by other fly species that normally would not be attracted to the sheep.

SARCOPHAGIDAE, FLESH FLIES. This is also a large, cosmopolitan family that is closely related to the Calliphoridae; however, most members are viviparous. Some larvae are parasites of food stored in wasps' nests, but most live in carrion or decaying organic matter. A few are important in attacking the skin of living animals and humans. In one genus, *Blaesoxiphia*, the larvae are entirely endoparasitic in the grasshopper family Acrididae.

OESTRIDAE, WARBLE FLIES. These (Fig. 8-157) comprise a small group of fast-flying fuzzy bumblebee-like flies, parasites of mammals; some of them have a life cycle as specialized as that of the Gasterophilidae. Many species have a simple life cycle. For example, the sheep botfly *Oestrus ovis* deposits young larvae in the nostrils of sheep; the larvae migrate into the sinuses and horns, where they mature. The mature larvae escape to the ground through the sheep's nostrils and pupate. Much more complex is the life cycle of the northern cattle grub *Hypoderma bovis*, which attacks cattle. The fly lays eggs on the hairs of the hind legs or flank of the animal. The larvae soon hatch, crawl down the hair, and burrow beneath the skin, making their way slowly through connective tissue to the esophagus; this journey takes about 4 months. After 3 months' development in the esophagus the maggots (still fairly small) journey through connective tissue again and come to rest beneath the skin in the lumbar region. Here the larvae attain most of their growth, each causing a swelling called a warble, provided with a small hole through the skin. Through this the larva first obtains air while maturing and then escapes to the ground for pupation.

TACHINIDAE (LARVAEVORIDAE), TACHINA FLIES. This (Fig. 8-158) is one of the largest families of Diptera. All its members are parasitic on insects and attack larvae or adults of many orders, especially lepidopterous larvae. Many members of this family have been propagated and introduced into various parts of the world in an effort to hold in

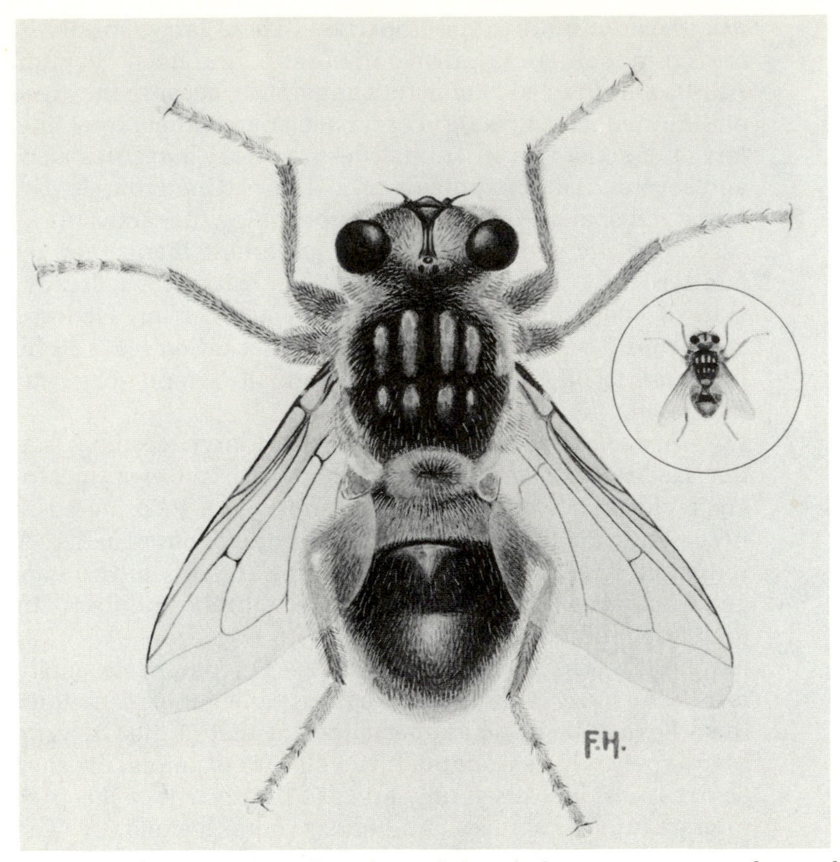

Fig. 8-157. A warble fly, the adult of the common cattle grub *Hypoderma lineatum*. (From Canadian Department of Agriculture)

check some injurious insect. The adult females of many species lay their eggs directly on the host; when they hatch, the maggots bore into the host body and feed on its tissues. In other species the eggs are laid on foliage; if this foliage is eaten by a larva of the host species, the parasite eggs hatch in the alimentary canal of the host, and from there the young larvae bore into the body cavity where development takes place. When mature, the parasite larvae either pupate in the host body or leave it and pupate in the ground.

SUPERFAMILY GLOSSINOIDEA

Here belong four small families living on blood meals or as ectoparasites on the bodies of birds and mammals. So far as is known, most members of the Glossinoidea have the feature, very unusual among insects, of uterine development of the young. One at a time larvae are retained in the body of the female in a special uterine pouch and nourished on glandular secretions.

GLOSSINIDAE, TSETSE FLY. This family includes only one genus, *Glossina*, of about 20 species that live in Africa. Tsetse flies, slightly larger than houseflies, transmit through their bites several serious trypanosome (protozoan) diseases, such as African sleeping sickness in humans and nagana disease in cattle. Tsetse flies are obligate blood feeders and usually take daily blood meals that are several times their own body weight. A single young larva is hatched within the female and for about 10 days the larva is retained and fed by secretions from highly modified glands in the mother's uterine wall. When adequately nourished, the mature larva weighs as much as the mother. Within 1 to 2 hours after leaving the mother, the larva buries itself in soil and pupates. About 30 days later the adult emerges.

HIPPOBOSCIDAE, LOUSE FLIES. This family includes greatly modified flies that suck blood and live as ectoparasites in the hair of mammals or the feathers of birds. These flies have flattened bodies and stout legs and generally appear louse- or ticklike. Most have well-developed wings with strong anterior veins. A few shed their wings once they become established in their host; some have reduced wings or wings may be lacking, as in *Melophagus* (Fig. 8-159). Usually the mature larvae are not deposited on the host.

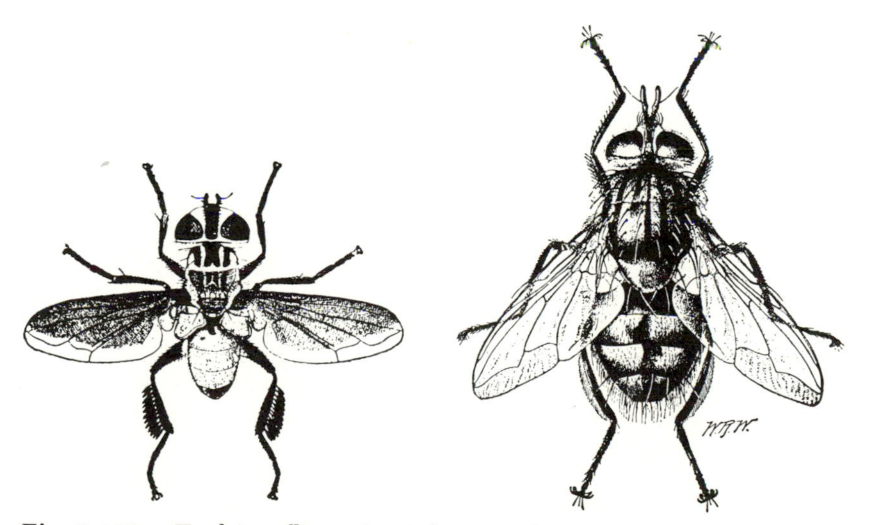

Fig. 8-158. Tachina flies. At right, *Winthemia quadripustulata*, a parasite of Lepidoptera larvae; at left, *Trichopoda pennipes*, a parasite of Hemiptera. (From U.S.D.A., E.R.B.)

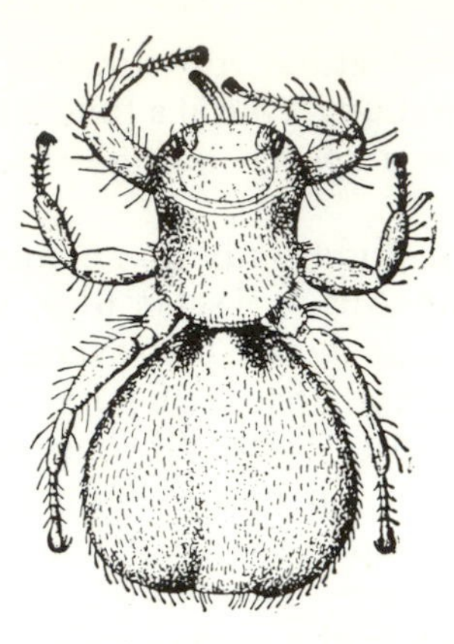

Fig. 8-159. The sheep tick *Melophagus ovinus.* (From Kentucky Agricultural Experiment Station)

The best known species in North America is the sheep tick, *Melophagus ovinus* (Fig. 8-159). This wingless ticklike insect is common on domestic sheep in many parts of the world, particularly in temperate latitudes. When its mature larvae are "born," they are glued to the host's wool where they immediately pupate. Two thirds of the species in the family are found only on birds.

OTHER GLOSSINOIDEA. The members of the Streblidae and their close relatives, the Nycteribiidae, are bloodsucking ectoparasites on bats. Some of these specialized species appear to have become host specific. Most of the species in these two families are tropical or subtropical in their distribution.

References

Berg, C. O., and L. Knutson, 1978. Biology and systematics of the Sciomyzidae. *Annu. Rev. Entomol.*, **23**:239–258.

Cole, F. R., 1969. The flies of western North America. Berkeley and Los Angeles: Univ. of California Press. 693 pp.

Crampton, G. C., et al., 1942. The Diptera or true flies of Connecticut. *Conn. State Geol. Nat. Hist. Surv.*, **64**: 1–509.

Curran, C. H., 1934. *The families and genera of North American Diptera.* New York: Privately published, The Ballou Press. (Reprinted, 1965, Woodhaven, N.Y.: Henry Tripp).

Demerec, M. (ed.), 1950. *Biology of Drosophila.* New York: Wiley. 632 pp.

Downes, J. A., 1971. The ecology of blood-sucking Diptera: An evolutionary perspective. In A. M. Fallis (Ed.), *Ecology and physiology of parasites.* Toronto: Univ. of Toronto Press. p. 232–258.

Frost, S. W., 1924. A study of the leaf-mining Diptera of North America. *Cornell Univ. Agr. Exp. Stn. Mem.* **78**:1–228.

Gillett, J. D., 1971. *Mosquitoes.* London: Weidenfield & Nicholson. 274 pp.

Herring, E. M., 1951. *Biology of the leaf miners.* The Hague: Junk. 420 pp.

Hull, F. M., 1962. Robberflies of the world: The genera of the family Asilidae. *U.S. Natl. Mus. Bull.* **224**:1–907 (2 vols.).

Knight, K. L., and A. Stone, 1977. A catalog of the mosquitoes of the world (Diptera:Culicidae). *Thomas Say Foundation, Entomol. Soc. Am.*, **6**:1–612.

Oldroyd, H., 1964. *The natural history of flies.* London: Weidenfield & Nicholson. 324 pp.

Oliver, D. R., 1971. Life history of the Chironomidae. *Annu. Rev. Entomol.*, **16**:211–230.

Roback, S. S., 1951. A classification of the muscoid calyptrate Diptera. *Ann. Entomol. Soc. Am.*, **44**:327–361.

Seguy, E., 1951. Ordre des Diptères (Diptera Linne, 1758). *Traité de Zool.*, **10**:449–744.

Stone, A., C. W. Sabrosky, W. W. Wirth, R. H. Foote, and J. R. Coulson (Eds.), 1965. *A catalogue of the Diptera of America north of Mexico.* Washington, D.C.: U.S. Govt. Printing Office. 1696 pp.

Wirth, W. W., and A. Stone, 1956. Aquatic Diptera. In R. L. Usinger (Ed.), *Aquatic insects of California.* Berkeley: Univ. of California Press. pp. 372–482.

Order TRICHOPTERA: Caddisflies

These are mothlike insects having aquatic larvae and pupae. The adults (Fig. 8-160) vary in length from 1.5 to 40 mm. They have chewing-type mouthparts, with all parts greatly reduced or subatrophied except for the two pairs of palpi; long multisegmented antennae; large compound eyes; and long legs. Except for the wingless or brachypterous females of a few species, the adults have two pairs of large membranous wings, with a fairly complete set of longitudinal veins, but a reduced number of crossveins. In most species the body and wings have hair but no scales; in a few species the antennae, palpi, legs, and wings may have patches of scales or a scattering of scales among the longer hair. The larvae (Figs. 8-161 to 8-164) are diverse in general appearance and habits. The eyes are each represented by a single facet, the antennae are small and one segmented, the mouthparts are of the chewing type and well developed. The thorax has three pairs of strong legs, the abdomen a pair of strong legs bearing hooks and frequently a set of finger-like gills.

Most of the adults are somber in color or tawny, but a few have wings that are marked with yellow or orange, or have silvery streaks.

Caddisflies are divided into two suborders, the Annulipalpia and Integripalpia (Fig. 8-165). Annulipalpia consist of one superfamily: the Hydropsychoidea (Fig. 8-165, 2), whose larvae are net makers or retreat makers. Integripalpia include two superfamilies: the Limnephiloidea (Fig. 8-165, 8), whose larvae are tube-case makers

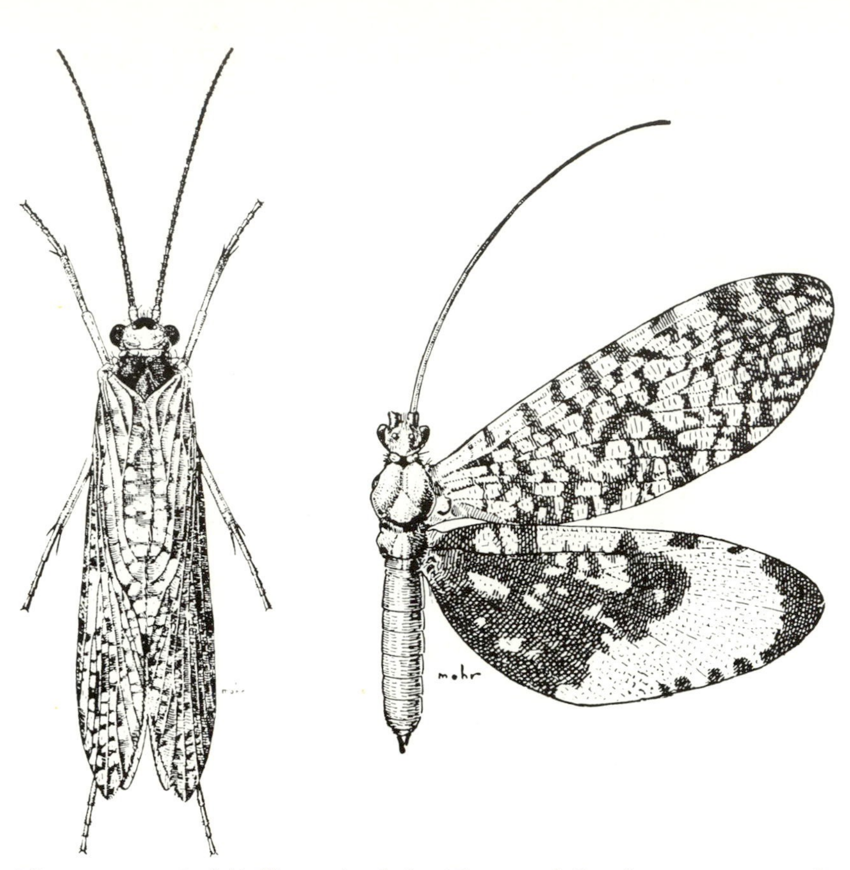

Fig. 8-160. Caddisflies. At left, *Rhyacophila fenestra*; at right, *Eubasilissa pardalis*. (From Illinois Natural History Survey)

(Fig. 8-164); and the Rhyacophiloidea (Fig. 8-165, 5 to 7). The Rhyacophiloidea include three families that have relatively primitive larval features: Rhyacophilidae have free-living larvae, Glossosomatidae are saddle-case makers (both the head and abdomen protrude), and Hydroptilidae are purse-case makers (Fig. 8-163). Limnephiloidean caddisflies are the most diverse and include about two dozen families (Fig. 8-165, 8 to 18). In all, about 30 families of caddisflies, several hundred genera, and more than 7000 species are recognized. A large number of the families are worldwide, however, some are restricted to either the northern hemisphere or to parts of the southern hemisphere. The nearctic caddisfly fauna contains over 800 species belonging to 17 families.

Case making has been developed to a high degree by the larvae of most families of these insects. The cases, which are portable "houses" built by the larvae, have fascinated observers of freshwater

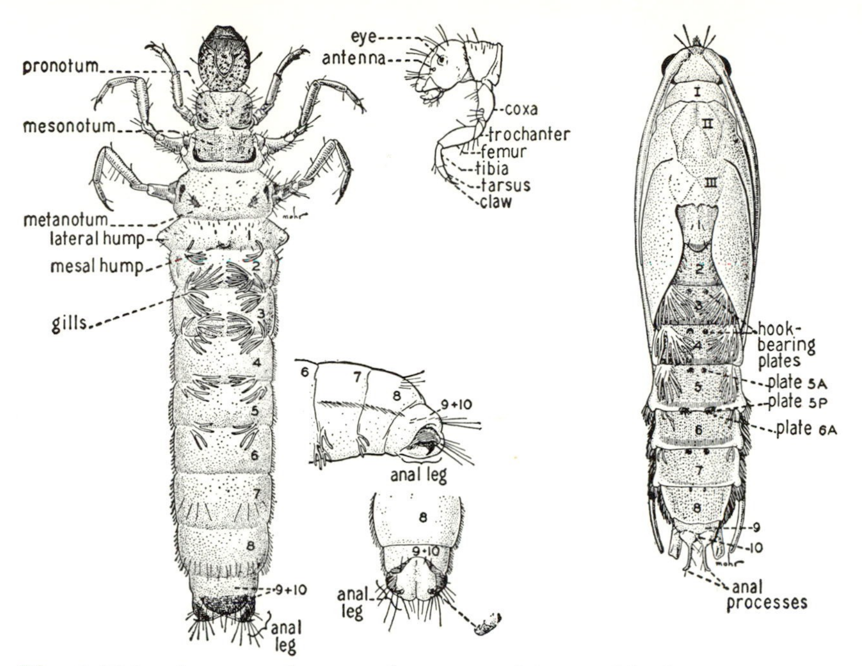

Fig. 8-161. Larva and pupa of a case-making caddisfly *Limnephilus submonilifer*. (From Illinois Natural History Survey)

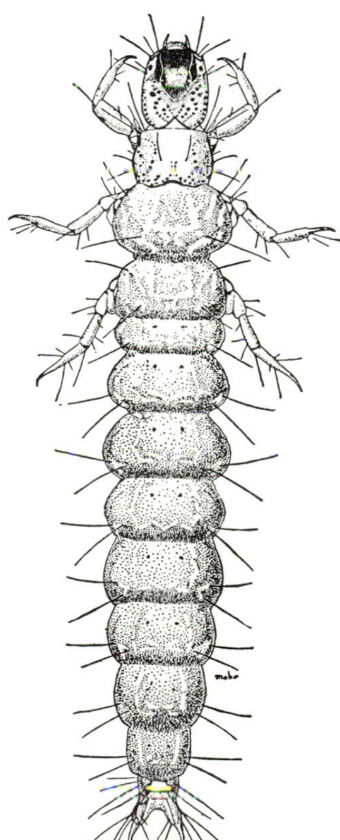

Fig. 8-162. Free-living caddisfly larva *Rhyacophila fenestra*. (From Illinois Natural History Survey)

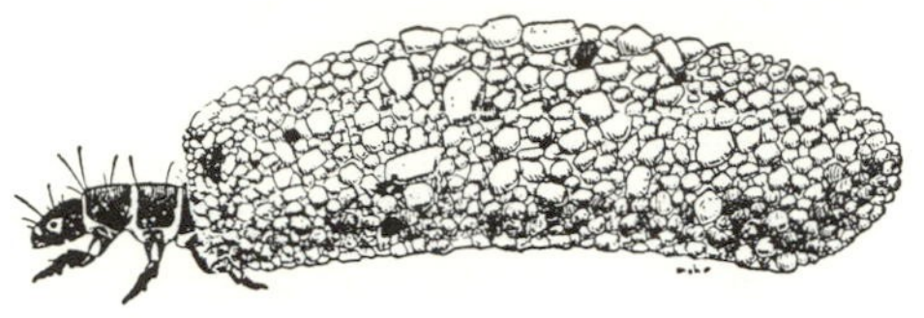

Fig. 8-163. Purselike caddisfly case and larva, *Ochrotrichia unio*. (From Illinois Natural History Survey)

Fig. 8-164. Caddisfly larva in cylindrical case, *Limnephilus rhombicus*. (From Illinois Natural History Survey)

insect life. Cases (Figs. 8-163, 8-164) are of varied shapes, ranging from a straight tube to the coiled case of *Helicopsyche*, which resembles a snail shell. Many types of materials are used in the construction of these cases. Small stones, sand grains, bits of leaves, sticks, conifer needles, and frequently small snail shells may be utilized. In most instances a given genus or species constructs a case of characteristic shape, but genera in different families often make very similar cases.

Caddisfly females lay from 300 to 1000 eggs each. In some species the eggs are discharged in strings (Fig. 8-166*B*); the female enters the

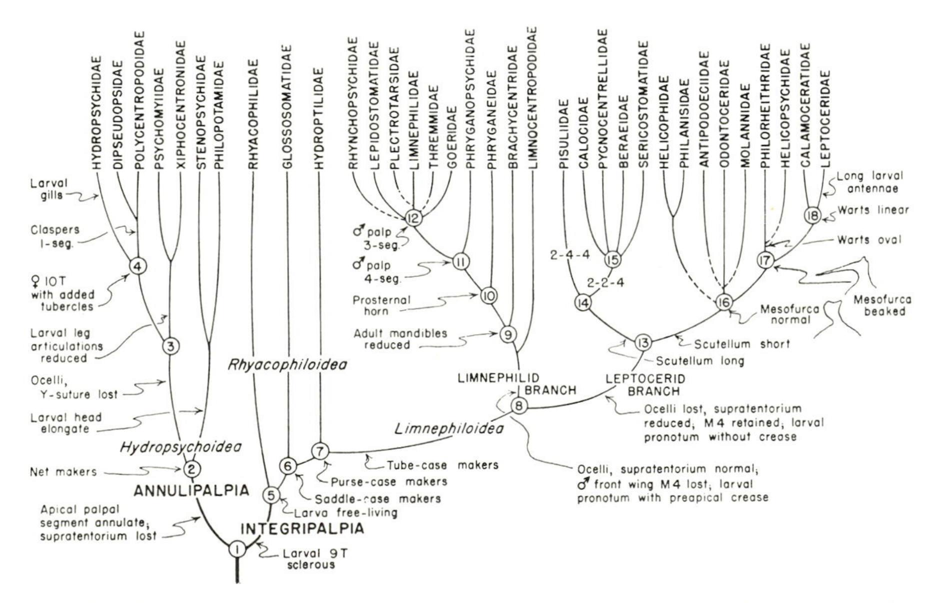

Fig. 8-165. Phylogenetic diagram of the Trichoptera families. (Reproduced with permission from the *Annual Reviews of Entomology*, Vol. 12, © 1967 by Annual Reviews, Inc.)

water and lays the eggs on stones or other objects, grouping the strings into irregular masses containing up to 800 eggs. Females of other groups extrude the eggs and form them into a large mass at the tip of the abdomen before depositing them (Fig. 8-166*A*). The eggs are encased in a gelatinous matrix that swells on absorbing moisture. These masses are attached to sticks or stones that are submerged in, adjacent to, or overhanging water.

Females of the family Limnephilidae frequently deposit egg masses on branches overhanging water. On numerous occasions observers have reported that these egg masses swell and liquefy, and the eggs hatch during rain. The gelatinous drops formed by this process run down the twigs and drop into the water, carrying the larvae along.

The larvae are all active, most of them feeding chiefly on small aquatic animals or microorganisms that encrust decayed organic matter in the water. A few genera, notably *Rhyacophila* (Rhyacophilidae) and *Oecetis* (Leptoceridae), are predominantly predaceous, feeding on small insect larvae. The Hydropsychoidea have larvae that weave a fixed net and shelter, the net being used to trap small aquatic organisms for food. In nearly all other families the larvae construct portable cases of various types. The larva uses these to protect the greater part of the body, which has thin integument. Only the heavily sclerotized head, legs, and anterior portion of the thorax are extruded from the case when the larva is actively moving about (Figs. 8-163, 8-164).

Prior to pupation, the larvae of all caddisflies spin a cocoon. The case makers form this very simply by spinning a silken lining inside

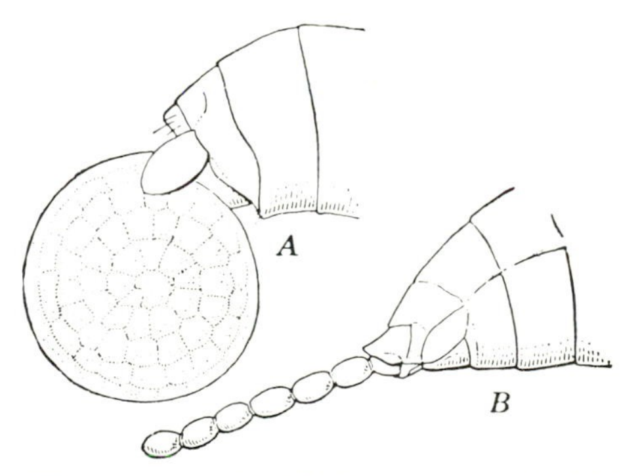

Fig. 8-166. Caddisfly eggs. (*A*) *Triaenodes tarda*; (*B*) *Cyrnellus marginalis*. (From Illinois Natural History Survey)

the case and closing the ends of the case with a barred or slit membrane. The free-living and net-making species spin an ovoid cocoon of silk and sand, stones, or bits of debris, which is firmly attached to a stone, log, or other rigid support. The pupae develop in the cocoon until the adult structures (except wings) are completely formed and fully sclerotized. The pupae (Fig. 8-161) are unusual in possessing strong mandibles. With these, the mature pupa cuts its way out of the cocoon. It then swims to the surface, crawls on a log or stone, and transforms into an adult.

Usually the complete life cycle requires a year, most of it spent in the larval stage. The egg stage lasts only a short time; the pupal stage requires 2 to 3 weeks; and the adults live about a month.

Of interest among caddisflies are the larval habits of "micro" caddisflies, or Hydroptilidae. This family contains only small individuals, ranging from 1.5 to 6 mm in length. The first instars are minute free-living forms, with small abdomens; the last instar builds a portable case and develops a swollen abdomen. Information is available only on two North American genera, *Mayatrichia* and *Ochrotrichia*. In the latter there are differences in the structure of the claws between the free-living and casemaking instars. The dimorphism is similar in many respects to typical examples of hypermetamorphosis.

Caddisfly larvae live in both lakes and streams, showing a definite preference for colder and unpolluted water. Their species are good indicators of levels of stream pollution and individual species have relatively narrow ranges of ecological tolerances. As a group they have a wider ecological tolerance but are more restricted than midge larvae (Chironomidae) in relation to pollution. Caddisflies are important in stream ecosystems because they form a major level in the food chain and are a primary source of food for many species of fishes.

Well-illustrated keys to larvae, pupae, and adults of the families of North American Trichoptera are given by Wiggins in Merritt and Cummins (1978).

References

Anderson, N. H., 1976. The distribution and biology of the Oregon Trichoptera. *Oreg. Agric. Exp. Stn. Tech. Bull.*, **131**:1–152.

Betten, C., 1934. The caddisflies or Trichoptera of New York State. *N.Y. State Mus. Bull.*, **292**:1–576.

Crichton, M. I. (Ed.), 1978. *Proceedings of the 2nd International Symposium on Trichoptera*. The Hague: Junk. 360 pp.

Denning, D. G., 1956. Trichoptera. In R. L. Usinger (Ed.), *Aquatic insects of California*. Berkeley: Univ. of California Press. pp. 237–270.

Nielson, A., 1948. Postembryonic development and biology of the Hydroptilidae. *Kgl. Danske Videnskab., Biol. Skr.*, **5**(1):1–200.

Ross, H. H., 1944. The caddisflies or Trichoptera of Illinois. *Bull. Ill. Nat. Hist. Surv.*, **23**(1):1–326.

——— 1956. *Evolution and classification of the mountain caddisflies.* Urbana: Univ. of Illinois Press, 213 pp.

——— 1964. Evolution of caddisworm cases and nets. *Am. Zool.*, **4**:209–220.

——— 1967. The evolution and past dispersal of Trichoptera. *Annu. Rev. Entomol.*, **12**:169–206.

Wiggins, G. B., 1977. *Larvae of the North American caddisfly genera (Trichoptera).* Toronto: Toronto Univ. Press. 401 pp.

——— 1978. Trichoptera. In R. W. Merritt and K. W. Cummins (Eds.), *An introduction to aquatic insects.* Dubuque, Iowa: Kendall/Hunt. pp. 147–185.

Order LEPIDOPTERA: Moths and Butterflies

These insects have two pairs of wings (Fig. 8-167), except for a few species that have apterous females. The body, wings, and other appendages are covered with scales that are often brilliant in color and arranged in showy patterns. Adult mouthparts are greatly reduced; in most forms only the maxillae are well developed. These are fused and elongated to form a coiled tube for sucking up liquid food. The large compound eyes, the long antennae, and the legs are all well developed. The species have complete metamorphosis. The larvae

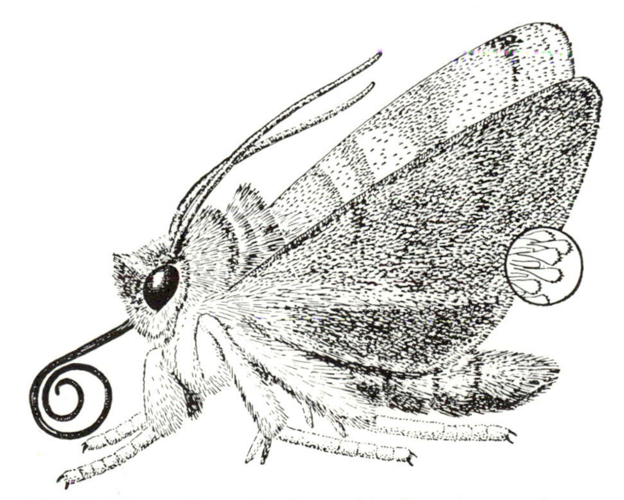

Fig. 8-167. A typical moth, showing scales on wings and body, and sucking tube, which is coiled under head when not in use. (From Illinois Natural History Survey)

are cylindrical (the familiar caterpillars); most of them have a definite head, thoracic legs, and five pairs of larvapods (including the end pair). The pupae are usually hard and brown, and the appendages appear cemented onto the body.

The order is a large one, embracing about 10,000 North American species. The adults of these vary greatly in size from minute forms a millimeter or two long to large species with a wing span of 15 cm. or more. The wings are distinctively patterned, and the order as a whole presents an attractive array of color and beauty.

The larvae of the Lepidoptera are plant feeders except for a scattering of species that are predaceous, scavengers, or feed on stored products. The great bulk of the species feed externally on foliage; a large number of the minute species mine inside leaves or leaf petioles; and another large group, including both large and small species, bore inside stems, trunks, or roots. A great number of these species attack cultivated plants and cause a high annual loss of crops and stored products. The order is therefore one of great economic importance.

In addition to its destructive species, the order Lepidoptera contains one of great commercial value: the silkworm *Bombyx mori*. This insect is the sole source of natural silk. The propagation of the species and the harvesting of the silk, known as sericulture, is an important industry in many parts of the world, with an annual harvest in oriental countries of many million dollars' worth of silk. At the turn of the century a sericulture industry was established in California. This failed because of high labor costs. In 1945 a new venture was started in Texas in which a large part of the work was done mechanically.

Phylogeny. In three families of Lepidoptera (Fig. 8-168), the front and hind wings are similar in size and venation and the front wing has a lobelike coupling device called the *jugum* (Fig. 8-169*A*: j) at the base of the hind margin. These characters resemble the condition in Trichoptera, and indicate that these three families, called the suborder Jugatae, are the most primitive group of the Lepidoptera. In the family Micropterygidae the larvae have legs on eight abdominal segments (reminiscent of some Mecoptera larvae) and the adults have functional mandibles. These conditions indicate that the Micropterygidae are descendants of the most archaic known branch of the suborder, indeed, of the entire order. In the other line of the Jugatae the adult mandibles were reduced, then lost, and the larval abdomen lost at least one pair of legs. This branch of the Jugatae gave rise to two typical jugate families, the Hepialidae (swifts) and the small moths of the family Eriocraniidae.

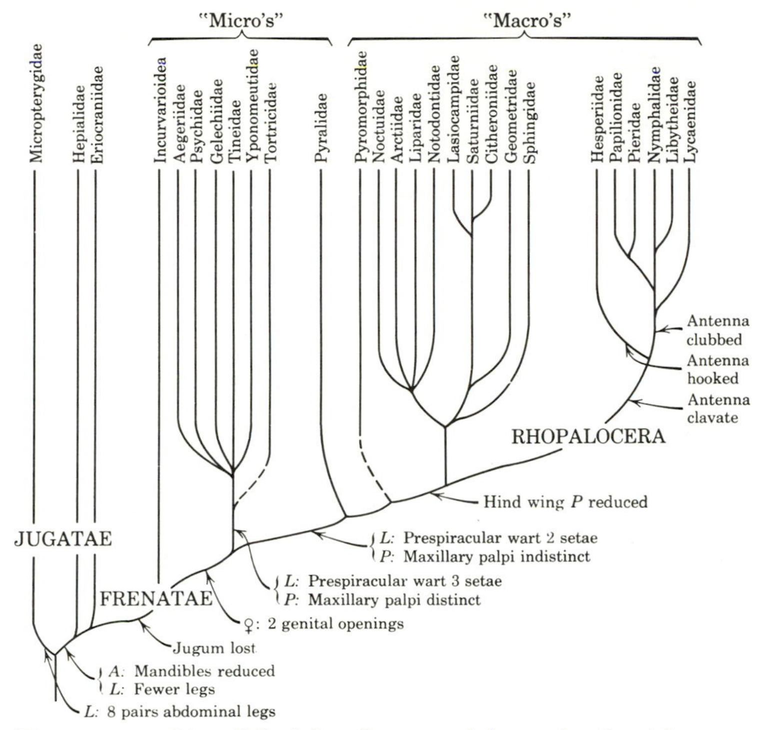

Fig. 8-168. Simplified family tree of the order Lepidoptera (based chiefly on Forbes, 1923). *A*, Adult; *L*, larva; *P*, pupa.

From the base of the Eriocraniidae line arose a lineage in which the jugum was lost and a long hair or *frenulum* (Fig. 8-169E: *f*) at the base of the hind wing was the chief wing-coupling structure. In this lineage the front and hind wings became dissimilar in both shape and venation. This was the beginning of the suborder Frenatae. Our only living representatives of the early Frenatae are a group of seldom-collected minute moths comprising the superfamily Incurvarioidea.

From an early ancestral Frenatae evolved a form having two genital openings in the female, one for insemination and one for egg laying. This form gave rise to two major branches. One branch that changed little gave rise to many families of small or minute moths, such as the Tineidae and Tortricidae, and to certain highly distinctive families such as the Psychidae (bagworms) and Aegeriidae (clearwings).

In the other branch certain body setae of the larvae atrophied and the maxillary palpi of the pupae became indistinct. The more primi-

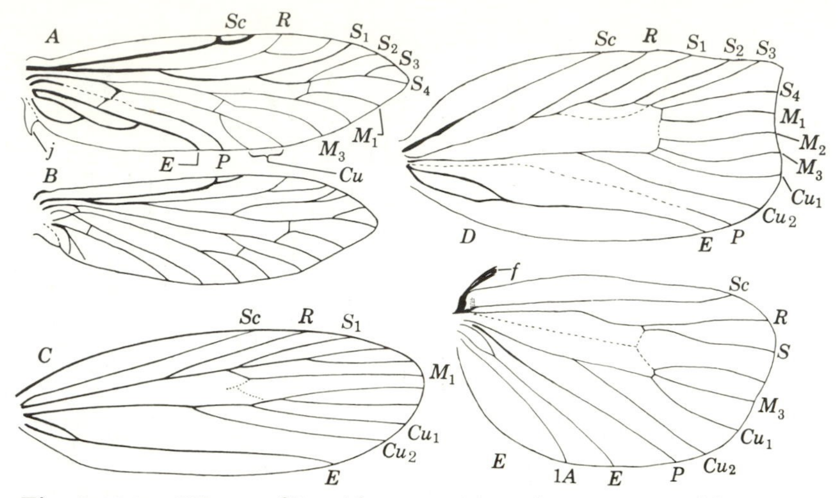

Fig. 8-169. Wings of Lepidoptera. (*A* and *B*) Front and hind wings of *Mnemonica*, Eriocraniidae; (*C*) front wing of *Achroia*, Pyralidae; (*D* and *E*) front and hind wings of *Archips*, Tortricidae. *f*, Frenulum; *j*, jugum. (*C* loaned by Dr. Kathryn M. Sommerman)

tive members of this branch may be the Pyralidae and the rarer Pyromorphidae, in which *P* of the hind wing remained distinct. More specialized members appear to be the larger moths and the butterflies, in which *P* of the hind wing became reduced and indistinct. The large-moth lineage evolved into the abundant families Noctuidae, Arctiidae, Saturniidae, and many others.

From the base of the large-moth branch evolved a lepidopteran whose adult was diurnal and had antennae enlarged at the tip. This lineage gave rise to the suborder Rhopalocera, which includes the skippers (Hesperiidae), having the adult antennae hooked at the tip; and the butterflies (Papilionidae and allies), having the antennae clavate.

KEY TO COMMON FAMILIES

1. Wings reduced to small pads or entirely lacking (Fig. 8-182*b*) 2
   Wings well developed, at least nearly as long as abdomen (Fig. 8-167) 4
2. Legs lacking or reduced to short stubs; usually associated with a bag-like case (Fig. 8-174). ..... **Psychidae**, p. 487
   Legs elongate and normal in appearance (Fig. 8-182*b*) 3

3. Abdomen having closely set scales or spines, or bristling dark gray hair; usually not found near cocoon .... **Geometridae**, p. 494
   Abdomen smoothly clothed with fine light wooly hair; moth usually found clinging to cocoon .... **Liparidae**, p. 497
4. Front wing having a *jugum*, a lobe at the base of the posterior margin for use in wing coupling (Fig. 8-169*A*); front and hind wings similar in venation and shape (**Jugatae**) .... **Hepialidae**, p. 486
   Front wing without a jugum; either anterior margin of hind wing having an enlarged lobe at base (Fig. 8-171*D*, *H*, *I*), or with a long basal spine, or *frenulum* (Fig. 8-171*E*), both used for wing coupling; or front and hind wings markedly different in shape and venation 5
5. Hind wing without a frenulum, and antenna clubbed or hooked at apex (Fig. 8-170*E*, *F*) (**Rhopalocera**) 6
   Either hind wing having a frenulum, or antenna not clubbed or hooked; instead either threadlike, or serrate, or pectinate (**Frenatae**) 11

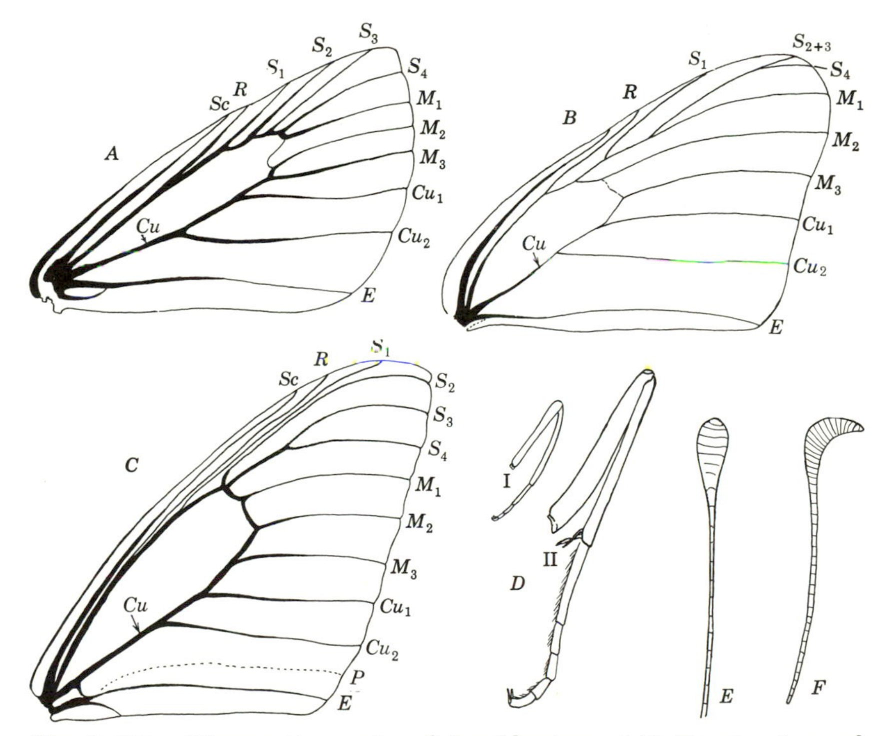

Fig. 8-170. Diagnostic parts of Lepidoptera. (*A*) Front wing of *Pamphila*, Hesperiidae; (*B*) front wing of *Pieris*, Pieridae; (C) front wing of *Papilio*, Papilionidae; (*D*) front (small) and middle legs of *Brenthis*, Nymphalidae; (*E*) tip of antenna of same; (F) tip of antenna of *Thanaos*, Hesperiidae.

6. Front wing having each of the 4 branches of sector and 3 of media arising from the discal cell (Fig. 8-170*A*); antennae usually hooked at apex (Fig. 8-170*F*) .... **Hesperiidae**, p. 500
   Front wing having some of these fused at base, branching beyond discal cell (Fig. 8-170*B*, *C*) 7
7. Front wing having *Cu* appearing 4-branched (Fig. 8-170*C*), because both M_2 and M_3 are more closely associated with it than with S_4 .... **Papilionidae**, p. 500
   Front wing having *Cu* appearing 3-branched (Fig. 8-170*B*) because M_2 is more closely associated with S_4 8
8. Labial palps longer than thorax, thickly hairy, and extending forward .... **Libytheidae**
   Labial palps shorter than thorax 9
9. Front legs reduced, and much shorter than the other legs (Fig. 8-170*D*) .... **Nymphalidae**
   Front legs larger in proportion to the others 10
10. Front wing having M_1 fused for a considerable distance with *S* (Fig. 8-170*B*); colors white, yellow, or orange, plus black marks .... **Pieridae**, p. 501
    Front wing having M_1 either not fused with *S*, or only slightly so, thus arising from discal cell or very near it; colors coppery, blue, or brown .... **Lycaenidae**
11. Hind wing having a posterior fringe as long as wing is wide (Figs. 8-172, 8-176); wing usually lanceolate. A large number of families of small moths difficult to identify, including Yponomeutidae, Gelechiidae, and Tineidae. .... Not keyed further here
    Hind wing markedly wider than its fringe (Figs. 8-177, 8-178) 12
12. Front wing narrow, more than four times as long as wide; hind wing and sometimes front wing having transparent areas devoid of scales (Fig. 8-177) .... **Aegeriidae**, p. 490
    Front wing wider, hind wings usually entirely covered with scales 13
13. Hind wing having 3 veins posterior to Cu_2 (Fig. 8-169*E*) 14
    Hind wing having 1 or 2 veins posterior to Cu_2 (Fig. 8-171*E*) 15
14. Front wing with *P* fairly well developed, at least toward apex (Fig. 8-169*D*) .... **Tortricidae**, p. 490
    Front wing with *P* atrophied (Fig. 8-169*C*) .... **Pyralidae**, p. 491
15. Front wing with *E* evenly bowed anteriorly, the vein coming close to central portion of *Cu* or apex of Cu_2 (Fig. 8-171*F*) .... **Sphingidae**, p. 494
    Front wing with *E* straight or only slightly sinuate (Fig. 8-171*C*) 16
16. Front wing having both M_2 and M_3 associated closely with *Cu*, which therefore appears 4-branched (Fig. 8-171*C*) 17
    Front wing having M_2 either midway between M_3 and M_1, or closer to M_1, so that *Cu* appears 3-branched (Fig. 8-171*A*) 20
17. Hind wing without a frenulum, the base of the front margin greatly expanded (Fig. 8-171*D*) .... **Lasiocampidae**, p. 493

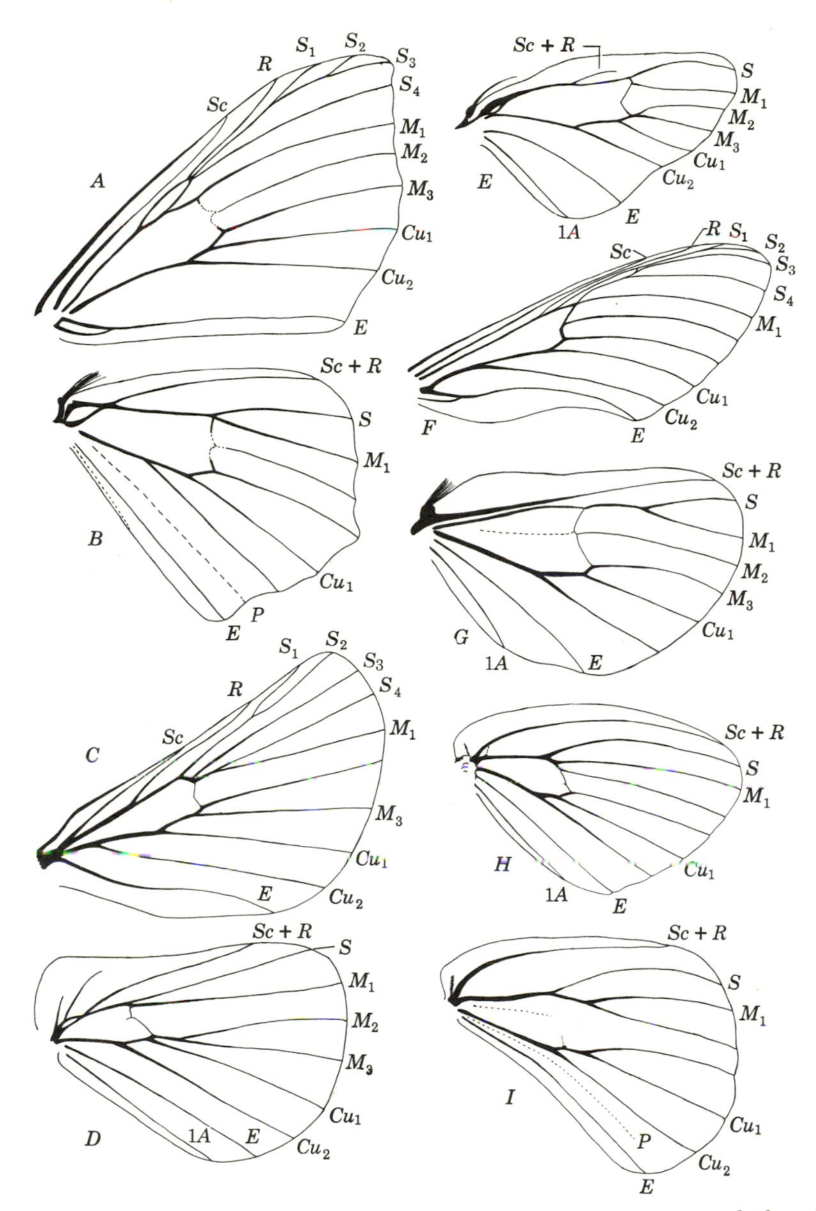

Fig. 8-171. Wings of Lepidoptera. (*A*, *B*) Front and hind, *Acidalia*, Geometridae; (*C*, *D*) front and hind, *Malacosoma*, Lasiocampidae; (*E*) hind, *Halisidota*, Arctiidae; (*F*) front, *Protoparce*, Sphingidae; (*G*) hind, *Hyperaeschra*, Notodontidae; (*H*) hind, *Citheronia*, Citheroniidae; (*I*) hind, *Samia*, Saturniidae.

Hind wing either with a frenulum or base of wing not expanded 18

18. Hind wing having *Sc* + *R* fused with S for about half the length of discal cell, then the 2 veins separating (Fig. 8-171*E*) . . . . **Arctiidae**, p. 499

Hind wing having *Sc* + *R* fused with S for only a short distance, or the veins not at all fused 19

19. Head having two ocelli . . . . **Noctuidae**, p. 497

Head without ocelli . . . . **Liparidae**, p. 497

20. Hind wing having *Sc* + *R* arcuate, curving forward from its base and well separated from S for its entire distance (Fig. 8-171*H*, *I*); frenulum obsolete 21

Hind wing having *Sc* + *R* either fused for a distance with S (Fig. 8-171*B*), or running close to it (Fig. 8-171*G*); frenulum present, often tuftlike 22

21. Hind wing having two veins after cubitus (Fig. 8-171*H*) . . . . **Citheroniidae**, p. 497

Hind wing having only one vein after cubitus (Fig. 8-171*I*) . . . . **Saturniidae**, p. 495

22. Hind wing with *Sc* making a short sharp angulation at the base of the wing (Fig. 8-171*B*) . . . . **Geometridae**, p. 494

Hind wing with *Sc* not angulate at base (Fig. 8-171*G*) . . . . **Notodontidae**

SUBORDER JUGATAE

The Jugatae are the most primitive present-day Lepidoptera. In addition to possessing a jugum, adults of this suborder have front and hind wings of very similar shape and venation. In one group the pupae have long stout mandibles used for cutting an exit from the cocoon; the adult mouthparts include well-developed mandibles and lobelike maxillae and labium, which are not elongated and appressed to form the usual sucking tube. These characters are of great interest, because they are also found in the Trichoptera, and they demonstrate the close relationship between primitive members of the Trichoptera and of the Lepidoptera.

The Jugatae are represented in North America by only a few species. The family Micropterygidae has mandibulate adults of small size; its larvae feed on moss. The Eriocraniidae also includes small species but these have vestigial mandibles and a sucking tube; the larvae are leaf miners. The Hepialidae includes a few larger species (with a wing expanse up to 5 cm) called swifts because of their rapid flights; their larvae are wood borers or root feeders.

SUBORDER FRENATAE

Here belong the greater number of the Lepidoptera, representing a great variety of shapes, sizes, and habits. The adult, in addition to lacking a jugum, has hind wings that differ in shape from the front ones (Fig. 8-172), and that also have the sector usually reduced to a single vein, in the front wings the sector usually has three or four branches. The pupa is *obtect*; that is, the appendages appear embedded in the body of the pupa and are incapable of movement.

In North America we have 50 families represented, many of them containing species of great economic importance. The diagnosis and identification of these families are difficult and require making special preparations of the wings after their scales have been removed, a process known as denuding.

The "Micros." Many of the more primitive families of the Frenatae comprise small slender moths having the hind wings with vein *P* present, or bearing a long posterior fringe, or extremely narrow and pointed in outline. These families are frequently termed the Microlepidoptera or the Microfrenatae. A few members of the Microfrenatae are quite large, so that size is only an average characteristic for the group. Although it is impractical to give defining characters for them here, a few families are mentioned that contain common species of importance.

TINEIDAE. The larvae feed chiefly on fungi or fabrics, or as scavengers. Included in the Tineidae are our common clothes moths (Fig. 8-173). The webbing clothes moth *Tineola bisselliella* has a larva that makes an indefinite web; in addition to fabrics, it has a marked liking for old feathers. The larva of the case-making clothes moth *Tinea pellionella* makes a portable case of silk and fabric fragments. Both species are common pests of fabrics of animal origin, such as wool, furs, and feathers.

PSYCHIDAE, THE BAGWORMS. This family has relatively large species, some attaining a wing spread of 30 mm. The larvae construct a case or bag of silken fabric and bits of leaf or bark (Fig. 8-174). Only the

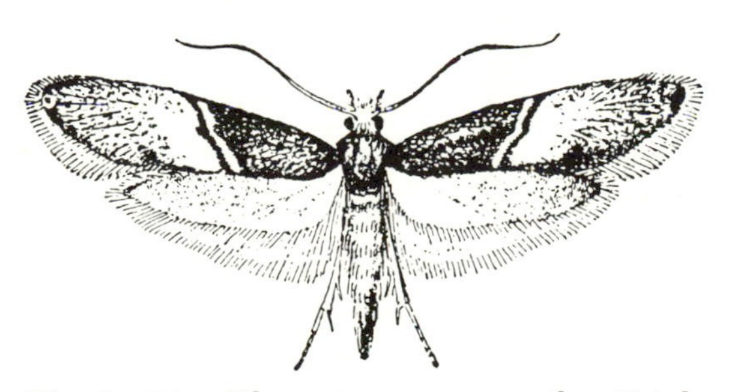

Fig. 8-172. The tapestry moth *Trichophaga tapetzella*, one of the Microlepidoptera. (From U.S.D.A., E.R.B.)

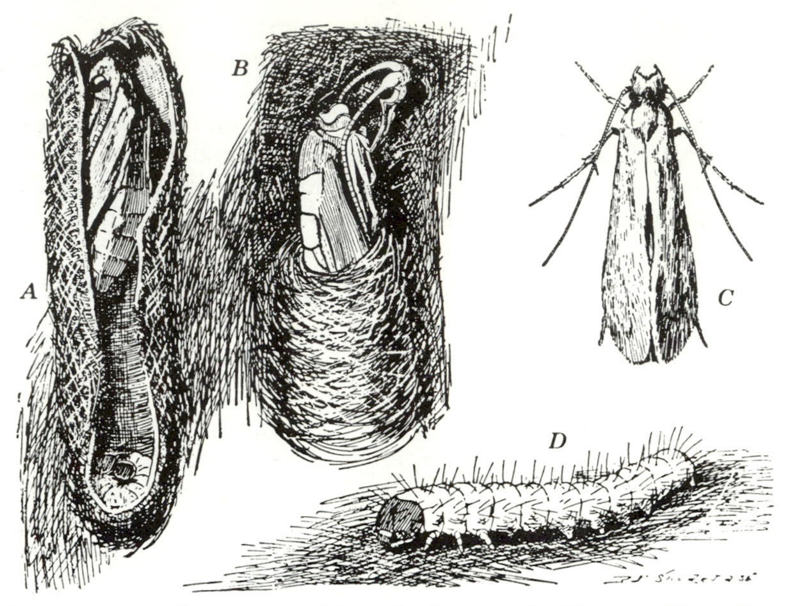

Fig. 8-173. The case-making clothes moth *Tinea pellionella*. (*A*) Cocoon cut open showing fully formed pupa within; (*B*) empty pupal skin projecting from door of cocoon after the moth has emerged; (*C*) adult moth; (*D*) the larva that does the damage to clothes. (After Snodgrass)

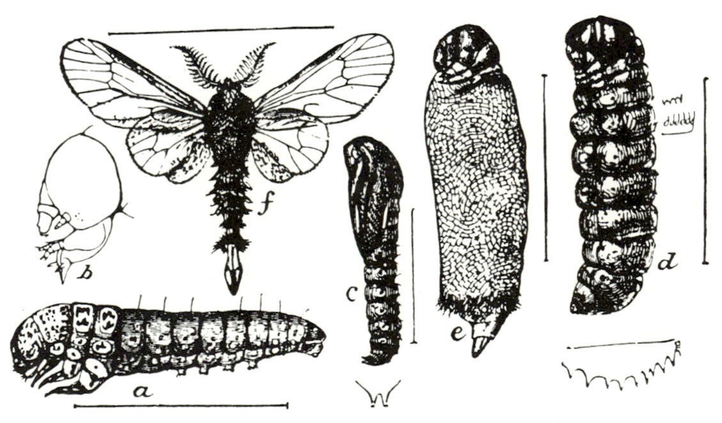

Fig. 8-174. Stages in the life cycle of the bagworm *Thyridopteryx ephemeraeformis*. (*a*) Full-grown larva; (*b*) head of same; (*c*) male pupa; (*d*) female pupa; (*e*) adult female; (*f*) adult male. (From U.S.D.A., E.R.B.)

heavily sclerotized head and thorax project from the case. These bags are a familiar sight hanging from the twigs or leaves of many species of coniferous and deciduous trees. The common bagworm in the eastern states is *Thyridopteryx ephemeraeformis* (Fig. 8-175). The males have fuzzy dark bodies, pectinate antennae, and usually clear wings. The females are larviform and have almost lost the power of locomotion. The life history has some peculiarities. By late summer the larva is full grown, whereupon it fastens its bag to a twig and pupates inside the bag. When mature, the male pupa emerges partially out of the bag, and the adult male emerges from it in this position. The female pupa stays within the bag; the female adult works itself partway out of the pupal skin and awaits fertilization. The males fly about looking for bags containing mature females and mate with them by means of an elongate extensile copulatory apparatus that can be extended deep into the bag containing the female. Soon after fertilization the female lays eggs, simply allowing them to fall into her old pupal skin, which becomes half filled wih them. This completed, the spent female crawls out of the bag, falls to the ground, and dies. The eggs lie dormant over winter and hatch the next spring. The newly hatched larvae crawl out of the old bag, disperse, and begin feeding.

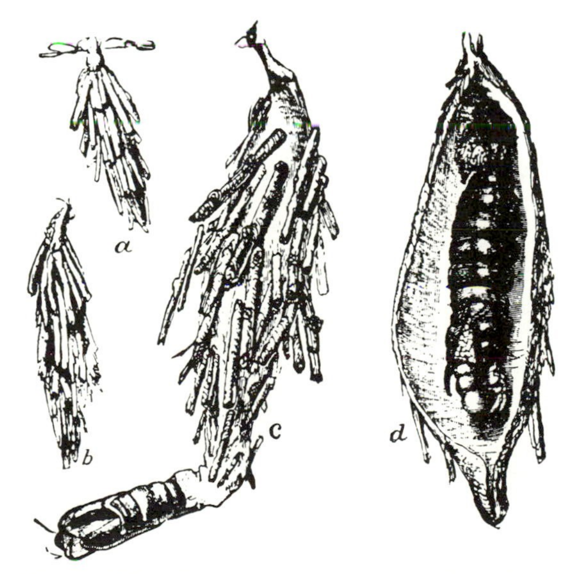

Fig. 8-175. Bagworms at successive stages (*a*, *b*, *c*). (*c*) Male bag; (*d*) female bag. (From U.S.D.A., E.R.B.)

GELECHIDAE. The larvae of this family exhibit a wide diversity of food and hosts. Several eastern species of *Gnorimoschema* make large stem galls on goldenrod and related Compositae; the larva feeds within the galls and when full grown pupates there. The pink bollworm *Pectinophora gossypiella* tunnels through the developing cotton boll and feeds on the seeds. It is a native of Asia that has spread to all the cotton-growing regions of the world and is one of the worst pests of this crop. Also included in the family are species that mine the needles of pine and other conifers and attack potatoes (Fig. 8-176) and tomatoes, and some that bore in twigs and fruits of certain trees. One species is a worldwide pest of stored grain, the Angoumois grain moth *Sitotroga cerealella*.

AEGERIIDAE, THE CLEARWINGS. These are moderate-sized narrow-winged forms in which the front and hind wings are coupled by a series of interlocking spines situated near the middle of the wing margins. In most species the wings have definite window-like areas free from scales; hence the name "clearwing" moths. The adults are diurnal and extremely rapid in flight, and in many the body and wings are banded with purple, red, and yellow, apparently mimicking some of the common wasps. The larvae are stem borers, attacking herbs, shrubs, and trees. Two of notable economic importance are the peach tree borer *Sanninoidea exitiosa* (Fig. 8-177) and the squash vine borer *Melittia cucurbitae*.

TORTRICIDAE. A very large family of small to medium-sized moths, having wide wings, the front pair with the apical margin truncate or even concave or excised. The larvae feed on nuts, fruits, and leaves, and in stems. In many species the larvae make nests by rolling and tying leaves, from which they get their common name, leaf rollers.

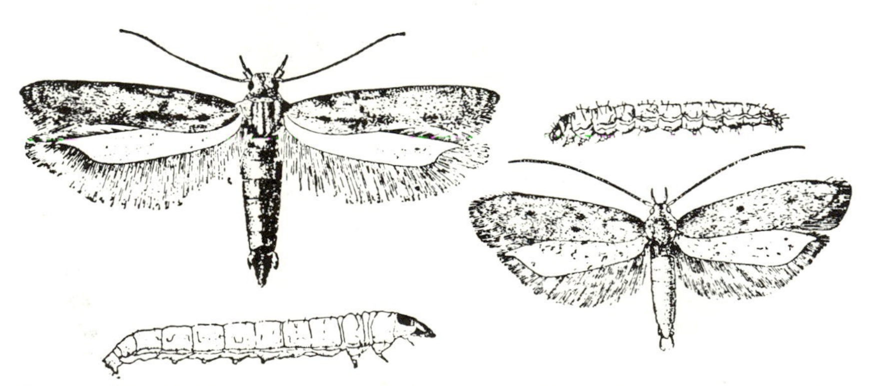

Fig. 8-176. Left, the potato tuberworm *Phthorimaea operculella*; right, the eggplant leaf miner *Keiferia inconspicuella*. Both belong to the Gelechiidae. (From U.S.D.A., E.R.B.)

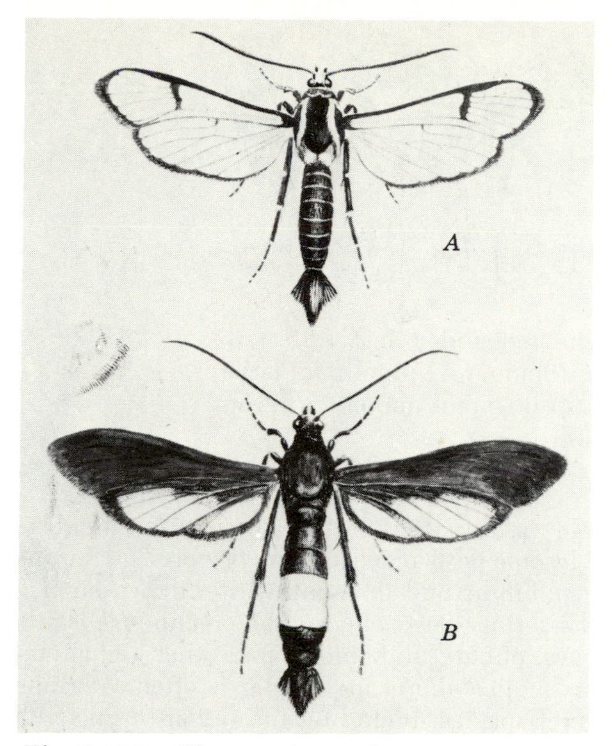

Fig. 8-177. The peach tree borer *Sanninoidea exitiosa*. (*A*) Male; (*B*) female. (From U.S.D.A., E.R.B.)

The group includes the redbanded leaf roller *Argyrotaenia velutinana* (Fig. 8-178), which feeds on many wild and cultivated trees and shrubs. Some other economic species in the family are the oriental fruit moth *Grapholitha molesta*, which feeds in the twigs and fruits of peaches, plums, and related fruit trees; the pine shoot moths, several species of *Rhyacionia*, which destroy the terminal buds of pine; and the grape berry moth *Polychrosis viteana*, which feeds on the leaves and in the fruits of grapes.

PYRALIDAE. This is a large family, economically one of the most important in the order. In taxonomic position the pyralids are intermediate in many characters between the "micros" and the "macros." The moths vary greatly in size and shape, but are usually delicate, trim, and have a rather detailed and soft coloration pattern. The larvae exhibit a wide range of food habits: feeding on leaves, fruits, and flowers; boring in stems or stalks; some being saprophagous and a few predaceous. Many spin an extensive web over their food and surroundings and are called webworms, such as the garden web-

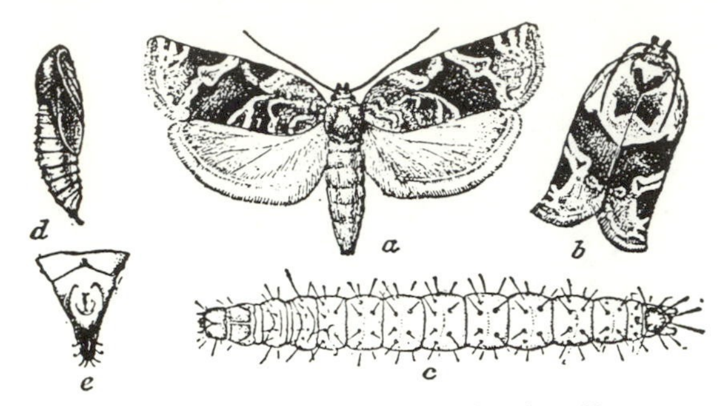

Fig. 8-178. The redbanded leaf roller *Argyrotaenia velutinana*. (*a*, *b*) Moth; (*c*) larva; (*d*) pupa; (*e*) tip of pupal abdomen. (From U.S.D.A., E.R.B.)

worm *Loxostege rantalis* (Fig. 8-179). The family includes some of the most troublesome pests of agricultural crops. The European corn borer *Ostrinia nubilalis* and the southwestern corn borer *Diatraea grandiosella* are serious pests of corn; the greenhouse leaf tier *Udea rubigalis* is a pest of chrysanthemums and other greenhouse crops; the greater wax moth *Galleria mellonella* is often a serious pest of beehives. Several species, including the Indian meal moth *Plodia interpunctella*, attack stored grain and prepared foods, and *Ephestia elutella* feeds on stored tobacco and other dried vegetable products. Many other species attack a wide assortment of crops.

The "Macros." In the Frenatae, the more specialized moth families containing the larger species, such as the miller moths, the various native "silkworm" moths, and the hawk moths, are usually referred to as the Macrofrenatae, or "macros." In these the hind wing usually is broad, has only a short fringe, and lacks vein, *E*. As is the case with the "micros," the "macros" contain many families that are difficult for the nonspecialist to identify.

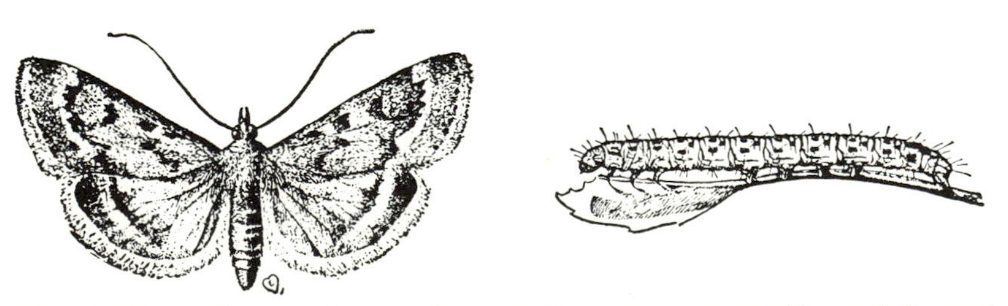

Fig. 8-179. The garden webworm *Loxostege rantalis*, adult and larva. (From U.S.D.A., E.R.B.)

LASIOCAMPIDAE. To this small family belong a few species of moderate size that have hairy larvae and velvety large-bodied adults. Tent caterpillars of the genus *Malacosoma* (Fig. 8-180) are the best known of these. The larvae feed on a variety of deciduous trees, including fruit trees, and periodically occur in outbreak numbers, the hordes of caterpillars defoliating thousands of acres of trees.

In the genus *Malacosoma*, winter is passed in the egg stage. The eggs hatch in late spring, and the young caterpillars of an egg mass spin a colonial webbed nest, usually around the fork of a branch. The caterpillars leave this to feed on foliage, each individual returning via a silken thread left in its wake. The larvae leave the nest and pupate singly in a protected spot under bark or debris where the cocoons are constructed. Adults emerge in late summer, and, after mating, females deposit eggs in bands around small twigs. Each egg mass contains several hundred eggs, the whole encased in a secretion that hardens and becomes impervious to the elements.

During an outbreak year the adults may be attracted in huge numbers to the light of towns. On a summer night in 1925 a tremendous flight in Edmonton, Alberta, covered the entire business section with moths about 15 cm (6 in.) deep, their greasy bodies completely stopping streetcar and automobile traffic, and making it difficult to walk. Under each streetlight and shop window the moths formed piles reaching an apex from 60 to 90 cm (2 to 3 ft) high. The great majority died from suffocation by their fellows. Their rotting bodies gave a distinctive odor to the city streets for some time.

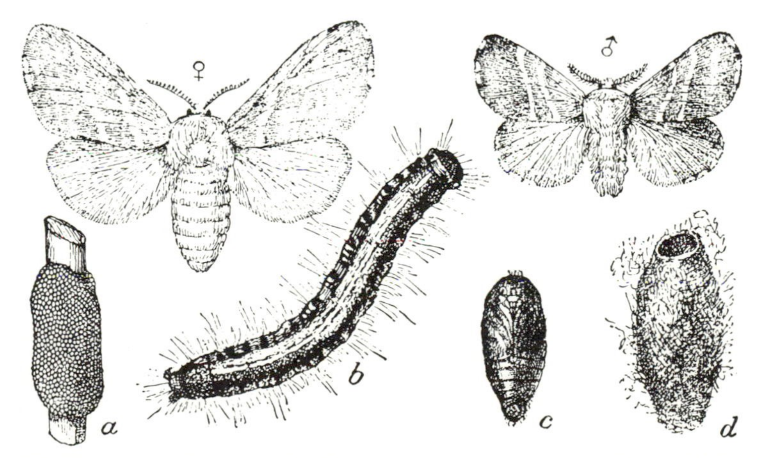

Fig. 8-180. Stages of the eastern tent caterpillar *Malacosoma americanum.* (*a*) Egg mass; (*b*) larva; (*c*) pupa; (*d*) cocoon; (♀) female moth; (♂) male moth. (From U.S.D.A., E.R.B.)

SPHINGIDAE, HAWK OR SPHINX MOTHS. These moths are all large, most of them having a wing spread of over 65 mm (2½ in.). The body is stout and spindle-shaped, frequently tapering to a sharp point at the posterior end. When spread, the posterior margins of the hind wings are seldom back as far as the middle of the abdomen; the front wings are long and proportionately narrow. The antennae are long and simple, frequently slightly thickened toward the tip. The moths are extremely rapid fliers and feed on nectar. The larvae are leaf feeders; most forms have a sharp horn on the eighth abdominal segment and are commonly called hornworms. In the main our species feed on a wide variety of herbs, vines, and trees. A few are of economic importance, particularly the tomato hornworm *Manduca quinquemaculata* and the tobacco hornworm *Manduca sexta* (Fig. 8-181).

GEOMETRIDAE, MEASURING WORMS, GEOMETERS. These are fragile moths, with slender or pectinate antennae, large delicate wings, and slender bodies (Fig. 8-182). The larvae are well known for their peculiar walking habit, consisting of a series of looping movements. They have long slender bodies but only two or three pairs of large abdominal legs, located near the end of the body. When they are walking, the abdomen is raised in a high loop, and the hind legs are brought forward to grasp a supporting position close to the thoracic legs;

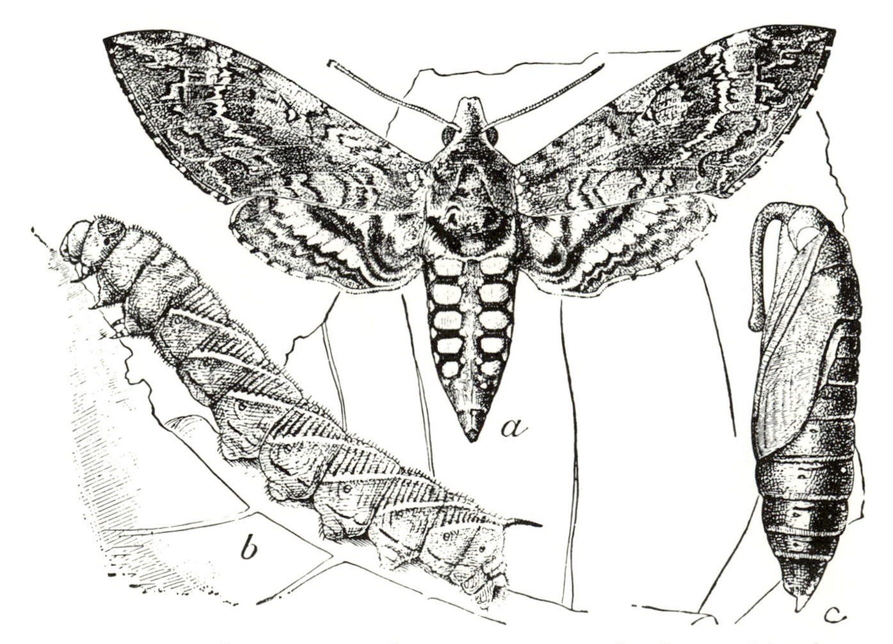

Fig. 8-181. A hornworm infesting tomato and tobacco, *Manduca sexta.* (*a*) Adult; (*b*) larva; (*c*) pupa. (From U.S.D.A., E.R.B.)

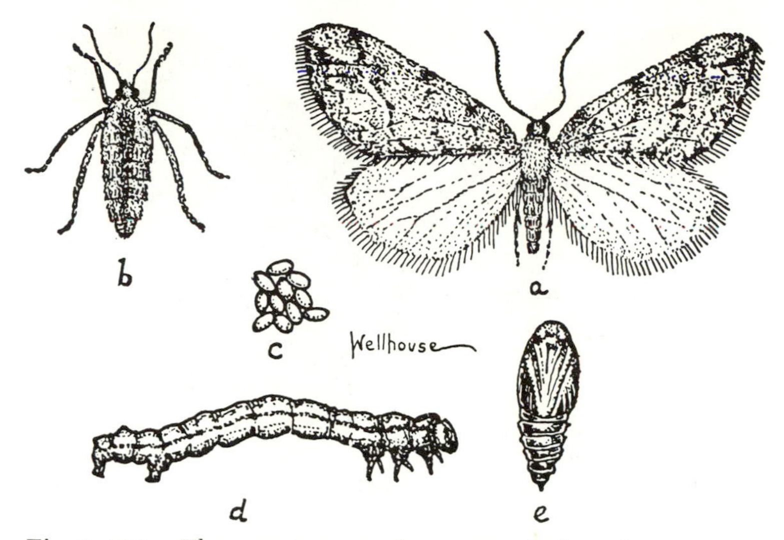

Fig. 8-182. The spring cankerworm *Paleacrita vernata*. (*a*) Male; (*b*) female; (*c*) eggs; (*d*) larva; (*e*) pupa. (From Wellhouse, *How insects live*, by permission of Macmillan Co.)

these latter are then released, the body stretched forward, and a new grip taken by the thoracic legs at the end of the reach. The hind legs then let go, the body arches, and the operation is repeated. Geometers make up a large family; their host list includes many plant families and genera. Certain species are common defoliators of deciduous and coniferous trees. Some species, known as cankerworms (Fig. 8-182), are locally very destructive to shade and fruit trees; these species have normal winged males but completely apterous females.

SATURNIIDAE, GIANT SILKWORM MOTHS. Here belong the largest North American moths and caterpillars. The adults are velvety or woolly, with broad wings having a showy pattern. The hind wings have no trace of a frenulum; instead the basal portion of the anterior margin is enlarged and projects under the front wing in the flight position, thus synchronizing the two pairs. The antennae are feathery in the males and frequently in the females also. The caterpillars are leaf feeders; the larger ones have a voracious appetite and will consume an entire large leaf with astonishing speed. The larvae are stout bodied, bear spiny tubercles or tufts of stiff hairs, and may attain a length of 100 mm (nearly 4 in.). The full-grown larvae spin large brown cocoons on branches or twigs near the ground, pupate, and pass the winter in this stage. A common eastern species is the promethea moth *Callosamia promethea* (Fig. 8-183), which has a red-

Fig. 8-183. The promethea moth *Callosamia promethea*. (From Comstock, 1936, *Introduction to entomology*, by permission of the Comstock Publishing Co.)

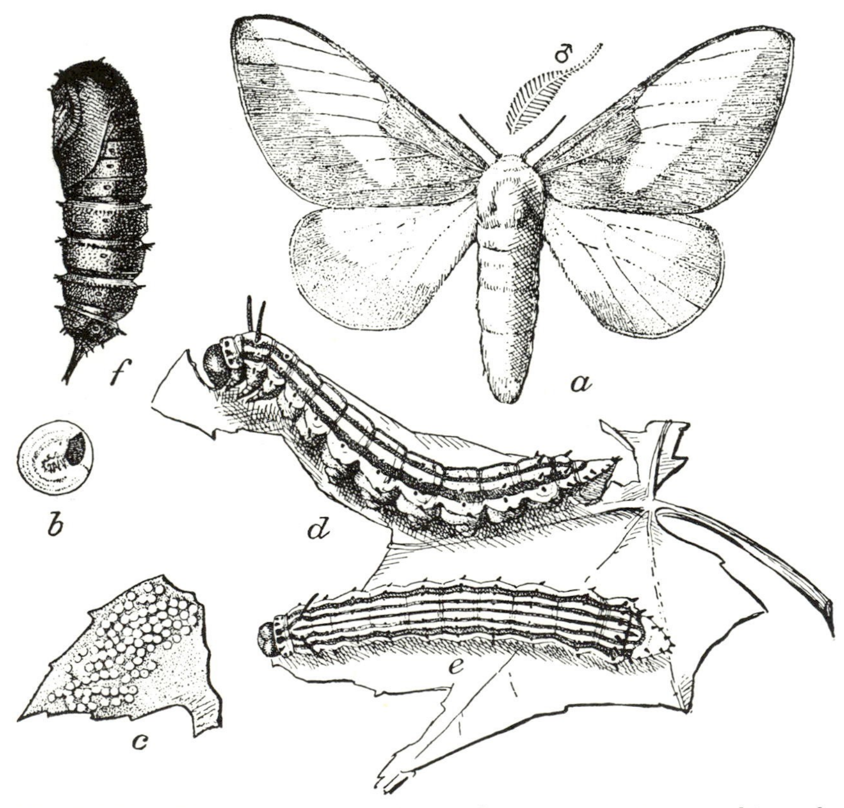

Fig. 8-184. The green-striped mapleworm *Anisota rubicunda*. (*a*) Female moth and antenna of male; (*b*) egg showing embryo within; (*c*) egg mass; (*d*, *e*) larva; (*f*) pupa. (From U.S.D.A., E.R.B.)

dish or brown adult; the larva is green with yellow, red, and blue spiny tubercles.

The related family Citheroniidae contains the green striped mapleworm *Anisota rubicunda* (Fig. 8-184), a widespread species east of the Rocky Mountains.

LIPARIDAE. This also is a small family containing species of only moderate size. The caterpillars are hairy, the adults frequently fuzzy. In several genera the females cannot fly, having only small padlike wings. To this family belong a few species of extreme economic importance, including the gypsy moth *Porthetria dispar* and the brown-tail moth *Nygmia phaeorrhoea* (Fig. 8-185). Both of these gained entrance to the United States from Europe and are destructive enemies of shade trees in northeastern United States. Another species of Liparidae attacking a wide variety of deciduous trees is the white-marked tussock moth *Hemerocampa leucostigma* (Fig. 8-186). The tussock moth larva has tufts of long nettling hairs at each end of the body, and "pencils" of white hairs on some of the central segments. It makes a cocoon on the bark of trees. The female is grublike; she lays a large group of eggs encased in a foamy white secretion that forms a protective covering for them.

NOCTUIDAE (PHALAENIDAE), OWLET OR MILLER MOTHS. This is the largest family of the Lepidoptera. The adults vary greatly in size, shape, and color; the structural characters are also diverse, so that the family can be distinguished from its relatives only on the basis of a combination

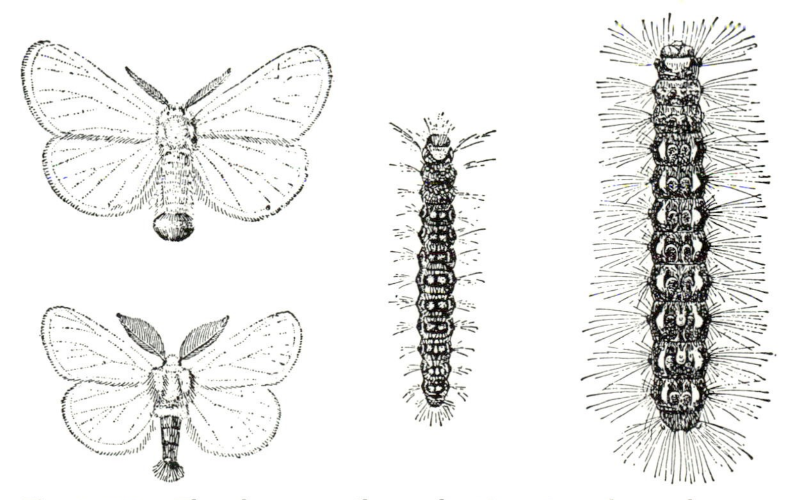

Fig. 8-185. The brown-tail moth *Nygmia phaeorrhoea*. Female above; male below; larva in center; enlarged larva to right. (From U.S.D.A., E.R.B.)

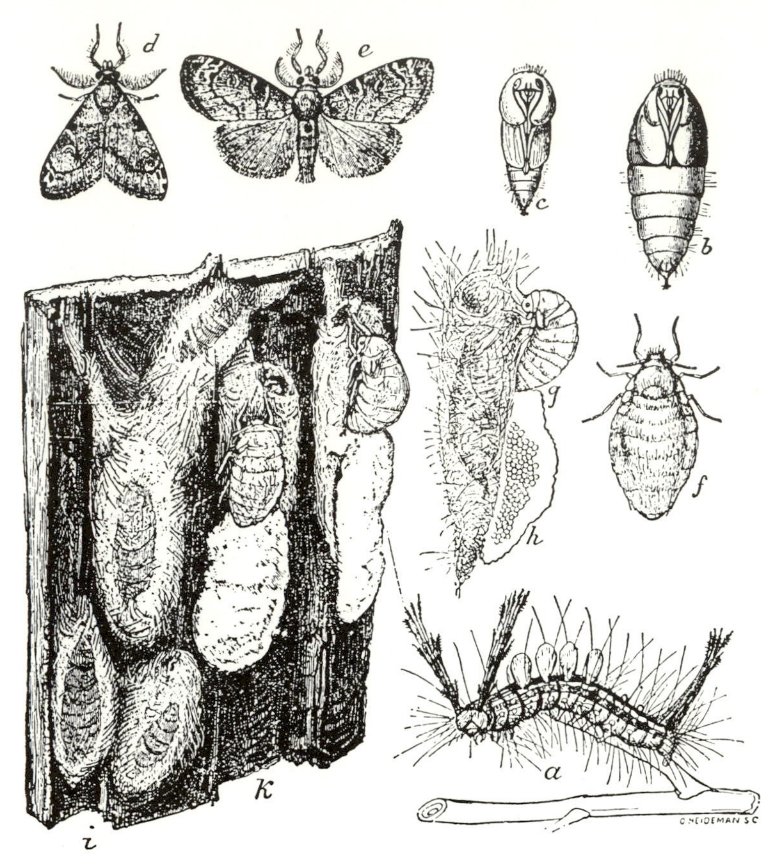

Fig. 8-186. The white-marked tussock moth *Hemerocampa leucostigma*. (*a*) Larva; (*b*) female pupa; (*c*) male pupa; (*d*, *e*) male moth; (*f*) female moth; (*g*) same, ovipositing; (*h*) egg mass; (*i*) male cocoons; (*k*) female cocoons, with moths laying eggs. (From U.S.D.A., E.R.B.)

of several critical differences. The larvae are usually leaf feeders or stem or root borers and for the most part are unadorned with horns or conspicuous processes. From the standpoint of agriculture the family is an important one. It includes many species whose larvae, called cutworms, attack a wide variety of grain, truck, and field crops. Other economic species are the armyworm *Pseudaletia unipuncta* and the fall armyworm *Spodoptera frugiperda* (Fig. 8-187), which attack pasture grasses, corn, and small grains; the corn earworm or cotton bollworm *Heliothis zea*, which attacks cotton, corn, tomatoes, and other crops; and the cabbage looper *Trichoplusia ni*, which feeds on cruciferous crops and "loops" like a geometrid larva,

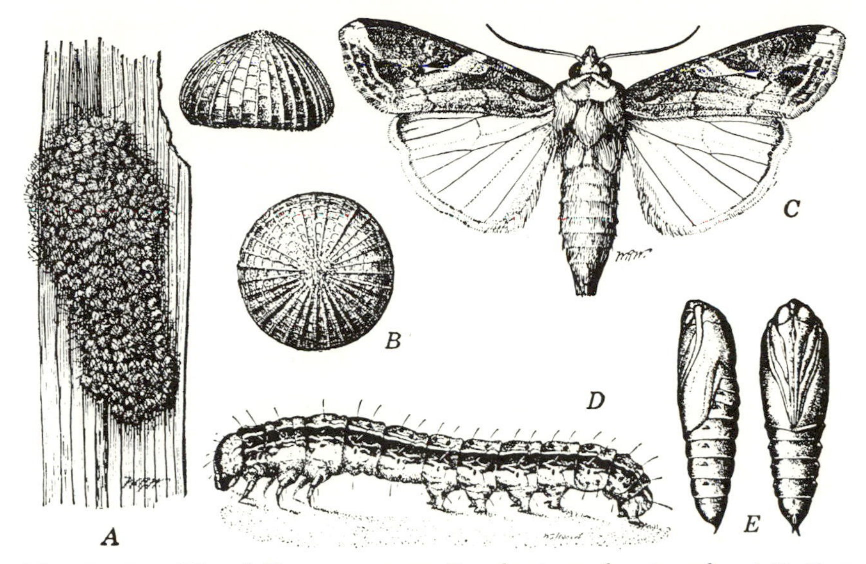

Fig. 8-187. The fall armyworm *Spodoptera frugiperda*. (*A*) Egg mass; (*B*) eggs; (*C*) adult; (*D*) larva; (*E*) pupa. (From U.S.D.A., E.R.B.)

but has only two middle pairs of abdominal legs. In all these the larvae do the damage.

ARCTIIDAE, THE ARCTIID MOTHS. This family is a close relative of the Noctuidae, differing from it mainly in that the adults are usually white or yellow or have intricate bright or yellow patterns. The larvae have thick tufts of hairs, and hence many of them are called woolly bears. These larvae are a common sight in late summer, hurrying along the ground looking for a sheltered place to make a cocoon and pupate.

OTHER MOTHS. There are many other moths, some of striking appearance, others similar in general characteristics to the few just mentioned. To identify these the student is referred to the works listed at the end of the section on Lepidoptera.

SUBORDER RHOPALOCERA

BUTTERFLIES AND SKIPPERS. In this suborder the hind wings have no frenulum, and in both front and hind wings the stem of media has been lost, resulting in a large central discal cell. Most of the species are brightly patterned. The adults are diurnal in habit and lovers of sunshine, in marked contrast to the crepuscular or nocturnal habit of most of the moths. When at rest, the Rhopalocera hold their wings upright over the body instead of folding them flat on the body as do the moths. The suborder is divided into two well-marked groups: the skippers and the butterflies.

Skippers. These are very rapid on the wing, able to fly in a straight line like a wasp or hawk moth. Most North American species belong to the Hesperiidae; they are dull colored, with yellow and brown predominating, and are less than 30 mm from wing tip to wing tip. Except for a few species, the larva has a large head accentuated by a small necklike prothorax. Most of the species live in a nest made by sewing together a few leaves of the host plant. A common eastern species is the silver-spotted skipper *Epargyreus tityrus*, which feeds on several legumes: *Wisteria* is a favorite.

Butterflies. The inversion "flutterbys" is descriptive of the flight of most members of this group. They have a very slow rate of wing stroke and hence fly in a series of up-and-down movements producing an erratic course. The butterfly group is represented in North America by seven families, most of which have members well known to the naturalist.

PAPILIONIDAE, THE SWALLOWTAIL BUTTERFLIES. In this family (Fig. 8-188), the margin of the hind wing is usually notched, and the vein M_3 ends in a finger-like projection or tail. These are large butterflies, many of them gaudily spotted or striped with many colors, yellow predominating in several species. The larvae are leaf feeders and may be as conspicuously marked as the adults. They are unique in having a

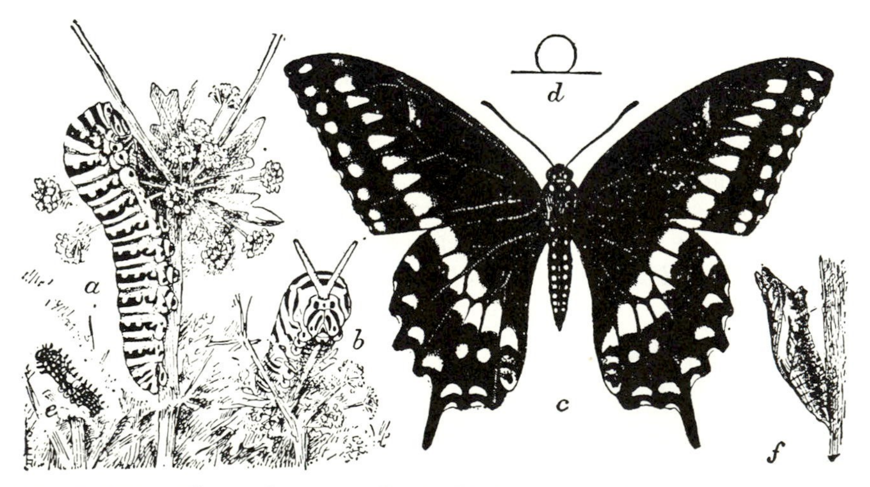

Fig. 8-188. The celery swallowtail *Papilio polyxenes*. (*a*) Larva from side; (*b*) larva showing head with odoriferous appendages; (*c*) male butterfly; (*d*) outline of egg; (*e*) young larva; (*f*) chrysalis. (From U.S.D.A., E.R.B.)

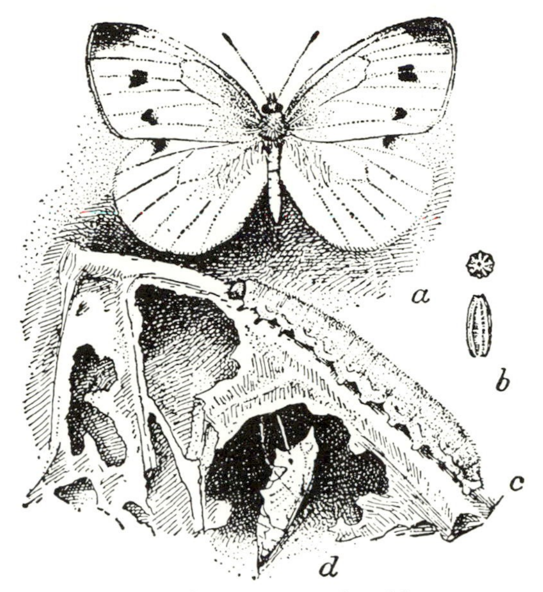

Fig. 8-189. The imported cabbageworm *Pieris rapae*. (*a*) Female butterfly; (*b*) egg (above as seen from above, below as seen from side); (*c*) larva, or worm, in natural position on cabbage leaf; (*d*) suspended chrysalis. (From U.S.D.A., E.R.B.)

forked eversible stink gland, or *osmeterium* (Fig. 8-188*b*) on the dorsum of the pronotum. This is shot out when the larva is alarmed; it is usually bright orange and emits a pronounced odor.

PIERIDAE, THE WHITES AND SULPHURS. These are predominantly white, yellow, or orange butterflies, some with extensive black markings. The larvae have an abundant supply of stiff hairs and look bristly. The imported cabbageworm *Pieris rapae* (Fig. 8-189), whose adult is white marked with black dots, occurs commonly in the central and northern states. The larvae are green and are pests of cabbage and related plants. This is an introduced species from Europe. A native species of the same genus, *P. protodice*, the southern cabbageworm, is a pest of cruciferous crops in the southern states. Another pierid that is often a pest locally is *Colias eurytheme*, the alfalfa caterpillar or orange sulphur. It is a highly variable species in color and occurs over most of the continent. The larva feeds on alfalfa, clover, and certain other legumes.

OTHER BUTTERFLIES. In North America there are many species of butterflies in addition to the few just listed. Every locality on the continent has a selection from strikingly colored to somber forms, many of them abundant locally. For more information regarding their identification characters, hosts, and range, the student should consult the references listed below.

References

Brock, J. P., 1971. A contribution towards an understanding of the morphology and phylogeny of the Ditrysian Lepidoptera. *J. Nat. Hist.*, **5**:29–102.

Common, I. F. B., 1975. Evolution and classification of the Lepidoptera. *Annu. Rev. Entomol.*, **20**:183–203.

Dominick, R. B. (Ed.), 1971–to date. *Moths of America north of Mexico.* London: E. W. Classey & R. B. D. Publishers. (published as separate fascicles for each family.)

Ehrlich, P. R., 1958. The comparative morphology, phylogeny and higher classification of butterflies (Lepidoptera: Papilionoidea). *Univ. Kans. Sci. Bull.*, **39**:305–370.

Ehrlich, P. R., and A. H. Ehrlich, 1961. *How to know the butterflies.* Dubuque, Iowa: W. C. Brown. 262 pp.

Holland, W. J., 1913. *The moth book.* New York: Doubleday. 479 pp. (Reprinted 1968. New York: Dover.)

——— 1931. *The butterfly book.* New York: Doubleday. 424 pp.

Howe, W. H., 1975. *The butterflies of North America.* Garden City, N.Y.: Doubleday. 633 pp.

Klots, A. B., 1951. *A field guide to the butterflies.* Boston: Houghton-Mifflin. 349 pp.

McDunnough, J., 1938–1939. Check list of the Lepidoptera of Canada and the United States of America. Pts. I and II. *Mem. So. Calif. Acad. Sci.*, **1**:1–272; **2**:1–171.

9
The Past History of Insects

The insects that we see around us belong to orders that have gradually evolved during a long period of time. Primitive relatives of insects extend back in time at least 350 million years, Devonian Period, and probably much longer (Fig. 9-1). Their history and development is closely interwoven with the history and development of other terrestrial animals and plants. These groups evolved into more and more terrestrial habitats and paved the way for insects also to adapt to these new habitats. As insects gradually evolved morphological, physiological, and behavioral adaptations to meet new ecological conditions, they, perhaps more than any other terrestrial group of animals, successfully subdivided their habitats into very small ecological niches. The phylogeny and interrelationships of existing insect orders are complex and reflect a series of interesting and significant evolutionary steps that form the general topics for this chapter.

FORMS OF EVIDENCE

Several types of evidence help us decipher the long evolutionary history of insects.

Fossils. The most direct evidence comes from fossils. Although far from complete, the insect fossil record provides very useful information for determining many important steps in the evolutionary development of insects because they provide fixed points in time that aid in unraveling the ancestry of present-day insect groups.

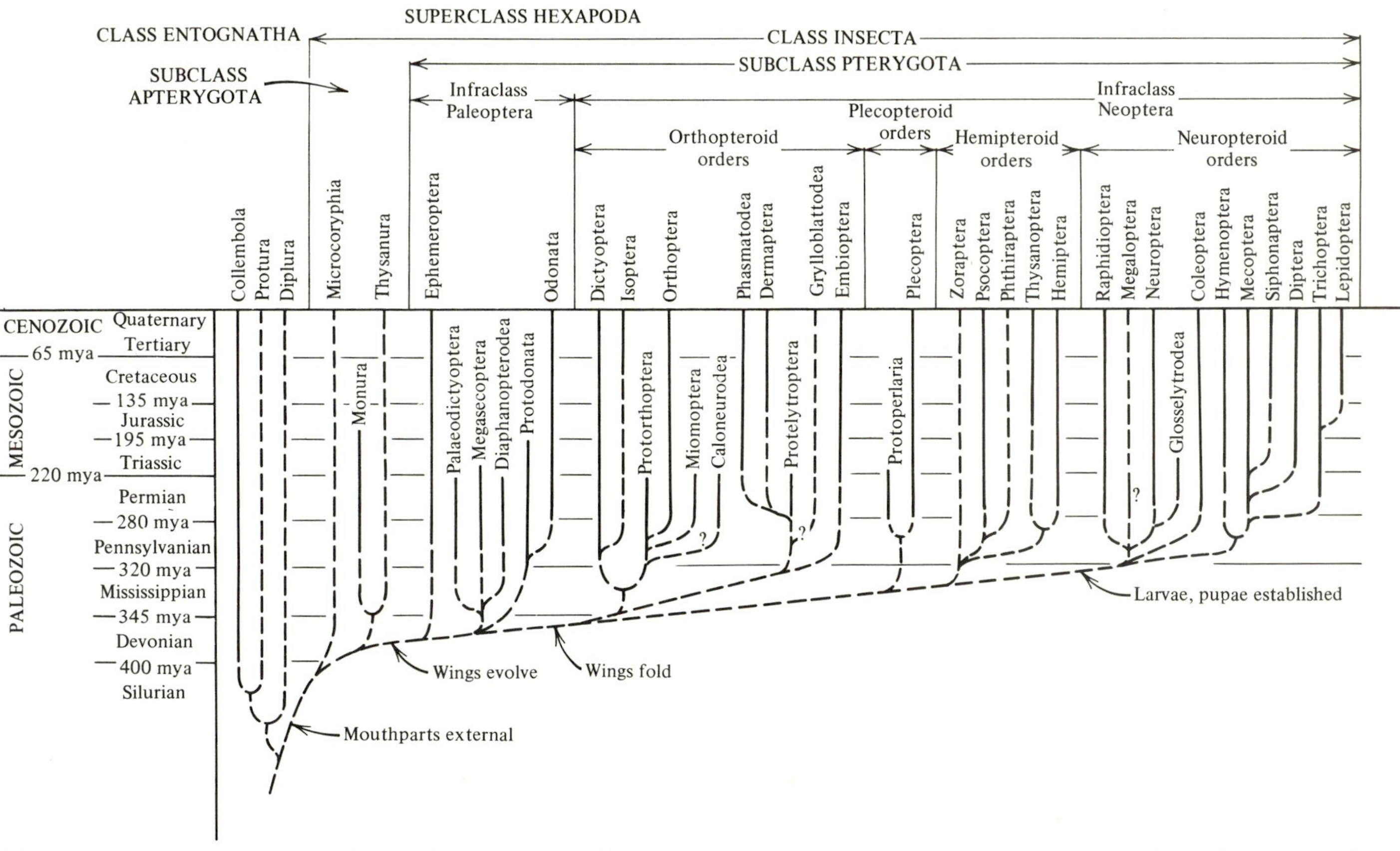

Fig. 9-1. Phylogenetic relations of the insect orders and their geological record. Solid lines show branches after they first appear in fossil record; broken lines show probable geological ranges based on comparisons of morphology and other indirect evidence. A number of proposed, but poorly known, extinct fossil orders are not included. Mya, Millions of years ago.

Fossil insects, despite their relatively small size and fragile structures, are known from a surprisingly large number of localities. Most localities are associated with coal deposits or with very fine-grained sediments deposited in quiet, anaerobic (lacking free oxygen) conditions. Concretions (Fig. 9-2) may form around buried insect parts and preserve them from later destruction. Terrestrial and lake deposits of shale, volcanic ash, peat, and similar materials may contain insect fossils. Most insect fossils usually are incompletely pre-

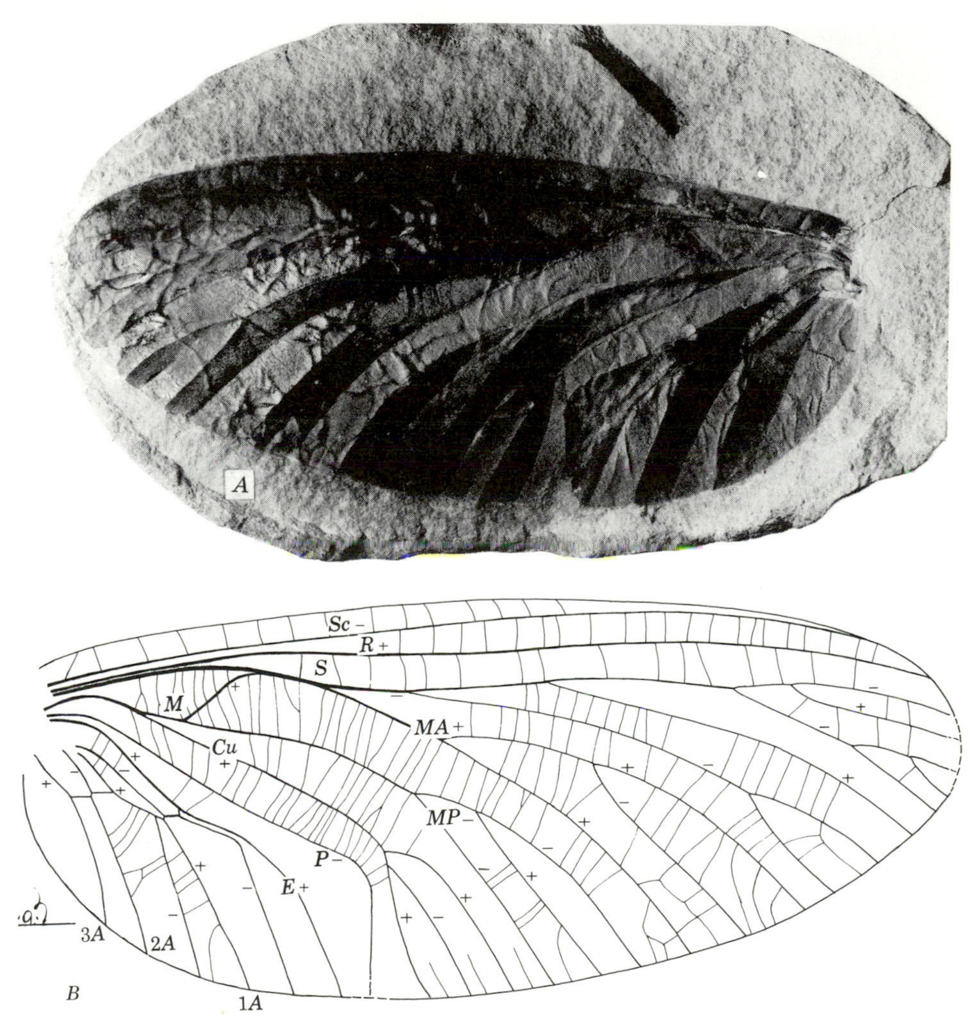

Fig. 9-2. A fossil from an iron nodule, Mazon Creek, Illinois; hind wing of an ancestral mayfly *Lithoneura mirifica*. (*A*) Photograph of fossil impression; (*B*) diagram of venation. (After Carpenter)

served; wings are most frequently encountered and form the basis for identification of most fossil specimens. Pennsylvanian and Lower Permian deposits in the Midwest and Midcontinent region of the United States have a long history of yielding insect fossils, and the Tertiary lake deposits of the Rocky Mountain region are also well known for their fossil insect faunas.

Coals, particularly those that are Cretaceous (120 million years old) or younger in age, commonly have amber nodules, formed from altered pitch and resin, that may include well-preserved, complete insects (Fig. 9-3). A large number of amber deposits are known, the most famous being the early Tertiary Baltic amber of northern Europe. Others have been discovered in Chiapas, Mexico, in Manitoba and western Canada, and in Alaska, Burma, and Siberia. The amber record of insects extends from the early part of the Cretaceous Period.

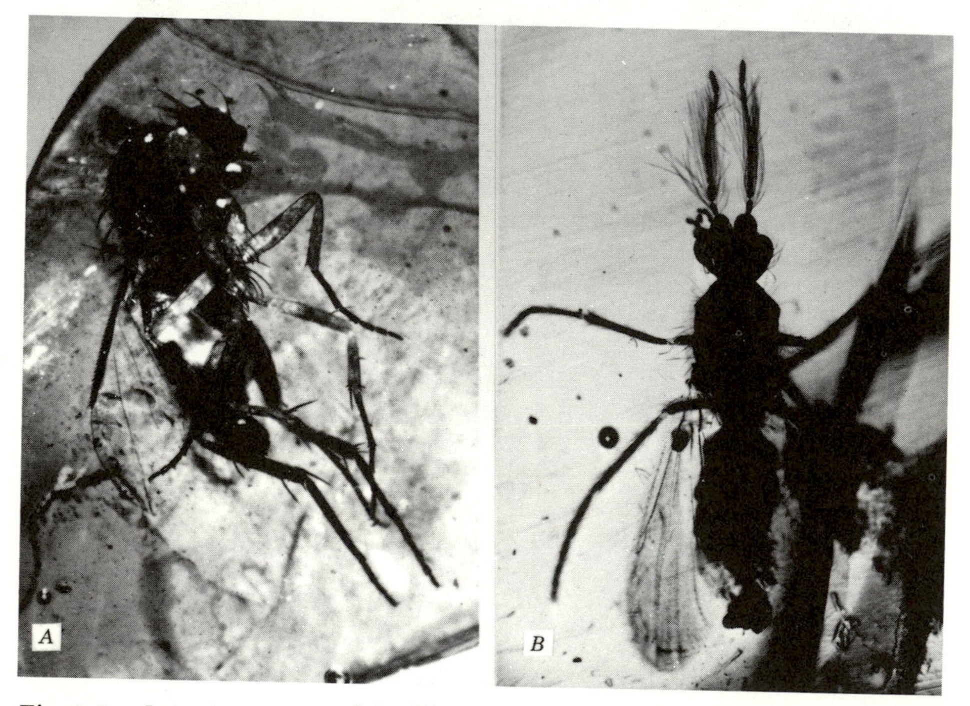

Fig. 9-3. Insects preserved in Upper Cretaceous Canadian amber. (*A*) A sciadocerid fly (Diptera), a group now known only from two living species from southern Chile and southeastern Australia and New Zealand; (*B*) a biting midge (Diptera, Chironomidae). (From McAlpine and Martin, 1969, *The Canadian Entomologist*, courtesy of F. McAlpine)

Phylogenetic Relations. A second, less direct type of evidence relies on analyses of similarities and differences in the morphology of living insects to establish phylogenetic relations; the broad aspects of this we examined in Chapter 8. This line of analysis attempts to identify primitive features and to establish a logical order by which newer adaptations and features were acquired by different lineages of insects. These powerful tools can greatly aid and supplement interpretations of the actual fossil record.

Geographical Distribution. A third type of evidence focuses on the present geographical distribution of groups of insects and interprets their phylogenetic relationships in the context of past geographical dispersals. At least three approaches to biogeographical studies are important in understanding insect distributions. Some insects have dispersed long distances, and overpassed areas for which they are not adapted, to form a number of nearly isolated, disjunct ranges. Island distributions in the Pacific Ocean represent an extreme of this form of distribution. A second approach considers that many present insect distributions are the result of fragmentation of once-continuous ranges. Plate tectonics and seafloor spreading commonly are invoked as mechanisms to fragment and move pieces of former continents during Mesozoic and Cenozoic time. The third approach attributes much of the present distributions to ecological factors, such as climatic fluctuations during the Quaternary Period, that displaced and isolated many fragments of insect populations.

Paleogeographical Distribution. Another type of evidence comes from studies of changes in past paleogeography and in paleogeographical distribution of fossil insects in the geological record. An example is a study of the distribution of Pennsylvanian insects by Durden (1974) that shows evidence for an insect distribution that had a strong latitudinal diversity gradient away from the Pennsylvanian paleoequator. Such distributional studies may give us a considerable understanding of the origin of the various insect groups.

PALEOGEOGRAPHY

Our perception of the world's past geography changed dramatically during the 1960s. Geologists learned a great deal more about the structure and age of ocean basins and they proposed two theories that are now widely accepted: seafloor spreading and plate tectonics. These theories unshackled the continental land masses from earlier concepts that had them firmly fixed in their present geographical positions. Most phylogenetic and dispersal interpretations previous to 1965 had been made on the assumption that the continents had remained in their relative geographical positions for most of geological time; thus these new concepts significantly changed some of those interpretations. These two geological theories have renewed interest in the earlier hypothesis of *continental drift* proposed by Wegener in 1912. The new lines of evidence give both a mechanism and a time framework in which the geographical and biological data that Wegener and others carefully compiled can be further interpreted.

Seafloor Spreading. The theory of seafloor spreading establishes that oceanic crust along the world's midoceanic ridges is very young and is forming by the intrusion of basaltic magmas (Fig. 9-4). On either side of the crests of these midoceanic ridges, the basaltic crustal rocks become progressively and systematically older as one goes away from the crests. The oldest oceanic crust in the Atlantic Ocean is of Middle Jurassic age (150 million years ago); the oldest oceanic crust in any of the oceans lies in the western Pacific Ocean east of the Philippines, and is of latest Triassic or earliest Jurassic age (195 to 180 million years ago).

Plate Tectonics. The theory of plate tectonics explains seafloor spreading, because if the ocean basins are not older than early or middle Mesozoic, what happened to the older rocks that made up the earlier oceanic crust? The theory of plate tectonics holds that the earth's crust and the upper part of its mantle form a set of 10 major

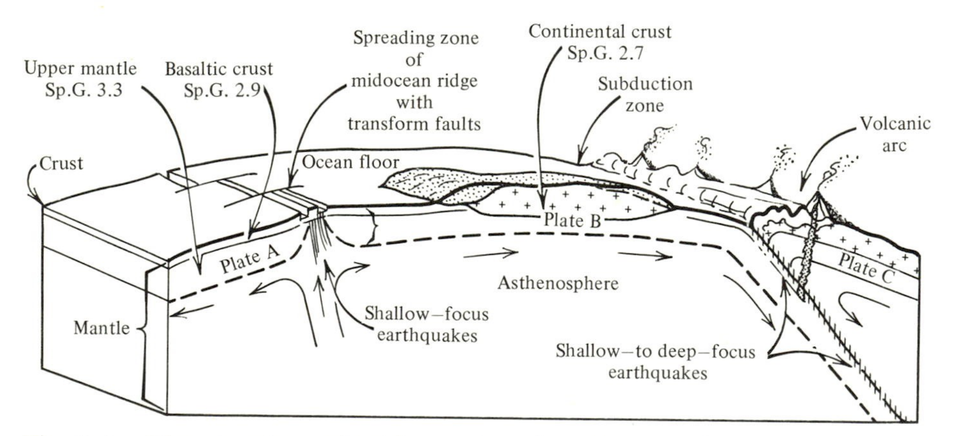

Fig. 9-4. The structure of midoceanic ridges as spreading zones and their relations to subduction zones and certain types of transform faults. Sp.G., Specific gravity.

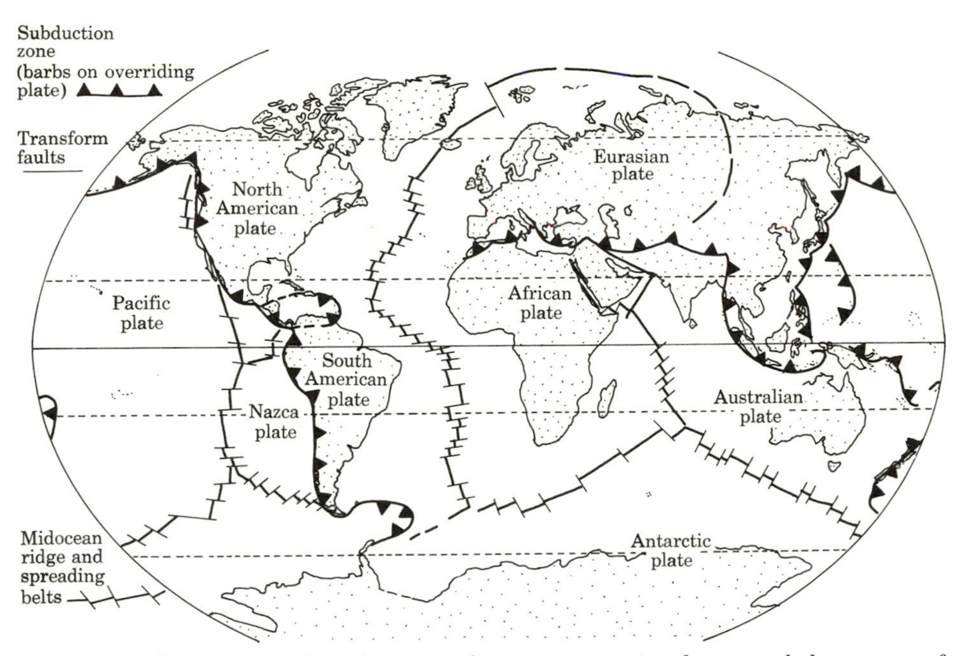

Fig. 9-5. The present distribution of major tectonic plates and the types of boundaries between them. The margins are characterized by earthquake activity of various kinds, and where these plate margins include a continent there is generally mountain-building activity. Subduction zones (barbs on overriding plate), Midocean ridge and spreading belts, and Transform faults. (From various sources)

rigid (lithospheric) plates and a number of smaller remnants of other plates (Fig. 9-5). The major plates are bound on one of their sides by midoceanic ridges, which are spreading strips. The other boundaries of the plates are located either at subduction zones or at major transform faults. A *subduction zone* is where one plate overrides another. In general, where less-dense continental crustal material is carried on one plate, that plate overrides the more dense (basaltic) oceanic crustal material carried on the other plate. Where both plates are composed of oceanic crust, either plate may be subducted. The depressed side of the convergence is marked by a deep trench on the ocean floor. These are well demonstrated around the margins of the Pacific Ocean (Fig. 9-5). Where two lithosphere plates each carrying continental crustal materials converge, active mountain building and elevation of large parts of continents occur, such as in the Himalayas and other parts of southern Asia and southern Europe (Fig. 9-5).

Paleogeographical Reconstruction. Figures 9-6, 9-7, and 9-8 show reconstructions of the positions of continents during Paleocene (60 million years ago), Early Cretaceous (120 million years ago), and Early Triassic (220 million years ago) epochs. A reconstruction (Fig. 9-9) for the Late Pennsylvanian or Early Permian (280 million years ago) shows the continents were rotated so that the paleoequator extended from southern California through northern Spain. North American and the northwest corner of Africa had begun to converge (Fig. 9-10) in Late Devonian time (360 million years ago) and were completely joined during the Late Pennsylva-

Fig. 9-6. Reconstruction of the earth's geography during the Paleocene Epoch (60 million years ago). (Modified from Smith and Briden, 1977)

Fig. 9-7. Reconstruction of the earth's geography during Early Cretaceous time (120 million years ago). (Modified from Smith and Briden, 1977)

nian and Early Permian to form a combined Euramerican-Gondwana continent called *Pangaea*.

EXTINCT ORDERS

As presently understood, the fossil record shows that insects contain 11, and possibly more, extinct orders, in addition to the many orders that are still living (Fig. 9-1). Most of the extinct orders first appeared in the fossil record during Pennsylvanian and earliest Permian time

Fig. 9-8. Reconstruction of the earth's geography during Early Triassic time (220 million years ago). (Modified from Smith and Briden, 1977)

and most became extinct before the end of the Permian Period. These extinct orders include the early ancestors of a number of existing orders. Most of these early fossil ancestors have primitive and less specialized morphological features compared to their descendants; hence the marked differences that we see between the living orders is much less sharply defined in their late Paleozoic representatives. On the other hand, these late Paleozoic orders clearly show that by the end of the Pennsylvanian, the basic steps in insect evolution were completed. These include development of wings, folding of the

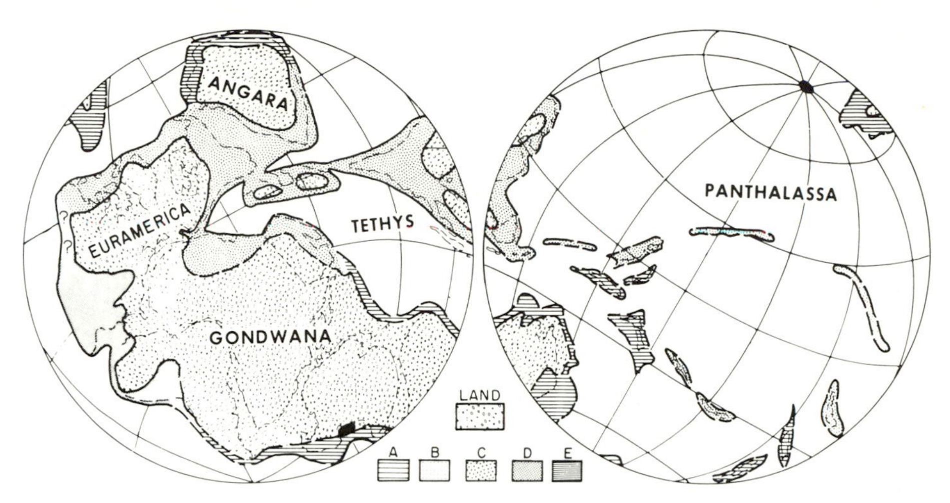

Fig. 9-9. Reconstruction of the earth's geography during Late Pennsylvanian and Early Permian time (280 million years ago). (After Ross and Ross)

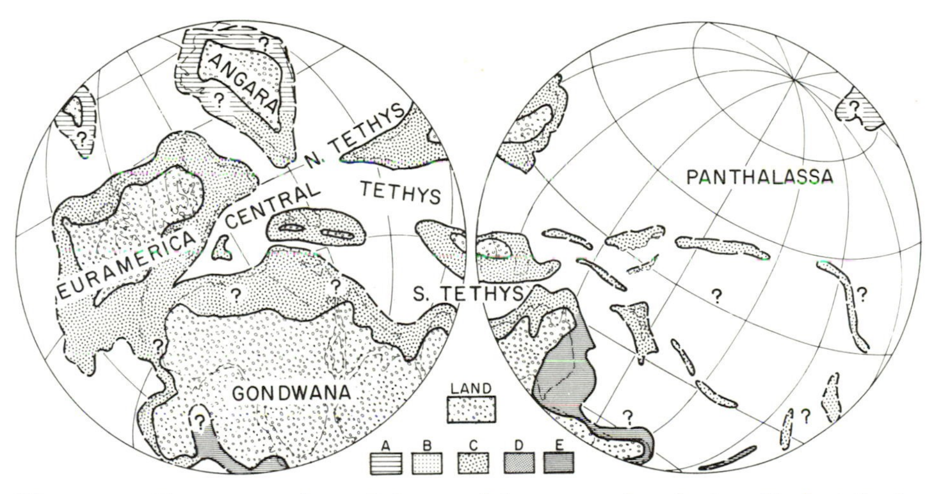

Fig. 9-10. Reconstruction of the earth's geography during Early Mississippian time (330 million years ago). (After Ross and Ross)

wing, diversification of feeding mechanisms, and complete metamorphosis in the life cycle.

The following briefly summarizes the extinct orders that are well established. Their relation to living orders is shown in Figure 9-1. A number of additional extinct orders have been proposed based on incomplete specimens and are not included here.

Subclass Apterygota

Order Monura

These are primitive apterygotes that retain tergal rudiments in the labial-mandibular series of head appendages (Fig. 9-11). The monurans are similar to the Microcoryphia (or Archeognatha) in having 10 complete segments with paired stylets in the abdomen. They differ in that the monurans lack or have only rudimentary cerci, whereas in the Microcoryphia an eleventh segment gives rise to cerci. Monura are known from the genus *Dasyleptus*; specimens are from the Upper Carboniferous (France, New Mexico, and Illinois), and Lower Permian (U.S.S.R., Czechoslovakia, Kansas, and Oklahoma).

Subclass Pterygota

Infraclass Paleoptera

Order PALAEODICTYOPTERA

This order makes up a large and quite diverse group that was widely distributed during Pennsylvanian and Permian time. These insects (Figs. 9-2, 9-12) were large and solidly constructed, many reaching wing spans of 20 cm and some more than 50 cm, and include some of the larger known insects. The head was small with slender antennae formed by many small segments. Eyes were large and prominent. Mouthparts were large, haustellate, and beaklike, and adapted to sucking liquids for food, probably from plants.

Legs were short and stout and comparable to those in Ephemeroptera. The prothorax had a pair of lobes, which commonly were cordate, and were attached to the pronotum. The lobes were heavily sclerotized and, in many, they were very large and overlapped the basal part of the front wings. They contained a venation system that is homologous to wing venation.

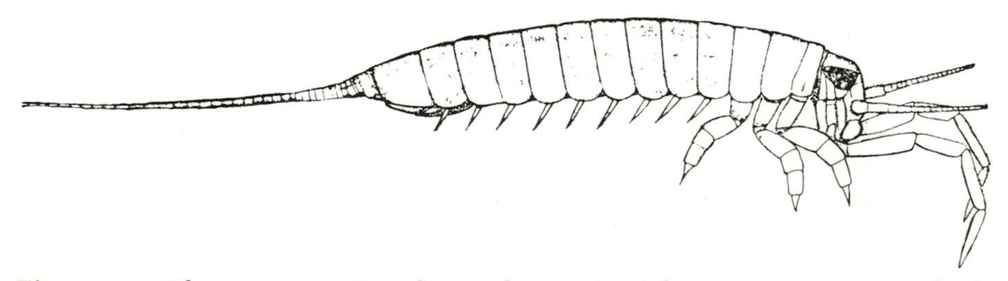

Fig. 9-11. The monuran *Dasyleptus brongniarti* from Lower Permian beds in the Kuznets Basin, U.S.S.R. (After Sharov in Rodendorf, 1962)

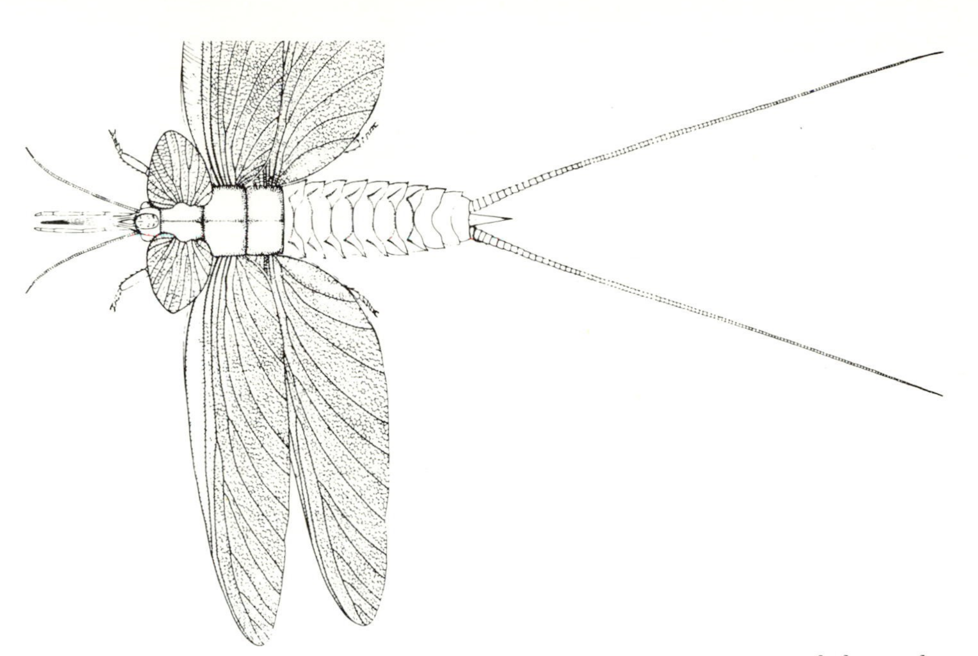

Fig. 9-12. *Stenodictya*, a Late Pennsylvanian representative of the order Paleodictyoptera, from France. (From Kukalova, 1969–1970, in *Psyche*)

Palaeodictyopteran wings included a wide variety of sizes and shapes (Fig. 9-12). The basal plates of the wing were similar to those in Protodonata and Odonata. The wings also commonly had color banding and circular spots, and most had hair, particularly along the veins near the wing base, along the wing margin, and on the wing membrane.

The abdomen was relatively short and had the general construction found in ephemeropteran nymphs. Multisegmented cerci were about twice the length of the abdomen. Females had a long, stout ovipositor, and the males had short, segmented claspers and paired aedeagus. Members of this order were abundant and mainly inhabitants of low, moist, tropical and subtropical forests and swamps.

Order MEGASECOPTERA

Among the insect orders from the early Pennsylvanian is a distinctive group, the Megasecoptera, that had specialized wing venation and falcate, or hooded, shaped wings (Fig. 9-13). These insects were medium to large (maximum 12 cm) in size with long cerci. Their bodies were slender and their mouthparts were formed into sucking beaks. The extinct order Diaphanopterodea was probably a de-

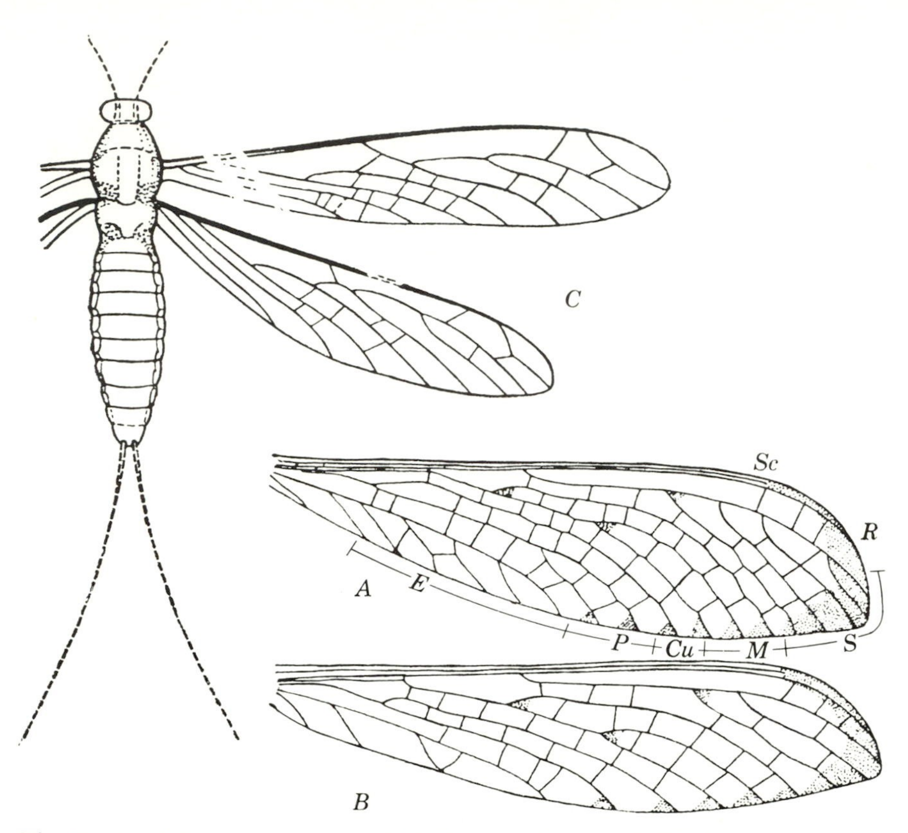

Fig. 9-13. Megasecopterans (*A*) Front wings and (*B*) hind wings of *Aspidothorax triangularis* from Upper Carboniferous of Europe; (*C*) *Pseudohymen angustipennis* from Lower Permian of the Ural region. (From Rodendorf and others; *A*, *B* after Carpenter)

scendent. Megasecopterans had similar habitat requirements to the Palaeodictyoptera. They are not known in strata younger than Triassic.

Order DIAPHANOPTERODEA

The Diaphanopterodea are known only from strata of Pennsylvanian and Permian age and had sucking beaks formed by much-modified mouthparts. Although their wing venation is basically similar to that of the Palaeodictyoptera and Megasecoptera, the Diaphanopterodea were able to fold their wings over their bodies. The folding mechanism is not clear, but it was not the same mechanism used by the Neoptera. Permian diaphanopterans show greatly reduced venation.

Order PROTODONATA (or MEGANISOPTERA)

This Pennsylvanian to Triassic order is characterized by large to gigantic insects that generally had wing spans between 12 and 75 cm and were the largest known insects (Fig. 9-14). They closely resembled dragonflies in general appearance with large eyes, long slender bodies, and strong mandibles, but had more primitive wing venation in which all the major veins except the subnodus were present, and veins *P* and *E* were separate at their bases. Protodonata gave rise to the Odonata during the Pennsylvanian. Apparently the protodonates were agressive and agile predators.

Infraclass Neoptera

ORTHOPTEROID ORDERS

Among the extinct late Paleozoic orthopterans at least four orders, Protorthoptera, Miomoptera, Calonerodea, and Protelytroptera, appear clearly established.

Order PROTORTHOPTERA

This Early Pennsylvanian to Permian order is one of the most common in the fossil record and contains a large diversity of groups. They are the earliest known fossilized neopterans. Protorthopterans may have coriaceous front wings and expanded anal areas on their

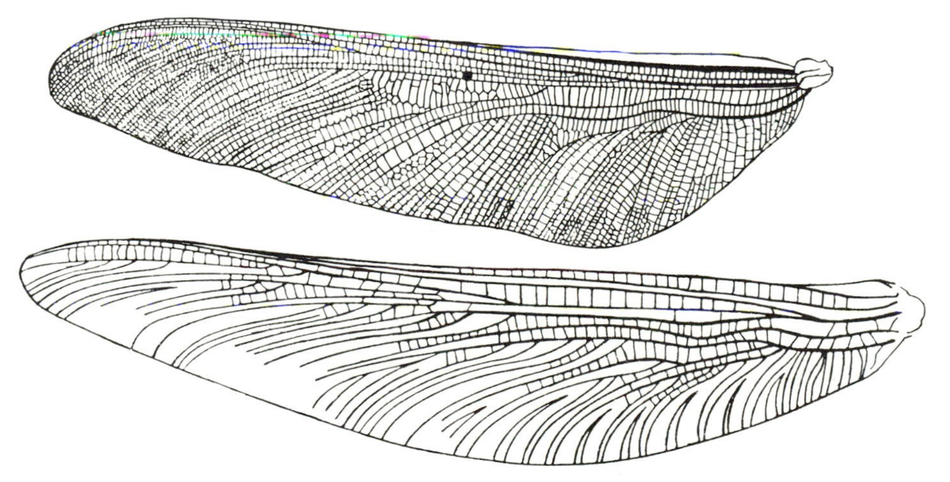

Fig. 9-14. Representatives of the order Protodonata. (*A*) *Meganeura moyni* from the Upper Pennsylvanian of France; (*B*) *Arctotypus sinuatus* from the Upper Permian of the U.S.S.R. (From Rodendorf and others; *A* after Handlirsch, 1906–1908; *B* after Martynov)

hind wings (Fig. 9-15*A*). Chewing mouthparts (mandibulate) were well developed.

One large group of protorthopterans, the Protoblattoidea, superficially resemble cockroaches, although they differed significantly in detail, particularly wing venation (Fig. 9-15*B*). An Early Pennsylvanian group of these retained a primitive archedictyon venation pattern with seven anal veins on both pairs of wings.

Order MIOMOPTERA

This order includes small to very small insects with relatively long wings (Fig. 9-16). They are rare in Early Pennsylvanian and common in Permian beds. Their wing venation is similar to the Pscoptera. Miomoptera had mandibulate mouthparts and short cerci.

Order CALONEURODEA

This Pennsylvanian–Permian order has long, nearly identical front and hind wings (Fig. 9-17) with the basal parts of veins *M* and *Cu* fused as in some protorthopterans. The mouthparts were mandibulate and cerci were present, but extremely short.

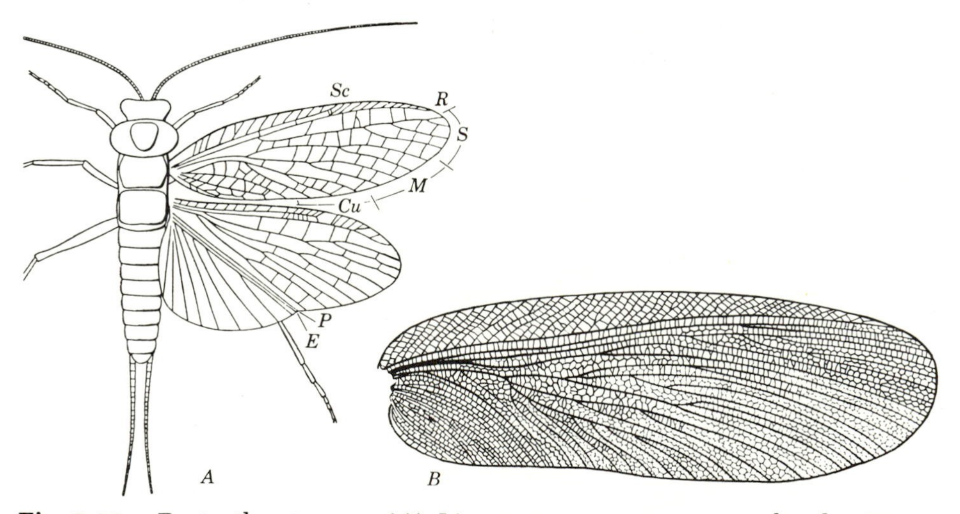

Fig. 9-15. Protorthopterans. (*A*) *Liomopterum ornatum*, suborder Paraplecoptera, Lower Permian of North America; (*B*) wing of *Stenoneura fayoli*, suborder Protoblattodea, from the Pennsylvanian of western Europe. (From Rodendorf and others; *A* after Carpenter, *B* after Handlirsch, 1906–1908)

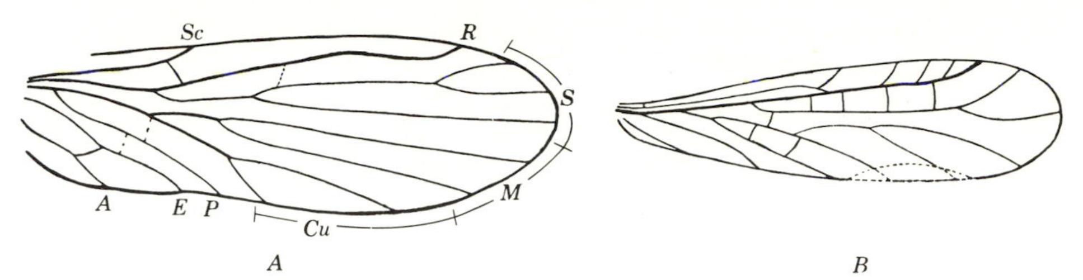

Fig. 9-16. Miomopteran front wings. (*A*) *Delopterum incertum* from Upper Permian of the Ural region, U.S.S.R.; (*B*) *Permembria delicatula* from Lower Permian of North America (after Tillyard). (From Rodendorf and others)

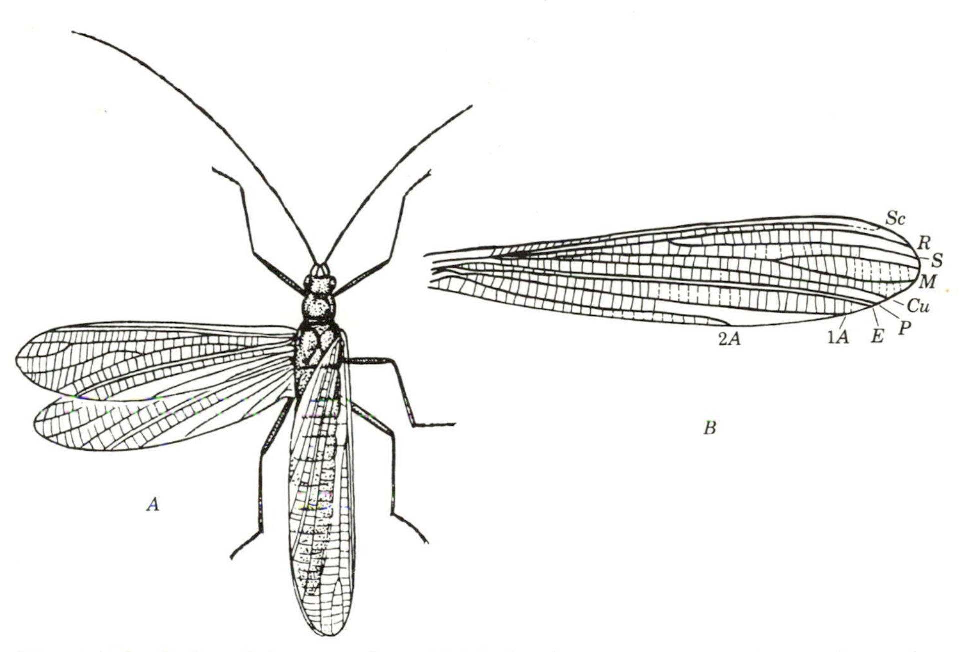

Fig. 9-17. Order Caloneurodea. (*A*) *Paleuthygramma tenuicorne* from the Lower Permian of the Ural region, U.S.S.R.; (*B*) front wing of *Euthygramma parallelum* from the Permian of Arkhangel region, U.S.S.R. (From Rodendorf and others)

Order PROTELYTROPTERA (or PROTOCOLEOPTERA)

Another order of small insects, the Protelytroptera, are known from Permian sediments. Typically their front wings were well-developed elytra and their hind wings were broad with both longitudinal and transverse folds in an enlarged anal portion (Fig. 9-18). They had short, stout antennae, broad bodies, strong legs, and short cerci.

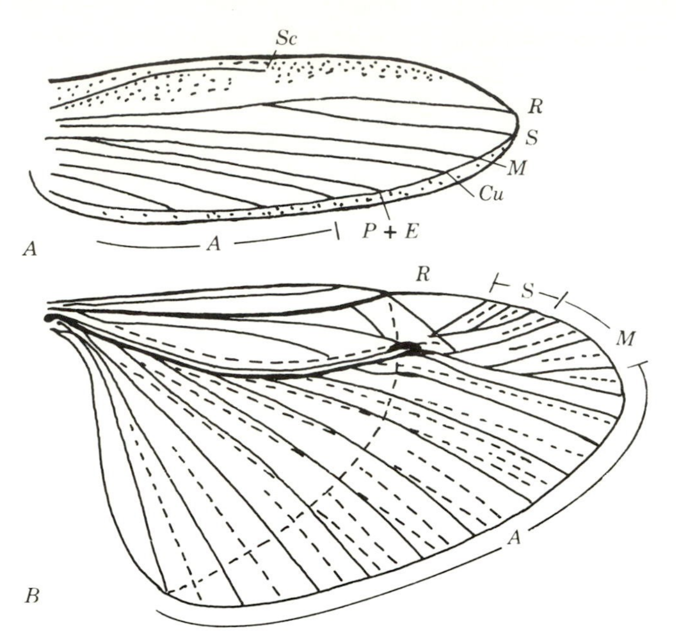

Fig. 9-18. Order Protelytroptera; wings of *Protelytron permianum* from the Lower Permian of North America. (*A*) Fore wing; (*B*) hind wing. (*A* Redrawn from Carpenter, *B* redrawn from Tillyard)

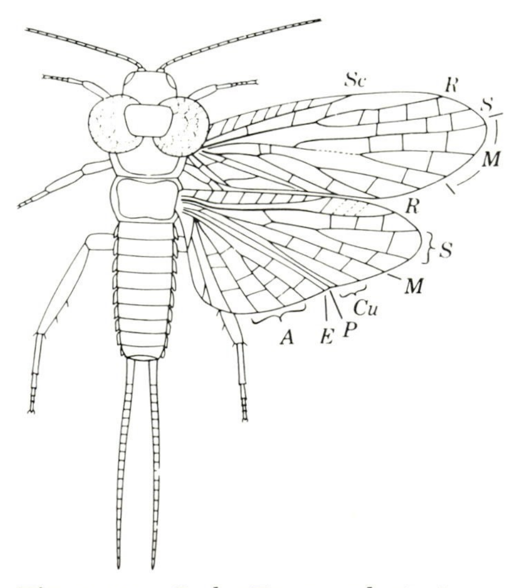

Fig. 9-19. Order Protoperlaria, *Lemmatophora typa*, from Lower Permian strata of Kansas. (Modified from Riek, 1970, after Carpenter)

Based on wing venation, these insects are probably ancestral to the Dermaptera, and the possession of elytra is considered independent of the formation of elytra in the Coleoptera.

PLECOPTEROID ORDERS

Order PROTOPERLARIA

This order, also known only from Permian sediments, is closely related to the Plecoptera (Fig. 9-19). The front wings were more elongate than the hind wings, which retained five anal veins arranged like those of modern Plecoptera. Some Protoperlaria had vestiges of a pair of flat lobes that projected horizontally from the prothorax. Cerci were long and the hind legs were stout and longer than the front and middle legs. The Protoperlaria apparently gave rise to the Plecoptera before the Permian.

NEUROPTEROID ORDERS

Order GLOSSELYTRODEA

This group, the only extinct order of the endopterygote Neoptera, ranged from the Permian into the Jurassic Period. Wing venation (Fig. 9-20) was more specialized than in either the Neuroptera or Mecoptera and the front wings were also elytra. In many features the wings, head, and thorax of Glosselytrodea are generally similar to

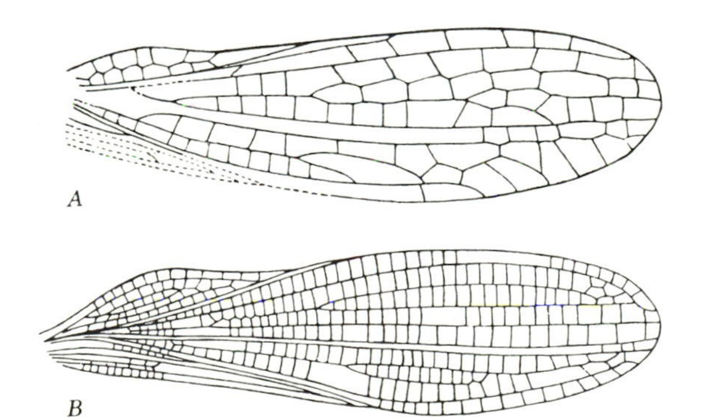

Fig. 9-20. Order Glosselytrodea. (*A*) Front wing of *Archoglossopterum shoricum* from Permian strata of the Kuznets Basin, U.S.S.R.; (*B*) front wing of *Jurina marginata* from the Permian of Arkhangel region, U.S.S.R. (From Rodendorf and others)

those of the Neuroptera. These two orders probably had a close common ancestor in Late Carboniferous or earliest Permian time.

FOSSIL INSECTS AND THEIR RELATION TO LIVING ORDERS

Paleopterans

The fossil record of four extinct orders and two living orders of Paleoptera show that the two surviving orders, the Ephemeroptera (mayflies: Fig. 9-21*A*, *B*) and Odonata (dragonflies) are only remnants of an early and very broad diversification of Pennsylvanian winged insects. They present two quite divergent branches within the Paleoptera. In the Ephemeroptera, retention of the ability to fly in both the preimaginal and imaginal stages and the post-Paleozoic reduction in mouthparts of these stages indicate they have been for a long time a separate lineage from the other paleopterans.

The Odonata have changed little since they first appeared and probably always have had predatory nymphal and imaginal stages.

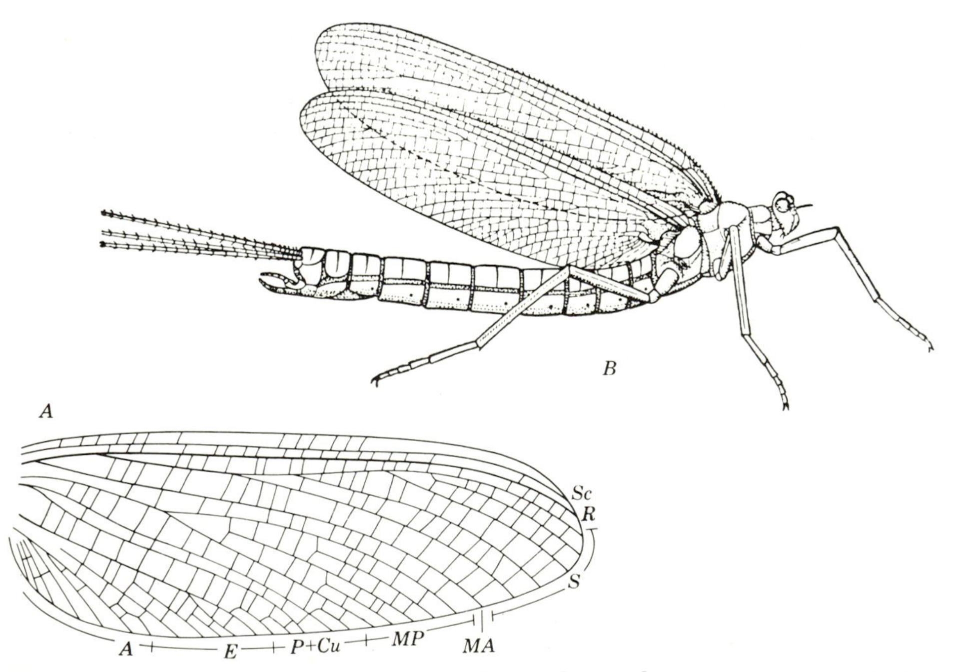

Fig. 9-21. Early representatives of the order Ephemeroptera. (*A*) *Triplosoba pulchella*, Upper Pennsylvanian of Europe. (Modified from Riek, 1970, after Edmonds and Traver); (*B*) *Protereisma permianum*, Lower Permian of North America. (From Rodendorf and others, after Tillyard)

The mouthparts of the Protodonata were similar to those of the active, agile Odonata; however, they had much more robust bodies. Fossilized nymphs of Palaeodictyoptera and Megasecoptera suggest they were terrestrial and the adults were plant feeders. The Paleoptera are important because they gave rise to one and probably to two groups that could fold their wings over their abdomen. This protected the wings when they were not used for flying and significantly increased the mobility of individuals.

Neopterans

The most important group to evolve from the early paleopterans is the Neoptera. In this group the wings articulate using sets of aligned sclerites (Fig. 3-42, 3-46) that permit the wings to be folded back along the body. This contrasts with the sclerite arrangements in the two living orders of the Paleoptera (Fig. 3-45).

Orthopteroid orders

Among the orthopteroid orders, the Dictyoptera (cockroaches) and the Isoptera (termites) are considered very primitive and both have generally similar wing venation. Dictyoptera were well represented during Pennsylvanian time by true cockroaches (suborder Blattaria); however, the Isoptera are first known in the fossil record from the Cretaceous. Wings of the Protorthoptera from Pennsylvanian and Permian rocks show side variation in the number of anal veins, differential change between front and hind wings, and changes in folding patterns. The most primitive protorthopterans, earliest Pennsylvanian, are the oldest known winged insects and had a true archedictyon venation. The Protelytroptera appear to have shared a common ancestor with the Dermaptera (earwigs). The Dermapterans are known from Middle Jurassic.

The Orthoptera (grasshoppers), and probably also the Phasmatodea (walking sticks), Miomoptera, and Caloneurodea, represent another set of early lineages from one or more other subgroups within the Protorthoptera. Phasmatodea are known from Triassic beds of Australia. Both the Miomoptera and Caloneurodea were specialized sufficiently by the time they first appeared in the fossil record (Pennsylvanian) that their affinities are difficult to determine. Their body and wing features link them to the orthopteroids. Orthopterans first appeared in Pennsylvanian beds, and since then their basic wing venation has been only slightly modified.

The other orthopteroid order having a fossil record is the Embioptera (embiids). One genus is recorded from the Permian of east-

Fig. 9-22. Embioptera, *Sheimia sojanensis*, from Permian strata of Arkhangel region, U.S.S.R. (After Rodendorf and others)

ern Europe (Fig. 9-22); the rest are Cenozoic, mostly from Baltic amber and mostly similar to modern forms. The specialized order Grylloblattodea lacks a fossil record.

Plecopteroid Orders

The Protoperlaria and Plecoptera (stoneflies) have well-established Permian fossil records. Both had aquatic nymphs and represent a distinctive, small, compact branch arising from the Protorthoptera. The Plecoptera are known from Permian, Jurassic, Cretaceous, and Cenozoic beds.

Hemipteroid Orders

The fossil hemipteroids all belong to living orders and only the Zoroptera and Phthiraptera (lice) lack a fossil record. Hemiptera (bugs) and Psocoptera (psocids) are well represented in Permian strata, and Thysanoptera (thrips) appear with certainty in Jurassic strata. All Permian Hemiptera (Fig. 9-23) were Homopterans and had sucking beaks characteristic of that suborder. Permian Hemiptera included a wide variety of extinct families, some with long ovipositors indicating their relationship to some early orthopteroids. They successfully adapted to liquid foods and they probably gradually displaced the palaeodictyopterans from this set of ecological niches.

The Early Permian psocopteran suborder Permopsocida (Fig. 9-24*A*) had subequal wings with a generalized wing venation. By the Late Permian time, at least two psocopteran families had evolved reduced hind wings and modified venation in the front wings (Fig. 9-24*B*) and are strikingly similar to modern psocids. Thysanoptera have a scant fossil record, the best known being from Jurassic strata of Turkistan. An earlier thriplike fossil is known from Upper Permian deposits in European U.S.S.R.

Neuropteroid Orders

The fossil record of neuropteroid orders is remarkably good. Both major branches within the neuropteroid orders first appeared during

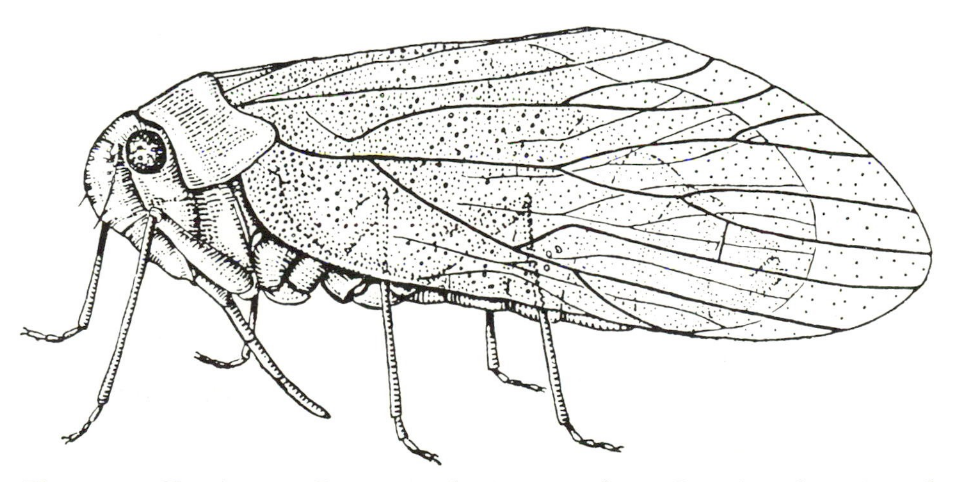

Fig. 9-23. Hemiptera, *Permocicada integra*, from Permian deposits of Arkhangel region, U.S.S.R. (After Rodendorf and others)

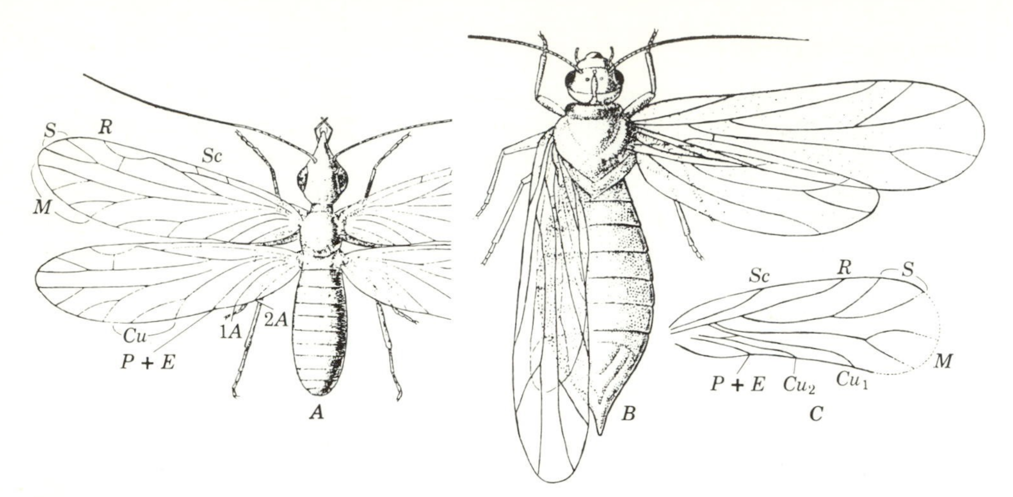

Fig. 9-24. Psocoptera. (*A*) Permopsocida, *Dichentomum tinctum*, Lower Permian of Kansas. (After Carpenter); (*B*) Parapsocida, *Zoropsocus tomiensis*, from Permian deposits in the Arkhangel region, U.S.S.R.; (C) Parapsocida, *Lophioneurodes sarbalensis*, Lower Permian, Kuznets Basin, U.S.S.R. (After Rodendorf and others)

the Pennsylvanian and in considerable abundance during the Permian.

Primitive Neuropteroids. Among the primitive neuropteroids (Fig. 9-1), the orders Megaloptera (alderflies), Neuroptera (lacewings) (Fig. 9-25), Glosselytrodea, and Coleoptera (beetles), occur in Permian beds. Raphidioptera (snakeflies) (Fig. 9-26) are reported with certainty from Jurassic beds. These orders were much more diverse and abundant in the Permian, Triassic, or Jurassic than at present.

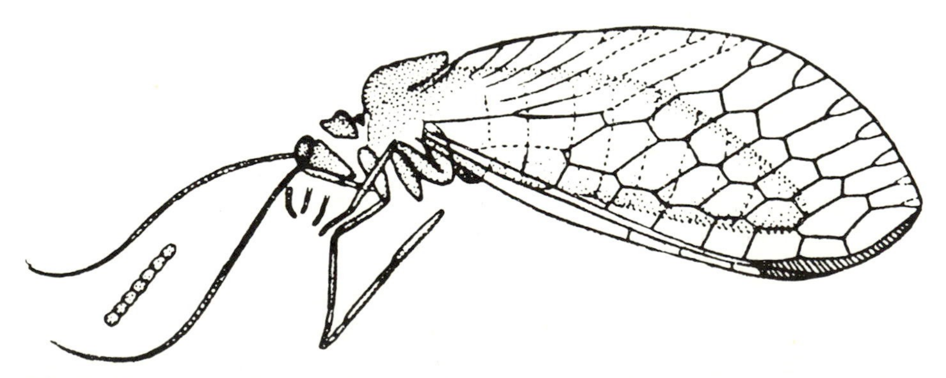

Fig. 9-25. Neuroptera, *Mesypochrysa latipennis*, Jurassic deposits of Kazakhstan, U.S.S.R. (From Rodendorf and others)

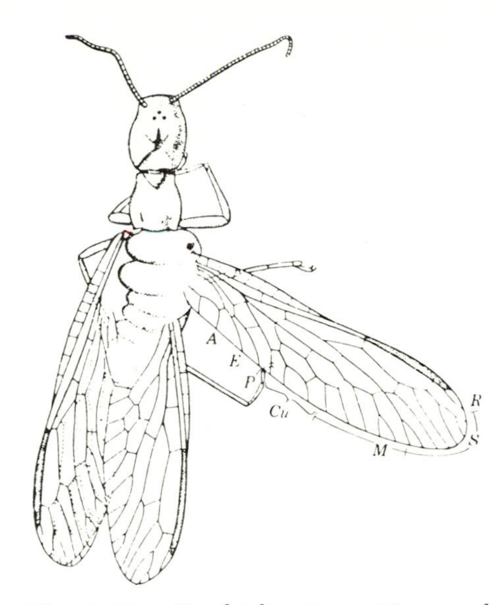

Fig. 9-26. Raphidioptera, *Mesoraphidia pterostigmalis*, from Jurassic deposits in Kazakhstan, U.S.S.R. (From Rodendorf and others, after Handlirsch, 1906–1908)

Most families in these orders had evolved before the Jurassic Period ended. The Coleoptera (beetles) are the only order of the primitive neuropteroids to gain and retain a major portion of terrestrial habitats and, at present, include nearly 40% of the known species of insects. The Permian fossil Coleoptera have the beginnings of elytrous venation and were members of the suborder Archostemmata. In the fossil record, the Archostemmata remain dominant until the Jurassic Period, at which time a large number of modern suborders and families appeared.

Advanced Neuropteroids. The advanced neuropteroid order include the Hymenoptera (wasps and bees), which forms an independent branch; the Mecoptera (scorpionflies), Siphonaptera (fleas), and Diptera (flies), which form a second branch; and the Trichoptera (caddisflies) and Lepidoptera (moths and butterflies), which form a third branch. Trichoptera-like forms and Mecoptera are both known from Permian rocks, suggesting that the separation of these two branches was even earlier, possibly Pennsylvanian. Hymenoptera first appeared in Triassic strata.

Mecopterans were well differentiated in the Permian Period and included at least three suborders. One of these, the Protomecoptera, had wing venations that were similar to early megalopteran-like fossils of about the same age, and its descendents, the family Meropeidae, today is a relict lineage. The suborder Eumecoptera is represented in Permian sediments by eight or more families (Fig. 9-27), most of which became extinct by the end of the Mesozoic or have been modified sufficiently to form modern families. The two lineages represented (1) by the surviving family Bittacidae, and (2) by the Panorpidae and Panorpodidae, evolved from two eumecopteran families that were distinct and separate lineages during all of the Mesozoic. A Mesozoic suborder, Paratrichoptera, appeared in the Triassic but did not survive into Cenozoic times. The Permian and Mesozoic Mecoptera were considerably more diverse than modern Mecoptera and gave rise to two specialized orders, the Siphonaptera and Diptera.

The Siphonaptera (fleas) have a meager fossil record, including an occurrence in Lower Cretaceous beds of Australia and another in early Tertiary amber (Fig. 9-28) of the Baltic region. Their lack of wings and specialized adaptations for parasitic bloodsucking obscure their origins within the Mecoptera.

The Diptera (flies) are better known as fossils (Fig. 9-29). Diptera-like wings are known from Permian deposits in Australia and possibly from the Triassic of the U.S.S.R. Diptera of nearly modern ap-

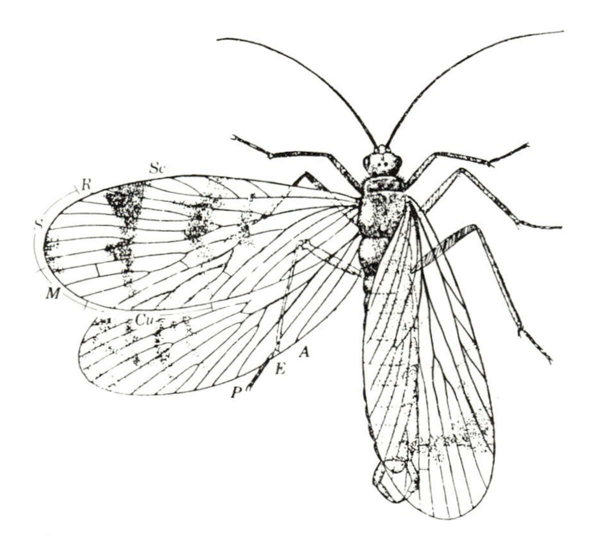

Fig. 9-27. A Permian mecopteran *Agetochorista tillyardi*, suborder Paramecoptera, Ural region, U.S.S.R. (From Rodendorf and others, after Handlirsch, 1906–1908)

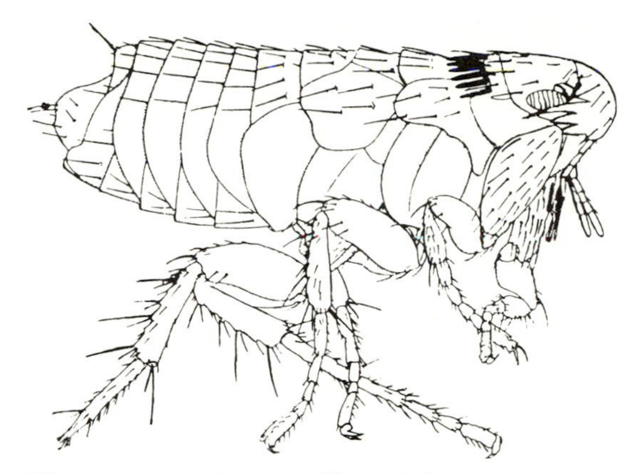

Fig. 9-28. *Palaeopsylla clebsiana*, a siphonapteran from late Eocene Baltic amber of Europe. (From Rodendorf and others)

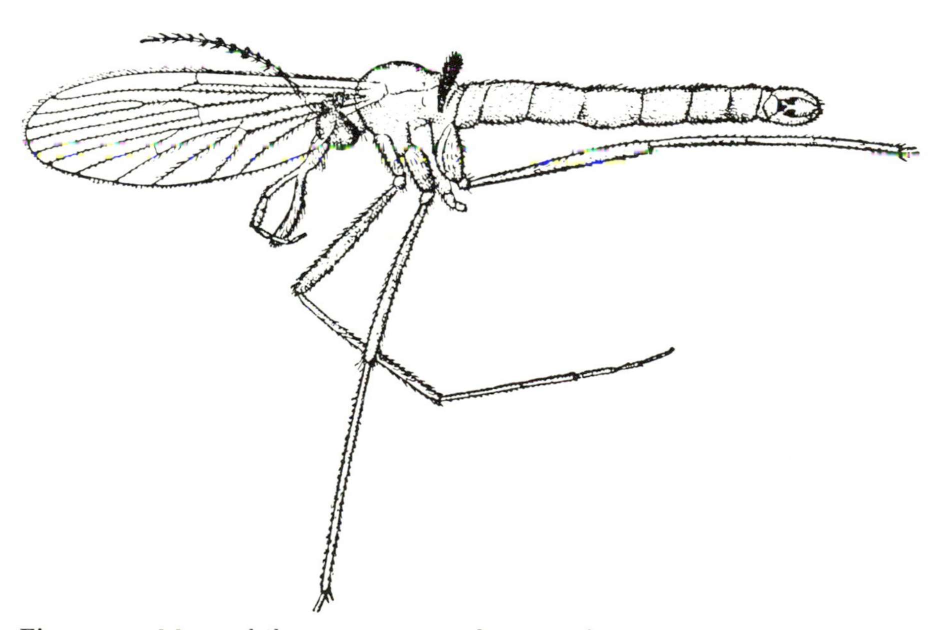

Fig. 9-29. *Macrochile spectrum*, a dipteran from the late Eocene Baltic amber of Europe. (From Rodendorf and others)

pearance are known by Early Jurassic time when Nematocera and several Brachycera families were abundantly represented. Fossils of most of the families of Diptera appeared in Eocene or earlier beds. The more advanced schizophoran families appear in Middle or Upper Cenozoic deposits.

Hymenoptera (sawflies, ants, bees, and wasps) have an extensive Mesozoic and Cenozoic fossil record. Primitive sawflies, family Xyelidae, suborder Symphyta, are the earliest fossils, appearing in the Lower Triassic of Central Asia, Upper Triassic of Australia, and Jurassic of Asia (Fig. 9-30*A*). The suborder Apocrita is first reported from Jurassic beds, and the Ichneumonoidea (Fig. 9-30*B*) and Chalcidoidea parasites are well represented in Cretaceous and younger deposits (Fig. 9-30C). A Sphecoidea is reported from the Lower Cretaceous of Australia. Formicidae (ants) (Fig. 9-31) and Vespidae (wasps) were abundant by the beginning of the Cenozoic and

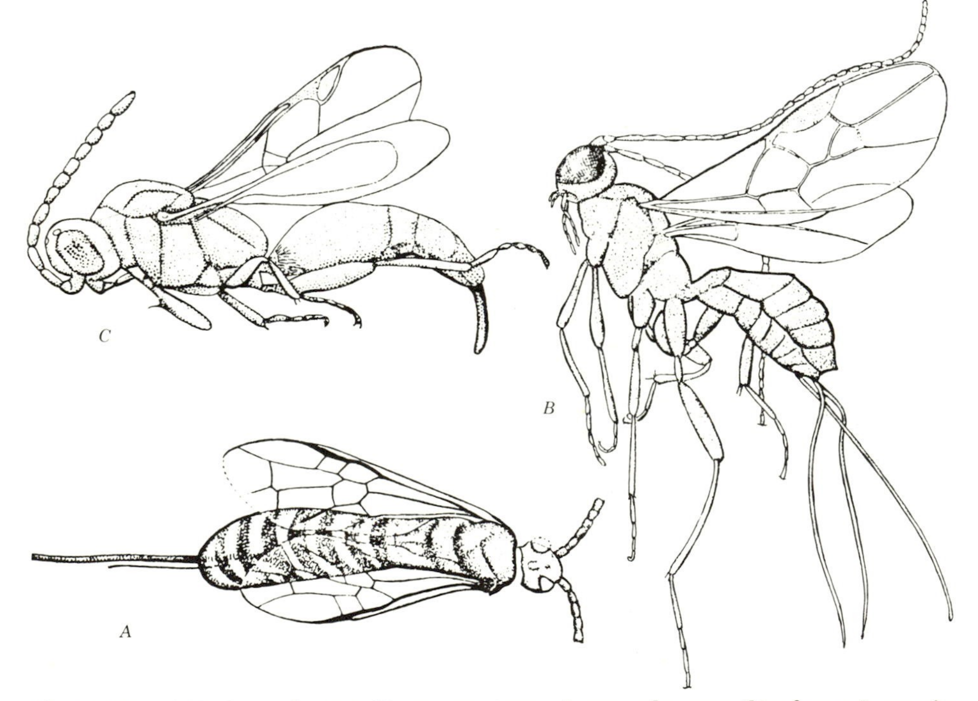

Fig. 9-30. (*A*) Symphytan Hymenoptera, *Anaxyela gracilis* from Jurassic deposits of Kazakhstan, U.S.S.R.; (*B*) ichneumonid Apocrita, *Microtypus longicornis* from Eocene Baltic amber of Europe; (C) proctotrupid *Cryptoserphus pinorus* from Eocene Baltic amber of Europe. (From Rodendorf and others)

Fig. 9-31. Oldest known ant, *Sphecomyrma freyi*, from Upper Cretaceous deposits from New Jersey. (From Wilson, Carpenter, and Brown; photograph by F. M. Carpenter)

Apoidea (bees) were well developed by the Oligocene. The several stages in Hymenoptera evolution appear to have general parallels in their order of fossil occurrence.

The Trichoptera (caddisflies) and Lepidoptera (moths and butterflies) form another highly advanced group of neuropteroid orders. Trichoptera, or their immediate ancestors, appeared during the Early Permian in North America and in the Late Permian of Australia. Their wing venation showed similarities to the early lineages of the Mecoptera, indicating their common ancestry. Permian trichopterans had reduced cross venation and terminal branches and represented intermediate stages leading to the Mesozoic trichopteran families. The Upper Triassic of Australia has the family Mesopsychidae, and the Upper Triassic and Lower Jurassic of northwestern Europe has the family Necrotaulidae. These families still retained separate bases for veins M and Cu_1. In most younger representatives, these two vein bases are fused. Cretaceous and Oligocene fossils provide most of the remainder of the trichopteran fossils. Caddisfly larval cases are commonly preserved in the fossil record. Case construction appears to have evolved prior to the Triassic Period.

The Lepidoptera, a group renowned for its great diversity of large showy species and many families, is a remarkably late evolutionary arrival and relatively scarce as fossils. Several primitive families are reported from Cretaceous deposits; however, most of the fossil forms are from the Eocene–Oligocene (Baltic amber) of northern Europe and a few from the Oligocene of North America. The Jugatae (Fig. 9-32*A*) were well established by the beginning of the Cenozoic. They had similar wing venation to some Trichoptera and presumably were derived from them in the mid-Mesozoic. In the more advanced Frenatae (Fig. 9-32*B*), known from Baltic amber, their front and hind wings were no longer closely similar. The most advanced suborder, Rhopalocera, is not reported in the fossil record.

INSECTS AND THE HISTORY OF LIFE

Early Life and Colonization of Terrestrial Habitats

Terrestrial floras and faunas appeared for the first time in the geological record near the middle of the Paleozoic Era, only about 400

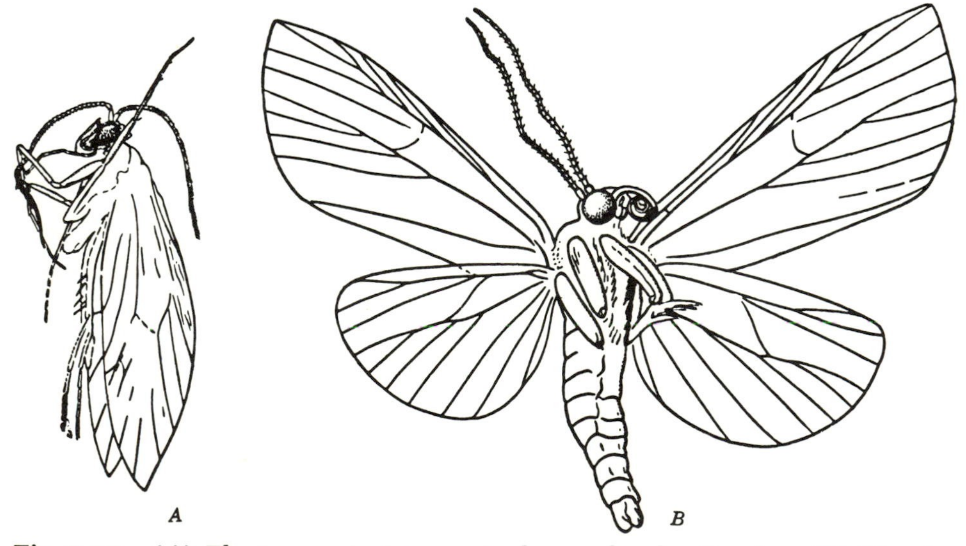

Fig. 9-32. (*A*) *Electrocrania immensipalpa*, suborder Jugatae; (*B*) *Glendotricha olgae*, suborder Frenatae; both from Eocene Baltic amber of Europe. (From Rodendorf and others)

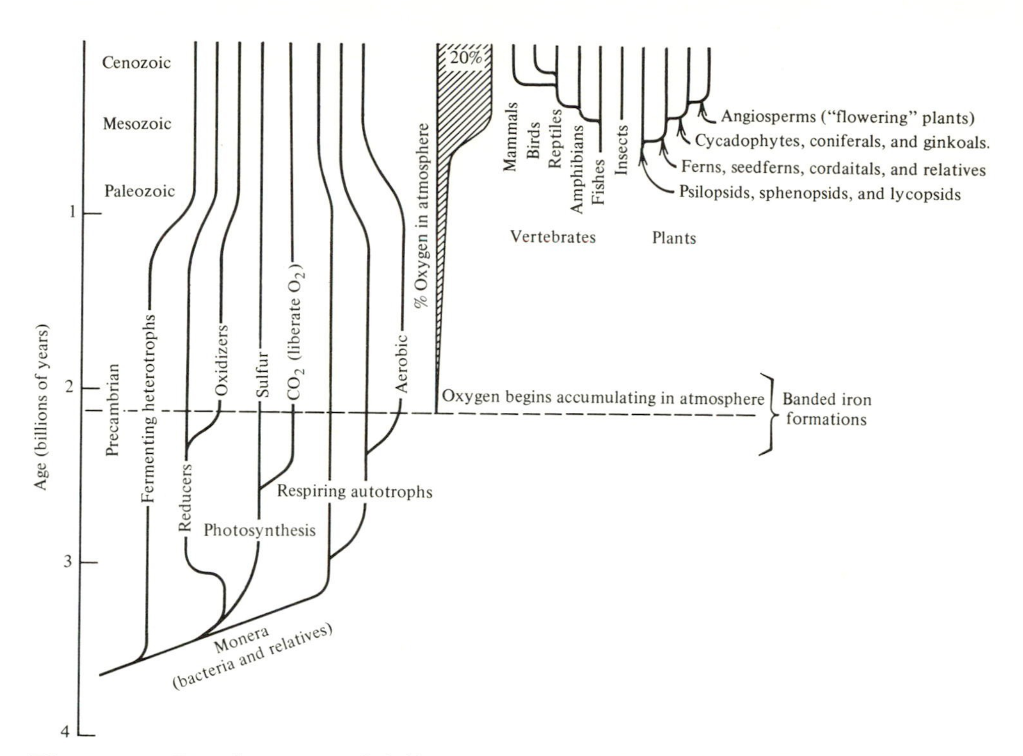

Fig. 9-33. Development of different stages of life and its effect on the earth's atmosphere.

million years ago. Prior to that time, organisms had evolved in aquatic environments, starting 3500 to 4000 million years ago (Fig. 9-33). During the Cambrian and Ordovician Periods all of the major animal phyla appeared, and many marine classes that survive today originated during these early Paleozoic periods. From the evidence of the geological record, we know surprisingly little about terrestrial deposits of these early Paleozoic periods.

Colonization of terrestrial areas by plants was well started by the later part of the Silurian Period (410 million years ago). Bryophytes (mosses) appeared before the end of that period and, in the Early Devonian, swamp-dwelling psilophytes were reasonably abundant and diverse. In addition to these early plants, Lower Devonian Rhynie Chert of Scotland has an assortment of small fossil Crustacea, Arachnida, and Collembola (*Rhyniella praecursor*, Fig. 9-34). Lower Devonian rocks near Alken in the Mosel Valley, Germany, have a wide variety of freshwater, brackish, and marine faunas and floras that were deposited in a lagoon.

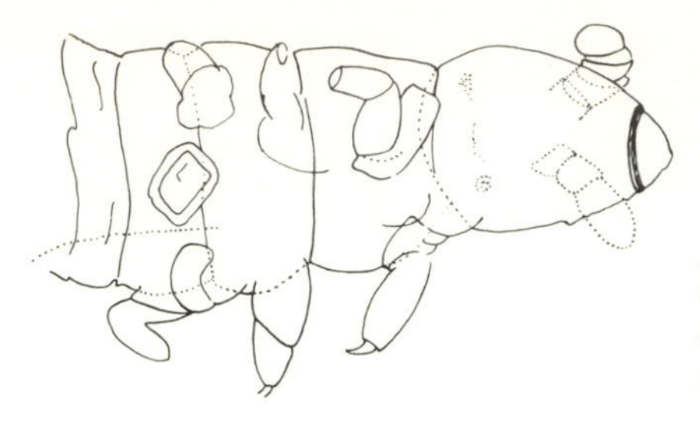

Fig. 9-34. *Rhyniella praecursor*, a Devonian fossil belonging to the order Collembola. (After Aubert)

By Middle Devonian time (375 million years ago), many low, wet terrestrial swamps were forested by tree ferns. Some had stumps up to half a meter in diameter and represented medium to large trees. Insect fossils have not been directly associated with these Middle Devonian forests; however, the height and concentration of these tree ferns indicate that complex, multistoried environments were rapidly becoming available to insects during this time, and flight could have been an advantage to insects in such environments.

Early Winged Insects

Mississippian and Pennsylvanian Periods. Fossil insects so far have not been found in Upper Devonian or Mississippian rocks. The rapid evolution of the amphibians and their evolution into four orders during the Mississippian Period may have been based on the amphibians' active dietary interest in aquatic freshwater arthropods, particularly insect nymphs.

The shallow continental shelves were covered by broad river deltas that were densely forested (Fig. 9-35) by psilopsids, lycopsids (scale trees), sphenopsids (horsetails and relatives), pteropsids (ferns), pteridosperms (seed ferns), and early conifers (Cordaitales and early Coniferales). These forests formed our oldest economic coal deposits and many excellent insect fossil localities. After mid-Pennsylvanian time, many swamp forests formed in broad intermontane lake basins.

Pennsylvanian insects include at least 12 insect orders (Fig. 9-1), and additional fragmentary fossils suggest that other orders were present as well. Fossil localities include about 35 extensive fossil assemblages and show the paleopterans, particularly the orders Paleodictyoptera, Megasecoptera, and Diaphanoptera, to be abundant and diverse. Neopterans also were very diverse and some were very abundant, particularly the order Protorthoptera. The orders Miomoptera, Caloneurodea, and Orthoptera also were broadly diversified. Cockroach-like insects, including the suborder Protoblat-

Fig. 9-35. Pennsylvanian coal swamp. (From R. Zallinger, mural *Age of Reptiles*, copyright 1975, Peabody Museum of Natural History, Yale Univ., reproduced with permission)

toidea (Protorthoptera), formed another major Pennsylvanian group composed of more than two dozen families. So many of these are fossilized that the Pennsylvanian Period has been called the "Age of Cockroaches." Of paleogeographical interest is the gradual expansion during the Pennsylvanian Period of insect fossil localities northward and southward from an initially narrow equatorial belt.

Permian Period. The supercontinent Pangaea extended from northern polar regions across the temperate, subtropical, and tropical regions to the southern polar region. Permian coal swamps yielded many coal deposits and at least 75 major fossil insect localities.

Nearly all the Pennsylvanian insect orders extended into Permian deposits and continued to expand in diversity (Fig. 9-36); however, 10 or more new orders appeared that dominated the assemblages (Fig. 9-1). The Psocoptera (psocids), Hemiptera (homopteran bugs), Mecoptera (scorpionflies), and Coleoptera (beetles) were well represented. True Odonata (dragonflies), Protelytroptera, Protoperlaria, Plecoptera (stoneflies), Neuroptera (lacewings), Glosselytrodea, Trichoptera (caddisflies), and possibly Megaloptera (alderflies), first appeared during the Permian Period. Although many of these early orders became extinct before the end of the Permian, 10 or more formed the primitive stocks from which modern orders evolved.

Two other terrestrial groups also evolved rapidly during the Permian: reptiles and plants. The reptiles, which appeared late in the Pennsylvanian Period, evolved rapidly during the Permian. Some were herbivores, others carnivores, and many others insectivores. Plants also continued to diversify. Ferns and seed ferns dominated

Fig. 9-36. Photograph of a fossil insect *Dunbaria fasciipennis*, from the Lower Permian of Kansas. (From Mavor, *General biology*, by permission of Macmillan Co.)

the moist environments; however, a broad range of early coniferophytes spread into progressively drier areas.

Diversification of Terrestrial Communities

Triassic. This was a time of major changes in terrestrial conditions and these changes are reflected in the history of insect orders. Large parts of the supercontinent Pangaea became warm, dry, and covered by red alluvial plains. Vegetation changed rapidly to adapt to these new conditions. The coniferophytes included many Cordaitales, Coniferales, and Taxales. Ginkgoales, and perhaps Gnetales, were becoming widespread in warm temperate latitudes, and Cycadeoidales and the flowering Cycadales in tropical and subtropical latitudes. Insects experienced major evolutionary modification because of these changes in their habitats.

Triassic insect fossil localities are considerably less numerous than those of either older or younger periods. There are about a dozen localities, principally in Australia and central Asia, but only some of these have yielded abundant, well-preserved fossils. At least nine orders that appeared in the Permian did not survive beyond the Triassic Period. Primitive Hymenoptera (sawflies), Phasmatodea (walking sticks), and Diptera (flies) first appeared during the Triassic. They were well differentiated, strongly suggesting an earlier origin. Feeding tunnels typical of modern beetles of the Buprestidae and Scolytidae (Fig. 9-37) are preserved in petrified Triassic trees.

Modernization of Insect Faunas

Jurassic. Pangaea remained more or less intact during the Jurassic Period. Along the western margin of North America, an elongate, north-trending mountain system extended from north central Mexico into east central Alaska. Sediments and wind shadows from this north-south mountain system and the transgressions and regressions of interior seas greatly influenced the climates of North America.

Plant evolution continued rapidly during Jurassic time. Dicotyledonous angiosperm-like pollen appeared in sediments before the end of the period. Pteridosperms (seed ferns) were rare and did not survive the Jurassic Period. Coniferophytes, particularly the Ginkgoales and cycadophytes, were widespread.

Jurassic insects are known from a number of well-studied localities, including those in the Lower Jurassic of northwestern Europe, several Middle Jurassic localities near Karatau in Asia, and others in the Irkutsk Region of Asia. Within the order Neuroptera, Mecoptera, and Diptera, many "modern" new suborders and superfamilies appeared in these Jurassic faunas. Insect feeding was well established on the varied floras. Some plant-flowering bodies were conspicuous, and it is likely that many insects visited and pollinated them. The Dermaptera (earwigs) and true Thysanoptera (thrips) first appeared in Jurassic time and, in general, the faunas had a modern appearance. The appearance of mammals, flying reptiles, and birds during the Jurassic suggests that both the Phthiraptera (lice) and Siphonaptera (fleas) may have become established on their host lineages at this time.

Coevolution of Insects and Flowering Plants

Cretaceous. The Cretaceous Period represents a long period of time, 60 to 70 million years. During this time the North Atlantic Ocean basin was spreading, gradually separating the southern parts of Europe away from North America, and also separating these two areas from both northern Africa and northern South America (Fig. 9-7). The South Atlantic Ocean basin started spreading within the Cretaceous Period so that Africa and South America did not start to separate until about the middle of the period. Throughout the Cretaceous, North America and Europe retained a northern land

Fig. 9-37. Feeding tunnels of (*A*) a buprestid beetle, and (*B*) a scolytid beetle in petrified Triassic trees, Petrified Forest National Monument, Arizona. (After Walker)

connection across northern Europe, Barents Shelf, and Greenland. Large, linear mountains in the Cordilleran region continued to have strong influences on local climates.

During the Cretaceous (Fig. 9-38), angiosperm plants, particularly the Dicotyledonae, evolved rapidly. Nearly all of the modern dicotyledonous plant families are known from the fossil record before the end of the period. Sedges and grasses appeared. This increase in flowering plants, many of which are insect pollinated, is paralleled by an increase in insect groups that had adapted to flowering plants as sources of food, particularly pollen- and nectar-feeding families in the Lepidoptera (butterflies). Insect fossils of Cretaceous age (Figure 9-39) are not common; however, Alaska, Canada, Siberia, and the Middle East have several Cretaceous fossil sites. These insect faunas are varied in taxonomic composition and include stoneflies, dragonflies, cockroaches, midges, aphids, caddisflies, parasitic hymenopterans, termites, and primitive lepidopteran families. Many of these are close relatives, or actually members of existing genera; some others represent genera of a primitive nature. A few are of exceptional interest because they represent families intermediate between existing families. An example is the fossil family Jascopidae, known from a nymph in Canadian Cretaceous amber, which is intermediate between the homopteran families Ceropidae and Cicadellidae and represents an early stage in the evolution of the Cicadellidae from the cercopid-like ancestor.

Fig. 9-38. Cretaceous landscape. (From R. Zallinger, mural *Age of Reptiles*, copyright 1975, Peabody Museum of Natural History, Yale Univ., reproduced with permission)

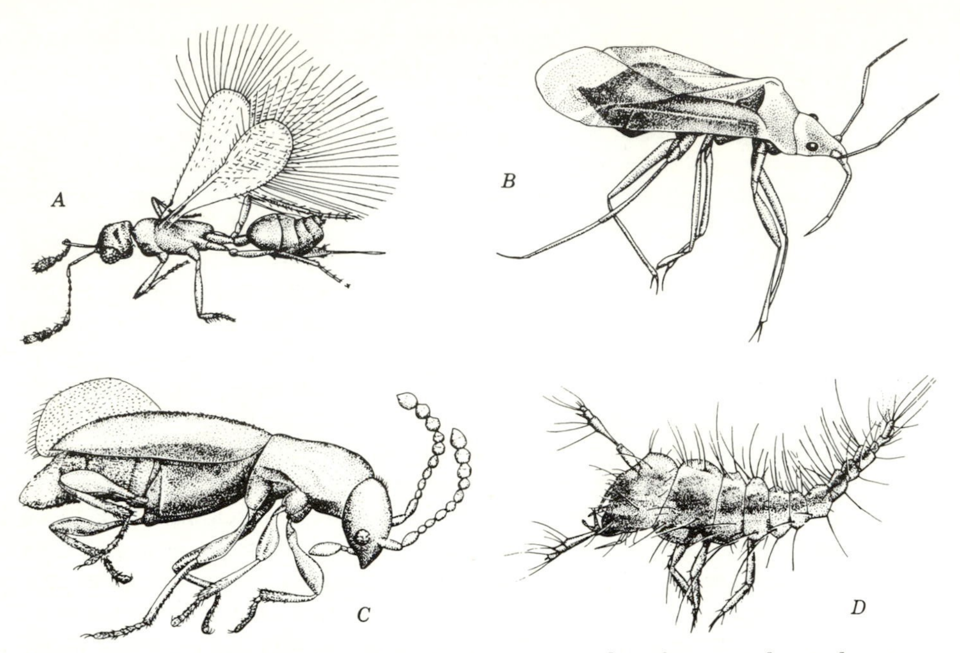

Fig. 9-39. Fossils preserved in Cretaceous amber from Cedar Lake, Manitoba, Canada. (*A*) Fairyfly (Hymenoptera: Mymaridae); (*B*) a small pirate bug (Hemiptera: Anthocoridae); (*C*) stone beetle (Coleoptera: Scydmaenidae); and (*D*) a larva of a rove beetle (Coleoptera: Staphylinoidea). (From McAlpine and Martin, 1969, in *The Beaver*, drawings by K. A. Hamilton, Agriculture Canada)

Development of Modern Faunal Realms

Cenozoic. Although the Cenozoic Era is short, about 60 to 70 million years, it has perhaps the most complete fossil insect record available.

Climatic Changes. During the Cenozoic most of the earth's present physical surface features formed. Also, climatic cooling during the Cenozoic followed by strong fluctuations in climatic temperatures in the Quaternary strongly modified the present distribution of insects. These events caused frequent and marked expansions and contractions in environmental conditions, shifts in climatic belts toward and away from the equator, and in insects the repeated development of disjunct ranges and relict ranges. Although similar extreme climatic events may have occurred earlier during the Pennsylvanian and early parts of the Permian Periods, the Mesozoic periods did not have strongly developed climatic fluctuations.

Geographical Changes. The Cenozoic was also a time of wider separation and increased isolation of most continents. North America and northern Europe continued to separate and by the middle of the Eocene Epoch, their connection at the north was severed so that a terrestrial dispersal path between them was terminated (Fig. 9-6). In late Eocene and Oligocene time, the warm coastal plain along the southern margins of central Europe, central Asia, and China was deformed as Africa, India, and the Middle East continental blocks pushed northward. By Miocene time, the Himalayan mountain system was nearly continuous from Europe into southeast Asia, and by Pliocene time it had reached lofty elevations. A series of large depressions formed on the northern side of this mountain chain, and many were sites of nonmarine lakes (the Black Sea, Caspian Sea, and Sea of Azov are remnants). The Himalayan mountains are climatic and natural environmental barriers that today separate the Palearctic and Oriental faunal realms (see Fig. 8-2).

Australia and New Guinea also moved northward to a position close to one segment of the Oriental faunal realm. A combination of climatic and environmental contrasts and relatively recent movement have limited the intermixing of the Oriental and Australian animals and plants.

From the Jurassic Period until the Pliocene Epoch, North America and South America were separated. South America separated from Africa and Antarctica in the later part of the Cretaceous Period. The present Isthmus of Panama connection between North and South America dates from middle Pliocene time, 4 to 6 million years ago. Although the exchange of animals and plants across this narrow isthmus has been significant, both continents have retained a number of distinct biotic features. The greatest changes have been in the mammals, where the North American placentate mammals have largely out-competed their marsupial and placentate South American counterparts.

North America and northern Asia were connected frequently across the Bering Straits during much of Cenozoic time and this has resulted in close similarities in the Nearctic and Palearctic parts of the Holarctic faunal realm (Fig. 8-2). In North America significant increases in the height of the Cordillera and Rocky Mountains during the Miocene, Pliocene, and Quaternary partitioned the basin and range region into small, ecologically distinctive areas. The Basin and Ranges, Rocky Mountains, and Great Plains were uplifted 2 to 4 km as a broad regional arch and subsequent erosion shaped the present mountain ranges. This uplift caused a rain shadow, resulting in expansion of grasslands in the Great Plains. The mountains became cooler and received more localized precipitation.

Insect fossils from the Upper Eocene–Lower Oligocene amber in the Baltic region of northern Europe contain several orders or suborders not seen in older deposits: the Mantodea, Embioptera, and the specialized parasitic coleopteran Strepsiptera, in addition to many other insects. In northern Europe (Germany and France), North America, Caucasus, and Central Asia, Oligocene insects are well known and have formed the basis for several classical studies. Miocene localities are known in northern Europe and Northern Caucasus and in the Mojave Desert of California. Pliocene localities are rarer.

Most of the insects known from Baltic amber belong to surviving genera. Several advanced families of hymenopterans and lepidopterans are not present in the Baltic amber collections and probably have evolved since Eocene time. At Florissant, Colorado, large numbers of Oligocene insect fossils are preserved in fine sediments of volcanic ash that settled in ponds at the foot of the old Rocky Mountains. These fossilized insects came from an extremely varied assortment of habitats, ranging from tropical to cool temperate, and they show that the Oligocene insect fauna contained fully as many genera as does that of the present day. Some of these genera no longer occur in North America, but are known now only from other regions, such as the genus *Glossina*, tsetse flies, which at present occurs only in equatorial Africa. Fossils of the curious lacewing family Osmylidae occur in the Florissant Oligocene (Fig. 9-40); living species are restricted to tropical areas of the world.

Ice Ages—Quaternary, Pleistocene. After the Pliocene, cooler temperatures repeatedly fluctuated with slightly warmer temperatures over the whole world. Glaciers grew and expanded and then melted and diminished in many parts of the globe, especially Antarctica, northern North America, and northern Eurasia. The northern glaciers spread southward and covered large areas (Fig. 9-41). Four main glacial episodes are recognized, and between each was a long interval with a climate warmer than our present one. At least the last two, and possibly earlier glaciations, internally included more than one glacial–interglacial cycle. We are living at present in the receding portion of the last glacial period, and remnants of the northern ice cap are now found only as isolated glaciers or as island masses such as that on Greenland.

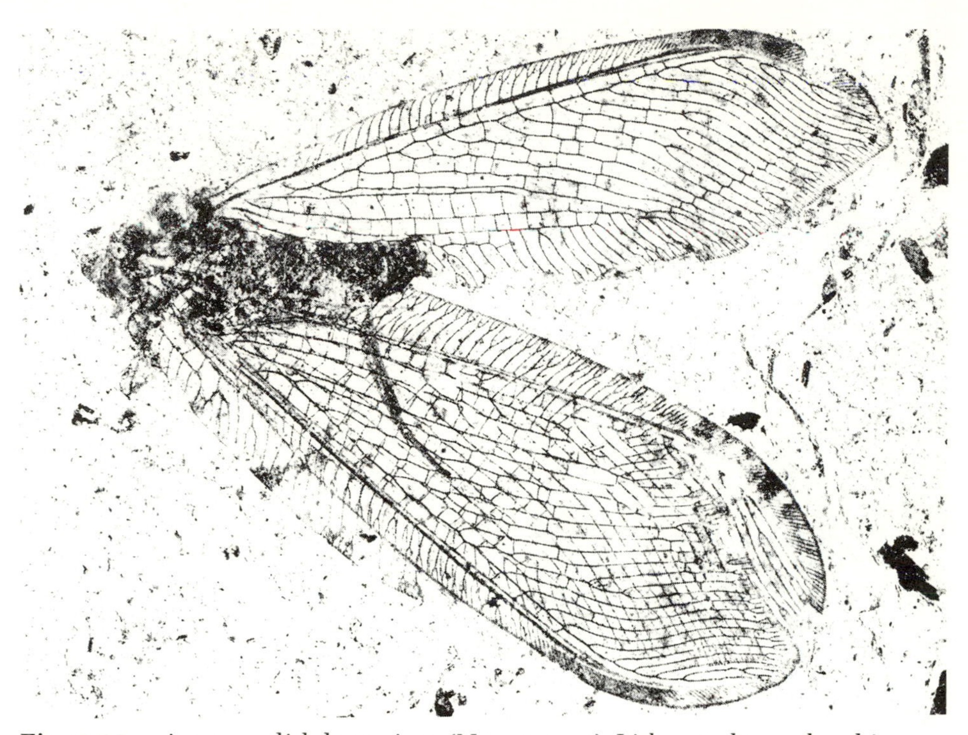

Fig. 9-40. An osmylid lacewing (Neuroptera) *Lithosmylus columbianus*, from the Oligocene shales of Florissant, Colorado. (Photograph loaned by F. M. Carpenter)

Dispersals, Disjunct Ranges, and Speciation. The glacial–interglacial fluctuations had a tremendous impact on the North American biota, especially its more temperate, northern elements. Although North America was faunally separated from western Europe, northwestern North America and northeastern Asia were connected by the Bering land bridge which provided a land dispersal route. As a result, when northern climates became as warm as or warmer than those of today, many cool temperate species of insects became holarctic in distribution, spreading between North America and Asia, some going in one direction, others in the opposite. During the next cooling episode when glaciers again expanded, the ranges of these species spread south as the southern edges of their ranges became cooler and better suited for their existence. As the northern edges of ranges were displaced south of the Bering land bridge, the range of each species was divided into two separate segments without genetic interchange; if the separation was long enough, the

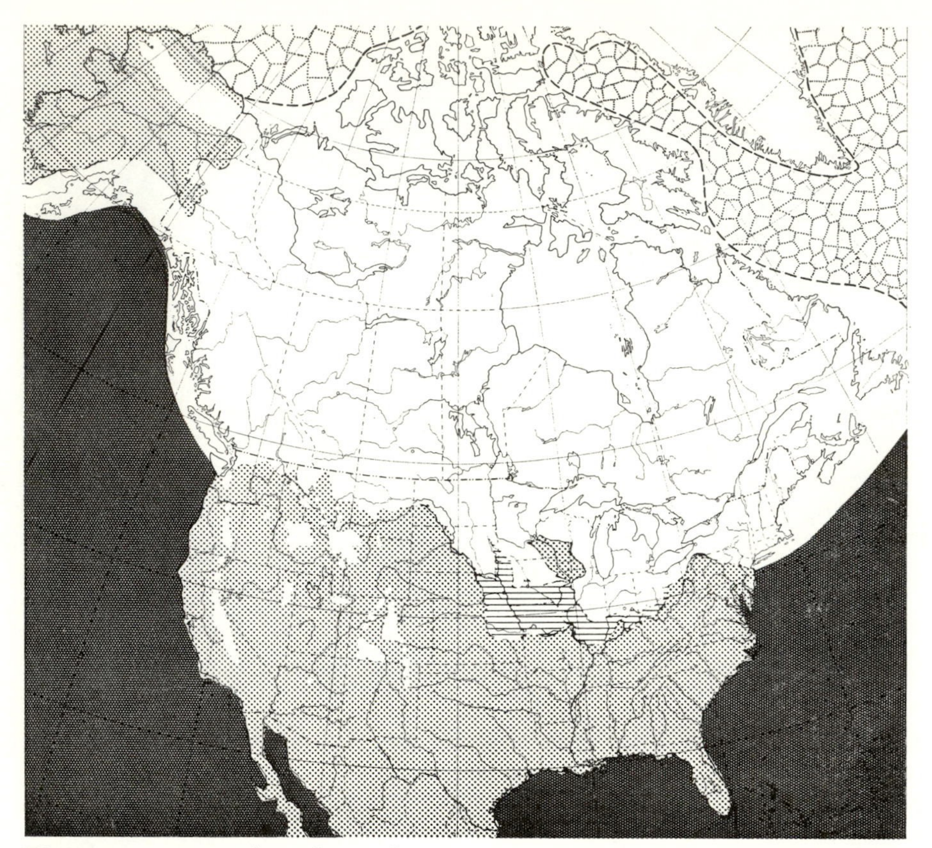

Fig. 9-41. General outline of maximum glacial ice cover during the Quaternary in North America. White area shows extent during last major glaciation (Wisconsinan); horizontal pattern shows additional area covered during preceding glaciations (Illinoisan, Kansasan, and Nebraskan). (From Illinois Natural History Survey)

populations in each segment potentially could evolve into separate species. When the glaciers dissipated and the next warm interglacial interval arrived, the ranges of the species moved northward. This resulted in another mixing of Asian and North American species, many of which were closely related but now genetically isolated, to form a new holarctic assemblage, and the process is ready to be repeated—a long-term climatic species-generating mechanism.

The same mechanism has operated between mountainous areas within the same continent. For example, the winter stonefly genus *Allocapnia*, restricted to cool streams of eastern North America, has evolved into a cluster of about 38 species, primarily through glacial–interglacial dispersals and isolation of geographically separated populations of parental species between the northern and southern Appalachian highlands, the Cumberland plateau, and the Ozark-Ouachita moutains (in Missouri, Arkansas, and Oklahoma). Current evidence suggests that all, or nearly all, of these species evolved within the 3 million years of the Pleistocene.

The northern, cool-adapted caddisfly *Glossosoma intermedium* now occurs from western Europe, across Asia, and eastward to Minnesota and Missouri. The species is remarkably similar throughout this wide range, suggesting that it attained this holarctic distribution during the last glacial episode, possibly only about 12,000 to 30,000 years ago. The range, however, is fragmented into a number of isolated populations, setting the stage for future speciation.

In the southwestern United States, climatic changes during the Quaternary have produced another mechanism favoring speciation. The glacial periods produced cool, rainy or pluvial periods; the interglacials produced hot, dry periods. Many insects of that area, including many species of beetles, wasps, and bees, are especially adapted to the desert. During the pluvial periods, their ranges were restricted to small, isolated, low-elevation, desert or semidesert areas between cool, wet mountain ranges. This was a time of speciation for these isolated, dry-adapted populations. During the following interglacial, the dry, hot areas expanded at the expense of the wet, cool areas and the resulting connection provided a period for dispersal and mixing of species.

Quaternary fossil insects are becoming increasingly better known, particularly from the study of peat macerals and organic-rich silts. In a number of orders, the Coleoptera being a good example, the exoskeletal parts are sufficiently durable that they can be wet-sieved from unconsolidated sediments. Most recent studies of Quaternary insects have focused on Coleoptera from the Wisconsinan, or latest Pleistocene desposits. Most of these insect fossils belong to present-day species of temperate-adapted coleopteran families, and suggest relatively slow evolutionary rates. The displacements, compressions, and expansions of climatic belts during the Quaternary had many ramifications for insects, particularly changes in their ranges and dispersal patterns, and not all groups of insects show the same evolutionary rates.

Faunal Mixing

Many components of the present North American insect fauna probably did not evolve here, but they or their ancestors dispersed into North America from various sources. An excellent example is the

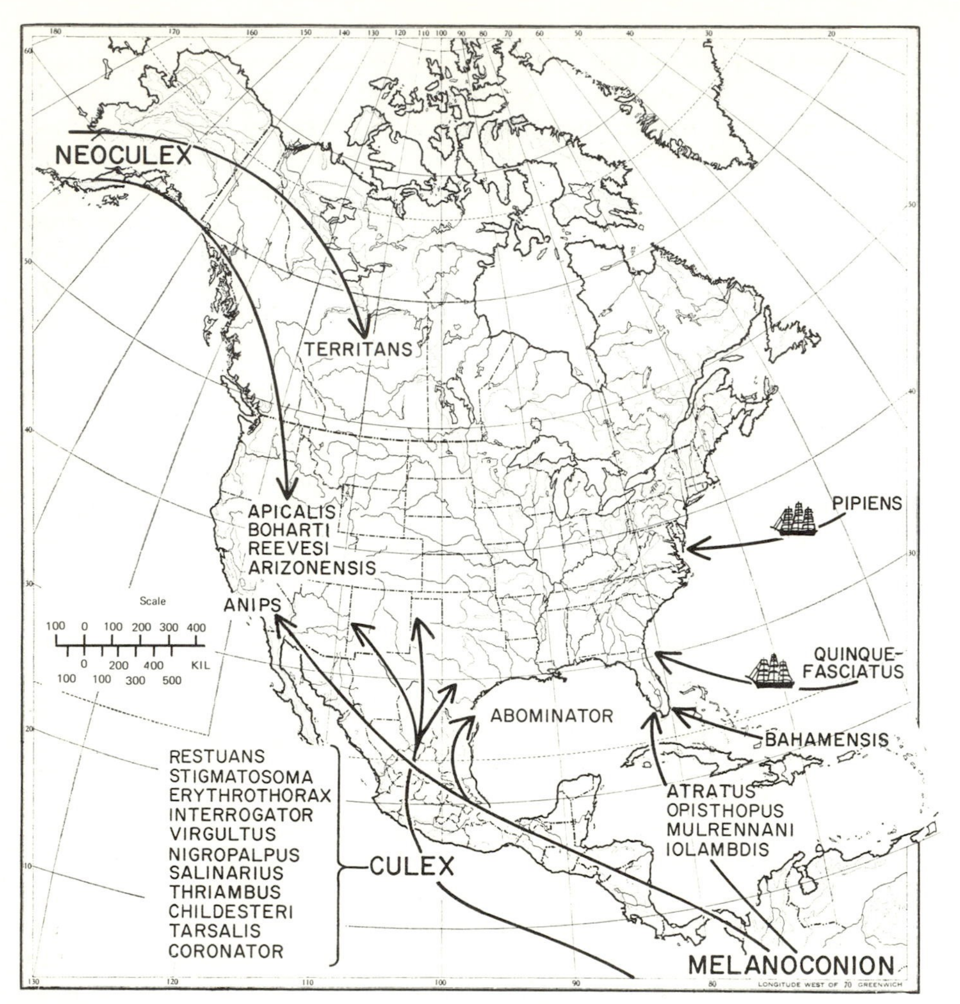

Fig. 9-42. Many insect lineages, such as the stonefly genus *Allocapnia*, evolved in North America. In other insect genera, the North American fauna reached this continent at different times and from various directions, as shown in this map for many of the nearctic species of the mosquito genus *Culex*. *Culex*, *Melanoconion*, and *Neoculex* are subgenera of *Culex*. (Courtesy of *Mosquito News* and the Illinois Natural History Survey)

mosquito genus *Culex* (Fig. 9-42). As noted in this figure, man brought in, either by accident or design, a large number of species from other continents. These dispersed and transported species, together with those that evolved here, constitute the mixture of insects that inhabit our present ecological communites.

Old Relict Faunas

Among the modern insects that live in the former parts of Gondwana, a surprisingly large number belong to families that are known only from Australia, New Zealand, and South America. In addition, some other insect groups are known from these areas and also southern Africa. Most of these insects are adapted to the southern temperate zone and most are sedentary with little means of dispersing across the intervening oceans. Examples include the flightless moss bugs of the family Peloridiidae (order Hemiptera), the elongate weevils of the family Belidae, the flightless cylindrical crickets of the family Cylindrachetidae, and three families and one subfamily of stoneflies of the Plecoptera. In the order Mecoptera, the primitive pygmy scorpionflies in the family Nannochoristidae form part of this southern distribution and they are the only surviving members of the Mecoptera having cool-adapted aquatic larvae. This same family is known also from the Permian of Australia. Dipteran flies having this distribution include the genus *Pelecorhynchus*, which is related to March flies and horseflies (family Tabanidae), the genus *Trichophthalma* or tangle-vein flies (family Nemestrinidae), the subfamily Ceratomerinae of the dance flies (family Empididae), and the primitive midge family Blephariceridae. Additional examples are known and emphasize the common pattern in these distributions. Most of these are survivors of old and primitive lineages and strongly suggest that these once had continuous geographical ranges prior to the breakup of Gondwana. In a sense, these insects are relict, endemic, and disjunct all at the same time!

REFERENCES

Carpenter, F. M., 1971. Adaptations among Paleozoic insects. *Proc. N. Am. Paleont. Conv., 1969*, Pt. I, pp. 1236–1251.

——— 1976. Geological history and evolution of the insects. *Proc. 15th Cong. Entomol.*, Washington, D.C., pp. 63–70.

Coope, G. R., 1970. Interpretations of Quaternary insect fossils. *Annu. Rev. Entomol.*, **15**:97–120.

Cox, C. B., I. N. Healey, and P. D. Moore, 1973. *Biogeography: An ecological and evolutionary approach*, 2nd ed. New York: Halsted (Wiley). 194 pp.

Dillon, L. S., 1956. Wisconsin climate and life zones in North America. *Science*, **123**:167–176.

Durden, C. J., 1969. Pennsylvanian correlations using blattoid insects. *Can. J. Earth Sci.*, **6**:1159–1177.

——— 1974. Biomerization: An ecologic theory of provincial differentiation. In C. A. Ross (Ed.), Paleogeographic provinces and provinciality. *Soc. Econ. Paleon. Mineral.*, Spec. Pub., **21**:18–53.

Emerson, A. E., 1971. Tertiary fossil species of the Rhinotermitidae (Isoptera), phylogeny of genera, and reciprocal phylogeny of associated Flagellata (Protozoa) and the Staphylinidae (Coleoptera). *Bull. Am. Mus. Nat. Hist.*, **143**(3):247–303.

Handlirsch, A., 1906–1908. *Die Fossilen insekten und die Phylogenie der rezenten Formen*. Leipzig: W. Engelmann. 1430 pp.

Kukalova, J., 1969–1970. Revisional study of the Order Palaeodictyoptera in the Upper Carboniferous shales of Commentry, France. Pts. I–III. *Psyche*, **76**:163–215; **76**:439–486; **77**:1–44.

Larson, S. G., 1978. *Baltic amber: A palaeobiological study*, Entomonograph. Klampenborg: Scand. Sci. Press. Vol. 1, pp. 1–192.

Löve, A., and D. Löve (Eds.), 1963. *North Atlantic biota and their history*. Oxford: Pergamon. 430 pp.

McAlpine, D. K., 1972. Insects and continental drift. *Aust. Nat. Mus.*, **17**(8):274–278.

McAlpine, J. F., and J. E. H. Martin, 1969. Canadian amber. Hudson Bay Co., *The Beaver*, Summer, pp. 28–37.

Riek, E. F., 1970. Fossil history. In D. F. Waterhouse, et al., (Eds.), *The insects of Australia*. Carlton, Victoria: Melbourne Univ. Press. pp. 168–186. Suppl. 1974, pp. 28–29.

Rodendorf, B. B. (Ed.), 1962. *Chlenistonogie; trakheynye i khlitserovye. Osnovy Paleontologie*. Moscow: Izdatel'stovo Akademii Nauk SSSR. 560 pp. (In Russian).

——— (Ed.), 1968, *Jurskie Nasekomye Karatan*. Moscow: Otdelenie Obshchey Biologii. Akademiya Nauk SSSR, Izdatel' svo "Nauka." 252 pp.

1974. *The historical development of Diptera* (Trans. from Russian, J. E. Moore and I. Thiele). Edmonton: Univ. of Alberta Press. 360 pp.

Ross, C. A. (Ed.), 1974. Paleogeographic provinces and provinciality. *Soc. Econ. Paleont. Mineral.* Spec. Pub. **21**:1–233.

——— (Ed.), 1976. *Paleobiogeography*. Benchmark Papers in Geology 131. Stroudsburg: Dowden, Hutchinson & Ross. 429 pp.

Ross, H. H., 1953. On the origin and composition of the nearctic insect fauna. *Evolution*, **7**:145–158.

——— 1956. *The evolution and classification of the mountain caddisflies*. Urbana: Univ. of Illinois Press. 213 pp.

Scudder, S. H., 1890. The Tertiary insects of North America: *Rep. U. S. Geol. Surv. of the Territories* (Hayden), **13**:103–116, 624.

Sharov, A. G., 1968. Phylogeny of the Orthopteroidea. *Trudy Paleont. Inst. Akad. Nauk SSSR*, **118**:1–221. (Trans., 1971. Israel Prog. Sci. Trans.,

Jerusalem, for U.S. Dept. of Commerce, Natl. Tech. Inform. Svc., Springfield, Va.)

Smith, A. G., and J. C. Briden, 1977. *Mesozoic and Cenozoic paleocontinental maps.* Cambridge and New York: Cambridge Univ. Press, 63 pp.

Wolfe, J. A., 1978. A paleobotanical interpretation of Tertiary climates in the Northern Hemisphere: *Am. Scientist,* **66**:694–703.

Wootton, R. J., 1972. The evolution of insects in freshwater ecosystems. In R. B. Clark and R. J. Wootton (Eds.), *Essays in hydrobiology.* Exeter: Univ. of Exeter Press. pp. 69–82.

——— 1976. The fossil record and insect flight. In R. C. Rainey (Ed.), Insect flight. *Symp. R. Entomol. Soc. London*, **7**:236–254.

10 Ecological Considerations

Because insects are so numerous, have such a great number of species, are so widely distributed, and include so many species that are of applied importance to us, it is extremely important to learn how they survive, how they function as populations, and how they interact with other populations, as well as with their physical environment. These inquiries lead us into the field of ecology, a complex subject to study because of the multiplicity of interrelated causes and effects, many with long chains of subsequent side effects.

Ecology is usually broadly divided into three levels of concentration. Each level is complete with its own problems, complexities, aims, and goals. The first level studies the *ecology of individuals*. This level includes studies of how individuals survive, feed, and behave in their environment. We examined some of the general features that relate to this level of inquiry in earlier chapters, such as in Chapter 5 on life processes, Chapter 6 on response and behavior, and Chapter 7 on life cycles, growth, and reproduction.

The second level studies the *ecology of populations*. This includes two aspects, both of which are particularly important to entomologists. One investigates population growth and the other the interactions of populations of different species. Both deal with population structure, abundance, and relative stability of populations. It is at this level that competitive and predatory imbalances in the insect world, which result in economically disasterous population outbreaks, were first studied in detail.

The third and most integrated level studies *communities*, *biomes*, and their *ecosystems*. The *community* includes all the organisms living together in the same region. The *biome* includes a number of similar communities and their organisms and also the physical features of the environments. Communities or biomes may be analyzed in terms of the processes and amounts of materials and energy that flow through them; that is, their *ecosystem*.

BIOMES AND COMMUNITIES

Life in natural surroundings is broken up into different types of interlocking landscapes, such as prairies, forests, deserts, lakes, and streams. The type of vegetation—that is, whether it is desert, prairie, or forest—is determined by climatic factors of temperature, rainfall, and evaporation (Fig. 10-1). In the main, forests occur in regions with a high rainfall, prairies in regions having lower rainfall, and deserts where rain is scant and evaporation is high. Different types of freshwater aquatic habitats depend on slope, rainfall, and a large variety of local factors including acidity and leaching qualities of the soil, drainage, seepage, and temperature. The vegetation type of the landscape is therefore a reflection of the climate, and widely separated areas having similar climate have the same kind of landscape aspects. These principal landscape aspect areas are the most inclusive ecological units and are called *biomes* (Fig. 10-2). Each of these is divided into smaller units. A forest, for instance, has an edge area and may have small open areas or glades scattered through it; in one

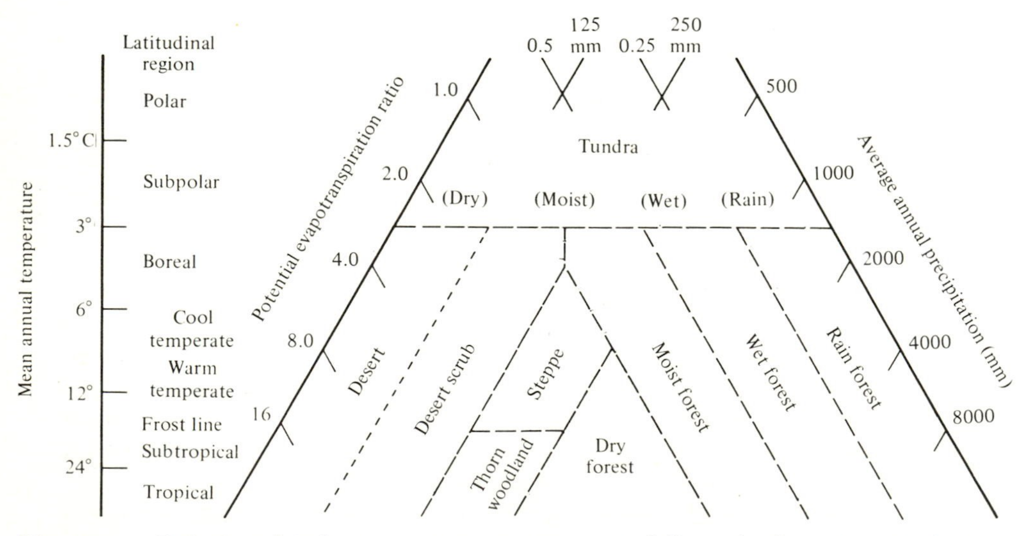

Fig. 10-1. Relationship between temperature, rainfall, and other climatic factors and vegetation types. (Data from Pianka, 1978; Holdridge, 1967)

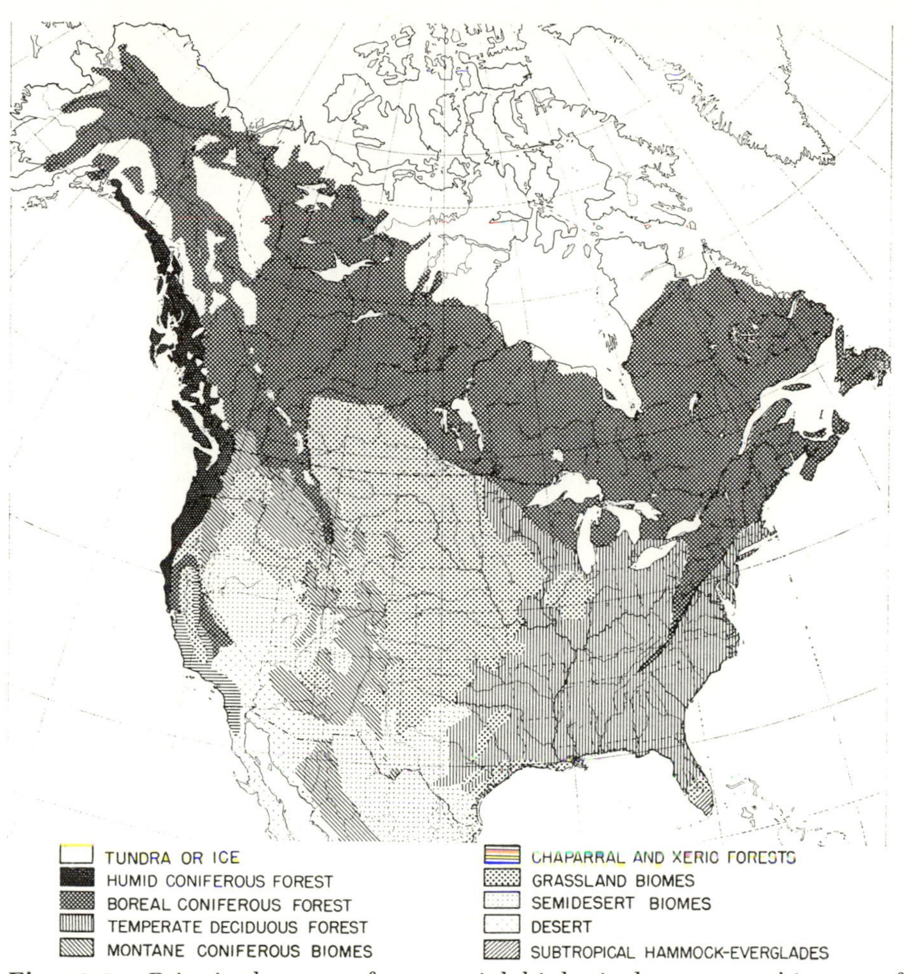

Fig. 10-2. Principal areas of terrestrial biological zones or biomes of North America. (From Illinois Natural History Survey)

place the forest may be well drained and high, with a preponderance of oaks and hickories or pines; in another place it may be low and swampy, having elms, gums, or cypress and other trees different from those in the better-drained areas. Each of these fairly uniform areas is considered by the ecologists as the biological unit of natural areas and is called a *community*. Each community has a definite set of animal species living in it, a set that persists year after year with only minor change. The animal species living in similar communities are practically the same. Thus oak-hickory communities in Wisconsin, Indiana, Missouri, and Oklahoma are each populated by very nearly the same species of animals.

Although an elm or gum forest community contains a fair proportion of the species found in an oak-hickory community, it lacks many species found there but possesses in addition species distinctive to itself. If we go further afield, a prairie community has a species makeup differing greatly from that of a forest community, and neither has much in common with aquatic communities.

Examining communities more closely, we see that the animal species in each are stratified in various ways. In terrestrial communities some of the animals live in the soil, some on the herbs, and some in the trees, if the habitat is a forest. There is a vital relationship between various organisms in the community, as between herbivorous animals and the plants they eat, or between predatory animals and their prey. Altogether these coordinated relations make a network of dependency that binds all the diverse individuals of a community into a biological whole.

ECOSYSTEMS

Ecosystems ecologists attempt to show how each of the living components (all the organisms) interact with one another and with environmental components to produce a biological-physical system. This system is studied in terms of the flow of food and nutrient energy through structured food chains (trophic levels), biological diversity, and exchange of materials between the organisms and the physical environments. No size limits are placed on an ecosystem—the largest one, of course, is the world's biosphere—but smaller sizes, such as a square kilometer of forest, a square meter of grasslands, or a pond, or even a pile of dung, are more practical for individual study. Ecosystems vary greatly in geographical extent and in internal size and complexity. Furthermore, the boundaries between adjacent ecosystems overlap and merge, such as at the edge of a pond or a river floodplain, and form *ecotones*.

Biological Components

Two major biological components (Fig. 10-3) are present in all ecosystems.

1. *Autotrophs,* those organisms that provide their own nourishment from the physical environment. In terrestrial ecosystems these are mainly green plants that use photosynthesis to convert atmospheric carbon dioxide and water to form complex organic substances.
2. *Heterotrophs,* those organisms that rely on other sources for food in which complex organic substances are produced to form other

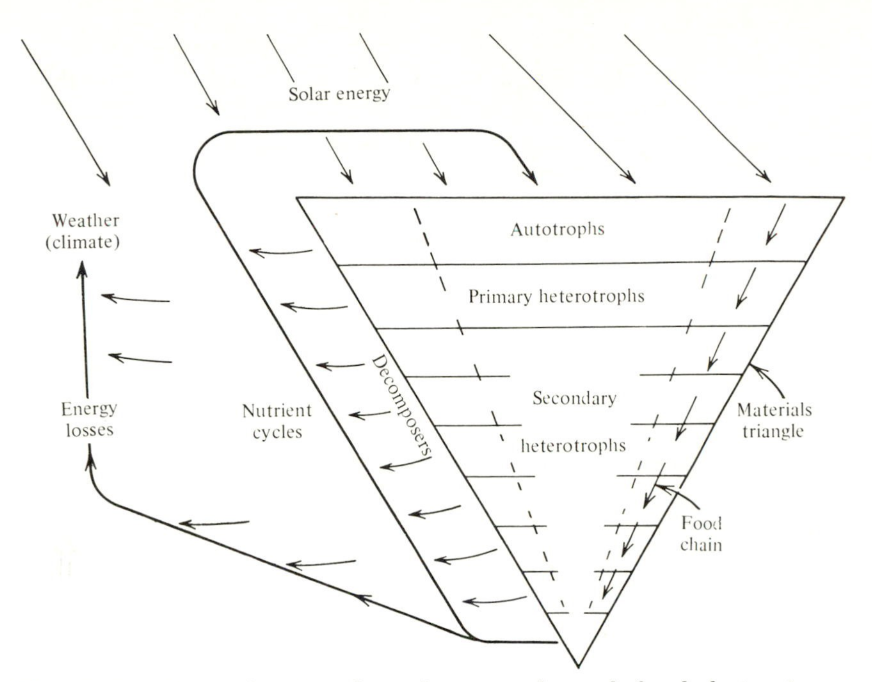

Fig. 10-3. Flow of materials and energy through food chains in an ecosystem.

complex substances. Heterotrophs are further subdivided into *primary consumers*, those animals that use the autotrophs directly for nutrients (herbivores); and *secondary consumers* (carnivores and insectivores), those animals that prey on the primary consumers or other secondary consumers and hence derive nutrients indirectly from the autotrophs. Usually several levels of secondary consumers are present in an ecosystem.

Decomposers. In addition to these highly visible biological components, decomposers (Fig. 10-3) make up a *third important biological component*. These are principally microorganisms, such as fungi, bacteria, protozoans, and small invertebrates (including many insects), that break down complex organic substances and release products that are recycled by plants and to a minor extent by heterotrophs.

Physical Components

There are three types of physical components.

1. Climate. This includes temperature, rainfall, seasons, and similar factors (Fig. 10-1). In large part the interaction of the different

aspects of the climatic component determines the rate (as a function of temperature) and amount (as a function of available water from rainfall) of soluble nutrients in the ecosystem.

2. Inorganic substances. These are chemical substances involved in complex recycling mechanisms, such as the water cycle (Fig. 10-4).
3. Organic compounds. These are products relating to the chemical and biological parts of the ecosystem. They are closely regulated by the climatic components, such as temperature and availability of water.

Processes

The various components of an ecosystem interact with one another through a series of processes. These processes include energy flow, water and mineral (nutrient) cycles, food chains, changes in diversity patterns, developmental and evolutionary changes, and control mechanisms between species.

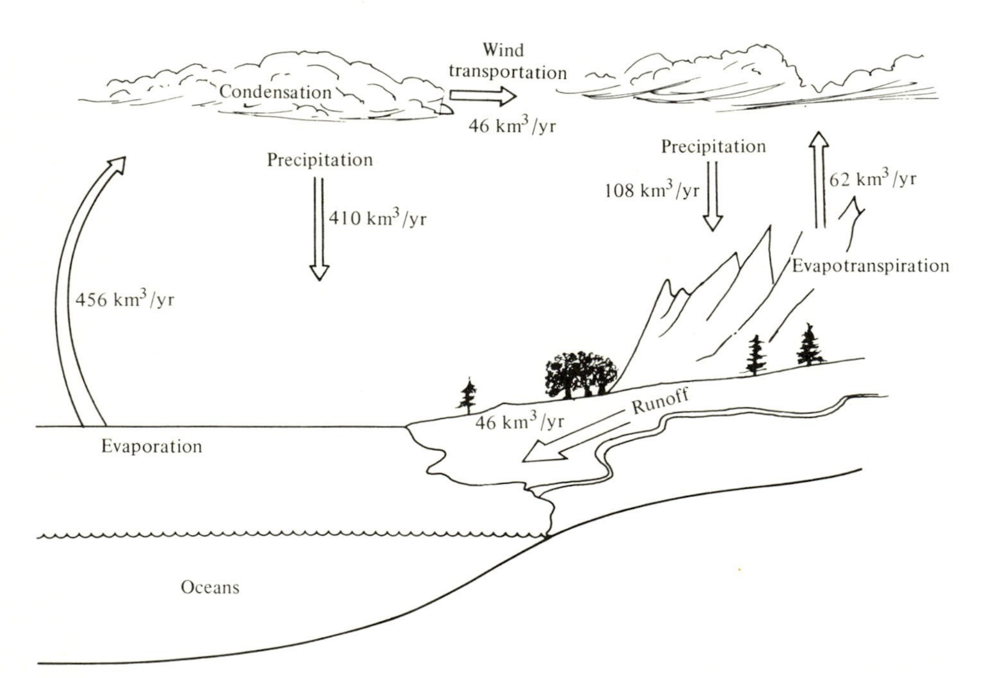

Fig. 10-4. The water cycle (hydrological cycle) involves the recycling of about 1000 km³ of water each year. (Modified from Ehrlich, Ehrlich, and Holden, 1977, and Budyko, 1974)

Energy flow. This originates from solar radiation of which only a fraction (up to 5%) is used to fix carbon in organic compounds by autotrophs. Most solar radiation (90 to 95%) is used up in evaporation and maintenance of climatic zonation. Primary consumers, such as the leaf-eating caterpillar, are able to transfer to their bodies only about 10 to 20% of the biochemical energy stored by the leaves (autotrophs). Each level of secondary consumption has about the same efficiency, 10 to 20%. These relations may be outlined as a trophic level pyramid (Fig. 10-3).

Water and Mineral Cycles. These also are powered by solar radiation. They are important in the hydrological cycle (Fig. 10-4) and in establishing many physical and chemical conditions of the environment. In addition, solution, transportation, and deposition of dissolved carbon dioxide and carbonate minerals are largely controlled by these processes, as well as the interchange of carbon dioxide between the atmosphere and water.

Food Chains. These are the basis for establishing trophic levels in an ecosystem and insects are primary and secondary consumers and decomposers (Table 10-1). As discussed later in this chapter under the section on autoecology, insects, more than any other group, have subdivided their food sources into extremely small divisions, so that commonly one insect species is a primary consumer on one preferred plant species, and only marginally successful on closely related plant species. In fact, many species in some leafhopper groups are consumers of the same species of tree, and each of the insect species is successfully adapted to different parts of the same tree. Such fine partitioning of autotrophic food sources is not common in other animal groups and, in part, this has contributed to the very high number of insect species. Likewise, species of insects that are secondary consumers also may prey on one or only a few other species of insects. Insect food chains are characterized also by changes in food sources as the insect develops and matures (Table 10-2). In many insects, the larval stages and adult stage have different food sources. Some insects have specific food needs in order to survive. Adult mosquitoes and many flies need to feed on the blood of vertebrates in order to complete their life cycle. Insects also are food for a great number of other animals, including many birds, voles, fishes, amphibians, reptiles, and mammals.

Evolution. In an ecosystem evolution is an important long-range process because through long intervals of time the relations between species gradually change. Some species may develop more defensive mechanisms against predators, others may become the prey of

Table 10-1. General Trophic Relations of Insect Orders in Consumption of Food.

PRIMARY CONSUMERS (HERBIVORES)	SECONDARY CONSUMERS (INSECTIVORES AND CARNIVORES)	DECOMPOSERS
Phasmatodea	Ephemeroptera (larvae)	Ephemeroptera (larvae)
Orthoptera	Odonata	Isoptera[c]
Grylloblattodea	Dictyoptera (Mantodea)	
Embioptera	Plecoptera (some)	Plecoptera (some)
Thysanoptera	Phthiraptera	Psocoptera
Hemiptera (Homoptera)[b]	Hemiptera (Heteroptera)[b]	
Hemiptera (Heteroptera)[c]	Coleoptera[a,b]	Coleoptera[a,b]
Coleoptera[a,b]	Hymenoptera (parasitic and provisioning wasps)	
Hymenoptera (sawflies and horntails)[b]	Neuroptera	
Hymenoptera (provisioning bees)	Megaloptera	
Mecoptera[a]	Rhaphidioptera	
Lepidoptera[a]	Siphonaptera	
Diptera[a,b]	Diptera[a,b]	Diptera[a,b]
	Trichoptera	

[a]Many orders have families that are restricted to certain food. In others larval and adult food are significantly different.
[b]Food sources mixed by stages of growth.
[c]Food sources mixed by specialization at lower taxonomic level.
[d]Omnivores include the Dictyoptera (Blattaria), Dermaptera, in North America.

Table 10-2. Examples of Shifts in Food Sources in Holometabolous Insects Between Larval and Adult Stages

	LARVAL FOOD	ADULT FOOD
Diptera		
Mosquitoes	Aquatic algae and microorganisms	Blood of vertebrates and invertebrates
Hymenoptera		
Wasps	Parasitic on insect host	Flower nectar
Lepidoptera		
Butterflies	Vegetation	Flower nectar

different species. Thus most ecosystems are considered relatively stable systems in which the species are gradually evolving together and the possibility of abrupt changes, although not common, is ever present. The evolution of ecological succession, in which different plant and animal assemblages appear in regular stages, has greatly increased the rate at which an ecosystem is able to recover from destruction by fire or other disasters.

Diversity Patterns. In ecosystems, diversity patterns are generally related to two factors—time and geographical location (Fig. 10-1)—and these are both partially interrelated. High species diversity is the rule in tropical ecosystems, and a decrease in the number of species to a very low diversity is found in subarctic and arctic ecosystems. In part, this is a function of annual climatic fluctuations and the number of species that are adapted to strongly different climatic seasons. It is also a result of evolutionary diversification during the relatively long and continuous geological time that tropical ecosystems have existed, in contrast to the short and intermittent history of arctic conditions.

POPULATIONS

Population Dynamics

Entomologists determined that insect populations had two distinctive characteristics.

1. Many populations were extremely stable in size, generation after generation.
2. Others could be extremely variable in size, with numbers rising rapidly in one or two generations to extremely high levels and then falling just as rapidly to very low numbers.

Stability in many insect populations has been well documented by light trap collections such as those carried out over many successive years in England. These collections show that the number of individuals of a species is remarkably constant. Studies in population size of a tropical rainforest butterfly, *Heliconius ethilla*, in Trinidad shows great stability through at least 26 generations.

In contrast, many other insect populations experience such outbreaks of high population levels as to constitute serious and disastrous economic pests. Locusts and tent caterpillars are frequently cited examples that have an unpredictable history of reaching epidemic levels in some years but not in others.

These two types of population dynamics have led to two nearly diametrically opposite viewpoints in explaining population dynamics. One viewpoint proposes that as a population increases its size, it becomes subject to a gradually increasing number of ecological processes that combine to reduce the rate of reproduction to a stable population-replacement level. The processes were identified as density dependent. The other viewpoint maintains that population sizes fluctuate widely and are limited by environmental fluctuations (for example, growing season, seasonal and geographical variations in rainfall), and only ultimately by the limitations of food resources (Fig. 10-5). In order to evaluate these viewpoints we need to look at the structure of populations and then their interactions.

Population Structure. When we examine the structure of a population of insects, we find that it is helpful to construct a life table for

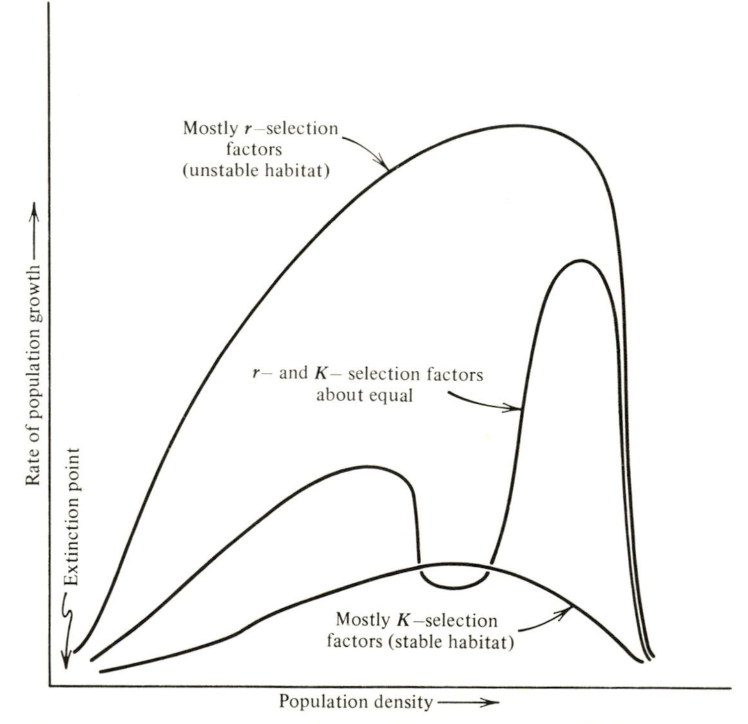

Fig. 10-5. Models for three different types of population growth and population densities. The deep trough, where r- and K-selection are approximately equal, is the result of predator–prey interactions. (Modified from Southwood, 1977)

Table 10-3. A Life Table for the Spruce Budworm, Green River, New Brunswick, Canada[a,b]

STAGES OF LIFE CYCLE	NUMBER OF INDIVIDUALS AT BEGINNING OF STAGE	NUMBER DYING BEFORE END OF STAGE	CAUSE OF DEATH	PERCENT CHANGE DURING STAGE
Egg	200	10	Parasites	15
		20	Other	
Early larva	170	136	Dispersal	80
Late larva	34	13.6	Parasites	
		6.8	Disease	
		10.2	Other	90
Pupa	3.4	0.3	Parasites	
		0.5	Other	25
Moth	2.5	0.5	Miscellaneous	20

[a] Modified from Morris, 1957.
[b] Generation survival, 2 individuals (1%); generation mortality, 198 individuals (99%). Sex ratio 1:1 (equal numbers of both sexes).

each species (e.g., Table 10-3), which shows the number of individuals in each stage of growth. From this table it can be seen that the spruce budworm initially has great potential for a population to increase dramatically, starting with 200 eggs in a single mass. This reflects the *intrinsic rate of increase* of the population. If all these eggs hatch and the resulting individuals reach maturity and produce a similar number of eggs, which in turn also reach maturity, the population would have grown from one individual (that one that laid the first mass of eggs) to 40,000 individuals in two generations. Fortunately, other processes and factors usually intervene to prevent such an increase in spruce budworm numbers. These include parasites, disease, predators, dispersals, and climatic factors, so that usually only two or three individuals reach maturity; that is, mortality rates are very high.

Spruce budworm populations, as shown in Table 10-3, are characterized by an age structure. In many insects, eggs do not hatch at exactly the same time, and also many insects have several generations per year. These populations commonly have eggs, larvae of many stages, pupae, and adults all existing together before the end of the growing season.

Sex ratios are also important. In many insects, males and females are reproduced in nearly equal numbers, but in many others females significantly outnumber males. In some, such as the social insects, males are a minor part of a population. In some weevils, such as the white fringed weevil and alfalfa root weevil, males are not known.

Other characteristics of populations include genetic differences. These include chromosomal differences brought about by different alleles at particular loci on the chromosomes, by rearrangements of chromosomal chains, and, perhaps most important, the mixing or recombinations of the large number of allelic combinations in successive generations. These genetic differences give each population a certain variation of morphological (phenotypic variation) and physiological expression that becomes extremely important through natural selection for the evolution of the population and of the species as a whole.

Changes in Populations. The contrasting types of insect populations mentioned earlier can be explained as the result of different controlling processes (Table 10-4). The stable population situation is one in which one or more predators, also at nearly stable population levels, are the main cause for high mortality rates in the population. A contributing factor is that environments in which these populations live are predictably stable also. A less-stable population also

Table 10-4. Relationship of *r*- and *K*-Selection Characteristics in Insects to Certain Ecosystem and Population Features[a]

	r-SELECTION	*K*-SELECTION
Climate	Uncertain, variable, unpredictable	Fairly consistent, predictable, reasonably certain
Life cycle	Short, a few weeks	Longer, usually several months
	Rapid development and growth rate	Slower development and growth rate
	Early reproduction and single reproduction	Delayed or extended reproduction period
	Small body size	Larger body size
	Potentially many offspring	Fewer and larger offspring
	Mortality catastrophic, usually density independent	Mortality more evenly distributed, density dependent
Population size	Variable in time, nonequilibrium, usually significantly below carrying capacity, periodic recolonization	Fairly constant, equilibrium, at or near carrying capacity, no need for recolonization
Competition	Variable, commonly low	Usually high

[a] Modified from E. R. Pianka, 1978.

has predators and environmental limitations that usually keep it partially in check, but the environment is less predictable and less stable. These less-stable populations may survive in greater proportions and reach an adult level that is limited mainly by competition with one another of their own species for the available food resource. Although these populations show more fluctuations in numbers, they infrequently reach epidemic proportions in nature. Very unstable populations, whose population numbers fluctuate wildly, frequently escape, at least temporarily, from their parasite and physical environmental constraints and overshoot their normal food resources. Often at these population levels they will extend their eating to include food that they normally do not feed on.

Controls or ecosystem balance (*cybernetics*) is an intergral part of natural ecosystems and is built into the interaction between processes and components. Usually these are feedback mechanisms and they are commonly visualized as mathematical models that form the science of *systems ecology*. A simple example is the interaction between population densities of predator and prey species. As the prey population density increases, there is a lag in the predator population density increase. As the prey population expands, the predators are able to find the prey more readily and this allows the predator population to gradually build up also. A point is reached where the population density of the predator is sufficient to search out and devour a very high proportion of the prey. Consequently, the prey population is rapidly reduced and this leads to greatly increased predation pressure, which continues to lower the prey population density. When the prey population density reaches a very low level, the predator population has difficulty locating the prey and its population density suddenly declines. The prey population is then able to expand gradually once again, and the cycle is repeated. The prey is thus limited by its *natural enemies*.

Outbreaks are those occasions when a species population can escape from this predator–prey cycle for a short time. In these cases, the prey population expands more rapidly than the predator population and reaches a threshold level beyond which the predator population growth cannot keep it in check. The prey species reaches outbreak, or epidemic proportions, and eats nearly all its available food resources (Fig. 10-5). At this level of population density, the carrying capacity of the prey population's food source is the limiting factor. Once the carrying capacity limit is exceeded, few individuals in the population are able to obtain sufficient food and the population density declines rapidly. These types of high population levels are *resource limited*.

Environmental stability has an important effect on insects and their adaptation. In general, all animals, particularly insects, inhabiting environments that have short-term stability (such as strongly contrasting seasons or unpredictable weather) are usually small, very mobile, and have high reproductive potential and short generation time; these are called *r-strategists*. On the other hand, animals, including insects, in highly stable environments tend to be larger, more territorial, have lower reproduction potential, and have longer generation time; these are called *K-strategists*. The *r*- and *K*- strategists form the ends of a spectrum that is referred to as the *r–K continuum*. It is possible to compare population density of various types of populations in the *r–K* continuum (Fig. 10-5) and their interaction in the predator–prey model. The desert locust, *Schistocerca gregaria*, and the housefly, *Musca domestica*, are *r*-selected strategists in that their physical environments are highly unstable and they have few natural predators and disperse rapidly and widely. At the other end of this spectrum, the codling moth, *Cydia pomonella*, usually lives in stable environments, has low reproductive rates and little fluctuation in population density and is a *K*-selected strategist. Most insects lie between these extremes of *r*- and *K*-selected strategists. Those nearest the *r* end of the continuum (such as aphids) are the ones most likely to escape control by natural enemies and become epidemic pests.

For many groups of insects, a wide range of habitats in an area may result in high diversity of species. Such an area may also be rich in number of species of predators, hence complex predator–prey relations may develop. Acidic grassland leafhoppers (Hemiptera) near Silwood, Berkshire, in southern England, include a rich fauna of species. In comparative studies of two habitats there, the more diverse habitat included 63 species of leafhoppers. Of these, 30 Cicadellidae, 10 Delphacidae, and 2 Cercopidae bred there regularly. The second habitat was environmentally more uniform and the regularly breeding species included 9 Cicadellidae and 1 Delphacidae.

In the more diverse habitat, the population size of the leafhoppers was controlled in large part by numerous species of parasitoids belonging to the broad-headed wasps Dryinidae (Hymenoptera) and the flies Pipunculidae (Diptera) (Fig. 10-6*F*). The more uniform habitat had a much smaller number of parasitoid species. The relationships between several of the common leafhoppers and their parasitoid predators is shown in Fig. 10-6. The patterns of parasitism appear localized. In those species populations, which remained in the same immediate habitat for a series of generations, the established ratios of parasitism (predator to prey) were maintained within fairly narrow limits. These ratios were maintained even when the frequency and species of parasitoids consistently changed between

Fig. 10-6. Parasitoid infestations by small broad-headed wasps (Hymenoptera, Dryinidae) and flies (Diptera, Pipunculidae) in grassland leafhop-

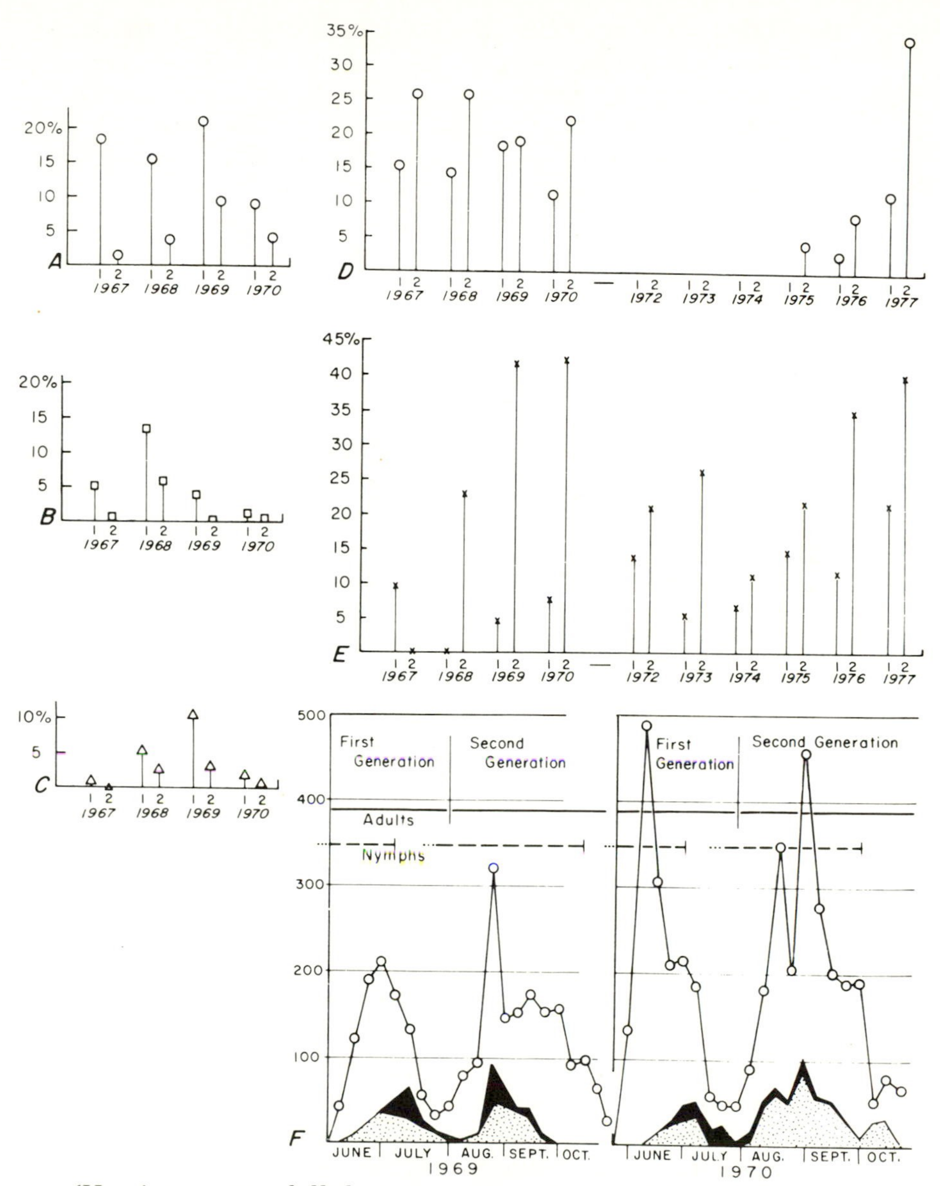

pers (Hemiptera, Cicadellidae) near Silwood, Berkshire, southern England. (*A*, *B*, *C*) Percent parasitism by Dryinidae on the three most abundant species of cicadellids in the more diverse habitat with high parasitism. (*A*) *Psammotettix confinis*; (*B*) *Arthaldeus pascuellus*; (*C*) *Jassargus pseudocellaris*. (*D*) *P. confinis*; (*E*) *Errastunus ocellaris*. In *A* to *E*, 1 represents first generation; 2, second generation. (*F*) Number of adult leafhoppers of *P. confinis* from weekly sampling data at the more diverse habitat for 1967–1970. Circles, total number of *P. confinis* adults; black areas, proportion of *P. confinis* parasitized by dryinid wasps; stippled areas, proportion parasitized by pipunculid flies. (Data from Waloff, 1975, 1980) the leafhopper host's first and second generations (Fig. 10-6*A* to *E*). In the example in Fig. 10-6*A* to *C*, the parasitoid dryinid species that attacked the first generation of leafhopper populations were different dryinid species from those that preyed on the second generation. In addition, the attack by dryinid species on the first generation of leafhopper was heavier than on the second leafhopper generation. Because some parasitoids will attack and kill a number of different host species (that is, they are polyphagous), the population levels of a number of species of both hosts and parasitoids commonly are interrelated. In a Cicadellidae species that newly established itself in the more uniform habitat, the initial proportion of parasitism fluctuated much more widely than in populations of the same leafhopper species in an established, but more diverse, habitat (Fig. 10-6*D*, *E*).

In *highly modified ecosystems*, such as crop cultivation by humans, many of the natural controls are short-circuited. The result has been frequent epidemic outbreaks of insect pests that have destroyed many crops. These have had catastrophic effects on human societies and have been reported since the beginning of written records. As the world human population increases and the production demands on tillable land increase, humans seek to develop effective controls that replace those natural controls lost in the change to cultivation. At the present time about one in six humans suffers from a serious disease transmitted by insects, and one is five is seriously malnourished as a result of insect damage to crops. The ecosystem approach to problems of balancing disturbed insect-population dynamics offers many valuable insights into possible solutions to a difficult task.

ECOLOGY OF INDIVIDUALS

Each insect species is specially adapted to live in a particular "niche" in the community. It is restricted to its particular habitat by various environmental factors, such as weather and food. The limits of these factors within which a species can exist are called its *ecological tolerance*, which differs for each species. Individuals of a species tend to move to sites that afford optimal conditions for their success. Ecological tolerance is distinctive for each species; therefore accurate identification of a species is important.

Most ecological studies of an insect species investigate the following.

1. Environmentally limiting factors that determine where the species may live.
2. Interrelationships of the species with competitors or enemies.

3. The effect of all these on the distribution and abundance of the species.

Environmental Factors

The most important environmental factors concerning the distribution and abundance of insects are weather, physical and chemical conditions of the environment, food, enemies, and competition.

Weather

Weather forms a blanket over the entire community and directly or indirectly affects conditions and organisms in practically all parts of the community. Weather is a composite condition of which light, temperature, relative humidity, precipitation, and wind are the most important ecological components. It is not the annual averages of these components (climate) that affect the species populations, but conditions from day to day. A single night's frost, for instance, may decimate a population of a subtropical insect, although the average temperatures for that year may be high. Similarly, conditions may differ tremendously and may result in great population differences within a short distance. In hilly country a single night's frost could be most severe in the valleys and might not affect the portion of the insect population living on the hill crests.

Light. Light is an extremely important factor in insect behavior and is considered in the sections on taxes and photoperiod. A great number of insects that are normally diurnal have been reared successfully for many generations, either in artificial light deficient in many wavelengths or in total darkness. Therefore, the effect of light on most insects is indirect and is expressed through quality of food caused by plant reactions to light.

Temperature. To insects, temperature is one of the most critical factors. Insects are cold-blooded, so at rest their body temperature, within narrow limits, is the same as that of the surrounding environment. Insects usually are unable to control the temperature of their surrounding environment; instead, each species has physiological adjustments that enable it to survive temperature extremes normally occurring in its ecological niche.

The honeybee is the best-studied example of an insect that regulates the temperature of its surrounding medium; in this case, the air within the hive. In summer the hive is maintained at about 35°C. If the temperature rises above this point, bees at the hive entrance set up ventilating currents by fanning their wings, and other bees may bring water and put it on the comb to obtain the cooling effect of its evaporation. In winter the bees keep the hive a safe temperature by heat obtained through oxidation of foods in the insects' bodies. Other social bees and ants also exercise a certain amount of control over nest temperatures.

Effects of temperature are shown in two ways: the effect on rate of development, and the effect on mortality.

Effect on Development. The temperature of insects is not constant and the chemical reactions of metabolism automatically speed up with an increase in temperature. As a result temperature has a marked effect on insect development and activities. Not all chemical reactions do respond at the same rate to temperature increase, and certain physical factors, such as the solubility of gases in liquids, tend to produce unfavorable metabolic conditions as temperature increases. As a result, insect development is not equally responsive to changes over the entire temperature scale. Development stops at a definite low point, called the *threshold temperature*; this point may be 5 to 30°C above the actual lethal low temperature. Development for each species stops also at a definite high point that is usually very close to that of the lethal high temperature.

Between these two points, rate of development changes with temperature. But the response is not uniform throughout the insect world. Each species has its own individual rate of development. Figure 10–7 illustrates differences in rate of development for four species of grasshoppers. Within a species, each developmental stage may have a different rate of development (Fig. 10–8); the eggs and pupae of the Japanese beetle have a much higher rate than do the larval stages, at identical temperatures.

An interesting example of dissimilar rates of development is shown by eggs and nymphs of the red-legged grasshoppers (Fig. 10-7). For nymphs the developmental rate increases steadily with increase in temperature to a point close to the lethal high temperature. The rate for the eggs increases with the lower range of increased temperature and then decreases with additional temperature increase. Temperature of decreased development of eggs is reached far below the lethal temperature. These examples show that it is necessary to study the various stages of the life history in order to obtain accurate information on the developmental phase of the species.

SEASONAL COORDINATION. The different rates of growth of species feeding on plants or cold-blooded animals are correlated extremely closely with the growth rate of their hosts. The result is that when the host has reached a point favorable for a certain insect to attack it, the attacker has reached the proper stage to attack. Let us examine

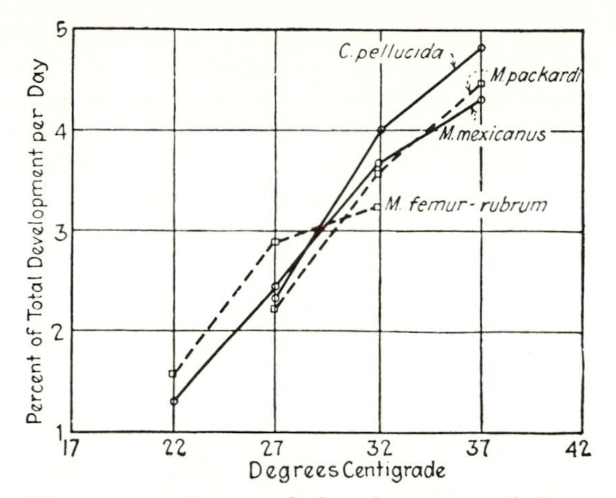

Fig. 10-7. Rates of development of four species of grasshoppers at constant temperatures from 22° to 37°C (71 to 98°F). (After Chapman, 1931, *Animal ecology*, by permission of McGraw-Hill Book Co.)

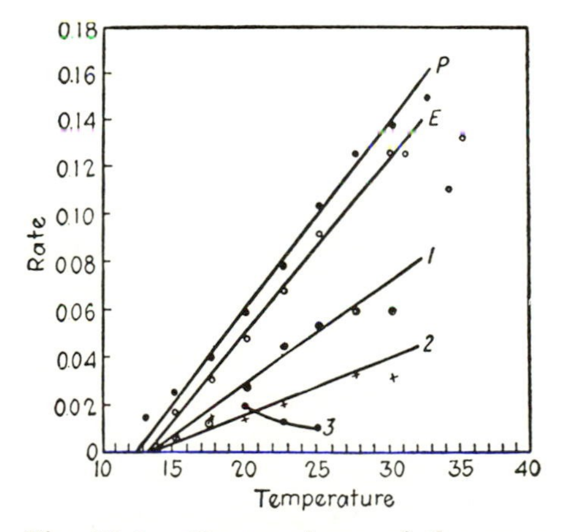

Fig. 10-8. Comparison of the rates of development of each stage of the Japanese beetle. Temperature scale in centigrade, 10 to 40°C (50 to 104°F). Numbers 1, 2, and 3 refer to the three larval instars; *E*, egg; *P*, pupae. (After Ludwig)

this relation in two species of Hymenoptera, a sawfly (Tenthredinidae) and its ichneumonid wasp parasite (Ichneumonidae). The sawfly adults emerge first early in spring at a time when the host plants have young leaves suitable for oviposition. The eggs hatch 1 or 2 weeks later when the plant is in the midst of vigorous growth and is providing a bountiful supply of food for the larvae. The ichneumon wasp has either a slower development, or one that starts at a higher temperature, so that the adult ichneumonid emerges about 3 or 4 weeks after the adult sawfly. At this time the sawfly larva is nearly full grown and at the right stage for the ichneumon adult to lay eggs on it.

Another example is a group of aphids or plant lice (Aphididae) feeding in the spring on apple. The developmental rate of the overwintering eggs is such that the young aphids hatch at almost the exact time the apple buds first begin to open in spring. The aphids feed immediately on the minute leaves of the opening buds.

So constant is this coincidence of certain insect events with definite plant events that the plant phenomena (which are easy to see) are used as guides in many control programs. There are "bud sprays" for early aphid control, "petal-fall sprays," "calyx sprays," and so on, in which plant development is taken as a criterion for insect development.

Effect on Mortality. The temperature range that insects can withstand varies tremendously with the species. The most heat-resistant insects die at temperatures of 47 to 52°C. Most have a high lethal temperature of from 38 to 44°C. Species that live in cool places have correspondingly lower heat tolerances, such as the mountain genus *Grylloblatta* (Grylloblattodea), whose optimum is about 3°C, and is normally active between −2° and 16°C; heat prostration occurs at about 28°C.

Lethal low temperatures vary as much as lethal high temperatures. Tropical insects usually succumb as the temperature drops near freezing (Fig. 10-9). The confused flour beetle, for example, will die in a few weeks at 7°C. Many insects die at temperatures only a few degrees below 0°C (32°F). Hibernating stages of most northern insects are remarkably resistant to cold. The hibernating pupa of the promethea moth, for example, can survive continued exposure to −35°C, and some other insects are known to survive −50°C.

Insects occurring in regions having freezing winters almost invariably exhibit a different temperature tolerance in each stage of their life cycles, with the exception of parasites of warm-blooded animals. At least one stage is resistant to low temperatures, and in this stage the species is able to withstand the winter temperatures. The resis-

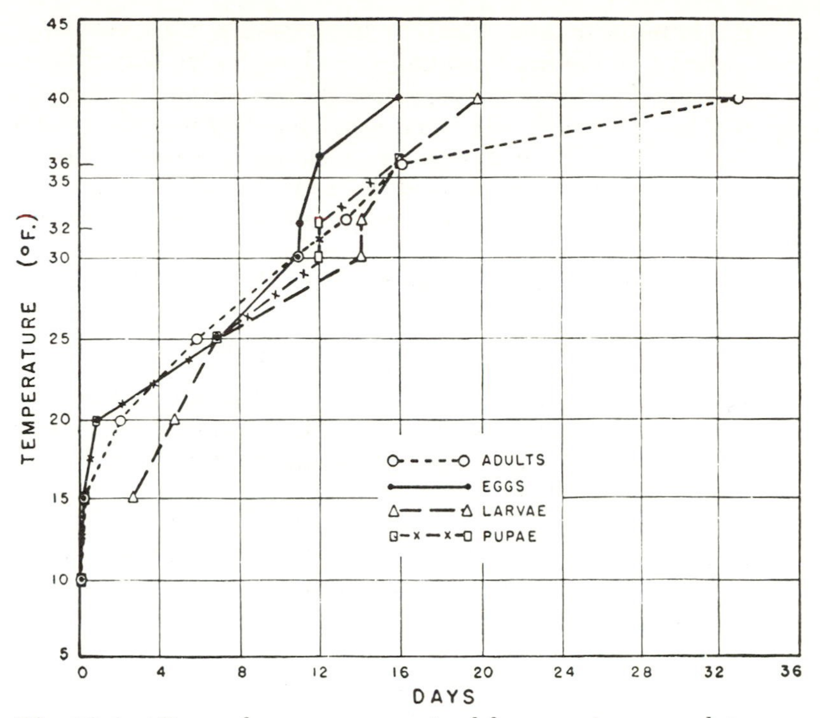

Fig. 10-9. Days of exposure required for assuring complete mortality of eggs, larvae, pupae, and adults of the cigarette beetle at various temperatures ranging from −10 to 5°C (15 to 40°F). (After Swingle)

tant form may be the egg, nymph, larva, pupa, or adult. In most cases only a single stage is cold resistant; when winter arrives, the resistant form lives, and individuals in any other stage die. Thus in chinch bugs only the adults are cold resistant; when extremely low winter temperatures occur, the adults live, and any nymphs still remaining die.

In their natural environment insects are well adjusted to prevailing usual temperatures. Temperature operates as a restricting factor in unusual or unseasonable periods of hot or cold weather. Generally unusual temperatures modify or control the range of a species along some frontier. The southern house mosquito *Culex quinquefasciatus* may expand northward and extend its range during years with mild winters, but is contracted southward during severe winters.

Unseasonable temperatures, such as early or late frosts, may be as effective as temperature extremes in this action, because unfavorable conditions may occur before a species has entered the stage at which it is immune to them, or after it has passed to a susceptible stage. For instance, in the north-central states hibernating chinch bugs cannot withstand many alternate periods of freezing and thawing. A winter that has a number of unusual warm thawing periods, each followed by a −20°C or lower period, produces an alternation of freezing and thawing that is extremely destructive to chinch-bug populations. Unseasonable temperatures would affect a species more frequently at some periphery of its range, and thus contribute to restricting its distribution. Such temperatures may also occur hit-and-miss anywhere over the range of the species, and affect abundance in local areas throughout the main body of the range.

Precipitation. Insects are ordinarily not affected directly by normal precipitation, but indirectly through the effect of precipitation on humidity, soil moisture, and plant food supply. Snow has an unusually important effect on soil temperatures. Bare soil is responsive to temperature changes to a depth of 0.6 m (24 in.); when covered with snow, even surface soil is remarkably insulated from changes in air temperature (Fig. 10-10). Thus a snow cover has a marked effect on both the extremes of temperature and average temperature to which insects in the soil are subjected.

Certain expressions of precipitation, however, have a direct effect on insects. Excessive precipitation may inflict severe physical damage to insects. Two centimeters of rain coming as a gentle sustained rain in one area may cause no harm, but coming as a sudden pelting downpour in another area may beat into the ground and kill most of the aphids or early-stage chinch bug nymphs. Hail inflicts the same type of physical damage.

Humidity and Evaporation. It is difficult to separate the factors of humidity and evaporation in their effect on insects, either experimentally or geographically. Humidity pertains to the amount of moisture in the air, and evaporation to the actual water loss of a surface. In experimental work, if insects are subjected to low humidities, the evaporation from their bodies increases. Because of their small size, increased evaporation quickly depletes the water content of insects' bodies so that insects need to replenish their water supply by feeding. Unless this precaution is taken, desiccation may be attributed to humidity conditions of the medium. The graph in Fig. 10-11 delineates the relationship between evaporation and humidity for a common grasshopper under conditions of starvation.

Available data indicate that, in general, humidity is important but not so critical a factor as is temperature, and that each species has an optimum, which may be different for various stages of the life cycle. In the bean weevil the larvae develop faster at higher humidities, but

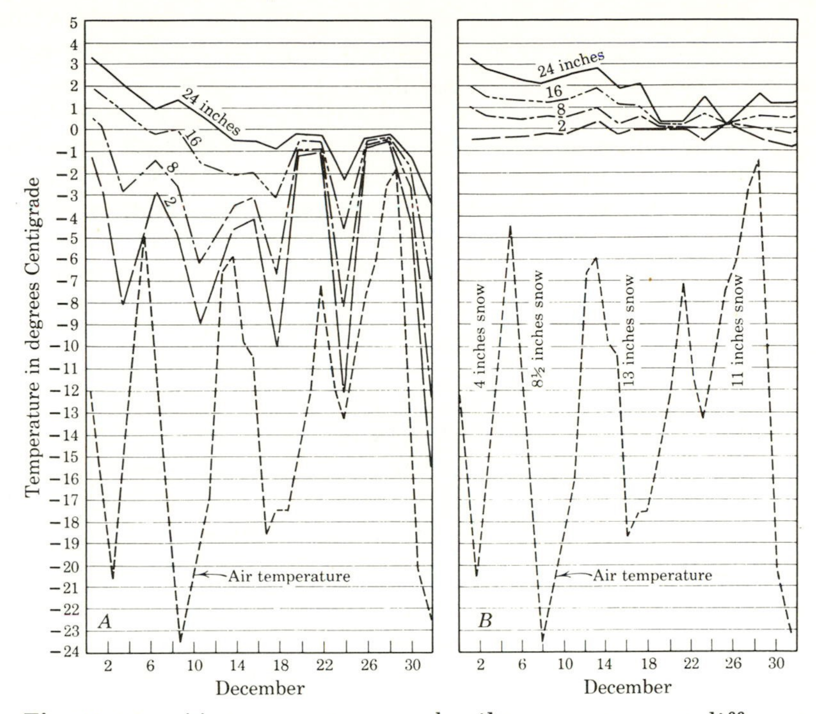

Fig. 10-10. Air temperatures and soil temperatures at different depths in bare (*A*) and snow-covered (*B*) ground for December in Montana. (After Mail)

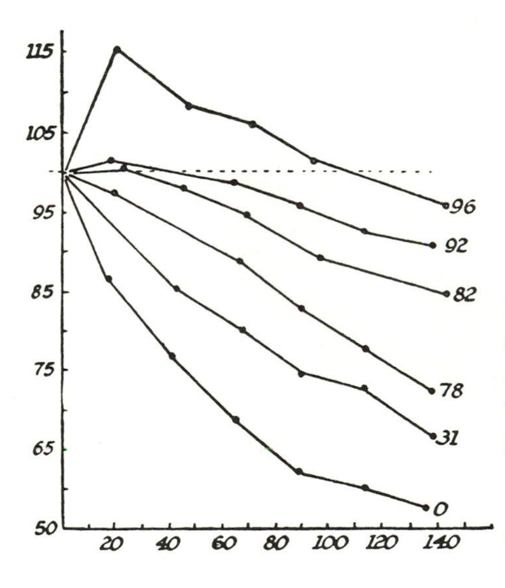

Fig. 10-11. Rate of loss of weight in *Chortophaga viridifasciata* at different relative humidities indicated at the end of each curve. Vertical figures represent weight as percentage of original weight; horizontal figures indicate time in hours. (From Wigglesworth, after Ludwig)

the eggs and pupae develop more rapidly at low humidities. In many cases, however, the rate of growth has been found to be practically constant over a wide range of humidity conditions.

Humidity also affects mortality rate. Low humidity has been found to increase mortality of *Drosophila*, and high humidity interferes with hatching and molting in some species of aphids. High humidities apparently reduce the resistance of some species to fungus attack.

Humidity and evaporation constitute two of the main barriers that restrict the geographical range of many species of insects along parts of their periphery. Many species occur in eastern North America whose ranges extend westward to about the Mississippi River. The less-humid conditions to the west appear to be the factor that prevents further extensive spread of these species in that direction (Fig. 10-12). Conversely, other species occur in the Great Plains area that do not extend much further eastward, probably because their optimal requirements are for low humidity.

Temperature and Humidity. Together these two have a marked effect on both general development and distribution of insect species. Their action is frequently critical on different phases of a species and at different times of the year (Fig. 10-13). Critical cold temperatures, for instance, might operate in winter against the hibernating mature larvae, whereas adverse humidities might operate during the summer against eggs or actively feeding larvae.

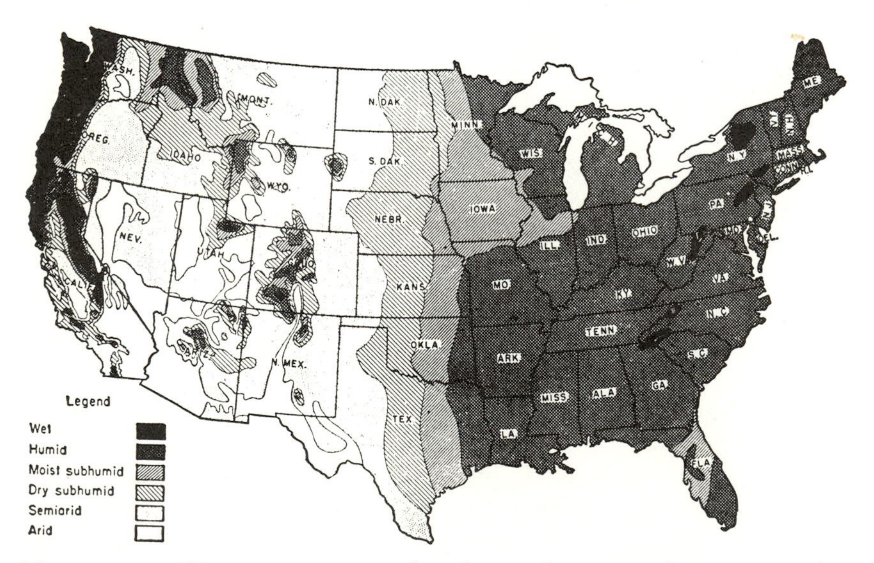

Fig 10-12. Climatic moisture bands in the United States. (Adapted from U.S.D.A.)

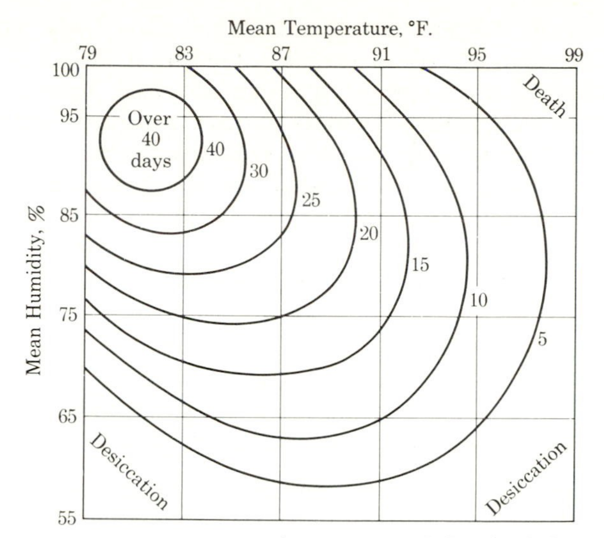

Fig. 10-13. Zones of maximum life of adult cerambycid beetles *Hoplocerambyx spinicornis* at different combinations of temperature and relative humidity. (From Linsley, after Beeson and Bhatia)

Investigations on different species of flour beetles of the genus *Tribolium* indicate that various combinations of temperature and humidity may have a profound effect on the competitive survival of individual species. Thus differences in abundance or ranges of competing species may reflect, not the empirical responses of these species occuring singly to the two factors, but the result of a threefold interaction between each species and the two weather factors, plus the additional competitive species in the environment.

Daily Rhythm. During the 24-hour cycle of day and night there is a daily rhythm of temperature and humidity characteristic of each area. Except during diapause, activities of most insects are related very definitely to this rhythm. The most conspicuous example is found in areas having hot days. During the heat of a summer day, when the humidity is depressed, many insects will be relatively inactive, locating cooler and moister niches. Toward dusk, as the temperature drops, there is a rapid increase in humidity. During this period a great number of insects emerge from daytime hiding and swarm over the ground and foliage, and in the air.

Air Movement. Air movement has little direct effect on insect physiology. It acts indirectly by influencing evaporation and humidity and, by causing evaporation, it aids in reducing body temperature. As drafts or wind, it plays a remarkable role in insect dissemination. The upward drafts caused by dawn and dusk air-convection currents carry an astonishing diversity of insects hundreds of meters in the air. Insects caught up by these currents include not only a large array of winged insects but also small wingless forms, such as springtails (Collembola). It is principally on these airborne insects that the swifts and nighthawks feed.

Many record exist of strong-winged insects, such as the *Erebus* moth, being blown by storms several thousand kilometers or more north from their tropical homes. Occasional specimens of the *Erebus* moth have been found, still alive, in Canada, the end of a journey started in Mexico, the West Indies, or southward. Shorter wind dispersals of large numbers of butterflies and moths are fairly common.

Periodic Dispersal. Several widespread species of insects that become abundant each year in northern states either die out during the winter or become greatly reduced in the northern part of their range. Overwintering populations persist in the extreme southern states, and the species move back into the northern states each spring. The armyworm *Pseudaletia unipuncta*, the aster leafhopper *Macrosteles fascifrons*, the potato leafhopper *Empoasca fabae*, and the potato psyllid *Paratrioza cockerelli* (Fig. 10-14), are examples of species that behave in this manner. Evidence to date indicates that the spring northern dispersal is windborne. Periodic windborne dispersal probably occurs in many more species than the few for which it is known.

Physical and Chemical Conditions of the Habitat

The habitat in which insects live may either temper or accentuate environmental conditions. Three general habitats are of paramount importance: surface, soils, and aquatic.

Surface Habitats. Surface habitats include the land surface and everything above it. This includes the aerial and arboreal regions because so many insects move from one to another. Conditions of terrestrial habitats differ from those of the atmosphere because of vegetative cover. In exposed areas such as treetops, upper foliage of desert and grassland plants, and bare unshaded soil or rock, the sun temperatures rather than shade temperatures (which are official weather temperatures) tend to prevail. Insects under leaves or other cover enjoy a moderation of temperature extremes, as do insects living within and under the tree canopy of a forest. The diurnal and

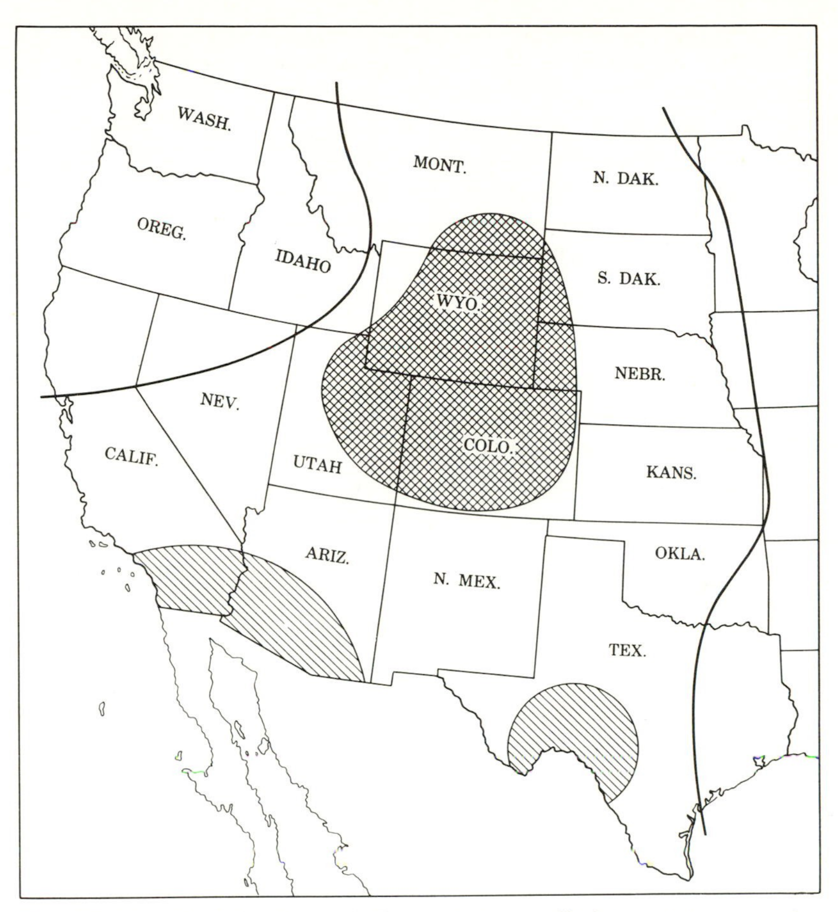

Fig. 10-14. Distribution of the potato psyllid *Paratrioza cockerelli* in the United States. The crosshatching indicates the area of greatest injury, the diagonal shading the approximate overwintering areas, and the heavy lines the eastern and western limits of summer occurrence. (After Wallis)

seasonal rhythm of temperature and humidity is greatest in the more exposed areas and progressively less in the more shaded or protected areas. At or near ground level, rotting organic matter produces heat of fermentation that adds to temperature. Insects in fungi, plant galls, leaf mines, and tunnels in living trees enjoy a high humidity that approaches an aquatic environment. Insects in rotten logs find a moderation of extreme temperatures because of insulation, and their habitat approaches that of soils.

Soils. Soils reflect the general climate but temper its extremes. At the same time they possess several important characteristics of their own.

Tempering of Climate. Depending on circumstances, soils act as sponges, insulators, and radiators. Soils store rain, giving it up slowly so that humidity, or moisture content, fluctuates over a much narrower range than does that of the air. The surface layers of soils soak up heat and insulate the parts beneath; the absorbed heat is also given up slowly, so that diurnal rhythm and temperature extremes are greatly moderated in comparison to those of the surface habitats.

Soil Properties. Many soils have properties that are not superimposed on them by the immediate climate, such as texture, moisture, drainage, chemical composition, and physiography. They reflect a development over a long period of time under varying climatic conditions.

Cultivation and urbanization have disturbed natural soil conditions more than any other element of the ecological pattern. Not only does cultivation change the original condition of the soil, but also plowing and tilling keep changing it at various intervals; drainage or irrigation decreases or increases moisture content; and methods of farming can increase or decrease chemical constituents and organic content. These influence soil texture profoundly. Although these changes are detrimental to many insect species, they allow others to increase and become major crop pests.

Soil texture varies from hard-packed clays to loose sands. Few insects are able to push or dig their way through hard-packed soils. Loams are probably the most common soils for digging and burrowing insects. Loams also are usually favorable in other charcteristics, such as moisture content, drainage, and organic content.

The critical effect of soil texture on species abundance was demonstrated in the pale western cutworm *Agrotis orthogonia*. A species of the northern prairies, it was a collectors' rarity in the early days of collecting in central North America. After extensive breaking of the prairie sod and cultivation of the land in northern United States and Canada, this cutworm increased sharply in numbers, and the larvae became extremely destructive to grain crops. Investigation revealed that the larvae live only in soil of fairly light texture because they move in response to daily or seasonal temperature and moisture changes. In the primeval northern prairie they occurred only in local sandy areas having loose soil, but not in the unbroken prairie sod. Cultivation transformed the sod into soil of optimal texture for the cutworm larvae and opened thousands of square kilometers of new habitat to them.

The *moisture* content of a soil is affected greatly by *drainage*. Impervious layers of substrata, such as clay or rock, may retard natural drainage, resulting in permanent or temporary semimarsh conditions or wet soils. In such situations only those insects occur that are at least partially modified for aquatic existence, such as many larvae of Diptera. More open types of subsoil, such as sand, gravel, or shale, allow free drainage, better aeration, and more rapid restoration of normal moisture content after rains. Soil insects in well-aerated soil do not need modifications for aquatic or semiaquatic existence.

An interesting demonstration of the effect of soil moisture on an insect species is shown in the western corn rootworm *Diabrotica virgifera*. Collecting records indicate that until irrigation became widespread, this species was fairly rare, occurring in the arid regions of New Mexico, Colorado, and Nebraska. The larvae feed on corn roots and, if present in large numbers, may destroy the root system of the plant and cause great reduction in yield. Since about 1890 the species has been a constant pest in south-central Nebraska, but practically disappeared during the drought years of the 1930s. Since then the species has become of major importance in the irrigated portion of the Platte River Valley, because of the increase in soil moisture caused by widespread irrigation.

Wireworms in the Pacific Northwest are another striking example of changes in species composition and abundance because of changes in soil moisture. There, four species of wireworms, *Limonius californicus*, *L. infuscatus*, *L. canus*, and *L. subauratus*, are wetland pests, normally restricted to swamp and river-bottom areas. When arid land was irrigated and used for farming, these wetland wireworms became important pests of potatoes, corn, lettuce, onions, and many other irrigated crops. The high soil moisture maintained by irrigation allows the wireworms to increase in great numbers (Fig. 10-15).

Drainage and texture together exert considerable influence on the distribution of some insects, such as the destructive Texas leafcutting ant *Atta texana*, whose range extends into Louisiana where it nests only in fine sandy loams having light subsoils and excellent drainage.

Physiography has few direct effects on insect distribution, but it has a marked influence on several soil factors. Flat country has slow rain runoff and must have adequate subsurface drainage to maintain good soil aeration. Hilly or mountainous country has rapid water

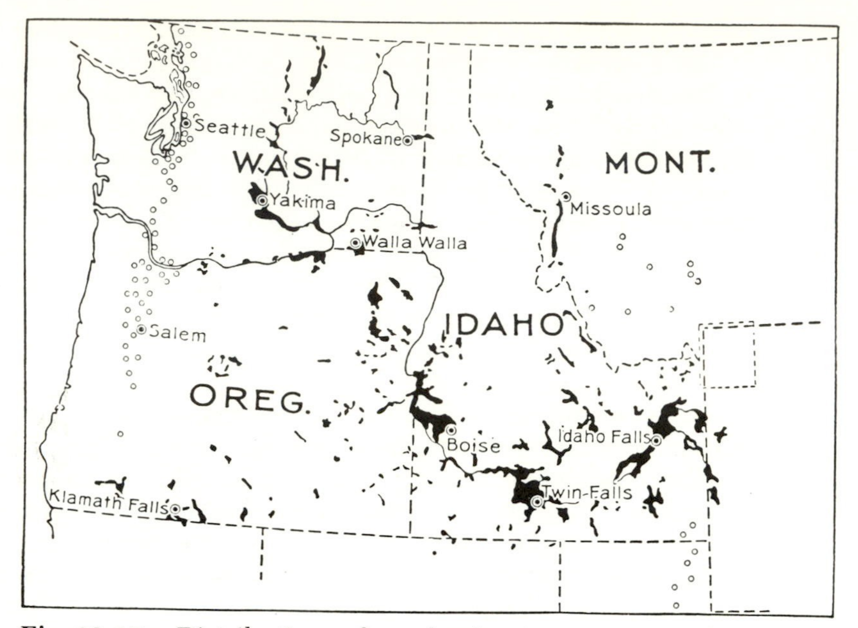

Fig. 10-15. Distribution of wetland wireworms in the Pacific Northwest. Black areas represent irrigation projects on which one or more species of wireworms cause serious injury to crops annually. Circles represent localities where wireworms are known to occur naturally, without benefit of irrigation, and to cause occasional injury. (From U.S.D.A., E.R.B.)

runoff, ensuring good general soil aeration. In addition, in the northern temperate zone, south, east, and west exposures have a higher soil temperature, greater evaporation, and, as a result, differences in the biota.

Chemical Composition. Soil chemistry may determine the species of plants that grow naturally in an area and in this way determines the distribution of many host-specific plant feeders. Deficiencies of mineral elements, resulting in similar plant deficiencies, inhibit the growth, abundance, and distribution of some insects but seemingly not that of others. Nitrogen deficiency lowers the productivity of some species of insects, but seems to contribute to outbreak numbers of others.

Aquatic Habitats. In water, respiration is critical in aquatic insects, and many characteristics of aquatic habitats are important because they have a direct bearing on this function. Of great importance to many aquatic insects are the diffusion of excess carbon dioxide out of the water and, more significantly, the diffusion or solution of oxygen into it. Most of the oxygen comes from the air, and any stirring movement that brings more water into direct contact with the air increases the oxygen supply. In lakes and ponds wind action is the chief agent. Waterfalls, rapids, or movement of current are stirring agents in streams.

Temperature has a direct bearing on aeration, because the colder the water, the greater is the amount of gases (including oxygen) that may be dissolved in a given volume of water. High temperatures greatly decrease the solubility of gases in water.

An important distinction must be made regarding aeration among aquatic insects. Many groups, such as mosquito larvae, horsefly larvae, and certain aquatic bugs, have extensile respiratory tubes that reach the surface, or the individuals periodically come to the surface to breathe; others, such as the water boatmen or adult diving beetles, take a bubble or film of air into the water with them, coming to the surface to replenish it from time to time (see p. 159). These groups are almost independent of the aeration factor in water, and many live in water almost devoid of oxygen.

Aquatic insects without modifications for obtaining direct contact with air are dependent for respiration on oxygen dissolved in the water. As with other ecological factors, various insects have different aeration requirements and are limited in distribution by it. Certain dragonfly nymphs and midge larvae are examples of forms able to tolerate very poorly aerated or even stagnant ponds. On the other hand, larvae of the midge family Blepharoceratidae have extremely high oxygen requirements and occur only in rapid mountain cascades.

Temperature. Aquatic temperatures do not have the same range as air temperatures and insect species usually show a definite restriction to water of a certain temperature range. For many species this is undoubtedly correlated with aeration, but in some temperature is probably the key factor. Certain mosquito larvae, for instance, live and transform normally in water at 18°C, but die during molting in water at 27°C. Since mosquitoes are not dependent on water for oxygen, it appears that this is a temperature factor.

A few aquatic insects have been found in hot springs with temperatures ranging from 43 to 49°C. These are chiefly aquatic beetles and fly larvae, which obtain oxygen directly from the air. For the most part development and activity increase as water temperatures rise, resulting in the production of great swarms of adults through late spring and summer.

In the winter stoneflies the reverse is true. The nymphs live in streams throughout the north-central and eastern states. During summer, nymphal growth seems to be retarded but increases with the advent of cool autumn weather. As a result the adults emerge during the winter months, beginning during late November and continuing until February or March (Fig. 10-16).

Depth. In lakes and ponds, depth has a marked influence on oxygen, temperature, and light; in running water the effect of depth is less marked owing to the stirring action of the current.

Water absorbs heat and light passing through it, converting the latter to heat. Heat rays and the red end of the light spectrum are absorbed first; at greater depths the other wavelengths are gradually absorbed until, even in very clear lake water, almost all light is absorbed at about 33m (100 ft). This has a profound effect on plant life. Practically none exists below the 20-m level, and most of it is concentrated in the first 2 to 3 m of water where there is a good supply of light for photosynthesis. This produces a zonation of food supply that in turn limits the distribution of many insects.

Depth in deep lakes (30 m or more deep) is accompanied by another phenomenon of great biological interest: the thermocline. In summer the surface waters of a large lake are appreciably warmer than the water at the bottom, which remains near the temperature of greatest density, 4 to 5°C (Fig. 10-17). This bottom layer being the heaviest, the upper warmer waters "float" on it, rather than mixing with it. Between the two is a relatively narrow dividing area called the *thermocline*, intermediate in general conditions between the fairly uniform upper and lower strata. The upper stratum, the *epilimnion*, is churned and agitated by wind action so that it is

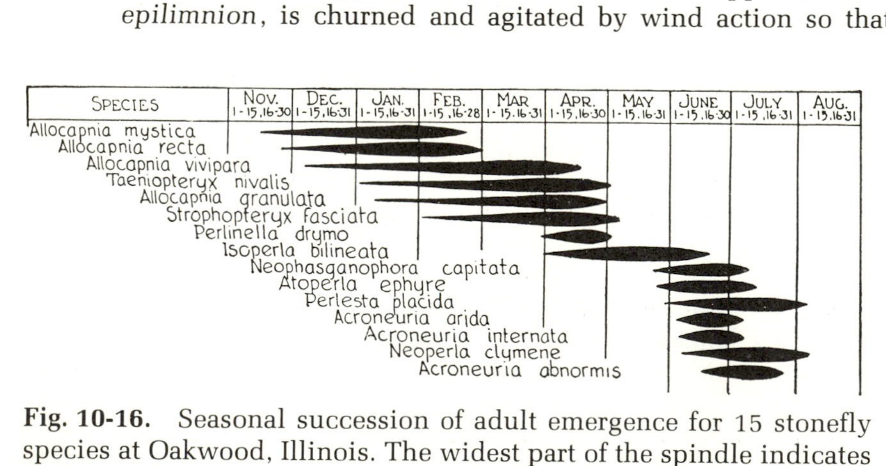

Fig. 10-16. Seasonal succession of adult emergence for 15 stonefly species at Oakwood, Illinois. The widest part of the spindle indicates the time of maximum abundance of adults for each species. (From Illinois Natural History Survey)

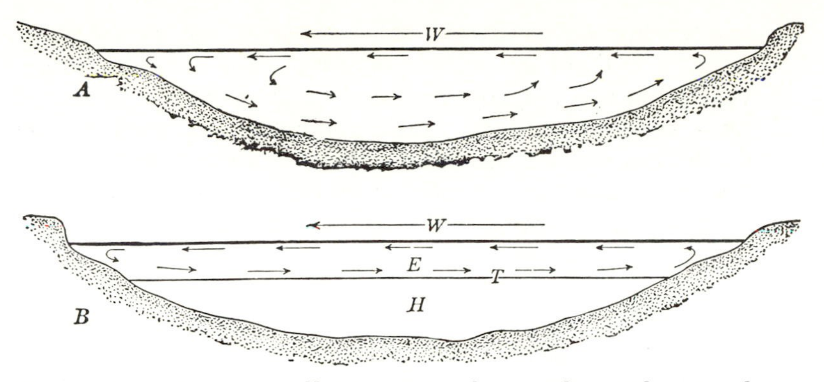

Fig. 10-17. Diagrams illustrating a thermocline. The circulation of the water (*A*) in a lake of equal temperature; (*B*) in a lake of unequal temperature. *W*, Direction of the wind; *T*, thermocline; *H*, hypolimnion. (From Ward and Whipple, after Birge)

almost uniform in temperature and well aerated. The bottom stratum, the *hypolimnion*, is stagnant, and its oxygen is gradually used up by organic oxidation. Almost all the life in a deep lake is the epilimnion; the hypolimnion is practically a biological desert.

Turbidity. Minute particles of clay and silt, "blooms" of algae or other organisms usually cloud water to some extent. The clouding or turbidity has an indirect effect on insects, because it reduces light penetration and therefore plant production on the bottom. Under conditions of continuous high turbidity, there is persistent settling of suspended material on the bottom, thus modifying its character and its fauna.

Substrate. Many aquatic insects live on or in the substrate, or bottom, and most species will live only where the substrate is of a particular type. The most useful categories for purposes of classifying substrates are based on sediment size: namely, clayey, silty, sandy, and gravelly (rocky). Mud (composed of both clay and silt) substrates are highest in organic material that may serve as food; sandy and rocky substrates have the least. Mud substrates in streams, however, are usually associated with slow current, lower oxygen content of the water, and higher temperatures. Sandy substrates are relatively unstable and usually have a relatively small fauna. Rocky substrates afford the most stable footing and are the favorite habitat for a large number of groups.

Vegetation. To some insects aquatic vegetation is primarily food; to others it is a haven. Vegetation beneath the water provides shelter and footing, especially valuable to species that are the prey of other animals or that have no special adaptation for swimming. Aquatic vegetation is especially abundant in lakes, and it is there that it is most useful. Relatively few lake-inhabiting insects frequent open water; they stay on the bottom and in the weed beds most of the time. Those that move about freely in the open water usually do so only at dusk or night, hiding on the bottom during the daytime.

Food

As a Factor in Distribution. The availability of suitable and sufficient food is one of the most important factors influencing the distribution and abundance of insects. Many insects are usually present wherever a suitable food is present and are prevented from extending their ranges because of the food factor. For many insect species this factor has been changed radically by human agriculture, travel, and transportation; for example, many plant lice have become successfully established on agricultural crops far beyond the range of the native hosts of the insects. On the other hand, the range of the host plant for a given species of insect may extend much beyond that of the insect, demonstrating that some other factor such as temperature or soil condition is the actual factor limiting the insect's success and distribution.

General Feeders. Many insects have a wide assortment of acceptable hosts or prey or feed on material such as decayed or dead organic matter that is widely distributed through most biotic communities. In these, food is only infrequently a limiting factor of distribution. Areas in which such species are absent usually have food material, but other factors such as temperature or moisture may be intolerable for the species.

Specific Feeders. A great many insect species—chiefly predators, parasites, and plant feeders—feed on only a small number of host species, or are restricted to a group of closely related host species, or may be restricted to a single host species. Species that have the most limited host tolerance are the ones that are most likely to have their distribution limited by food. The overall range of the Hepatica sawfly *Pseudodineura parvula* appears to cover all the north-central and northeastern states; yet in the north-central states the sawfly occurs only in the scattered localities in which its host, *Hepatica*, is found. The black locust sawfly *Nematus tibialis* is normally confined by the restricted eastern range of its host, *Robinia pseudoacacia*. Where black locust have been transplanted, the sawfly has ultimately been found, even in England, indicating that the sawfly has a wider ecological tolerance than the natural range of its host.

Host Crossover. Insects may defoliate their particular species of plant host, so that the insect species may attempt to adopt a new host or else have its numbers reduced to the carrying capacity of the original host. Some species, such as the forest tent caterpillar, make the change to closely related hosts with ease and without evident ill effects. Chinch bugs, for instance, will feed on oats or wheat until nearly full grown and can transfer to corn readily and without noticeable mortality from food reactions.

Other species will attempt a change from one host to a close relative with the greatest difficulty, either after compulsion of a period of starvation (in the case of immature larvae) or under circumstances of extreme necessity (in the case of ovipositing females). Usually host transfer cannot be made by advanced larvae, which may eat the new food but develop symptoms of intestinal disturbance, such as diarrhea, and die. If first-instar larvae are put on the new host, they eat but suffer an extremely high mortality. During succeeding instars the mortality rate decreases until the pupal stage is reached, and here the mortality is again high. But of thousands of first-instar larvae, only a few reach the adult stage. These prefer to oviposit on the new host, and the resulting larvae feed on it readily and without ill effects.

An interesting example is the satin moth *Stilpnotia salicis*. A European pest of Lombardy poplar, it was introduced on the Pacific Coast about 1922 and in a few years became very abundant, completely defoliating Lombardy and other introduced poplars in each locality to which it spread. In British Columbia by the late 1920s it had spread up the valley of the Lower Fraser River. There it built up a large population and practically exterminated the Lombardy poplars. When Lombardy poplars were no longer available, the satin moths began laying eggs on the native cottonwood *Populus trichocarpa hastata*, an abundant tree in this region. At first only sporadic colonies became established on this new host, which is also a member of the poplar genus; but in a few years the satin moth was the most serious insect enemy of cottonwood in this area. Laboratory rearings demonstrated that in making this host crossover the satin moth went through very high initial mortality.

The gypsy moth *Porthetria dispar* is an example that does not follow this model. Its larvae are plant feeders and will feed on over 450 species of plants. For normal development, however, it is necessary for the first two larval instars to feed on one of the 42 favored species, including willows, birches, and oaks. Later instars of the larvae may switch to one of the other host species (one of the favorites being white pine), suffer no ill effects, and develop into maturity.

Host crossovers have a great effect on the distribution and abundance of the insects involved and represent an important type of ecological tolerance. There is always the possibility of this happening when new plants are brought into the national agricultural economy, as, for instance, the soybean. When the soybean was introduced into the central United States it was virtually unattacked by any of the native insect species. Gradually some of these began using it as a host, and now the soybean in this country has several serious insect enemies, including grasshoppers, root maggots, and white grubs.

As a Factor in Abundance and Size. Amount of available food is an important factor affecting the population of a species in a given community. It is not uncommon for a species to utilize its entire available food supply, with a resulting sharp reduction in population because of starvation.

Although there may be considerable variation in size between different individuals of the same species, a definite minimum amount of food is required for the normal development of an individual. Housefly maggots, for instance, die during pupation if the larvae are removed prematurely from their food supply. Most sawfly larvae will die without further development if removed from their food only two or three feedings prior to completing their full food intake. If, therefore, an excessive number of individuals are feeding on an insufficient amount of food, those with a head start complete their normal food intake and mature, and many of the remainder run out of food and fail to develop.

Certain parasitic species having a wide range of hosts are notable exceptions. In these, the size of the individual parasite is determined by the size of the respective host species. An excellent example of this is the mutillid wasp *Dasymutilla bioculata*; larvae feeding on small prey species develop into small individuals, those feeding on large prey species develop into large individuals (Fig. 10-18).

Agriculture has changed the insect food factor in several ways: (1) by providing suitable food when or where it would not be present under natural conditions; (2) by providing better food; and (3) by simply providing a greater food supply, and modifying environmental conditions.

Extension of Food supply. In the late 1930s and early 1940s, in many states, especially in the corn belt, a tenebrionid beetle *Cynaeus*

Fig. 10-18. Correlation in size between *Dasymutilla bioculata* and its hosts *Microbembix monodonata* (left) and *Bembix pruinosa* (right). Each vertical row has *Dasymutilla* male and female above, host wasp at bottom. (Material, courtesy of C. E. Mickel, photo by W. E. Clark)

angustus became extremely abundant and widespread and developed into a major pest of stored grain. This beetle was first described in 1852 from California, and prior to 1938 remained a collector's rarity, occurring at the base of yucca plants. In 1938 the beetle was encountered as a stored-grain pest in Washington, Kansas, and Iowa. By 1941 it was known to be widely distributed through the corn belt. In the man-made conditions of grain storage the species reached population peaks that are in astonishing contrast to the scarcity of the species in its natural habitat.

Better Food. Some introduced agricultural crops have increased the fecundity and thereby the abundance of native insect species. The two grasshoppers, *Melanoplus differentialis* and *M. bivittatus*, showed marked increases in fecundity on a diet of soybeans and alfalfa, respectively. This correlates with field observations of an increase in population of the two species following the planting of large acreages of the two crops.

Additional Host Material. Practically every crop supplies at least a few insects with additional food. The Colorado potato beetle, the corn aphids, and other pests on potatoes and corn have flourished on the thousands of acres of planted host crops. They form huge populations that dwarf the scatterd colonies that existed before agriculture when their hosts were relatively sparse and the individual plants not so luxuriant as those of improved agricultural varieties.

Enemies

A wide array of organisms prey on or parasitize insects. Some parasites, such as the malarial organisms *Plasmodium* sp., seem to do the insect no harm, but the majority have a harmful effect on the insect host. These enemies constitute an environmental factor having a definite effect on the abundance, and sometimes the distribution, of the host species. Each stage of the host species may be subject to attack by a different set of enemies, or several stages may be attacked by the same one. Usually predaceous enemies and plant enemies, such as fungi, are more general in their attack on various stages, and internal parasites are restricted regarding the stage they attack.

Internal Parasites. Insects are attacked by several groups of internal parasites, of which the most important are certain groups of insects, parasitic worms, bacteria, and fungi.

Insects. The larvae of many families of Hymenoptera (Ichneumonidae, Chalcididae, Scelionidae, and many others) and a few families of Diptera (Pyrgotidae, Tachinidae) are entirely endoparasitic on insects or closely allied arthropods. A few Lepidoptera larvae and several Coleoptera larvae, including the Stylopoidea, are endoparasitic. There are more than 11,000 species of parasitic insects known at present in North America. Most are fairly specific as to the group they attack. Some will attack a wide variety of lepidopterous caterpillars; others will attack only certain primary parasites in these caterpillars (see p. 423). For more discussion of certain of these see Chapter 11.

Other Animals. Some species of Protozoa and invertebrate parasites pass one stage of their life cycles in insects. Examples of such protozoan parasites are the malarial organisms *Plasmodium* sp. and sleeping-sickness organisms *Trypanosoma* sp. Among the parasitic worms that spend part of their life cycle in insects are trematodes, nematodes (for example, *Filaria*), and Acanthocephala (for example,

Macracanthorhynchus hirudinaceus, the thorny-headed worm of swine). In each case only one of the early stages of development is passed in the insect, which is an intermediate host for the parasite. This group of parasites does not appear to have a deleterious effect on the insect, at least not the fatal effect of the insectan parasites. On the other hand, certain Protozoa, especially the microsporidians, have insects as their only host; these parasites frequently kill the host.

Fungi, Bacteria, and Viruses. Species of these groups attack insects in various stages and may be destructive to their hosts. Cultures for rearing insects are very susceptible to attack by fungi and bacteria because the best development of both types of these parasitic organisms is attained under conditions of relatively high humidity and temperature, which are frequently increased to an unnatural degree in caged experiments.

Among common fungus diseases of insects is *Empusa muscae*, the housefly fungus (Fig. 10-19). Other members of the same genus attack a large variety of insects, including grasshoppers, aphids, and chinch bugs. A famous fungus disease is *Beauveria globulifera*, often referred to as *Sporotrichum globulifera*, the white fungus of the

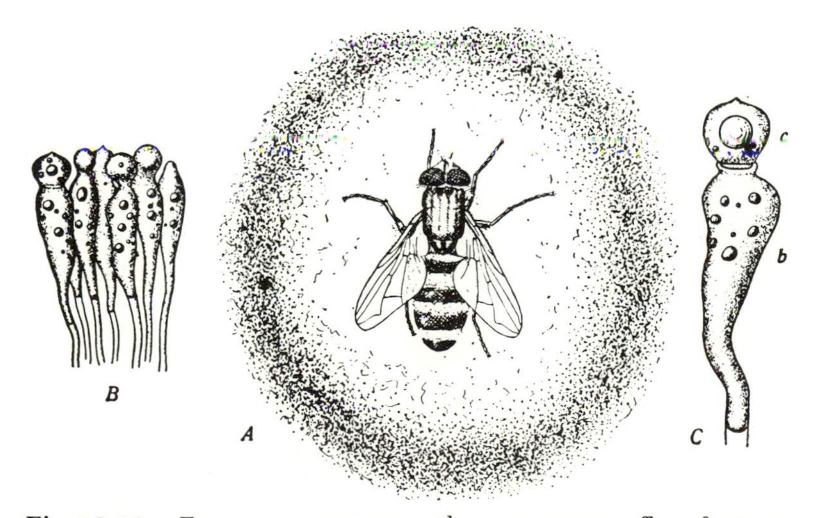

Fig. 10-19. *Empusa muscae*, the common fly fungus. (*A*) Housefly (*Musca domestica*), surrounded by fungus spores (conidia); (*B*) group of conidia in several stages of development; (*C*) basidium (*b*) bearing conidium (*c*) before discharge. (From Folsom and Wardle, 1934, *Entomology*, by permission of P. Blakiston's Son & Co.)

chinch bug. During warm and humid seasons this fungus kills large numbers of chinch bugs and other insects in late spring and early summer and at times has controlled the bugs to the point of local extermination. Entomophagous fungi of the genus *Isaria* are the chief species attacking insects under artificial conditions.

Of unusual interest is the fungus family Laboulbenaceae. Most of the species are entomophagous and produce elongate or ornate fruiting structures outside the body of the host insect. A species occasionally encountered in the eastern states is *Cordyceps ravenelii*, a parasite of white grubs (Fig. 10-20).

Bacterial diseases are less numerous than fungi but at times are strikingly devastating. Flacherie, an infectious and highly fatal disease of silkworms, is caused by a bacterium. Grasshoppers and chinch bugs also are attacked by bacteria, and attempts have been made to control these insect pests by propagating and disseminating the bacterial disease. *Bacillus popilliae*, called milkly disease, has been used successfully to control Japanese beetle larvae. The bacterial spores are mixed with an inert dust and the mixture applied on top of the soil in grub-infested areas. Rain washes the spores into the ground and into contact with the grubs.

Virus diseases are extremely toxic to susceptible insect species. Polyhedrosis viruses in particular have proved sufficiently virulent against certain sawfly and lepidopterous larvae to be employed successfully as control agents.

External Parasites. Insects have few ectoparasites, such as lice or fleas, in which the adult stage or both immature and adult stage use the body of the host as a home. A few mites infest various insects. An unusual ectoparasite is the bee louse *Braula caeca*, a curious minute fly that is ectoparasitic in the adult stage on honeybees.

In the Hymenoptera, some families whose larvae are mostly endoparasites, such as the Braconidae, contain genera whose larvae are attached externally to their host larvae. These parasites have the same host relation as their endoparasitic allies, in that normally only one parasite individual lives on one host individual, the latter almost always dying when or before the parasite is mature.

Fig. 10-20. Fruiting structures of a fungus *Cordyceps ravenelii*, arising from the body of a white grub *Phyllophaga*. (After Riley)

Predators. As in the case of internal parasites, insects are their own worst enemies. Staphylinidae and some Carabidae, two very large beetle families that feed in both adult and larval stages, are almost exclusively predators on other insects. Many families of wasps are predaceous, as are larvae of Tabanidae, Dolichopodidae, and some other large families of Mecoptera and Diptera. Odonata (damselflies and dragonflies) are predaceous as both nymphs and adults. The same is true of certain families of Hemiptera such as Pentatomidae (stinkbugs), Reduviidae (assassin bugs), and Phymatidae (ambush bugs); in some other families of Hemiptera, such as the Miridae (plant bugs), most genera are phytophagous, but some are predaceous. There are many other small groups of predaceous forms, including many of the aquatic Hemiptera.

Noninsectan predators of insects include members of several large groups. Spiders are primarily insectivorous; there are about 3000 species of spiders in North America, and each spider population takes its toll of insects. Centipedes feed on insects to a large extent also.

Vertebrates contain many groups that are insectivorous. Among the fish, perch, sunfish, crappies, bass, and sheepshead use insects for a large share of their diet. Reptiles and amphibians are largely insectivorous, as are bats and moles; other mammals such as mice, skunks, shrews, and raccoons eat large numbers of insects.

Birds are voracious insect eaters. Swifts, nighthawks, and flycatchers feed entirely on insects caught on the wing. Robins, wrens, chickadees, cuckoos, quail, and prairie chickens live almost entirely on insects when the latter are abundant. During insect outbreaks many birds of omnivorous food habits switch temporarily to an insect diet. Crows, blackbirds, gulls, owls, and small hawks are in this group and have been noted especially feeding on grasshoppers during periods of abundance.

All these animals are abundant and, being comparatively large individuals, eat proportionately large numbers of insects. In doing so they exert a steady ecological force against insect populations.

Predaceous Plants. A list of insect predators would not be complete without mention of those curious plants that trap animal prey and digest them. Most of these plants belong to the orders Nepenthales (sundews and Old World pitcher plants) and Sarraceniales (New World pitcher plants), and commonly live in nutrient-deficient soils. Bladderworts (*Utricularia*) are aquatic plants that trap small organisms in bladder-like pouches; sundews (*Drosera*) are bog plants having sticky tentacle hairs on their leaves that encompass prey; and pitcher plants (*Sarracenia*) have leaves in the shape of pitchers, partially filled with water, with stiff hairs pointing to the water. The hairs allow insects to get to the bottom of the pitcher but prevent their escape. Apparently the nitrogen derived from the digestion of insects and other small invertebrates aids growth in these plants. Interestingly, a few other insects, such as pitcher-plant moths (*Exyra*), larvae of certain flies (*Sarcophaga*), and some others, have developed an immunity and antidigestive enzyme so that they live in the walls of the pitchers and prey on other insects that become trapped by these plants. None of these plants is sufficiently abundant to be of importance ecologically by reducing insect numbers.

Protection Against Enemies

Camouflage. We have all been surprised at one time or another to discover a "stick" come to life in the net, or, in examining a tree trunk, to see what appeared to be a piece of bark take wings and fly away. Camouflage is widely used by insects to blend in with their environments so as to escape detection from predators. The walking sticks (Phasmatodea, Fig. 10-21) resemble the sticks and twigs on which they live, even to the extent that the young are green in the spring and the adults change to brown by autumn. Many moths at rest resemble bark. Some of the larger forms, the underwing or *Catocola* moths, are perfect bark mimics with wings folded, but often conspicuous in flight owing to brightly colored underwings (Fig. 10-21). Grasshoppers resemble lichens, various types of soil, dried leaves, or grass, depending on the species and its food. Psocids have similar protective color patterns, especially those that feed on algae or lichens on bark or rock bluffs. A few larvae at rest curl up and resemble a fresh bird dropping, notably the sawfly *Megaxyela aviingrata*. In general most of the leaf-feeding larvae are green or are mottled so that they blend into the foliage on which they feed. Flattid bugs (cicadas) in East Africa come to rest in considerable numbers on stems and arrange themselves to look like flowers of the lupines. In at least one of these cicada species, the females are green and the males yellow. When a group of these are at rest, the females rest above the males, resulting in a close parallel to the sequence by which flowers lower on an influorescence open before the ones above.

Cryptic camouflage (Fig. 10-21*A*, *B*) is very important for insect survival, as is shown by the following well-known example. In the English Midlands, extensive insect collections have been made since before the middle of the last century. In collections made about 1850, the peppered moth *Biston betularia* was common on the light-gray trunks of birch trees. This moth was normally white or very light

Fig. 10-21. Protective mimicry and coloration. At left, a walking stick insect on a twig; at right, an underwing (*Catocala*) with wings spread (*A*) and at rest on bark (*B*). (From Folsom and Wardle, 1934, *Entomology*, by permission of P. Blakiston's Son & Co.)

gray with black spots and narrow stripes. There also was an uncommon form that was black. This darkly colored form occurred at a very low frequency in the populations. As the industrial revolution in this part of England continued and expanded in the later part of the nineteenth century, the trees of the area became increasingly dirty from soot and pollution. By 1895 the black (or melanin) forms of the peppered moth in the vicinity of the major industrial towns had increased to comprise nearly 95% of the moth's local populations. Predators, mostly birds, could easily see the light-colored forms on the soot-covered tree trunks, but not the black forms. In areas away from the industrial centers, where the tree trunks were still light-colored, the black forms were easily seen and there the light forms constituted the great bulk of the peppered moth population. This tendency for black-colored prey to be selectively spared because they blend better into their industrialized community's environment is frequently referred to as industrial melanism.

Another form of camouflage mimicry is less passive. The mantid *Hymenopus coronatus* has the same red coloration as the red flowers of certain Malaysian orchids among which it lives. It so closely resembles the orchid flowers that insects in search of orchid nectar also examine the mantid for nectar but find instead that they are the mantid's lunch.

Scare Tactics. Certain insects try to bully their predators into leaving them alone. Many butterflies have eyespots on their hind wings that are usually hidden when the individual is at rest. When disturbed by a predaceous bird, these butterflies suddenly move their front wings away to expose these "eyes" and to frighten the bird. Caterpillars also may have conspicuous eyespots, and when attacked they assume a pose that makes the predator believe it sees two large eyes staring back.

Flowers That Mimic Insects. Even insects can be deceived by mimicry. Some of the most striking mimics are a group of orchids, particularly in the genus *Ophrys*. These have evolved a part of the flower that resembles the female of certain species of bees and wasps. These flowers also produce a mimic of the sexual attractant odors usually associated with the females of the particular insect species they mimic. In some, the mimic's odor is more attractive to the male than the female insect. The result of this mimicry is that the male insect is misled into attempting to copulate with the flower's "dummy female" and in doing so the male insect becomes covered with pollen, which the insect in its bliss proceeds to transfer to other blossoms.

Building Protective Structures. Certain larvae build cases, houses, or canopies that give the occupant some physical protection and, in addition, may resemble the host or surroundings and result in protective resemblance. Bagworms (Lepidoptera, Fig. 8-175) and also many caddisfly larvae (Fig. 8-163) are common examples. Cases of the latter may be very difficult to see unless the insect is in motion. The larvae of several leaf-feeding beetles construct an urnlike case that is difficult to see in its natural surroundings.

Poisons, Bites, and Stings. Certain insects gain protection by inflicting pain on their assailants. Several caterpillars have sharp hairs, containing a poisonous fluid that causes a rash and extreme pain. So delicate are these hairs that one has only to brush against them lightly to feel the excruciating nettling sensation that they produce. Tussock moth larvae and eucleid moth or saddleback larvae (Fig. 10-22) are protected by these. Other insects bite the aggressor, as, for instance, ants. Bees and wasps, are provided with painful

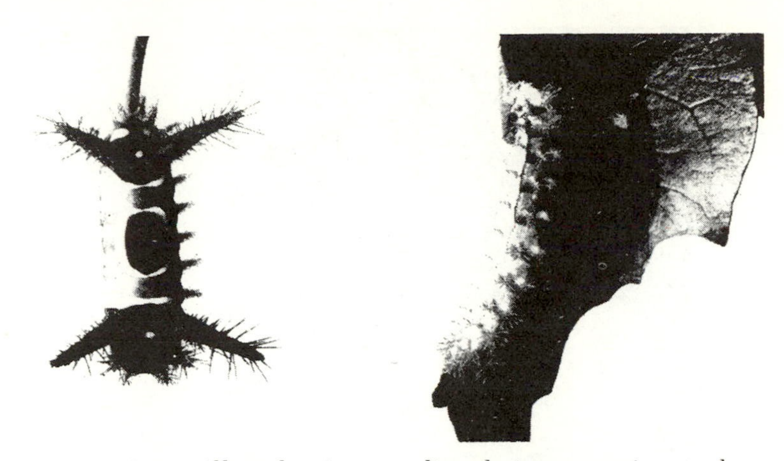

Fig. 10-22. Caterpillars having nettling hairs annoying to humans. Left, the saddleback *Sibine stimulea*; right, *Automerus io*. (From U.S.D.A., E.R.B.)

stings; bees use these only for protection. Solitary wasps primarily use their stings for paralyzing their prey and are generally nonaggressive toward intruders. On the other hand, social wasps primarily use their stings for defense and are highly aggressive to intruders, but inject their prey with paralyzing fluids through their mandibles rather than their sting.

Noxious Secretions. A large number of insects have mechanisms for producing and ejecting noxious smelly substances. Swallowtail butterfly larvae have an eversible pair of horns, the *osmeterium*, on the pronotum, that give off an odor thought to be repellent to some animals. Nymphal stages of most Heteroptera have stink glands on the dorsum of the abdomen. Many beetles, such as the blister beetle, cockroaches, and others have less conspicuous repellent-producing glands (see p. 205). Other insects, without such definite glands, apparently have a disagreeable taste, because birds especially refuse them as food. Swallowtail butterfly and milkweed butterfly adults, and both larvae and adults of many brightly colored leaf-feeding beetles are in this category.

A large number of species possessing protective devices of the types discussed in the two preceding paragraphs are strikingly marked or gaudily colored (Fig. 10-23). This striking ornamentation may be the display of warning colors, to aid the memory of an assailant who has attacked a protected species and became aware of its defense. Birds and other vertebrates have few instincts to avoid protected species, hence each individual must learn for itself.

Fig. 10-23. A brightly colored mal-tasting butterfly *Papilio ajax*. (From Folsom and Wardle, 1934, *Entomology*, by permission of P. Blakiston's Son & Co.)

Mimicry. This phenomenon is relatively common in insects as a means of protection from certain predators (Fig. 10-24). Two, three, or more species, usually distantly related, closely resemble each other for the advantage of one or more of the species. Several types of mimicry are known. One of the first forms of mimicry was recognized by H. W. Bates in 1862 as a result of studying tropical butterflies from South America. Members of the butterfly family Heliconiidae are bright, conspicuous, and very unpalatable to birds. Members of the butterfly family Pieridae have a similar color pattern

Fig. 10-24. Mimicry in bees and wasps. (*A*) Drone bee *Apis mellifera*; (*B*) dronefly *Eristalis tenax*. (From Folsom and Wardle, 1934, *Entomology*, by permission of P. Blakiston's Son & Co.)

and closely resemble the Heliconiidae, but are edible. This form of mimicry, *Batesian mimicry*, makes use of deceptive coloration and patterns by an edible mimic species to protect it from a predator that has learned to recognize by experience its distasteful model.

Another form of Batesian mimicry makes use of similarities in morphology and in behavior. The darkling beetles (*Eleodes*) stand on their two front pairs of legs and raise their abdomen to spray a noxious defensive fluid on their attacker. *Megasida* is a similar-appearing beetle that adopts the same pose when attacked, but it lacks abdominal glands to secrete defensive fluids.

Mullerian mimicry occurs where a number of distantly related species, all of whom are distasteful, resemble one another. These species groups are referred to as *mimic rings*. This phenomenon, also first recognized in tropical butterflies, was explained by Fritz Muller in 1878 as an important means by which each of the species lets the predator learn it is inedible because the predator needs to sample only one individual from all those having similar color patterns to learn its lesson, rather than one individual from each distasteful species. This results in a conservation of individuals for all the species of the ring.

Some members of the same species gain advantages by mimicking others of the same species. This is called *automimicry* and may take many forms. In many bees and wasps, the males lack stingers and are defenseless, whereas the females are well endowed with defensive stingers. By closely resembling females, the males may thus gain protection from predators. A somewhat different form of automimicry is common in the danaine butterflies (Fig. 10-25), which include the monarch (*Danaus plexippus*) and queen (*D. gilippus*) butterflies. The larvae of these butterflies feed on a number of plants, but their principal food is milkweed (*Asclepias*) which contains cardiac glycosides or cardenolides. These chemicals are very toxic to vertebrate predators such as birds, and a small amount triggers vomiting (below lethal levels). The danaine larvae are able to tolerate the cardenolides and retain them in significant amounts through to the adult stage, which gives the butterfly protection from predators. Larvae of these species may feed on other plants but they enjoy equal protection from predation also because they cannot be visually separated from those that fed on milkweed. The distantly related, edible viceroy butterfly (*Limenitis archippus*) is a Batesian mimic and also enjoys protection from predators.

More complex forms of mimicry exist, and some mimic rings include mixtures of these different types of mimic relationships. One of the most interesting is a series of ring groups that include the soft-bodied beetles, subfamily Lycinae, relatives of the fireflies. In

Fig. 10-25. The viceroy, below, and the milkweed butterfly, above. (From Folsom and Wardle, 1934, *Entomology*, by permission of P. Blakiston's Son & Co.)

southwestern North America, the lycids *Licus loripes* and *L. fernandezi* live together with their mimics, the longicorn beetles *Elytroleptus ignitus* and *E. apicalis*. *L. loripes* and *E. ignitus* are a bright light orange, and *L. fernandezi* and *E. apicalis* are bright light orange with brown-tipped elytra. The lycids are inedible to most predators and outnumber the longicorn beetles by a ratio of 100 to 1. Investigations have shown that not only are the longicorn beetles a Batesian mimic, but they are also occasional predators on the lycids, and in this sense are *aggressive mimics*. By eating fluids from the lycids, the longicorn beetles ingest the same toxic substances that the bird predators have learned to reject, and thus the longicorn beetle gains an additional, but perhaps temporary, form of protection. Other parallel series of mimic groups based on lycids and longicorn beetles are known from other parts of North America, South America, Borneo, and Cuba, suggesting this mimic relationship is very ancient. Most of these rings have additional species, including butterflies, hemipterans (some of which also may be inedible), and others even have wasps (with their own independent protection supplied by a stinger). In Cuba no less than 15 species form a very complex mimic ring.

Much more remains to be learned about mimicry in insects. For example, a great many insects (and spiders) look like ants. The reasons for mimicking ants are not always clear because many of the relationships do not entirely fit the models that we have just discussed.

Competition

Among insects, competition is chiefly for food. This competition may be between either individuals of the same species or individuals of different species. Critical competition commonly produces no reaction, and all individuals may starve. If, for instance, sawfly larvae overpopulate a host, they feed quietly until the entire host is stripped; then all of them wander until exhaustion and death if additional food is not found. In the case of critical competition involving two or more insect species, their different requirements may mitigate in favor of one of them. An interesting example is found in two hymenopterous genera *Opius* and *Tetrastichus*, that parasitize the Mediterranean fruit fly larvae in Hawaii. Within a single fruit fly larva, only a single larva of a species of the braconid *Opius* can develop to maturity, but as many as 10 to 30 individuals of the minute chalcid *Tetrastichus* (Fig. 10-26) can. If both oviposit in the

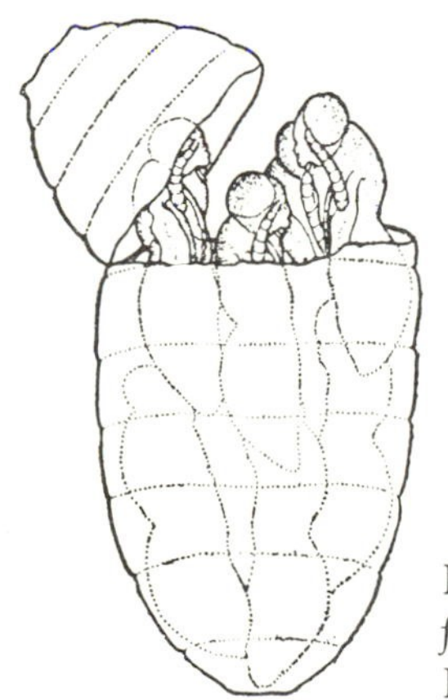

Fig. 10-26. Parasitic wasps *Tetrastichus giffardianus*, in puparium of fruit fly. (After Pemberton and Willard, 1918)

same fruit fly larva, the *Opius* larva kills most of the *Tetrastichus* larvae, but a few of the latter escape destruction. These develop more rapidly than the *Opius* larva and reach maturity, but leave too little food for the larger braconid larva, so that the *Opius* larva invariably dies.

Competition for food is frequently active and aggressive. In Hawaii three wasp species, *Opius tryoni*, *O. fullawayi*, and *O. humilis*, parasitize fruit fly larvae. The female wasps lay their eggs in the fly larvae, and several individual wasps of all three species may oviposit in the same larva. Only one survives, and it is the victor of a battle among the newly hatched larvae. The first instar of each *Opius* larva has a relatively large hard head bearing a pair of long sharp mandibles that can be opened and shut with great force and speed (Fig. 10-27). These larvae are pugnacious and attack any other parasite larva within the fly larva, using these sharp mandibles to pierce and lacerate their antagonist's body. Whether the struggle is between individuals of the same species or of different species, only one *Opius* larva remains after the struggle is over. *O. tryoni* is almost invariably the victor over the other two species, because of its greater agility, reaction time, mandible force, and other combative advantages.

Cannibalistic tendencies occur in many insect groups, and are invariably accentuated by crowding. The confused flour beetle lives and feeds on a variety of stored-grain products; generations are continuous, and adults, eggs, and all stages of larvae occur together in the food. Large larvae and adults may feed on eggs and small larvae of their own species but apparently make no effort to hunt them out.

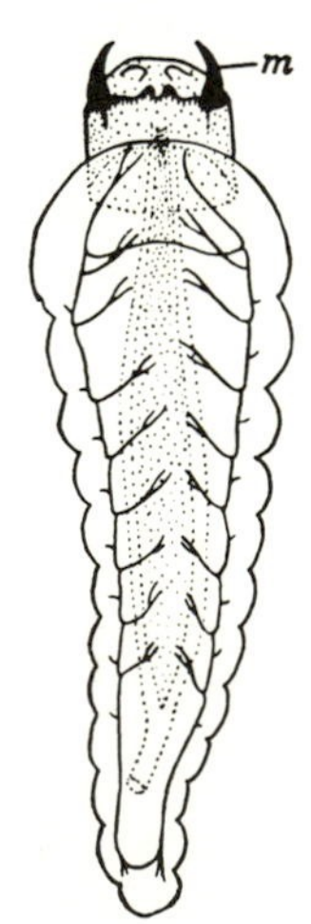

Fig. 10-27. First-instar larva of *Opius humilis*. Note the sharp heavily sclerotized mandibles, *m*. (Redrawn from Pemberton and Willard, 1918)

When the infestation of these beetles is small in relation to the volume of their food medium, the older individuals encounter the younger stages less frequently. As the infestation increases per unit volume of food medium, these encounters are more common and cannibalism increases. By this mechanism a population level is reached where the losses attributable to cannibalism are equal to the reproductivity of the adults, and overcrowding beyond this point is prevented.

REFERENCES

Allee, W. C., A. E. Emerson, O. Park, T. Park, and K. P. Schmidt, 1949. *Principles of animal ecology*. Philadelphia: Saunders. 837 pp.

Allee, W. C., and K. P. Schmidt, 1951. *Ecological animal geography*, 2nd ed. New York: Wiley. 597 pp.

Anderson, R. M., B. D. Turner, and L. R. Taylor (Eds.), 1979. *Population dynamics*. 20th Symposium British Ecological Society. Oxford: Blackwell. 434 pp.

Andrewartha, H. G., and L. C. Birch, 1954. *The distribution and abundance of animals*. Chicago: Univ. of Chicago Press. 782 pp.

Breymeyer, A. I., and G. M. Dyne, 1980. *Grasslands, systems analysis and man*. New York: Cambridge Univ. Press. 950 pp.

Chapman, R. N., 1931. *Animal ecology*. New York: McGraw-Hill. 464 pp.

Clark. L. R., P. W. Geier, R. D. Hughes, and R. E. Morris, 1967. *The ecology of insect populations in theory and practice*. London: Methuen. 232 pp.

Clements, F. E., and V. E. Shelford, 1939. *Bio-Ecology*. New York: Wiley. 425 pp.

Elton, C., 1927. *Animal ecology*. New York: Macmillan. 207 pp.

Folsom, J. W., and R. H. Wardle, 1934. *Entomology with special reference to its ecological aspects*, 4th ed. Philadelphia: P. Blakiston's Son & Co. 605 pp.

Harcourt, D. G., 1969. The development and use of life tables in the study of natural insect populations. *Annu. Rev. Entomol.*, **14**:175–196.

Harper, J., 1977. *Population biology of plants*. London: Academic Press. 892 pp.

Hassell, M. P., 1978. *The dynamics of arthropod predator-prey systems*. Princeton: Princeton Univ. Press. 328 pp.

May, R. M. (Ed.), 1976. *Theoretical ecology*. Philadelphia and Toronto: Saunders. 317 pp.

Mound, L. A., and N. Waloff (Eds.), 1978. *Diversity of Insect Faunas*. Symposia Royal Entomological Society. Oxford: Blackwell. 204 pp.

Parker, J. R., R. C. Newton, and R. L. Shotwell, 1955. Observations on mass flights and other activities of the migratory grasshopper. *USDA Tech. Bull.*, **1109**:1–46.

Pianka, E. R., 1978. *Evolutionary ecology*, 2nd ed. New York: Harper & Row. 397 pp.

Price, P. W., 1975. *Insect ecology*. New York: Wiley. 514 pp.

Roughgarden, J., 1979. *Theory of population genetics and evolutionary ecology: An introduction*. New York: Macmillan. 634 pp.

Southwood, T. R. E. (Ed.), 1968. *Insect abundance*. Symposia Royal Entomological Society. Oxford: Blackwell. 160 pp.

——— 1977. Entomology and mankind. *Am. Scientist*, **65**:30–39.

Spurr, S. H., and B. V. Barnes, 1980. *Forest ecology*, 3rd ed. New York: Wiley. 688 pp.

Waloff, N., 1980. Studies on grassland leafhoppers (Auchenorrhyncha, Homoptera) and their natural enemies. *Adv. Ecol. Res.*, **11**:82–215.

Wickler, W., 1968. *Mimicry in plants and animals* (Trans. R. D. Martin). New York: McGraw-Hill. 255 pp.

11 Useful Insects

Because insects are so diverse in numbers of species and in their adaptation to so many habitats,they commonly are both a help and a hindrance to humans. We frequently are reminded of insects' abilities to be nuisances, pests, and obvious competitors for food and fiber products; however, less apparent are their important beneficial roles. These beneficial aspects, especially as they occur in North America, are explored in this chapter. Important areas in which insects are particularly beneficial are (1) in helping to maintain balanced levels of insect populations through parasitism and predation; (2) attacking weeds and reducing the plants' ability to spread and to withstand diseases; (3) pollination of flowering plants, including many that humans use as foods (especially fruits and berries); and (4) the production of useful products including honey, beeswax, silk, shellac, dyes, and drugs. Several other insect benefits include insects as human food, as laboratory study animals, and as part of recreational activities by nature enthusiasts. Another insect benefit commonly unrecognized is their importance as decomposers in recycling much of the chemical materials of the ecosystem.

PREDATORY AND PARASITIC INSECTS

As discussed in the previous chapter, predation by many insects helps to keep the population size and density of other insects within reasonable limits. More than half of the world's insect species feed on other insects, most of which in turn feed on material that is not of direct economic importance to humans, such as poison ivy foliage or rotten wood in the ground cover. Some of these insect parasites and predators feed on insects that in turn feed on something useful to humans or harmful to their health or well-being. This latter group of insect predators and parasites thus become part of our arsenal in combating noxious or harmful insects, and may be manipulated to a minor extent to increase their effectiveness in managing the populations of the harmful insects involved.

These relations among insects have an increasing importance to humans and human food resources for several reasons. First, the use of pesticide chemicals to control insect pests has been shown to have many side effects. There are some hazards to humans and animals in their use. Some insects, particularly the Diptera, are showing increasing resistance to these chemicals so that greater concentrations are needed to effect the desired drop in population sizes. Other critical insect population relations, besides that involving the targeted pest, are usually affected and disturbed. Many of the chemicals, such as DDT, are long-lived in the natural environment and readily carried by runoff into streams, lakes, rivers, and the oceans, where they are commonly concentrated by other organisms. Second, the use of chemical pesticides to control target populations of pest insects is at best a temporary and expensive solution to a more complicated problem of insect and plant interactions. Thus the large group of insect predators and parasites of other insects has become an exciting part of the field of entomology known as *biological control* (see also Chapter 12 on harmful insects).

Predatory insects differ from parasitic insects in several important ways. Predators are generally active in seeking out their prey and they usually consume many individuals during their lifetime. A parasitic insect, on the other hand, generally is attached to (or is in) the body of a single host individual and generally does not kill its host. A more extreme form of parasitism, common in many insects, is where the parasitizing insect is attached to or within the body of a host and grows along with the host to a certain point, eventually killing the host. This relationship is called *parasitoid* and is nearly always the larval stage of a free-living adult; the female adult usually can search out and deposit its eggs on the host's egg or larvae under difficult circumstances. For example, the ichneumonid wasp *Megarhyssa lunator* (Fig. 11-1) has a long ovipositor with which it actually drills through wood to lay eggs in the larval tunnel of its wood-boring host, the pigeon tremex, or horntail, *Tremex columba* (Fig. 8-118).

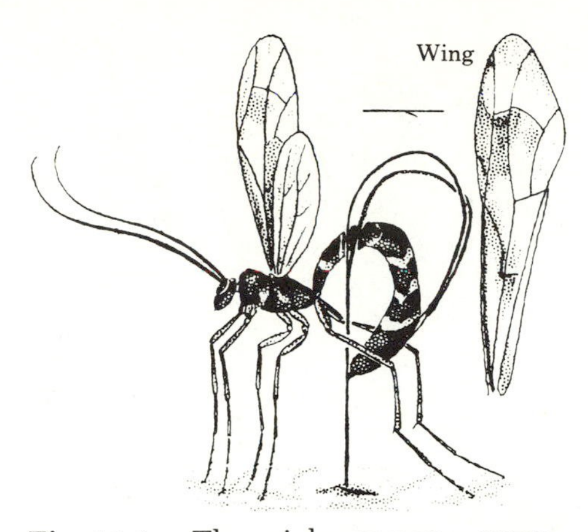

Fig. 11-1. The ichneumon wasp *Megarhyssa lunator* drilling into dead wood to locate larval tunnel of prey. [From *The common insects of North America* by Lester A. Swan and Charles S. Papp: fig. 1155 (p. 527). Copyright © 1972 by Lester A. Swan and Charles S. Papp; reprinted by permission of Harper and Row, Publishers Inc.]

INSECT PREDATORS

Coleoptera

Family Coccinellidae. Among the Coleoptera, the family Coccinellidae (ladybird beetles) (Fig. 8-95) contains many helpful predators of other insects. Some, such as *Rodolia cardinalis* (Fig. 11-2*A*), feed on scale insects and have the capability of a complete life cycle in less than 2 weeks during the summer. The larvae feed on eggs and nymphs of the prey and the adults on all the prey's life stages. *Coccinella californica* (Fig. 11-2*B*) has a passion for aphids and will consume nearly 500 aphids before reaching its adult stage. As an adult it will continue to eat aphids at the rate of 30 to 40 per day. Others of the family Coccinellidae have equally large appetites. In California the convergent ladybird beetle, *Hippodamia convergens*, is usually used in aphid-control projects because large natural populations of it hibernate in stumps and forest litter, and under rocks. These can be easily collected and kept under refrigeration until

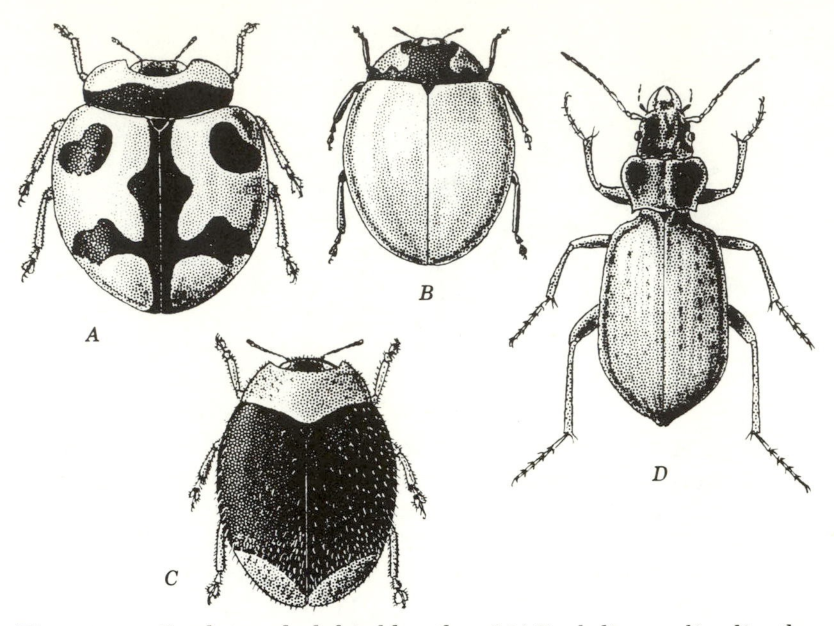

Fig. 11-2. Predatory ladybird beetles. (*A*) *Rodolia cardinalis*, the vedalia, a predator on citrus cottony-cushion scale; (*B*) the California ladybird beetle *Coccinella californica*, an active predator on aphids; and (*C*) the mealybug destroyer *Cryptolaemus montrouzieri*. (*D*) The ground beetle *Calosoma sycophanta*. [From *The common insects of North America* by Lester A. Swan and Charles S. Papp; *A*, fig. 805 (p. 407); *B*, fig. 826 (p. 412); *C*, fig. 804 (p. 407); *D*, fig. 542 (p. 337). Copyright © 1972 by Lester A. Swan and Charles S. Papp; reprinted by permission of Harper and Row, Publishers Inc.]

shipped. There are more than 125,000 individuals per gallon. In practice, release of large numbers of *H. convergens* early in the season is not entirely successful because a certain population level of aphids is needed to sustain them.

One of the early successful attempts at biological control of an insect pest includes the coccinellid *Rodolia cardinalis*, the vedalia, which was imported from Australia in 1888 to help control the citrus cottony-cushion scale (*Icerya purchasi*; family Coccidae). This scale, which was apparently introduced into California prior to 1872, nearly destroyed the citrus groves in the next 15 years. At that time, the cottony-cushion scale was known only from Australia, where it was not a serious problem because it was held in check by several natural predators, including a small Diptera (*Cryptochaetum iceryne*) and the coccinellid *R. cardinalis*. Both predators were introduced to California and the coccinellid quickly became well established. With the introduction of about 500 individuals, this beetle soon became well established in all the citrus areas of California, and within only 2 years they had brought the cottony-cushion scale under control. After the Second World War, chlorinated hydrocarbons and organophosphates were widely used in California. These, particularly DDT, tend to persist on foliage for months and, in addition to eliminating other citrus pests, they also locally eliminated the coccinellid *R. cardinalis*. The cottony-cushion scale quickly reappeared in outbreak numbers and *R. cardinalis* had to be reestablished after the insecticide residue had dropped to low levels.

Another coccinellid beetle, *Cryptolaemus montrouzieri* (Fig. 11-2*C*) is a predator of the citrus mealybug, *Pseudococcus citri*. *Cryptolaemus montrouzieri* was introduced in California in 1891 and became a well-established and extremely effective predator. Its survival through the winters in inland California is chancy and since 1917 it has been reared at mass-production levels and distributed to California citrus groves as needed.

Coccinellid beetles include a number of other species that are important aphid and scale insect predators. In addition to those already mentioned, the black lady beetle, *Rhizobius ventralis*, a related species, *R. debilis*, and *Lindorus lophantae* were imported to aid in controlling the citrus black scale insect, *Saissetia oleae*. Locally, these have been effective, and they often have resorted to preying on other scales with surprising success. Nearly a dozen species of coccinellid beetles have been imported as predators and have become successfully established. In the Caribbean and Mexico the citrus blackfly (of Asiatic origins) is preyed on by the coccinellid *Catana clausensi* and a parasitic wasp *Eretmocerus serius*, both of which were introduced from Malaya in 1930. Between these two predators, principally because of the effectiveness of the wasp, the citrus blackfly is not a serious pest.

Family Carabidae. This large family, the ground beetles (Fig. 8-85), are predominantly insect predators and rank close to the coccinellids for the most valuable predator's award. With few exceptions, larvae of this family are predaceous on other insects. Carabids include nearly 2000 North American species. They eat larvae and pupae of Lepidoptera and adults of most other insect orders. The European carabid *Calosoma sycophanta* (Fig. 11-2*D*) was introduced into New England in 1905 to help control the gypsy moth, *Porthetria dispar*, which earlier had become a serious introduced pest. The larva of this carabid consumes upwards of 50 gypsy moth caterpillars during its 2 weeks as a larva. Adult *C. sycophanta* live for several years and will eat several hundred more caterpillars. The number of gypsy moth caterpillars eaten by these beetles substantially helps in keeping the moth infestations under control.

Family Staphylinidae. This family, the rove beetles, also are mainly predators of other insects and include more than a thousand species in the contiguous United States. Many, such as the small *Aleochara bimaculata* (Fig. 11-3*A*), have aggressive larvae that are host specific and must depend on a single prey, such as in this case the maggot of the cabbage fly, *Hylemya brassicae*. Because of this dependence staphylinids may be quite effective in reducing a specific pest to tolerable limits.

Family Cleridae. The checkered beetles are mainly predators on the larvae and, in some, the adults of bark and wood-boring beetles. The bark beetles of the western forests are much sought after by the clerid *Enoclerus sphegeus* (Fig. 11-3*B*) and the powder-post beetles (Lyctidae, *Lyctus* and *Xylobiops*) are the prey of the clerid *Tarostenus univittatus*. Other clerids, including *Thanasimus*, *Tillus*, and *Thaneroclerus*, are important predators of boring beetles.

Other predaceous beetles. Several other groups of beetles are predominantly predaceous. In the Lampyridae, or lightning bugs, most species are predaceous on other insects, as well as snails and slugs, and they help in keeping those garden pests under a degree of control. The Cantharidae, or soldier beetles (Fig. 11-3*C*), are mostly predaceous on the eggs and larvae of other insects. Some cantharid species of *Podabrus* and *Canthris* feed on aphids, and several species are fond of eating grasshopper eggs and larvae of moths and beetles. All of the Cicindelidae, or tiger beetles (Fig 11-3*D*), apparently are predaceous on other insects —with no apparent preference for prey. The cicindelids in natural populations are not abundant and their effectiveness in insect control has not been thoroughly explored. The Meloidae, or blister beetles, are a mixed blessing. They are common and their larvae are predaceous or parasitic on other insects; however, the adult is herbivorous and several are crop pests of considerable importance, particular to the potato family.

Other beetle families include some species and genera that are predaceous, although many of their relatives are not. Scavenging beetles such as the Histeridae, Silphidae, Nitidulidae, and Tenebrionidae include some species with mixed feeding habits and the Elateridae have at least several species that are predatory on larvae of

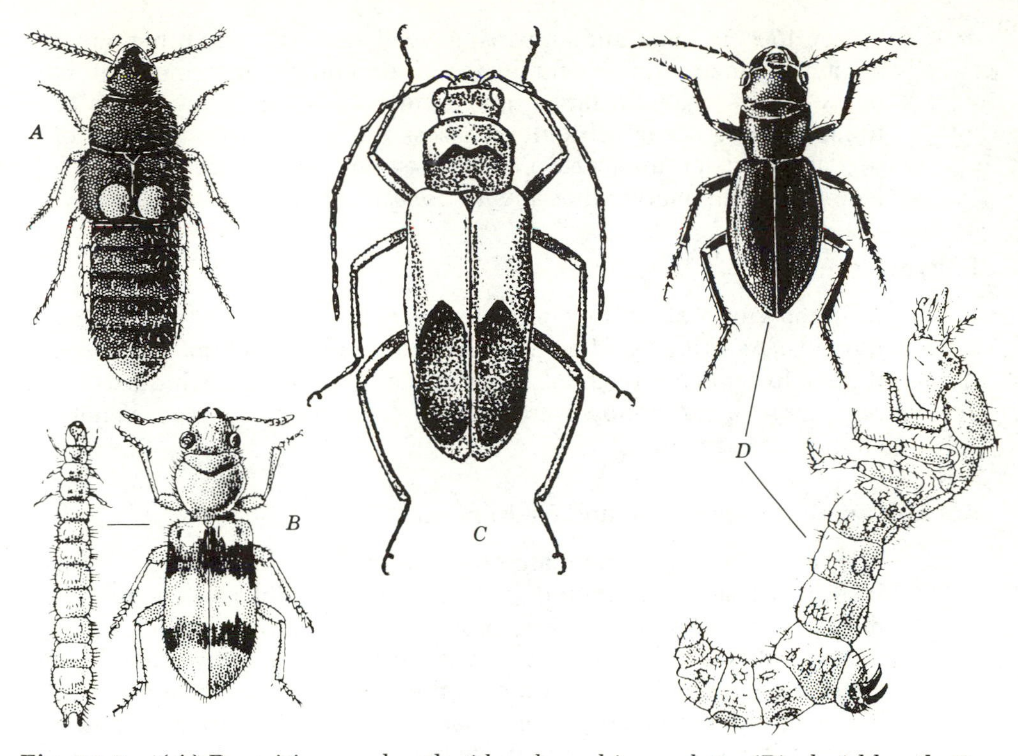

Fig. 11-3. (*A*) Parasitic rove beetle *Aleochara bimaculata*; (*B*) clerid beetle *Enoclerus sphegeus*, a predator of *Dendroctonus* bark beetles; (*C*) soldier beetle *Chauliognathus pennsylvanicus*, Cantharidae; and (*D*) adult and larva of California black tiger beetle *Omus californicus*. [From *The common insects of North America* by Lester A. Swan and Charles S. Papp: *A*, fig. 653 (p. 364); *B*, fig. 691 (p. 375); *D*, fig. 524 (p. 332). Copyright © 1972 by Lester A. Swan and Charles S. Papp; reprinted by permission of Harper and Row, Publishers Inc.]

scarab beetles (Scarabaeidae). The Stylopidae, or twisted-wing beetles, are especially beneficial in some countries because of their predation on the pests of planthoppers (Fulgoridae) and leafhoppers (Cicadellidae). Most North American species attack species of Hymenoptera.

Odonata

These ancient predators have great flight speed and strength and large appetites. Larvae and adults of Odonata are instrumental in reducing the numbers of larvae and adults of Diptera, particularly mosquitoes, in lakes and streams, as well as nearly any other small creature. When experimentally fed, a dragonfly larva may eat as many as six mosquito larvae per minute for a sustained length of time. Although it is difficult to assess the tremendous amount of predation this group accomplishes, they obviously have a major effect in limiting many other aquatic insect populations.

Dictyoptera

Another group of ancient predators are the Mantodea, or praying mantids. As skilled and aggressive predators throughout their life, they help reduce unwanted insect pests. They have an indiscriminate, large appetite; however, they may capture a few other beneficial insects as well.

Raphidioptera, Megaloptera, and Neuroptera

The larvae of these orders are almost entirely predaceous on other insects. These orders include the large aquatic dobsonflies and alderflies. The lacewings mainly prey on aphids, thrips, scale insects, mites, and young grasshoppers. Antlion larvae are aptly named because they capture mostly ants in their pits. Because the egg and larvae of Neuroptera are not as sensitive to pesticides as some pests, it is possible in orchards with controlled spraying to disperse eggs of the lacewings, such as *Chrysopa californica*, to control the grape mealybug, *Pseudococcus maritimus*, where its natural predators have been reduced by pesticide application. This is one form of *integrated management*, and this and similar approaches to insect control have been used successfully for several years in a number of orchards.

Hemiptera

Most of the Hemiptera are plant pests. However, among this diverse and numerous order are a number of families that are predators of insects, or include genera that are. The Phymatidae, or ambush bugs (Fig. 11-4*A*), are cryptic predators, hiding among plant flowers and leaves to capture passing insects. Also, the Reduviidae, or assassin bugs (Fig. 11-4*B*) the Nabidae, or damsel bugs, and the Anthocoridae, or pirate bugs, are almost all insect predators. Among the Pentatomidae, or stinkbugs, and the Miridae, or plant bugs, several genera are predaceous. In general these Hemiptera feed on other Hemiptera, caterpillars, and beetle larvae, but some seem to specialize on aphids, ants, bedbugs, or other groups.

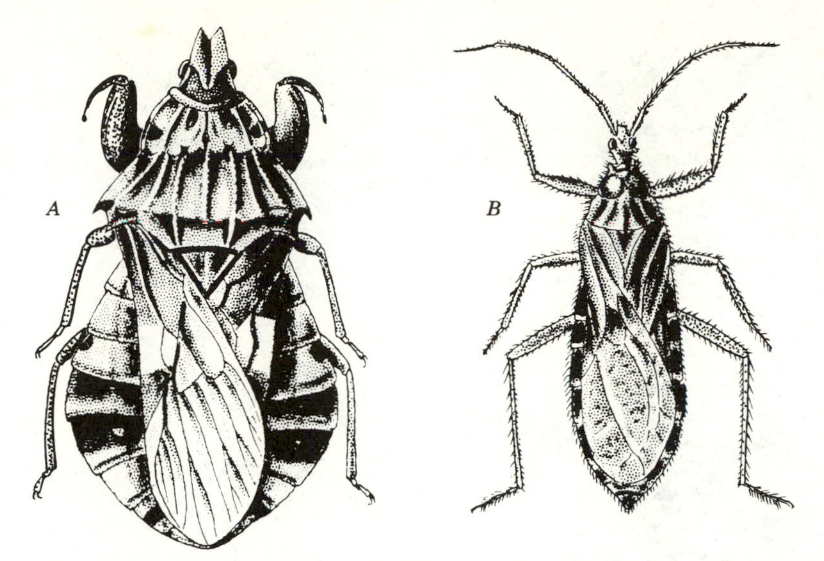

Fig. 11-4. Hemipteran predators. (*A*) Ambush bug *Phymata erosa*, a predator on insects that visit flowers; and (*B*) assassin bug, or kissing bug, *Reduvius personatus*. A predator of bedbugs and other insects, and will bite humans. [From *The common insects of North America* by Lester A. Swan and Charles A. Papp: *A*, fig. 103 (p. 120); *B*, fig. 105 (p. 122). Copyright © 1972 by Lester A. Swan and Charles S. Papp; reprinted by permission of Harper and Row, Publishers Inc.]

Diptera

Many of the flies are predaceous on other insects, particularly as larvae. Included as helpful predators are the larvae of Syrphidae, or flower flies (Fig. 11-5*A*), many of which feed on various homopteran plant pests. The syrphids have a 2- or 3-week larval stage, so their populations may build up rapidly during the growing season. Another family of predators, both as larvae and adults, are the Asilidae, or robber flies (Fig. 11-5*B*). The larvae are ground dwellers and feed on coleopteran larvae; the adults feed on other insects by capturing them in flight. Several other families have important predaceous members, such as the Bombyliidae, or bee flies (Fig. 11-5*C*), which are known to prey heavily on orthopteran eggs. Some Itonididae (or gnats) and some Culicidae (or mosquitoes) are predaceous as larvae, and local attempts have been made to establish populations of certain of these for pest control. Other predaceous flies include the Rhagionidae, or snipe flies, and the Empididae, or

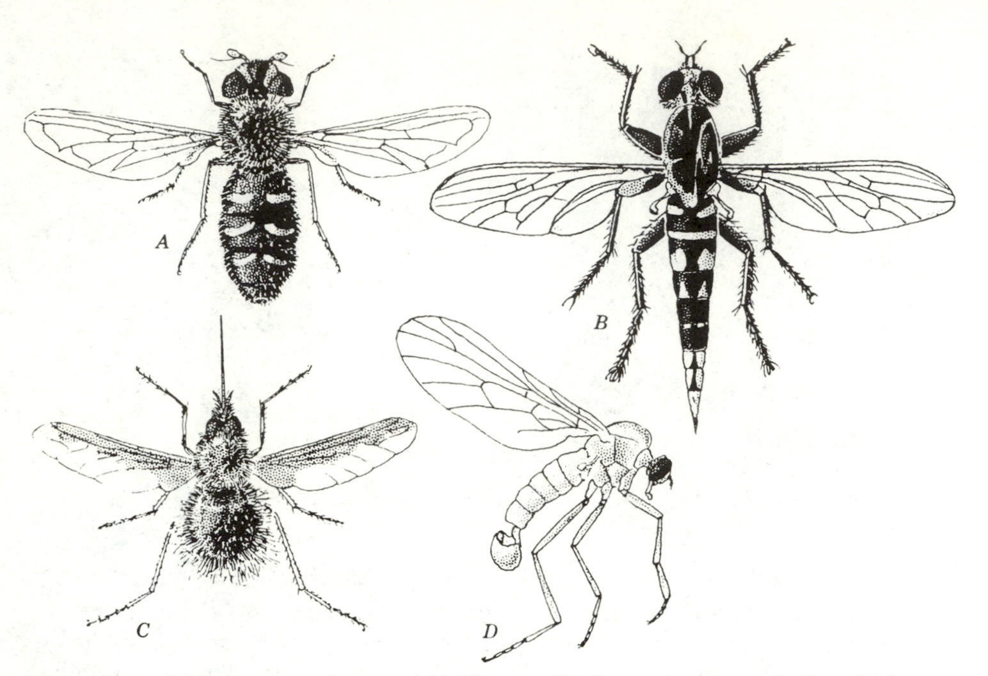

Fig. 11-5. Dipteran predators. (*A*) Flower fly *Scaeva pyrastri*, Syrphidae, a predator of aphids; (*B*) robber fly *Efferia interrupta*, Asilidae, whose larvae prey on grubs and adults on adult horseflies; (C) bee fly *Bombylius major*, a predator of larvae of several species of solitary bees; and (*D*) dance fly *Ramphomyia* sp., a predator of small insects. [From *The common insects of North America* by Lester A. Swan and Charles S. Papp: *A*, fig. 1325 (p. 617); *B*, fig. 1305B (p. 612); *C*, fig. 1312 (p. 614); *D*, fig. 1321 (p. 617). Copyright © 1972 by Lester A. Swan and Charles S. Papp; reprinted by permission of Harper and Row, Publishers Inc.]

dance flies (Fig. 11-5*D*), both of which are predaceous as larvae and adults.

Hymenoptera

Although many Hymenoptera are parasitic, a number of families include predatory members. The Vespidae capture mainly caterpillars to feed to their young. *Polistes exclamans* has proven important in keeping the tobacco hornworm, *Protoparce sexta*, at acceptable levels. Because *Polistes* (Fig. 8-124) has small nests and is subject itself to predation, small shelters can be built to help protect it. Hornets have similar food habits, although less docile. Sphecid wasps also provide their larvae with insects and each species appears to have some preferences as to the kind of insects it captures. One, *Bembix hinei*, uses horseflies principally and is common around livestock in the southern states. Another, *Sphecius speciosus* (Fig. 11-6), is a major cicada predator. Two forest pests—the spruce budworm, *Choristoneura fumiferana*, and the jackpine sawfly, *Neodiprion swainei*—are preyed on by several vespid wasps and these offer possible value for control.

Formicidae (ants) have a number of groups that prey on insect pests, including some other ants. Most of these predaceous ants appear to be surviving primitive lineages; they are widely distributed in the tropics and temperate zones and are important in limiting populations of the cotton boll weevil, termites, and many plant-feeding insects. Some species of ants are harmful in the sense they feed on the secretions of scale insects, aphids, and other plant-feeding insects; the ant actually protects these by discouraging parasites and predators.

Mecoptera

Scorpionflies (Panorpidae) and hanging flies (Bittacidae) include some forms that are predaceous on other insects or are at times. Hanging flies (Fig. 11-7) capture flies by hanging by their forelegs from plants and using the other two pairs for grasping prey.

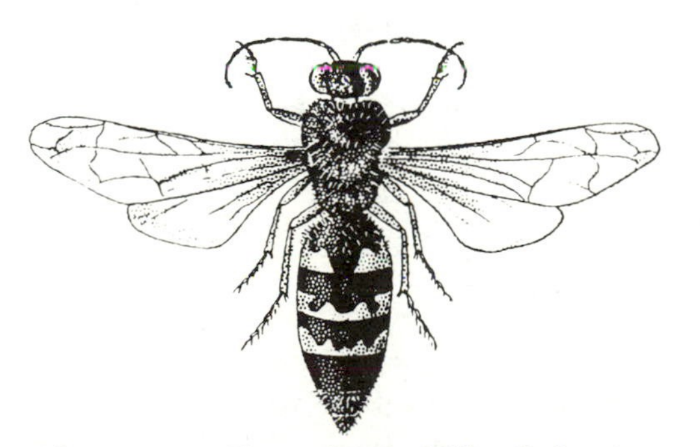

Fig. 11-6. Giant cicada killer *Sphecius speciosus*. [From *The common insects of North America* by Lester A. Swan and Charles S. Papp: fig. 1228 (p. 566). Copyright © 1972 by Lester A. Swan and Charles S. Papp; reprinted by permission of Harper and Row, Publishers Inc.]

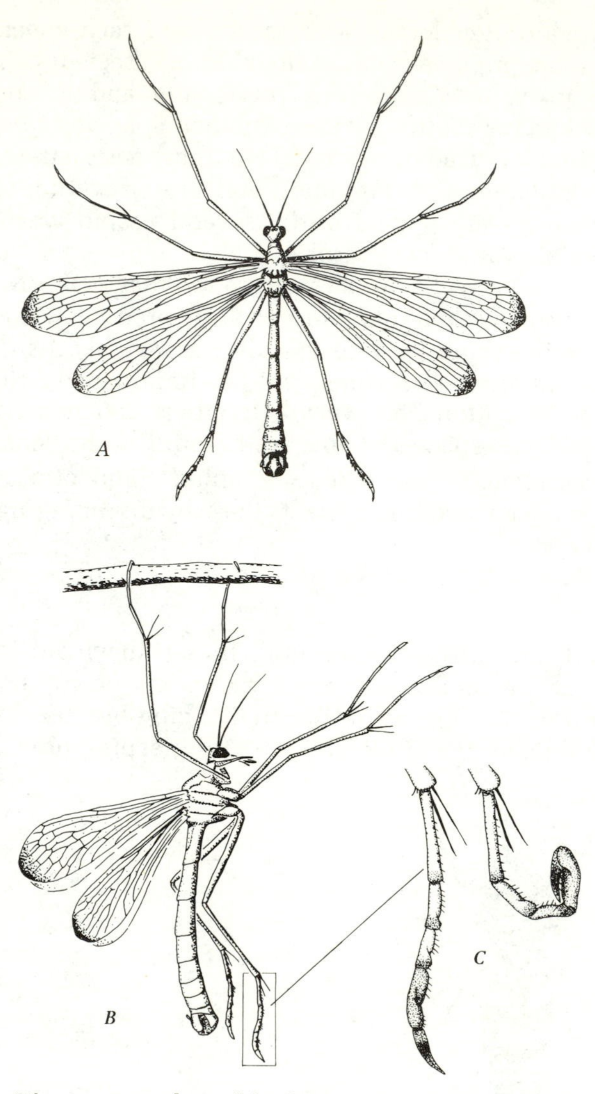

Fig. 11-7. The black-tipped hanging fly *Hylobittacus apicalis*, predator of flying insects: (*A*) dorsal view; (*B*) hanging position; and (C) enlargement of prehensile tarsi. (Redrawn from Thornhill, 1980).

INSECT PARASITES

The host–parasite relationships that are developed among insects are commonly very specific and precise. A parasite usually requires only one host during its preadult stages, and frequently it shares that host with one or more other parasites. Parasitoids allow the host to feed and to grow to a certain stage, at which point they kill the host and then enter a pupal stage. This relationship is based on intricate timing of the egg, larval, and adult stages of both host and parasitoid.

Hymenopteran Parasites

Two major lineages of Hymenoptera—the primitive parasites including the Ichneumonidae, Braconidae, and Gasteruptionidae, and the specialized parasites (Fig. 8-111)—are very important and economically significant in parasitizing other insects. In these, the female ovipositors are adapted to puncturing, boring, drilling, and injecting fluids and eggs, and serving as sense organs to locate and taste prey.

The small braconids are parasites on Coleoptera, Lepidoptera, Diptera, and Hemiptera (Homoptera) and a few on other groups. Some are external and others are internal parasites. The tomato hornworm, *Protoparce quinquemaculata*, is host for the braconid *Apanteles congregatus* (Fig. 11-8*B*), and the cabbageworm, *Pieris protodice*, is host to the braconid *Apanteles glomeratus*. Parasitism by these wasps generally renders the host caterpillar inactive or unable to complete its life cycle.

The braconids are important also in efforts to control a number of pests that did not have native predators or parasites subsequent to being introduced into North America. Among the imported pests are *Stilpnotia salicis* (the satin moth), *Nygmia phaeorrhoea* (the browntail moth, Fig. 8-185), and *Porthetria dispar* (the gypsy moth). These moths came from Europe, so a number of European parasites have been imported to try to reduce these pests' population levels. The imported braconids, *Apanteles solitarius, A. lacteicolor, A. melanoscelus* (Fig. 11-8*A*), and *Meteorus versicolor*, are parasites on one or more of these moths, and have aided in reducing their populations to tolerable limits prior to the widespread application of DDT during the late 1940's and early 1950's. They have been reestablished since abandonment of DDT in these control programs. Among apple pests, the codling moth, *Cydia pomonella*, is a serious pest. This moth has a braconid parasite, *Ascogaster quadridentata carpocapsae*, which has sometimes been effective in restricting the codling moth's populations.

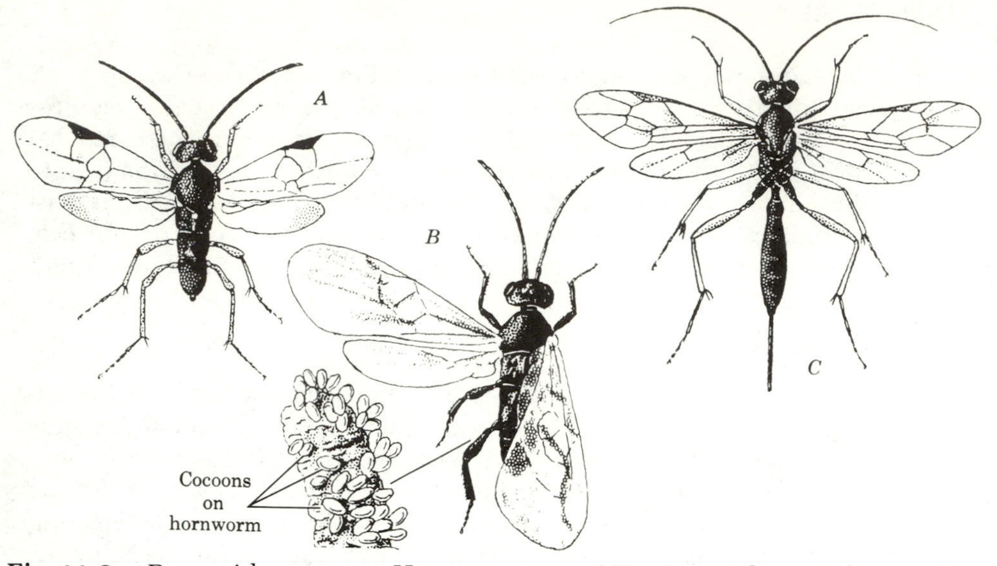

Fig. 11-8. Braconid wasps, Hymenoptera. (*A*) *Apanteles melanoscelus*; (*B*) *Apanteles congregatus*; and (*C*) ichneumon wasp *Horogenes punctorius*, a parasite of the European corn borer. [From *The common insects of North America* by Lester A. Swan and Charles S. Papp: *A*, fig. 1150 (p. 524); *B*, fig. 1151 (p. 524); *C*, fig. 1161 (p. 531). Copyright © 1972 by Lester A. Swan and Charles S. Papp; reprinted by permission of Harper and Row, Publishers Inc.]

In addition to the imported braconids that help control accidentally introduced pests, a great number of potential North American pests are also controlled by native braconids. The wheat stem sawfly, *Cephus cinctus*, which has extended its range from prairie grasses to the grainfields of the Canadian prairies, is a serious pest. Most of the species of wasps, which control this sawfly in natural grasslands, are unable to parasitize the pest population effectively in large wheatfields. Only *Bracon cephi* has become an effective parasite in large fields.

In the case of the introduced oriental fruit moth *Grapholitha molesta*, a number of North American parasites have used it as a host, but these relationships have not become selectively or firmly established. Therefore, no one parasite has become effective as a control, although the overall number of parasitic species that attacks this moth is important and substantially reduces its population. The North American braconid *Macrocentrus ancylivorus* frequently includes the oriental fruit moth as a host. Other hosts in vegetation near orchards act as natural reservoirs for the parasite. Several successful efforts established *M. ancylivorus* throughout much of the peach-growing region of the eastern states in the 1930's. After the oriental fruit moth was discovered in California in the mid-1940's, *M. ancylivorus* was mass cultivated there using an alternate host of the larvae of the potato tuber moth, *Gnorimoschema operculella*, which were a by-product of a lady beetle mass culture. The introduction of DDT and related pesticides in the late 1940's reduced the dependence on these mass-produced parasites, and their cultivation has been curtailed.

Ichneumonid wasps, likewise, have proven extremely beneficial parasites to a wide range of insect pests. This group of wasps are almost entirely parasites, mostly of Lepidoptera larvae and Hymenoptera. Ichneumonid wasps were used in attempts to control the European spruce sawfly, *Diprion hercyniae*, which reached high infestation levels in eastern Canada and northeastern New England in the 1930's and early 1940's. The ichneumonids *Aptesis basizonia*, *Exenterus abruptorius*, *E. amictorius*, *E. tricolor*, *E. claripennis*, and *E. confusus*, were among the introduced predators that became established on this sawfly. Hundreds of millions of these parasites were cultured or imported and released in an extensive control program prior to 1942. Subsequently a deadly virus became established in the sawfly population and since 1942 only small population levels of the spruce sawfly remain.

Dahlbominus fuscipennis is another introduced ichneumonid wasp that has been effectively used locally as a parasitic control against the larch sawfly in the northeastern states. The European pine-shoot moth, *Rhyacionia bouliana*, is preyed on by several wasps that have been imported, but with limited success.

The European corn borer first appeared in North America about 1917 and became the target for an extensive parasitic program. Of the more than 22 species introduced as European corn borer parasites, 6 have become established and 2 of these—*Horogenes punctorius* (Fig. 11-8C) and *Phaeogenes nigridens*—are ichneumonid wasps. These commonly parasitize 50 to 60% of the borer larvae.

Chalcid wasps are specialized parasites that belong to several families. Most are very small (1 to 2 mm) and are very effective as internal parasites of Coleoptera, Lepidoptera, Hemiptera (Homoptera), and some Diptera. As with several other hymenopteran parasites, a number of chalcids have been introduced to help control introduced insect pests because native predators and parasites did not prove effective. *Aphytis chrysomphali* is a chalcid that parasitizes the red scale *Aonidiella aurantii* and was probably introduced at the same time the red scale was introduced. *Aphytis lingnamensis* and *A. melinus*, imported chalcids, combine with *A. chrysomphali* to control red scale effectively in most coastal California citrus groves. *Comperiella bifasciata*, another introduced chalcid, parasitizes and feeds on the citrus yellow scale *Aonidiella citrina* and has controlled the scale in much of the pest's range. Other *Aphelinus* chalcid imports have helped reduce the black-scale pest *Saissetia oleae*, in southern coastal California, although additional control is needed elsewhere. The wooly apple aphid, *Eriosoma lanigerum*, a northeastern North American species, escaped its parasitic predator, *Aphelinus mali*, when the aphid was inadvertently spread to the West Coast. Later introductions of *A. mali* became established in Washington and Oregon, but not farther south.

Chalcid wasps (Fig. 11-9) include two families that have very small individuals and are adapted to parasitizing eggs. These are the

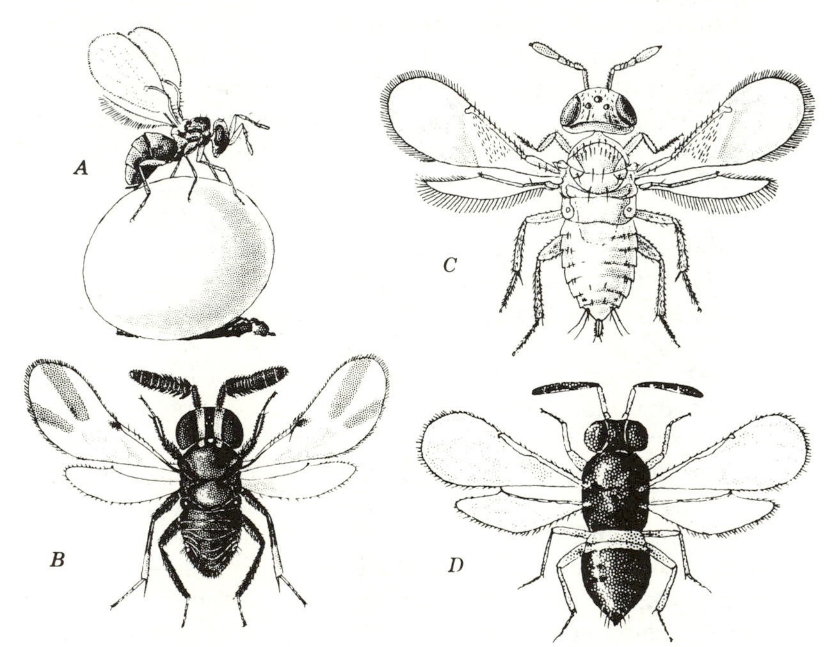

Fig. 11-9. Chalcid wasps. (*A*) *Trichogramma minutum*, an egg parasite of several hundred species of insects; (*B*) *Comperiella bifasciata*, a larval parasite on yellow and several red-scale insects; (*C*) golden chalcid *Aphytis chrysomphali*, a parasite on numerous scale insects; (*D*) *Aphelinus mali*, a parasite on many harmful aphids. [From *The common insects of North America* by Lester A. Swan and Charles S. Papp: *A*, fig. 1165 (p. 532); *B*, fig. 1172 (p. 536); *C*, fig. 1167 (p. 534); *D*, fig. 1169 (p. 534). Copyright © 1972 by Lester A. Swan and Charles S. Papp; reprinted by permission of Harper and Row, Publishers Inc.]

Mymaridae and Trichogrammatidae, including one species less than a quarter of a millimeter long. These are effective in controlling many insects. The alfalfa weevil *Hypera postica* has been brought under control in Utah with the introduction of *Mymar pratensis* (Mymaridae) from southern Europe. The Trichogrammatidae are parasites on a large number of insect eggs and have been shown to be important in controlling the earworm *Heliothis zea*, and the spruce budworm *Choristoneura fumiferana*. The families Encyrtidae and Eupelmidae also have important egg parasites of economic significance.

The paralyzing wasps are also important in the general maintenance of many insect populations. Some are fairly selective in choosing their prey, such as the Scoliidae (p. 425); however, others are not, such as the Vespidae.

Diptera

Although many families of Diptera are beneficial to humans, the Tachinidae are flies of sturdy build and rapid flight that parasitize many other insects, including a number of pests that we have already mentioned.

Tachinids generally parasitize the larval stages of their hosts, which usually are beetles and moths. Several species have been used in the tropics; the tachinid *Microceromasia sphenophori* was introduced and has controlled the sugar-cane borer *Rhabdoscelus obscura* in Hawaii. The North American tachinid *Winthemia quadripustulata* (Fig. 11-10*A*) is a parasite on a number of crop caterpillars, and *Lydella thompsoni* (Fig. 11-10*B*) was brought from Europe to eastern North America to help control the European corn borer. Several other tachinids are parasites on the introduced European brown-tail and gypsy moths. In addition to their role in these and other specific pest-control problems, tachinid flies have a broad spectrum of hosts and generally aid significantly in the "balance of nature."

WEED MANAGEMENT

One of the most beneficial groups of insects are those that feed on weed plants that are undesirable from our point of view. The species of greatest interest in control by insects are not the common farm or garden weeds that otherwise can be controlled, but those weeds that outcompete other vegetation in large areas that are rendered unfit for agricultural development. Classical examples of this latter type of weed are (1) the introduction of the American prickly pear cactus

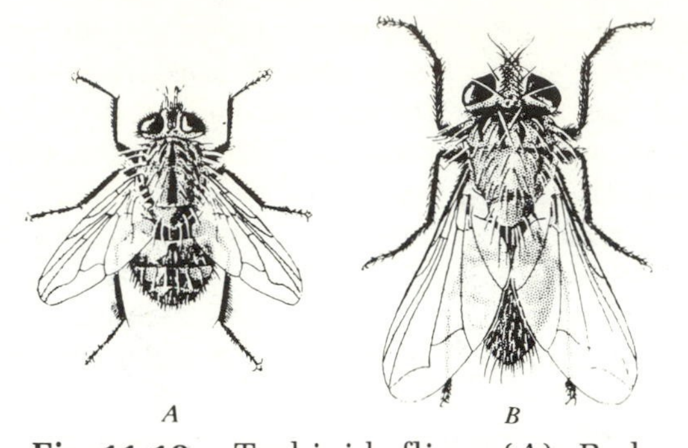

Fig. 11-10. Tachinid flies. (*A*) Red-tailed tachinid fly *Winthemia quadripustulata*, a parasite of armyworms, corn earworms, and many other caterpillars; (*B*) *Lydella thompsoni*, a parasite of the European corn borer. [From *The common insects of North America* by Lester A. Swan and Charles S. Papp: *A*, fig. 1395 (p. 648); *B*, fig. 1396 (p. 648). Copyright © 1972 by Lester A. Swan and Charles S. Papp; reprinted by permission of Harper and Row, Publishers Inc.]

(*Opuntia* sp.) into Australia, where it spread and ruined millions of acres of rangeland for cattle production; (2) the introduction of the European perennial herb, Saint-John's-wort *Hypericum perforatum* into the arid grazing lands of Australia and California, where it outcompeted and replaced the range grasses over thousands of square miles; and (3) the introduction of the tropical American plant pamakani *Eupatorium adenophorum* into Hawaii, where it thrived and formed impenetrable thickets, commonly 3 m high, that killed the grass and ruined large areas as rangeland. In each of these and many comparable examples, acreages of national importance were effectively removed from agricultural production.

When these events occurred, it was suggested that insects deleterious to the plants in question might be a determining factor in reducing the weed populations. Initially this idea was received with less than an enthusiastic response. But by 1925 the tremendous spread of prickly pear in Australia led to a search for cactus-feeding insects that might reduce the pest. The South American moth *Cactoblastis cactorum* proved to be hightly effective and when introduced into Australia reduced the dense cactus areas by about 99%, and the rangeland is again agriculturally productive. Later, the leaf-feeding beetle *Chrysolina gemellata* greatly reduced the Saint-Johns-wort in both Australia and North America, and a tephritid fly *Procecidochares utilis* reduced pamakani in Hawaii. About a dozen pest plant species have now been controlled by insects in various parts of the world. The successes have established weed control by insects as an important field of study.

Recent extension of this method of weed control has been the use of insects in controlling water plants that clog streams, ponds, and lakes, thereby ruining boating, fishing, and swimming. One of these is alligator weed *Alternanthera philoxeroides*, originally introduced from South America into Florida and now common throughout the Southeast and occurring westward to California. Several insects (Fig. 11-11) that feed on it in South America were introduced into the southern United States and one of them, a flea beetle of the genus *Agasicles*, seems to provide good control of alligator weed.

In using introduced insects for weed control, extreme caution must be used to avoid bringing in an insect that might transfer to another host that was a desirable plant species. For this reason, each

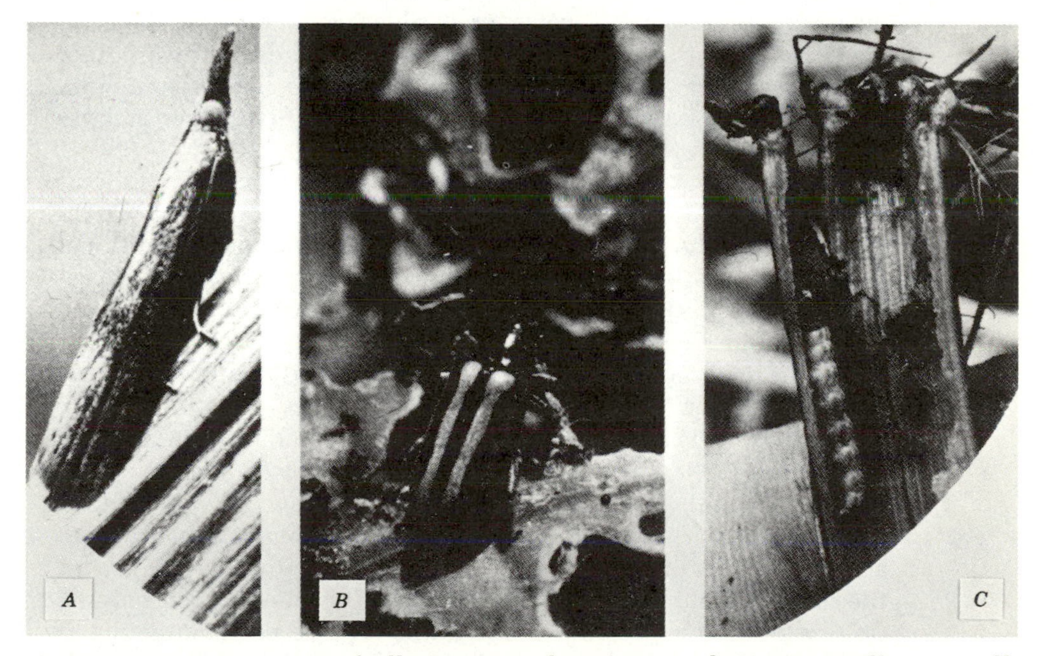

Fig. 11-11. Insect pests of alligator weed. (*A*) A moth *Vogtia malloi* is well camouflaged on a dead stem; (*B*) South American flea beetle *Agasicles hygrophilia* on its natural host, alligator weed; (*C*) larva of *V. malloi* in stem of alligator weed. (From *USDA Tech. Bull.* 1547)

introduction of an insect species into a new country is preceded by a great deal of study concerning its host plants and climatic adaptations prior to release.

INSECT POLLINATION

Were it not for the existence of insect-pollinated plants, human diet would be greatly different from what it is. All our fruits of the rose, strawberry, apple, and cherry family (the Rosaceae) are dependent on insect pollination, as are also the melons and squashes (the Cucurbitaceae), and the protein-rich peas, beans, and their allies (the Leguminosae). Although we eat primarily vegetative or prepollinated fruiting structures of the carrot family (the Umbelliferae) and the cabbage family (the Cruciferae), we depend on insect pollination for seeds from which to grow the edible plants. The seeds themselves also produce oil and condiments such as various peppers and celery seed. These are but a few of the insect-pollinated plant families that produce food, condiments, and beverages used by humans. In a most informative and lucid compendium, McGregor (1976) lists over 200 species of cultivated crop plants belonging to about 50 plant families, grown in the United States (including Hawaii), that are pollinated chiefly by insects. Flies, butterflies, beetles, and other plant-dwelling insects pollinate many of these plants, particularly those lacking showy flowers.

Insects pollinate another vast and varied assortment of agricultural plants: most of the flowers used as garden ornamentals. Included are all the plant families mentioned above and an additional long list. A smaller list includes certain plants grown for drug production (poppies) and others grown for insecticides (such as the pyrethrum-producing daisies).

Few of the species used in tree farming in this country are pollinated by insects. The few exceptions are species grown as nursery stock, such as the native dogwoods, cherries, elderberries, tulip poplar, and magnolias, and an assortment of introduced species.

The bulk of the pollination of plants with showy flowers is done by honeybees. If the grower does not have enough bees locally to achieve satisfactory pollination, he contracts with a beekeeper for the rental of enough hives of bees to accomplish the needed pollination. In many parts of the country this type of bee contracting is highly organized and is an important part of the agricultural economy.

Many wild bees are excellent crop pollinators if present in sufficient numbers. A great deal of study has been carried out seeking methods to increase the local populations of the efficient wild bee pollinators. To date, excellent results have been achieved with the alkali bees *Nomia cockerelli* in some of the west-central states, and with the leaf-cutter bee *Megachile pacifica*. In tropical regions, the stingless bees of the genera *Melipona* and *Trigona* are excellent pollinators and are readily "domesticated," but they cannot endure freezing weather and are unusable in temperate climates. Although only limited success has been achieved using other native bees, they are frequently sufficiently numerous to pollinate small holdings quite satisfactorily.

MAKERS OF USEFUL PRODUCTS

Throughout history, insects have produced several compounds that continue to be of great use to humans.

Honey and Beeswax. In North America, these two products are made by the honeybee, *Apis mellifera*. Human's earliest sweetener was honey, and this led to the discovery of beeswax, which was a product of many uses. Both are a commercial commodity today, but beeswax is overshadowed in comparison with modern synthetic compounds that produce similar but cheaper results.

Silk. For centuries, silk, spun by larvae of the silkworm moth, was the "ultrasheer" fabric, used primarily for hose and a wide array of garment materials. In the last few decades in North America and Europe, nylon and other tougher synthetic compounds have largely supplanted silk fabrics except for those who are allergic to some of the synthetic textile products. Worldwide, however, silk is still an important product; about 70 million pounds a year are produced annually.

Shellac. The Asiatic lac insect, *Laccifer lacca*, one of the scale insects, produces a secretion that encrusts the twigs of its host plant. This secretion is separated from the host twigs by being melted and is then processed into shellac. Many of its former roles are now filled by synthetic products but in the Orient a large amount of shellac is still used as a lacquer.

Other Products. In the past a few species of insects have been the source of dyes and drugs, but synthetic products have essentially replaced them. A southwestern scale insect *Dactylopius coccus* is the source of the crimson dye cochineal. The best-known drug of insect origin is cantharidin, made from the dried bodies of the European blister beetle *Lytta vesicatoria*, commonly called the Spanish fly.

OTHER USES

Insects as Human Food

Prehistorically insects were a common source of human food; however, now they are consumed only on a limited scale. Among the commonest are grasshoppers, which may be stewed or fried. In Mexico, fried insects, including the highly prized caterpillars that bore in agave plants, the *gusanos de maguey*, are considered delicacies and are canned and exported. In Japan, similarly treated caddisworms and stonefly nymphs and silkworm pupae are also exported.

The high fat and protein content of insects makes them an ideal food additive for chiefly carbohydrate diets. In Africa, such meals may be supplemented with insects, whose protein partially staves off the protein-deficiency disease of children, kwashiorkor.

Subjects for Scientific Study

Insects have made or are making significant contributions as subjects of study to most general fields of scientific inquiry. In the fifteenth and sixteenth centuries, they contributed to the early beginnings of comparative anatomy. At the beginning of this century, through studies of the fruit fly *Drosophila*, they contributed to the development of the first comprehensive studies in genetics. Later these studies led to some of the best of the early ideas of speciation and evolution. Today, studies of *Drosophila* genetics continue to contribute to a better understanding of evolutionary theory. Many insect groups have contributed to a better understanding of the principles of phylogeny, ecology, population dynamics, and biogeography.

The reasons for this are severalfold: (1) the tremendous numbers of known insect species, both fossil and living, and the remarkable number of living species that have changed little from ancient ancestral forms and that therefore are essentially "living fossils"; and (2) the short life cycle and high reproductive potential of many insects compared with those of ecologically comparable forms of life such as birds or mammals.

Because of their small size, insects were not prominent in early studies of physiology and biochemistry. However, since the advent of ultra-microtechniques in chemical analysis and techniques for radio-labeling elements, insects have become increasingly important as study subjects in both physiology and biochemistry. This is especially true in the fields of endocrinology and behavior, for in insects it is possible not only to perform remarkably intricate surgical operations, but also to correlate biochemical data with effects on the internal chemistry and external behavior of test animals. These same reasons have made insects ideal research organisms with regard to ecological problems of many types.

Recreational and Aesthetic Values

The nature-loving peoples of the world have recently become very much aware of the broad spectrum of recreational values to be found in natural areas. These values range from sportfishing and hunting, to hiking, and through a variety of activities to simply enjoyment of the wild as an admixture of observation, reflection, and enjoyment. The recent awareness of these values has been graphically demonstrated in the United States by the tremendous number of vacationers literally jamming the national parks, state parks, and other areas that have camping, picnicking, and trailer accommodations.

Although not immediately apparent to many who traverse the native areas, it is true that many objects of enjoyment owe their existence to insects. Almost all plants with showy flowers—from tiny chickweeds, to daisies, to hollies and dogwoods, and to tulip trees and magnolias—owe their existence to a great variety of insect pollinators. Entire families of birds, such as swifts, swallows, woodpeckers, warblers, wrens, flycatchers, and kingbirds, are completely dependent on insects for food, and many other birds feed chiefly on insects. Even the seed-loving cardinals will utilize insects when feeding their young. Snakes, lizards, frogs, and toads feed largely on insects, as do bats, shrews and some of the other mammals. When other food is scarce, foxes, skunks, raccoons, and bears turn to insects as a food source. Thus a surprising proportion of the aesthetic elements of the land biota is completely or partially dependent on the presence of insects. Insects themselves are aesthetic elements in their own right, as is evidenced by the many persons who study them as a hobby.

The importance of insects applies equally to the aesthetics of freshwater life. Most of our game fish rely for much of their food either on insects or on smaller fish that in turn feed on insects. Some of this feeding is spectacular. When a brood of the large stoneflies is emerging on the Gunnison River, the news spreads by radio and TV, and fishermen from Colorado and nearby states converge on the area for an orgy of trout fishing. Similar events occur in many parts of the world, for example, the mass emergence of snowflies (white caddisflies) on the trout streams of Tasmania. No greater tribute to the role of insects in sportfishing can be found than the hundreds of fish lures that mimic particular insects.

With regard to freshwater life, insects are assuming another important role concerning aesthetics. Many if not all bodies of fresh water are threatened with pollutants that may have a detrimental effect on the life in them. Aquatic insects are proving to be among the most sensitive organisms for detecting changes in freshwater ecosystems, and are therefore being used more and more extensively in monitoring programs designed to establish pollution indices.

DECOMPOSERS AND NUTRIENT RECYCLING

Insects are important in the annual turnover of key nutrients in an ecosystem as discussed in the previous chapter. Each year green plants produce, through photosynthesis and other chemical processes using water and carbon dioxide, a supply of carbohydrates, vitamins, and other chemicals needed for their growth. But plants cannot manufacture from these raw materials certain necessities for growth, especially those containing elements such as nitrogen, potassium, phosphorus, sulfur, and many others. These must be obtained from the environment. Because they occur in some soluble form in terrestrial situations, these necessities for growth are gradually leached out of the soil and may reach critically low levels. At this stage the chief source of these necessities may be from the various parts of the plants that drop to the ground and form the layer of litter and duff on the soil surface. This layer may be in the form of old leaves, dead branches, or whole herbs, shrubs, or trees that die and fall to the ground or into streams. This layer of dead plant remains also includes the bodies of dead animals and the feces of living ones.

This layer of detritus is broken down by a variety of microorganisms into its elemental parts, thus releasing elements necessary for new green plant growth. But a number of critical studies have demonstrated that insects and certain allied forms such as mites and millipeds that ingest detritus hasten this process of chemical release by disrupting the plant tissues mechanically and allowing faster penetration of microorganisms into the dead plant material. The dead leaf material or duff is attacked primarily by collembolans, proturans, diplurans, saprophytic mites, millipeds, and many saprophytic insect larvae. Dead logs and branches are attacked by termites, boring beetle larvae, and an assortment of other insects, millipeds, and assorted invertebrate relatives.

The nutrient recycling by insects often takes a more direct role. Many insects feed on the living leaves, twigs, or stems of plants, and their feces return to the detrital layer, shredded plant material that is ready to be worked over by the invertebrates or microorganisms that ingest such materials.

Again, insects play an especially important role in the degradation and ultimate consumption of plant material falling into headwater streams. Many of the larger stonefly and caddisfly larvae shred and ingest the plant material just as do terrestrial leaf eaters. Their feces contribute to the continuing flow of nutrients downstream. Many types of mayfly and caddisfly larvae have evolved mechanisms for filtering this smaller particulate material out of the water currents and using it, or microorganisms living on it, as food.

The total effect of these processes in both the terrestrial and

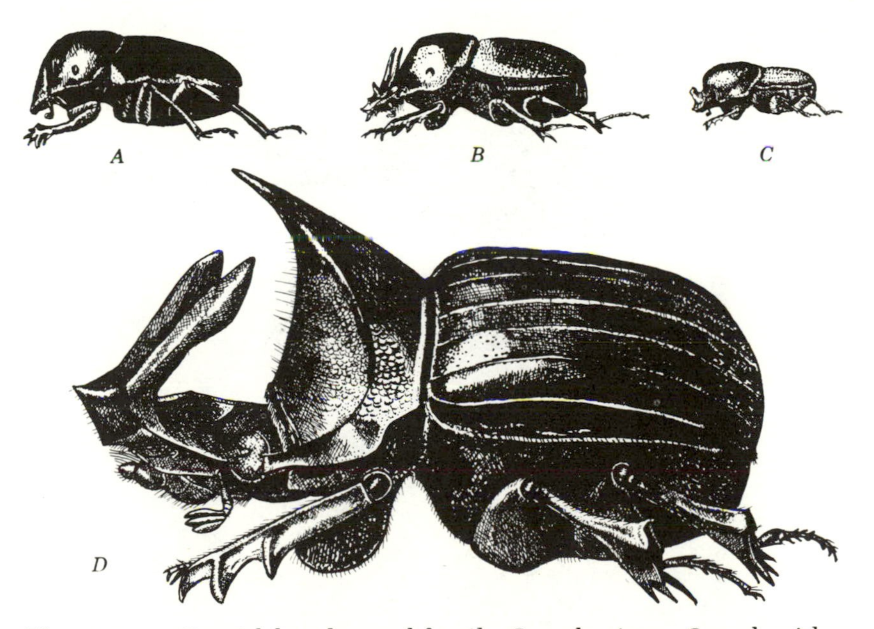

Fig. 11-12. Coprid beetles, subfamily Scarabaeinae, Scarabaeidae. (*A*) *Garreta nitens*, a ball-rolling species; (*B*) *Onthophagus gazella* and (*C*) *Euoniticellus intermedius* are two tunneling species; and (*D*) *Heliocopris gigas*, a large species able to resist attack by small vertebrates, such as toads. (Redrawn from Waterhouse, 1974).

aquatic environments is to utilize as much as possible of the recycled nutrient sources in its area of origin and before it is swept downstream and into the ocean.

The importance of insect decomposers was recognized many thousands of years ago by the early Egyptians, who believed the scarab dungroller *Scarabaeus sacer* was sacred. This species and many others of the Scarabaeidae (Fig. 11-12) consume great amounts of dung, up to 10 or 12 times their own weight per day, and also roll their eggs in balls of dung, so that the emerging larvae can feed on it. Usually the egg-bearing ball is rolled to a suitable location and buried by the scarab (Fig. 11-13). These beetles are instrumental in cleaning up livestock pastures and in removing dung so it is not available as breeding grounds for flies and other insect pests.

Where dung beetles are lacking or where the native beetles are ineffective, the dung of livestock accumulates and decomposes very slowly. In Australia, where much of the agriculture is pastoral, dung produced by the imported European cattle posed a serious problem because the native dung beetles are not able to decompose it rapidly.

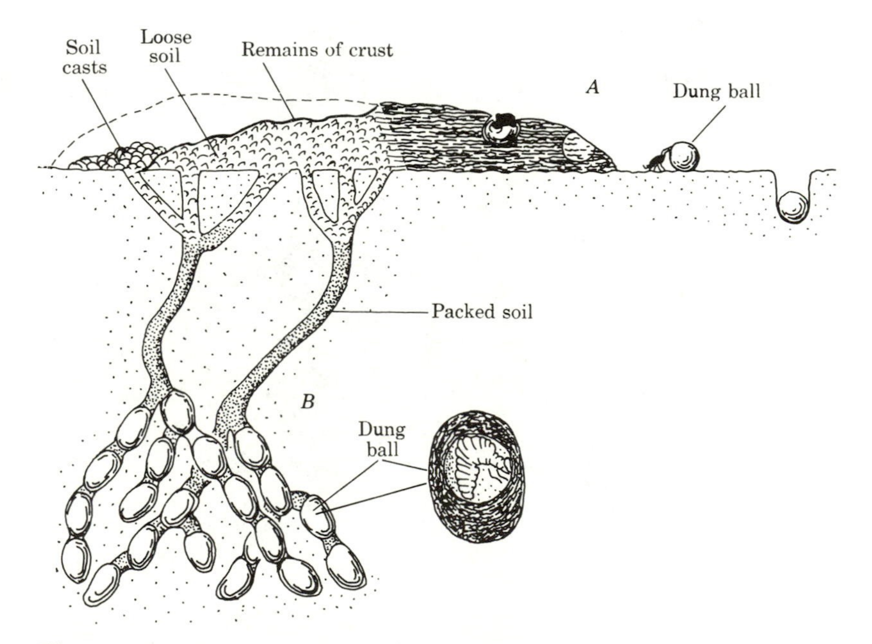

Fig. 11-13. Disintegration of dung by coprid beetles: *A*, ball-rolling beetles, such as *Garreta nitens*; *B*, tunneling beetles, such as *Onthophagus gazella*. (Redrawn from Waterhouse, 1974).

Each animal in a herd was taking a fifth of an acre out of pasture each year by its production of dung. The dung takes 2 to 3 years to decompose. Flies and other dung-inhabiting pests were also particularly troublesome. Beginning in 1967, the Australian government started to import various dung beetles in an attempt to reduce these problems. Within 6 years after this project started, one species, *Onthophagus gazella* had become established across northern Australia and as far south as Newcastle on the central eastern coastline. During the moist seasons this beetle is very effective. Several other species that decompose dung also have become established. The results have been a great increase in available pasture, better nitrogen recycling, and healthier forage. The Australians are now searching for an effective predator for dung-dwelling fly maggots, which have continued to be a problem.

In North America, the common dung beetles are *Canthon pilularis*, *Aphodius fimetarius*, and *Geotrupes splendidus*. These, along with others, have been effective in their role of decomposers of dung.

REFERENCES

Apple, J. L., and R. F. Smith (Eds.), 1976. *Integrated pest management*. New York: Plenum. 214 pp.

Arnaud, P. H., Jr., 1978. *A host-parasite catalog of North American Tachinidae (Diptera)*. USDA Misc. Publ. 1319. 860 pp.

Bohart, G. E., 1972. Management of wild bees for the pollination of crops. *Annu. Rev. Entomol.*, **17**:287–312.

Chapman, R. F., and E. H. Bernays (Eds.), 1978. *Insect and host plant*. Amsterdam: Nederlandse Entomologische Vereniging. pp. 201–766. (Reprinted from Entomologic Experimentalis et applicata, Vol. 24, No. 3.)

Daly, H. V., J. T. Doyen, and P. R. Ehrlich, 1978. *Introduction to insect biology and diversity*. New York: McGraw-Hill. 564 pp.

Debach, P., 1974. *Biological control by natural enemies*. London and New York: Cambridge Univ. Press 323 pp.

Free, J. B. 1970 *Insect pollination of crops*. London and New York: Academic Press. 544 pp.

Heinrich, B. 1979. *Bumblebee economics*. Cambridge, Mass.: Harvard Univ. Press. 246 pp.

Heinrich, B., and G. A. Bartholomew, 1979. The ecology of the African dung beetle. *Sci. Am.* **241**:146–156.

Huffaker, C. B., and P. S. Messenger (Eds.), 1976. *Theory and practice of biological control*. New York: Academic Press. 788 pp.

McGregory, S. E., 1976. *Insect pollination of cultivated crop plants*. USDA,Agric. Handbook 496. 411 pp.

Ridgeway, R.L., and S. B. Vinson (Eds.), 1977. *Biological control by augmentation of natural enemies*. New York: Plenum. 480 pp.

Swan, L. A., 1964. *Beneficial insects*. New York: Harper & Row. 429 pp.

Swan, L. A., and Papp, C. S., 1972. *The common insects of North America* New York: Harper & Row. 750 pp.

Waterhouse, D. F., 1974. The biological control of dung. *Sci. Am.*, **236**: 100–107.

12
Insect Pests and Their Control

In North America, many thousands of different species of insects and arachnoids are in some way of economic importance. These include a wide variety of plant-feeding groups, others attacking stored products, and others feeding on humans and their domestic animals. The great majority of these pests occur only sporadically and locally, such as the sawfly that attacks deciduous azaleas or some of the rare mosquitoes that become abundant only during periods of unusual weather. Only about a thousand species are the persistent pests that cause the great proportion of our insect and mite damage.

Damage occurs in myriad ways. Most conspicuous is defoliation of plants caused by chewing insects, by which large quantities of leaves or twigs are removed; equally severe may be root removal or mines in leaves, stems, or roots. Sucking insects remove plant juices or digest the contents of cells; some secrete toxins that poison part or all of the plant, or transmit various disease organisms from plant to plant. In some instances, damage is caused by oviposition (Fig. 12-1). Damage to humans and domestic animals is caused chiefly by "bites" inflicted by species with sucking mouthparts, such as flies, fleas, lice, or mites. A number of these transmit organisms causing some of the world's worst diseases, such as malaria and bubonic plague.

Reducing the numbers of these pest species to a point of eliminating or preventing material losses has been and still is one of the primary objectives of applied entomology. Achieving this result continues to be a more and more difficult task, as we shall see later in this chapter. The expense of controlling pests at a certain acceptable

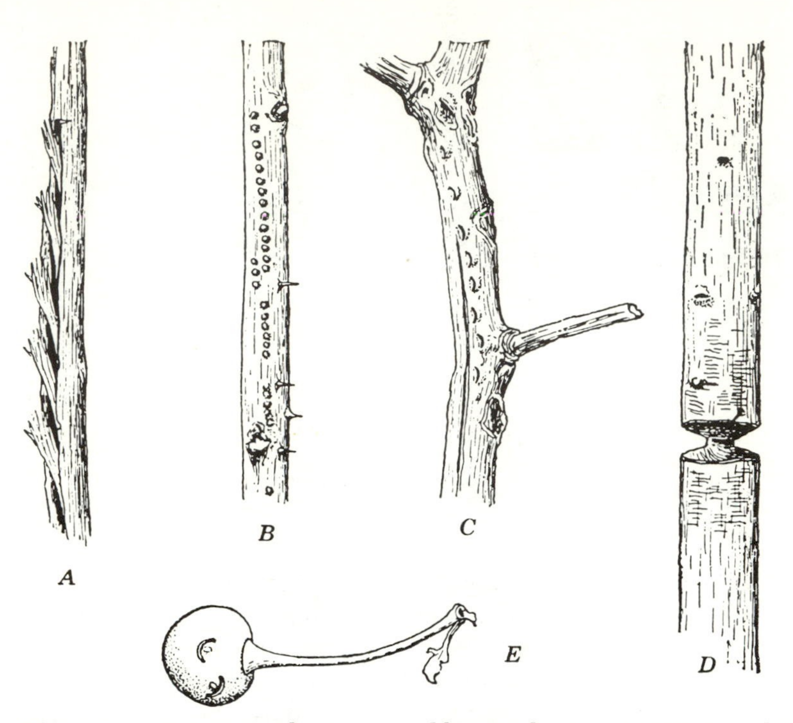

Fig. 12-1. Injury to plants caused by egg laying. (*A*) Twig split by periodical cicada; (*B*) holes in raspberry cane made by tree cricket; (*C*) slits in bark of apple twig made by treehopper; (*D*) twig of pecan nearly cut in two by twig girdler; (*E*) cherry showing two egg punctures of plum curculio. (After Metcalf and Flint *Destructive and useful insects*, by permission of McGraw-Hill Book Co.)

level is often examined as a problem of comparing the costs of different treatments against a cost/benefit ratio.

Presented here is a brief survey of the most destructive species or groups in relation to the crops or commodities they attack or other types of damage they cause.

INSECT PESTS OF PLANTS

By far the greatest number of important insect pests attack farm crops and animals. Injury is of many types, and insect species of widely different habits are involved.

Field-Crop Insects. All the major field crops suffer losses from insect attack. Serious enemies of cotton (Fig. 12-2) are the boll weevil and bollworm, which feed inside the boll, destroying the developing cotton fiber; the cotton leafworm, whose larvae eat the foliage; and the cotton aphid, which sucks juices from the leaves and stems. Corn may be almost completely destroyed by grasshoppers feeding on the foliage or by chinch bugs sucking the plant juices. The corn yield is annually reduced by several species of borers in cobs and stalks, including the European corn borer and the southwestern corn borer, and the corn rootworm. Wheat and other small grains are injured extensively by various species of cutworms, wireworms, aphids, and grasshoppers, depending on climatic conditions and region. Larvae of the Hessian fly attack the stems and crowns of grains; and this species is the most destructive single pest attacking wheat. Field enemies of tobacco, a high cash-value crop, are chiefly leaf feeders, such as hornworms and flea beetle adults; cutworms and tobacco

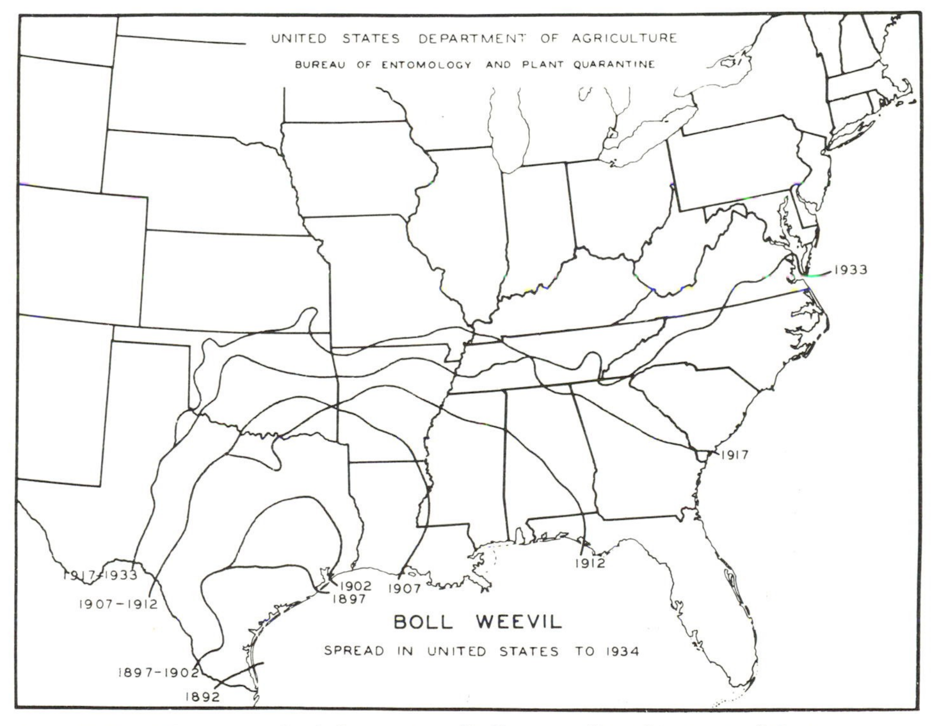

Fig. 12-2. The spread of the cotton boll weevil in the United States, up to 1933. The limit of distribution has been more or less stable in the Gulf coastal area since 1933. (From *USDA Handbook* **515**, 1978)

budworms also cause serious damage. The potato beetle feeds on potato foliage, and various leafhoppers suck the plant juices. Potato tubers are injured by soil-inhabiting larvae such as wireworms and flea beetle larvae.

Truck Crop and Garden Insects. Each plant species grown in truck farm or garden is subject to ravages from one or more insects specific in their food preference. These include such insects as cabbage loopers, cabbage butterflies, and cabbage aphids, that feed on cabbage, cauliflower, and other cruciferous crops; the carrot rust fly, specific on carrots; melonworms, asparagus beetles, and the Mexican bean beetle. In addition to pests specific to each crop, there are many general feeding insects that may attack almost any of these crops. Garden webworms, grasshoppers, blister beetle adults, cutworms, and fall armyworms are among the groups most likely to occur occasionally in destructive numbers.

Greenhouse Insects. In the greenhouse, warm humid conditions are maintained throughout the winter months. As a result we find in them many insect species that are normally tropical and subtropical in distribution. Most troublesome of these are several species of thrips, mealybugs, whiteflies, and scale insects. In addition, several species that are outdoor in habit during the summer invade greenhouses and continue active all winter, instead of becoming dormant. The melon aphid, green peach aphid, and the greenhouse leaf tier are examples of this type.

The different kinds of plants grown under glass are legion, and few are not attacked either by general feeding insects like larvae of leaf tiers or by specific pests such as the chrysanthemum midge, whose larvae make galls on leaves and stems (Fig. 12-3). Normally 40 to 50 species of potentially destructive insect species are found during the winter season in greenhouses. When one considers the variety of hosts involved and the fact that these may all occur in a range of glass of only a few thousand square feet, it poses a serious control problem and demands constant alertness on the part of the operator.

Fruit Insects. All classes of fruit—citrus, deciduous, and small—suffer heavily from insect damage, and in each group the major pests are different.

Citrus fruit trees are injured mostly by scale insects, mealybugs, whiteflies, thrips, and mites. The purple scale, California red scale, and black scale are especially important, damaging fruit and trees or producing honeydew on which grows a black sooty fungus that discolors the fruit. Many of the scales on citrus have a wide host range,

Fig. 12-3. Injury caused by the chrysanthemum midge *Diarthronomyia hypogaea*. About natural size. (From Metcalf and Flint, *Destructive and useful insects*, by permission of McGraw-Hill Book Co.)

but, being subtropical in distribution, they are pests of other fruits only in the citrus belt, in Florida, southern Texas, and southern California.

Deciduous fruits, including apple, pear, cherry, peach, plum, and their allies, have many destructive pests. Apple fruit is attacked chiefly by larvae of the codling moth. This insect is the most important species on the apple control calendar. Peaches, cherries, and other soft fruits are entered by larvae of the plum curculio, which also attacks apples. The branches and foliage of the entire group suffer from San Jose scale, oriental fruit moth, aphids, red spiders, a host of leaf-feeding insects, and many that bore in the tree or deform the fruit.

Small fruits are a group of wide taxonomic composition and have more specific insect pests. Grapes are attacked by the grape berry moth, many aphids and leafhoppers, and leaf-eating beetles that eat roots and leaves. Currants, raspberries, and strawberries are attacked by a variety of aphids, leaf-feeding larvae, and stem or crown borers.

Shade Tree and Forest Insects. Trees in general support thousands of insect species, which may defoliate, girdle, or bore into the tree, or suck its juices. Many of these species have only a slight effect on the host tree, but some damage the tree severely or may even kill it. As a result there is a high annual loss in both shade and forest trees.

Shade trees in the northeastern states are attacked especially by the gypsy and brown-tail moths. Elms suffer most from the elm leaf beetle and from Dutch elm disease, carried from tree to tree by the small European elm bark beetle. Direct injury by bark beetles and wood borers weakens and kills trees of many species.

Forest trees are visited periodically with insect outbreaks that kill huge tracts of timber. This is a loss of natural resources that in past years was given little attention, but, now that our forests are dwindling, increased efforts are being made to find means of checking losses. Larvae of forest tent caterpillars, gypsy and brown-tail moths, hemlock loopers, budworms, and tip moths are perennial defoliators of various deciduous and evergreen trees. Bark beetles are the greatest single enemy of conifers. Sawflies feeding on conifers occasionally appear in outbreak numbers and may cause tremendous damage. A severe sawfly outbreak was caused by the introduced European spruce sawfly, which in 1938 defoliated about 12,000 square miles of spruce timber, chiefly in the eastern provinces of Canada (Fig. 12-4).

Insects as Carriers of Plant Pathogens. Insects affect certain plants seriously by disseminating disease-producing organisms. Commonly the diseases are much more destructive than the injury by insect feeding. Under these circumstances control of the disease may resolve itself into a problem of very thorough control of the insect, because even a few insects are able to inflict, indirectly, staggering losses.

A number of plant pathogens, not actually carried by insects, gain entrance to the plant through insect feeding or oviposition punctures. Brown rot of peach commonly enters through feeding punctures of plum curculio adults, and bacterial rot of cotton through feeding and oviposition punctures of various insects.

Fig. 12-4. Distribution of the European spruce sawfly in North America during the epidemic of 1938. (Modified after Balch, 1952)

Insects assist in the dissemination of some plant diseases by transporting them on the body or in the digestive tract. Fire blight bacteria are carried on the legs and body of bees, beetles, and some other insects, as well as by birds and other animals. Spores of certain fungus diseases, such as apple canker, are eaten by insects and pass through the digestive tract in healthy conditions. In these cases insects are only one of many ways by which the disease is spread.

More important, insects are the principal or sole transmitters or vectors of a disease organism from one plant to another. The insects become infected with the pathogen, usually either bacterial or viral, by feeding on an infected plant; then, the pathogen is injected either mechanically or with the saliva into the tissue of the next plant on which the insects feed. Various species of leafhopper transmit aster yellow, and the beet leafhopper transmits curly top of sugar beets, both viral diseases. Many other viral diseases are transmitted by other insects. Bacterial diseases, such as cucurbit wilt disease, are carried by insects. In cucurbit wilt, the bacteria pass the winter in the digestive tract of the hibernating vectors, the cucumber beetles, which start the next year's infections. Many of these viral and bacterial diseases are exceedingly destructive.

Insect Pests of Stored Food. Grain and meat, flour, grain meals, and other highly nutritious foodstuffs are eaten by many insects. When in storage, these commodities suffer a heavy loss from insect ravages and necessitate constant preventive and remedial measures to keep them to a minimum.

In North America the chief pests of stored grains and grain products are the adults and larvae of the sawtooth grain beetle, the confused flour beetle, cadelle, mealworms, and the granary and rice weevils; and the larvae of the Indian meal moth and the Mediterranean flour moth. Peas and beans in storage are eaten by various pea weevils (Bruchidae). Meats and cheeses are eaten by larder beetles and maggots of the cheese skipper.

Large quantities of stored foods are attacked first by the group of insects just listed. After a certain amount of damage is done molds enter, followed rapidly by a host of other insect species, and soon the entire mass of food may be reduced to a small percentage of the original.

INSECT PESTS OF HUMANS AND ANIMALS

Human Habitation. Some insect species have become almost "domesticated," especially north of the frost line, in that they are found almost entirely in human habitations. In the case of ectoparasitic species, the relationship antedates civilization and is due to the parasites staying with the warm-blooded host. With other species, however, the relationship is more recent and is due to the relatively high temperatures at which houses and buildings are maintained even through severe winters. Thus some species, originally semitropical, are now found much farther north and are able to maintain themselves in human habitations.

Ectoparasites and pests of stored foods are of prime importance in human habitations. In addition, larvae of clothes moths and carpet beetles eat anything containing animal fibers, such as woolen garments, upholstery, and carpets. Silverfish and cockroaches are general feeders that eat starchy foods such as bookbindings and are an unsightly nuisance. Cockroaches drop excrement promiscuously, and spot and taint food and quarters; when very abundant, they will give a house, store, or restaurant a disagreeable and penetrating odor. Ants frequently invade buildings and may become a serious nuisance in the kitchen and food-storage rooms.

Termites are the most destructive pests of buildings. They eat the wood in foundations, flooring, and walls, necessitating extensive repairs. Other insects live in wood in dwellings, such as Lyctidae beetles, and carpenter ants may eat out extensive galleries in wood of buildings to use for nests.

Insects Parasitic on Humans and Domestic Animals. Both humans and domestic animals suffer annoyance and exposure to disease from the activities of insects. Certain of these insects, such as the Anoplura, confine their attacks to one or two closely related species of animals. Others, such as mosquitoes, are general feeders on a wide variety of warm-blooded vertebrates.

Domestic fowl are attacked chiefly by Mallophaga (chewing lice) and mites, several of which live on hens, ducks, turkeys, and geese. On young fowl, infestations of lice often cause death; on older birds the lice cause lack of condition and lower egg production. Mites sometimes become very injurious by reducing the general health of the flock. Blackflies are vectors of at least one duck disease similar in many respects to malaria. Hens suffer also from attacks of specific fleas, of which the southern sticktight flea is the most persistent.

Domestic animals and humans have a variety of specific parasites, including Anoplura (sucking lice), fleas, bedbugs, a few Mallophaga, and several kinds of mites. These latter include such annoying forms as itch mites, chiggers, and ticks. Sheep have in addition "sheep ticks"; these are odd wingless flies of the family Hippoboscidae. Attacks by ectoparasites result in irritation and loss of condition, but seldom in death unless disease transmission is involved.

Several vertebrates are attacked internally by larvae of botflies and warble flies. In the horse the larvae attach to various regions of the digestive tract and cause severe loss of weight and condition. Certain warble flies develop in the sinuses of sheep, and other species along the back of cattle where they form a pocket just beneath the hide. Larvae of the screwworm fly enter wounds, feed on the flesh beneath the skin, and annually cause large losses to all kinds of livestock.

In addition to these and other specific pests, all warm-blooded vertebrates are attacked by a great number of bloodsucking flies: mosquitoes, horseflies, blackflies, *Symphoromyia* flies, stable flies, and horn flies. Some of the fleas, ticks, mites, and bedbugs are also general feeders. The annoyance these cause is often severe. Blackflies especially may be destructive and occasionally cause the death of large numbers of horses and mules in local areas. Mosquitoes, blackflies, and horseflies are at times abundant enough to cause an exodus of tourists from an area, to reduce land values near suburban centers, or to retard settling of large tracts, as in the extreme northern part of Canada and in Alaska. The effect of these attacks on livestock in general may result in a loss of condition equal to or greater than that caused by specific parasites.

Stings and Other Irritants. There are some protective devices of insects that cause injury, allergic reactions, or extreme irritation, such as bee and wasp stings, ant bites, and nettling or poison hairs of certain caterpillars. Although very unpleasant and painful, these are only a negligible part of insect injury as a whole. Some persons have a great fear of insects that may cause them anguish, and others are strongly allergic to insect stings.

Insects as Carriers of Animal Pathogens. Insects are vectors of some of the most important diseases of humans and other vertebrates. As with plant diseases, in some cases insect transmission is only one of several ways by which the pathogen is spread, and in other cases the insect vector is the only known agent by which the pathogen is disseminated from one host individual to another (Fig. 12-5).

In the first category are typhoid fever, summer diarrhea, and some kinds of dysentery, all caused by species of the bacterial genus *Bacillus*. Houseflies get the pathogen on their feet or mouthparts through contact with sewage, saliva, or other infected material and then contaminate food or other items on which they alight later. These pathogens are transmitted in a variety of ways, but under some particular conditions flies may be the principal effective method of dispersal.

Bubonic plague (the black death), caused by *Bacillus pestis*, is another contagious disease in this first category. Rats and small

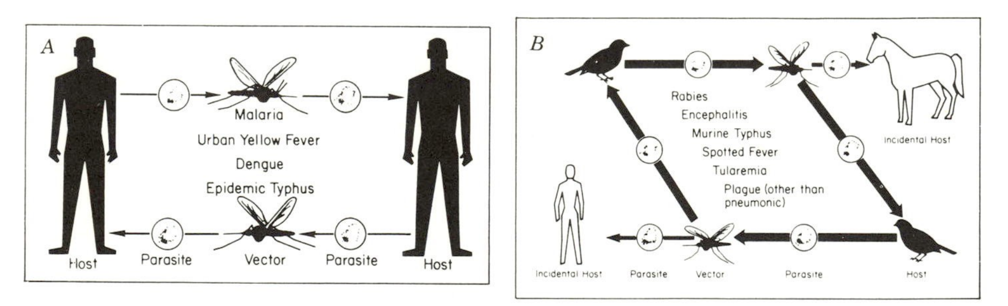

Fig. 12-5. (*A*) Disease cycle with three primary living factors (host, parasite, and vector), in which humans are the principal or only host. (*B*) Disease cycle with three primary hosts in which humans are an incidental host. (From U.S. Dept. of Health, Education, & Welfare, Public Health Service)

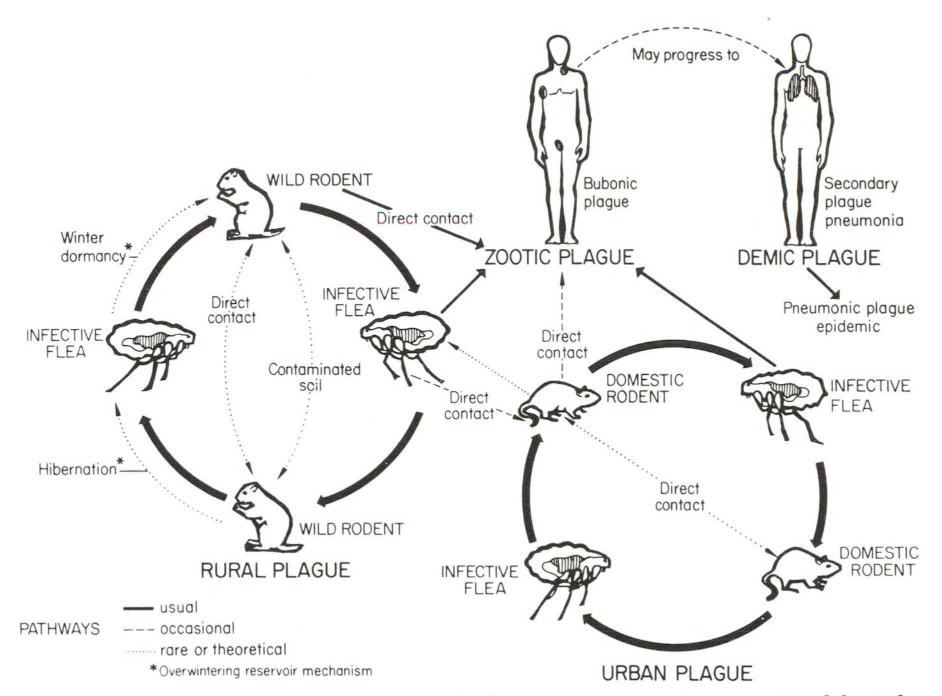

Fig. 12-6. Epidemiological cycle of plague. (From U.S. Dept. Health, Education, & Welfare, Public Health Service)

mammals serve as the reservoir of the disease, and rat fleas carry the bacteria from rat to rat or from rat to human (Fig. 12-6).

Insects are the sole vectors of several important human diseases. Malaria is caused by species of the protozoan genus *Plasmodium*, transmitted from one person to another by some species of mosquitoes belonging to the genus *Anopheles*; yellow fever and dengue (breakbone fever) are caused by viruses carried by several species of mosquitoes of which *Aedes aegypti* is the chief vector in North America; African sleeping sickness is caused by protozoans of the genus *Trypanosoma*, carried by flies of the genus *Glossina*; elephantiasis (filariasis) is caused by nematode worms of the genus *Filaria* transmitted by several species of mosquitoes. In all these instances the mosquito or fly, when feeding on a person carrying the pathogen, draws up into its buccal cavity or digestive system some of the pathogen; some of the pathogen is discharged during feeding at a later date into the tissues of another person. In this manner healthy persons are inoculated with the disease.

Typhus is caused by an almost ultramicroscopic organism called

Rickettsia, which is carried by body lice or cooties. These take up the pathogen when feeding and then later expel it in the feces. Scratching on the part of the bitten person works the pathogen into the skin and effects inoculation.

Ticks and mites are the only known vectors of several important diseases, of which three are of special interest. Texas fever, a lethal disease of cattle, is caused by a species of Sporozoa, *Babesia bigemina*. The pathogen is transmitted by the cattle tick *Margaropus annulatus*. Rocky Mountain spotted fever, a highly fatal human disease of increasing incidence, caused by a *Rickettsia* organism, is maintained in some of the small wild rodents, and a few species of ticks of the genus *Dermacentor* effect the transfer of the pathogen by feeding on infected rodents during nymphal development, and afterwards, when adult, biting humans. A third disease is scrub typhus, an oriental disease caused by another *Rickettsia* organism. This is transmitted from wild rodents to humans by chiggers (early instars of the mite family Trombidiidae) and was a serious hazard to humans in both the Burma and Pacific theaters during World War II.

Under most circumstances the practical control of these diseases is obtained by control of the vectors. This has been particularly effective in the case of Texas fever; control of the cattle tick has virtually eliminated the disease from the United States. Extremely satisfactory results have been obtained also in reducing outbreaks of typhus by controlling body lice. Mosquitoes, flies, and *Dermacentor* ticks are more difficult to control, and the species involved have such a wide dispersal range that measures aimed at control of these vectors have not always produced such remarkable results as those obtained with Texas fever and typhus.

INSECT PEST CONTROL

When a pest species reaches a certain density, it will cause damage or there are other consequences that make it desirable to reduce its numbers. In the case of damage caused solely by the activities of the pest species (for example, caterpillars feeding on cabbage, cockroaches despoiling stored food), the pest density may be fairly high before it is economically profitable to attempt control methods. In the case of damage from pathogens transmitted by an insect vector from plant to plant or animal to animal, only a single pest individual may cause extensive damage; in these situations the density must be kept as close to zero as possible.

Most cases of insect control involve arbitrary decisions as to how much damage is tolerable. Current standards of blemish-free fruits and noncontaminated vegetables impose almost pest-free standards of production, a level very unnatural and costly to attain. Were we to relax such high "purity" standards, many insects that compete with us for a mutually desired food would not be categorized as pests, and consequently much applied control could be reduced or eliminated.

In certain instances inimical factors of the environment may drastically reduce populations of pest species. A famous example is the outbreak of the Mormon cricket that threatened to destroy the crops in the vicinity of Salt Lake City, Utah, in the pioneer days of that settlement. A large flight of sea gulls checked the outbreak effectively. In 1935 the chinch bug was reduced to the status of a rarity in many cornbelt states by adverse winter conditions plus a fungal disease. Less dramatic examples include phytophagous insects such as the red-banded leafroller on apple that was formerly kept to noneconomic densities by native predators and parasites.

Biological Control

Present types of biological control fall into three categories: use of insect parasites, parasitoids or predators (see also Chapter 11); use of pathogenic organisms; and genetic control. Biological control commonly is used in a more limited sense for pest control by parasites, parasitoids, predators, and pathogens.

Insect Parasites, Parasitoids, and Predators. The possibility of propagating and distributing natural enemies for the control of destructive insects has kindled the imagination of entomologists for many decades. It has been found, however, that with destructive insects endemic to the United States we can do little to improve on existing natural control. It is difficult and costly to improve on existing natural control unless there is a high probability that great destruction of crops will occur. This situation points up that biological control methods are primarily preventive and not corrective.

With introduced pests the situation is entirely different, as seen in the previous chapter. The particular species may have an abundance of parasites or predators holding its numbers in check in its native land. When it is accidentally introduced into another country, usually only the pest without its parasites is transported. Freed from enemies, the pest in the new land is able to flourish at an unimpeded rate.

The ideal control for such an introduced pest would be to establish efficient enemies of it so that they would reduce the numbers of the pest to insignificant proportions. This might result in a permanent control that would obviate the necessity for an annual program of more expensive measures.

This ideal has been achieved only rarely. The most outstanding example has been the control of introduced cottony-cushion scale by the importation and establishment of the Australian vedalia ladybird beetle. So effective are the beetle and its larvae in controlling the scale in California that only occasionally and locally does the scale become important as a pest. Many parasites, especially of introduced pests such as the Japanese beetle, gypsy moth, and European corn borer, are imported by the U. S. Department of Agriculture and released in the United States. Numerous imported parasites fail to maintain themselves in the United States under natural conditions, because either they do not adjust to climate or suitable hosts are lacking at the right time. Some species have become successfully established and aid in controlling the pest species. It is hoped that eventually sufficient introduced parasite populations will be built up so that the populations of many pest species will drop well below their present destructive level. In some areas this result has already been achieved for the satin moth by introduced hymenopterous parasites, especially in Washington State and British Columbia. Propogation and dispersal of bacterial diseases of Japanese beetles have also given promise of being effective. Considerable success has been achieved in islands, such as Hawaii, by using a variety of parasites against insects attacking many crops.

Sufficient work has been done in biological control to show that a number of factors influence its success or failure. A few of these factors are the ecological requirements of the parasites, their effect on each other (see under Competition, p. 599), their host specificity, their rate of increase, and the character of their dispersal. To be effective, well-trained personnel and a great amount of specialized equipment are necessary, together with an organization for gathering parasite material in foreign countries and returning it to the United States alive and healthy.

Because of these factors, the work on biological control in the United States is done chiefly by the federal government; a notable exception is much intensive work done by the state of California. The final distribution and liberation of parasites are often performed cooperatively by scientists of the federal government and interested state agencies. The Canadian Government also is extremely active in biological control efforts.

Pathogenic Organisms. Spectacular natural outbreaks of fungal or bacterial diseases of chinch bugs and various other insects led entomologists to hope for insect pest control through dissemination of pathogenic organisms. Early efforts failed, because the years that did not naturally favor the spread of the known diseases also were ecologically unsuited for their artificial propagation. Since about 1930, several groups of organisms have been found that are suitable for artificial control, including viruses, bacteria, protozoa, fungi, and rickettsia. The milky disease of the Japanese beetle has proven useful in some parts of the American range of the beetle. Several kinds of polyhedrosis viruses are proving excellent controls for certain sawflies destructive to conifers. In California both the polyhedrosis virus and the Thuringian bacterium have proved effective as a field control for the alfalfa caterpillar. These successes have led to the discovery of more than 1000 other similar insect diseases.

Genetic Control. During the 1950's an entirely new kind of biological control was undertaken by entomologists in the U. S. Department of Agriculture. The target species was the destructive screwworm fly, which attacks warm-blooded mammals, including domestic animals. The method consisted of two steps: (1) rendering active and sexually aggressive males of the screwworm fly sterile through exposure of the fly pupae to gamma radiation; and (2) rearing and releasing such sterile males in great numbers into the natural populations. Females mating with these sterile males received abnormal spermatozoa at copulation and their eggs were inviable. In this type of control program, release of sterile males continues until the species is annihilated locally. By this method the screwworm was eliminated from the 440 km^2 (170-square mile) island of Curacao in 1955, and from Florida and the southeastern states in 1958. Since then the screwworm fly has been essentially eliminated as an economic pest of cattle. An indirect result of this is the marked increase in deer populations of the southwestern United States, a bonus feature of this pest-control program.

Another approach to genetic control involves efforts to manipulate the sex-determining apparatus of a species so that few females are produced. Genetic strains producing such results have been isolated in mosquitoes and described in certain moths. How these genetic strains can be used effectively and maintained in field-control operations present considerable complexity.

Cultural and Management Control. Some insect pests of agricultural or forest crops may be kept below the damage level by various cultural or management practices. An important general approach is keeping crops healthy by proper fertilizing, drainage, irrigation, and cultivation, and by planting crops that are well adjusted physiologically to the climate and soil.

Against certain pests specific cultural methods are of value, such as clean cultivation, crop rotation, certain times of harvesting or planting, and the use of insect-resistant or tolerant varieties.

Clean cultivation. This practice eliminates weeds that may serve as host to insects that attack the crop. The buffalo treehopper breeds on many herbaceous weeds; the adult hoppers fly into adjacent fruit trees, cut slits in the twigs, and in them lay their eggs. Clean cultivation of an orchard prevents this injury by eliminating the primary host. Weeds and soil debris also serve as hibernating or pupating quarters for a wide variety of harmful insects, and clean cultivation tends to discourage a buildup of population in that area.

Crop rotation. Rotation of crops has been found especially effective against some insects whose larvae feed on roots. Some species of *Diabrotica* rootworms can be controlled by crop rotation. These beetle larvae feed primarily on corn roots. If corn is grown continuously on the same ground for over 3 years in localities favoring these insects, they build up large populations and cause severe damage to corn. If, however, corn is eliminated and wheat or legumes substituted for a year, the rootworms starve. For this reason a rotation of corn with wheat or other crops arranged so that corn follows corn for no more that 2 or 3 years eliminates rootworm damage almost completely.

Choice of time of planting crops. This method is useful as a control measure for certain insects. The Hessian fly, a serious wheat pest whose larvae feed in the leaf sheath of wheat, can be partially controlled by regulating the time of planting winter wheat. The entire fall generation of adult midges normally emerges within a short period, following late summer rains. The adults live only 3 or 4 days, laying their eggs in grooves of wheat leaves. If winter wheat is planted after this generation is past, the plants will have no eggs laid on them and consequently will be entirely free from attack until spring. To take advantage of these conditions entomologists in wheat-growing areas annually establish dates for sowing winter wheat that will (1) allow the plants enough good weather to attain satisfactory growth before winter, and yet (2) be late enough to avoid all but a light infestation of Hessian flies (Fig. 12-7). The spring generation of Hessian flies attacks only late shoots (tillers) of winter wheat and does little damage. Early or late planting of corn and other crops is sometimes of assistance in reducing infestation and damage by such pests as rootworms and European corn borers.

Forest Management. Appropriate practices tend to keep down the abundance of insect pests that change the ecological structure of the forest community. For example, some of the most destructive species of bark beetles build up outbreak populations in over-age stands of pine. By cutting the tress for timber before they reach old age, this

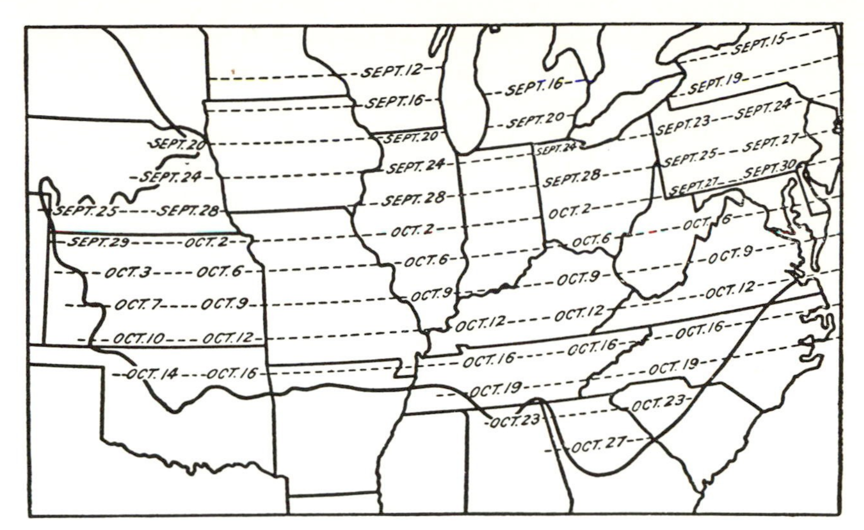

Fig. 12-7. Sample of a chart showing safe dates for sowing wheat in several north-central and eastern states to escape injury by the Hessian fly. (From U.S.D.A., E.R.B.)

beetle population increase is prevented, and younger trees in the stand are given a better chance for development. By selective cutting and logging, diversity of the forest can be achieved in regard to both species of trees and age groups represented by these trees. In general, the greater the diversity, the fewer the insect outbreaks that will occur.

Tree selection is used as a remedy for the gypsy moth. For its successful early development, the gypsy moth needs a favored host, such as willow, birch, or oak, but its later instars move to pine and other less-favored species, which they may also defoliate. To decrease infestation by this pest, the favored tree host is kept to a minimum and less-favored species of trees are replanted.

Quarantine. The most obvious way to avoid damage by an insect is to prevent its becoming established in a country if it is not already there. There are hundreds of insects in other parts of the world, especially in temperate areas of Europe and Asia, which we believe might become pests of great economic consequence if established in North America. To prevent their entrance, the United States federal government maintains an inspection of imports into the country, especially living plants or animals or packing material that is likely to harbor pests and serve as a carrier for them. Most or all of this material is fumigated before being allowed into the country. In addition, states may have restrictive regulations regarding the movement of critical materials within the state or into the state. The Canadian Government maintains a similar service.

It is admittedly impossible to prevent indefinitely the entrance of all potential new pests into the country, but quarantine records show that hundreds and sometimes thousands of new importations are prevented each year. It is impossible also to estimate how much we gain by this. Experience with such destructive importations as the cotton boll weevil (Fig. 12-2), the European corn borer, and the Japanese beetle, however, emphasizes that we cannot afford to take the chance of allowing free entry to every insect species.

PLANT RESISTANCE

In some species of crop plants, one variety may be highly susceptible to a particular insect pest's attack, whereas another variety may be highly resistant. This resistance in a plant is mediated by several factors, some of which result from the *direct* interaction of the plant and insect pest species and others that are *indirect* or *ecological* and relate to such variables as time of year, severe climate, and natural enemies.

Natural Resistance. Most resistance is generally controlled by the genetic composition of a plant. The different varieties of a plant species are an expression of the genetic variation within the species. In some species of insect pests, local populations, called *races* or *biotypes*, also vary genetically from region to region. For example, in two regions such as Canada and Kansas, biotypes of particular strains of pea aphid have quite different levels of attack on peas.

Different levels of resistance in crop plants and their effectiveness are seen in the pea aphid (*Acythosiphon pisum*), which feeds on alfalfa (*Medicago sativa*). A young pea aphid on a common crop variety of alfalfa may produce about 60 offspring in 10 days, whereas the same aphid on a resistant variety of alfalfa may produce no aphid offspring or one offspring in the same period of time. Likewise, on a susceptible commercial strain of sorghum (*Sorghum vulgare*), the chinch bug (*Blissus leucopterus*) oviposits about 100 eggs per generation. In contrast, on a resistant strain of sorghum, less than one egg per generation will be oviposited by the chinch bug.

Resistance of some soybean varieties to leafhoppers is apparently the result of hairy foliage of the resistant types, in contrast to the almost hairless foliage of susceptible strains. In another example, certain structural features of a plant may negate in part the effects of feeding injury by an insect pest species. In some varieties of corn, shorter, broader stalks increase the strength of the plant and reduce the wind breakage of the plant and also crop losses from insect borers.

In an example of more complex interactions, plant resistance to the European corn borer is produced by (1) a reduction in the amount of ascorbic acid (vitamin C) in the corn plant tissue and (2) by certain toxic substances produced by the plant at different periods of growth and in different tissues. This toxic reaction is called *antibiosis*.

Ecological Adaptations. Indirect interaction of a plant to pests is commonly the result of ecological adaptations. For example, in some crop species, different varieties mature at different times. As a consequence, certain varieties mature before the pest species becomes abundant and these varieties assure a good crop.

Resistance by Plant Breeding. At times, control of an insect pest may be accomplished by introducing a resistant variety of a species or by breeding new varieties that mix the characteristics of a resistant strain with the high yields of a susceptible strain. By this means genetic factors that give resistance are selected for and combined with those that produce commercial characteristics of the crop. This promises to be one of the very important endeavors in insect control. Resistant plant strains have now been developed through breeding experiments for wheat against the Hessian fly and the wheat stem sawfly, alfalfa against the blue alfalfa and spotted alfalfa aphids, and a number of others. Various degrees of resistance have been achieved for a considerable number of other crops.

In developing new genetic varieties of resistant crops, other difficulties may arise. A plant's resistance to other pests, commonly related to a high carbohydrate to nitrogen ratio, may be lowered so that the plant is attacked more readily by fungal and bacterial pathogens. In addition, plant resistance may be only temporary because the insect pest may develop immunity or otherwise overcome the resistance. An example of this breakdown of resistance is the "biotype E" greenbug which now readily feeds on "greenbug resistant" varieties of grain sorghum.

Physical Control

To obtain control of an insect pest, several direct methods have been used. The simplest was physical control, which included removing insects by hand or using mechanical devices to trap or kill them. Hand picking was practiced on large caterpillars, such as tobacco or tomato hornworms. In this procedure the number of insects was usually not excessive, and the individuals were large in size and easy to see. Nests of larvae can be cut out of trees and destroyed. A number of mechanical devices are used with good effect against a limited number of pests. One of the most common is the use of both screen doors and window screens to keep insects out of buildings. Various traps of the maze type are used to catch flies. Bands of burlap or paper are fastened around trunks of fruit trees to provide hibernating or pupating quarters for codling moths; periodically these bands are examined and the insect occupants killed. Bands of screen, gauze, or sticky substances are put around trees to prevent ascent of wingless female moths and larvae (Fig. 12-8).

Various physical barriers formerly were used, especially furrows in the soil or wooden or paper barriers, against migrating wingless insects, such as Mormon crickets or chinch bugs, attacking field crops.

Insects can endure only limited extremes of heat, cold, and other physical phenomena. This limited endurance is utilized to kill insect pests. Such physical factors are difficult to control over a large area, so that with a few exceptions their use is restricted to buildings and tight enclosures. Infrared radiation is employed in many mills and grain elevators as a control measure. During hot weather in summer, the building temperature is raised to about 60° C (140° F)

Fig. 12-8. Banding traps used to prevent ascent of larvae and wingless female moths. (From U.S.D.A., E.R.B.)

for several hours, and this kills all the insects in the building. Heat transmitted by irradiation is used to control insects in stored products.

Cooling is used extensively in storage for insect control. Furs, tapestries, and other valuable articles of animal origin are kept in lockers below 5°C (40°F). This does not kill all the insects, but at this temperature they are completely inactive and do no damage.

Electricity is used to some extent to kill insects. Screens and lights can be fitted with electrically charged grills that electrocute insects coming between the elements. Limited control of some insect species has been obtained using light traps to attract and kill adult insects, especially Lepidoptera.

Chemical Control

Synthetic insecticides have been used extensively since World War II to reduce attack, damage, or destruction of crops and livestock, and epidemics of insect-borne diseases. The loss of about one half of the world's food supply to pests has intensified the search for those insecticides that are effective and at the same time do not induce adverse side effects.

Insect Formulations. An insecticide consists of a toxicant (a poisonous substance) and one or more inert nonpoisonous materials that function to dissolve the poison or act as a carrier or as an emulsifier, dispersant, or spreader-sticker. Preparation of chemicals for a particular method of application is called *formulating*, and the product sold for such specialized use is a *formulation*. The type and quality of a toxicant's formulation greatly influences its effectiveness as an insecticide. Insecticides are commonly formulated as dusts, granules, insecticide-fertilizer mixtures, wettable powders, emulsifiable concentrates, flowables, solutions, soluble powders, slurries, aerosols, fumigants, and specialized formulations.

Factors Influencing the Effectiveness of Insecticides. Certain chemical insecticides more readily penetrate or permeate the body structure of an insect than others. Those entering through the tracheae or the cuticle have a rapid effect. For example, fumigant poisons enter the tracheal system in the form of a gas. Organic solvents such as oils may penetrate readily the outer integument, so an oil in a spray mixture commonly increases toxicity. The early developmental stages of insects are the most susceptible to insecticides. Just before hatching, eggs are very susceptible and larvae and nymphs are more susceptible than pupae and adults.

Various wetting agents are used in the formulation of insecticides so that the insecticide can penetrate deeply into heavy foliage or spread around curved and irregular surfaces.

Environmental conditions also affect the effective toxicity of insecticides. With most insecticides, a sequence of high temperature that permits rapid penetration into the insect, and then a low temperature that slows the rate of breakdown of the toxic substance, is the most effective. In contrast, some insecticides, such as DDT and methoxychlor, are most effective at relatively low temperatures.

Applying an insecticide most effectively presents many problems. First, the insecticide must be applied at a selected time and in some cases (as with mosquito larvae) within a period of only a few days. Second, weather conditions are important because sprays and dusts cannot be applied during rains or in high winds, and some crops are susceptible to insecticide burning during periods of high temperature and humidity.

Specialized sets of machinery are available for applying sprays, dusts, and aerosols (Fig. 12-9), for dipping cattle, or for administering fumigants. A careful choice of these must be made for each control project, the area and local conditions being taken into consideration, such as topography, height and spacing of the crop, and labor conditions.

One promising new development is the ultra-low-volume technique that, under certain conditions, has produced comparable reductions in pest populations using only 75 to 90% of the amount of insecticide used by more conventional equipment. The efficiency of ultra-low-volume spraying (ULV) depends on the production of a spray with a high proportion of exceedingly fine droplets, in the range of 10 to 60 microns, which are distributed by a turbulent airstream so that the droplets reach all sides of the infested plant.

The combined application of different insecticides on a crop must

Fig. 12-9. Application of insecticides; the use of airplane equipment has been of great aid in many situations. (Courtesy of Piper Aircraft Corp.)

be carefully analyzed. Two different chemicals when brought together may react to form different compounds that may be ineffective. A mixture of chemicals may injury plants even though each chemical separately has no effect. Some chemicals when brought together may change their physical characteristics and one of them may become unstable and even hazardous in application. Other chemicals greatly increase the toxicity of certain chemicals so that the resultant action is considerably greater than the sum of the two chemicals used separately. This is called a *synergistic action*. Insecticide synergist agents include butoxide, sesamex, MGK 264, piperonyl, and sulfoxide.

Precautions in Using Insecticides. Some insecticides are highly toxic to humans and many are potentially toxic to humans, especially to the personnel handling them, who may get large amounts on their skin. In general, great care should be taken to avoid getting insecticides on your skin and clothing. Insecticides should be used in open and adequately ventilated areas and insecticide fumes should not be breathed. Every year there are an estimated 200 deaths from the misuse of insecticides and many more instances of nonfatal poisoning. Each user should read the instructions about the insecticide and follow them carefully. From the standpoint of human safety, every effort should be made to *decrease*, not *increase*, the use of insecticides, and to use them safely.

Insecticides must be used with caution, because they may damage the host as well as the insects, or leave a residue that is toxic to humans or other animals. Some plants are highly sensitive to one insecticide but not to others. Lists outlining these "dos" and "don'ts" can be obtained from the label on the insecticide and from the local agricultural extension offices. For each insecticide there are legal limits to the amount of residue that is permitted on the marketed product. These limits must be taken into consideration in planning the control program for each crop.

An important characteristic of insecticides is the length of time they remain toxic after being applied. Many poisons remain toxic for long periods and are spoken of as having a high residual action. Some have practically no residual action, losing their potency almost immediately after they are applied, owing to chemical deterioration. Some of the synthetic contact poisons, such as DDT, have a long residual action; others, such as some of the organophosphorus insecticides, have a short residual life. The length of residual action governs to a large extent the frequency with which the insecticide must be reapplied.

Mode of Action of Insecticides. The way in which an insecticide acts on an insect to cause its death can be classified into one of four methods.

1. *Physical poisons.* These kill the insect by some physical action; for example, the mineral oils exclude air, and silica aerogel dust leads to a loss in water.
2. *Protoplasmic poisons.* These kill the insect by precipitating protein from the tissues. This group includes the arsenicals.
3. *Metabolic inhibitors.* These inhibit metabolism of an insect and include a variety of compounds. For example, respiratory poisons, such as hydrogen cyanide, dinitrophenols, and rotenone, inhibit respiratory enzymes; sodium fluoracetate inhibits carbohydrate metabolism.
4. *Nerve poisons.* This group includes many of the current insecticides. They are of three types:
   a. Anticholinesterase compounds, such as organophosphorus and carbamate insecticides, which increase excitation of the nervous system;
   b. Chemicals that affect the ionic permeability of nerve membranes, such as DDT, chlordane, and pyrethrins; and
   c. Compounds that affect the nerve receptors or synaptic ganglia, such as nicotine.

Another group of organic compounds that indirectly lead to the death of an insect pest are the *chemosterilants*. These induce varying degrees of sterilization in insects. When used as field sprays, these chemosterilants produce sterilized males in sufficient numbers to swamp out normal males over large areas.

Classification of Insecticides. On the basis of their chemical structure, insecticides are classified into *inorganics* and *organics*. The organics are further divided into oils, botanicals, and synthetics.

Inorganic insecticides have for the most part been replaced by organic insecticides. Inorganics include (1) lead arsenate, which is used to control chewing insects on shrubs and to safeguard fruit trees and forest and shade trees; (2) sodium fluosilicate, which is used as a bait for ants, cockroaches, and grasshoppers; and (3) sulfur, finely ground to a dust, which has been used to control mites and certain fungi.

Oils are petroleum products, such as mineral oil, used to safeguard fruit trees and to serve as solvents or carriers. Diesel fuel serves as a carrier for insecticides in aerial application and over water being treated to control mosquitoes.

Oils may serve as insecticides themselves. *Summer oils*, highly refined oils that are not so toxic to plants, are applied to trees in leaf, such as citrus trees, to control mites and scale insects. *Dormant oils*, relatively unrefined oils, are applied when no leaves are present.

Botanicals are plant products used in insect control. These include geraniol and eugenol, which attract insects; citronella and oil of cedar, which repel insects; and cottonseed oil, which serves as a solvent or extender. However, most organics in this group are insect toxicants of which nicotine, pyrethrum, and rotenone are the most important.

Synthetic Organic Insecticides. Synthetic organic chemicals comprise an exceptionally long list.

1. Chlorinated Hydrocarbons. Known as organochlorine insecticides, these include DDT (dichloro diphenyl trichloroethane), a highly effective chemical in controlling insect vectors of human diseases and many crop pests. Because DDT is a stable chemical, it tends to remain in the environment and is not easily leached from the soil. DDT accumulates in fatty tissue in animals, where it is stored. Other chemicals in this group are benzene hexachloride, methoxychlor, perthane, toxaphene, and the cyclodienes (aldrin, dieldrin, chlordane, endosulfan, endrin, and heptachlor). Methoxychlor is chemically similar to DDT but is far less toxic to mammals. It is safer to use on plants than DDT and does not accumulate as readily in fatty tissues of animals. Toxaphene has a wide use against numerous species of insects. However, it is considerably more toxic to mammals than DDT and is very toxic to fish. On the other hand, its residues do not accumulate in fatty tissues as readily. Toxaphene has been used to control grasshoppers, cutworms, webworms, cotton insect pests, legume insects, and insect pests of livestock. Cyclodienes are effective against a wide range of insects but are more toxic to mammals than DDT. Their use has been greatly restricted because their residues accumulate in soils.

2. Organophosphorus Insecticides. These comprise the largest number of synthetic insecticides. Some of them are absorbed by plants, whose sap becomes toxic to insects, or by animals, whose blood becomes toxic to other organisms. These chemicals are called *systemic* insecticides. Plant systemics include demeton, dicrotophos, dimethoate, disulfton, mevinphos, monocrotophos, phorate, and phosphamidon. Animal systemics include coumaphos, crufomate, ronnel, famphur, phosmet, and trichlorfon. Chemicals that are generally nonsystemic include diazinon, malathion, parathion, and tepp. Parathion is highly toxic to mammals and is used extensively on fruit, vegetables, ornamentals, and field crops. Its high toxicity requires that it be kept away from humans and livestock. It is applied in the form of capsules. Methyl parathion, a less hazardous chemical, is used to control cotton insect pests and as a substitute for DDT. Various commercial modifications of parathion are used on crops close to harvest and to control insects, such as mosquitoes and flies, that affect public health. A single spring applications of carbobenthion will replace a number of insecticidal spray applications for control of the pine tip moth.

3. Carbamate Insecticides. These chemicals break down readily and leave no harmful residue. Their mammalian toxicity varies from low to high. Cabaryl is a well-known compound with low mammalian toxicity and is used in home gardens and on crops against blister beetles, Japanese beetle, grasshoppers, scale insects, Colorado potato beetle, and Mexican bean beetle. Other carbamates are carbofuran, metalkamate, methomyl, and propoxur.

4. Synthetic Pyrethroids. Synthesized from petroleum-based chemicals, these are related to the natural pyrethrins. Allethrin is used to determine the existence and extent of cockroach infestations.

5. Insect Hormones. The discovery that chemical substances similar to the juvenile hormone could be used as insecticides has led to the synthesis of many hundreds of substances with biological activity comparable to the corpora allata hormone. These nontoxic or low-toxic substances disrupt the course of morphogenesis and thus prevent normal development and indirectly cause death in many diverse groups of insects. Many of the synthesized juvenile hormones act specifically on a given group of insects and they show considerable promise for use in insect-control programs because they are harmless or have low toxicity to other very useful insects, humans, and many other animals; and they are effective in extremely low dosages. This new group of chemicals, which came onto the market in 1975 and which interfere with the normal growth patterns of insects and indirectly cause their death, are called *insect growth regulators* and are referred to as IGRs. Methoprene is an IGR used to control floodwater mosquitoes.

So far no IGR has been developed that interferes specifically with the various processes associated with the insect-moulting hormones ecdysone and 20-hydroxyecdysone. A synthetic diflubenzuron has been developed that interferes with moulting by inhibiting chitin synthesis. The major drawbacks in developing a successful moulting IGR are the high cost of production from natural products and the poor penetration of the chemical into the cuticle. There are a number of genera of plants with ecdysteroids, that is compounds with an insect moulting hormone activity. These ecdysteroids are most common in ferns and in one family of gymnosperms and one family of angiosperms. Their occurrence in ferns may explain why few insect species live on ferns.

*6. **Insect Repellants.*** These chemicals are primarily used to keep insects away from humans and animals and are not necessarily toxic. Naphthalene and camphor have been used for centuries in homes for keeping insects out of stored clothing. Creosote is used to keep termites out of wood.

Of the many compounds tested as repellants of mosquitoes, other biting flies, ticks, fleas, and chiggers, some of the most effective are dimethyl phthalate, Indalone, and DET (diethyl toluamide). Each of these compounds is effective against only a certain number of species, but various mixtures of them applied to clothing and skin give fair protection to the user for a short period. Benzyl benzoate is proving especially effective as protection against mites and ticks. Pyrethrum is used to repel bloodsucking insects. Dimethyl phthalate has been used to repel bloodsucking insects including anopheles mosquitoes, ticks, and chiggers.

*7. **Insect Attractants.*** These comprise natural substances, such as sugar, molasses, yeast extract, fatty acids, geraniol, and eugenol. Synthetic chemicals include phenethyl propionate and eugenol, which together as a mixture are a very effective attractant. Other synthetic attractants are medlure and trimedlure, which are used to attract the male Mediterranean fruit fly. Attractants are used in various types of traps to catch insects.

Since the late 1950s, ways of using pheromones to control insect pest species have been extensively investigated, particularly for those pest species that appear to be highly dependent on pheromone communication for their survival. Control of insect pest populations by manipulation with pheromones include changing aggregation patterns (particularly at breeding sites), mating behavior, and feeding patterns. Examples of successful control methods utilizing pheromones are the release of the synthetic sex attractants looplure in cabbage and gossyplure in cotton fields, to disrupt the behavioral patterns in male cabbage loopers and pink bollworm moths, respectively, causing a high proportion of unfertilized females.

Pheromones have been used also as lures to attract insect pests to traps with insecticide or on a sticky board. Female moths of certain pest species emit pheromones that strongly attract males from the surrounding area. These females are placed in traps in orchards and one unmated female may attract thousands of males to the trap. Synthetic pheromones are used to attract males of various moths and include disparlure for the gypsy moth, and codlure for the codling moth.

8. Microbial Insecticides. These have been developed from some of the many insect pathogens, such as viruses, bacteria, fungi, and protozoans. As an emerging area of pest control, they offer a safe and effective method to control insects because they are specific against the target pest, are not toxic or pathogenic to humans, do not attack predators and parasites, and are not deleterious to the environment. An example is the bacterial insecticide *Bacillus popilliae*, which causes milky disease in the larvae of the Japanese beetle.

Of the many different types of viruses, the world's first naturally occurring virus that was registered in 1975 as a pesticide is a *nuclear polyhedrosis virus*. It is used against two serious cotton pests, the bollworm and the tobacco budworm. A second nuclear polyhedrosis virus is used to control outbreaks of the Douglas-fir tussock moth.

9. Fumigants. These chemicals are volatile substances that are toxic and repellant to insects. They are used as fumigants to control insects, nematodes, and fungi in soil and to rid houses, warehouses, stores, granaries, greenhouses, mills, and other buildings, as well as citrus trees, of insects. They are usually applied in an enclosure such as a box, boxcar, drum, bin, building, or tent. Fumigation must be handled with care during the application, and the structure must be ventilated properly when the treatment is completed. Chemicals in use are hydrogen cyanide, paradichlorobenzene (PDB), carbon disulfide, ethylene dichloride, carbon tetrachloride, methyl bromide, naphthalene, and dichloropropene. Two serious hazards are the flammability of certain gases and their danger to humans. Hydrogen cyanide is deadly to humans and especially dangerous because it is nearly odorless.

INSECT PEST MANAGEMENT

The advent of the synthetic insecticides following the Second World War appeared to usher in an era of almost complete insect control. These early synthetic insecticides were cheap and effective, returning profits many times the cost of both insecticide and application. Within a few years, however, problems appeared, soon followed by others, that necessitated complete restructuring of insect-control programs. Three main problems appeared: the pests developed insecticide *resistance*; the insecticide *residues* were persistent; and insecticide *costs rose* because of the need for increasing amounts and frequency of insecticide application.

Insecticidal Resistance. Within a year or two of the widespread use of DDT for housefly control in the Mediterranean area, strains of houseflies appeared that were resistant to DDT. Since then the housefly has evolved strains resistant to almost every useful insecticide. The German cockroach has done the same. A surprising array of other insect pests also have evolved strains resistant to almost every insecticide used for their control. These insects include many species of anopheline mosquitoes, certain rootworms (*Diabrotica*), several phytophagous mites, the cotton boll weevil, and several Lepidoptera. When a population is treated with an insecticide, some individuals are more resistant and may survive, because genes conferring resistance to a certain insecticide exist in the pest species at low levels before its exposure to the insecticide. Most of the other individuals do not have these particular genes and do not survive. In the next generation a greater proportion of the surviving individuals will carry the resistance gene and if the population was again treated with the same insecticide, the process would repeat itself, leading to greater resistance in successive populations. If carried on long enough, this leads to a population almost entirely resistant to the insecticide. Most past and some current practices of insecticide application lend themselves perfectly to the rapid buildup of insecticidal resistance by this model. A spray schedule is followed in which the insecticide is applied at regular intervals whether the pest species is scarce or abundant. This regularity of application sustains a selection pressure, year in and year out, for individuals bearing the resistance genes. If there is even a trace of resistance genes in the population, such a prophylactic or precautionary spray schedule is eventually self-defeating from the standpoint of control of the pest species.

Residues and Environmental Contamination. Soon after their widespread use, persistent synthetic insecticides encountered trouble on an entirely different front. These insecticides accumulate on and in human food and in the environment, where they are ingested by invertebrates, such as earthworms in the soil and midge larvae in ponds and lakes. These small animals concentrate the insecticide in their tissues and are eaten by larger ones, such as robins and small fish. At each step the concentration of the insecticide within the organism increases until it eventually becomes lethal to the larger animals higher in the food chain. A well-studied example is the attempted control of the Clear Lake gnat *Chaoborus astictopus* in Clear Lake, California, where the gnat is a nuisance when it emerges in tremendous swarms. DDD was applied periodically at 20 ppb, and depressed larval populations. In the food chain of midge larvae to small fish to larger fish to western grebes, the DDD was accumulated at every step, reaching 2000 ppm in the grebes, which died. In another example, runoff from forests that had been sprayed with DDT killed the aquatic insects in many streams and greatly reduced the game fish, which relied to a large extent on the insects for food. Many of these insecticides decompose so slowly that these deleterious events continue for years, long after applications are discontinued.

Effect on Parasites and Predators. Nearly all insecticides currently in use are broad-spectrum and many are persistent compounds that kill not only individuals of the pest species, but also nontarget species. Notable among their victims are parasites and predators that may actually be keeping the targeted pest or other potential pest species at nonpest population levels. When this happens, the potential pests may become actual pests (called *secondary pests*), and these require additional control measures. Thus the use of DDT for citrus pest control in California was followed by a resurgence of the cottony-cushion scale, previously kept under control by the predaceous vedalia beetle; the outbreak was caused by a great reduction of the predatory beetle's population by DDT. Many other examples have been documented.

Rising Costs. The switch to short-lived insecticides with limited residual effect has greatly increased the cost of controlling pest insects on many crops and of controlling insect vectors of human diseases, because more insecticide applications are needed per year. In cotton fields, where pest resistance also is involved, growers may use 10 to 20 and commonly more applications during the growing season, representing a massive increase in production cost.

Reappraisal of Insect Control. In view of these circumstances, a new look at the entire approach to insect control was needed. The continued use of some long-lived insecticides would lead to worldwide environmental contamination that would be a serious hazard to all animal life including humans. As a result, several countries have banned the use of DDT and almost all have imposed rigorous restrictions on the use of DDT and several other persistent insecticides. Such a change in thinking began in the late 1950s with suggested programs termed *integrated control*, combining the use of parasites and insecticides. In the early 1960s control programs were planned that took into account all of the critical biological requirements of the particular pest species. With this change of direction, a switch was made from "insect control" to "pest management" or "integrated pest management," with its broader connotations.

INTEGRATED PEST MANAGEMENT

The present practice of pest management is strongly ecological in its approach. It involves selection from a great diversity of possible control methods so as to develop a system of techniques that are mutually reinforcing in particular situations. For this reason, it is usually called *integrated pest management* or IPM. The general aims of IPM are (1) to understand thoroughly the interactions of organisms and their environment in those ecosystems, such as a pasture, cultivated field, or orchard; (2) to determine the level, or threshold, of economic injury that will necessitate control measures; and (3) to develop a program or series of treatments that will not upset other highly desirable interactions between other organisms of the ecosystem.

Because most ecosystems in IPM programs are agricultural, sometimes they are referred to as *agroecosystems*. These are very highly specialized ecosystems and they have one or only a few types of plants or domesticated animals and usually only a few species of insects or other threatening organisms that pose a threat of rapidly becoming extremely abundant and serious pests. A larger number of other species are potential pests, but they are relatively rare. Only occasionally and under unusual circumstances will these rare species reach pest levels.

The amount or level of pest damage that can be economically tolerated has to be determined for individual crops and consumers. This level may vary from crop to crop and locality to locality, and is known as the *economic-injury level*. In many situations it is possible to establish an *economic threshold*—that is, an early warning that the population density of a potential pest has reached a level where one or more steps in the management program should be initiated to prevent the pest from eventually reaching a population-density level that will cause damage to the harvest in excess of the economic-injury level.

Many crops can sustain considerable damage from insects with little or no decrease in crop productivity. Commonly an economic threshold for a pest species is exceeded only in certain critical local populations. This means it is necessary to apply insecticides only in the particular areas where and when these population levels are reached.

The "when" has two important aspects. The population might be at a low density for an early generation but reach a high density during a subsequent generation during the same year. Monitoring will be necessary from generation to generation. The other important aspect concerns the exact time in the pest's life cycle when it can be controlled most effectively. In the Sudan, for example, egg laying of the bollworm *Heliothis armigera* occurs in a 10-day synchronous period over a large cotton-growing area. By monitoring the onset of this period, spray applications can be timed to kill the hatching larvae, which provides the most efficient control. Sampling and monitoring the pest species populations on a regular basis makes it possible to predict population trends before, during, and after the growing season.

Integrated pest-management programs plan to use the best selective control methods that disrupt the rest of the ecosystem as little as possible. They depend on the time during the growing season, the relative abundance of other insects, and many other factors. A pest-management program may use a selective spray at a certain time to control a pest species when it is most vulnerable. Yet at the same time the spray will not eliminate the predator and parasite populations, which can continue to control the much smaller pest population that may remain. Some management programs have found that a combination of pest-resistant crop varieties, predators, and parasitoids can reduce or eliminate the need for insecticide sprays. Another type of program combines a number of cultivation practices, predators and parasitoids (which may be local, natural populations or augmented by bringing into the area additional ones), and application of minor amounts of carefully selected insecticides. Every effort is made to keep the release and the amount of insecticides, especially the persistent types, to a minimum in the ecosystem.

Other methods include crop rotation, manipulation of seeding dates, use of native and introduced parasites and predators, use of trap crops, use of resistant crop strains, or any of the other methods available.

Integrated pest management also considers the best use of application equipment and techniques that will eliminate the most pest species per unit of insecticide released.

Operational Programs. The widely recognized need for integrated approaches to pest management meant that programs could be developed, tested, improved, and then adopted and put into practice. Research into these stages of pest management is an ongoing effort at federal, state, university, and commercial laboratories. Many of these laboratories have a long history of investigations into agricultural and other pest problems in many parts of North America.

Research is centered on the more pressing problems associated with protecting major crops that are planted in large acreages and assuring an adequate harvest. The shift in emphasis from pest destruction to crop protection recognizes that pest-management systems are only one part of the total, complex crop-production system. The development of operational programs rely on effectively organized teams of research groups, excellent exchange of information between them, and well-established priorities for suitable working facilities. Researchers having broad, but different backgrounds, on the same teams tend to speed the development of effective control programs.

Because ecosystems involve complex interactions, models are widely used to simplify planning and to permit more rapid predictions. In different parts of its range the same crop may have quite different pest species, thus a single ecosystem model does not always apply to all areas. Computerization of data for rapid handling and analysis in these models is frequently an aid. However, because models are simplifications, it may still be difficult to transfer model results to actual field crop situations.

Many areas of agriculture, such as the small-plot grower, the home gardener, and the more labor-intensive agriculturist, are only beginning to have available to them these types of programs. It will be a number of years before the full potential of integrated pest-management research and its application reaches all agricultural crops.

The dissemination of pest-management information to farmers is accomplished chiefly by the extension services of the federal and state governments. These also distribute information concerning fertilizers, weed control, cultural practices, and other facets of farm management. All of this is a complicated matrix of information for one individual to follow. This situation has led to the establishment of various agricultural consultants, called *agrifieldmen*, who contract for various facets of farm management, including insect pest and weed control, irrigation, fertilizing, or any combination of these and other activities. Those specializing in insect pest management usually work out or learn management models for the important insects of their immediate area, monitor pest populations at prescribed intervals, and take appropriate action as indicated.

CHALLENGES OF THE IMMEDIATE FUTURE

In a very real sense, humans and insects are competing with each other for the available food resources of the world. The United Nations' World Health Organization estimates that nearly 80% of the world's human population are starving or are seriously undernourished. Although our attempts to increase food production through agricultural development, pest management, and plant improvement have been dramatic, our human population level has increased much more rapidly, so that the numbers of starving and undernourished people are increasing at an ever-increasing rate. In order for the food and health benefits gained through the effective management of insect pests to appreciably help the great majority of people living on this planet, we must also take steps to keep the size of the human population within the world's carrying capacity.

REFERENCES

Annu. Rev. Entomol., vols. **1–25** (1956–80). Articles about many topics discussed in this chapter.

Balch, R. E., 1952. The spruce budworm and aerial forest spraying. *Can. Geog. J.* **45**:201–209.

Berozoa, M. (Ed.), 1976. Pest management with insect sex attractants and other behavior-controlling chemicals. *Am. Chem. Soc. Symp., Ser. 23.* 192 pp.

Blum, M. S., and N. A. Blum (Eds.), 1979. *Sexual selection and reproductive competition in insects.* New York: Academic Press. 464 pp.

Carter, W., 1973. *Insects in relation to plant disease*, 2nd ed. New York: Wiley. 759 pp.

Caswell, R. L. (Ed.), 1977. *Pesticide handbook—Entoma. 1977–78*, 27th ed. College Park, Md.: Entomological Soc. of America. 248 pp.

Cherrett, J. M., and G. R. Sagar (Eds.), 1977. *Origins of pest, parasite, disease, and weed problems.* Oxford: Blackwell. 413 pp.

Chiang, H. C., 1978. Pest management of corn. *Annu. Rev. Entomol.*, **23**:101–123.

Cox, G. W., and M. D. Atkins, 1979. *Agricultural ecology.* San Francisco: W. H. Freeman. 722 pp.

DeBach, P., 1974. *Biological control by natural enemies.* Cambridge: Cambridge Univ. Press. 323 pp.

Ebeling, W., 1959. *Subtropical fruit pests.* Berkeley: Univ. of California Div. Agr. Sci. California. 436 pp.

Geier, P. W., L. R. Clark, D. J. Anderson, and H. A. Nix (Eds.), 1973. *Insects: Studies in population management.* Canberra: Ecological Society of Australia Memoirs, vol. 1, 295 pp.

Gibbs, A. J. (Ed.), 1973. *Viruses and invertebrates.* Amsterdam: North-Holland, 673 pp.

Harborne, J. B. (Ed.), 1978. *Biochemical aspects of plant and animal coevolution.* New York: Academic Press. 436 pp.

Harris, K. F., and K. Maramorosch (Eds.), 1977. *Aphids as virus vectors.* New York: Academic Press. 560 pp.

Harrison, G., 1978. *Mosquitoes, malaria and man.* New York: E. P. Dutton. 314 pp.

Headley, J. C., 1972. Economics of agricultural pest control. *Annu. Rev. Entomol.*, **17**:273–286.

Herms, W. B., 1950. *Medical entomology*, 4th ed. New York: Macmillan. 643 pp.

Horsfall, W. R., 1962. *Medical entomology.* New York: Ronald Press. 467 pp.

Huffaker, C. B. (Ed.), 1980. *New technology of pest control.* New York: Wiley. 500 pp.

Huffaker, C. B., and P. S. Messenger (Eds.), 1976. *Theory and practice of biological control.* New York: Academic Press. 788 pp.

Jones, D. P., and M. E. Solomon (Eds.), 1974. Biology in pest and disease control. *13th Symposium British Ecological Society.* New York: Wiley. 398 pp.

Maramorosch, K., and K. F. Harris, 1979. *Leafhopper vectors and plant disease agents.* New York: Academic Press. 654 pp.

McEwen, F. L., and G. R. Stephenson, 1979. *The use and significance of pesticides in the environment.* New York: Wiley. 538 pp.

Metcalf, C. L., W. P. Flint, and R. L. Metcalf, 1962. *Destructive and useful insects, their habits and control*, 4th ed. New York: McGraw-Hill. 1087 pp.

Metcalf, R. L., and W. H. Luckmann (Eds.), 1975. *Introduction to insect pest management.* New York: Wiley. 587 pp.

Mitchell, E. R., 1975. Disruption of pheromonal communication among coexistent pest insects with multichemical formulations. *BioScience*, **25**:493–499.

Narahashi, T. (Ed.), 1979. *Neurotoxicology of insects and pheromones.* New York: Plenum. 308 pp.

Neal, J. W., Jr. (Ed.), 1979. *Guidelines for the control of insect and mite pests of foods, fibers, feeds, ornamentals, livestock, forests, and forest products.* USDA, Agric. Handbook 554. 822 pp.

Pal, R., and M. J. Whitten, 1974. *The use of genetics in insect control.* Amsterdam: North-Holland. 241 pp.

Peairs, L. M., and R. H. Davidson, 1956. *Insect pests of farm, garden, and orchard*, 5th ed. New York: Wiley. 661 pp.

Pfadt, R. E. (Ed.), 1978. *Fundamentals of applied entomology*, 3rd ed. New York: Macmillan. 798 pp.

Smith, K. G. V. (Ed.), 1978. *Insects and other arthropods of medical importance.* New York: Wiley. 600 pp.

Stairs, G. R., 1972. Pathogenic microorganisms in the regulation of forest insect populations. *Annu. Rev. Entomol.*, **17**:355–372.

Theiler, M., and W. G. Downs, 1973. *The arthropod-borne viruses of vertebrates.* New Haven: Yale Univ. Press. 578 pp.

Tinsley, T. W., 1979. The potential of insect pathogenic viruses as pesticidal agents. *Annu. Rev. Entomol.*, **24**:63–88.

Wallace, J. W., and R. L. Mansell (Eds.), 1976. *Biochemical interaction between plants and insects.* New York: Plenum. 436 pp.

Watson, M. A., and R. T. Plumb, 1972. Transmission of plant-pathogenic viruses by aphids. *Annu. Rev. Entomol.*, **17**:425–452.

Whitten, M. J., and G. G. Foster, 1975. Genetical methods of pest control. *Annu. Rev. Entomol.*, **20**:461–476.

See also References listed under Growth of North American Entomology, Life Processes, Response and Behavior, Ecological Considerations, and Useful Insects.

Index

(In a series of entries, important references are indicated by **bold face** type)

Abbott, J., 10
Abdomen, 33, 99–105
Abductor muscle of coxa, 121
Absorption in digestion, 147
Academy of Natural Sciences of Philadelphia, 11
Acalymma vittata, 399, 400
Acalyptratae, 442, 444, **464**, 466
Acanthocephala, 360
Acanthoscelides, 382
Acarina, 40
Accessory glands, 123, 125
Acerentomon, 89
Acerentulus barberi, 281
Acetylcholine, 192
Acherontia atropos, 200
Achorutes armatus, 282, 283
Achroia, 482
Acidalia, 485
Acinus, 139
Acrididae, 305, 306
Acridium ornatum, 306
Acron, 63, 70
Activation hormone, 245, 261, 262
Aculeata, 412, 425
Acythosiphon pisum, 648
Adams, J.A., 285
Adephaga, Coleoptera, 374, 375, **385**–386
Adipohemocytes, 163
Adult, 234, 236–238
  see also Imago
Aedeagus, 104, 125
Aedes, 446, 458
  aegypti, 86, 187, 457, 641
Aegeriidae, 481, 484, **490**
Aeolothripidae, 334
Aeolothrips, 334
African driver ant, 215
African sleeping sickness, 641
Agaontidae, 423
Agasicles hygrophilia, 621
Agassiz, L., 15, 20
Age structure of populations, 561
Agetochorista tillyardi, 528
Aggregation in presocial insects, 218
Agrell, I.P.S., 263
Agricultural ants, 430
Agriotes, 392
Agroecosystems, 661
Agromyza, 450
Agromyzidae, 442, 450, 453, 464, **465**
Agrotis orthogonia, 578
Agulla, 365
Air movement, 575
Air receptors, 181
Air sacs, 114, 115, 153, 155
Air tubes, 160
Alaglossa, 75, 76, 84
Alary muscles, 121
Alcohols, 206
Aldehydes, 206
Alderflies, 366, 535, 610
Aldrin, 22
Aleiodes, 66
Aleochara bimaculata, 608, 609
  curtala, 388, 389
Aleurodidae, 337, 338, 343, 348
Alexander, R.D., 105
Alfalfa, 648
Alfalfa caterpillar, 501, 645
Alfalfa root weevil, 562, 619
Aliform muscles, 165
Alimentary canal, 107–110
Alinotum, 90, 91
Alkali bees, 623
Allantoic acid, 150
Allantoin, 150
Allee, W.C., 601
Allelochemics, 202, 205–206
Alligator weed, 621
Allocapnia, speciation, 544
Allomones, 205
Allothrips, 334
Alternanthera philoxeroides, alligator weed, 621
Alternation of generations, 251, 252
Alula, 464
Alydus, 342
Amber, fossil, 506, 542
  Baltic, 529, 532, 542
Amblycera, 325
Amblycerus, 382
Amblyplygida, 38, 40
Ambush bug, 591
American cockroach, 296, 298
Ametabolous, 235
Amino acids, 137, 169
Amitermes hastatus, 215
Amnion, 227, 230
Amphibolips confluens, 247
Amphicerus, 382
Anabolic phase, 168
Anabolism, 167
Anabrus simplex, 308
Anagasta kueniella, 206
Anal papillae, 153

Anal syphon, 153
Anal veins, 95
Anamorphosis, 281
Ananthakrishnan, T.N., 335
Anaphothrips, 334
Anasa tristis, 360, 361
Anaxyela gracilis, 530
Ancistrocerus, 411
Anderson, D.T., 56, 263
Anderson, N.H., 478
Anderson, R.M., 601
Andrena, 417
Andrenidae, 410, 415, 417
Andrewartha, H.G., 601
Andricus erinacei, 253
  seminator, 247
Angiosperms, 537, 539
Angoumois grain moth, 490
Anisoptera, 291, 292–295
Anisota rubicunda, 496, 497
Anisozygoptera, 291
Anlage (pl. anlagen), 229
Annelida, 27
Annulipalpia, 473, 476
Anobiidae, 376, 380, 385
Anopheles, 458, 641
  maculipennis, 224
Anoplura, **327,** 639
Antennae, 29, 65, 73, 229
Antennal flagellum, 71, 178
Antennal sclerite, 71
Antennal sockets, 67
Anterior arms, 66
Anterior body region, 62
Anthidium, 417
Anthocoridae, 337, 344, 345, 540, 610
Anthomyia, 449
Anthomyiidae, 442, 444, 449, 450, 451, 452, **466**
Anthonomus grandis, 402, 403
Anthrax, 446
Anthrenus, 380
  scrophulariae, 393, 394
Anthrophoridae, 410
Antibiosis, 649
Anticoagulin, 140
Antipodoeciidae, 476
Antireflective structure, 187
Antlions, 372, 610
Ants, 179, 213–215, 220, 408, 412, 420, **428–430,**
  531, 568, 594, 599, 610, 613
Anus, 100, 103, 107, 109, 229, 232
Aonidiella aurantii, 617
  citrina, 618
Aorta, 66, 111, 112, 165
Aortic diverticulum, 112
Apanteles congregatus, 615, 616
  glomeratus, 615
  lacteicolor, 615
  melanoscelus, 239, 422, 616
  solitarius, 615
Apex, body region, 63
Aphaenogaster, 213
Aphelinus mali, 618
Aphididae, 251, 339, 570
Aphidlions, 372
Aphidoidea, 337, 338, 343, **350**
Aphids, 172, 224, 251, 339, 346, 348–352, 539,
  564, 570, 606, 610, 612, 613, 633, 635, 636
Aphis, 339
  gossypii, 350
  pomi, 350
Aphodius fimetarius, 629
Aphrophora, 339
Aphytis chrysomphali, 617, 618
  lingnamensis, 617
  melinus, 617
Apidae, 410, 413, 416, **432**
Apis mellifera, 178, 201, 202, 433, 596, 623
Aplopus mayeri, 303
Apocrita, 410, 411, 413, **420,** 421, 530
Apodeme, 31, 62, 128, 129
Apoidea, 410, 412, **432,** 531
Apolysis, 135
Apophysis, 31, 128, 129
Appendages: abdominal, 100, 101–105
  ancestry of, 29
  embryonic, 226
  head, 70–85
  intercalary, 64, 71, 72, 229
  muscles of, 120
  reproductive types, 101–105
Apple, J.L., 629
Apple aphid, 350
Apple canker, 637
Apple maggot, 465
Apple mirid, 224
Apple tree borer, 401
Apterobittacus apterus, 435
Apterygota, 52, 235, 266, **267,** 284, 504, 514
Aptesis basizonia, 617
Aquatic environments, 580–584, 625–626
  adaptation for, 159–161, 626
Arachnida, 37, 38, 39, 113
Aradidae, 337, 344, 345, 360, 361
Aradus acutus, 361
Araneae, 38, 40
Arborization, 191, 192
Archicerebrum, 29
Archilestes californica, 224, 290, 291, 292
Archips, 482
Archoglossopterum shoricum, 521
Archostemata, 374, 375, 387, 527
Arctiidae, 481, 482, 485, 486, **499**
Arctiid moths, 499
Arctocorixa, 358
Arctotypus sinuatus, 517
Arge, 411
Argentine ant, 430

Argidae, 409, 410, 411, 413
Argyrotaenia velutinana, 491
Arista, 185
Arixeniidae, 312
Armyworm, 498, 499, 576, 620, 634
Arnaud, P.H., Jr., 629
Arnett, R.H., Jr., 407
Arnold, J.W., 173
Arrhenotoky, 244
Arsenicals, 12
Arthaldeus pascuellus, 565
Arthroplea, 78
Arthropleona, 282
Arthropoda: ancestor of, 27–29
  characteristics, of, 27–36
  classes of, 37–58
Arthropod appendages, 70–76
Articulation, 60, 61
Ascalaphidae, 369
Aschiza, 442, 444, 462–464
Asclepias, 597
Ascogaster quadridentata carpocapsae, 615
Ascorbic acid, 649
Asilidae, 442, 443, 446, 447, 448, **460,** 611
Asilomoidea, 442
Asilua, 446
Askew, R.R., 263
Asparagus beetle, 399, 634
Aspidothorax triangularis, 516
Assassin bugs, 359, 591, 610, 611
Aster leafhoppers, 576
Aster yellow, 637
Atheas exiguus, 355
ATP, 167, 168
Attagenus, 380
  megatona, 393
Atta texana, 430, 579
Attractants, insect, 657
Augomonoctenus, 413
Australian bulldog ant, 214
Australian cockroach, 298
Australian faunal realm, 273
Autecology, 557, 566
Autotrophs, 554
Axillary sclerites, 94
Axon, 191

Babesia bigemina, 642
Bacillus pestis, 640
  popilliae, 590, 658
Backswimmers, 160, 359
Bacteria, 589
Bacterial disease, 637, 644
Bacterial rot of cotton, 636
Bagworms, 481, 487, 488, 489, 594
Baits, 657
Baker, E.W., 57
Balch, R. E., 664
Bald-faced hornet, 408
Ball, E.D., 310
Banks, N., 303
Bark beetles, 206, 403, 608, 609, 636, 646
Barnes, R.D., 57
Barrington, E.J.W., 173
Baryodma, 388
Basal sclerites, 94
Basipodite, 73
Batesian mimicry, 597
Bauman, R. W., 318
Beak, 85
Bean weevils, 400
Beauveria globulifera, 589
Bed bugs, 355, 357, 610, 611, 639
Bedford, G. O., 304
Bee flies, 611, 612
Bee louse, 181, 466, 590
Bees, 114, 140, 170, 172, 173, 178, 236, 271, 408, 412, 413, 417, 420, 430, **432–433,** 530, 531, 594, 596, 597, 622
Bees wax, 169, 623
Beet leafhopper, 348, 637
Beetles, 139, 140, 150, 160, 171, 172, 196, 220, 236, 237, 373–407, 526, 535, 608, 636
Behavior, 175
  learned, 206, 212–213
  stereotyped, 206, 207–212
Beier, M., 304
Belidae, 547
Belostoma, 340
Belostomatidae, 337, 340, 343, 359
Bembix hinei, 613
  pruinosa, 587
Benzene hexachloride (BHC), 22
Benzyl benzoate, 657
Beraeidae, 476
Berg, C.O., 472
Berner, L., 288
Berothidae, 369, 370
Beroza, M., 221, 664
Berytidae, 346
Bethune, C.J.S., 17
Bethylidae, 410, 412, **425**
Bethyloidea, 410, 412, 425
Betten, C., 478
Bibio, 446, 451
Bibionidae, 442, 443, 446, 447, 451
Biological control, 23, 422, 604–621, 644
Bioluminescence, 196
Biome, 551–553
Biotype, 648
"Biotype E" greenbug, 649
Birch leaf miners, 418
Birch, M. C., 221
Biston betularia, 592–593
Biting sheep louse, 327
Bittacidae, 436, 528, 613
Black ant, 428
Black carpet beetle, 393

Black death, the, 640
Black flies, 68, 79, 80, 81, 443, 453–455, 639
Black lady beetle, 607
Black locust sawfly, 584
Black scale, 634
Black scale pest, 618
Bladderworts, 591
Blaesoxiphia, 469
Blastoderm, 225, 226
Blastophaga psenes, 424
Blatchley, W. S., 310, 363, 407
Blatta orientalis, 102, 296, 298
Blattaria, 296, 523
Blattella, 296–298
  germanica, 296, 297, 298
Blattopteriforma, 295
Blattopteroidea, 295
Blephariceridae, 547
Blissus, 342
  leucopterus, 361, 648
Blister beetle, 240, 395–398, 595, 608, 634
Blood, 111, 162–167
  function of, 164
  properties, 162
Blood cells, 111, 163
Blood pressure, 164–166
Blood sugar, 167
Blowfly, 150, 469
Blue alfalfa aphid, 649
Bluebottles, 469
Blum, M. S., 664
Body louse, 329, 330, 641, 642
Body regions, 28, 62–105
Body wall, 128–137
Bohart, G. E., 629
Boll weevil, 402, 403
Bollworm pest, 490, 633, 658, 662
Bombardier beetle, 206
Bombus, 432
Bombyliidae, 442, 446, 448, 611
Bombylius major, 612
Bombyx, 245
  mori, 480
Booklice, 321–323
Borboridae, 442, 453, 464, **465**
Boreidae, 435, 436
Boreus, 435, 436
Boring beetle, 627
Borror, D.J., 272, 294
Bostrichidae, 375, 376, 382, 385
Bostrichoids, 375, 376, 393–394
Botflies, 467, 468, 639
Brachycentridae, 476
Brachycera, 442, 443, 444, 459–473, 530
Brachypanorpa, 436
Bracon cephi, 616
Braconid, 599, 616
Braconidae, 77, 209, 233, 410, 416, **422,** 590, 615
Bradley, J. C., 407
Brain, 117–119
Braula caeca, 181, 466, 590
Braulidae, 442, 445, 466
Breakbone fever, 641
Brenthis, 483
Brevicoryne brassicae, 252
Breymeyer, A. I., 601
Brine maggots, 464
Bristletails, 220, 284
Britton, E. B., 407
Britton, W. E., 363
Broad-headed wasps, 564, 565
Brock, J. P., 502
Brood care, 218–219
Brooks, A. F., 363
Brown rot of peach, 636
Brown-tail moth, 497, 615, 636
Bruchidae, 375, 381, 382, 383, **400,** 638
Bruchophagus platyptera, 423
Bruchus, 382
  pisorum, 400
Brues, C. T., 272
Bubonic plague, 21, 438, 640
Budworm, 613, 636
Buffalo gnats , 453
Buffalo treehopper, 646
Bugs, 171, 172, 182, 236, 270, 335–363
Bulb mites, 40
Bumblebee, 173, 413, 432
Buprestidae, 375, 376, 384, **392**, 537
Buprestid beetle, 182, 392, 538
Burks, B. D., 288
Bursa copulatrix, 260
Bushcrickets, 183
Butterflies, 151, 170, 196, 197, 220, 479, 482, 499–502, 527, 531, 539, 559, 596, 599
Byers, G. W., 436

Cabbage aphid, 252, 634
Cabbage butterfly, 173, 634
Cabbage fly, 608
Cabbage looper, 498, 634
Cabbage maggot, 466
Cabbageworm, 501, 615
Cactoblastis cactorum, 620
Caddisflies, 89, 104, 105, 160, 240, 271, 473–479, 527, 532, 535, 539, 545, 594, 627
Cadelle, 638
Calamoceratidae, 476
California black tiger beetle, 609
California ladybird beetle, 606
California red scale, 634
Caliroa cerasi, 244
Callibaetis flunctuans, 287
Calliphoridae, 442, 451, 466, 469
Callosamia promethea, 495, 496
Calocidae, 476
Caloneurodea, 504, 517, **518,** 519, 523, 534
Calosoma sycophanta, 606, 607, 608

Calypters, 464
Calyptratae, 442, 444, 464, 466
Calyx, 123
Cambarus, 358
Camel crickets, 307
Camnula pellucida, 306
Camouflage mimicry, 594
Campaniform sensilla, 178
Campodea folsomi, 280
Campodeidae, 279
Campylenchia latipes, 346
Cankerworms, 495
Cannabalism, 600
Cantharidae, 375, 376, 378, 384, 608, 609
Cantharidin, 624
Canthon, 389
pilularis, 629
Canthris, 608
Cantrell, I.J., 312
Cap cell, 179
Capillary tubes, 113
Capitate antennae, 71
Capnia lacrusta, 317
Capniidae, 318
Carabidae, 374, 375, 382, 384, **385**, 591, 607
Carbamates, 12, 656
Carbohydrates, 168
Cardenolides, 597
Cardia, 110, 144, **145**
Cardines, 75
Cardo, 74, 75, 76
Carnivores, predation, 558
Carotenoids, 133
Carpenter, F.M., 365, 372, 436
Carpenter ants, 639
Carpet beetles, 393, 394, 638
Carrion beetles, 376, 386
Carrot rust fly, 634
Carrying capacity, 563–566, 664
Carson, R., 23
Carter, W., 664
Casemaking clothes moth, 488
Castes, 215, 301
Caswell, R. L., 664
Catabolism, 167
Catacola moths, 592, 593
Catana clausensi, 607
Catesby, M., 10
Cattle grub, 469, 470
Cattle tick, 642
Cave crickets, 307
Cecidomyia poculum, 247
Cecidomyiidae, 255, 441, 442, 444, 447, **453**
Cecropia moth, 245
Celerio lineata, 185
Cell body, 191
Cement layer, 130, 132
Centipedes, 47, 69, 591
Cephalic fan, 80
Cephalothorax, 33, 37
Cephidae, 409–411, 414
Cephus cinctus, 616
Cerambycidae, 196, 375, 381, 382, 383, **400**, 575
Cerambycid beetles, 575
Ceratitis captitata, 465
Ceratomerinae, 547
Ceratophyllidae, 438
Ceratopogonidae, 444, 447, 451
Cerci, 100, 101, 102, 103, 104, 229
Cercopidae, 337, 339, 343, 347, 539, 564
Ceresa bubalis, 346
Ceropales, 417
Cervical sclerites, 86
Cervix, 86
Ceuthophilus, 308
maculatus, 307
Chalcididae, 411, 415, 588
Chalcidoidea, 410, 412, 414, 423, 530
Chalcid wasps, 233, 423, 599, 617, 618
Chalcis, 411
Chamberlin, J.C., 57
Chaoboridae, 442, 445
Chaoborus astictopus, 659
Chapman, P.J., 323
Chapman, R.F., 105, 125, 173, 221, 263, 629
Chauliognathus pennsylvanicus, 378, 609
Checkered beetle, 608
Chelicerata, 32, 33, 36–42
Chelifer, 38
Chelymorpha, 381
Chemical control, 651–658
Chemoreceptors, 176, 181–182
Chemosterilants, 654
Chemotaxis, 209, 210
Cherrett, J.M., 664
Cherry fruit fly, 465
Chewing lice, 325, 327
Chiang, H.C., 664
Chicken body louse, 326, 327
Chicken head louse, 326, 327
Chiggers, 639, 642
Chigoe, 438, 439
Chilopoda, 43, 46, 48, 88, 113
China, W.E., 363
Chinch bugs, 235, 361, 571, 585, 590, 643, 644, 648, 650
Chironomidae, 184, 442, 443, 446, 447, 457, **458**, 459
Chironomus, 117, 163, 446, 451
ferrugineorittatus, 459
tentans, 459
Chitin, 130, 168
Chlorinated hydrocarbons, 22, 607, **655**
Chloropidae, 442, 453, 464, 465–466
Choerocampinae, 184
Cholesterol, 138
Chordotonal organs, 178–180
Chorion, 224, 225

Choristoneura fumiferana, 613, 619
Christiansen, K., 283
Chrysanthemum midge, 634, 635
Chrysididae, 410, 411, 412, 416, 425, 430
Chrysis, 411
Chrysobothris femorata, 392
Chrysolina gamellata, 621
Chrysomelidae, 375, 376, 381, 382, 383, 387, 399
Chrysopa, 368, 370
  californica, 610
  oculata, 224
Chrysopidae, 206, 224, 368, 369, 370, 371, 372
Cibarium, 141
Cicada, 141, 146, 200, 346–348, 592, 613, 632
Cicada killer, 431, 613
Cicadellidae, 337, 339, 342, 346–348, 539, 564, 565, 566, 609
Cicadidae, 197, 198, 337, 339, 346–348
Cicindela, 382
Cicindelidae, 375, 382, 384, 385, 608
Cigarette beetle, 571
Cimbex americana, 418, 419
Cimbicidae, 409, 410, 413, 414
Cimex, 259
  lectularius, 357
Cimcidae, 337, 344, 357, 410
Cimicomorpha, 337, 341, 355
Circulatory system, 110–113, 162, 165
Circulifer tenellus, 348
Circumesophageal connective nerves, 118
Citheronia, 485
Citheroniidae, 481, 485, 486, 497
Citrus blackfly, 607
Citrus black scale, 607
Citrus cottony-cushion scale, 606
Citrus fruit insects, 634
Citrus mealybug, 607
Citrus yellow scale, 618
Claassen, P.W., 318
Claduis, 413
Clark, L.R., 601
Clasper, 104
Clausen, C.P., 263, 407
Clavate antennae, 71
Clay, T., 330
Clean cultivation, 646
Clearwings, 481, 490
Cleavage, 224–226
  holoblastic, 226
  superficial, 225
Clements, F. E., 601
Cleridae, 608
Clerid beetle, 609
Click beetles, 196, 376, **391**–392
Climacia, 370
Climate, 555–556
Climatic changes, 540
  moisture bands, 574
Cloeon, 287
Clothes moths, 638
Cloudsley-Thompson, J. L., 25, 57
Cloverseed chalcid, 423
Clypeus, 67, 69, 141
Coarctate form, 396
Cobben, R. H., 363
Coccidae, 149, 351, 606
Coccinella californica, 605
Coccinellidae, 375, 376, 385, **387**, 394–395, 605–607
Coccinellid beetle, 607
Coccoidea, 270, 337, 338, 343, **350**–352
Cochineal, 623
Cockroaches, 74, 109, 112, 121, 139, 142, 143, 151, 159, 177, 236, 269, **295–298**, 523, 535, 595, 638, 659
Codling moth, 564, 615, 635
Coelopidae, 442, 452, 464
Coevolution, insects and flowering plants, 537–539
Cold hardiness, 570–572
Cole, F. R., 472
Coleoptera, 52, 131, 138, 150, 172, 196, 198, 243, 266, 271, 275, 364, **373–407**, 504, 526, 527, 535, 540, 545, 588, 605, 611, 615, 617
  tarsi, 381
Colias eurytheme, 501
Collateral branch of axon, 192
Collembola, 52, 69, 92, 100, 108, 151, 196, 224, 226, 279, **282–283**, 504, 533, 534, 576, 626
Colleterial glands, 123
Colletes, 417
Colletidae, 410, 416, 417
Colonization of terrestrial habits, 432
Color, 133–135
  pigment, 133–135
  protection, 595
  receptors, 190
Colorado potato beetle, 151, 399, 588
Common, I.F.B., 502
Common booklouse, 323
Communal societies, 216
Communication, 196–206
Community, 551–554
Comperiella bifasciata, 618
Competition, 562, 566, 599–601
Complete metamorphosis, 236, 479
Compound eyes, 66–69, 133, 184–191
Comstock, J.H., 16, 18, 19, 20, 57
Cone-headed grasshopper, 307
Confused flour beetle, 398, 399, 570, 600, 638
Coniopterygidae, 369, 370
Conocephalus strictus, 307
Conopidae, 442, 448, 460
Conotrachelus nenuphar, 403
Consumers, 555–558
Continental drift, 508
Control methods, 642–663
Cook, A. J., 20

Cook, D. R., 57
Coope, G. R., 547
Coordination, nervous system, 193
Cootie, 329, 330, 642
Copeognatha, 321
Copepoda, 44, 45
Coprid beetles, 627
Copris minutus, 373
Copulatory organ, 102, 104
Coquillet, D. W., 18
Corbet, P. S., 294, 295
Coreidae, 337, 340, 345, **360**
Corimelaena, 339
Corium, 335
Corixa, 340
Corixidae, 160, 337, 340, 341, **358**–359
Corn, insects on, 633
Corn aphids, 588
Cornea, 186–189, 191
Corneagenous cells, 186, 191
Corn earworm, 498, 620
Corn rootworm, 633
Corpora allata, 119, 125, 194, 195, 261
Corpora cardiaca, 119, 166, 194
Corporotentorium, 66, 67, 70
Corpotendon, 66
Corpus luteum, 194, 195
Corrodentia, 321
Corydalidae, 367
Corydalus, 367
Corythuca floridanus, 355
Costa, 95
Costal crossveins, 96
Cotton, insects on, 633
Cotton aphid, 633
Cotton boll weevil, 22, 613, 633, 648, 659
Cotton bollworm, 498
Cotton fleahopper, 356
Cotton leafworm, 633
Cottony cushion scale, 18, 644
Counce, S. J., 263
Cox, C. B., 548
Cox, G. W., 664
Coxa, **88,** 92, 93, 122
Coxal process, 91
Coxopodite, 70, 71, 72, 73, 88, 101, 104
Crab louse, 329
Crampton, G. C., 472
Crane flies, 145, 441, 443, 453, **456**
Cresson, E. T., 13, 18
Crichton, M. I., 478
Crickets, 100, 183, 197, 198, 269, 547
Cricotopus trifasciatus, 459
Crioceris asparagi, 399
Crop, 108, 109, 110
  damage to, 398, 399, 631–638
  rotation of, 646
Crossveins, 94, **96**
Crown borers, 636
Crowson, R.A., 407
Crozetia crozetensis, 80
Crozier, R.H., 433
Crustacea, 34, 35, 43, 44–46, 76, 113
Cryptic camouflage, 592
Cryptocercus punctulatus, 298
Cryptochaetum iceryne, 60
Cryptolaemus montrouzieri, 606, 607
Cryptoseruphus pinorus, 530
Crystalline cone, 186–191
Ctenocephalides canis, 224, 438
  felis, 438
Cubital plate, 94
Cubitus, 95
Cuckoo wasps, 412, 425, 430
Cuclotogaster heterographus, 326, 327
Cucujidae, 375, 385
Cucujoids, 375, 376, **394**–404
Cucumber beetles, 637
Cucurbit wilt disease, 637
Culex quinquefasciatus, 546, 571
Culicidae, 184, 442, 443, 445, 446, **456–458,** 611
Cultural control of insects, 645
Cuneus, 356
Cupedidae, 375, 387
Curculionidae, 375, 376, 377, 379, 380, **402–403**
Curly top of sugar beets, 637
Curran, C. H., 472
Currant worm, 418, 419
Cursoria, 295
Cutaneous respiration, 160–161
Cuticle, 128–130
Cuticulin, 130
Cuticulum layer, 132
Cut worms, 633, 634
Cuvier, 14
Cyclorrhapha, 442, 444, **462**
Cydia pomonella, 564, 615
Cydninae, 344, 363
Cylindrachetidae, 547
Cynaeus angustus, 586, 587
Cynipidae, 253, 415, 424
Cynipoidea, 410, 412, 416, 424
Cyrnellus marginalis, 477
Cystocytes, 163
Cytoplasm, 225

Dactylopius coccus, 623
Dahlbominus fuscipennis, 617
Daily rhythm of insects, 575
Daly, H. V., 629
Damage, relation to mouthparts, 76–85
Damsel bugs, 610
Damselflies, 161, 170, 224, 268, 289–**292,** 591
Danaine butterflies, 597
Danaus gilippus, 1966, 597
  plexippus, 597
Dance flies, 547, 612
Dance of the honey bee, 201–202, 203

Darkling beetles, 376, 398, 597
Darwin, C., 14
Dasyleptus, 514
  brongniarti, 514
Dasymutilla bioculata, 426, 586, 587
Davis, C., 315
Day, W. C., 288
DDT, 12, 22, 604, 607, 615, 617, 653, 655
Dean, G. A., 20
Deathwatch beetle, 197
Debach, P., 629, 664
Deciduous fruit insects, 632, 634–636
Decker, G. C., 25
Decomposers, 555, 558, 626
DeCoursey, R. M., 363
Deerflies, 461
Defoliation: coniferous forests, 420, 637
  by insects, 631–637
DeGeer, 8, 10
Delany, M. J., 285
de la Torre-Bueno, J. R., 273, 364
DeLong, D. M., 363
Delopterum incertum, 519
Demerec, M., 472
Demodicidae, 40
Dendrites, 191
Dendroctonus, 380, 609
Dengue, 21, 641
Denning, D.G., 478
Dermacentor ticks, 642
Dermal glands, 128, 132
Dermal light receptors, 184
Dermaptera, 52, 144, 266, 269, 275, 277, 310–312, 504, 523, 537
Dermestes, 393
  lardarius, 393
Dermestidae, 376, 380, 385, **393**–394
Desert, adaptation to, 545
Desert locust, 564
Deutocerebrum, 118
Deutonymph, 281
Development: physiology of, 223–240
  temperature effect, 574–575
Developmental changes, 223–246
De Wilde, J., 263
Diabrotica, 646, 659
  virgifera, 579
Diachasma tryoni, 224
Diapause, 244–246
Diaphanopterodea, 504, 516
Diapheromera femorata, 303
Diarrhea, 640
Diaspididae, 351
Diastraea grandiosella, 492
Diastrophus nebulosus, 247
Diathronomyia hypogaea, 635
Dictyoptera, 52, 266, 269, 275, 276, 278, 295–300, 504, 523, 610
Didea fasciata, 463
Dieldrin, 22
Diffusion, in respiration, 155, 156
Digestion, 139–148
  extraintestinal, 140
Digestive enzymes, 140, 143
Digestive system, 107, 146, 152, 232
Digger wasps, 430
Dilaridae, 369, 370
Dillon, E. S., 407
Dillon, L. S., 548
Dineutes, 380
Diploglossata, 312
Diplopoda, 44, 46, 48, 113
Diplura, 52, 69, 266, 267, 277, **279**, 280, 504, 627
Diprion hercyniae, 420, 617
Diprionidae, 409, 410, 413, 414, 420
Dipseudopsidae, 476
Dipsocoridae, 337
Dipsocoromorpha, 337, 338
Diptera, 52, 102, 119, 138, 139, 144, 146, 150, 171, 172, 238, 266, 275, 278, **440–472**, 504, 527, 528, 530, 536, 537, 547, 564, 565, 579, 588, 591, 604, 606, 609, 611, 612, 615, 617, 619
Disease: distribution by housefly, 467
  insect carriers, 458, 566, 636–642
  of insects, 588–590, 644–645
Dispersal: past geographical, 507, 543–547
  periodic, 576
Distal nerve fiber, 177
Diversification of terrestrial colonies, 536
Diversity patterns, 556, 558
Diving air stores, 160, 581
Diving beetles, adult, 160, 581
Dixon, A. F. G., 263, 363
Dobsonflies, 237, 366, 610
Dog fleas, 224
Dolichopodidae, 442, 451, 452, 460, 591
Dolichopsyllidae, 438
Dolichopus, 451
Dome receptor, 177
Domestic animals, insects affecting, 639–642
Dominick, R.B., 502
Donacia, 383
Doodlebug, 372
Dorilaidae, 464
Dormant oils, 655
Dorsal diaphragm, 112, 165
Dorsal sinus, 112
Dorsal vessel, 110, 111
Dorsum, 62
Douglas-fir tussock moth, 658
Downes, J. A., 472
Draeculacephala mollipes, 348
Dragonflies, 161, 170, 173, 234, 236, 268, 289–294, 522, 535, 539, 581, 591, 610
Drainage, in soil, 579
Driver, E. C., 57
Drone bee, 596

Drone fly, 596
Drosera, sundews, 591
Drosophila, 179, 182, 441, 574, 624
  melanogaster, 82, 465, 466
Drosophilidae, 442, 444, 453, **465**
Dryinidae, 410, 564, 565
Dryopidae, 375, 385, 392
Dryopoid beetles, 392
Dunbaria fasciipennis, 536
Dung beetles, 627, 628–629
Dung flies, 464, 465, 629
Dunnet, G. M., 439
Dupuis, C., 25
Durden, C. J., 548
Dutch elm disease, 636
Dynastes tityus, 389, 390
Dysentery, 640
Dytiscidae, 160, 374, 375, 382, 383, 385
Dytiscus, 382

Earwigs, 209, 269, **310**–312, 537
Earworm, 619
Ebling, W., 303, 664
Ecdysis, 135–137, 234
Ecdysone, 195, 245, 262
Echidnophaga gallinacea, 438, 439
Echinophthiriidae, 327
Eclosion, 232
Eclosion hormone, 194
Ecological tolerance, 566
Ecology, 551–602
Economic-injury level, 661
Economic threshold, 661
Ecosystem, 551, 552, 554–559, 663
  balance of, 563
  modification of, 566
Ecotones, 554
Ectoderm, 107, 229
Ectoparasites, 563–566, 590, 615–619
Edmondson, W. T., 57
Edmunds, G. F., Jr., 105, 288
Edwards, H., 18
Edwards, J. G., 407
Edwards, Milne, 14
Efferia interrupta, 612
Egg, 223–233
Egg burster, 233
Egg-laying, 241
  see also Oviposition
Eggplant leaf miner, 490
Ehrlich, P. R., 502
Eisner, T., 174, 221
Ejaculatory duct, 124
Elateridae, 188, 196, 375, 380, 384, **391,** 608
Electrocrania immensipala, 532
Elephantiasis, 641
Eleodes, 398, 597
Elm leaf beetle, 636
Elmidae, 375, 384
Elton, C., 601
Elytra, 198, 373, 519
Elytroleptus apicalis, 598
  ignitus, 598
Embaphion muricatum, 398
Emerson, A. E., 303
Emerson, K. C., 330
Embiids, 313, 523
Embioptera, 52, 218, 266, 269, **313**–315, 504, 523, 524, 542
Embryo, 226–233
Embryology, 64, 223–233
Empididae, 442, 443, 448, 449, 452, 460, 547, 611
Empis, 117, 449
Empoasca fabae, 576
Empusal vein, 95, 96
Empus muscae, 589
Encyrtidae, 619
Endelomyia aethiops, 244, 418
Endites, 70, 71, 72, 73, 74
Endocrine system, 175, 193–195, 260–263
Endocuticle, 129, 130, 135
Endoderm, 107, 229
Endoparasites, 588, 615–619
Enemies of insects, 562–566, 588–592, 603–619
  fungi, bacteria, viruses, 589–590
  other animals, 588–589
  other insects, 588, 603–619
Energy flow, 552, 554–559
Engelmann, F., 263
Enicocephalidae, 337
Enicocephalomorpha, 337
Enoclerus sphegeus, 608, 609
Entognatha, 44, 50, 52, 72, 226, 265–267, 279–284, 504
Entomobryidae, 283
Entomology: history of, 4–24
  literature of, 24–25
  relations to humans, 1–4
Entotrophi, 279
Environment: effect of mortality, 570–572
  effects on development, 568–570
  humidity and evaporation, 552, 572
  light, 567
  precipitation, 552, 572
  stability of, 563
  temperature, 552, 567, 573
Environmental factors, 552, 567–584
Enzymes, 140, 143, 167
Epargyreus tityrus, 500
Ephemera, 286
Ephemerella grandis, 287
Ephemerida, 286
Ephemeridae, 288
Ephemeroptera, 52, 76, 96, 234, 236, 266, 268, 275, **286**–288, 504, 522
Ephestia elutella, 492
Ephydridae, 442, 453, 464, 465

Epicauta, 396
pennsylvanica, 396, 397
vittata, 396, 398
Epicranial arms, 67
Epicranial stem, 67
Epicuticle, 129, 130–132, 135
Epidemic outbreak of insect pests, 566
Epidermis, 128, 129, 135
Epilachna, 381
varivestis, 395
Epilimnion, 582
Epimeron, 90, 91
Epipharynx, 72, 83
Epiphragma fascipennis, 456
Epiproct, 100
Episternum, 90, 91
Epithelium, 115
Epitrix, 339
hirtipennis, 400
Erebus, 576
Eremotylus, 411
Eretmocerus serius, 607
Eriococcidae, 352
Eriocraniidae, 480–482, 486
Eriophyidae, 40, 41
Erisoma lanigerum, 618
Eristalis, 462
aeneus, 463
tenax, 463, 596
Errastunus ocellaris, 565
Esophagus, 108, 109, 110, 141
Esig, E.O., 25
Ethiopian faunal realm, 273
Etiolation, 356
Eubasilissa pardalis, 474
Eucleid moth, 594
Eumecoptera, 528
Eupatorium adenophorum, 620
Eupelmidae, 619
Euplexoptera, 310
European blister beetle, 624
European brown-tail moth, 619
European carabid, 607
European corn borer, 22, 492, 616, 617, 619, 620, 633, 644, 646, 648, 649
European earwig, 312
European elm bark beetle, 636
European spruce sawfly, 420, 617, 636, 637
Eurosta solidaginis, 247, 440
Eurypauropodidae, 49
Eusternum, 91
Euthycera, 450
Euthygramma parallelum, 519
Euoniticellus intermedius, 627
Euura, 418
hoppingi, 248
salicisnodus, 248
Evaniidae, 410, 412
Evans, H. E., 221, 433
Evans, J. W., 363
Evaporation, 158, 572, 574
Evolution, 265–272, 522–547, 557
Ewing, H. E., 281, 330, 439
Excretion, 148, 149–153
Excretory system, 110, 149–153
Exenterus abruptorius, 617
amictorius, 617
claripennis, 617
confusus, 617
tricolor, 617
Exites, 72, 73
Exocrine glands, 202
Exocuticle, 129, 130, 132, 135
Exoskeleton, 59, 128
Extension services, 663
External parasites, 615–622
External respiration, 113
Extraintestinal digestion, 140
Exuviae, 234
Exyra, pitcher plants, 592
Eyes, 65, 66–69, 117, 185–191, 229
apposition image, 189
superposition image, 189

Fabricius, 9, 10
Face of head, 65
Facet, eye, 186
Fairyfly, 540
Fall armyworm, 498
Fall webworm, 218
False wireworm, 398
Fanniidae, 452
Fat body, 125, 168, 169
Fatty acids, 167, 168
Faunal mixing, 545
Faunal realms, modern, 273, 540
Felt, E.P., 18, 263
Femur (pl. femora), 92, 93, 121, 122
Fernald, C.H., 20
Ferris, G.F., 330
Fertilization, 259
Fibrils, 170
Field-crop insects, 633
Figs, 423
Fig wasp, 424
Filaria, 641
Filariasis, 21, 641
Filiform antennae, 71
Filter chamber, 146, 147
Fire ants, 430
Fire blight bacteria, 637
Firebrats, 97, 284, 285
Fireflies, 196, 376, 390
Fitch, A., 13, 15
Flacherie, 590
Flagellum, 71
Flannagan, J.F., 288
Flat bugs, 360

Flathead apple tree borer, 392
Flat-headed borers, 376
Flathead wood borer, 392
Flattid bugs, 592
Flea beetles, 621, 633
Fleas, 142, 224, 238, 271, 436–440, 527, 528, 537, 639
Flesh flies, 466, 469
Fletcher, J., 17
Flicker vision, 190
Flies, 171, 172, 180, 236, 440–473, 527, 536, 557, 564, 565, 611, 642
Flight, 170–173
Flightless moss bug, 547
Flour beetle, 398, 399, 575
Flour moth, 206
Flower, J.W., 97, 105
Flower flies, 444, 462, 463, 611, 612
Flowers, insect mimics, 594
Flower thrips, 332
Flow receptors, 181
Follicle mites, 40
Follicles, 124
Follicular epithelial cells, 258
Folsom, J.W., 105
Food, 584
  habits, 76–87, 251–255, 554–556, 603–619
  insects as human food, 624
Food chains, 554–557
Food sources, shifts in, 251–255, 558
Food storage, 83, 148, 215–218
Foote, R.H., 273
Foot pad, 181
Forbes, S.A., 18, 20
Foregut, 107, 108, 119, 142, 232
Forest insects, 636
Forest management, 646
Forest tent caterpillar, 585
Forficula, 209
  auricularia, 312
Formica polyctena, 179
Formicidae, 410, 412, 414, 428–430, 530, 613
Formulation, 651
Fossils, insect, 503–506, 511–542
Fox, I., 439
Fraenkel, G.S., 221
Frankliniella tritici, 332
Free, J.B., 629
Frenatae, 481, 483, 486–502, 532
Frenulum, 481, 482
Frison, T.H., 318
Froeschner, R.C., 363–372
Frontal connective nerve, 118, 119
Frontal ganglion, 118, 119, 120
Frontal suture, 67, 69
Frontoclypeal suture, 67, 69
Frost, S.W., 472
Fruit flies, 82, 103, 110, 465
Fruit insects, 634
Fulgoridae, 337, 339, 341, 346–347, 609
Fumigants, 658
Fungi, 588, 590
Fungus beetles, 375
Fungus diseases, 637, 644
Fungus gnats, 441
Furca, 91, 92
Furcula, 282

Galea, 74, 75
Galium, 210
Galleria mellonella, 492
Gall fly, 247
Gall gnats, 453
Gall makers, 247
Gall midges, 441
Galls, 247–248, 424
Gall wasp, 247, 424
Ganglia, 117–120, 192, 193
Garden fleahopper, 356
Garden insects, 634
Garden webworms, 492
Gargaphia, 342
Garreta nitens, 627, 628
Gaster, 410, 421
Gasterophilidae, 442, 451, 452, **467**–468
Gasterophilus, 451, 467
  haemorrhoidalis, 468
  intestinalis, 467, 468
Gasteruptionidae, 410, 411, 615
Gastric caeca, 107, 108, 109
Gastric mill, 108
Gastrulation, 228, 229, 230
Geier, P.W., 664
Gelastocoridae, 337, 339, 341, 343
Gelastocoris, 339
Gelechiidae, 481, 490
Gena, 67, 69
Genal suture, 67, 69
Genetic control, 643, 645
Genetic strains, 645
Genetic studies, 624
Genitalia: female, 101–102, 103
  male, 102, 104
Geographical changes in faunal realms, 541–542
Geometridae, 481, 483, 485, 486, 494
George, R.S., 439
Geosargus, 451
Geotaxis, 208
Geotrupes splendidus, 629
German cockroach, 296, 298
Germarium, 258
Germ band, 226
Germ cells, 225
Germ disc, 226
Germ layers, 228
Gerridae, 337, 338, 340, 344, 354, 355

Gerris, 340
  rufomaculata, 354
Gerromorpha, 337, 338, 355
Gertsch, W.J., 57
Giant water bugs, 359
Gibbs, A.J., 665
Gigantostraca, 37
Gilbert, P., 26
Giles, E.T., 312
Gill, T., 25
Gillett, J.D., 472
Gill respiration, 161
Gills: rectal, 117, 161
  segmental, 101
Gland cell, 129
Glands, cephalic tubular, 151
Glendotricha olgae, 532
Glossae, 75, 84
Glosselytrodea, 504, 521, 526, 535
Glossina, 243, 471, 542, 641
Glossinidae, 442, 471
Glossinoidea, 441, 442, 444, 466, **470–472**
Glossosoma intermedium, 545
Glossosomatidae, 474, 476
Glover, T., 12
Glowworms, 196
Glycogen, 168
Glycoprotein, 130
Glypta rufiscutellaris, 422
Gnathal segments, 33
Gnathochilarium, 48
Gnats, 511, 659
Gnorimoschema, 490
  operculella, 617
Goeridae, 476
Gomphocerus rufus, 194
Gonads, 122–125
Gondwana, 511, 547
Gonopods, 104
Gonopophysis, 104
Gonopore, 124
Gradual metamorphosis, 235–236
Granary weevil, 403, 638
Granulocytophagous cells, 163
Grape berry moth, 491, 636
Grape mealybugs, 610
Grape phylloxera, 254
Grapholitha molesta, 491, 616
Grasshoppers, 65, 73, 118, 137, 139, 194, 197, 224, 233, 236, 269, **304–307**, 523, 572, 587, 590, 592, 610, 633, 634
Gravity, reactions to, 208
Greater wax moth, 492
Greenhouse insects, 634
Greenhouse leaf tier, 492, 634
Greenhouse thrips, 335
Green peach aphid, 350, 634
Green striped mapleworm, 497
Grote, A.R., 13, 18
Ground beetles, 100, 378, 385, 386, 607
Growth, 127, 223–241
Grumulus, 227
Gryllacrididae, 308
Gryllidae, 305, 308–309
Gryllobatta, 570
Grylloblattids, 312–313
Grylloblattodea, 52, 266, 269, 278, 312–313, 504, 570
Grylloblatta campodeiformis, 313
Gryllotalpa, 218
  hexadactyla, 309
Gryllotalpidae, 305, 309
Gryllus, 199
Guinea pig lice, 325
Gupta, A.P., 57
Gurney, A.B., 300, 313, 321, 323, 372
Guthrie, D.M., 300
Gypona, 339
Gypsy moth, 21, 422, **497**, 585, 607, 615, 619, 636, 644, 647
Gyrinidae, 374, 375, 377, 380, 385, 386
Gyropidae, 325

Habituation, 212
Hackman, R.H., 174
Haddow, J.R., 439
Haematomyzus elephantis, 325
Haematopinidae, 328
Haematopinus asini, 328
Hagan, H.R., 263
Hagen, H.A., 20
Hair, 62, 179
Halictidae, 410, 416, 417
Halictus, 417
Haliplidae, 375, 382, 383, 385
Haliplus, 382
Halteres, 180, 440
Haltica, 400
Halticus bracteatus, 356
Hamilton, K.G.A., 96, 105, 273
Handlirsch, A., 548
Hanging flies, 613, 614
Haplodiploidy, 219–220, 244
Haplorhynchites aeneus, 131
Harborne, J.B., 665
Harcourt, D.G., 601
Hardy, D.E., 273
Harlequin bug, 353, 362
Harmolita grandis, 423
  tritici, 423
Harpalus, 378, 381
Harper, J., 601
Harper, P.P., 318
Harris, K.F., 665
Harris, T.W., 11, 12
Harrison, G., 665
Harvestmen, 41
Harvey, 5

Hassell, M.P., 601
Hatch, M.H., 407
Hatch Act, 17
Hatching, 232
Hawk moths, 112, 184, 185, 200, 492, **494**
Head, 63–85, 229
Head capsule, 63, 65, 66, 67
Headley, J.C., 665
Hearing organs, 178–180, 182–185, 196–201
Heart, 110–113, 165
Hebard, M., 300, 312
Hectopsyllidae, 438
Heidemann, O., 18
Heinrich, B., 221, 629
Heleomyzidae, 442, 452, 464
Helfer, J.R., 310
Helichus, 382
Heliconiidae, 596, 597
Heliconis ethilla, 559
Helicophidae, 476
Helicopsychidae, 476
Heliocopris gigas, 627
Heliothis armigera, 662
  zea, 498, 619
Heliothrips haemorrhoidalis, 335
  rubrocinctus, 333
Hellgramite, 367
Helobia, 451
Hemerobiidae, 369, 370, 372
Hemerobius, 370
Hemerocampa leucostigma, 497, 498
Hemimeridae, 312
Hemimetabolous development, 236
Hemiptera, 52, 146, 172, 180, 197, 206, 233, 236, 251, 253, 266, 270, 274, 275, 278, 335–368, 504, 525, 535, 540, 547, 564, 565, 591, 599, 610, 611, 615, 617
Hemipteroid orders, 266, 319, 504, 525
Hemlock loopers, 636
Hemocoel, 111
Hemocytes, 111, 163–164
Hemolymph, 111, 162
Hennig, W., 273
Henous, 396
  confertus, 398
Hepatica, 584
Hepburn, H.R., 174
Hepialidae, 480, 481, 483, 486
Herbridae, 337
Hercules beetle, 389
Hermann, H.R., 221
Hermaphroditic insects, 122
Herms, W.B., 665
Herring, E.M., 472
Herring, J.L., 363
Hesperiidae, 481, 482, 483, 484
Hesperoboreus, 436
Hessian fly, 453, 454, 633, 646, 647, 649
Heterarthrus nemorata, 418
Heteroceridae, 375, 380
Heterocerus, 380
Heterojapyx gallardi, 123
Heteromera, 396
Heteroptera, 150, 198, 337, 352–363, 595
Heterothripidae, 334
Heterothrips, 334
Heterotrophs, 554–555
Hexagenia, 286
  limbata, 286
Hexapoda, 35, 50, 52, 265–272, 504
Hindgut, 107, 108, 109–110, 145, 232
Hinks, W.D., 312
Hinton, H.E., 407, 436
Hippoboscidae, 442, 444, 471, 639
Hippodamia convergens, 395, 605, 606
Histeridae, 375, 376, 384, 608
Histogenesis, 240
Histolysis, 240
Hitchcock, S. W., 318
Hoff, C. C., 57
Holland, G. P., 439
Holland, W. J., 502
Holoarctic assemblage, 544
Holoblastic cleavage, 226
Holocrine secretion, 143
Holometabola, 364
Holometabolous development, 236
  insects, 67, 364, 558
Homogenous layer, 132
Homoptera, 146, 147, 337, **346–352**, 525, 535, 611, 615, 617
Honey, 623
Honeybee, 84, 149, 156, 170, 173, 201, 413, **432–433**, 567, 590
Honey dew, 430
Hopkins, G.H.E., 331, 439
Hoplocerambyx spinicornis, 575
Horistonotus, 391, 392
Hormones, 125, 193–195, 245–246, 260–263, 656
  activation, 261, 262
  ecdysone, 195, 262
  eclosion, 194–195
  endogeneous, 260
  exogenous, 260
  glandular, 194–195, 261
  juvenile, 262–263
  molting, 262
  neotenin, 262–263
  neurohormones, 194–195, 261
  releaser, 194, 195
  tissue, 194–195, 261
Horn, D.C., 273
Horn, G.H., 18
Hornet, 408, 426, 612
Horn flies, 639
Horntails, 409, 410, 418–420, 421, 604
Hornworms, 633
Horogenes punctorius, 616, 617

Horridge, G.A., 221
Horse botfly, 467–468
Horseflies, 79, 173, 443, 459, **461–462**, 547, 581, 612, 613, 639
Horseshoe crab, 37
Horse sucking louse, 329
Horsfall, W.R., 665
Host crossover, 585
Hosts of insects, 247–250
House, H.L., 174
House ant, 430
Houseflies, 83, 114, 173, 224, 466–467, 589, 640, 659
Housefly fungus, 589
Hover fly, 173, 462, 464
Howard, L.O., 18, 25
Howe, W.H., 502
Hubbard, C.A., 440
Huffaker, C.B., 629, 665
Hull, F.M., 473
Human flea, 436
Human habitation, pest of, 638, 642
Humeral crossvein, 96
Humidity, 182, 572, 574, 575
Hungerford, H.D., 363
Hyalophora cecropia, 245
Hydraulic pressure of blood, 162, 164–166
Hydrological cycle, 556
Hydrometridae, 337, 343, 354
Hydrophilidae, 160, 375, 377, 380
Hydrophobic hairs, 160
Hydropsychidae, 476
Hydropsychoidea, 473, 476, 477
Hydroptilidae, 474, 476, 478
Hydroscaphidae, 376, 387
Hygrotaxis, 210
Hylemya, 449
  antiqua, 466, 467
  brassicae, 466, 608
Hylobittacus apicalis, 614
Hymenoptera, 52, 75, 76, 79, 109, 112, 137, 139, 146, 150, 172, 214, 218, 236, 266, 270, 271, 277, 407–434, 504, 527, 530, 531, 536, 539, 540, 564, 565, 570, 588, 590, 612–613, 615–619
Hymenopus coronatus, 594
Hynes, H.B., 318
Hyperaeschra, 485
Hypera postica, 619
Hyperaspis binotata, 224
Hypericum perforatum, 620
Hypermetamorphosis, 238, 239, 422
Hyperparasites, 250, 423
Hyphantria, 218
Hypoderma bovis, 469
  lineatum, 470
Hypognathous, 63
Hypolimnion, 583
Hypopharynx, 65, 75, 77, 79, 83, 109
Hystrichopsyllidae, 438

Ice age, 542, 544
Icerya purchasi, 122, 606
Ichneumonidae, 247, 410, 411, 416, **421**–422, 570, 588, 615
Ichneumonoidea, 530
Ichneumon wasp, 224, 421, 422, 570, 604, 605, 616, 617
Identification by pheromones, 205
Illies, J., 318
Imaginal disc, 237, 238
Imago, 234, 268, 286
Imported cabbageworm, 501
Imported currantworm, 419
Impulse, 176–177
Incurvarioidea, 481
Indian meal moth, 492, 638
Industrial melanism, 592–593
Ingestion, 140
Inocellia, 365
Inocelliidae, 365
Inquilines, 220
Insect: carriers: of animal pathogens, 640–642
    of plant pathogens, 636–637
  control: by bacteria, 590
    by viruses, 590
  crop losses, 2–4
  development, 223–240, 568–570, 571, 582
  dispersal, 576
  diversity, 51–56, 272
  galls, 247–248, 424
  growth regulators, 656
  irritants to humans, 640
  literature on, 24–25
  losses, 2–4, 21–23
  parasites, 604, 615–619, 643
  pest control, 604, 642, 661–662
  pests, 585, 631–643
    of plants, 632–638
    of stored food, 637–638
  physical control of pests, 649–651
  pollination, 433, 539, 622–623
  predators, 370–373, 591, 604–614
  table of orders, 52, 266, 504
Insecta, 44, 50–56, 266–272, 504
  evolution of, 33–36, 63–66, 504, 522–545
Insecticides, 12, 22–24, **651–658**
  precautions in using, 653
Insectivores, 558
Insects: colonial, 213—220, 300–303, 425–433
  social, 203, 213–220, 300–303, 425–433
  southern temperate zone, 547
  temperature range, 570
Instar, 234
Integrated control of insects, 24, 660–661
Integrated pest management, 610, 661–663
Integripalpia, 473, 476
Integument, 128–137

Intercalary segment, 29, 64
Intermediate metabolism, 168–169
Internal parasites of insects, 423, 588–590
Internal skeleton, 92
International Rules of Zoological Nomenclature, 19
Interneurons, 192, 193
Intestine, 107–110, 140–147
Intima, 115, 116
Ips, 206
Iridomyrmex humilis, 430
Iron longimanus, 287
Irritability, 175
Isaria, 590
Ischnocera, 325
Ischnopsyillidae, 438
Isoperla confusa, 316
Isopoda, 44, 45
Isoptera, 52, 197, 214, 266, 269, 276, 278, **300–303**, 504, 523
Isotoma andrei, 282
  cinerea, 231
Itch mites, 639
Itonididae, 453, 611
Iwata, K., 221
Ixodidae, 40

Jackpine sawfly, 613
Jacobson, M., 221
Jalysus, 340
Janus, 411
Japanese beetle, 389, 426, 568, 590, 644, 648, 658
Japygidae, 124, 279
Japygids, 112, 279
Japyx diversiunguis, 280
Jascopidae, 539
Juargus pseudocellaris, 565
Jefferson, T. 10
Jewett, S.G., Jr., 318
Johannsen, O.A., 263
Johnson, C.G., 221
Johnston's organ, 179, 180, 184
Joint: dicondylic, 170
  monocondylic, 170
Jones, D.P., 665
Jones, J.C., 163, 174
Joose, E.N.G., 283
Jugal bar, 96
Jugal veins, 96
Jugatae, 480, 481, 486, 532
Jugum, 480, 482
Jumping plant lice, 348
June beetles, 373, 389
Jurina marginata, 521
Juvenile hormone, 194, 246, 262–263

K-strategists, 560, 562–566
Kairomones, 205, 206
Kaston, B.J., 57
Katydids, 196, 307
Ked flies, 243
Keiferia inconspicuella, 490
Kennedy, J.S., 363
Key, K.H.L., 304, 310, 313
Kim, K.C., 331
Kineses, 207–208
King, P.E., 57
Kissing bug, 611
Klots, A.B., 502
Knight, K.L., 473
Kormondy, E.J., 295
Koss, R.W., 289
Krantz, G.W., 57
Krishna, K., 221, 303
Kristensen, N.P., 57, 273
Krombein, K.V., 433
Kukalova, J., 548

Labella, 83
Labial glands, 109, 140, 151
Labial palpi, 75, 184
Labia minor, 310
Labium, 65, 66, 71, **74**–75, 86, 109, 229
Laboulbenaceae, 590
Labral apodeme, 64
Labrum, 29, 65, 67, 69, 71 **72**, 141, 229
Laccifer lacca, 623
Lace bugs, 355
Lacewings, 89, 170, 224, 368, **369**–373, 535, 543, 610
Lacinia, 74, 75, 86
Ladybird beetles, 224, 374, 376, 394–395, 605, 606
Laemobothriidae, 327
Lamarck, 14
Lamellate antennae, 71
Lamellicorn beetles, 389
Laminate cuticle, 130
Lampyridae, 196, 376, 384, **390–391**, 608
Lampyroids, 375, 376, 390–393
Languria, 380
Languriidae, 375, 380, 383
Lanternaria phosphorea, 347
Lantern fly, 347
Larch sawfly, 418
Larder beetles, 376, 638
Larson, S.G., 548
Larva, 233–240
  aquatic, 161
Larvaevoridae, 469
Larval nests, 109
Lasiocampidae, 481, 484, 485, 493
Lateral process, 70, 72
Lateral tracheal trunks, 115
Lauterback, K.-E., 273
Lauxaniidae, 442, 452, 464

Lawrence, P.A., 263
Layne, J.N., 440
Leaf beetles, 399
Leaf-cutter bee, 623
Leaf-cutting ant, 430
Leafhoppers, 64, 236, 346–348, 564, 565, 566, 609, 637, 648
Leaf miners, 464, 465
Leaf roller, 491, 492
Learning, 212–213
  associative, 212
  explorative, 212
  insight, 213
  latent, 212
LeBaron, W., 16
Lecanium, 147
LeConte, J.L., 18
Lee, K.E., 303
Leech, H.B., 407
Leeuwenhoek, 5
Leftwich, A.W., 25
Leg joint, 60–61
Legs, 92–93, 229
Lemmatophora typa, 520
Lens, 186, 190, 191
Lepidoptera, 52, 77, 87, 102, 109, 112, 133, 137, 140, 150, 180, 196, 197, 218, 266, 275, 278, **479–502**, 504, 527, 531, 532, 539, 588, 594, 607, 615, 617, 659
Lepidosaphes ulmi, 352
Lepidostomatidae, 476
Lepisma saccharina, 285
Lepismatidae, 285
Leptinotarsa, 245, 381
  decemlineata, 399
Leptoceridae, 476, 477
Leptura, 382
Lethocerus, 359
  americanus, 359
Leucopterine, 151
Leuctridae, 318
Levi, H.W., 57
Lewis, T., 335
Libytheidae, 481, 484
Lice, 224, 324–330, 537, 639
Licus fernandezi, 598
  loripes, 598
Life, early, 532
Life cycles, 223–241, 405
Life tables, 560, 561
Light, 208, 567
  reaction to, 184–191, 208
Lightning bugs, 196, 391, 608
Limenitis archippus, 597
Limnephilidae, 476, 477
Limnephiloidea, 473, 474, 476
Limnephilus rhombicus, 476
  submonilifer, 475
Limnocentropodidae, 476
Limonius californicus, 579
  canus, 579
  infuscatus, 579
  subauratus, 579
Limulus, 37
Lindorus lophantae, 607
Linguatulids, 31, 32
Lingula, 75
Linnaeus, C., 8, 9, 10
Linsenmaier, W., 273
Liomopterum ornatum, 518
Liparidae, 481, 483, 486, 497
Lipid layer, 132
Lipids, 167, 168
Lipke, H., 221
Liposcelis divinatorius, 323
Liquid diet, 145–146
Lithoneura mirifica, 505
Lithosmylus columbianus, 543
Litomastix truncatellus, 233
Locke, M., 174
Locust borer, 401
Locusta, 180
Locusts, 180, 197, 305–306, 559
Locy, W.A., 25
Lomamyia, 370
Longevity of adults, 243
Longhorn beetles, 400, 401
Longhorned cricket, 183, 308
Long-horned grasshoppers, 307–308
Louse flies, 471
Löve, A., 548
Loxostege rantalis, 492
Ludius inflatus, 391
Lycaenidae, 481, 484
Lycinae, 597
Lycids, 598
Lyctidae, 608, 638
Lyctocoris, 340
Lyctus, 608
Lydella thompsoni, 619, 620
Lygaeid bugs, 361–362
Lygaeidae, 337, 340, 344, 346, 361–362
Lygaeus, 340
Lygus, 339, 342
  lineolaris, 356
Lymantria monacha, 103
Lyonet, 7
Lytta vesicatoria, 624

McAlpine, D.K., 548
McAlpine, J.F., 548
McDunnough, J., 502
McEwen, F.L., 665
McGregor, E.A., 58
McGregory, S.E., 630
Machilidae, 267
Machilis, 97
  variabilis, 104
McKenzie, H.L., 364
Mackerras, I.M., 273

McKittrick, F.A., 300
MacLeod, E.G., 372
Macracanthorhynchus hirudinaceus, 589
Macrocentrus ancylivorus, 616, 617
  gifuensis, 233
Macrochile spectrum, 529
Macrofrenatae, 492
Macroglossa, 209, 210
Macromia magnifica, 293, 294
Macropsis trimaculata, 348
"Macros," 492
Macrosteles fascifrons, 576
  focifrons, 87
Macroxyela, 66
MacSwain, J. W., 57
Maddrell, S. H. P., 174
Magicicada, 200
  septendecim, 347
Malacosoma, 218, 485, 492
  americana, 109, 493
Malaria, 21, 641
Malathion, 655
Male glands, accessory, 151
Mallis, A., 25
Mallophaga, **325–327,** 639
Malpighi, 5
Malpighian tubules, 107, 108, 109, 110, 149–152
Malyshev, S.I., 434
Management practices, 645–648
Mandibles, 65, 67, 71, **72**–73, 87, 229
Mandibular muscles, 120, 122
Mandibulata, 32, **33**–36, 43–56
Manduca quinquemaculata, 494
  sexta, 494
Manson, P., 21
Mantidae, 298
Mantids, 269, 295, **298**–300, 610
Mantispa brunnea, 371
Mantispidae, 369, **370**–371
Mantispids, 369, 370, 371
Mantis religiosa, 298
Mantodea, 296, **298**–300, 542, 610
Manton, S.M., 27, 35–36, 57
Maple worm, 495, 496
Maramorosch, K., 665
March fly, 443, 547
Margaropus annulatus, 642
Masarinae, 426
Mating, 241, 259–260
Matsuda, R., 105, 263
Matthews, R.W., 221, 434
Maxillae, 65, 67, 71, **73**–75, 81, 229
Maxillipeds, 44
May, R.M., 601
Mayatrichia, 478
May beetle, 378
Mayetiola destructor, 453, 454
Mayflies, 6, 78, 161, 234, 236, 268, **286**–288, 505, 522
Maynard, E.A., 283
Meadow grasshoppers, 307
Mealworms, 638
Mealybugs, 352, 353, 634
Measuring worms, 494
Mechanoreceptors, 176–181
Mechilis maritima, 101
Mecoptera, 52, 266, 270, 271, 274, 276, 278, 407, 408, 430, **434**–436, 504, 527, 528, 535, 537, 547, 591, 613
Medfly, 22, 465, 599, 657
Media, 95
Median Plate, 94
Medicago sativa, 648
Medical entomology, 21
Mediterranean flour moth, 638
Mediterranean fruit fly, 22, 465, 599, 657
Megachile pacifica, 623
Megachilidae, 410, 416, 417
Megacyllene robinae, 401
Megaloptera, 52, 236, 237, 266, 271, 276, 364, **366**–367, 504, 526, 535, 610
Megalothorax, 282
Meganeura moyni, 517
Meganisoptera, 517
Megaphasma dentricus, 304
Megarhyssa lunator, 604, 605
Megasecoptera, 504, 515–516, 523, 534
Megaselia, 449
Megasida, 597
Megaxyela aviingrata, 592
Melanin, 133
Melanism, 593
Melanoconion, 546
Melanolestes picipes, 360
Melanophila, 182
Melanoplus bivittatus, 306, 587
  differentialis, 306, 587
  femur rubrum, 306
  mexicanus, 306
  spretus, 306
Melanotus, 380, 392
Melipona, 623
Melittia cucurbitae, 490
Melittidae, 410, 416
Meloe angusticollis, 398
Meloidae, 240, 375, 379, **395**–398, 608
Meloids, 239, 398
Melon aphid, 350, 634
Melon worms, 634
Melophagus, 471
  ovinus, 472
Melsheimer, F. V., 11
Membracidae, 337, 342, 346
Menacanthus stramineus, 326, 327
Mengeidae, 404, 406
Menke, A.S., 364
Menopon gallinae, 327
Menoponidae, 327
Mentum, 75
Merocrine secretion, 143

Meromyza americana, 44(
Meron, 92
Meropeidae, 435, 528
Merope tuber, 435
Merostomata, 37
Merritt, R.W., 274
Mesenteron, 107, 108, 109, 143
Mesoderm, 229, 230
Mesomachilis, 284
Meson, 62
Mesoraphidia pterostigmalis, 527
Mesosoma, 421
Mesothorax, 89, 90
Mesoveliidae, 337
Mesypochrysa latipennis, 526
Metabolism, 164, 167–169
Metallic wood borers, 392
Metamorphosis, 125, 169, **223**, **233**–241
  complete, 236–238
  gradual, 235–236
  hypermetamorphosis, 238
  indirect, 236
Metasoma, 421
Metathorax, 89, 90
Metcalf, C.L., 665
Metcalf, R.L., 665
Meteorus versicolor, 615
Mexican bean beetle, 395, 634
Miastor, 241, 255
Michelbacher, A.E., 58
Michener, C.D., 221, 434
Mickel, C.E., 25
Microbembix monodonata, 587
Microbracon, 209
Microcentrum rhombifolium, 307
Microceromasia sphenophori, 619
Microchorista philpoti, 79
Microcoryphia, 52, 104, 105, 266, 267, 277, **284**–285, 504
Microfrenatae, 487
Microlepidoptera, 487
Micromalthidae, 375, 387
Micromalthus debilis, 241, 255
Micropterygidae, 480, 481, 486
Micropyle, 224, 225
"Micros," 487
Microstelus focifrons, 64
Microthorax, 86
Microtypus longicornis, 530
Midges, 160, 184, 256, 443, 453, **458**, 539, 547, 581
Midgut, 107, 108, 109, 111, 143, 232
Milkweed, 597
Milkweed bug, 361
Milkweed butterfly, 595
Milky disease, 590, 645, 658
Miller, N.C.E., 364
Miller moths, 492, 497
Millipedes, 48
Mills, H.B., 283
Milne, L.J., 364
Mimicry, 196, 596–599
  aggressive, 598
  automimicry, 597
  Batesian, 597
  Mullerian, 597
  mimic rings, 597, 599
Miomoptera, 504, 517, **518**, 519, 523, 534
Miridae, 337, 338, 344, 345, **356**–357, 591, 610
Mitchell, E.R., 665
Mitchell, T.B., 434
Mites, 42, 610, 627, 639, 642, 659
Mnemonica, 482
Mockford, E.L., 323, 324
Modifier effects of hormones, 194
Molannidae, 476
Mole crickets, 219, 309
Molting, 135–137, 195, 234
  fluid, 135–136
Monarch butterflies, 597
Moniliform antennae, 71
Monocrepidius, 392
Monoctenus, 413
Monomorium minimum, 428
  pharaonis, 430
Monura 267, 504, 514
Mordellidae, 375, 379, 380
Mordellistena, 380
Mormon cricket, 308, 643, 650
Morpho, 135
Morrill Act, 15
Mortality, 561, 570–572, 574
Moscoidea, 442
Mosquitoes, 85, 86, 138, 153, 172, 180, 184, 187, 224, 238, 443, 453, **456**–458, 557, 610, 611, 639, 641, 642, 645, 659
Mosquito larvae, 581
Moth fly, 443
Moths, 170, 173, 196, 197, 236, **479**–**499**, 527, 531
  pests, 490, 491, 492, 493, 494, 497, 498
Mound, L.A., 335, 364
Mouth, 29–35, 65, 107, 108, 141, 229, 232
Mouthparts, 72–87
  chewing, 77–79
  chewing-lapping, 83–84
  cutting-sponging, 79, 81
  filtering, 79, 80
  piercing-sucking, 84–86, 87, 270
  siphoning-tube, 85, 87
  sponging, 83
Mud dauber wasps, 431
Mudge, 20
Mud wasps, 431
Muesebeck, C.F.W., 434
Müller, 14
Mullerian mimicry, 597
Murgantia histrionica, 353, 363
Musca, 449
  domestica, 224, 467, 564, 589
Muscidae, 442, 449, 450, **466**–467

Muscle, 120–122, 169–173
appendages of, 120–122
depressor, 170
diaphragm, 165
direct, 171
flight, 170–173
indirect, 171
levator, 170
reaction, 192–193
segmental bands, 120
visceral, 120
wing-folding, 171
Muscoidea, 422, **466–470**
Musculature, 120–122
Musical organs, 196–202
Mutillidae, 410, 412, 414, 418, **426**, 586
Mutillid wasp, 426, 586
Mycetophilidae, 441, 442, 447
Mydaidae, 442, 447, 451, **461**
Mydas, 451
clavatus, 461
Mydas flies, 461
Mymaridae, 540, 619
Mymar pratensis, 619
Myofibrils, 169
Myriapoda, 34, 46
Myrientomata, 281
Myrmeleontidae, 369, 372
Mystax, 460
Myxophaga, 375, 376, 387
Myzine, 415, 417
Myzus persicae, 350

Nabidae, 337, 340, 344, 345, 610
Nabis, 340
Nachtigall, W., 221
Naiads, 236
Nallachius, 370
Nannochoristidae, 79, 547
Narahashi, T., 665
Nasutes, 302
Natural control, 23, 559–566, 603–622, 648–649
Neal, J.W., Jr., 665
Neck, 86
Necrotaulidae, 531
Needham, J.G., 289, 295
Needham, J.H., 318
Neididae, 340
Nematocera, 441, 442, 453–459, 530
Nematus ribesi, 418, 419
tibialis, 584
Nemestrinidae, 547
Nemobius, 308
fasciatus, 308
Nemopoda, 450
Nemouridae, 318
Neoclytus, 382
Neoculex, 546
Neodiprion lecontei, 63, 420
swainei, 613
Neomecoptera, 435
Neoptera, 52, 98–**99**, 266, **268–272**, 295, 504, 517–522, 523–532
Neosminthurus clavatus, 282
Neotenin, 246, 262
Nepa, 340, 358
Nepenthales, 591
Nephrocytes, 151
Nepidae, 337, 340, 343
Nepomorpha, 337, 341, **358–359**
Nerve cells, 117–120, 176–193
Nerve cord, 66, 118–120
Nervous integration, 191
Nervous system, 117–120, 176–193
Nest-building, 218
Neuroctenus simplex, 361
Neurohormones, 261
Neurons, 192
Neuroptera, 52, 146, 150, 266, 271, 276, 364, **368–373**, 504, 534, 535, 537, 543, 610
Neuroptera-Coleoptera branch, 266, 270, 271, **364–407**
Neuropteroid orders 266, 270, 271, 364, 525–533
Neurosecretion, 192, 193–195
New, T.R., 372
Nezara, 112
Nicoletiidae, 285
Nicrophorus, 380
marginatus, 386
Nidi cells, 143
Nielson, A., 479
Nigronia, 367
Nitidulidae, 375, 385, 608
Noctuidae, 481, 482, 486, **407**
Nocturnal insects, 189
Nomenclature, binomial, 8–9
Nomia cockerelli, 623
Northern cattle grub, 469
Norton, E., 18
Nosopsyllus fasciatus, 436, 438
Notodontidae, 481, 485, 486
Notonectidae, 160, 337, 341, 359
Notum, 88, 89, 90–91
Novák, V.J.A., 221, 264
Noxious secretions, 595
Nuclear polyhedrosis virus, 658
Number of species, 54–55
Nurse cells, 258
Nutrition, 137–139
Nutritive cells, 258
Nycteribiidae, 472
Nygmia phaeorrhoea, 497, 615
Nymph, 236
aquatic, 161
Nymphalidae, 481, 483, 484

Obtect, 487
Occipital condyle, 67, 69, 86
Occipital foramen, 66, 67

Occipital ganglion, 118, 119
Occipital suture, 67, 69
Occiput, 67, 69
Ocellar pedicel, 118
Ocelli, 65, 67, 69, 117, 184, 190
  dorsal, 184, 190
  lateral, 69, 184, 190
Ocellus, median, 69
Ochrotrichia, 478
  unlo, 476
Ocularium, 67, 68
Ocular sclerite, 67
Odonata, 52, 96, 236, 266, 268, 276, **289-295**, 504, 517, 522, 535, 591, 609
Odontoceridae, 476
Odorous house ant, 430
Odors, response to, 202-206, 209-210, 657-658
Oecanthus, 308
  niveus, 308
Oecetis, 477
Oecophylla, 213
Oedipoda, 197
Oedipodinae, 306
Oenocytes, 195
Oenocytoids, 163
Oestridae, 442, 450, 469
Oestrus ovis, 469
Oil emulsion, 654
Oils, 654
Oldroyd, H., 473
Olfactory receptors, 181
Oligarces, 241
  paradoxus, 255, 256
Oligonyx mexicanus, 299
Oligotoma onigra, 315
  saundersii, 314, 315
Oliver, D.R., 473
Ommatidia, 133, 186-190
Omnivorous insects, 139, 557, 558
Omnochromes, 133
Omus californicus, 609
Oncopeltus fasciatus, 361
Onion maggot, 466, 467
Onion thrips, 335
Onthophagus gazella, 627, 628, 629
Onychophora, 30, 31, 113
Oocytes, 258, 259
Oogonia, 258
Ootheca, 297
Opilionida, 40, 41
Opius, 599, 600
  fullawayi, 600
  humilis, 600
  tryoni, 600
Opuntia, 620
Orange sulphur butterfly, 501
Organophosphates, 12, 607, 655-656
Organ systems, 107-125
Oriental cockroach, 296, 298
Oriental fruit fly, 465
Oriental fruit moth, 491, 616, 635
Orthoptera, 52, 102, 112, 123, 142, 149, 150, 180, 183, 198, 200, 266, 269, 275, 276, 278, **304-410**, 504, 523, 534, 611
Orthopteroid orders, 266, 295, 523
Orthorrhapha, 442, 444, 459-462
Orussidae, 409, 410, 411, 418
Osborn, H., 18, 20, 25
Osmeterium, 595
Osmylidae, 542
Ostia, 112, 165
Ostrinia nubilalis, 492
Otitidae, 442, 444, 452, 464
Outbreak of population numbers, 563
Ovarioles, 123, 124, 258
Ovary, 123
Oviduct, 102, 123, 124
Oviparous, 241
Oviposition, 101-103, 241, 604-605, 615
  injury by, 632
Ovipositor, 101-103
Ovum, 223-225
Owen, 14
Owlet moths, 497
Oystershell scale, 352

Pacarina, 339
Pacemaker neuron, 166
Packard, A.S., Jr., 18, 20, 21
Paedogenesis, 241, 255
Pal, R., 665
Palaeopsylla clebsiana, 529
Paleacrita vernata, 495
Paleodictyoptera, 504, 514-515, 523, 534
Paleogeography, 507-513, 541-547
Paleoptera, 52, 99, 266, **268**, **286-295**, 504, 514-517, 522-523
Paleuthygramma tenuicorne, 519
Pale western cutworm, 578
Palpigrada, 38, 40
Palpomyia, 451
Palpus, 74, 75
Pamakami, 620, 621
Pamphiliidae, 409, 410
Pangaea, 535, 536, 537
Pangaeus, 339
Panoistic ovarioles, 258
Panorpa, 436
  chelata, 435
Panorpidae, 434, 435, 436, 528, 613
Panorpodidae, 436, 528
Papilio, 483
  ajax, 596
Papilionidae, 481, 482, 483, 484, 500, 596
Paracalocoris colon, 224
Paracantha culta, 103
Paraglossae, 75, 77
Paraleptophlebia packii, 287
Paralyzing wasps, 424-428
Paramecoptera, 528

Paranotal lobes, 97
Paraproct, 100
Paraquinones, 206
Parasitic insects, 248–250, 423–424, 564–566, 586–590, 603–604, 615–619
primary, 250, 421
secondary, 250
Parasitoids, 249, 423–424, 564–566, 586–587, 588, 604, 615–619, 643
Paratenodera sinensis, 298, 299
Paratrichoptera, 528
Paratrioza cockerelli, 576, 577
Paraulacizes, 339
Parfin, S., 367, 372
Paris green, 18
Parker, J.R., 601
Parthenogenesis, 244, 251–255
viviparous generations, 252
Parthogenetic species, 251–255
Partial metamorphosis, 236
Passalus, 199
Pathogenic organisms, 643, 644–645
Paurometabolous development, 236, 270
Pauropoda, 44, 46, 49
Pea aphid, 648
Peach bark beetle, 403, 404
Peach tree borer, 490, 491
Peacock butterfly, 200
Peairs, L.M., 665
Peanut bug, 347
Pear psylla, 349
Pear thrips, 335
Pea weevils, 400
Peck, W.D., 10
Pectinate antennae, 71
Pectinophora gossypiella, 490
Pedicel, 71
Pediculidae, 328
Pediculus humanus, 328, 329, 330
Pedipalpi, 38
Pegomyia, 450
Pelecinidae, 410, 412, 416, 417
Pelecinus, 417
polyturator, 416
Pelecorhynchus, 547
Peloridiidae, 337, 547
Peloridium, 336
Pemphigus, 253
Pemphredoninae, 431
Penis, 124
Pennak, R.W., 57
Penny, N.D., 436
Pentastomida, 30–32
Pentatomidae, 337, 339, 353, 362–363
Pentatominae, 344, 591
Pentatomorpha, 337, 341, 359–363, 591, 610
Peppered moth, 592
Peregrinus maidis, 346
Pericardial cells, 151
Pericardial glands, 194, 195
Perikaryon, 191
Perillus bioculatus, 353
Periodic cicada, 347
Periodic dispersal, 576
Peripatus, 31
Periplaneta, 297
americana, 296, 298
australasiae, 298
Periplasm, 224, 225
Peristaltic movements, 165
Peritrophic membrane, 144
Perla, 89
Permembria delicatula, 519
Permocicada integra, 525
Permopsocida, 525
Pesticides, 604, 651–658
Pest insects, 401, 632–642
introduced, 643
management, 658–663
Pests of stored grain, 490, 587, 637–638
Petersen, A., 274
Pfadt, R.E., 665
Phaeogenes nigridens, 617
Phalaenidae, 497
Phallus, 124
Pharaoh ant, 430
Pharate condition, 135, 137
Pharyngeal pump, 140
Pharynx, 108, 141
Phasic reception, 177
Phasmatodea, 52, 266, 269, 278, **303–304**, 504, 523, 536, 592
Pheromones, 202, 203–206, 657–658
alarm and alert, 205
sex, 204
Philanisidae, 476
Philonix prinoides, 247
Philopotamidae, 476
Philopteridae, 327
Philorheithridae, 476
Phlaeothripidae, 334
Phloeotribus liminaris, 403
Phoridae, 442, 448, 449, 464
Photinus pyralis, 196
Photoperiod, 194–195, 245–246
hormone control, 245–246
releaser effect, 194–195
Photoreceptor cells, 176, 184–191
Phototaxis, 208
Photurus pennsylvanicus, 390
Phragma, 90, 91
Phryganeidae, 476
Phryganopsychidae, 476
Phthiraptera, 52, 266, 270, 279, **324–331**, 504, 525, 537
Phthorimaea operculella, 490
Phyllophaga, 237, 378, 389
Phyllotreta, 399
Phylloxeridae, 253

Phymata, 340
erosa, 611
Phymatidae, 337, 340, 344, 591
Physical conditions, 552, 566–584
Physical control, 649–651
Physiography, 579
Physopoda, 331
Phytophagous, 246
Pianka, E.R., 602
Picture-winged flies, 444, 464
Pieridae, 151, 481, 483, 484, **501–502**, 596
Pieris, 483, 501
protodice, 501, 615
rapae, 501
Piesmatidae, 337, 345
Pigeon tremex, 604
Pigment cells, 189
Pigments, 133, 151, 184, 185
Pilifer, 184, 185
Pine shoot moth, 491, 617
Pink bollworm, 490
Pipunculidae, 442, 464, 564, 565
Pirate bug, 540, 610
Pisuliidae, 476
Pitcher plants, 591
Placodes, 178
Plant bugs, **356–357**, 591, 610
Plant hopper, 87, 609
Plant lice, 140, 348–350, 570. *See also* Aphids
Plant resistance, 648
Plasma, 111
Plasmocytes, 163
Plasmodium, 588, 641
Platygaster hiemalis, 233
Plecoptera, 52, 96, 266, 269, 276, 277, 279, 316–318, 504, 534, 536, 547
Plecopteroid orders, 266, 315, 504, 521, 524
Plectrotarsidae, 476
Pleistocene dispersals, 542–545
Pleroneura, 413
Pleura, 89, 90
Pleural region, 88, 89
sclerites, 100
suture, 90
Pleurodema, 90, 91, 92
Plical vein, 95
Plodia interpunctella, 492
Plum curculio, 403, 632, 635, 636
Plum leafhopper, 348
Poblicia, 339
Podabrus, 608
Podocytes, 163
Pogonomyrmex, 430
Poison hairs, 133, 564–565, 640
Poisons, 564–565, 594
Polarized light, 190
Polistes, 427, 612
Pollenia, 451
Pollination, 433, 622–623
Pollution, indicators of, 478, 626
Polycentropodidae, 476
Polychrosis, 245
viteana, 491
Polyctenidae, 352
Polyembryony, 233, 234
Polyhedrosis viruses, 590, 645
Polyphaga, 375, 376, 387–407
Polystoechotes, 370
Polystoechotidae, 369, 370
Pompilidae, 410, 416, 417
Popham, E.J., 312
gpopillia japonica, 389
Populations, 559–565, 661–662
density in, 560
dynamics of, 558
genetic differences in, 562
growth in, 560
intrinsic rate of increase of, 561
management of, 661–662
stability of, 563
structure of, 560
Pore canals, 129, 130, 132
Porthetria dispar, 497, 585, 607, 615
Postembryonic development, 223, 233–241
Posterior, body region, 62
Postgena, 67, 69
Postlabium, 75
Postnotum, 90, 91
Postoccipital suture, 67
Postocciput, 67, 69
Potato beetle, 634
Potato psyllid, 576, 577
Potato tuberworm, 490, 617
Poultry insects, 639
Powder-post beetles, 376, 608
Praying mantids, 298–300, 610
Precipitation, 552, 556, 557, 567, 572
Predaceous plants, 591–592
Predator-prey interaction, 560, 563–566
Predators, 560, 563–566, 591, 603–614, 643, 660
birds, 591
cannibalism, 600
centipedes, 591
cryptic, 610
mantids, 370–371
spiders, 591
vertebrates, 591
Preisner, H., 335
Prelabium, 75
Prementum, 75, 76, 77
Preoral cavity, 107, 109
Pretarsus, 92
Prey population, 562–566
Price, P.W., 602
Prickly pear cactus, 619
Primary germ cells, 186
Pringle, J.W.S., 174
Pristiphora erichsoni, 418
Proboscis, 83, 85, 184
Procecidochares utilis, 621

Processes: external, 62
internal, 62
Proctodeal valve, 107, 108
Proctodeum, 107, 145
Proctotrupes, 411
Proctotrupidae, 411
Proctotrupiodea, 410, 412, 414
Prohemocytes, 163
Promachus vertebratus, 460
Promethea moth, 495, 496
Proprioceptors, 176–180
Propupa, 332
Prosopistoma foliaceum, 287
Prostomium, 29, 63, 229
Protection against enemies, 592–599
camouflage, 592–594
mimicry, 596–599
protective structures, 594
scare tactics, 594
Protective resemblances: camouflage, 304, 592–594
mimicry, 596–599
Proteins, 137, 169
Protelytron permianum, 520
Protelytroptera, 504, 517, **519–521**, 523, 535
Proteolytic enzymes, 140, 143–144
Protereisma permianum, 522
Prothoracic glands, 194, 195, 261
hormone, 261
Prothorax, 89
Protoblattoidea, 268, 518, 535
Protocerebrum, 117, 118, 194
Protocoleoptera, 519
Protodonata, 504, 517, 523
Protomecoptera, 527
Protonymph, 281
Protoparce, 485
quinquemaculata, 615
sexta, 612
Protoperlaria, 504, 520, **521**, 524, 535
Protorthoptera, 268–269, 504, 517, 518, 523, 534, 535
Protura, 52, 69, 88, 89, 92, 99, 266, 267, 278, **280–281**, 504, 627
Provancher, L., 18
Proventriculus, 108, 142
Provisioning wasps and bees, 425–428, 430–432
Psammotettix confinis, 565
Pselaphidae, 220
Pselliopus, 342
barberi, 360
Psephenidae, 375, 376, 384
Pseudaletia unipuncta, 498, 576
Pseudatomoscelis seriatus, 356
Pseudococcus citri, 607
maritimus, 610
Pseudodineura, 413
parvula, 584
Pseudohymen angustipennis, 516
Pseudopods, 458
Pseudoscorpionida, 38, 40
Pseudoscorpions, 38, 40, 101
Psocids, 243, 321–323, 535, 592
Psocoptera, 52, 243, 266, 269, 270, 279, **321–323**, 504, 525, 535, 592
Psychidae, 481, 482, 487
Psychoda, 446
Psychodidae, 442, 443, 444, 445, 446
Psychodid line, 443
Psylla, 339
pyricola, 349
Psyllidae, 270, 337, 339, 343, 348
Pteridines, 133, 151
Pterostichus substriatus, 386
Pterygota, 52, 93, 235, 266, 268–272, 286, 504, 514–532
Pthirius, 328
pubis, 329
Pthirudae, 328
Ptilinum, 444
Pulex irritans, 436
Pulicidae, 438
Pulsating organs, 112, 165
Pulvilli, 181
Punkies, 447
Pupa, 236–238
Pupal cells, 110
Puparium, 444
Purple scale, 634
Pycnocentrellidae, 476
Pycnogonida, 37, 42
Pycnoscelus, 297
surinamensis, 297
Pygmy grasshopper, 306
Pygmy mole crickets, 309
Pygmy scorpionflies, 547
Pyralidae, 482, 484, 491
Pyrethrum, 12
Pyrgotidae, 588
Pyromorphidae, 481, 482
Pyrrhocoridae, 337, 345

Quadraspidiotus perniciosus, 351, 352
Quarantine, 647
Queen, 204, 215–217, 219–220, 300–303, 428–430, 432–433
Queen butterflies, 597
Queen substance, 204
Quiescent period, 244

r- and *K*-selection factors, 560, 562–564
Rabbit flea, 260–261
Rabbit hormone, 260–261
Races, 648
Radial sector, 96
Radius, 95–96
Rainey, R.C., 221
Ramphocorixa, 358
Ramphomyia, 612

Raphidia ratzeburgi, 365
Raphidiidae, 365
Raphidioptera, 52, 236, 266, 271, 276, 364, **365**, 504, 526, 527, 610
Rat flea, 436
Rat-tailed maggots, 160, 462, 463
Ray, J., 8
Reaumur, 9, 10
Receptors, 175–192
Recruitment of workers, 204
Rectal gills, 117, 161
Rectum, 108, 109, 110, 149–152
Recurrent nerve, 118, 119, 120
Red-banded leaf roller, 49, 643
Red-headed pine sawfly, 420
Redi, 5
Red-legged grasshopper, 568
Red scale, 617
Red spiders, 635
Reduviidae, 337, 345, 359–360, 610
Reduvius personatus, 611
Reed, W., 21
Reflexes, 207
  arc, 193, 207
  phasic, 207
  tonic, 207
Regenerative cells, 143
Regions, body, 62–63
Rehn, J.A.G., 310
Releaser effect, 194
Remington, C.L. 285
Rempel, J.G., 105
Repellents, insect, 657
Repetitious generations, 251
Reproduction, 100, 127, 255–260
Reproductive caste, 214–220, 300–302, 428–430 432–433
Reproductive opening, 102–103, 123–125
Reproductive system, 101–103, 122–125
Resilin, 171
Resistance to insecticides, 22, 659
Resistant plant strains, 649
Resources, limited, 563
Respiration, 153–161, 164, 168
  blood, 157
  control, 159
  external, 153
  internal, 153, 168
  metabolism, 168
  movements, 158
Response to contact, 209
Response to odors, 209
Retina, 186
Retinular cells, 186, 189, 191
Rhabdom, 186, **187**, 189, 191
Rhabdoscelus obscura, 619
Rhagio, 446
Rhagionidae, 442, 444, 446, 448, 462, 611
Rhagoletis, 449
  cingulata, 465
  pomonella, 110, 465
Rhizobius debilis, 607
  ventralis, 607
Rhodopsins, 187
Rhopalocera, 481, 482, 483, 499–502, 532
Rhyacionia, 491
  bouliana, 617
Rhyacophila, 477
  fenestra, 474, 475
Rhyacophilidae, 476, 477
Rhyacophiloidea, 474, 476
Rhyncophthirina, 325
Rhyniella praecursor, 533, 534
Rhysodidae, 374
Rice weevil, 403, 638
Richards, O.W., 105, 125, 274, 283, 434
Richards, W.R., 283
Ricinidae, 327
Ricinuleida, 38, 40
Ricker, W.E., 318
Rickettsia, 642
Ridgeway, R.L., 630
Riek, E.F., 318, 372
Riley, C.V., 13, 16, 17
Riley, W.A., 331
Roback, S.S., 473
Robber flies, 443, **460–461**, 611, 612
Rocky Mountain spotted fever, 21, 642
Rodendorf, B.B., 548
Rodolia cardinalis, 395, 605, 606, 607
Roesel, 9
Romoser, W.S., 221, 274
Rootworms, 399, 646, 659
Rose slug, 418
Ross, C.A., 548
Ross, E.S., 315
Ross, H.H., 57, 274, 319, 367, 434, 479, 548
Ross, R., 21
Roth, L.M., 300
Rothschild, M., 440
Roughgarden, J., 602
Round dance, 201, 202
Roundheaded apple tree borer, 401
Roundheaded borers, 401
Rove beetles, 376, **388–389**, 540, 608, 609
Royal Entomological Society of London, 24
Russell, L.M., 25

Sacken, O., 18
Saint John's-wort, 620, 621
Saissetia oleae, 607, 618
Salda, 342
Saldidae, 337, 345
Salivary channel, 141
Salivary duct, 141
Salivary glands, 65, 66, 75, 107, 109, 110, 139–141
Salivation, 139–141
Salmon, J.T., 284
Salt and water, balance of, 148–153

Samia, 485
San Jose scale, 22, 351, 352, 635
Sanninoidea exitiosa, 490, 491
Saperda, 382
candida, 401
Saprophagous, 246
Sarcolemma, 170
Sarcophaga, 117, 449, 592
Sarcophagidae, 242, 442, 444, 449, 450, 466, **469**
Sarcoplasm, 168, 170
Sarcosomes, 170
Sarcostyles, 170
Sarraceniales, 591
Satin moth, 585, 644
Saturniidae, 481, 482, 485, 486, 495–496
Saunders, D.S., 221
Saunders, W., 17
Sawflies, 63, 77, 103, 136, 408, 409, 410, 418–420, 530, 536, 570, 584, 592, 636
Sawtooth grain beetle, 638
Say, T., 11, 12
Scaeva, 446
Scale insects, 62, 133, 146, 236, 346, 350–352, 610, 613, 623, 634
Scape, 71
Scaphidiidae, 387
Scarab beetle, 145, 170, 373, 376, 389, 609, 628
Scarabaeidae, 375, 376, 377, 378, 380, **389**, 609, 627, 628
Scarabaeus sacer, 628
Scare tactics, 594
Scelionidae, 588
Sceliphron, 411
servillei, 431
Schistocerca, 197
americana americana, 305
gregaria, 195, 564
Schizophora, 442, 444, 462, 464–472, 530
Schleiden, 14
Schmitt, J.B., 105
Schneirla, T.C., 221
Scholl-Rempel-Church hypothesis, 63
Schwann, 14
Sciomyzidae, 442, 450, 452, **464**
Sclerites, 59, **61–62**
Scolia, 417
dubia, 425
Scoliidae, 410, 412, 417, 418, 425–426, 619
Scoliid wasps, 425
Scolopale cap, 180
Scolopale cell, 180
Scolopidium, 179, 180
Scolops, 347
Scolytidae, 375, 376, 379, 380, **403–404**, 537, 538
Scopeuma, 450
Scopeumatidae, 450, 452
Scorpionflies, 103, 434–436, 527, 535, 613
Scorpionida, 38–40
Scorpions, 38
Scotopic image, 189
Scott, H.G., 284
Screw worm, 639
fly, 645
Scrub typhus, 642
Scudder, G.G.E., 105
Scudder, S.H., 18, 548
Scutellerinae, 344, 363
Scydmaenidae, 540
Seafloor spreading, 507, 508
Seasonal coordinations, 568–570
Seasonal cycles, 250–251
Secondary pests, 660
Sector, 95, 96
Segmental gills, 101
Segmental muscle bands, 120–121
Segmentation, 27–35, 50–51, 62–65, 88–92, 99–101, 226–228
Seguy, E., 473
Semidalis, 370
Semiochemicals, 202
Seminal vesicle, 124
Sense cell, 176, 177
Sense organs, 128, 175–191, 196–201
Sensilla, 175–191
Sensory neurons, 181, 192–193
Sensory setae, 133
Sepsidae, 452, 464
Sericostomatidae, 476
Serosa, 230
Serrate antennae, 71
Seta (pl. setae), 129, 133, 176
Setaceous antennae, 71
Seventeen-year locust, 347
Sexual maturity, 255–263
Shade tree insects, 636
Shaft lice, 327
Sharov, A.G., 548
Sheep botfly, 469
Sheep lice, 327
Sheeptick, 472
Sheimia sojanensis, 524
Shellac, 623
Shore flies, 464, 465
Shorey, H.H., 221
Short-horned bugs, 358–359
Sialidae, 367
Sialoidea, 366
Sickle dance, 201
Silk, 109, 623
Silkworm, 480
Silkworm moth, 194, 492, 495
Silpha, 380
Silphidae, 375, 376, 377, 380, 384, 608
Silverfish, 100, 267, 284–285, 638
Silver-spotted skipper, 500
Simple metamorphosis, 236
Simuliidae, 79, 442, 443, 444, 447, 453–455
Simulium, 68, 446
johannseni, 455
rittatum, 455

Simulium *Continued*
tahitiensis, 80
Siphlonurus occidentalis, 287
Siphonaptera, 52, 236, 238, 271, 278, **436–439**, 504, 527, 528, 537
Siricidae, 409, 410, 414, 420
Sisyra, 368
Sisyridae, 369, 370
Sitophilus granarius, 403
oryza, 403
Sitotroga cerealella, 490
Skippers, 482, 499, 500
Slabaugh, R.E., 285
later, J.A., 364
Slave raiding, 205
Sleeping sickness, 21, 471, 588
Slingerland, M.V., 20
Sminthuridae, 283
Sminthurus viridis, 224
Smith, A.G., 549
Smith, C., 372
Smith, C.F., 364
Smith, D.S., 174
Smith, E.H., 26
Smith, E.L., 285
Smith, J., 18
Smith, K.G.V., 665
Smith, L.M., 279
Smith, R.C., 274
Smith, R.F., 26, 295
Smithers, C.N., 321, 324
Smoke fly, 443
Snakeflies, 365, 526
Snipe flies, 462, 610
Snodgrass, R.E., 58, 105, 174, 264
Snow cover, 572, 573
Snow flies, 626
Snowy tree cricket, 242
Snyder, T.E., 303
Social bees, 432–433, 568
Social insects, 201–205, 213–220, 412, 425
Social wasps, 427
Socket, 129
Soft bodied beetles, 597
Sofus, 413
Soil: chemical composition, 580
moisture, 579
physiography, 579
properties, 578
texture, 578
Solar radiation, 555, 557
Soldier beetles, 376, 378, 608, 609
Soldier caste, 215
termites, 197, 301
Soldier flies, 462
Solenopotes capillatus, 224
Solenopsis invicta, 430
molesta, 430
Solitary wasps, 412, 430–432, 612
Solpugida, 40, 41
Solubia, 339
Sommerman, K.M., 324
Sorghum, 648
Sorghum vulgare, 648
Sound receptors, 176, 182–184
Southern cabbage worm, 501
Southern house mosquito, 571
Southern sticktight flea, 639
Southwestern corn borer, 492, 633
Southwood, T.R.E., 602
Sowbugs, 44
Spacing, 205
Spanish fly, 624
Specific feeders, 584
Speed, flight, 173
Spencer, G.J., 26
Spermatheca, 123
Spermatogonia, 256, 257
Spermatophore, 260
Spermatozoa, 256–258, 259, 260
Sperm transfer, 259–260
Sperm tube, 124
Sphacophilus, 413
Sphaeriidae, 376
Sphecidae, 410, 411, 412, 415, **430–432**
Sphecid wasps, 430–432, 613
Sphecius speciosus, 431, 613
Sphecoidea, 410, 430, 530
Sphecomyrma freyi, 531
Sphenophorous phoeniciensis, 224
Spherule cells, 163
Sphingidae, 173, 481, 484, 485, **494–495**
Sphinx convolvuli, 111
Sphinx moths, 173, 494–495
Spiders, 38–40, 591
Spilopsyllus cuniculi, 261
Spina, 92
Spinasternum, 91, 92
Spiracle control, 153–154
Spiracles, 88, 100, 113–116, 153–154
Spiracular adaptations, 116, 158
Spiracular branches, 114
Spittle bugs, 347
Spodoptera frugiperda, 498, 499
Sponge feeders, 370
Spongeflies, 370
Spongillaflies, 370
Sporotrichum globulifera, 589
Spotted alfalfa aphid, 649
Spradberry, J.P., 434
Sprays: calyx, 570
petal-fall, 570
Springtails, 196, 231, 282–283, 576
Spruce budworm, 561, 613, 619
Spurious vein, 446
Spurr, S.H., 602
Spurs, 62
Squamae, 444, 464, 466
Squash blister beetle, 398
Squash bugs, 360, 361

Squash vine borer, 490
Stability of populations, 559-560, 562-566
Stable fly, 467, 639
Stadium, 234
Stagmomantis carolina, 299
Stain moth, 615
Stairs, G.R., 665
Stannard, L.J., Jr., 335
Staphylinidae, 220, 375, 376, 377, **388-389**, 591, 608
taphylinoids, 375, 376, 388-389
Stegobium, 380
Stemmata, 67, 184, **190**
Stem miners, 465
Stenobothrus, 198
Stenocranus, 339
Stenodictya, 515
Stenoneura fayoli, 518
Stenopelmatus, 309
Stenopsychidae, 476
Stenus, 388
Sterile workers, 215-216, 300, 432
Sternites, 100, 104
Sternum, 88, 89, 90, **91**-92, 100
Sticktight flea, 438, 439
Stilpnotia salicis, 585, 615
Stimuli, 175, 176, 191, 192, 194
Stinging organ, 424
Stinging wasps, 424
Stingless bees, 623
Stings, 594
Stink bugs, 206, **362**-363, 591, 610
Stipes, 74, 75
Stipites, 76
Stipulae, 75
Stiretrus, 339
Stobbart, R.H., 174
Stomach reaction, 146
Stomodeal nervous system, 118-119
Stomodeal valve, 107, 108
Stomodeum, 107, 108, 119, 141
Stomoxys calcitrans, 467
Stone, A., 473
Stone beetle, 540
Stoneflies, 89, 228, 233, 236, 269, 316-319, 524, 535, 539, 544, 625, 627
Stored food pests, 637-638
Storer, T.I., 58
Størmer, L., 58
Stratiomyidae, 442, 444, 446, 447, 448, 451, 462
Stratiomys, 446
Streblidae, 442, 445, 472
Strepsiptera, 404, 406, 542
Stress receptors, 178-181
Striated fibers, muscles, 169
Stridulation, 197
Stridulation organ, 199
Striped cucumber beetle, 400
Structural colors, 133-135
Styli, 102, 104
Stylopidae, 404, 406, 609
Stylopoidea, 171, 239, 243, 376, **404-407**, 588
Stylopoids, 375, 376, 404-407
Subcosta, 94, 95
Subcoxa, 88
Subcoxal sclerites, 88, 89
Subduction zone, 509
Subesophageal ganglion, 118, 195
Subimago, 235, 268, 286
Submentum, 75, 76
Sucking lice, 324, **327**-330, 639
Sugar-cane borer, 619
Sulfurs, butterflies, 501
Summer oils, 655
Sundews, 591
Supella longipalpa, 298
Superficial cleavage, 225
Superior arms, 66
Superlinguae, 76
Superposition type image, 189
Supraesophageal ganglion, 117
Sutures, 61, 67, 69, 130
Swallowtail butterflies, 500-501, 595
Swammerdam, 5, 6
Swan, L.A., 630
Swarming flights, 428
Swifts, 480, 576
Symbiosis, 144, 145, 218, 220
Sympathetic nervous system, 119
Symphoromyia, 81, 462, 639
Symphyla, 43, 46, 48, 76
Symphypleona, 282
Symphyta, 409, 410, 413, 418-420, 530
Synapse, 192, 193
Synergistic action, 653
Synthetic insecticide, 22, 655-657
Synthetic pyrethroids, 656
Syrphidae, 173, 220, 442, 444, 446, 448, **462-464**, 611, 612
Syrphid flies, 462
Systema Naturae, 8
Systemic poisons, 655
Systems ecology, 563
Systolic pressure, 165

Tabanidae, 79, 117, 442, 443, 444, 446, 447, 451, 459, **461-462**, 547, 591
Tabanomoidea, 442
Tabanus, 173, 446, 451
lineolus, 461
Tachina flies, 469, 471, 620
Tachinidae, 442, 444, 449, 450, 451, 469-470, 588, 619
Tactile receptors, 176
Tactile stimuli, 177
Taenidia, 115, 116, 157
Taeniopterygidae, 318
Taeniothrips inconsequens, 224, 335
Tangle-vein flies, 547
Tanytarsus, 241

Tapestry moth, 487
Tapinoma sessile, 430
Tardigrada, 30, 31
Tarnished plant bug, 356
Tarostenus univittatus, 608
Tarsus, 92, 93, 122
Tauber, C.A., 372
Tauber, M.J., 264
Taxes, 207–212
Telopodite, 70, 72, 73, 101
Temperature, effect of, 182, 567–572, 574–575
581–582
Tenaculum, 282
Tendipedidae, 458
Tenebrio, 149, 399
Tenebrionidae, 375, 376, 379, **398–399**, 608
Tenebrionid beetle, 586
Tenorio, J.M., 273
Tent caterpillar, 218, **493**, 585, 636
Tenthredinidae, 409, 410, 413, 414, **418,** 570
Tenthredo, 413
Tentorial pit, 67, 70
Tentorium, 66, **69**–70
Tephrita, 451
Tephritidae, 442, 449, 451, 453, **465**
Terebrantia, 333
Tergite, 100
Tergum, **88**, 89, 90, 100
Termites, 145, 214, 215, 269, 300–303, 523, 539,
613, 627
Testes, 124
Tetrapyrrole pigments, 133
Tetrastichus, 415, 599, 600
giffardianus, 599
Tetrigidae, 305, 306
Tettigoniidae, 305, 307–308
Texas fever of cattle, 642
Texas leafcutting ant, 579
Thanaos, 483
Thanasimus, 608
Thaneroclerus, 608
Theiler, M., 666
Thelytoky, 244
Thereva, 446
Therevidae, 442, 446, 448
Thermobia, 97
domestica, 285
Thermocline, 582, 583
Thermoreceptors, 176, 182
Thermotaxis, 210
Thief ant, 430
Thiele, H.-U., 407
Thigmotaxis, 208, 209
Thin walled peg, 177
Thomas, C., 16
Thorax, 50–51, 89–99
Three-striped blister beetle, 398
Thremmidae, 476
Threshold temperature, 568
Thripidae, 334
Thrips, 224, 236, 237, 270, 331–335, 537, 610,
623
Thrips tabaci, 335
Thuringian bacterium, 645
Thyanta, 339
Thyce, 380
Thyreocorinae, 344, 363
Thyridopteryx ephemeraeformis, 488,
489
Thysanoptera, 52, 266, 270, 275, 278, **331–335**,
504, 525, 537
Thysanura, 52, 72, 76, 151, 220, 266, 267, 277,
284–285, 504
Tibia, 92, 93, 122
Tibicen linnei, 431
Tibiotarsus, 92
Ticks, 40, 639, 642
Tietz, H.M., 274
Tiger beetles, 385, 608
Tillus, 608
Timema, 303
Timemidae, 303
Tinea pellionella, 487, 488
Tineidae, 481, 487
Tineola bisselliella, 224, 487
Tingidae, 337, 355–356
Tinsley, T.W., 666
Tiphia, 426
Tip moths, 636
Tiphiidae, 410, 412, 415, 417, 418, **426**
Tipulidae, 442, 443, 445, 451, **456**
Tissue determination, 228
Tissue hormones, 260
Tobacco budworm, 633, 658
Tobacco flea beetle, 400
Tobacco hornworm, 494, 612, 649
Tobacco insects, 612, 633–634, 649
Tobacco thrips, 335
Tomato hornworm, 494, 615, 649
Tomocerus niger, 108
Tonic reception, 177
Tormogen cell, 129, 133, 176, 177
Tortricidae, 481, 482, 484, **490–491**
Toschi, G.A., 372
Totoglossa, 75, 76
Townes, H., 434
Toxoptera graminum, 224
Toxotrypania curvicauda, 103
Tracheae, 66, 98, **113–117, 153–161**
Tracheal system, **113–117,** 153–161
branches, 113
closed, 116
open, 115
trunks, 113–114
ventilation of, 157
Tracheoblast, 154
Tracheole cell, 154
Tracheole fluid, 155
Tracheoles, 113, 153, 154
Tree cricket, 632

Tree hopper, 219, 346, 632
Trehalase, 167
Trehalose, 162, 167, 168
Treherne, J.E., 222
Tremex columba, 76, 420, 421, 604
Trench fever, 21, 329
Triaenodes tarda, 477
Trialeurodes vaporariorum, 349
Triatoma sanguisuga, 360
Tribolium, 575
  castaneum, 398
  confusum, 398, 399
Trichiosoma, 413
  triangulum, 76
Trichodectes, 327
  ovis, 327
Trichodectidae, 327
Trichogen cell, 129, 133, 176, 177
Trichogramma minutum, 618
Trichogrammatidae, 619
Trichoid sensilla, 176–178
Trichophaga tapetzella, 487
Trichophthalma, 547
Trichoplusia ni, 498
Trichopoda pennipes, 471
Trichoptera, 52, 137, 240, 266, 271, 277, **472–479**, 504, 527, 532, 535
  case making, 474
Trichoptera-Lepidoptera, 271
Tricobothria, 341
Tridactylidae, 305, 309
Tridactylus minutus, 309
Trigona, 623
Trigonalidae, 410, 412
Trilobita, 32, 34, 36
  leg, 70
Triphelps, 342
Triplosoba pulchella, 522
Triticerebral commissure, 118, 119
Tritocerebrum, 118, 120
Tritonymph, 281
Triungulin, 396
Triungulinid, 406
Trochanter, 92, 93, 122
Trochantin, 91
Trogium pulsatorium, 323
Trombiculidae, 40
Trombidiidae, 642
Trophic levels, 554–559
  of insect orders, 558
Trophocytes, 258
Tropisternus, 380
Truck crop insects, 634
Truman, J.W., 222
Trypanosoma, 588, 641
Trypetidae, 465
Trypoxylinae, 431
Tsetse fly, 167, 243, **471**
Tubulifera, 333
Tunga penetrans, 439
Tungidae, 438
Turpaneidae, 465
Tussock moth, 594
Tuxen, S.L., 274, 281
Twig girdler, 632
Twisted-wing beetles, 171, 609
Tympanal membrane, 183–184
Tympanal organs, 180
Tympanum, 183
Typhoid fever, 640
Typhus, 21, 329, 641–642
Tyroglyphidae, 40

Udea rubigalis, 492
Uhler, P.R., 18
Umbonia, 219
Underwing moth, 592, 593
Urate cells, 151
Uric acid, 150, 151
Urine, 150
Urogomphi, 376
Uropygida, 38, 40
Usinger, R.L., 274, 364
Uterine development of young, 471
Utricularia, 591
Utriculi majores, 151
Uvarov, B.P., 310

Vagina, 123
Valve: cardiac, 107, 108
  pyloric, 108
Valvifer, 101, 102, 103
Valvula, 101, 102, 103
van der Weele, W.H., 367
Van Duzee, E.P., 364
Vanes, 133, 134
Vannal fold, 98
Vas deferens, 124, 258
Vas efferens, 258
Vedalia ladybird beetle, 18, 395, 606, 644
Vegetation, aquatic, 583
Veliidae, 337, 344
Veins of wings, 94–99
Velvet ants, 412, 426
Venation, 94, 95
Venter of body, 62
Ventilation: control of, 158
  of tracheal system, 157
Ventral diaphragm, 112, 165
Ventral gland, 194, 195
Ventral nerve cord, 118, 119
Ventricular ganglion, 119
Ventriculus, 109
Vermiform cells, 163
Vertex, 67
Vespa, 415
Vespidae, 410, 411, 412, 415, 417, **426–428,** 530, 612, 619
Vespoidea, 410, 425
Vespoid wasps, 426–428, 612–613

Vespula maculata, 408, 427
Viceroy butterflies, 597, 598
Viereck, 434
Vinegar gnats, 444, 465, 466
Virus disease, 589-590, 636-637. 641
Visceral muscles, 120
Vitamin C, 649
Vitamins, 138
Vitelline membrane, 224, 225
Viviparity, 242-243
Vogtia malloi, 621
Voltinism, 243-244
Von Baer, 14

Waggle dance, 201, 202
Walker, E.M., 295
Walking sticks, 269, 303-304, 523, 536, 592
Wallace, A.R., 14
Wallace, J.W., 666
Waloff, N., 602
Walsh, 16
Warble flies, **469,** 470, 639
Wasps, 170, 172, 271, 408, 410, 412, 413, 417, **420-427,** 430-432, 530, 531, 594, 596, 597, 599, 600, 607
  social, 427
  solitary, 430-432
Water absorption, 145
Water beetles, 376
Water boatmen, 160, 183, 358, 581
Water bug, 359
Water current receptors, 181
Water cycle, 556, 557
Water depth, 582
Waterhouse, D.F., 58, 105, 174, 274, 630
Water requirements, 139
Water striders, 354, 355
Watson, J.R., 335
Watson, M.A., 666
Wax canal filament, 132
Wax layer, 130, 132
Wax moth, greater, 492
Weather, 567-576
Webbing clothes moth, 224, 487
Weber, H., 106
Web spinners, 218, 269, **313-315**
Webworms, 491, 492, 634
Weesner, F.M., 303
Weevils, 224, 373, **402-404**, 547, 562
Weiss, H.B., 26
Western corn rootworm, 579
Westfall, M.J., Jr., 289, 295
Wharton, G.W., 57, 58
Wheat jointworm, 423
Wheat stem sawfly, 616, 649
Wheat strawworm, 423
Whirligig beetles, 386
White, M.J.D., 264
White ants, 300-303
Whiteflies, 348, 349, 634
White fringed weevils, 562
White-marked tussock moth, 497, 498
Whites, The, butterflies, 50
Whitten, M.J., 666
Wickler, W., 602
Wiggins, G.B., 479
Wigglesworth, V.B., 125, 174
Williams, S.R., 58
Williston, S.W., 18
Wilson, E.O., 222
Winged segment, 90-92
Wing pad, internal, 238
Wings, 53, 91, 93-99
  asynchronous, 171, 172
  basal folds, 98
  synchronous, 171-172
Winter stoneflies, 317, 582
Winthemia quadripustulata, 471, 619, 620
Wireworms, 188, 391-392, 579, 633, 634
Wirth, W.W., 472
Withycombe, C.L., 373
Woglum, R.S., 365
Wolfe, J A., 549
Wood borers, 392, 636
Wood cockroach, 145, 298
Woodward, T.E., 364
Wool sower gall, 247
Wooly apple aphid, 618
Wootton, R.J., 549
Worker caste, 215, 300-302, 427, 432-433
Wright, M., 295
Wygodzinsky, P., 285, 364

Xanthopterine, 151
Xenopsylla cheopis, 438
Xiphocentronidae, 476
Xiphosura, 37
Xiphydriidae, 409, 410
Xyelidae, 409, 410, 413, 530
Xylobiops, 608

Yellow fever, 21, 641
Yellow fever mosquito, 457
Yellow jackets, 426-428
Yponomeutidae, 481

Zenillia, 449, 451
Zimmerman, E.C., 273
Zoogeographic regions, 272-273
Zoophagous food, 246-247
Zoraptera, 52, 266, 270, 277, **319-321**, 504, 525
Zorotypidae, 319
Zorotypus, 320
  hubbardi, 320
  snyderi, 320
Zwick, P., 319
Zygoptera, 289, 291, **292**
Zygote, 225, 233